Handbook of
NATURALLY OCCURRING
COMPOUNDS

Volume I
Acetogenins, Shikimates,
and Carbohydrates

Handbook of
NATURALLY OCCURRING COMPOUNDS

Volume I
Acetogenins, Shikimates, and Carbohydrates

T. K. DEVON

**PFIZER CENTRAL RESEARCH LABORATORIES
SANDWICH, ENGLAND**

A. I. SCOTT

**STERLING CHEMISTRY LABORATORY
YALE UNIVERSITY, NEW HAVEN, CONNECTICUT**

Academic Press Inc.,   New York   San Francisco   London   1975

A Subsidiary of Harcourt Brace Jovanovich, Publishers

Science

ACADEMIC PRESS, INC.
111 Fifth Avenue, New York, New York 10003

United Kingdom Edition published by
ACADEMIC PRESS, INC. (LONDON) LTD.
24/28 Oval Road, London NW1

Library of Congress Cataloging in Publication Data

Devon, T   K
    Handbook of naturally occurring compounds.

      CONTENTS: v. 1.   Acetogenins, shikimates, and
carbohydrates.   v. 2.   Terpenes.
     1.   Natural products—Handbooks, manuals, etc.
I.   Scott, Alastair L., joint author.   II.   Title.
III.   Title: Naturally occurring compounds.
QD415.7.D48   547'.7'0212   76-187258
ISBN 0–12–213601–2 (v. 1)

PRINTED IN THE UNITED STATES OF AMERICA

CONTENTS

CONTENTS

PREFACE

With the advent of spectroscopy as a tool in the structural elucidation of natural products, there has been a rapid increase in our knowledge of the structures and variety of naturally occurring compounds. It has consequently become an increasingly difficult task to maintain current awareness in the natural products field, and the acquisition of retrospective data is a problem only partly alleviated at present by literature reviews, monographs, and compendia. It was in an attempt to pool the chemical and biochemical data of natural products that the Card Index File of Naturally Occurring Compounds was conceived and initiated at the University of Sussex in 1966. This project, the first essential step in the construction of a more comprehensive system, undertook to search for, list, and classify all reported naturally occurring compounds the structures of which had been determined (with the exclusion of polymeric and macromolecular compounds). The interest so often expressed in this file led, in 1970, to the reorganization of part of the data into the publishable format now presented.

It is somewhat inevitable, when attempting so wide-ranging a project, that the selection criteria should be controversial; thus, the decision to include steroidal aglycones as "natural products" is likely to be frowned upon by the purists. Others, however, may consider the absence of antibiotic degradation products a severe limitation. Similarly, the lack of inclusion, at this stage, of spectroscopic and botanical data may appear limiting; however, an attempt to be so comprehensive so rapidly would have greatly delayed the availability of the present material. It was felt that the present format and material could well stand alone in satisfying particular needs in this field; the provision of a literature reference for each compound should assist in the further acquisition of data.

No pretence of complete comprehensiveness is being claimed for this handbook, nor will it likely be found to be error free; however, considerable care has been taken to abstract the literature as deeply and thoroughly as our resources have permitted. A certain degree of editing of material for structural and sterochemical correlations has been undertaken and has hopefully removed the bulk of errors and ambiguities from the handbook. It would, of course, be greatly appreciated if diligent users who spot errors or omissions would forward them to the authors so that subsequent yearbooks can set the records straight.

We would like to indicate our debt to the numerous authors of earlier compendia, monographs, and reviews from whose works the original Card Index File of Naturally Occurring Compounds gained much of its foundation. For the facilities enabling the construction of the File our appreciation is tendered to Chemical Laboratory, University of Sussex. At Yale University we gratefully received assistance and encouragement from many colleagues in the Sterling Chemistry Laboratory. Dr. Paul Reichardt (University of Alaska) has become a valuable participant in this undertaking, and we look forward to his active collaboration with us in the production of the supplementary volumes. The final preparation of material for publication was ably performed by Mrs. Diane Devon with contributions to the indexing from Jeffery and Ronald Burns. Finally our appreciation is extended to Pfizer Central Research for the facilities and atmosphere which enable this endeavor to continue to flourish.

GUIDE TO HANDBOOK USAGE

Contents

This Handbook contains at present most of the known naturally occurring compounds to which structures have been assigned.

Limitations

Excluded from the Handbook are polymeric compounds, such as proteins and polysaccharides, synthetic derivatives of natural products, and degradation products (artifacts).

Data

Each structure is stored in the Handbook with its name, molecular formula, molecular weight, optical rotation (α_D), melting point, literature reference (usually the latest), and classification number.

DATA FORMAT

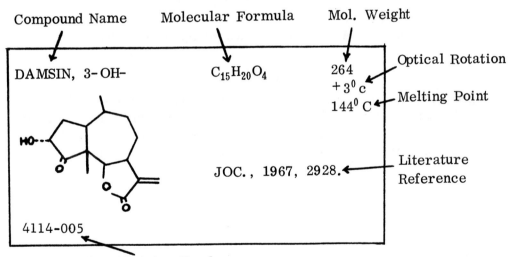

Compound Name   Molecular Formula   Mol. Weight

DAMSIN, 3-OH- $C_{15}H_{20}O_4$ 264

Optical Rotation — $+3^0$ c

Melting Point — 144^0 C

JOC., 1967, 2928. — Literature Reference

4114-005

Classification Number

Indices

Compounds can be retrieved by name using the ALPHABETICAL INDEX or by structural type using the STRUCTURAL CLASSIFICATION GUIDES at the beginning of each class. MOLECULAR WEIGHT AND MOLECULAR FORMULA INDICES are also provided at the end of the book.

Code Numbers

CLASSIFICATION CODE NUMBERS are used to specify structural types, and the key to these is held within the individual Structural Guides. The first two digits represent the major class and the second two the subclass. Each individual compound also possesses a COMPOUND SEQUENCE NUMBER which, in combination with the Classification Code Number, supplies a unique address for that compound. The compounds are stored in the Handbook in ascending sequence, new compounds being simply inserted at the end of its appropriate section.

Classification Guide

The Catalogue of each primary class of compounds will be preceded by its corresponding Classification Guide which in general consists of:

    Introduction
    Biogenetic Chart
    Main Skeleton Index
    Less Common Skeletons Index

Main Volumes

The wide range of structural types of naturally occurring compounds have been classified and then collected into three groups, each of which will be issued as a separate volume:

    Volume I   Acetogenins, Shikimates, and Carbohydrates
    Volume II   Terpenes
    Volume III Alkaloids and Related Nitrogenous Compounds

Supplements

Each April an annual supplement for the preceding year will be published for each volume which will contain new compounds, structure changes in earlier reported compounds, and any additional data and indices. At suitable intervals these supplements will be cumulated and merged into the main volumes.

ABBREVIATIONS

Nonstandard Journal Abbreviations

Aust. J. C.	Australian Journal of Chemistry
Can. J. C.	Canadian Journal of Chemistry
Helv.	Helvetica Chimica Acta
Ind. J. C.	Indian Journal of Chemistry
JACS	Journal of the American Chemical Society
JCS	Journal of the Chemical Society
JOC	Journal of Organic Chemistry
J. Ind. C.S.	Journal of the Indian Chemical Society

Symbols Used for Solvents (optical rotations)

a	Acetone	m	Methanol
b	Benzene	n	HCl
c	Chloroform	p	Pyridine
d	Dioxan	r	Diethyl ether
e	Ethanol	t	Carbon tetrachloride
h	Hexane	w	Water

Other Abbreviations

ac	Acetate	Glu	Glucosyl
Ac	Acetyl	Mann	Mannosyl
AMB	Alpha-methyl-butyryl	Me	Methyl
Amorph	Amorphous	m.e.	Methyl ester
Ang	Angeloyl	OI	Optically Inactive
Ara	Arabinosyl	Ph	Phenyl
Bu	Butyl	pic	Picrate
Bz	Benzoyl	Pr	Propyl
Caff	Caffeoyl	Rac	Racemic
Cinn	Cinnamoyl	Rha	Rhamnosyl
Coum	Coumaroyl	Sen	Senecioyl
DMA	Dimethylallyl	Tig	Tigloyl
Et	Ethyl	Val	Valeryl
Fruct	Fructosyl	Van	Vanilloyl
Ger	Geranyl	Ver	Veratroyl

INTRODUCTION

Of the three volumes that constitute this handbook, this particular compilation presents the greatest challenge to introduction. It would be a simple matter to state the contents of the other volumes and contend that this volume provides coverage of all other classes of natural products. Alas, such an abdication cannot be. In biogenetic terms the compounds included in this volume are derived from one (or more) of four major pathways:

> shikimic acid pathway
> acetate/malonate pathway
> carbohydrate metabolism
> tricarboxylic acid cycle

Thus, Volume I coverage can be clearly differentiated from that of Volume II (terpenes derived from the mevalonate pathway). Volume III material (alkaloids, amino acids, etc.) cannot be so readily distinguished in these terms, for the aromatic amino acids are products of the shikimic acid pathway, and some classes of alkaloids are of acetogenic origin. Suffice it to say here that the majority of nitrogenous compounds are located in Volume III, notable exceptions being the tetracyclines and aminosaccharides, which have been placed in Volume I, and the terpene alkaloids located in Volume II. This volume also has the distinction of embracing the largest collection of compounds, some 4800 of the 12,000 natural products in our files (as of early 1972). These structurally defined compounds have been assigned a two-digit Primary Classification Code Number(PCCN) that ranges from 01 to 35 plus a "miscellaneous" class, PCCN = 39, which has been provided for complex or ambiguous structures. The primary classification is given in the table overleaf, secondary subclassifications being outlined at the beginning of the appropriate primary sections of the handbook. Because of some of the classification problems associated with compounds such as the phenols, some effort has been taken to provide introductions to the biogenetic schemes utilized in this volume. It should be borne in mind that such structural/biogenetic schemes are frequently hypothetical simplifications, and the inclusion of a compound in a particular class is in no way a comment on its actual biosynthetic origin. Thus, while the anthraquinone class is considered to be a member of the acetogenic group, it is now well known that several plant anthraquinones are derived via the shikimic acid pathway.

THE PRIMARY CLASSES IN VOLUME I

Shikimate Aromatics

01	$C_6 - C_1$
02	$C_6 - C_2$
03	$C_6 - C_3$
04	$C_6 - C_n$ ($n \geqslant 4$)
05	Coumarins
06	Lignans
07	Terphenyls
19	Prearomatics

Acetate/Shikimate Aromatics

09	Neoflavonoids
10	$C_6 - C_3 - C_6$
11	Flavones
12	Flavanones
13	Flavans
14	Isoflavonoids
15	Biflavonoids
16	Pterocarpans/Coumestans
17	Anthocyanins
18	Rotenoids/Peltogynans/ Aurones
28	Benzophenones/Stilbenes

Acetate/Malonate Aromatics

21	Orcinols
22	Phloroglucinols
23	Depsides/Depsidones
24	Xanthones
25	Naphthoquinones

26	Anthraquinones
27	Polycyclics
28	Biphenyls/Grisans

Acyclic and Heterocyclic Acetogenins

31	Hydrocarbons, Fatty Acids, etc.
32	Oxyheterocyclics and Phenylpolyynes
33	Thio Compounds
34	Macrolides

Carbohydrates

35	Saccharides/Cyclitols/ Aminoglycosides

Complex Classes

08	Phenols (not classified elsewhere)
19	Phenol Epoxides/ Shikimate Prearomatics
20	Benzoquinones
28	$C_6 - C_n - C_6$
29	Cyclopentanoids
30	Tetronic Acids/ Tropolones/ Pyrones/ Furans
39	Miscellaneous

SHIKIMATE-DERIVED AROMATICS

Aromatic compounds of shikimic acid origin in general are characterized by their oxygenation patterns, lack of nuclear alkylation, and limited structural modification. The scheme provided on the next page summarizes the shikimic acid pathway and the relationships between the aromatics derived via shikimic acid. From this scheme it should be evident how the structural characteristics of shikimates originate and serve to distinguish them from acetogenic aromatics. It is of course important to realize that reliance on such features for biogenetic analysis is a double-edged tool and can lead to "misplacing" shikimate-looking aromatics of acetogenic origin. The overall utility of the classification does, however, significantly outweigh such reservations.

Chart 1: The Shikimic Acid Pathway

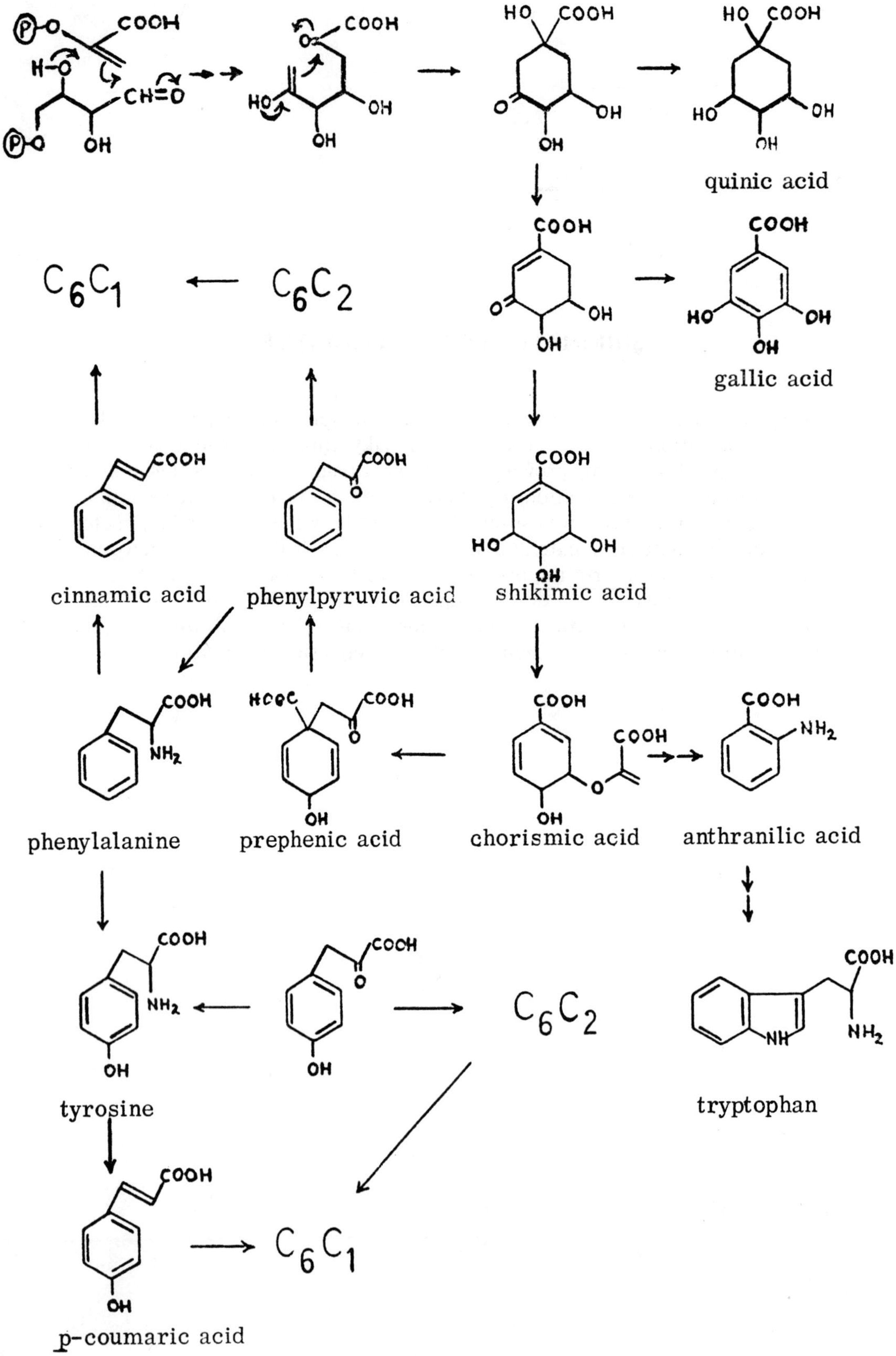

quinic acid

gallic acid

C_6C_1 ← C_6C_2

cinnamic acid

phenylpyruvic acid

shikimic acid

phenylalanine

prephenic acid

chorismic acid

anthranilic acid

tyrosine

C_6C_2

tryptophan

p-coumaric acid → C_6C_1

Oxygenation Patterns of Shikimate-Derived Aromatics

As has already been intimated, aromatic compounds derived from shikimic acid frequently display characteristic oxygenation patterns. The origin of these patterns can be rationalized quite simply by a consideration of the biogenetic scheme illustrated in Chart 1 and the invocation of an *ortho*-oxygenation process. Thus from Chart 1 the key intermediate prephenic acid can lead to either the nonoxygenated aromatic series (e.g., cinnamic acid) or the *para*-hydroxy series exemplified by *p*-coumaric acid; *ortho*-hydroxylation of the latter leads to the familiar 3,4-dihydroxy series (e.g., caffeic acid) and further to the 3,4,5-trihydroxy derivatives. It should be mentioned that the 3,4-di- and 3,4,5-trioxygenated benzoic acids can also be derived from 5-dehydroshikimic acid directly. An aspect of oxygenation which is particularly significant in coumarin biosynthesis is the introduction of an oxygen function at the *ortho* position relative to the side chain. Such free phenols are not common but there can be some ambiguity in biogenetic classification as a result of this process. Salicylic acid, for example, with its single *ortho*-hydroxyl function is difficult to rationalize biogenetically except as an oxygenated shikimate-derived benzoic acid. It has been a general policy in dealing with the phenols to place those in the acetogenic classes which most obviously belong there, and to put the remainder into the shikimate classes. Thus, some unusual oxygenation patterns will be found which cannot necessarily be rationalized as discussed above. In the absence of biosynthetic evidence, such ambiguities have to be accepted, although from time to time biogenetic models are presented which shed some light on the origins of such patterns—e.g., the rearrangement mechanism leading to gentisic acid:

SHIKIMATE CLASSES

$$Ar^* = \text{(benzene ring)} \quad or \quad \text{(HO-phenyl ring)} \quad or \quad \text{(HO, OH dihydroxy ring)} \quad or \quad \text{(HO, HO, OH trihydroxy ring)}$$

* See page 5

01	C_6C_1	$Ar-C$
02	C_6C_2	$Ar-C-C$
03	C_6C_3	$Ar-C-C-C$
04	C_6C_n	$Ar-[C]_n \quad n \geqslant 4$
05	Coumarins	(coumarin structure)
06	Lignans	$Ar-C-C-C$ $Ar-C-C-C$
07	Terphenyls	(benzoquinone with two Ar groups)
1902	Prearomatics	e.g. (shikimic acid structure: HO, COOH, HO, OH)

01	$C_6 C_1$	Ar—C

0101	Ar—CH$_3$
0102	Ar—CH$_2$O·
0103	Ar—CHO
0104	Ar—COO·
0105	
0199	MISC.

CRESOL, p-	C_7H_8O	108 OI 36^0C

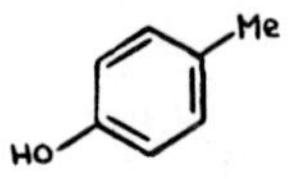

Chem. Abstr., 1956 (50) 9693.

0101-001

PYROGALLOL-1,3-DI-ME ETHER, -5-Me-	$C_9H_{12}O_3$	168 OI 41^0C

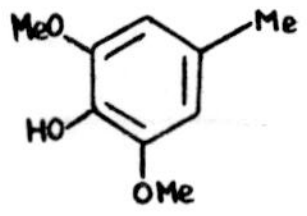

Ber., 1936 (69) 1870.

0101-006

CRESOL ME ETHER, p-	$C_8H_{10}O$	122 OI Oil

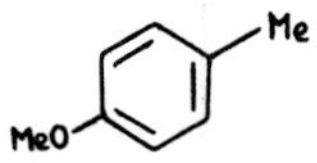

Helv., 1964, 111.

0101-002

CRESOL, ortho-	C_7H_8O	108 OI Oil

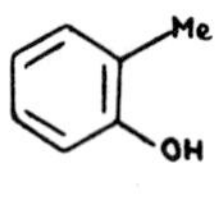

Phytochem., 1968, 278.

0101-007

HOMOCATECHOL	$C_7H_8O_2$	124 OI 65^0C

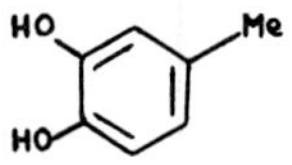

Chem. Zentr., 1931 (II) 586.

0101-003

BENZYL ALCOHOL	C_7H_8O	108 OI Oil

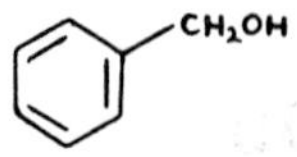

Arch. Pharm., 1932 (270) 249.

0102-001

CREOSOL	$C_8H_{10}O_2$	138 OI Oil

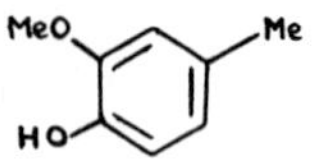

Phytochemistry, 1968, 278.

0101-004

ANISYL ALCOHOL	$C_8H_{10}O_2$	138 OI 25^0C

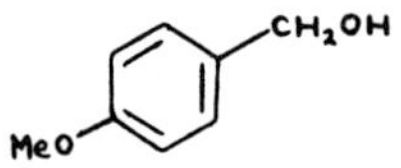

Chem. Abstr., 1951, 3124.

0102-002

PYROGALLOL-1-ME ETHER, -5-Me-	$C_8H_{10}O_3$	154 OI 68^0C

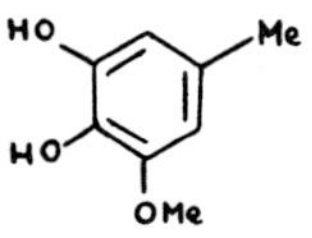

Ber., 1936, 1870.

0101-005

BENZYL ALCOHOL, 3,4-diOH-	$C_7H_8O_3$	140 OI

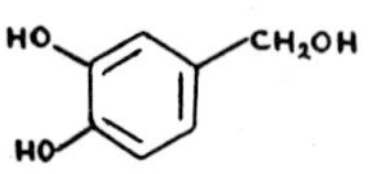

Phytochem., 1968, 119.

0102-003

VANILLYL ALCOHOL	$C_8H_{10}O_3$	154
		OI
		115^0C

Chem. Abstr., 1947 (41) 7681.

R = H

0102-004

GENTISYL ALCOHOL	$C_7H_8O_3$	140
		OI
		100^0C

R = H

Experientia, 1948 (4) 277.

0102-009

VANILLOLOSIDE	$C_{14}H_{20}O_8$	316
		120^0C

R = Glucosyl

Chem. Abstr., 1947 (41) 7681.

0102-005

SALIREPIN	$C_{13}H_{18}O_8$	302
		166^0C

R = Glucosyl

Tet. Lett., 1967, 1869.

0102-010

BENZYL ALCOHOL ME ETHER, 2,3-di-Br-4,5-di-OH-	$C_8H_8O_3Br_2$	310
		OI
		130^0C

Tet., 1967, 1185.

0102-006

GENTISYL ALCOHOL, 3-Cl-	$C_7H_7O_3Cl$	174
		OI

Agr. Biol. Chem., 1971, 105 & 445.

0102-011

SALIGENIN	$C_7H_8O_2$	124
		OI
		86^0C

Ann., 1845 (56) 35.

0102-007

BENZALDEHYDE, p-OH-	$C_7H_6O_2$	122
		OI
		116^0C

Helv., 1945, 722.

0103-001

SALICIN	$C_{13}H_{18}O_7$	286
		-63^0w
		206^0C

Natwiss., 1963 (50) 477.

0102-008

ANISALDEHYDE	$C_8H_8O_2$	136
		OI
		Oil

Biochem. J., 1955 (60) 255.

0103-002

PROTOCATECHUALDEHYDE	$C_7H_6O_3$	138 OI 154°C

Chem. Abstr., 1937 (31) 2638.

0103-003

BENZALDEHYDE, 3,4,5-triOH-	$C_7H_6O_5$	170 OI 212°C

JCS, 1967, 410.

0103-008

VANILLIN	$C_8H_8O_3$	152 OI 81°C

R = H

Compt. Rend., 1955 (241) 987.

0103-004

SYRINGIC ALDEHYDE	$C_9H_{10}O_4$	182 OI 112°C

R = H

Chem. Abstr., 1956 (50) 5100.

0103-009

VANILLOSIDE	$C_{14}H_{18}O_8$	314 -87°w 189°C

R = Glucosyl

Chem. Abstr., 1947 (41) 7681.

0103-005

SYRINGIC ALDEHYDE, gluco-	$C_{15}H_{20}O_9$	344

R = Glucosyl

Gazz., 1888 (18) 215.

0103-010

VERATRUMALDEHYDE	$C_9H_{10}O_3$	166 OI 44°C

R = Me

Compt. Rend., 1955 (241) 987.

0103-006

BENZALDEHYDE, 3,4,5-tri-OMe-	$C_{10}H_{12}O_4$	196 OI 75°C

R = Me

Chem. Abstr., 1965 (63) 5444.

0103-011

PIPERONAL	$C_8H_6O_3$	150 OI 37°C

Compt. Rend., 1955 (241) 987.

0103-007

PROTOCATECHUALDEHYDE, 5,6-diBr-	$C_7H_4O_3Br_2$	294 OI 203°C

Tet., 1967, 1185.

0103-012

10

SALICYLALDEHYDE	$C_7H_6O_2$	**12**2 OI 2^0C	BENZOIC ACID	$C_7H_6O_2$	122 OI 122^0C

SALICYLALDEHYDE $C_7H_6O_2$ **12**2 OI 2^0C

R = H

J. Am. Pharm. Ass., 1934 (23) 332.

0103-013

BENZOIC ACID $C_7H_6O_2$ 122 OI 122^0C

R = H

Ber., 1941 (74) 1617.

0104-001

ANISALDEHYDE, σ - $C_8H_8O_2$ 136 OI 35^0C

R = Me

Chem. Zentr., 1919 (I) 747.

0103-014

PERIPLANETIN $C_{13}H_{16}O_7$ 284 193^0C

R = Glucosyl (1-linked)

Tet., 1959 (5) 10.

0104-002

HELICIN $C_{13}H_{16}O_7$ 284 Amorph.

R = Glucosyl

Chem. Zentr., 1899 (II) 259.

0103-015

VACCINIIN $C_{13}H_{16}O_7$ 284 $+48^0e$ 105^0C

R = Glucosyl (6-linked)

Biochem. Z., 1922 (131) 611.

0104-003

ASARALDEHYDE $C_{10}H_{12}O_4$ 196 OI 114^0C

Chem. Abstr., 1930 (24) 2235.

0103-016

BENZOIC ACID, p-OH- $C_7H_6O_3$ 138 OI 211^0C

R = H

Natwiss., 1955 (42) 583.

0104-004

SALICYLALDEHYDE, 4-OMe- $C_8H_8O_3$ 152 OI 42^0C

Chem. Abstr., 1949 (43) 3567.

0103-017

BENZOIC ACID ME ESTER, -p-OH- $C_8H_8O_3$ 152 OI

R = Me

Chem. Abstr., 1971 (74) 72768.

0104-005

BENZOIC ACID GLUCOSE ESTER, -p-OH-	$C_{13}H_{16}O_8$	300

R = Glucosyl

Z. Nat., 1961 (16B) 249.

0104-006

PICRORHIZIN $C_{20}H_{28}O_{14}$ 492

R = Glucosyl

Chem. Abstr., 1955 (49) 6977.

0104-011

ANISIC ACID $C_8H_8O_3$ 152 OI 184^0C

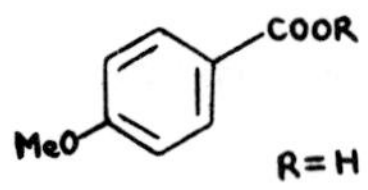

R=H

JCS, 1956, 4259.

0104-007

VERATRIC ACID $C_9H_{10}O_4$ 182 OI 181^0C

R = H

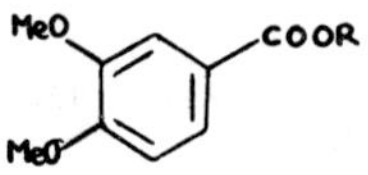

Ann., 1871, 282.

0104-012

ANISIC ACID ME ESTER $C_9H_{10}O_3$ 166 OI 48^0C

R = Me

Biochem. J., 1944 (38) 131.

0104-008

TECOMIN $C_{15}H_{20}O_9$ 344 -178^0p 219^0C

R = Glucosyl

Experientia, 1970, 1187.

0104-013

PROTOCATECHUIC ACID $C_7H_6O_4$ 154 OI 195^0C

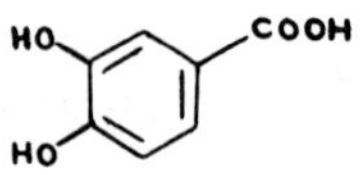

Arch. Pharm., 1955, 441.

0104-009

HESPERALIN $C_{14}H_{22}NO_4^+$ 268 OI

R = -$CH_2.CH_2.NMe_3^+$

Arch. Pharm., 1967 (300) 176.

0104-014

VANILLIC ACID $C_8H_8O_4$ 168 OI 205^0C

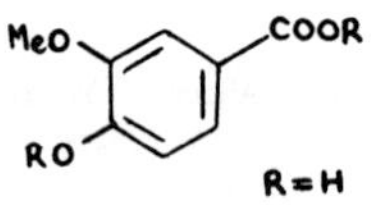

R=H

Chem. Abstr., 1956 (50) 7957.

0104-010

PIPERONYLIC ACID $C_8H_6O_4$ 166 OI 228^0C

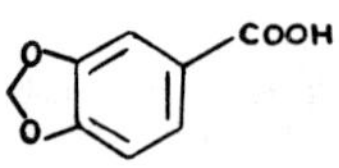

JCS, 1950, 2376.

0104-015

GALLIC ACID	$C_7H_6O_5$	170 OI 240^0C

R = R' = H

Arch. Pharm., 1955, 441.

0104-016

GRANDIFOLINE	$C_{36}H_{53}NO_{12}$	691 -7^0e Amorph.

Acta Chem. Scand., 1971, 1900.

0104-021

GALLIC ACID, 4-0-glucosyl-	$C_{13}H_{16}O_{10}$	332 233^0C

R = Glucosyl; R' = H

Ber., 1919 (52) 820.

0104-017

SALICYLIC ACID	$C_7H_6O_3$	138 OI 155^0C

R = H

Ann., 1843 (48) 60.

0104-022

GLUCOGALLIN	$C_{13}H_{16}O_{10}$	332 -24^0w 202^0C

R = H; R' = Glucosyl

Ber., 1918 (51) 1760.

0104-018

SALICYLIC ACID ME ETHER	$C_8H_8O_3$	152 OI 99^0C

R = Me

Rec. Trav., 1931 (50) 1046.

0104-023

SYRINGIC ACID	$C_9H_{10}O_5$	198 OI 198^0C

Chem. Zentr., 1937 (I) 1304.

0104-019

SALICYLIC ACID ME ESTER	$C_8H_8O_3$	152 OI Oil

R = H

JCS, 1910, 1742.

0104-024

EUDESMIC ACID	$C_{10}H_{12}O_5$	212 OI 168^0C

JCS, 1953, 355.

0104-020

GAULTHERIN	$C_{14}H_{18}O_8$	314 91^0C

R = Glucosyl

Compt. Rend., 1924 (179) 991.

0104-025

VIOLUTIN	$C_{19}H_{26}O_{12}$	446
		-36^0w
		170^0C

R = Vicianosyl

JCS, 1932, 2770.

0104-026

BENZOIC ACID ME ESTER, 2-OH-5-OMe- $C_9H_{10}O_4$ 182   OI   Oil

R = H

Bull. Soc., Chim. Biol., 1936 (18) 1405.

0104-031

BENZOIC ACID, 2,3-di-OH- $C_7H_6O_4$ 154   OI 204^0C

Bull. Soc. Chim. Fr., 1936, 1982.

0104-027

PRIMULAVEROSIDE $C_{20}H_{28}O_{13}$ 476 -20^0w 196^0C

R = primeverosyl

JCS, 1948, 2220.

0104-032

BENZOIC ACID, 3,5-diOH- $C_7H_6O_4$ 154   OI

Nature, 1960, 78.

0104-028

BENZOIC ACID, 2,4-diOH- $C_7H_6O_4$ 154   OI

Nature, 1960, 78.

0104-033

BENZOIC ACID, 2-OH-6-OMe- $C_8H_8O_4$ 168   OI 135^0C

Pharmazie, 1953, 1041.

0104-029

BENZOIC ACID ME ESTER, -2-OH-4-OMe- $C_9H_{10}O_4$ 182   OI 48^0C

R = H

Compt. Rend., 1945 (220) 790.

0104-034

GENTISIC ACID $C_7H_6O_4$ 154   OI 198^0C

Nature, 1960, 78.

0104-030

PRIMVEROSIDE $C_{20}H_{28}O_{13}$ 476 -71^0w 204^0C

R = primeverosyl

Compt. Rend., 1945 (220) 790.

0104-035

BENZOIC ACID, 2,3,4-triOH-	$C_7H_6O_5$	170 OI

Nature, 1960, 78.

0104-036

ELLAGIC ACID 4,4'-DI-ME ETHER	$C_{16}H_{10}O_8$	340 OI 338^0C

R'= H   R = Me

JCS, 1949, 2115.

0105-003

GENTISIC ACID ME ESTER, -5-ME-3-NO$_2$-	$C_9H_9NO_6$	227 OI 137^0C

Phytochem., 1971 (10) 679.

0104-037

NASUTIN-C	$C_{16}H_{10}O_8$	340 OI

R= H   R'= Me

Aust. J. C., 1964, 901.

0105-004

KURAMERINE	$C_{28}H_{44}O_8N^+$	522 Cl: -20^0m pic: 106^0C

Tet., 1969, 2727.

0104-038

NASUTIN-C, glucosyl-	$C_{22}H_{20}O_{13}$	492 $+79^0$m 215^0C

Tet., 1962, 357.

0105-005

ELLAGIC ACID	$C_{14}H_6O_8$	302 OI Sub

Ann., 1955 (591) 156.

0105-001

NASUTIN-B	$C_{17}H_{12}O_8$	354 OI 298^0C

Aust. J. C., 1964, 901.

0105-006

NASUTIN-A	$C_{14}H_{16}O_6$	270 OI $>300^0$C

Aust. J. C., 1964, 901.

0105-002

FLAVELLAGIC ACID, -3,4,3'-O-tri-Me-	$C_{17}H_{22}O_9$	370 OI 279^0C

Tet., 1967, 879.

0105-007

LUTEOIC ACID $C_{14}H_8O_9$ 320

OI

340^0C

Pharm. Acta Helv., 1938 (13) 277.

0105-008

CASTALIGIN $C_{41}H_{26}O_{26}$ 934

Ann., 1969 (721) 186.

0199-005

VALONEAIC ACID $C_{21}H_{14}O_{15}$ 506

Ann., 1955 (591) 156.

0199-001

VESCALIN $C_{27}H_{20}O_{18}$ 632

Ann., 1971 (747) 51.

0199-006

SALICORTIN $C_{20}H_{24}O_{10}$ 524

136^0C

R = Glucosyl

Tet. Lett., 1970, 3827.

0199-002

SALIREPOSIDE $C_{20}H_{22}O_9$ 406

-37^0

206^0C

Natwiss., 1964 (51) 291.

0199-007

TREMULACIN $C_{27}H_{28}O_{11}$ 528

122^0C

R = 2-0-Bz-Glucosyl

Tet. Lett., 1970, 3827.

0199-003

TRICHOCARPIN $C_{20}H_{22}O_9$ 406

-46^0m

135^0C

Natwiss., 1964 (51) 140.

0199-008

---

deleted

0199-004

CORILAGIN $C_{27}H_{22}O_{18}$ 538

-246^0e

205^0C

Ann., 1954 (587) 67.

0199-009

CHEBULINIC ACID — $C_{41}H_{34}O_{28}$ — 974
JCS, 1967, 2381.
0199-010

DIGALLIC ACID, 2,3'-0- — $C_{14}H_{10}O_{10}$ — 338 — OI — 240^0C
Ann., 1952 (578) 34.
0199-015

CHEBULAGIC ACID — $C_{41}H_{32}O_{28}$ — 972 — -57^0w — 240^0C
** linked
JCS, 1967, 2381.
0199-011

DELTOIDIN — $C_{20}H_{22}O_9$ — 406 — 220^0C
Can. J. Chem., 1971 (49) 49.
0199-016

BENZOIC ACID, BENZYL ESTER — $C_{14}H_{12}O_2$ — 212 — OI — 21^0C
Chem. Abstr., 1928 (22) 1597.
0199-012

SALICIN, 6'-SALICYLOYL- — $C_{20}H_{22}O_9$ — 406
Arch. Biochem. Biophys. 1963 (102) 33.
0199-017

PILORUBROSIN — $C_{20}H_{22}O_9$ — 406 — -66^0m — 168^0C
Chem. Comm., 1971, 597.
0199-013

POPULIN — $C_{20}H_{22}O_8$ — 390 — -2^0p — 179^0C
R = 6-Benzoyl-Glucosyl; R' = H
JACS, 1934, 2495.
0199-018

GALLIC ACID, 3-0-galloyl- — $C_{14}H_{10}O_9$ — 322 — OI — 286^0C
Aust. J.C., 1969, 597.
0199-014

POPULIN, salicoyl- — $C_{27}H_{26}O_{10}$ — 510 — -15^0e — 191^0C
R = 6-Benzoyl-Glucosyl
R' = Salicoyl
J. Pharm. Chim., 1942 (2) 289.
0199-019

GLYCOSMIN $C_{22}H_{26}O_{10}$ 450

                                        -36^0e

                                          169^0C

R = 6-Veratroyl-Glucosyl
R' = H   J. Proc. Inst. Chem.
                            (India), 1952 (24) 96.

0199-020

02	C_6C_2	Ar—C—C

0201	Ar—CH=CH$_2$; Ar—Et
0202	Ar—CH$_2$—CH$_2$O·
0203	Ar—CH$_2$CHO ; Ar—CO·Me
0204	Ar—CH$_2$COO·

STYRENE	C_8H_8	104
		Oil
	Chem. Zentr., 1932 (I) 750.	
0201-001		

PYROGALLOL -1, 3- DI- ME ETHER, 5- Et-	$C_{10}H_{14}O_3$	182
		OI
		70^0C
	Ber., 1936 (69) 1870.	
0201-006		

STYRENE, p-OH-	C_8H_8O	120
		OI
		74^0C
	Helv., 1945, 722.	
0201-002		

BARNOL	$C_{10}H_{14}O_3$	182
		OI
		146^0C
	Acta Scand., 1964, 638.	
0201-007		

PHLOROL, p-	$C_8H_{10}O$	122
		OI
	Phytochem., 1968, 278.	
0201-003		

PHLOROL, m-	$C_8H_{10}O$	122
		OI
R = H		
	Phytochem., 1968, 278.	
0201-008		

GUAIACOL, Et-	$C_9H_{12}O_2$	152
		OI
		Oil
	Ind. Soap J., 1946 (12) 77.	
0201-004		

PHLOROL ME ETHER, m-	$C_9H_{12}O$	136
		OI
R = Me		
	Ann., 1873 (170) 345.	
0201-009		

STYRENE, 3-OMe-4-OH-	$C_9H_{10}O_2$	150
		OI
		Oil
	Phytochem., 1968, 278.	
0201-005		

PHLOROL - ISOBUTYRYL ESTER, m-	$C_{12}H_{16}O_2$	192
		OI
R = CO. CHMe$_2$		
	Ann., 1873 (170) 345.	
0201-010		

SARISAN	$C_{10}H_{10}O_3$	178 OI Oil

Chem. Abstr., 1970 (73) 11351.

0201-011

TYROSOL	$C_8H_{10}O_2$	138 OI

R = H

Chem. Abstr., 1955, 16358.

0202-005

PHENETHYL ALCOHOL, β-	$C_8H_{10}O$	122 OI Oil

R = H

Ber., 1900 (33) 1720.

0202-001

SALIDROSIDE	$C_{14}H_{20}O_7$	300 -32^0w 160^0C

R = Glucosyl

Natwiss., 1964 (51) 360.

0202-006

PHENETHYL ALCOHOL ME ETHER	$C_9H_{12}O$	136 OI Oil

R = Me

J. Ind. C. S., 1938 (15) 509.

0202-002

PHENETHYL ALCOHOL, 3,4-di-OH-	$C_8H_{10}O_3$	154 OI 82^0C

Helv., 1950, 1877.

0202-007

PHENETHYL ALCOHOL AMB ESTER	$C_{13}H_{18}O_2$	206

R = ·CO

Crop Sci., 1970, 545.

0202-003

ACETOPHENONE, p-OH-	$C_8H_8O_2$	136 OI 109^0C

R = H

Helv., 1938, 1524.

0203-001

PHENETHYL ALCOHOL ISOVALERYL ESTER	$C_{13}H_{18}O_2$	206 OI

R = $COCH_2CHMe_2$

Crop Sci., 1970, 545.

0202-004

PICEIN	$C_{14}H_{18}O_7$	298 -86^0w 195^0C

R = Glucosyl

JACS, 1942, 693.

0203-002

ACETOPHENONE, p-OMe-	$C_9H_{10}O_2$	150 OI 38^{0}C		ACETOVERATRONE	$C_{10}H_{12}O_3$	180 OI 50^{0}C

ACETOPHENONE, p-OMe- $C_9H_{10}O_2$ 150
                                           OI
                                           38^{0}C

R = Me

Arch. Pharm., 1936, 126.

0203-003

ACETOVERATRONE $C_{10}H_{12}O_3$ 180
                                        OI
                                        50^{0}C

R = Me

Helv., 1949, 2171.

0203-008

ACETOPHENONE, 3,4-di-OH- $C_8H_8O_3$ 152
                                            OI
                                            118^{0}C

Chem. Abstr., 1959
(53) 7330.

0203-004

ACETOPHENONE C_8H_8O 120
                              OI
                              20^{0}C

Chem. Abstr., 1955,
11244.

0203-009

ACETOVANILLONE $C_9H_{10}O_3$ 166
                                     OI
                                     115^{0}C

R = H

Helv., 1949, 1351.

0203-005

ACETOPHENONE, 4-OH-3-
-(3,3-DMA)- $C_{13}H_{16}O_2$ 204
                                    OI
                                    94^{0}C

R =

Ber., 1970, 90.

0203-010

ANDROSIN $C_{15}H_{20}O_8$ 328
                                  224^{0}C

R = Glucosyl

JCS, 1909, 734.

0203-006

ACETOPHENONE, 4-OH-3-
-(3-OH-3,3-DMA)- $C_{14}H_{18}O_3$ 234
                                         OI
                                         124^{0}C

R = CH:CH. C(OH)Me$_2$

Ber., 1970, 90.

0203-011

NEOLLOYDOSIN $C_{20}H_{28}O_{12}$ 460
                                         -52^{0}w
                                         237^{0}C

R = Primeverosyl

Chem. Ind., 1967, 2147.

0203-007

ACETOPHENONE, 4-OH-3-
-isopentadienyl- $C_{13}H_{14}O_2$ 202
                                         OI
                                         138^{0}C

R = CH:CH. C(:CH$_2$)Me

Ber., 1970, 90.

0203-012

ACETOPHENONE, 3-isopentenyl--4-OMe-	$C_{14}H_{18}O_2$	218 OI

Ber., 1970, 90.

0203-013

TREMETONE, OMe-	$C_{14}H_{16}O_3$	232

Ber., 1970, 90.

0203-018

ENCEOLIN, des-OMe-	$C_{13}H_{14}O_2$	202 OI

Ber., 1970, 90.

0203-014

ACETOVANILLOCHROMENE	$C_{14}H_{16}O_3$	232 Oil

Phytochem., 1971 (10)
1665.

0203-019

TREMETONE, dehydro-	$C_{13}H_{12}O_2$	200 OI

Tet., 1964, 1419.

0203-015

PHENYLACETALDEHYDE	C_8H_8O	120 OI 34^0C

Chem. Zentr., 1931
(I) 2128.

0203-020

TREMETONE, (-)-	$C_{13}H_{14}O_2$	202

Tet., 1964, 1419.

0203-016

PHENYLACETIC ACID	$C_8H_8O_2$	136 OI 77^0C

JACS, 1948, 4238.

0204-001

TOXOL	$C_{13}H_{14}O_3$	218 -25^0m 52^0C

Tet., 1964, 1419.

0203-017

MANDELIC ACID, (-)-	$C_8H_8O_3$	152 -130^0w 133^0C

JACS, 1951, 3954.

0204-002

PHENYLACETIC ACID, 2-OH-	$C_8H_8O_3$	152
		OI
		146^0C

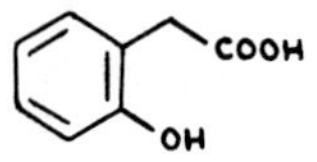

Nature, 1962 (194) 579.

0204-003

GENTISIC ACID, homo-	$C_8H_8O_4$	168
		OI
		153^0C

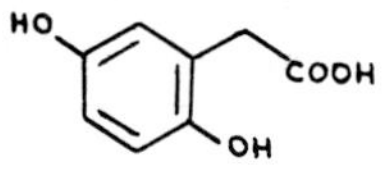

Chem. Abstr., 1951 (45) 216.

0204-006

PHENYLACETIC ACID, 4-OH-	$C_8H_8O_3$	152
		OI
		148^0C

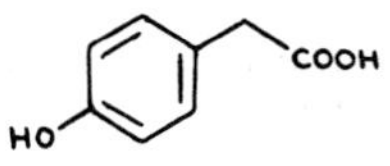

JOC, 1954, 784.

0204-004

PHENYLGLYOXYLIC ACID, -2,5-di-OH-	$C_8H_6O_5$	182
		OI
		141^0C

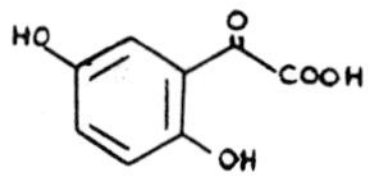

Chem. Ind., 1954, 1143.

0204-007

PHENYLACETIC ACID, 3,4-di-OH-	$C_8H_8O_4$	168
		OI
		129^0C

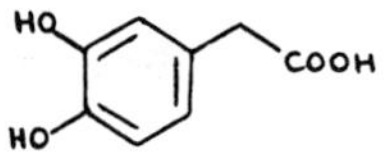

JCS, 1950, 3380.

0204-005

MALVONE	$C_{29}H_{36}O_{19}$	688

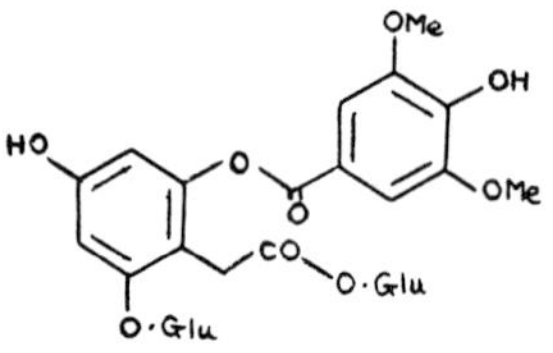

Phytochem., 1970, 1647.

0204-008

03	C_6C_3	Ar-C-C-C

0301	Ar-CH$_2$CH$_2$CH$_3$
	Ar-CH=CH·CH$_3$
	Ar-CH$_2$CH=CH$_2$
0302	Ar-CH$_2$CH$_2$CH$_2$O·
0303	Ar-CH$_2$CH$_2$CHO
0304	Ar-CH$_2$CH$_2$COO·
0399	MISC.

CHAVICOL	$C_9H_{10}O$	134
		OI
		17^0C
	JACS, 1933, 1556.	
0301-001		

ESDRAGOL	$C_{10}H_{12}O$	148	GUAIACOL, propyl-	$C_9H_{12}O_2$	152
		OI			OI
		Oil			Oil
	Compt. Rendu, 1893 (117) 1089.			Phytochem., 1968, 278.	
0301-002			0301-007		

ANETHOL	$C_{10}H_{12}O$	148	CATECHOL, allyl-	$C_9H_{10}O_2$	150
		OI			OI
		22^0C			49^0C
	Ann., 1867 (141) 260.			Chem. Zentr., 1934 (I) 2143.	
0301-003			0301-008		

FOENICULIN	$C_{14}H_{18}O$	202	CHAVIBETOL	$C_{10}H_{12}O_2$	164
		OI			OI
		24^0C			9^0C
	Ber., 1938 (71) 2708.			Chem. Abstr., 1949, 8617.	
0301-004			0301-009		

ANOL ISOVALERYL ESTER	$C_{14}H_{18}O_2$	218	EUGENOL	$C_{10}H_{12}O_2$	164
		OI			OI
		Oil			Oil
	Acta Chem. Scand., 1966, 992.			Ber., 1877 (10) 628. Ann., 1858 (108) 320.	
0301-005			0301-010		

EUGENOL ME ETHER C$_{11}$H$_{14}$O$_2$ 178 OI Oil Ber., 1888 (21) 1057. 0301-011	
   deleted 0301-012	SAFROL C$_{10}$H$_{10}$O$_2$ 162 OI 8^{0}C JOC, 1948, 443. 0301-017
EUGENOL, iso- C$_{10}$H$_{12}$O$_2$ 164 OI 33^{0}C Chem. Abstr., 1955, 6877. 0301-013	SAFROL, iso- C$_{10}$H$_{10}$O$_2$ 162 OI Oil Chem. Zentr., 1937 (II) 4050. 0301-018
EUGENOL ME ETHER, iso- C$_{11}$H$_{14}$O$_2$ 178 OI 17^{0}C Chem. Abstr., 1954, 12378. 0301-014	ACETEUGENOL C$_{12}$H$_{14}$O$_3$ 206 OI 30^{0}C R = Ac Chem. Zentr., 1899 (I) 825. 0301-019
	GEIN C$_{21}$H$_{30}$O$_{11}$ 458 -54^0 183^{0}C R = Vicianosyl JCS, 1949, 2054. 0301-020

EUGENOL, OMe-	$C_{11}H_{14}O_3$	194
		OI
		Oil

Natwiss., 1964 (51) 360.

0301-021

PYROGALLOL-1,3-DIME ETHER, 5-Pr-	$C_{11}H_{16}O_3$	196
		OI
		Oil

Ber., 1888 (21) 2020.

0301-026

ELEMICIN	$C_{12}H_{16}O_3$	208
		OI
		Oil

Aust. J. C., 1968, 3001.

0301-022

PYROGALLOL-1-ME ETHER, -5-Pr-	$C_{10}H_{14}O_3$	182
		OI
		Oil

Monatsh., 1883 (4) 182.

0301-027

ELEMICIN, iso-	$C_{12}H_{16}O_3$	208
		OI
		Oil

Chem. Abstr., 1954, 12378.

0301-023

BENZENE, 1-ALLYL-2,4-DI-OMe-	$C_{11}H_{14}O_2$	178
		OI
		Oil

Pharm. Weekbl., 1971 (106) 182.

0301-028

MYRISTICIN	$C_{11}H_{12}O_3$	192
		OI
		Oil

Bull. Soc. Chim. Fr., 1946, 361.

0301-024

CARPACIN	$C_{11}H_{12}O_3$	192
		OI
		47^0C

Aust. J. C., 1969, 1803.

0301-029

MYRISTICIN, iso-	$C_{11}H_{12}O_3$	192
		OI
		45^0C

Chem. Zentr., 1927 (II) 1518.

0301-025

CROWEACIN	$C_{11}H_{12}O_3$	192
		OI

JCS, 1939, 439.

0301-030

ASARONE, α-	$C_{12}H_{16}O_3$	208 OI 60^0C

Chem. Abstr., 1947, 2776.

0301-031

APIOL, iso-	$C_{12}H_{14}O_4$	222 OI 55^0C

Bull. Soc. Chim. Fr., 1926 (39) 1019.

0301-036

ASARONE, β-	$C_{12}H_{16}O_3$	208 OI Oil

Geometric Isomer

JCS, 1937, 1338.

0301-032

DILLAPIOL	$C_{12}H_{14}O_4$	222 OI Oil

JCS, 1934, 1681.

0301-037

ASARONE, γ -	$C_{12}H_{16}O_3$	208 OI Oil

Aust. J. C., 1968, 3001.

0301-033

CINNAMYL ALCOHOL, dihydro-	$C_9H_{12}O$	136 OI Oil

Arch. Pharm., 1893 (231) 55.

0302-001

ASARONE, 3-OMe-γ -	$C_{13}H_{18}O_4$	238 OI 25^0C

JCS, 1934, 1681.

0301-034

CINNAMYL ALCOHOL	$C_9H_{10}O$	134 OI 34^0C

R = H

JCS, 1926, 2763.

0302-002

APIOL	$C_{12}H_{14}O_4$	222 OI 30^0C

JCS, 1938, 1602.

0301-035

CINNAMYL ACETATE	$C_{11}H_{12}O_2$	176 OI Oil

R = Ac

JCS, 1897, 358.

0302-003

VIMALIN	$C_{16}H_{22}O_7$	326 -61^0m 144^0C

Natwiss., 1964 (51) 217.

0302-004

SYRINGIN	$C_{17}H_{24}O_9$	372 -18^0w 191^0C

R = Glucosyl

Compt. Rend., 1954 (238) 1835.

0302-009

CONIFERYL ALCOHOL	$C_{10}H_{12}O_3$	180 OI 75^0C

Monatsh., 1922 (42) 447.

R = H

0302-005

GUAIACYL-GLYCEROL	$C_{10}H_{14}O_5$	214 Rac

Chem. Ind., 1965, 180.

0302-010

CONIFERIN	$C_{16}H_{22}O_8$	342 -67^0w 185^0C

R = Glucosyl

Ber., 1927 (60) 1031.

0302-006

GLYCOL, C-Syringoyl-	$C_{12}H_{16}O_6$	256

Chem. Abstr., 1971 (74) 50590.

0302-011

DACRINIOL	$C_{15}H_{22}O_3$	250 OI 52^0C

Acta Chem. Scand., 1967 (21) 2247.

0302-007

ZANTHOXYLOL	$C_{14}H_{20}O_2$	220 OI Oil

JCS, 1968, 481.

0302-012

SYRINGENIN	$C_{11}H_{14}O_4$	210 OI 64^0C

Ber., 1951 (84) 67.

R = H

0302-008

CINNAMIC ALDEHYDE	C_9H_8O	132 OI Oil

Ann., 1835 (14) 50.

0303-001

CINNAMIC ALDEHYDE, dihydro-	$C_9H_{10}O$	134 OI Oil	CINNAMIC ALDEHYDE, 3,4-CH_2OH-

CINNAMIC ALDEHYDE, dihydro- $C_9H_{10}O$ 134 OI Oil

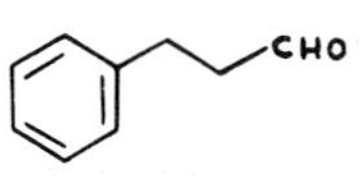

Chem. Zentr., 1902 (II) 1486.

0303-002

CINNAMIC ALDEHYDE, 3,4-CH_2OH- $C_{10}H_8O_3$ 176 QI 85°C

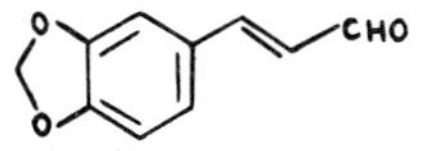

Chem. Abstr., 1942 (36) 6754.

0303-007

CINNAMIC ALDEHYDE, 2-OMe- $C_{10}H_{10}O_2$ 162 OI 46°C

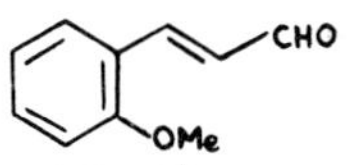

J. Am. Pharm. Ass., 1944 (33) 298.

0303-003

SINAPIC ALDEHYDE $C_{11}H_{12}O_4$ 208 OI 108°C

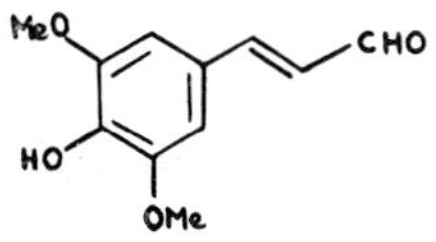

JACS, 1953, 5344.

0303-008

CINNAMIC ALDEHYDE, 4-OMe- $C_{10}H_{10}O_2$ 162 OI 58°C

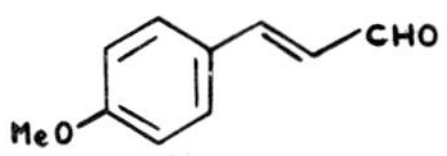

Chem. Zentr., 1908 (I) 1057.

0303-004

CINNAMIC ACID, cis- $C_9H_8O_2$ 148 OI 68°C

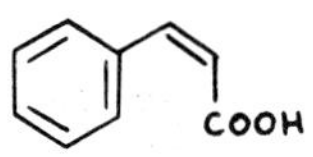

Ber., 1891 (24) 1101.

0304-001

FERULYL ALDEHYDE $C_{10}H_{10}O_3$ 178 OI 84°C

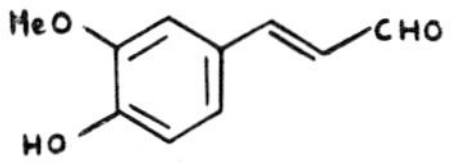

JACS, 1953, 5344.

0303-005

CINNAMIC ACID, trans- $C_9H_8O_2$ 148 OI 133°C

R = H

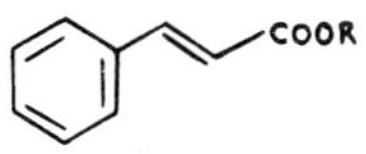

Ann., 1839 (31) 265.

0304-002

PARVIFLORAL $C_{15}H_{18}O_3$ 246 OI 125°C

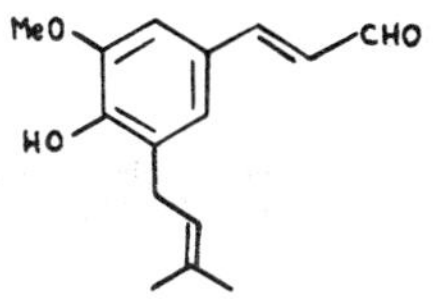

Aust. J.C., 1967, 565.

0303-006

CINNAMIC ACID ME ESTER $C_{10}H_{10}O_2$ 162 OI 36°C

R = Me

Biochem. J., 1940 (34) 82.

0304-003

CINNAMIC ACID AMIDE	C_9H_9NO	147
		OI
		148^0C
R = NH_2		
		J. Biochem. Jap., 1958 (45) 9.
0304-004		

PHLORETIC ACID	$C_9H_{10}O_3$	166
		130^0C
		JACS, 1940, 2422.
0304-009		

MELILOTIC ACID	$C_9H_{10}O_3$	166
		OI
		82^0C
		Ann., 1863 (126) 257.
0304-005		

PHENYPYRUVIC ACID, p-OH-	$C_9H_8O_4$	180
		OI
		Acta Chem. Scand., 1957 (11) 1431.
0304-010		

COUMARIC ACID, o-	$C_9H_8O_3$	164
		OI
		209^0C
R = H		Chem. Zentr., 1950 (II) 2453.
0304-006		

COUMARIC ACID, p-	$C_9H_8O_3$	164
		OI
		208^0C
R = H		JCS, 1909, 243.
0304-011		

MELILOTOSIDE	$C_{15}H_{18}O_8$	326
		-61^0e
		240^0C
R = Glucosyl		J. Pharm. Soc. Jap., 1934 (54) 617.
0304-007		

COUMARIC ACID ME ETHER, p-	$C_{10}H_{10}O_3$	178
		OI
		171^0C
R = Me		JCS, 1910 (97) 1944.
0304-012		

COUMARIC ACID GLUCOSIDE, -cis-ortho-	$C_{15}H_{18}O_8$	326
		216^0C
		Can. J. Biochem. Physiol., 1962 (607) 1763.
0304-008		

COUMARIC ACID ME ESTER, p-	$C_{10}H_{10}O_3$	178
		OI
		138^0C
R = H		Arch. Biochem. Biophys., 1958 (78) 263.
0304-013		

CINNAMIC ACID ME ESTER, p-OMe-	$C_{11}H_{12}O_3$	192 OI 88°C
R = Me		Biochem. J., 1940 (34) 82.
0304-014		

FERULIC ACID	$C_{10}H_{10}O_4$	194 OI 168°C
R = H		Nature, 1955 (176) 1016.
0304-019		

COUMARIC ACID ME ESTER, -4-0-Geranyl-	$C_{20}H_{26}O_3$	314 OI 66°C
R = Geranyl		Aust. J.C. 1966, 451.
0304-015		

FERULIC ACID ME ESTER, -4-0-Geranyl-	$C_{21}H_{28}O_4$	344 OI 61°C
R = Geranyl		Aust. J.C., 1966 (19) 451.
0304-020		

CAFFEIC ACID	$C_9H_8O_4$	180 OI 195°C
R = H		Arch. Biochem. Biophys., 1956 (60) 21.
0304-016		

FERULIC ACID, iso-	$C_{10}H_{10}O_4$	194 OI 228°C
		Chem. Zentr., 1940 (I) 1515.
0304-021		

PHASELIC ACID	$C_{13}H_{12}O_8$	296 +28°w
R =		Gazz. Chim. Ital., 1960 (90) 212.
0304-017		

FERULIC ACID CHOLINE ESTER, iso-	$C_{15}H_{22}NO_4$	280 OI
		Phytochem., 1970, 667.
0304-022		

CAFFEIC ACID, dihydro-	$C_9H_{10}O_4$	182 OI 139°C
		Helv., 1927 (10) 472.
0304-018		

CINNAMIC ACID, 3,4-di-OMe-	$C_{11}H_{12}O_4$	208 OI 180°C
		JCS, 1910 (97) 1944.
0304-023		

CINNAMIC ACID, 2,3-diOH- $C_9H_8O_4$ 180
OI

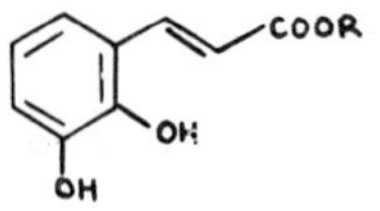

R = H

Planta Med. Phytother.,
1970 (4) 189.

0304-024

CINNAMIC ACID, 2-OH-3-OMe- $C_{10}H_{10}O_4$ 194
OI
184^0C

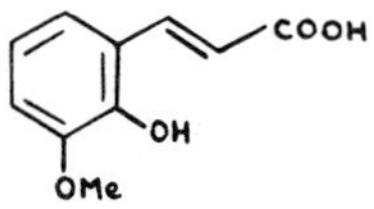

Can. J. Biochem.,
1964 (42) 493.

0304-029

CINNAMIC ACID GLUCOSE
ESTER, 2,3-DIOH- $C_{15}H_{18}O_9$ 342

R = Glucosyl

Planta Med. Phytother.,
1970 (4) 189.

0304-025

SINAPIC ACID $C_{11}H_{12}O_5$ 224
OI
192^0C

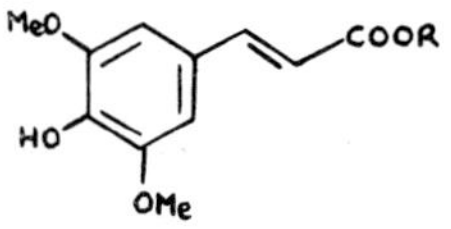

R = H

Chem. Ind., 1954, 1457.

0304-030

CINNAMIC ACID,
-2-0-Glucosyloxy-4-OMe- $C_{16}H_{20}O_9$ 356

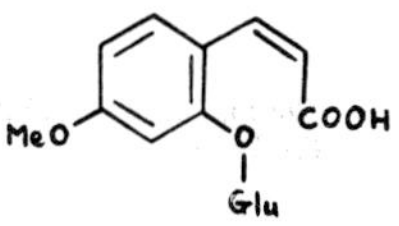

Phytochem., 1963 (2) 137.

0304-026

SINAPIN $C_{16}H_{25}NO_6$ 327

$R = \underset{}{\overset{+}{NMe_3}}$

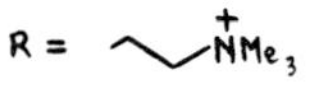

Z. Nat., 1953 (8B) 151.

0304-031

CINNAMIC ACID,
-2,4-di-Glucosyloxy- $C_{21}H_{28}O_{14}$ 504

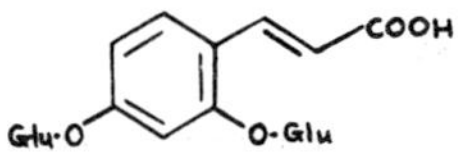

Phytochem., 1965 (4) 255.

0304-027

$C_{10}H_{10}O_4$ 194
OI
83^0C

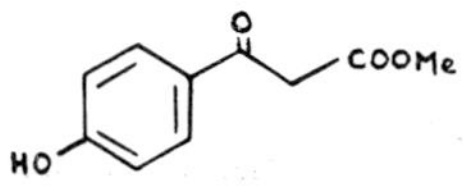

Phytochem., 1971 (10)
475.

0304-032

COUMARIC ACID, Glucosido-Furo- $C_{17}H_{18}O_9$ 366
-61^0w
125^0C

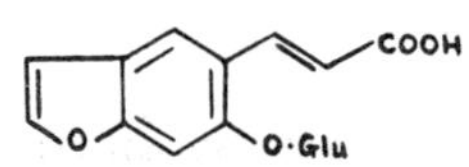

Helv., 1950, 1637.

0304-028

PAJANEELIN $C_{15}H_{18}O_8$ 326
-173^0
238^0C

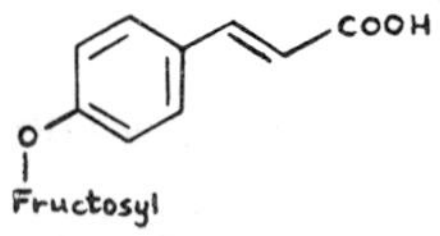

Chem. Abstr., 1947
(41) 4132.

0304-033

34

ANETHOL GLYCOL $C_{10}H_{14}O_3$ 182 115^{0}C Helv., 1939 (22) 382. 0399-001	LATIFOLONE $C_{11}H_{12}O_4$ 208 OI 88^{0}C Bull. Soc. Chim. Biol., 1963 (45) 1119. 0399-006
LASERINE $C_{21}H_{26}O_7$ 390 + 38^{0}c Oil Coll. Czech. Chem. Comm., 1968 (33) 2911. 0399-002	CINNAMYL CINNAMATE $C_{18}H_{16}O_2$ 264 OI 44^{0}C Chem. Zentr., 1910 (II) 1295. 0399-007
ANISYL KETONE $C_{10}H_{12}O_2$ 164 OI Oil Bull. Soc. Chim. Fr., 1902 (27) 990. 0399-003	CONIFERYL BENZOATE $C_{17}H_{16}O_4$ 284 OI 72^{0}C JACS, 1949, 2683. 0399-008
PROPIOVANILLONE, α-OH- $C_{10}H_{12}O_4$ 196 110^{0}C JACS, 1940, 3251. 0399-004	deleted 0399-009
PROPIOSYRINGONE, α-OH- $C_{11}H_{14}O_5$ 226 JACS, 1939, 516. 0399-005	KUTKIN $C_{24}H_{28}O_{11}$ 492 211^{0}C JOC, 1970, 3159. 0399-010

$C_{25}H_{34}O_{11}$ 510

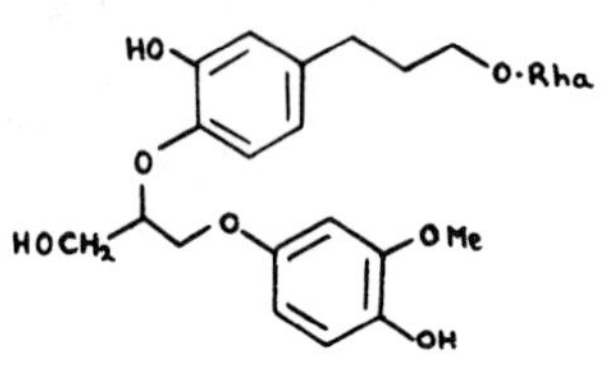

Can. J. C., 1971, 3607.

0399-011

POPULOSIDE $C_{22}H_{24}O_{10}$ 448
-38^0m
187^0C

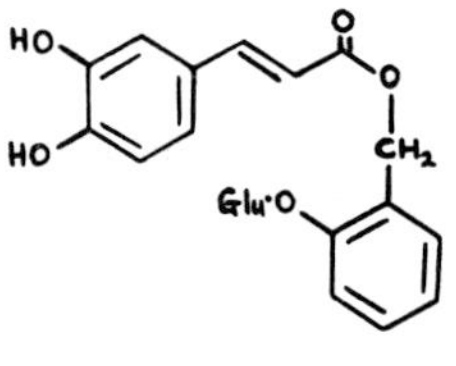

Phytochem., 1970, 857.

0399-016

CHICORIC ACID $C_{22}H_{18}O_{12}$ 474
$+383^0$m
206^0C

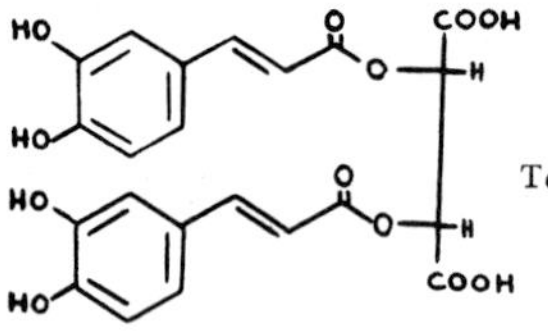

Tet., 1958 (4) 43.

0399-012

ANOLOXIDE $C_{14}H_{18}O_3$ 234

33^0C

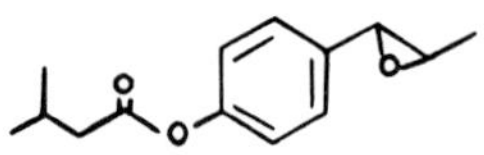

Ber., 1969, 1691.

0399-017

CINNAMIC ACID BENZYL ESTER $C_{16}H_{14}O_2$ 238
OI
39^0C

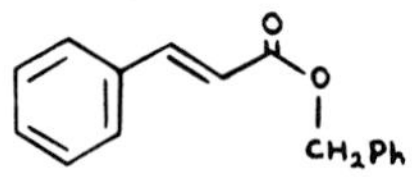

Chem. Zentr., 1902 (I) 445.

0399-013

EUGENOLOXIDE, iso- $C_{15}H_{20}O_4$ 264

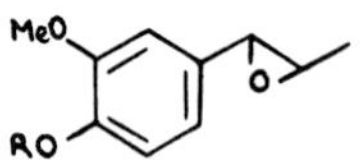

Ber., 1969, 1691.

R = i-Butyryl

0399-018

ROSMARINIC ACID $C_{18}H_{16}O_8$ 360
$+145^0$
204^0C

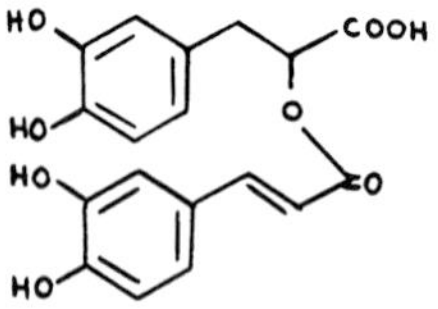

Ric. Sci., 1958 (28) 2329.

0399-014

EUGENOL-O-Angeloyl Ester, —epoxy-iso- $C_{15}H_{18}O_4$ 262
-48^0
Oil

R = Angeloyl

Ber., 1971 (104) 2033.

0399-019

RUBROPILOSIN $C_{22}H_{24}O_9$ 432
-69^0e
97^0C

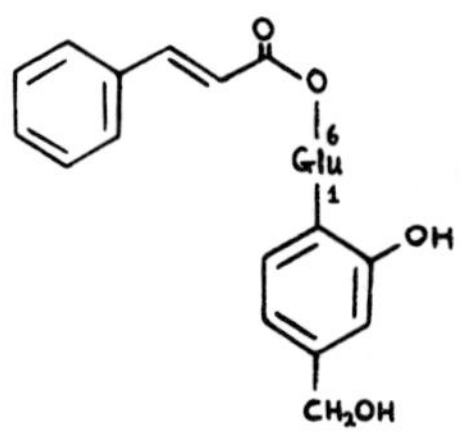

Chem. Comm., 1971, 597.

0399-015

0 4	C_6C_n	$Ar-(C)_n$

0401	$Ar-(C)_4$
0402	$Ar-(C)_5$
0403	$Ar-(C)_6$
0404	$Ar-(C)_7$
0499	$Ar-(C)_n \quad n > 7$

This class of compounds owes its biogenetic origin to the shikimic acid pathway (aromatic ring) and the "acetate" pathway (sidechain). The sidechain may acyclic, heterocyclic, saturated, unsaturated, or substituted.

BENZENE, 1-Bu-**3**, 4-methylene--dioxy-	$C_{11}H_{14}O_2$	178 OI Oil

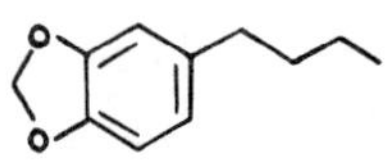

JCS, 1967, 2597.

0401-001

CINNAMYLIDENE ACETONE, -3, 4-METHYLENEDIOXY-	$C_{11}H_{10}O_3$	190 OI 89^0C

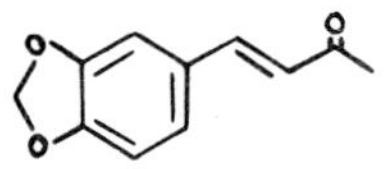

J. Chromatog., 1967, 375.

0401-006

BETULIGENOL	$C_{10}H_{14}O_2$	166 -18^0e 82^0C

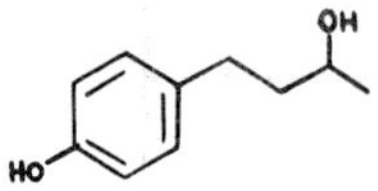

Chem. Abstr., 1951, 4222.

0401-002

BUTYRIC ACID, 4-OH-4-PH-	$C_{10}H_{12}O_3$	180

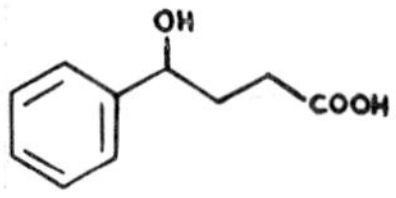

JCS, 1909, 1009.

0401-007

PHENETHYL KETONE, Me-	$C_{10}H_{12}O$	148 OI Oil

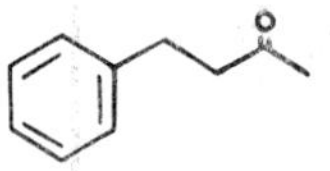

Chem. Abstr., 1939, 2281.

0401-003

deleted

0402-001

CINNAMYLIDENE ACETONE	$C_{10}H_{10}O$	146 OI 64^0C

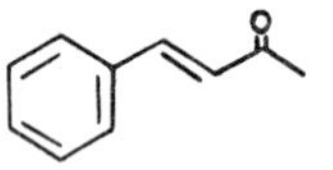

J. Chromatog., 1967, 375.

0401-004

CHAVICIC ACID	$C_{12}H_{10}O_4$	218 OI 201^0C

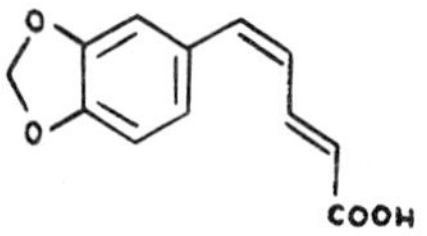

Ann., 1935 (517) 278.

0402-002

ZINGERONE	$C_{11}H_{14}O_3$	194 OI

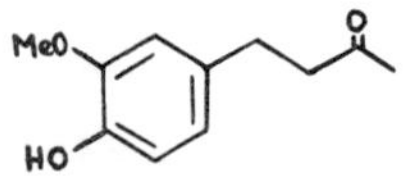

JCS, 1917, 769.

0401-005

CHAVICIC ACID, iso-	$C_{12}H_{10}O_4$	218 OI 202^0C

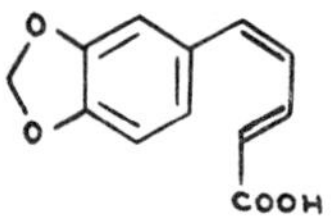

J. Prakt. Chem., 1928 (119) 271.

0402-003

PIPERIC ACID	$C_{12}H_{10}O_4$	218 OI 216^0C

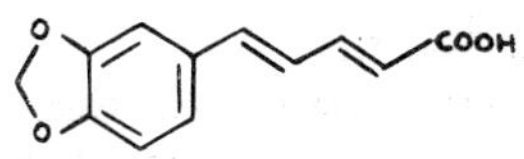

Ber., 1922 (55) 2653.

0402-004

PSILOTIN	$C_{17}H_{20}O_8$	352 -145^0w 130^0C

R = Glucosyl

Tet., 1965, 2939.

0402-009

PIPERIC ACID, tetrahydro-	$C_{12}H_{14}O_4$	222 OI 101^0C

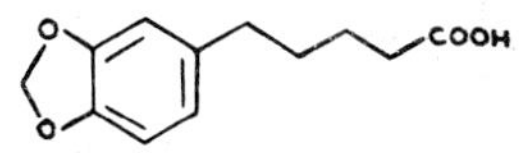

Chem. Abstr., 1949, 1085.

0402-005

PARACOTOIN	$C_{12}H_8O_4$	216 OI 149^0C

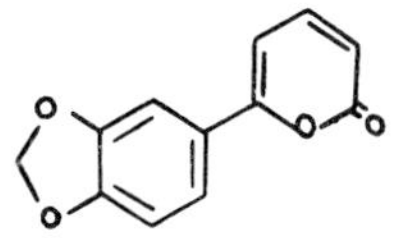

Bull. Soc. Chim. Fr., 1960, 23.

0402-010

COUMALIN, 6-PH-	$C_{11}H_8O_2$	172 OI 68^0C

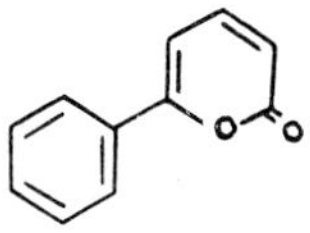

Rec. Trav. Chim., 1927 (46) 594.

0402-006

PARACOTOIN, 4-OMe-	$C_{13}H_{10}O_5$	246 OI 224^0C

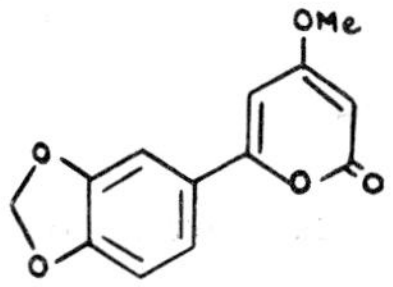

JACS, 1957, 4507.

0402-011

PYRONE, 2,3-DIHYDRO-3-ME- -6 - PH-GAMMA-	$C_{12}H_{12}O_2$	188 69^0C

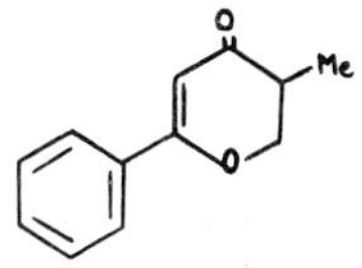

JCS, 1954, 3245.

0402-007

deleted

0404-001

PSILOTININ	$C_{11}H_{10}O_3$	190 Rac 152^0C

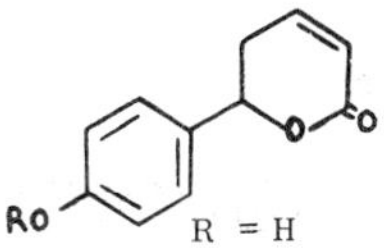

R = H

Tet., 1965, 2939.

0402-008

deleted

0404-002

deleted

0404-003

KAWAIN, des-OMe-dehydro- $C_{13}H_{10}O_2$ 198
OI
116^0C

Tet., 1971, 1043.

0404-008

---

deleted

0404-004

KAWAIN $C_{14}H_{14}O_3$ 230
$+105^0e$
106^0C

Tet. Lett., 1968, 1797.

0404-009

KAWAIC ACID $C_{14}H_{14}O_3$ 230
OI
185^0C

Ber., 1930, 2418.

0404-005

KAWAIN, dehydro- $C_{14}H_{12}O_3$ 228
OI
140^0C

JOC, 1959 (24) 1829.

0404-010

KAWAIC ACID, dihydro- $C_{14}H_{16}O_3$ 232
OI
140^0C

Ber., 1930, 2414.

0404-006

KAWAIN, dihydro-5,6-dehydro- $C_{14}H_{14}O_3$ 230
OI
96^0C

*   dihydro

Chem. Abstr., 1967, 21775.

0404-011

KAWAIN, des-OMe- $C_{13}H_{12}O_2$ 200
$+135^0m$
85^0C

Aust. J.C., 1967, 2199.

Stereochemistry (?)

0404-007

MARINDININ $C_{14}H_{16}O_3$ 232
$+30^0e$
60^0C

Rec. Trav. Chim., 1939 (58) 521.

0404-012

KAWAIN-5-OL, dihydro- $C_{14}H_{16}O_4$ 248
+73°c
92°C

Tet. Lett., 1970, 3259.

0404-013

YANGONIN, 11-OMe-Nor- $C_{15}H_{14}O_5$ 274
OI
218°C

Chem. Abstr., 1966,
15307.

0404-018

MUNDULEA LACTONE $C_{19}H_{20}O_3$ 296
OI
104°C

Chem. Comm., 1967,
577.

0404-014

METHYSTICIN $C_{15}H_{14}O_5$ 274
+9°a
136°C

JOC, 1959, 1829.

0404-019

YANGONIN $C_{15}H_{14}O_4$ 258
OI
152°C

R = Me

Tet., 1958 (4) 135.

0404-015

METHYSTICIN, dihydro- $C_{15}H_{16}O_5$ 276
+21°
116°C

*   dihydro

Ber., 1929 (62) 360.

0404-020

YANGONIN, bis-nor- $C_{13}H_{10}O_4$ 230
OI
242°C

R = H

Chem. Abstr., 1969
(70) 857.

0404-016

$C_{14}H_{12}O_4$ 244
OI
160°C

Tet., 1971, 1043.

0404-021

HISPIDIN $C_{13}H_{10}O_5$ 246
OI
259°C

JCS, 1962, 2085.

0404-017

$C_{14}H_{10}O_4$ 242
OI
174°C

Tet., 1971, 1043.

0404-022

PIPEROLIDE	$C_{15}H_{14}O_4$	258
		OI
		111^0C

Phytochem., 1971 (10) 1627.

0404-023

GINGEROL, 8-	$C_{19}H_{30}O_4$	322

R = H ; n = 7

Aust. J. C., 1969, 1033.

0499-004

YANGONIN, (+)-5,6,7,8- -tetrahydro-	$C_{15}H_{18}O_4$	262
		$+20^0c$
		90^0C

Ber., 1971 (104) 2688.

0404-024

GINGEROL, 10-	$C_{21}H_{34}O_4$	350

R = H ; n = 9

Aust. J. C., 1969, 1033.

0499-005

SHOGAOL	$C_{17}H_{24}O_3$	276
		OI
		Oil

Chem. Abstr., 1930 (24) 2445.

0499-001

BENZENE, 1-UNDECENYL-3, -4-METHYLENE-DIOXY-	$C_{18}H_{26}O_2$	274
		OI
		38^0C

Chem. Abstr., 1949, 1085.

0499-006

GINGEROL, 6-	$C_{17}H_{26}O_4$	294

R = H ; n = 5

Aust. J. C., 1969, 1033.

0499-002

PIPATALINE	$C_{19}H_{28}O_2$	288
		OI
		$37^0 C$

Chem. Ind., 1967, 2173.

0499-007

GINEROL, Me-6-	$C_{18}H_{28}O_4$	308

R = Me ; n = 5

Aust. J. C., 1969, 1033.

0499-003

ASPERYELLONE	$C_{20}H_{22}O$	278
		OI

Tet. Lett., 1969, 4049.

0499-008

CORTISALIN	$C_{21}H_{20}O_3$	320 OI 290°C	CRYPTOCARYALACTONE	$C_{17}H_{18}O_4$	286 +16°c 125°C

CORTISALIN — $C_{21}H_{20}O_3$ — 320 — OI — 290°C

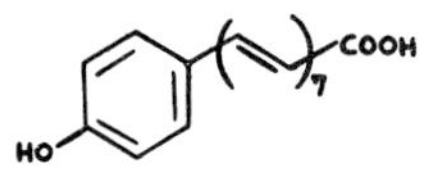

Acta Chem. Scand., 1952 (6) 580.

0499-009

CRYPTOCARYALACTONE — $C_{17}H_{18}O_4$ — 286 — +16°c — 125°C

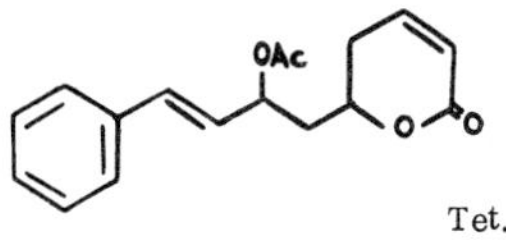

Tet. Lett., 1971, 3401.

0499-011

THITSIOL — $C_{23}H_{36}O_2$ — 344 — OI

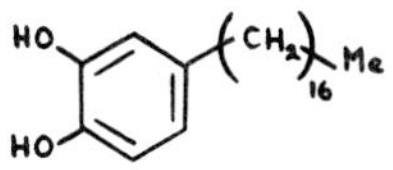

Ber., 1922 (55) 191.

0499-010

C_6C_n

0 5	COUMARINS		

0501		0507	
0502		0508	
0503		0509	
0504		0510	
0505		0511	
0506		0599	MISC.

45

UMBELLIFERONE R = H 0501-001	$C_9H_6O_3$ 162 OI 228^0C Nature, 1948 (161) 400.	AURAPTENE, oxo- R = 0501-006	$C_{19}H_{20}O_4$ 312 OI 64^0C Ber., 1970 (103) 3619.

HERNIARIN R = Me 0501-002	$C_{10}H_8O_3$ 176 OI 118^0C Helv., 1939 (22) 382.	MARMIN R = 0501-007	$C_{19}H_{24}O_5$ 332 $+25^0e$ 124^0C Tet., 1967, 2526.

SKIMMIN R = Glucosyl 0501-003	$C_{15}H_{16}O_8$ 342 -80^0p 220^0C Phytochemistry, 1965 (4) 255.	GEIPARVARIN R = -CH$_2$ 0501-008	$C_{19}H_{18}O_5$ 326 OI 160^0C Aust. J. C., 1967, 1943.

UMBELLIFERONE, (3,3-DMA)- R = 3,3-di-Me-Allyl 0501-004	$C_{14}H_{14}O_3$ 230 OI Tet. Lett., 1968, 3947.	R = 0501-009	$C_{19}H_{18}O_5$ 326 126^0C Ber., 1970 (103) 3619.

AURAPTENE R = Geranyl 0501-005	$C_{19}H_{22}O_3$ 298 OI 68^0C Ber., 1970 (103) 3619.	R = 0501-010	$C_{19}H_{18}O_6$ 342 125^0C Ber., 1970 (103) 3619.

UMBELLIPRENIN	$C_{24}H_{30}O_3$	366
		OI
		63^0C
R = farnesyl		
	Ber., 1938 (71) 1667.	
0501-011		

SAMARKANDONE	$C_{24}H_{30}O_5$	398
		$+25^0$e
		216^0C
R =		
	Chem. Abstr., 1968 (69) 59040.	
0501-016		

GALBANUM ACID	$C_{25}H_{32}O_5$	412
R =		
	Chem. Abstr., 1966 (64) 3461.	
0501-012		

KARATAVIC ACID	$C_{24}H_{28}O_5$	396
		-105^0e
		111^0C
R =		
	Khim. Priv. Soedin., 1968 (5) 283.	
0501-017		

GUMMOSINE	$C_{24}H_{30}O_4$	382
		-54^0
		167^0C
R =		
	Chem. Abstr., 1967 (67) 11394.	
0501-013		

HERNIARIN, 3-(1,1,-DMA)-	$C_{15}H_{16}O_3$	244
		127^0C
	Tet. Lett., 1968, 4395.	
0501-018		

BADRAKEMIN	$C_{24}H_{30}O_4$	382
		-64^0chf
		199^0C
R =		
	Chem. Abstr., 1968 (69) 35865.	
0501-014		

SUBEROSIN, des-Me-	$C_{14}H_{14}O_3$	230
		OI
		81^0C
R = 3,3-DMA		
	JCS, 1954, 1392.	
0501-019		

SAMARKANDINE	$C_{24}H_{32}O_5$	400
		$+30^0$
		176^0C
R =		
	Chem. Abstr., 1968 (69) 59040.	
0501-015		

PEUCEDANOL	$C_{14}H_{16}O_5$	264
		$+31^0$e
		174^0C
R =		
	Chem. Abstr., 1968 (69) 96521.	
0501-020		

ACETOGENINS, SHIKIMATES, AND CARBOHYDRATES

OSTRUTHIN	$C_{19}H_{22}O_3$	298 OI 119°C	**MICROMELIN**	$C_{15}H_{12}O_6$	288 218°C

R = Geranyl

Ber., 1942 (75) 1623.

0501-021

R =

Aust. J. C., 1967, 973.

0501-026

SUBEROSIN $C_{15}H_{16}O_3$ 244 OI 88°C

R = 3, 3-DMA

JCS, 1954, 1392.

0501-022

ANGELOL $C_{20}H_{24}O_7$ 376 -95°c 105°C

R =

Tet. Lett., 1965, 4559.

0501-027

GEIJERIN $C_{15}H_{16}O_4$ 260 OI 121°C

R = Isovaleryl

Aust. J. C., 1967, 1943.

0501-023

OSTHENOL $C_{14}H_{14}O_3$ 230 OI 124°C

R = H

Ber., 1937 (70) 1023.

0501-028

GEIJERIN, dehydro- $C_{15}H_{14}O_4$ 258 OI 130°C

R = Senecioyl

Aust. J. C., 1967, 1943.

0501-024

VELLEIN $C_{20}H_{24}O_8$ 392 188°C

R = Glucosyl

Aust. J. Sci. Res., 1951 (4A) 107.

0501-029

THAMNOSMIN $C_{15}H_{14}O_4$ 258 102°C

R =

Tet. Lett., 1969, 1845.

0501-025

OSTHOL $C_{15}H_{16}O_3$ 244 OI 84°C

R = 3, 3-DMA

Ann., 1932 (495) 187.

0501-030

MERANZIN	$C_{15}H_{16}O_4$	260 -33^0e 98^0C

R =

Arch. Pharm., 1941
(279) 213.

0501-031

ANGELICAL	$C_{11}H_8O_4$	204 OI 257^0C

R = CHO

J. Pharm. Soc. Jap.,
1956 (76) 666.

0501-036

MICROPUBESCIN	$C_{15}H_{14}O_4$	258 OI 130^0C

R =

Chem. Abstr., 1968
(69) 2887.

0501-032

ULOPTEROL	$C_{15}H_{18}O_5$	278

Khim. Prir. Soedin.,
1970 (6) 300.

0501-037

AURAPTENOL	$C_{15}H_{16}O_4$	260 $+14^0$e 109^0C

R =

Tet., 1965, 89.

0501-033

GRAVELLIFERONE	$C_{19}H_{22}O_3$	298 OI 166^0C

Experientia, 1968 (24)
992.

0501-038

PHEBALOSIN	$C_{15}H_{14}O_4$	258 120^0C

R =

Aust. J.C., 1966, 483.

0501-034

GRAVELLIFERONE ME ETHER	$C_{20}H_{24}O_3$	312 OI 71^0C

Tet. Lett., 1968, 4395.

0501-039

	$C_{15}H_{16}O_4$	260 OI 138^0C

R =

Tet., 1966, 1489.

0501-035

PRANFERIN	$C_{18}H_{22}O_5$	318

Khim. Prir. Soedin,
1971 (7) 255.

0501-040

NODAKENETIN R = H 0502-001	$C_{14}H_{14}O_4$	256 -25^0c 192^0C Ber., 1939 (72) 2089.
PRANTSCHIMGIN R = Senecioyl 0502-006	$C_{19}H_{22}O_5$	330 -23^0chf 139^0C Chem. Abstr., 1967 (66) 2438.
NODAKENIN R = Glucosyl 0502-002	$C_{20}H_{24}O_9$	418 $+57^0$w 216^0C Ber., 1937 (72) 2089.
FELAMEDIN R = Benzoyl 0502-007	$C_{21}H_{18}O_5$	350 -99^0chf 133^0C Z. Nat., 1967, 1231.
MARMESIN, (+)- R = H 0502-003	$C_{14}H_{14}O_4$	256 $+27^0$chf 190^0C Tet. Lett., 1968, 5463.
SESELIFLORIN 0502-008	$C_{18}H_{20}O_5S$	348 Khim. Prir. Soedin., 1970 (6) 522.
AMMAJIN R = Glucosyl 0502-004	$C_{20}H_{24}O_9$	418 -29^0c 216^0C JOC, 1961, 161.
PRANGOSINE 0502-009	$C_{14}H_{13}NO_3$	243 OI 131^0C Chem. Abstr., 1967 (67) 54284.
DELTOIN R = Angeloyl 0502-005	$C_{19}H_{22}O_5$	330 -38^0c 105^0C Acta Chem. Scand., 1970, 2863.
PEUCEDANIN 0502-010	$C_{15}H_{14}O_4$	258 OI 109^0C Ber., 1939 (72) 52.

SMIRNIORIN	$C_{18}H_{18}O_7$	346 −138° 144°C	RUTAMARIN	$C_{21}H_{24}O_5$	352 +14°c 108°C

R = Ac

R = Ac

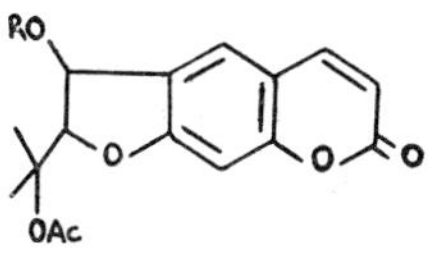

Khim. Prir. Soedin., 1969 (5) 592.

Lloydia, 1967, 73.

0502-011

0502-016

SMYRNIORIDIN $C_{21}H_{22}O_7$ 386 −229°c 127°C

CHALEPENSIN $C_{16}H_{14}O_3$ 254 OI 85°C

R = Angeloyl

Khim. Prir. Soedin., 1970 (6) 185.

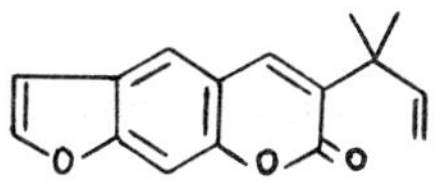

Tet. Lett., 1968, 4395.

0502-012

0502-017

PSORALENE $C_{11}H_6O_3$ 186 OI 163°C

PSORALENE, 4,5'8-tri-Me- $C_{14}H_{12}O_3$ 228 OI 227°C

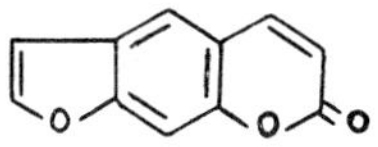

Helv., 1950 (33) 1637.

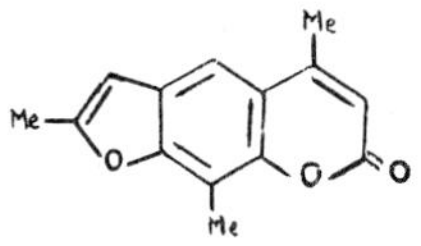

Biochemistry, 1963 (2) 1127.

0502-013

0502-018

CHALEPIN $C_{19}H_{22}O_4$ 314 +31°c 118°C

XANTHYLETIN $C_{14}H_{12}O_3$ 228 OI 130°C

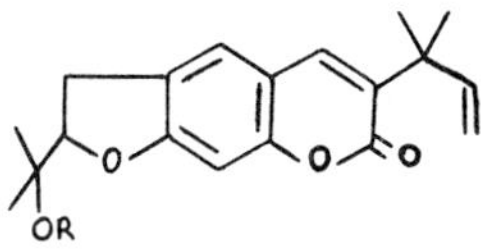

R = H

Lloydia, 1967, 73.

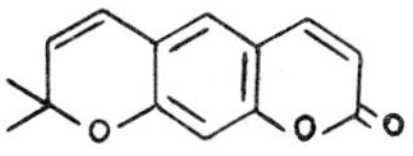

JCS, 1954, 1392.

0502-014

0503-001

HELIETTIN $C_{19}H_{22}O_4$ 314 0° 166°C

DECURSINOL $C_{14}H_{14}O_4$ 246 +104°p

R = H

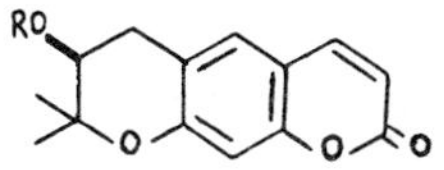

(Racemate ?)

R = H

Tet., 1967 (23) 1129.

Tet. Lett., 1969, 3.

0502-015

0503-002

DECURSIN R = Senecioyl Tet. Lett., 1966, 1461. 0503-003	$C_{19}H_{20}O_5$ 328 $+173^0 c$ $110^0 C$	$C_{14}H_{12}O_4$ 244 OI $178^0 C$ JCS, 1967, 2593. 0503-008
DECURSINOL, 3'-epi- R = H Tet. Lett., 1969, 3. 0503-004	$C_{14}H_{14}O_4$ 246 $-11^0 c$ $181^0 C$	MAJURIN Tet. Lett., 1971, 1657. 0504-001	$C_{14}H_{12}O_3$ 228 $96^0 C$
DECURSINOL, 0-angeloyl-3'-epi- R = Angeloyl Acta Chem. Scand., 1966, 2497. 0503-005	$C_{19}H_{20}O_5$ 328	DISCOPHORIDIN Aust. J. C., 1963 (16) 143. 0504-002	$C_{14}H_{12}O_4$ 244 $+20^0 e$ $135^0 C$
ANDELIN Khim. Prir. Soedin., 1970 (6) 190. 0503-006	$C_{24}H_{26}O_7$ 426 $-28^0 c$	OROSELOL Tet., 1964, 2605. 0504-003	$C_{14}H_{12}O_4$ 244 OI $150^0 C$
XANTHYLETIN, 3-(1,1-DMA)-8- -(3,3-DMA)- Tet. Lett., 1970, 4305. 0503-007	$C_{24}H_{28}O_3$ 364	COLUMBIANIN R = Glucosyl J. Pharm. Sci., 1968, 865. 0504-004	$C_{20}H_{24}O_9$ 408 $+118^0 w$ $276^0 C$

COLUMBIANIDIN	$C_{19}H_{20}O_5$	328 +227°c 121°C
R = Tigloyl		
	Acta Chem. Scand., 1964 (18) 2111.	
0504-005		

EDULTIN	$C_{21}H_{22}O_7$	386
R = Tigloyl R' = Ac		
	Tet. Lett., 1966, 4735.	
0504-010		

LIBANORIN	$C_{19}H_{20}O_5$	328 78°C
R = Senecioyl		
	Khim. Prir. Soedin., 1969, 222.	
0504-006		

OROSELOL, 9-OAc-O-Ac-dihydro-	$C_{18}H_{18}O_7$	346 123°C
R = R' = Ac		
	Ber., 1969 (102) 1673.	
0504-011		

COLUMBIANADIN, oxy-	$C_{19}H_{20}O_6$	344 +305°m 97°C
R = Epoxy-tigloyl		
	Acta Chem. Scand., 1964 (18) 2111.	
0504-007		

OROSELOL, 9-OAc-O-i-val-dihydro-	$C_{21}H_{24}O_7$	388 Oil
R = Ac R' = Isovaleryl		
	Ber., 1969, 1673.	
0504-012		

VAGINOL	$C_{14}H_{14}O_5$	262
R = R' = H		

	Ind. J. C., 1970 (8) 200.	
0504-008		

PEUCENIDIN	$C_{21}H_{22}O_7$	386 -48°c 126°C
R = Ac R' = Senecioyl		
	Acta Chem. Scand., 1970, 2893.	
0504-013		

VAGINIDIN	$C_{19}H_{22}O_6$	346 +172°m 139°C
R = Isovaleryl R' = H		
	Acta Chem. Scand., 1970, 2893.	
0504-009		

ATHAMANTIN	$C_{24}H_{30}O_7$	430 +96°m 60°C
R = R' = Isovaleryl		
	Tet. Lett., 1966, 4735.	
0504-014		

ARCHANGELICIN	$C_{24}H_{26}O_7$	426
		$+113^0m$
		101^0C
R = R' = Angeloyl		
	Acta Chem. Scand., 1964 (18) 932.	
0504-015		

SELINIDIN	$C_{19}H_{20}O_5$	328
		$+ 20^0d$
		98^0C
R = Angeloyl		
	Tet., 1967, 1883.	
0505-004		

ANGELICIN	$C_{11}H_6O_3$	186
		OI
		140^0C

	Tet., 1958 (2) 203; (4) 256.	
0504-016		

LOMATIN, hexanoyl-	$C_{20}H_{24}O_5$	344
		$+30^0t$
		Oil
R = Hexanoyl		
	Acta Chem. Scand., 1970, 2863.	
0505-005		

SESELIN	$C_{14}H_{12}O_3$	228
		OI
		120^0C

	Ber., 1939 (72) 963.	
0505-001		

LOMATIN, octanoyl-	$C_{22}H_{28}O_5$	372
		$+32^0t$
		Oil
R = Octanoyl		
	Acta Chem. Scand., 1970, 2863.	
0505-006		

LOMATIN	$C_{14}H_{14}O_4$	246
		$+52^0e$
		183^0C
R=H		

	Tet. Lett., 1969, 3.	
0505-002		

LOMATIN, cis-4-octenyl-	$C_{22}H_{26}O_5$	370
		$+31^0t$
		Oil
R = cis-4-octenyl		
	Acta Chem. Scand., 1970, 2863.	
0505-007		

NUTTALIN	$C_{19}H_{20}O_5$	328
		$+75^0c$
		59^0C
R = Senecioyl		
	J. Pharm. Sci., 1968, 865.	
0505-003		

KHELLACTONE, (+)-cis-	$C_{14}H_{14}O_5$	262
		$+81^0c$
		174^0C
R = R' = H		

	Acta Chem. Scand., 1971, 529.	
0505-008		

PTERYXIN, iso-	$C_{21}H_{22}O_7$	386		PTERYXIN, epoxy-	$C_{21}H_{22}O_8$	402

PTERYXIN, iso- $C_{21}H_{22}O_7$ 386

R = Angeloyl
R' = Ac

J. Pharm. Sci., 1967 (56) 184.

0505-009

PTERYXIN, epoxy- $C_{21}H_{22}O_8$ 402

Oil

R = Ac
R' = epoxy-angeloyl

Ber., 1970 (103) 3619.

0505-014

SAMIDIN $C_{21}H_{22}O_7$ 386
+49°c
138°C

R = Senecioyl
R' = Ac

Tet. Lett., 1969, 3365.

0505-010

KHELLACTONE, 3'-OAc, -4'-O-senecioyl-(+)-cis- $C_{21}H_{22}O_7$ 386
-12°e
121°C

R = Ac
R' = Senecioyl

Acta Chem. Scand., 1966, 2496.

0505-015

SAMIDIN, dihydro- $C_{21}H_{24}O_7$ 388
+19°c
118°C

R = Isovaleryl
R' = Ac

Ber., 1959 (92) 2338.

0505-011

SUKSDORFIN $C_{21}H_{24}O_7$ 388
+4°e
140°C

R = Ac
R' = Isovaleryl

Tet., 1967, 1235.

0505-016

VISNADIN $C_{21}H_{24}O_7$ 388
+9°e
85°C

R = AMB
R' = Ac

Tet. Lett., 1969, 3365.

0505-012

ANOMALIN, (-)- $C_{24}H_{26}O_7$ 426
-78°e
174°C

R = R' = Angeloyl

Chem. Pharm. Bull., 1966, 94.

0505-017

PTERYXIN $C_{21}H_{22}O_7$ 386
+10°e
82°C

R = Ac
R' = Angeloyl

J. Pharm. Sci., 1962 (51) 149.

0505-013

CALIPTERYXIN $C_{24}H_{26}O_7$ 426

R = Angeloyl
R' = Senecioyl

J. Pharm. Sci., 1967 (56) 184.

0505-018

SESELIN, 3'-senecioyloxy- R = Senecioyl R = Angeloyl 0505-019	$C_{24}H_{26}O_7$ 426 +18^{0}c 79^{0}C Acta Chem. Scand., 1971, 530.	$C_{24}H_{26}O_9$ 458 179^{0}C R, R' = epoxy-Angeloyl and/or epoxy-Tigloyl Tet. Lett., 1970, 3577. 0505-024
KHELLACTONE, 3', 4'-di-O- -senecioyl-(+)-cis- R = R' = Senecioyl 0505-020	$C_{24}H_{26}O_7$ 426 +16^{0}c 108^{0}C Acta Chem. Scand., 1966, 2497.	$C_{24}H_{26}O_9$ 458 211^{0}C isomer of 024 Tet. Lett., 1970, 3577. 0505-025
SESELIN, 3'-isovaleryloxy- R = Isovaleryl R' = Angeloyl 0505-021	$C_{24}H_{28}O_7$ 428 +21^{0}c Acta Chem. Scand., 1971. 529.	PELIFORMOSIN $C_{24}H_{26}O_7$ 426 +67^{0}c 155^{0}C R = Angeloyl ; R' = Senecioyl Tet. Lett., 1969, 3365. 0505-026
FLOROSELIN R = Tigloyl R' = . CO. CH = CH. SMe 0505-022	$C_{23}H_{24}O_7S$ 444 Khim. Prir. Soedin., 1970, 517.	ANOMALIN, (+)- $C_{24}H_{26}O_7$ 426 R = R' = Angeloyl J. Pharm. Soc. Jap., 1968 (88) 513. 0505-027
R = Angeloyl R' = epoxy-Angeloyl 0505-023	$C_{24}H_{26}O_8$ 442 Tet. Lett., 1970, 3577.	DAPHNETIN $C_9H_6O_4$ 178 OI 256^{0}C R = H Ber., 1884 (17) 933. 0506-001

DAPHNIN	$C_{15}H_{16}O_9$	340 -115^0m 215^0C

R = Glucosyl

Chem. Abstr., 1930
(24) 4787.

0506-002

GRAVELLIFERONE, 8-OMe-	$C_{20}H_{24}O_4$	328 OI 132^0C

Tet. Lett., 1970, 4305.

0506-007

COLLININ	$C_{20}H_{24}O_4$	328 OI 68^0C

Aust. J. Sci. Res.,
1949 (A2) 127.

0506-003

LUVANGETIN	$C_{15}H_{14}O_4$	258 OI 108^0C

Ber., 1939 (72) 1450.

0506-008

BROSIPARIN	$C_{15}H_{16}O_4$	260 OI

Chem. Abstr., 1971
(74) 72458.

0506-004

RUTARETIN	$C_{14}H_{14}O_5$	262 -34^0c 192^0C

Arch. Pharm., 1967
(300) 73.

0506-009

SABANDINONE	$C_{10}H_6O_4$	290 OI 154^0C

An. Quim., 1970 (66)
1017.

0506-005

XANTHOTOXOL	$C_{11}H_{16}O_4$	212 OI 250^0C

R = H

Ber., 1937 (70) 248.

0506-010

DAPHNETIN DI-ME ETHER, -3-(1,1-DMA)-	$C_{16}H_{18}O_4$	274 OI 86^0C

Tet. Lett., 1970, 4305.

0506-006

XANTHOTOXIN	$C_{12}H_8O_4$	226 OI 145^0C

R = Me

JACS, 1950, 4826.

0506-011

ACETOGENINS, SHIKIMATES, AND CARBOHYDRATES

PRANGENIN	$C_{15}H_{14}O_4$	258 OI 97°C
R = n-Bu		
	JACS, 1935, 2563.	
0506-012		

HERACLENOL	$C_{16}H_{16}O_6$	304 +16°p 118°C
R =		
	Natwiss., 1964 (51) 537.	
0506-017		

IMPERATORIN	$C_{16}H_{14}O_4$	270 OI 104°C
R = 3,3-DMA		
	Chem. Abstr., 1956, 11616.	
0506-013		

HERACLENOL, O-Me-	$C_{17}H_{18}O_6$	318
R =		
	Ind. J. C., 1970 (8) 855.	
0506-018		

HERACLENIN, (+)-	$C_{16}H_{14}O_5$	286 +25°p 108°C
R = epoxy-3,3-DMA		
	Tet., 1964, 87.	
0506-014		

HELACLENOL, Glucosyl-	$C_{22}H_{26}O_{11}$	466
R =		
	Ind. J. C., 1970 (8) 1146.	
0506-019		

HERACLENIN, (-)-	$C_{16}H_{14}O_5$	286 -24°p 107°C
R = epoxy-3,3-DMA		
	Aust. J. C., 1966, 483.	
0506-015		

HERACLENIN, Iso-	$C_{16}H_{14}O_4$	286 OI 100°C
R = $CH_2.CO.CMe_2$		
	Chem. Ind., 1970, 746. Ind. J. Chem., 1970 (8) 855.	
0506-020		

HERACLENIN, (±)-	$C_{16}H_{14}O_5$	286 OI 114°C
Racemate		
	Aust. J. C., 1966, 483.	
0506-016		

XANTHOTOXOL, O-geranyl-	$C_{21}H_{24}O_4$	340 OI 53°C
R = Geranyl		
	Tet., 1966, 3221.	
0506-021		

58

BENAHORIN	$C_{17}H_{16}O_4$	284

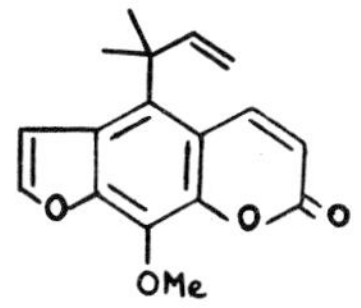

An. Quim., 1971 (67) 441.

0506-022

CICHORIIN	$C_{15}H_{16}O_9$	340
		215^0C

R = Glucosyl; R' = H

JCS, 1939, 1266.

0507-003

IMPERATORIN ME ETHER, - epoxy-allo- $C_{17}H_{16}O_5$ 300 104^0C

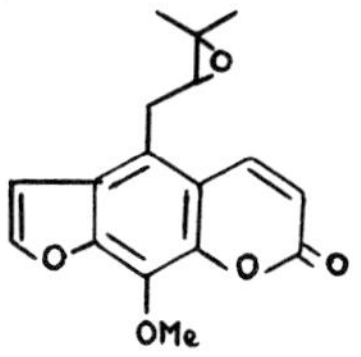

Tet. Lett., 1969, 1845.

0506-023

PRENYLETIN $C_{14}H_{14}O_4$ 246 OI 145^0C

R = 3, 3-DMA
R' = H

Tet. Lett., 1967, 2147.

0507-004

$C_{17}H_{18}O_6$ 318 -31^0e 175^0C

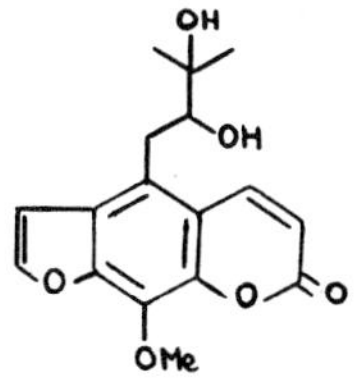

Tet., 1966, 2923.

0506-024

SCOPOLETIN $C_{10}H_8O_4$ 192 OI 205^0C

R = H

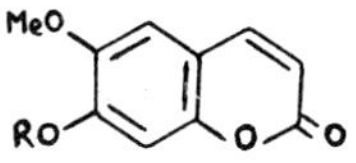

JOC, 1957 (22) 978.

0507-005

AESCULETIN $C_9H_6O_4$ 178 OI 270^0C

R = R' = H

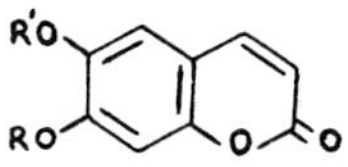

Ber., 1899 (32) 287.

0507-001

AESCULETIN DI-ME ETHER $C_{11}H_{10}O_4$ 206 OI 146^0C

R = Me

Aust. J. C., 1968. 3079.

0507-006

AESCULIN $C_{15}H_{16}O_9$ 340 -15^0m 205^0C

R = H; R' = Glucosyl

JCS, 1931, 1288.

0507-002

FABIATRIN $C_{16}H_{18}O_9$ 354 227^0C

R = Glucosyl

Can. J. C., 1962 (40) 256.

0507-007

SCOPOLETIN, O-(3,3-DMA)-	$C_{15}H_{16}O_4$	260 OI 82°C

R = 3,3-DMA

Phytochem., 1970, 894.

0507-008

NIESHOUTIN	$C_{15}H_{16}O_4$	260 126°C

Tet. Lett., 1967, 2147.
JCS, 1967, 145.

0507-013

SCOPOLETIN, O-geranyl-	$C_{20}H_{24}O_4$	328 OI 84°C

R = Geranyl

Acta Chem. Scand., 1970, 1113.

0507-009

CEDRELOPSIN	$C_{15}H_{16}O_4$	260 OI 172°C

R = H

JCS, 1968, 481.

0507-014

SCOPOLETIN, Iso-	$C_{10}H_8O_4$	192 OI 205°C

JCS, 1966, 1805.

0507-010

BRAYLEYANIN	$C_{20}H_{24}O_4$	328 OI 95°C

R = 3,3-DMA

Aust. J. Sci. Res., 1949 (A2) 608.

0507-015

AYAPIN	$C_{10}H_6O_4$	190 OI 232°C

Ber., 1937 (70) 702.

0507-011

OBLIQUETOL	$C_{14}H_{14}O_4$	246 OI 218°C

Tet. Lett., 1967, 2147.

0507-016

OBLIQUETIN	$C_{15}H_{16}O_4$	260 OI 138°C

Tet. Lett., 1967, 2147.

0507-012

OBLIQUIN	$C_{14}H_{12}O_4$	244 +76°c 162°C

R = CH_3

JCS, 1969, 526.

0507-017

OBLIQUOL	$C_{14}H_{12}O_5$	260
		$+77^0$c
		192^0C
R = CH_2OH		
	Tet. Lett., 1967, 2147.	
0507-018		

TODDACULINE	$C_{16}H_{18}O_4$	274
		OI
		95^0C
R = 3,3-DMA		
	Chem. Abstr., 1966, 10433.	
0508-002		

BRAYLIN	$C_{15}H_{14}O_4$	258
		OI
		150^0C

	Chem. Abstr., 1951 (45) 2938.	
0507-019		

ACULEATIN	$C_{16}H_{18}O_5$	290
		-17^0
		113^0C

R =

	J. Ind. Chem. Soc., 1942 (19) 425.	
0508-003		

SPHONDIN	$C_{12}H_8O_4$	216
		OI
		192^0C

	Chem. Abstr., 1956, 12999.	
0507-020		

TODDALOLACTONE	$C_{16}H_{20}O_6$	308
		$+56^0$c
		132^0C

R =

	Ber., 1939 (72) 53.	
0508-004		

RUTACULTIN	$C_{16}H_{18}O_4$	274
		OI
		101^0C

	Phytochem., 1971 (10) 191.	
0507-021		

ANGELICONE	$C_{16}H_{16}O_5$	288
		OI
		130^0C

R = Senecioyl

	J. Pharm. Soc. Jap., 1956 (76) 538.	
0508-005		

LIMETTIN	$C_{11}H_{10}O_4$	206
		OI
		147^0C

R = H

	J. Pharm. Soc. Jap., 1959 (79) 840.	
0508-001		

COUMURRAYIN	$C_{16}H_{18}O_4$	274
		OI
		158^0C

R = 3,3-DMA

	JOC, 1968, 3574.	
0508-006		

PINNARIN	$C_{16}H_{18}O_4$	274 OI 162°C	COUMARIN-5-geranyloxy-7-OMe-	$C_{20}H_{24}O_4$	328 OI 87°C
R = 1,1-DMA	Phytochem., 1970, 833.			JCS, 1945, 540.	
0508-007			0508-012		
SIBIRICIN	$C_{16}H_{18}O_5$	290 +60°d 152°C	ACULEATIN, 7-de-Me-7--(3,3-DMA)-iso-	$C_{20}H_{24}O_5$	344 105°C
R =	Ber., 1968, 2741. Tet., 1968, 3247.			Ber., 1968, 2741.	
0508-008			0508-013		
MEXOTICIN	$C_{16}H_{20}O_6$	308 +38°c 185°C	CLAUSENIN	$C_{14}H_{12}O_5$	260 156°C
R =	Tet., 1967, 3472.			Tet. Lett., 1966, 5767.	
0508-009			0508-014		
	$C_{16}H_{18}O_5$	290 OI 129°C	TRACHYPHYLLIN	$C_{19}H_{20}O_4$	312 OI 214°C
R =	Tet., 1967, 4613.			Aust. J.C., 1969, 2175.	
0508-010			0508-015		
GLABRALACTONE	$C_{16}H_{16}O_5$	288 OI 130°C	AVICENNIN	$C_{20}H_{20}O_4$	324 OI 141°C
R = Senecioyl	J. Pharm. Soc. Jap., 1956 (76) 649.			JCS, 1963, 3910.	
0508-011			0508-016		

XANTHOXYLETIN, allo-	C$_{15}$H$_{14}$O$_4$	258 OI 115^{0}C

XANTHOXYLETIN, allo-   C$_{15}$H$_{14}$O$_4$ 258 OI 115^{0}C   JCS, 1937, 1545.   0508-017

BERGAPTOL   C$_{11}$H$_6$O$_4$ 202 OI 278^{0}C   R = H   JCS, 1945, 540.   0508-022

DENTATIN, nor-   C$_{19}$H$_{20}$O$_4$ 312 OI 182^{0}C   R = H   Tet., 1968, 753.   0508-018

BERGAPTENE   C$_{12}$H$_8$O$_4$ 216 OI 191^{0}C   R = Me   Ber., 1912 (45) 3705.   0508-023

DENTATIN   C$_{20}$H$_{22}$O$_4$ 326 OI 95^{0}C   R = Me   Tet., 1968, 753.   0508-019

IMPERATORIN, Iso-   C$_{16}$H$_{14}$O$_4$ 270 OI 109^{0}C   R = 3,3-DMA   Ber., 1939 (72) 52.   0508-024

CLAUSENIDIN   C$_{19}$H$_{20}$O$_5$ 328 OI 136^{0}C   Tet. Lett., 1966, 5767.   0508-020

PRANGOLARIN   C$_{16}$H$_{14}$O$_5$ 286 +25^{0}c 104^{0}C   R =   Chem. Ind., 1963, 1430.   0508-025

XANTHOXYLETIN   C$_{15}$H$_{14}$O$_4$ 258 OI 133^{0}C   JCS, 1954, 1392.   0508-021

PEUCEDANIN, oxy-   C$_{16}$H$_{14}$O$_5$ 286 (-) 142^{0}C   Cf. Prangolarin (Stereoisomer)   Ber., 1939, 52.   0508-026

PEUCEDANIN, iso-oxy- R = 0508-027	$C_{16}H_{14}O_5$ Chem. Ind., 1970, 746.	286 OI 143^0C
BERGAMOTTIN R = Geranyl 0508-032	$C_{21}H_{22}O_4$ Ber., 1937 (70) 2272.	338 OI 60^0C
OSTRUTHOL R = 0508-028	$C_{21}H_{22}O_7$ Ber., 1933 (66) 1150.	386 -18^0p 136^0C
R = 0508-033	$C_{21}H_{22}O_5$ Tet., 1966, 1489.	354 OI 135^0C
PABULENOL R = 0508-029	$C_{16}H_{14}O_5$ Tet. Lett., 1971, 1977.	286 -4^0e 135^0C
PSORALIDIN 0508-034	$C_{16}H_{14}O_4$ Chem. Abstr., 1948 (42) 7492.	270 OI 315^0C
PANGELINE Cf. Pabulenol (Stereoisomer) 0508-030	$C_{16}H_{14}O_5$ Chem. Abstr., 1971 (74) 45611.	286 $+11^0$ 122^0C
PINNARIN, furo- 0508-035	$C_{17}H_{16}O_4$ Phytochem., 1970, 833.	284 OI 125^0C
PANGELINE, Angeloyl- R = 0508-031	$C_{21}H_{20}O_6$ Chem. Abstr., 1971 (74) 45611.	368 -20^0 88^0C
BERGAPTENE, iso- R = Me 0508-036	$C_{12}H_8O_4$ Ber., 1934 (67) 59.	216 OI 222^0C

ARCHANGELIN R = 0508-037	$C_{21}H_{22}O_4$ Chem. Abstr., 1969 (70) 37673.	338 $+14^0c$ 132^0C
FRAXIDIN, Iso- R = H MeO RO OMe 0509-003	$C_{11}H_{10}O_5$ Ber., 1939 (72) 52.	222 OI 149^0C
BRUCEOL HO 0508-038	$C_{19}H_{20}O_5$ Tet., 1963, 593.	328 -297^0c 201^0C
CALYCANTHOSIDE R = Glucosyl Bull. Soc. Chim. Biol., 1955 (37) 365. 0509-004	$C_{17}H_{20}O_{10}$	384 -42^0m 220^0C
BRUCEOL, deoxy- * deoxy Chem. Comm., 1968, 368. 0508-039	$C_{19}H_{20}O_4$	312
FRAXIDIN MeO MeO OH 0509-005	$C_{11}H_{10}O_5$ Ber., 1937 (70) 1019.	222 OI 197^0C
FRAXETIN R = H MeO HO OR 0509-001	$C_{10}H_8O_5$ Ber., 1929 (62) 120.	208 OI 228^0C
COUMARIN, 6,7,8-tri-OMe- MeO MeO OMe 0509-006	$C_{12}H_{12}O_5$ JCS, 1954, 1392.	236 OI 104^0C
FRAXIN R = Glucosyl Ber., 1938 (71) 1931. 0509-002	$C_{16}H_{18}O_{10}$	370 205^0C
COUMARIN, 6-OMe-7,8-methylene dioxy- MeO O O 0509-007	$C_{11}H_8O_5$ Phytochem., 1970, 891.	220 OI 220^0C

FRAXINOL	$C_{11}H_{10}O_5$	222
		OI
		173^0C
R = H		
		JOC., 1959 (24) 523.
0510-001		

PIMPINELLIN, iso-	$C_{13}H_{10}O_5$	246
		OI
		150^0C
R = Me		
		JCS, 1956, 4170.
0511-001		

MANDSHURIN	$C_{17}H_{20}O_{10}$	384
		-27^0m
		182^0C
R = Glucosyl		
		Chem. Abstr., 1971 (74) 108116.
0510-002		

PHELLOPTERIN	$C_{17}H_{16}O_5$	300
		OI
		102^0C
R = 3,3-DMA		
		Chem. Abstr.. 1942, 464.
0511-002		

NIESHOUTOL	$C_{15}H_{16}O_5$	276
		OI
		144^0C
		Tet., 1970 (26) 4473.
0510-003		

BYAK-ANGELICOL	$C_{17}H_{16}O_6$	316
		$+34^0p$
		106^0C
R =		
		Chem. Abstr., 1947, 4472.
0511-003		

PIMPINELLIN	$C_{13}H_{10}O_5$	246
		OI
		118^0C
R = Me		
		Chem. Abstr., 1956, 12999.
0510-004		

BYAK-ANGELICIN	$C_{17}H_{18}O_7$	334
		$+25^0p$
		118^0C
R =		
		Chem. Abstr., 1947. 5879.
0511-004		

BERGAPTEN, 6-(3,3-DMA)-iso-	$C_{17}H_{16}O_5$	300
		OI
		96^0C
R = 3,3-DMA		
		JOC., 1970, 2294.
0510-005		

	$C_{22}H_{26}O_8$	418
R =		
O-i-Val		
		Tet. Lett., 1968, 3947.
0511-005		

PSORALENE, 5-OMe-8-geranyloxy- $C_{22}H_{24}O_5$ 368 OI R = Geranyl JOC, 1968, 3577. 0511-006	COUMARIN $C_9H_6O_2$ 146 OI 68^0C R = H Chem. Abstr., 1952 (46) 6335. 0599-002	
KNIDILIN $C_{17}H_{16}O_5$ 300 OI 118^0C R = Me Chem. Abstr., 1967 (67) 11395. 0511-007	KARATAVIKINOL $C_{24}H_{32}O_4$ 384 -12^0e 52^0C R = Khim. Prir. Soed., 1969, 225. 0599-003	
KNIDITSIN $C_{21}H_{22}O_5$ 354 OI 78^0C R = 3,3-DMA Chem. Abstr., 1967 (67) 11395. 0511-008	COUMARIN, 4-OH- $C_9H_6O_3$ 162 OI R = H Tet., 1958 (4) 36 & 135. 0599-004	
SABANDININ $C_{11}H_8O_5$ 320 OI 193^0C Phytochem., 1971 (10) 1621. 0511-009	FERULENOL $C_{24}H_{30}O_3$ 366 OI 64^0C R = Farnesyl Tet. Lett., 1964, 2783. 0599-005	
MELILOTOL $C_9H_8O_2$ 148 OI 25^0C J. Pharm. Soc. Jap., 1934 (54) 107. 0599-001	AMMORESINOL $C_{24}H_{30}O_4$ 382 OI 108^0C Ber., 1937 (70) 1255 & 1679. 0599-006	

COUMARIN, 8-OMe-4-Me-

$C_{11}H_{10}O_3$ 190
OI
165^0C

Chem. Ind., 1965, 383.

0599-007

GLAUPALOL

$C_{15}H_{16}O_4$ 260
OI
203^0C

Chem. Comm., 1967, 547.

0599-012

HALKENDIN

$C_{13}H_{10}O_5$ 246
OI
173^0C

Aust. J. C., 1971, 209.

0599-008

PINNATERIN

$C_{16}H_{18}O_4$ 274
OI
131^0C

Phytochem., 1971 (10) 1971.

0599-013

HALFORDIN, iso-

$C_{14}H_{12}O_6$ 276
OI
152^0C

Tet. Lett., 1968, 447.

0599-009

BICOUMOL

$C_{18}H_{10}O_6$ 322
OI

Chem. Abstr., 1967 (67) 64177.

0599-014

HALFORDIN

R = Me

$C_{14}H_{12}O_6$ 276
OI
136^0C

Tet. Lett., 1968, 447.

0599-010

DICOUMAROL

$C_{19}H_{12}O_6$ 336
OI
289^0C

JACS, 1954, 1650.

0599-015

HALFORDININ

R = 1,1-DMA

$C_{18}H_{18}O_6$ 330
OI
110^0C

Tet. Lett., 1970, 3611.

0599-011

KOTANIN

R = Me

$C_{24}H_{22}O_8$ 438
$+33^0c$
$>315^0C$

JOC, 1971, 1143.

0599-016

KOTANIN, des-Me- R = H 0599-017	$C_{23}H_{20}O_8$ 424 -13^0c $>315^0$c JOC, 1971, 1143.	CANDICANIN 0599-021	$C_{32}H_{26}O_{10}$ 570 OI 153^0C Tet. Lett., 1971, 4221.
DAPHNORETIN R = H 0599-018	$C_{19}H_{12}O_7$ 352 OI 246^0C Ann., 1963 (662) 113.	PHEBALIN 0599-022	$C_{30}H_{28}O_6$ 484 0^0 176^0C Chem. Ind., 1971, 1020.
DAPHNORINE R = Glucosyl 0599-019	$C_{25}H_{22}O_{12}$ 514 -78^0w 203^0C Natwiss., 1963 (50) 521.	THAMNOSIN 0599-023	$C_{30}H_{28}O_6$ 484 JACS, 1968, 814.
LASIOCEPHALIN 0599-020	$C_{20}H_{14}O_6$ 350 OI 215^0C Chem. Ind., 1971, 855.	NOVOBIOCIN 0599-024	$C_{31}H_{36}N_2O_{11}$ 612 -63^0e 156^0C JACS, 1957, 3789.

69

The lignans are formed by the dimerization of C_6C_3 precursors (generally cinnamyl alcohols) linked through the α position:

Other positions of dimeric linkage (e.g., α-3′; 3-3′) are known, and examples will be found in Class 0699.

0601		0604	
0602		0605	
0603		0606	
0698	Irregular Oxygen-ation Patterns	0699	Misc. Structural Variations

GUAIARETIC ACID, nor-dihydro-	$C_{18}H_{22}O_4$	302 OI 185^0C

Chem. Abstr., 1953,
7126.

0601-001

GUAIARETIC ACID	$C_{20}H_{24}O_4$	328 -94^0e 100^0C

JCS, 1934, 1423.

0601-002

PHYLLANTHIN	$C_{24}H_{34}O_6$	418 $+12^0c$

Tet., 1967, 1915.

0601-003

LARICIRESINOL, seco-iso-	$C_{20}H_{26}O_6$	362 -36^0a 113^0C

Tet., 1959 (7) 262.

0601-004

LARICIRESINOL	$C_{20}H_{24}O_6$	360 $+20^0a$ 168^0C

JCS, 1939, 1237.

0601-005

GALBELGIN	$C_{22}H_{28}O_5$	372 -102^0c 138^0C

JCS, 1958, 4471.

0601-006

GALGRAVIN	$C_{22}H_{28}O_5$	372 OI 121^0C

JCS, 1958, 4471.

0601-007

VERAGUENSIN	$C_{20}H_{20}O_5$	340 $+34^0c$ 131^0C

JCS, 1962, 1459.

0601-008

GALBACIN	$C_{20}H_{20}O_5$	340 -114^0c 116^0C

Aust. J. C., 1954, 104.

0601-009

OLIVIL, (-)-	$C_{20}H_{24}O_7$	376 -127^0w 142^0C

Gazz. Chim. Ital.,
1938 (68) 87.

0601-010

MATAIRESINOL	$C_{20}H_{22}O_6$	358
		-49^0a
		119^0C
R = R' = H		
		JCS, 1950, 71.
0601-011		

PLUVIATOLIDE	$C_{20}H_{20}O_6$	356
		-36^0c
		160^0C
		Aust. J. C., 1970 (23) 133.
0601-016		

MATAIRESINOL MONO-GLUCOSIDE	$C_{26}H_{32}O_{11}$	520
		-58^0m
		92^0C
R, R' = H, Glucosyl (?)		
		Phytochem., 1970, 2407.
0601-012		

TAIWANIN-A	$C_{20}H_{14}O_6$	350
		OI
		Chem. Comm., 1965, 592.
0601-017		

ARCTIGENIN	$C_{21}H_{24}O_6$	372
		-29^0
		102^0C
R = H		
		JCS, 1937, 384.
0601-013		

SAVININ	$C_{20}H_{16}O_6$	352
		-88^0c
		141^0C
		JACS, 1953 (75) 235.
0601-018		

ARCTIIN	$C_{27}H_{34}O_{11}$	534
		-39^0e
		111^0C
R = Glucosyl		
		J. Pharm. Soc. Jap., 1935 (55) 816.
0601-014		

HINOKININ	$C_{20}H_{18}O_6$	354
		-40^0c
		65^0C
		JCS, 1950, 71.
0601-019		

HELIANTHOIDIN	$C_{26}H_{28}O_8$	468
		-145^0c
		135^0C
		JCS, 1969, 693.
0601-015		

PARABENZLACTONE, (-)-	$C_{20}H_{18}O_7$	370
		-11^0c
		160^0C
		Tet. Lett., 1970, 2016.
0601-020		

73

SVENTENIN ACETATE $C_{22}H_{20}O_8$ 412

OAc

Chem. Abstr., 1971
(74) 126624.

0601-021

EUDESMIN $C_{22}H_{26}O_6$ 386
-64°c
107°C

OMe
OMe

MeO
OMe

Aust. J. C., 1963, 147.

0601-026

CUBEBIN $C_{20}H_{20}O_6$ 356
-17°a
132°C

OH

Arch. Pharm., 1967
(300) 559.

0601-022

DIAEUDESMIN $C_{22}H_{26}O_6$ 386
+316°c
158°C

OMe
OMe

MeO
OMe

JCS, 1967, 2228.

0601-027

PINORESINOL $C_{20}H_{22}O_6$ 358
+84°c
121°C

OH
OMe

HO
OMe

Acta Chem. Scand.,
1960, 226.

0601-023

PLUVIATILOL $C_{20}H_{20}O_6$ 356
-136°c
162°C

OR
OMe

R = H

AJC, 1970 (23) 133.

0601-028

PHILLYGENOL $C_{21}H_{24}O_6$ 372
+122°e
135°C

OR
OMe

MeO
OMe

R = H

Acta Chem. Scand.,
1949, 898.

0601-024

FARGESIN $C_{21}H_{22}O_6$ 370
139°C

R = Me

Bull. Chem. Soc. Jap.,
1970, 3631.

0601-029

PHILLYRIN $C_{27}H_{34}O_{11}$ 534
+48°e
154°C

R = Glucosyl

Acta Chem. Scand.,
1949 (3) 898.

0601-025

ASARININ, (+)- $C_{20}H_{18}O_6$ 354
+119°c
123°C

Ber., 1961, 851.

0601-030

ASARININ, (-)- antipode	$C_{20}H_{18}O_6$ 354	
	J. Pharm. Soc. Jap., 1937 (57) 184; 289.	
0601-031		

PAULOWNIN	$C_{20}H_{18}O_7$ 370 $+29^0$ 105^0C	
	Chem. Pharm. Bull., 1966, 641.	
0601-036		

SESAMIN, (+)-	$C_{20}H_{18}O_6$ 354 $+68^0c$ 124^0C	
	Biochem. Z., 1928 (201) 454.	
0601-032		

PAULOWNIN, Iso-	$C_{20}H_{18}O_7$ 370 $+127^0c$ 132^0C	
* epimer	Chem. Pharm. Bull., 1970, 421.	
0601-037		

SESAMIN, (-)- antipode	$C_{20}H_{18}O_6$ 354	
	Khim. Prir. Soedin., 1966 (2) 149.	
0601-033		

CALOPIPTIN	$C_{21}H_{24}O_5$ 356	
	Aust. J. C., 1968, 2095.	
0601-038		

FAGAROL Racemate	$C_{20}H_{18}O_6$ 354 Rac. 130^0C	
	Acta Chem. Scand., 1955 (9) 1111.	
0601-034		

MATAIRESINOL, OH-	$C_{20}H_{22}O_7$ 374	
	Phytopathology, 1971 (61) 841.	
0601-039		

GMELINOL	$C_{22}H_{26}O_7$ 402 $+23^0c$ 124^0C	
	Aust. J. C., 1954, 83.	
0601-035		

THUJAPLICATENE, γ -	$C_{20}H_{20}O_7$ 372	
	Can. J. C., 1970, 3144.	
0602-001		

THUJAPLICATIN ME ETHER, -di-OH-	$C_{21}H_{24}O_9$	420

Can. J. C., 1967, 305.

0602-002

SYRINGARESINOL, DL-	$C_{22}H_{26}O_8$	418 Rac. 175°C

R = H (Racemate)

Chem. Abstr., 1967 (67) 32607.

0603-003

PODORHIZOL	$C_{22}H_{24}O_8$	416 -52°c 126°C

Tet. Lett., 1969, 885.

0602-003

LIRIODENDRIN	$C_{34}H_{46}O_{18}$	742 270°C

R = Glucosyl

JOC, 1958, 179.

0603-004

MAGNOLIN	$C_{23}H_{28}O_7$	416 +56° 97°C

Ar = 3,4,5-tri-OMe-phenyl
Ar' = 3,4-di-OMe-phenyl

Bull. Chem. Soc. Jap., 1970, 3631.

0602-004

LIRIORESINOL-C DI-ME ETHER	$C_{24}H_{30}O_8$	446 +284°c 145°C

JCS, 1968, 3042.

0603-005

LIRIORESINOL-A	$C_{22}H_{26}O_8$	418 127°c 211°C

JOC, 1958, 179.

0603-001

GALCATIN	$C_{21}H_{24}O_4$	340 -9°c 118°C

JCS, 1968, 74.

0604-001

LIRIORESINOL-B	$C_{22}H_{26}O_8$	418 +62°c 175°C

R = H

JACS, 1964, 1186.

0603-002

GALCATIN, Iso-	$C_{21}H_{24}O_4$	340 +5°c 107°C

JCS, 1966, 893.

0604-002

GALBULIN	$C_{22}H_{28}O_4$	356
		-8^0c
		135^0C

Acta Chem. Scand.,
1954, 1827.

0604-003

TAXIRESINOL, Iso-	$C_{19}H_{22}O_6$	346
		171^0C

R = H

Acta Chem. Scand.,
1969, 2021.

0604-008

OTOBAPHENOL	$C_{20}H_{22}O_4$	326
		$+40^0$c
		135^0C

JCS, 1966, 1775.

0604-004

TAXIRESINOL 6-ME-ETHER, Iso-	$C_{20}H_{24}O_6$	360
		$+35^0$a
		181^0C

R = Me

Acta Chem. Scand.,
1969, 2021.

0604-009

OTOBAENE	$C_{20}H_{18}O_4$	322
		$+47^0$c
		127^0C

JCS, 1966, 1775.

0604-005

LARICIRESINOL, Iso-	$C_{20}H_{24}O_6$	360
		$+44^0$e
		158^0C

Acta Chem. Scand.,
1969, 2021.

0604-010

OTOBAIN	$C_{20}H_{20}O_4$	324
		134^0C

R = H

JCS, 1966, 1775.

0604-006

OLIVIL, Iso-	$C_{20}H_{24}O_7$	376
		62^0e
		98^0C

JCS, 1940, 1321.

0604-011

OTOBAIN, OH-	$C_{20}H_{20}O_5$	340
		116^0C

R = OH

JCS, 1966, 1775.

0604-007

CONIDENDRIN	$C_{20}H_{20}O_6$	356
		-55^0a
		255^0C

JACS, 1955, 432.

0604-012

deleted 0604-013	TAIWANIN-C $\qquad$ $C_{20}H_{12}O_6$ 348 OI 276^0C Tet. Lett., 1967, 849. 0604-018
JUSTICIDIN-C $\qquad$ $C_{22}H_{18}O_7$ 394 OI 266^0C Tet. Lett., 1970, 923. 0604-014	DIPHYLLIN $\qquad$ $C_{21}H_{16}O_7$ 380 OI R = H $\qquad$ 291^0C Tet. Lett., 1967, 3517. Chem. Comm., 1968, 653. 0604-019
HELIOXANTHIN $\qquad$ $C_{20}H_{12}O_6$ 348 OI 240^0C JCS, 1969, 693. 0604-015	JUSTICIDIN-A $\qquad$ $C_{22}H_{18}O_7$ 394 OI 263^0C R = Me Tet., 1970, 4301. 0604-020
JUSTICIDIN-B $\qquad$ $C_{21}H_{16}O_6$ 364 OI 240^0C JOC, 1971, 3453. Tet., 1970, 4301. 0604-016	CLEISTANTHIN-B $\qquad$ $C_{27}H_{26}O_{12}$ 542 R = Glucosyl Curr. Sci., 1970 (39) 395. 0604-021
COLLINUSIN $\qquad$ $C_{21}H_{18}O_6$ 366 $+132^0c$ 196^0C Tet. Lett., 1967, 4183. 0604-017	CLEISTANTHIN $\qquad$ $C_{28}H_{28}O_{11}$ 540 -67^0c 136^0C R = 3, 4-di-O-Me-Xylosyl Tet., 1969, 2815. 0604-022

TAIWANIN- E	$C_{20}H_{12}O_7$	364 OI 265^{0}C

R = H

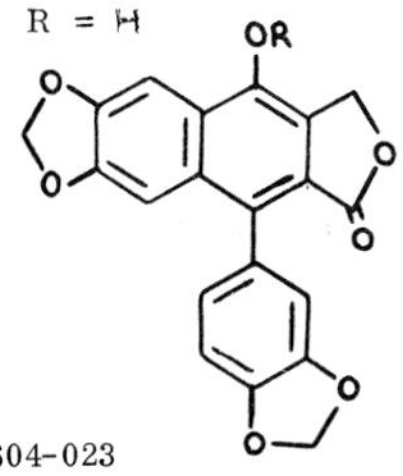

Tet. Lett., 1969, 1079.

0604-023

PODOPHYLLOTOXIN GLUCOSIDE, $C_{27}H_{30}O_{12}$ 546 -4'-desMe-deoxy- -77^{0}m 148^{0}C

R = Glucosyl

Helv., 1964 (47) 1203.

0605-003

TAIWANIN- E ME ETHER $C_{21}H_{14}O_7$ 378 OI 228^{0}C

R = Me

Tet., 1970, 4301.

0604-024

PICROPODOPHYLLIN, deoxy- $C_{22}H_{22}O_7$ 398 +32^{0}c 172^{0}C

R = H

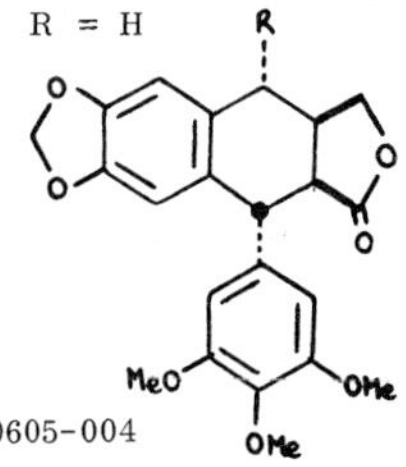

JACS, 1954, 4034.

0605-004

JUSTICIDIN- D $C_{21}H_{14}O_7$ 378 OI 272^{0}C

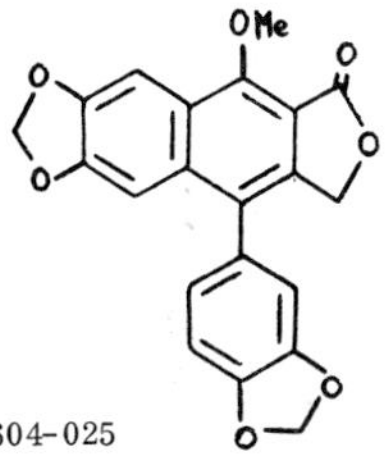

Tet., 1970, 4301.

0604-025

PICROPODOPHYLLIN $C_{22}H_{22}O_8$ 414 +5^{0}c 226^{0}C

R = OH

JOC, 1966, 3224.

0605-005

PODOPHYLLOTOXIN, 7-desOH- $C_{21}H_{20}O_7$ 384 -128^{0}c 246^{0}C -4'-desMe-

R = H

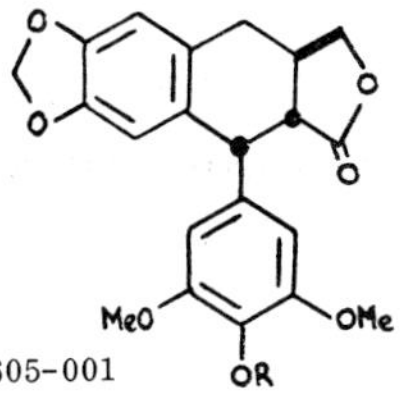

J. Pharm. Sci., 1971 (60) 649.

0605-001

PODOPHYLLOTOXIN, -4'-des-O-Me- $C_{21}H_{20}O_8$ 400 -130^{0}c 250^{0}C

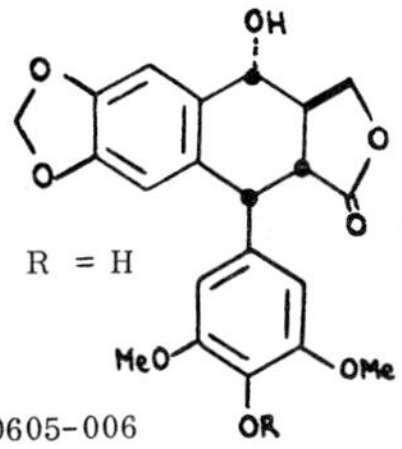

R = H

JACS, 1953, 1308.

0605-006

PODOPHYLLOTOXIN, deoxy- $C_{22}H_{22}O_7$ 398 -119^{0}c 172^{0}C

R = Me

JACS, 1954, 4034.

0605-002

PODOPHYLLOTOXIN $C_{22}H_{22}O_8$ 414 -101^{0}c

R = Me

JACS, 1953, 5916.

0605-007

PODOPHYLLOTOXIN, dehydro- $C_{22}H_{18}O_8$ 410
OI
273^0C

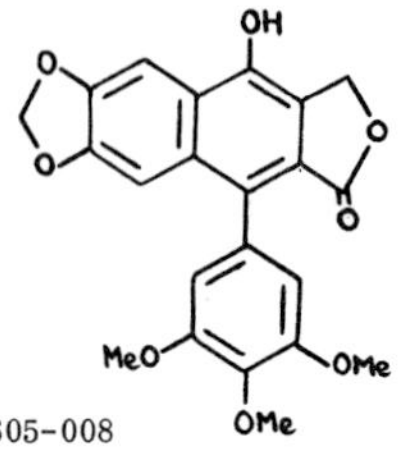

Acta Chem. Scand.,
1954, 1296.

0605-008

LYONIRESINOL, 2-O-Rhamnosyl- $C_{28}H_{38}O_{12}$ 666

amorph

R = Rhamnosyl

Tet., 1969, 2325.

0606-002

PODOPHYLLOTOXIN GLUCOSYL
ESTER, deoxy- $C_{28}H_{34}O_{13}$ 578
-152^0m
123^0C

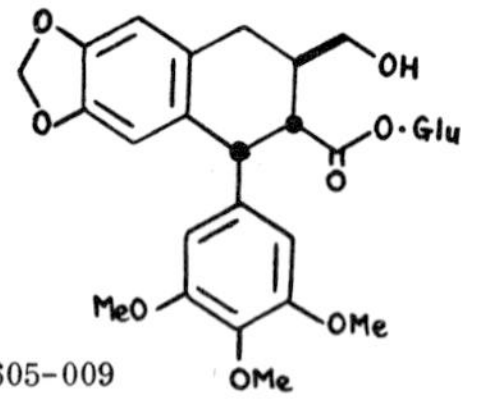

Helv., 1963 (46) 2127.

0605-009

THOMASIC ACID $C_{22}H_{24}O_9$ 432
Rac
233^0C

R = CH$_2$OH (Racemate)

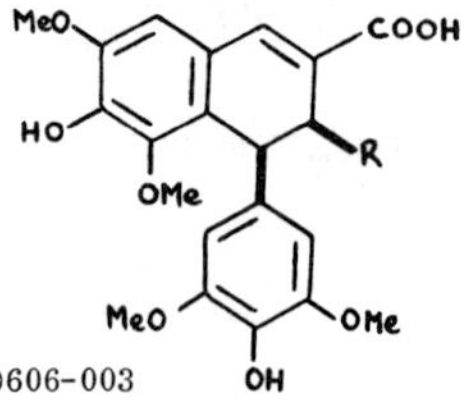

Tet., 1968, 1475.

0606-003

PLICATIC ACID $C_{20}H_{22}O_{10}$ 422
-10^0w
amorph

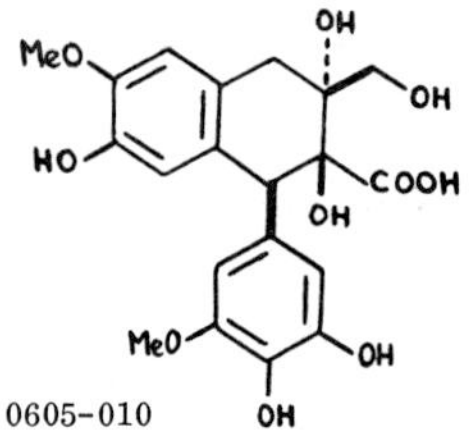

Can. J. C., 1967, 319.

0605-010

THOMASIDIOIC ACID $C_{22}H_{22}O_{10}$ 446

R = COOH (Racemate)

Tet., 1969, 2325.

0606-004

PLICATINAPHTHOL $C_{20}H_{16}O_8$ 384
260^0C

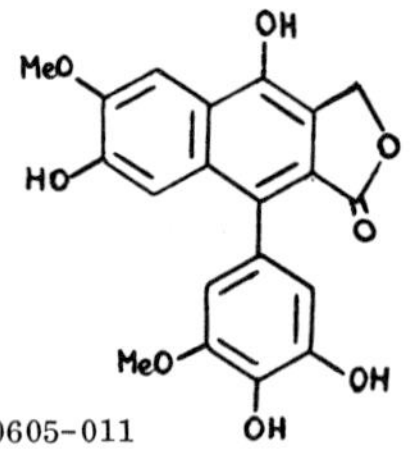

Can. J. C., 1969, 457.

0605-011

$C_{21}H_{22}O_5$ 354
OI

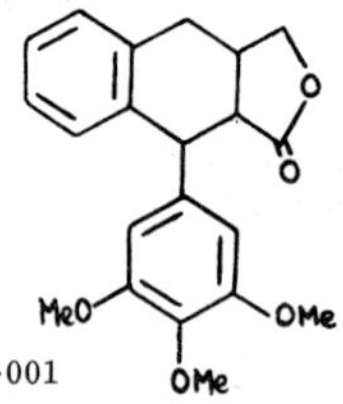

Chem. Abstr., 1947, 2917.

0698-001

LYONIRESINOL, DL- $C_{22}H_{26}O_8$ 418
Rac
130^0C

R = H (Racemate)

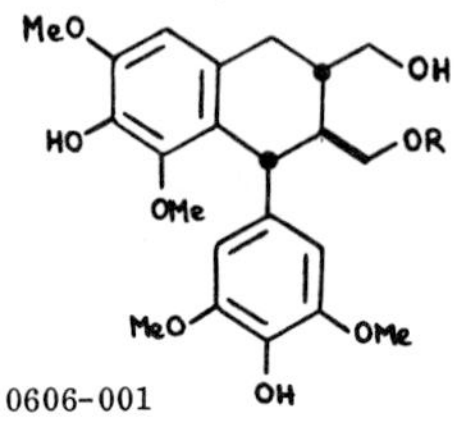

Tet., 1969, 2325.

0606-001

JUSTICIN, neo- $C_{21}H_{14}O_7$ 378
274^0C

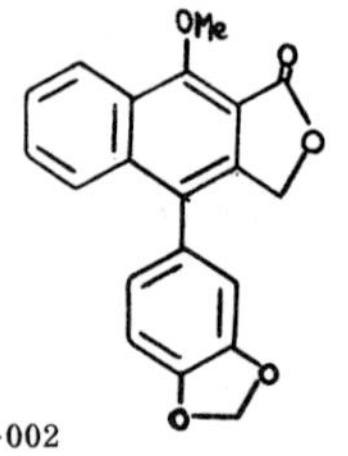

Chem. Pharm. Bull.,
1970, 862.

0698-002

PELTATIN, alpha-	$C_{21}H_{20}O_8$	400 -125^0c 244^0C

JACS, 1955, 1710.

0698-003

CARPANONE, Iso-	$C_{20}H_{18}O_6$	354 216^0C

Tet. Lett., 1969, 5159.

0698-008

PELTATIN, beta-	$C_{22}H_{22}O_8$	414 -115^0e 231^0C

R = H

JACS, 1952, 6285.

0698-004

CARPANANONE	$C_{20}H_{20}O_6$	356 228^0C

Tet. Lett., 1969, 5159.

0698-009

PELTATIN-A ME ETHER, beta-	$C_{23}H_{24}O_8$	428

R = Me

Tet. Lett., 1969, 2759.

0698-005

SESANGOLIN	$C_{21}H_{20}O_7$	384 $+48^0$c 88^0C

JCS, 1967, 1968.

0698-010

PELTATIN-A ME ETHER, -5'-des-OMe-β-	$C_{22}H_{22}O_7$	398 -146^0 142^0C

Tet. Lett., 1969, 2759.

0698-006

PHYLLANTHIN, hypo-	$C_{24}H_{30}O_7$	430 $+4^0$c 128^0C

Tet. Lett., 1971, 3175.
Tet., 1970, 3051.

0698-011

CARPANONE	$C_{20}H_{18}O_6$	354 OI 211^0C

Tet. Lett., 1969, 5159.

0698-007

FUTOQUINOL	$C_{21}H_{22}O_5$	354 98^0C

Chem. Pharm. Bull.,
1970, 100.

0699-001

FUTOENONE

$C_{20}H_{20}O_5$ 340
-58^0c
197^0C

Tet. Lett., 1968, 2003.

0699-002

TRUXILLIC ACID, epsilon-

$C_{18}H_{16}O_4$ 296

192^0C

Ber., 1935 (68) 2121.

0699-007

MAGNOLOL

$C_{18}H_{18}O_2$ 266
OI
103^0C

J. Pharm. Soc. Jap.,
1930 (50) 183.

0699-003

TRUXINIC ACID, beta-

$C_{18}H_{16}O_4$ 296

210^0C

Ber., 1937 (70) 483.

0699-008

EGONOL

$C_{19}H_{18}O_5$ 326
OI
117^0C

Ber., 1939, 1146.

0699-004

TRUXINIC ACID, delta-

$C_{18}H_{16}O_4$ 296
Rac
175^0C

Ber., 1937 (70) 483.

0699-009

EGONOL, homo-

$C_{20}H_{22}O_5$ 342
OI
121^0C

JCS, 1967, 2402.

0699-005

TRUXINIC ACID, neo-

$C_{18}H_{16}O_4$ 296

209^0C

Ber., 1937 (70) 483.

0699-010

TRUXILLIC ACID, alpha-

$C_{18}H_{16}O_4$ 296

285^0C

Ber., 1935 (68) 2108.

0699-006

deleted

0699-011

HINOKIRESINOL $C_{17}H_{16}O_2$ 252

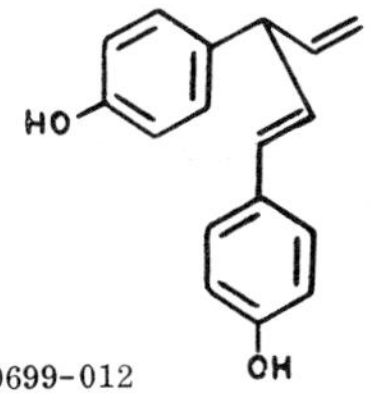

Tet. Lett., 1967, 793.

0699-012

TRACHELOSIDE $C_{27}H_{34}O_{12}$ 550

R = Me

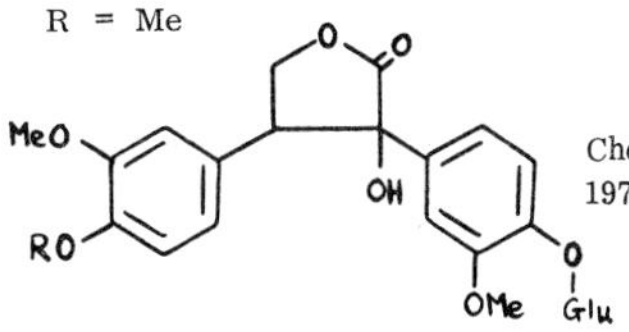

Chem. Pharm. Bull.,
1971, 866. **1972, 2075**

0699-017

AGATHARESINOL $C_{17}H_{18}O_4$ 286

R = H

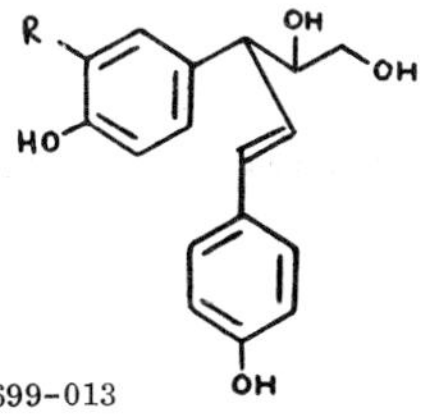

Tet. Lett., 1967, 793.

0699-013

TRACHELOSIDE, nor- $C_{26}H_{32}O_{12}$ 536
-48^0e
95^0C

R = H

Chem. Pharm. Bull.,
1971, 866.

0699-018

SEQUIRIN-C $C_{17}H_{18}O_5$ 302

R = OH

JCS, 1969, 1921.

0699-014

CONIFERYL ALCOHOL, dehydro- $C_{20}H_{22}O_4$ 358
$+11^0$a
140^0C

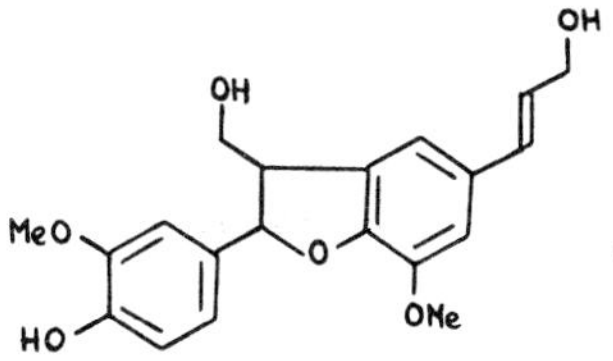

Ann., 1970 (736) 170.

0699-019

SUGIRESINOL $C_{17}H_{18}O_4$ 286
250^0C

R = H

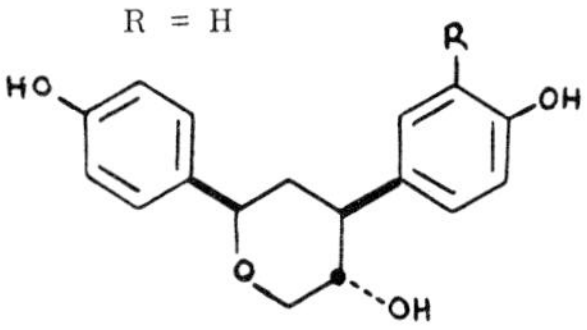

Tet. Lett., 1967, 793.

0699-015

EUPOMATENE $C_{20}H_{18}O_4$ 362
OI
155^0C

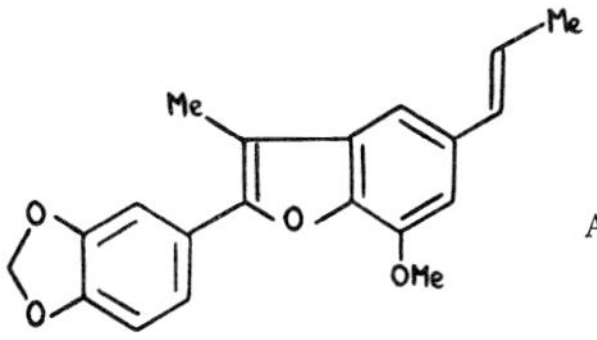

Aust. J. C., 1969, 1011.

0699-020

SEQUIRIN-B $C_{17}H_{18}O_5$ 302
212^0C

R = OH

JCS, 1969, 1921.

0699-016

HORDATINE-A $C_{28}H_{38}N_8O_4$ 550
$+69^0$w

R = H

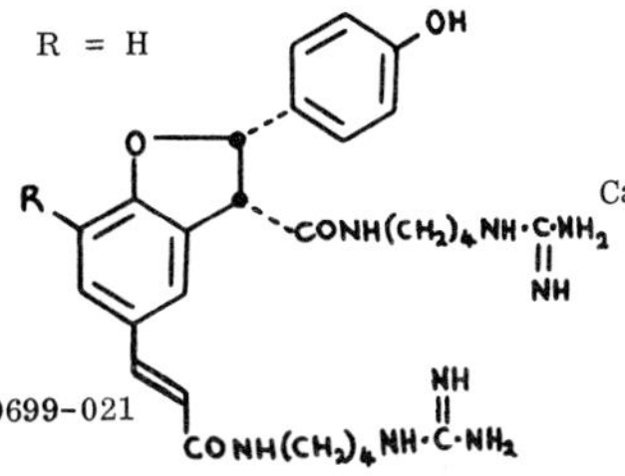

Can. J. C., 1967, 1745.

0699-021

HORDATINE-B

$C_{29}H_{40}N_8O_5$ 580
    $+54^0w$

R = OMe

Can. J. C., 1967, 1745.

0699-022

HINOKIOL

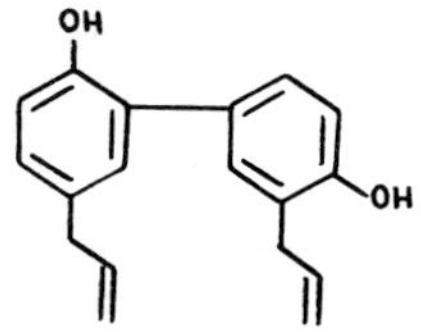

$C_{18}H_{18}O_2$ 266
    Ol
    87^0C

Chem. Pharm. Bull.,
1972, 210.

0699-025

---

deleted

0699-023

ASATONE

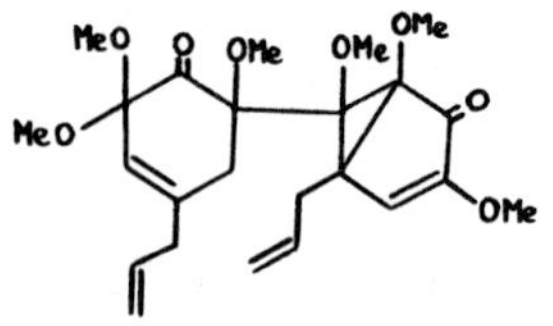

$C_{24}H_{32}O_8$ 448
    0^0m
    101^0C

Tet. Lett., 1972, 1607.

0699-026

SESAMOLIN

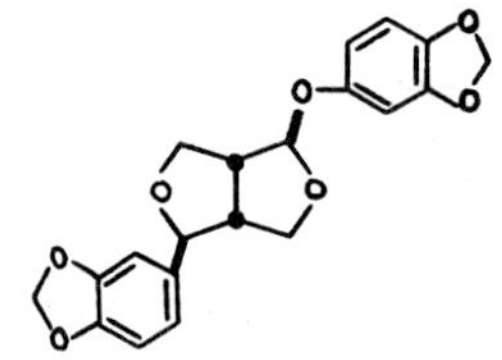

$C_{20}H_{18}O_7$ 370
    $+212^0c$
    94^0C

JCS, 1970, 2332.

0699-024

07	TERPHENYLS

The terphenyl class could well be regarded as a special type of lignan in that it too is derived biogenetically through the dimerization of a C_6C_3 precursor:

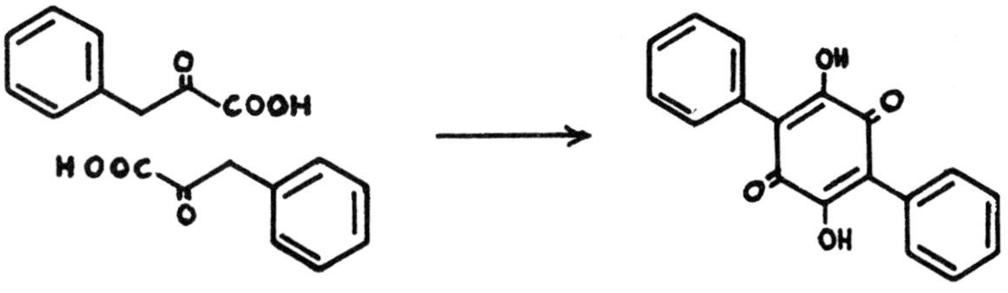

The oxidative cleavage of the middle aromatic ring provides a rationalization for the biogenesis of the series of acids and lactones known as the "lichen acids." The contraction of the middle ring to a cyclopentanoid provides the single representative of Class 0799.

0701	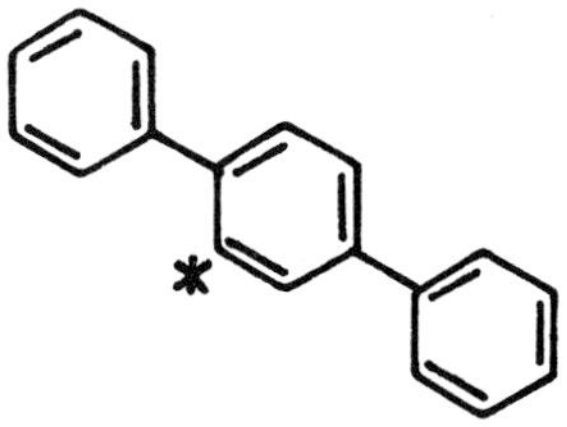 * This ring can be at various oxidation levels
0702	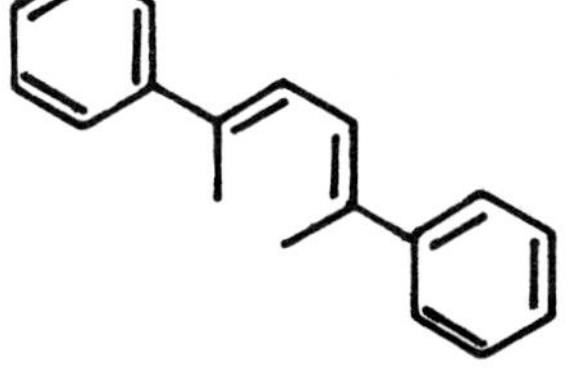 Includes the "lichen acids"
0799	MISC.

VOLUCRISPORIN

$C_{18}H_{12}O_4$ 292
OI
>300^{0}C

Chem. Ind., 1959, 731.

0701-001

LEUCOMELONE, proto-

$C_{32}H_{28}O_{14}$ 636
OI
204^{0}C

J. Pharm. Soc. Jap.,
1942 (62) 129.

0701-006

POLYPORIC ACID

$C_{18}H_{12}O_4$ 292
OI
306^{0}C

JACS, 1950, 1824.

0701-002

PHLEBIARUBRONE

$C_{19}H_{12}O_4$ 304
OI
249^{0}C

Tet., 1967, 3985.

0701-007

ATROMENTIN

$C_{18}H_{12}O_6$ 324
OI
-

Ann., 1928 (465) 243.

R = H

0701-003

THELEPHORIC ACID

$C_{20}H_{12}O_9$ 396
OI

Tet., 1960 (10) 135.

0701-008

AURANTIACIN

$C_{30}H_{20}O_8$ 532
OI
290^{0}C

R = Benzoyl

Acta Chem. Scand.,
1956 (10) 1111.

0701-004

MUSCARUFIN

$C_{25}H_{16}O_9$ 460
OI
275^{0}C

Ann., 1930 (479) 11.

0701-009

LEUCOMELONE

$C_{18}H_{12}O_7$ 340
OI
320^{0}C

J. Pharm. Soc. Jap.,
1942 (62) 129.

0701-005

XYLERYTHRIN

$C_{26}H_{16}O_5$ 408
OI
253^{0}C

R = H

Acta Chem. Scand.,
1970, 3445.

0701-010

86

XYLERYTHRIN, 5-O-Me-	$C_{27}H_{18}O_5$	422 OI
R = Me		
	Acta Chem. Scand., 1969, 2583.	
0701-011		

CALLOPISMIC ACID	$C_{20}H_{16}O_5$	336 OI 128^0C
R = Et		
	Ann., 1897 (297) 271.	
0702-003		

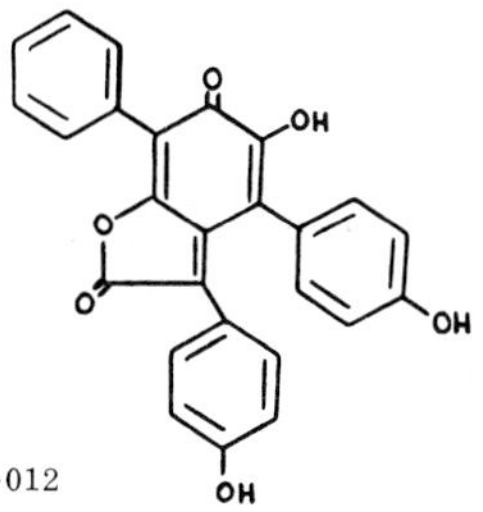

PENIOPHORIN $C_{26}H_{16}O_6$ 424 OI 300^0C

Acta Chem. Scand., 1970, 3444.

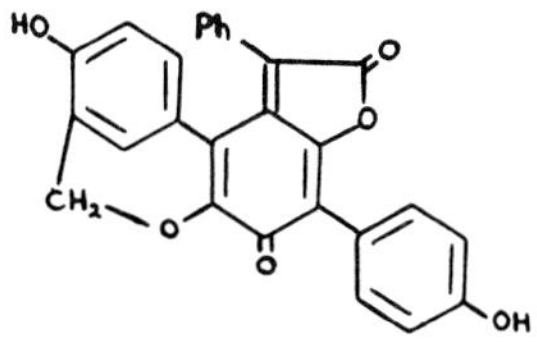

0701-012

PULVIC ACID LACTONE $C_{18}H_{10}O_4$ 290 OI 224^0C

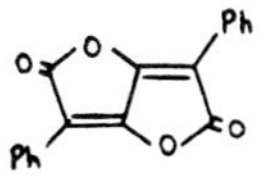

Ber., 1935 (68) 1569.

0702-004

PENIOPHORININ $C_{27}H_{16}O_6$ 436 OI $\sim 305^0$C

Acta Chem. Scand., 1970, 3449.

0701-013

CALYCIN $C_{18}H_{10}O_5$ 306 OI 245^0C

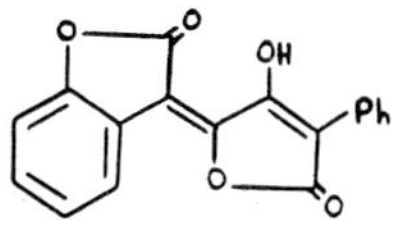

Tet. Lett., 1967, 3541.

0702-005

PULVIC ACID $C_{18}H_{12}O_5$ 308 OI 217^0C

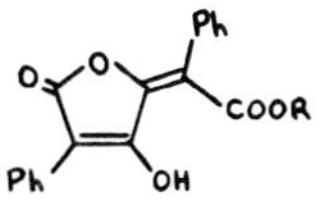

R = H

JCS, 1952, 1345.

0702-001

PULVINAMIDE $C_{18}H_{13}NO_4$ 307 OI 220^0C

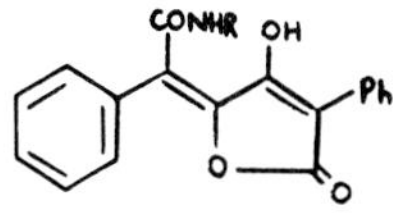

R = H

Phytochem., 1970, 2477.

0702-006

VULPINIC ACID $C_{19}H_{14}O_5$ 322 OI 148^0C

R = Me

Helv., 1926 (9) 446.

0702-002

RHIZOCARPIC ACID $C_{28}H_{23}NO_6$ 469 $+110^0$c 179^0C

R = .CH.COOMe
         |
       CH_2Ph

JACS, 1950, 4454.

0702-007

EPANORIN

$C_{25}H_{25}NO_6$ 435
-2^0c
136^0C

R = . CH . COOMe
       |
    CH_2CHMe_2

JACS, 1950, 4454.

0702-008

VARIEGATIC ACID

$C_{18}H_{12}O_9$ 372
OI
235^0C

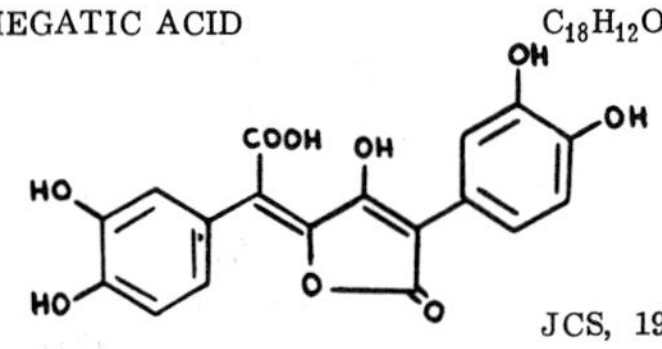

JCS, 1968, 2968.

0702-011

PINASTRIC ACID

$C_{20}H_{16}O_6$ 352
OI
205

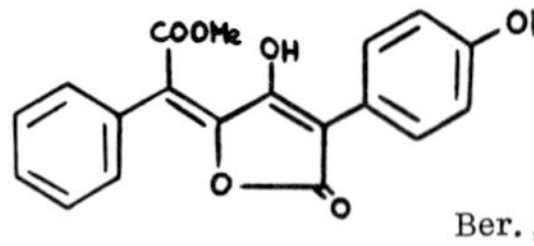

Ber., 1935 (68) 1565.

0702-009

VARIEGATORUBIN

$C_{18}H_{10}O_9$ 370
OI
$>320^0C$

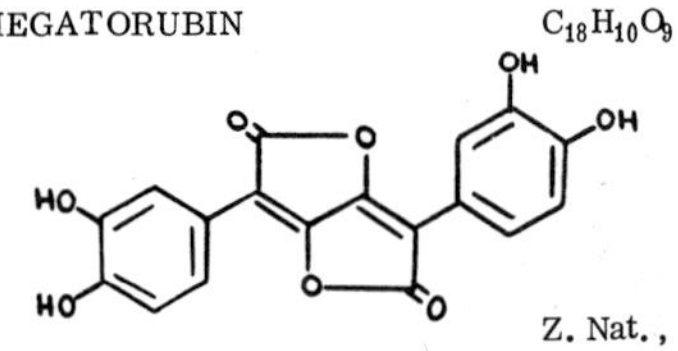

Z. Nat., 1971(26B) 376.

0702-012

LEPRAPINIC ACID

$C_{20}H_{16}O_6$ 352
164^0C

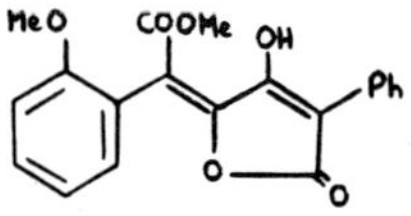

Tet., 1965, 3205.

0702-010

INVOLUTIN

$C_{17}H_{14}O_6$ 314
-23^0
172^0C

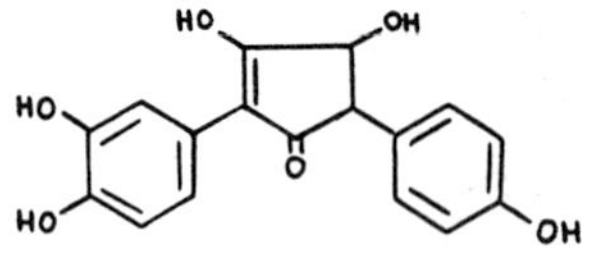

JCS, 1967, 405.

0799-001

08	MISCELLANEOUS PHENOLS

The absence of any carbon sidechain appended to the aromatic nucleus effectively prevents any meaningful biogenetic speculation. Thus, such "simple" phenols have been placed in this 08 class, which is subclassified according to the number of oxygen substituents on the aromatic ring. There are also some polyalkylated phenols the biogenetic origin of which are either ambiguous or complex (e.g., the xylenols and the tocopherols) and which are most conveniently deposited in the 0899 category.

0801	Mono-oxygenated
0802	Di-oxygenated
0803	Tri-oxygenated
0804	Tetra-oxygenated
0899	Complex and ambiguous (incl. polyalkylated and halogenated phenols)

PHENOL C_6H_6O 94
OI
42^0C

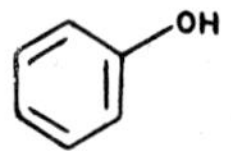

Phytochem., 1968, 278.

0801-001

PHENOL, 2-decaprenyl- $C_{56}H_{86}O$ 774
OI
Oil

R = (ip)$_{10}$

JACS, 1966, 5912.

0801-006

PHENOL, 2,6-di-Br- $C_6H_4OBr_2$ 250
OI
52^0C

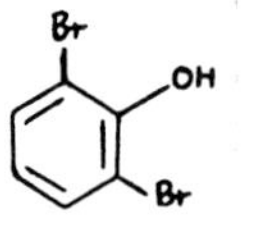

Science, 1967 (155) 1558.

0801-002

CATECHOL $C_6H_6O_2$ 110
OI
103^0C

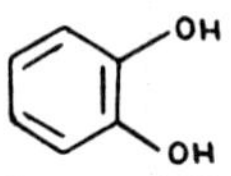

Compt. Rendu., 1955 (241) 48.

0802-001

PHENOL, 2,4-di-Cl- $C_6H_4OCl_2$ 162
OI
78^0C

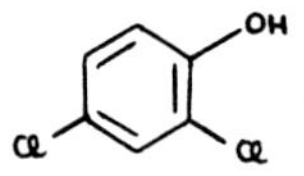

Biochem. Biophys. Res. Comm., 1970 (39) 1104.

0801-003

GUAIACOL $C_7H_8O_2$ 124
OI
30^0C

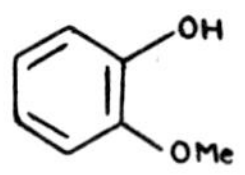

Helv., 1939 (22) 382.

0802-002

PHENOL, 2-tetraprenyl- $C_{26}H_{38}O$ 366
OI
Oil

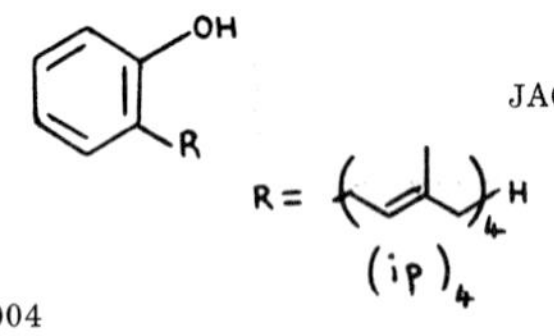

JACS, 1966, 5919.

$R = \{ \}_4$ (ip)$_4$

0801-004

PHENOL, 2-decaprenyl-6-OMe- $C_{57}H_{88}O_2$ 804
OI
Oil

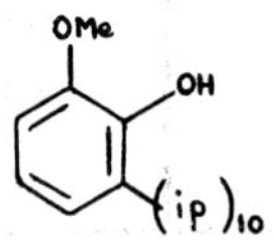

(ip)$_{10}$

JACS, 1966, 5919.

0802-003

PHENOL, 2-nonaprenyl- $C_{51}H_{78}O$ 706
OI
Oil

R = (ip)$_9$

Tet. Lett., 1967, 1237.

0801-005

PHENOL, m-OMe- $C_7H_8O_2$ 124
OI

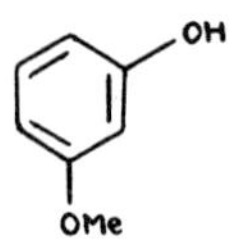

Phytochem., 1968, 278.

0802-004

90

HYDROQUINONE	$C_6H_6O_2$	110
		OI
		171^0C

RO — benzene ring — OH

R = H

J. Am. Pharm. Ass.,
1950 (39) 202.

0802-005

ARBUTIN, 6'-O-Galloyl-	$C_{19}H_{20}O_{11}$	424
		-31^0
		227^0C

R = 6-O-Galloyl-Glucosyl

JCS, 1965, 7312.

0802-010

ARBUTIN	$C_{12}H_{16}O_7$	272
		-60^0w
		142^0C

R = Glucosyl

JCS, 1952, 4740.

0802-006

ARBUTIN, 4-O-Galloyl-	$C_{19}H_{20}O_{11}$	424
		-20^0a/w
		139^0C

JCS, 1965, 7312.

0802-011

PYROSIDE	$C_{14}H_{18}O_8$	314
		-59^0w
		215^0C

R = 6-O-Ac-Glucosyl

JCS, 1964, 5649.

0802-007

HYDROQUINONE ME ETHER	$C_7H_8O_2$	124
		OI
		53^0C

JCS, 1968, 859.

0802-012

ARBUTIN, 2'-O-Caffeoyl-	$C_{21}H_{22}O_{10}$	434
		165^0C

R = 2-O-Caffeoyl-Glucosyl

JCS, 1964, 5649.

0802-008

HYDROQUINONE ETHYL ETHER	$C_8H_{10}O_2$	138
		OI
		66^0C

Arch. Pharm., 1891
(229) 84.

0802-013

ARBUTIN, 2'-O-Galloyl-	$C_{19}H_{20}O_{11}$	424
		-7.6^0
		165^0C

R = 2-O-Galloyl-Glucosyl

Tet., 1968 (24) 4015.

0802-009

HYDROQUINONE DIME ETHER	$C_8H_{10}O_2$	138
		OI
		56^0C

R = Me

JCS, 1968, 859.

0802-014

ARBUTIN, methyl-	$C_{13}H_{18}O_7$	286
		159^0C
R = Glucosyl		
JCS, 1952, 4740.		
0802-015		

PYROGALLOL 1-ME-ETHER	$C_7H_8O_3$	140
		OI
		40^0C
Ber., 1936 (69) 1870.		
0803-002		

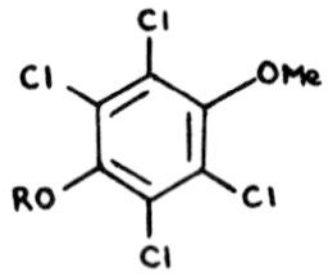

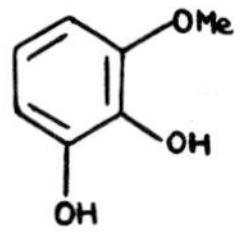

DROSOPHILIN-A $C_7H_4O_2Cl_4$ 260
OI
117^0C

JACS, 1952, 2943.

R = H

0802-016

PYROGALLOL 1,3-DI-ME ETHER $C_8H_{10}O_3$ 154
OI
56^0C

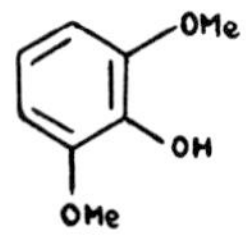

Bull. Soc. Chim. Fr.,
1904 (31) 478.

0803-003

DROSOPHILIN-A ME ETHER $C_8H_6O_2Cl_4$ 274
OI
165^0C

R = Me

Tet. Lett., 1966, 1229.

0802-017

HYDROQUINONE, OMe- $C_7H_8O_3$ 140
OI
83^0C

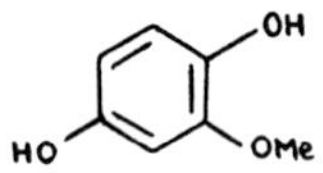

Ber., 1888 (21) 602;606.

0803-004

PLASTOHYDROQUINONE ME ETHER, $C_{29}H_{50}O_2$ 430
-phytyl-

(?)

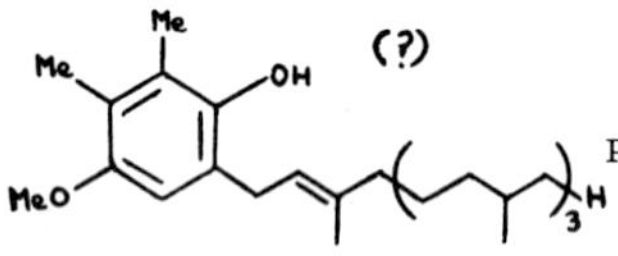

Phytochem., 1970, 213.

0802-018

SESAMOL $C_7H_6O_3$ 138
OI
66^0C

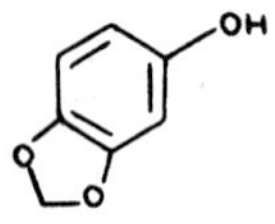

Chem. Abstr., 1954,
10652.

0803-005

PYROGALLOL $C_6H_6O_3$ 126
OI
133^0C

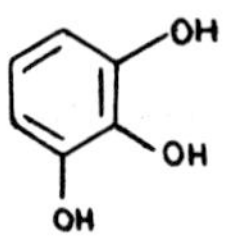

Chem. Ind., 1959, 1283.

0803-001

TAXICATIGENIN $C_8H_{10}O_3$ 154
OI
45^0C

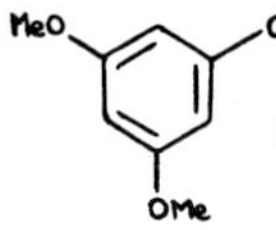

Monatsh., 1897 (18)736.

R = H

0803-006

TAXICATIN

$C_{14}H_{20}O_8$

-72^0

170^0C

R = Glucosyl

Arch. Pharm., 1907
(26) 241.

0803-007

TOCOPHEROL, delta-

$C_{27}H_{46}O_2$ 402

Oil

JACS, 1959, 3374.

0899-004

ANTIAROL

$C_9H_{12}O_4$ 184

OI

146^0C

Arch. Pharm., 1896
(234) 438.

0804-001

TOCOTRIENOL, alpha-

$C_{29}H_{44}O_2$ 424

Nature, 1965 (207) 521.

0899-005

TOCOPHEROL, alpha-

$C_{29}H_{50}O_2$ 430

$+1^0e$

3^0C

Chem. Comm., 1965, 40.

0899-001

TOCOTRIENOL, beta-

$C_{28}H_{42}O_2$ 410

Oil

Biochem. Biophys. Res.
Comm., 1964 (17) 542.

0899-006

TOCOPHEROL, beta-

$C_{28}H_{48}O_2$ 416

$+3^0e$

Oil

Helv., 1939 (22) 260.

0899-002

TOCOTRIENOL, gamma-

$C_{28}H_{42}O_2$ 410

Nature, 1965 (207) 521.

0899-007

TOCOPHEROL, gamma-

$C_{28}H_{48}O_2$ 416

$+2^0e$

-3^0C

Helv., 1939 (22) 260.

0899-003

TOCOTRIENOL, delta-

$C_{27}H_{40}O_2$ 396

Nature, 1965 (207) 521.
Biochem. J., 1966 (100) 138.

0899-008

PLASTOCHROMANOL-8 $C_{53}H_{82}O_2$ 750

Biochem. J., 1965 (96) 17 C.

0899-009

XYLENOL, 2,5- $C_8H_{10}O$ 122
OI

Phytochem., 1968, 278.

0899-014

SOLANACHROMENE $C_{53}H_{80}O_2$ 748

* dehydro

Chem. Abstr., 1967
(67) 73698.

0899-010

XYLENOL, 2,6- $C_8H_{10}O$ 122
OI

Phytochem., 1968, 278.

0899-015

TOCOPHEROL DIMER, α- $C_{58}H_{98}O_4$ 858

Chem. Abstr., 1971
(75) 140631.

0899-011

XYLENOL, 3,5- $C_8H_{10}O$ 122
OI

Phytochem., 1968, 278.

0899-016

XYLENOL, 2,3- $C_8H_{10}O$ 122
OI

Phytochem., 1968, 278.

0899-012

DYSIDEA SUBSTANCE-A $C_{12}H_5O_2Br_5$ (580)
OI
185^0C

Tet. Lett.,1972, 1715.

0899-017

XYLENOL, 2,4- $C_8H_{10}O$ 122
OI

Phytochem.,1968, 278.

0899-013

DYSIDEA SUBSTANCE-B $C_{12}H_8O_2Br_2$ (342)
OI
96^0C

Tet. Lett., 1972, 1715.

0899-018

ACETATE/SHIKIMATE AROMATICS

The extension of a shikimate-derived aroyl precursor by "acetate" has already been observed in the C_6C_n aromatics, but when the side chain itself is capable of cyclization, a new spectrum of aromatic compounds results. Thus the extension of a cinnamoyl precursor by three "acetate" units gives rise to a species capable of condensing to either a C_6-C_3-C_6 (e.g., chalcone) moiety or a stilbene. Similarly the benzoyl homologue can theoretically provide the benzophenone and biphenyl aromatic series (to date there is no clear-cut example of a biphenyl derived by this route). Modifications to these basis aromatic units in the form of heterocyclization, alkylation, and rearrangement account for the further diversification of structural types.

A class of compounds that has been uncovered only relatively recently is the neoflavonoid group. The neoflavonoids are probably derived by condensation of a shikimate moiety to a phenolic species derived from "acetate" (in many cases the phenol is a "phloroglucinol" type complete with acyl side chain). For convenience the 4-alkyl coumarins are included in the 09 class although the acyl precursor would undoubtedly be "acetate" derived.

Chart 2 overleaf provides a summary of the biogenetic relationships postulated for the acetate/shikimate aromatics.

09 Neoflavonoids		**16** Pterocarpans Coumestans (*oxo)	
10 $C_6-C_3-C_6$		**17** Anthocyanins	
11 Flavones		**1801** Rotenoids	
12 Flavanones		**1802** Peltogynans	
13 Flavans		**1803** Aurones	
14 Isoflavonoids		**2802** Benzophenones	
15 Biflavonoids		**2804** Stilbenes	

0901	
0902	
0903	 R = alkyl
0999	MISC.

OBTUSAQUINOL

$C_{16}H_{16}O_3$ 256

Chem. Comm., 1968, 1390.

0901-001

DALBERGIONE, 4-OMe-R-

$C_{16}H_{14}O_3$ 254
$+13^0$c
115^0C

Tet., 1965, 2683.

0901-006

LATIFOLIN

R = H

$C_{17}H_{18}O_4$ 386

Tet., 1965, 1495.

0901-002

DALBERGIONE, 4-OMe-S-

$C_{16}H_{14}O_3$ 254

S-Configuration

Tet., 1965, 2683.

0901-007

LATIFOLIN, 5-O-Me-

$C_{18}H_{20}O_4$ 300
-40^0m
106^0C

R = Me

Phytochem., 1968, 647.

0901-003

DALBERGIONE, 3,4-di-OMe-R-

$C_{17}H_{16}O_4$ 284
$+60^0$c
42^0C

Tet., 1965, 2697.

0901-008

DALBERGIONE QUINOL,
-R-3,4-diOMe-

$C_{17}H_{18}O_4$ 286

Tet., 1965, 2697.

0901-004

DALBERGIONE, 4'-OH-4-OMe-S-

R = H

$C_{16}H_{14}O_4$ 270
-52^0d
175^0C

Chem. Comm., 1968, 1390.

0901-009

KUHLMANNIQUINOL

$C_{18}H_{20}O_5$ 316

Oil

Chem. Comm., 1968, 1390.

0901-005

DALBERGIONE, 4,4'-di-OMe-S-

$C_{17}H_{16}O_4$ 284
-139^0c
110^0C

R = Me

Tet., 1965, 2683.

0901-010

CALOPHYLLIC ACID

$C_{25}H_{24}O_6$ 420
-58°c
215°C

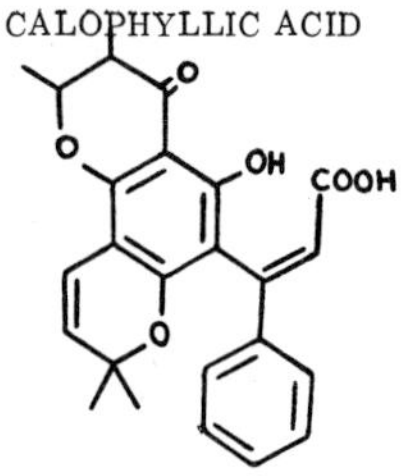

Bull. Soc. Chim. Fr.,
1957, 929.

0901-011

MAMMEA A/BA

$C_{25}H_{26}O_5$ 406
OI
125°C

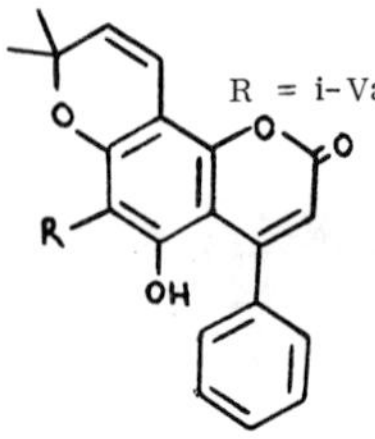

R = i-Valeryl

Tet. Lett., 1966, 145.

0902-005

MESUOL

R = i-Butyryl

$C_{24}H_{24}O_5$ 392
OI
154°C

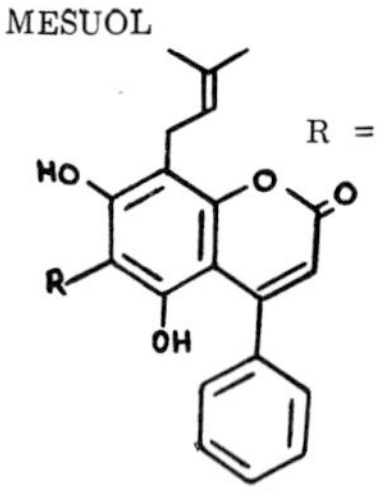

Tet. Lett., 1966, 5727.

0902-001

R = i-Valeryl

$C_{25}H_{26}O_6$ 422
149°C

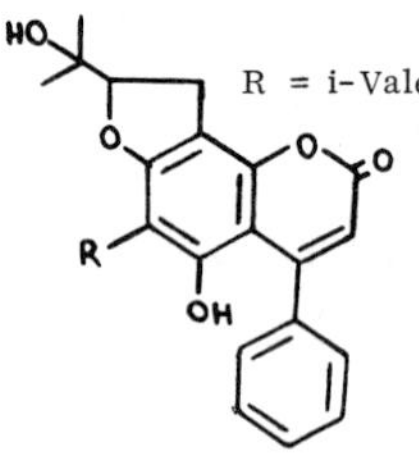

Tet. Lett., 1970, 3980.

0902-006

MAMMEISIN

R = i-Valeryl

$C_{25}H_{26}O_5$ 406
OI
100°C

Tet. Lett., 1966, 145.

0902-002

MAB-3

R = α-Me-Butyryl

$C_{25}H_{26}O_6$ 422
134°C

Tet. Lett., 1970, 3983.

0902-007

MAMMEA A/AB

R = α-Me-Butyryl

$C_{25}H_{26}O_5$ 406
108°C

Tet. Lett., 1966, 145.

0902-003

MEGUASIN

R = i-Butyryl

$C_{24}H_{22}O_5$ 390
OI
152°C

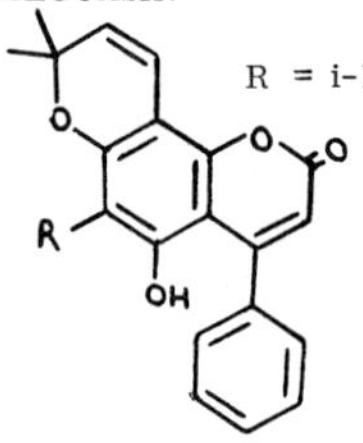

JOC, 1969, 3784.

0902-008

MAMMEA A/BB

R = α-Me-Butyryl

$C_{25}H_{26}O_5$ 406
124°C

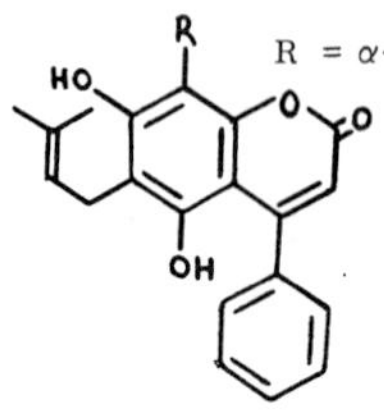

Tet. Lett., 1966, 145.

0902-004

MAMMEIGIN

R = i-Valeryl

$C_{25}H_{24}O_5$ 404
OI

Tet. Lett., 1966, 145.

0902-009

MAB-5 $C_{25}H_{24}O_5$ 404

                                                    79^0C

        R = α-Me-Butyryl

                              Tet. Lett., 1970, 3983.

0902-010

INOPHYLLOLIDE, dihydro- $C_{25}H_{24}O_5$ 404
                                                   $+43^0a$
                                                   201^0C

                              Tet. Lett., 1968, 2383.

0902-015

APETALOLIDE $C_{26}H_{24}O_5$ 416
                                                   OI
                                                   204^0C

                              Tet. Lett., 1967, 2633.

0902-011

TOMENTOLIDE-A $C_{25}H_{22}O_5$ 402
                                                   OI
                                                   203^0C

                              Tet. Lett., 1967, 2633.

0902-016

CALOPHYLLOLIDE $C_{26}H_{24}O_5$ 416

                                                   159^0C

                              Bull. Soc. Chim. Fr.,
                              1957, 929.

0902-012

DALBERGIN $C_{16}H_{12}O_4$ 268
                                                   OI
                                                   210^0C

                              Chem. Comm., 1968,
                                    1390.

0902-017

INOPHYLLOLIDE, cis- $C_{25}H_{22}O_5$ 402
                                                   $+70^0c$
                                                   150^0C

        *cis

                              Tet. Lett., 1968, 2383.

0902-013

DALBERGIN ME ETHER $C_{17}H_{14}O_4$ 282
                                                   OI

                              JCS, 1957, 970.

0902-018

INOPHYLLOLIDE, trans- $C_{25}H_{22}O_5$ 402
                                                   $+13^0c$
                                                   190^0C

        *trans

                              Tet. Lett., 1968, 2383.

0902-014

MELANNEIN $C_{17}H_{14}O_6$ 314
                                                   OI
                                                   222^0C

                              Tet., 1968, 2617.

0902-019

EXOSTEMIN

$C_{18}H_{16}O_6$ 328

OI

174^0C

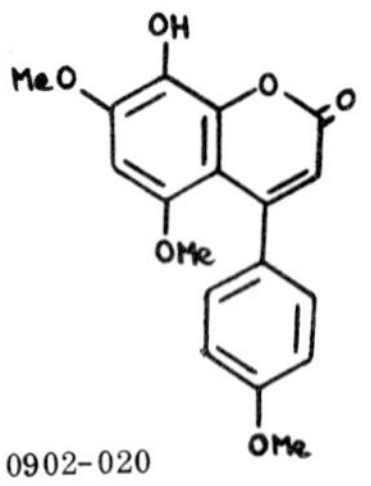

Chem. Abstr., 1967
(67) 108521.

0902-020

MAMMEA B/BB

$C_{22}H_{28}O_5$ 372

122^0C

R = α - Me - Butyryl

JCS, 1967, 2425.

0903-004

KUHLMANNIN

$C_{17}H_{14}O_5$ 298

OI

211^0C

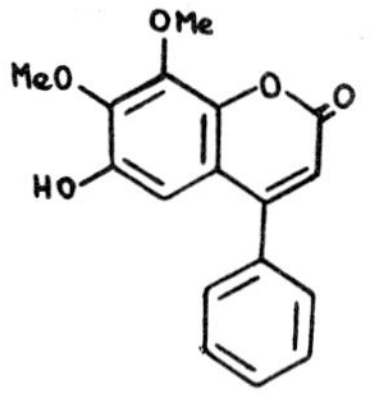

Chem. Comm., 1968,
1390.

0902-021

MAMMEA B/BA, OAc-

$C_{24}H_{30}O_7$ 430

R = i - Valeryl

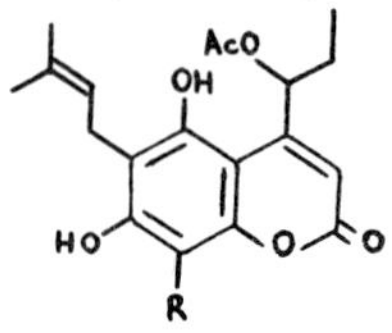

Tet. Lett., 1970, 251.

0903-005

MAMMEA B/AA

$C_{22}H_{28}O_5$ 372

109^0C

R = i - Valeryl

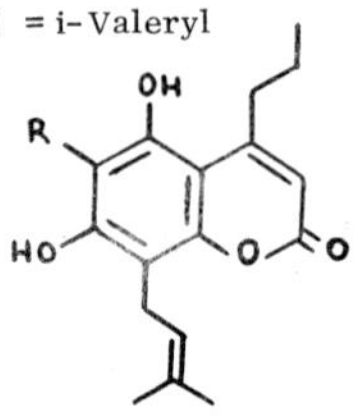

Tet. Lett., 1970, 3980.

0903-001

MAMMEA B/BB, OAc-

$C_{24}H_{30}O_7$ 430

R = α - Me - Butyryl

Tet. Lett., 1970, 251.

0903-006

MAMMEA B/AB

$C_{22}H_{28}O_5$ 372

114^0C

R = α - Me - Butyryl

Tet. Lett., 1970, 3980.

0903-002

MAMMEA B/BC

$C_{21}H_{26}O_5$ 358

OI

133^0C

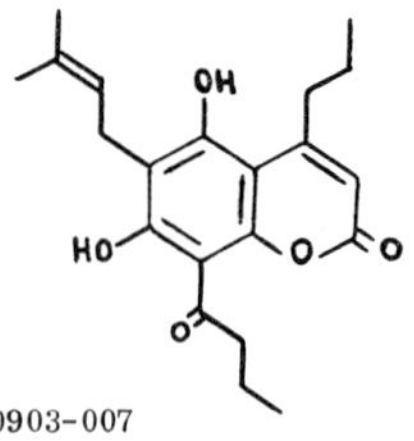

JCS, 1967, 2545.

0903-007

MAMMEA B/BA

$C_{22}H_{28}O_5$ 372

OI

127^0C

R = i - Valeryl

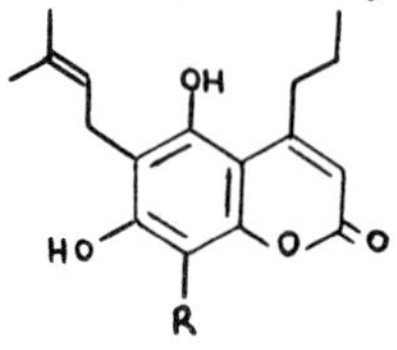

JCS, 1967, 2545.

0903-003

SURANGIN-A

$C_{27}H_{36}O_5$ 440

-2^0c

84^0C

R = H

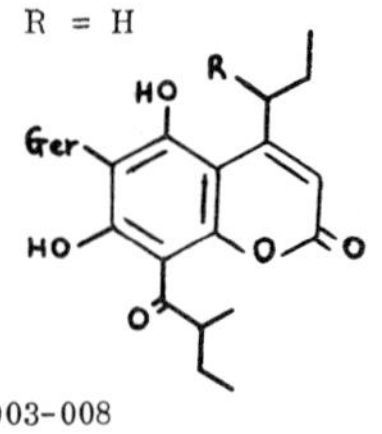

0903-008

SURANGIN-B R = OAc	$C_{29}H_{38}O_7$ 498 -30^0c 99^0C Tet., 1969, 1455.	$C_{22}H_{28}O_6$ 388 R = α-Me-Butyryl Tet. Lett., 1970, 3975.
	0903-009	0903-014

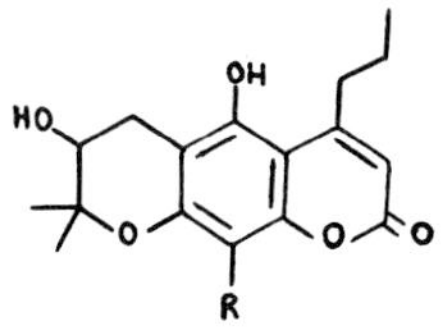

R = Butyryl

$C_{21}H_{26}O_6$ 374

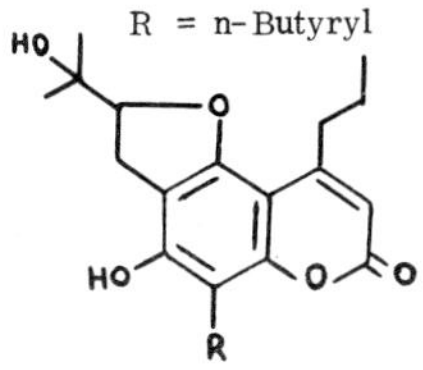

Tet. Lett., 1970, 3975.

0903-010

R = i-Valeryl

$C_{22}H_{28}O_6$ 388

Tet. Lett., 1970, 3975.

0903-015

R = α-Me-Butyryl

$C_{22}H_{28}O_6$ 388

Tet. Lett., 1970, 3975.

0903-011

$C_{22}H_{28}O_7$ 404

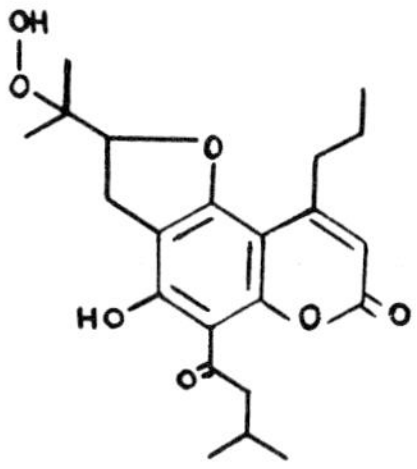

Tet. Lett., 1970, 3975.

0903-016

R = i-Valeryl

$C_{22}H_{28}O_6$ 388

Tet. Lett., 1970, 3975.

0903-012

R = i-Valeryl

$C_{22}H_{26}O_7$ 402

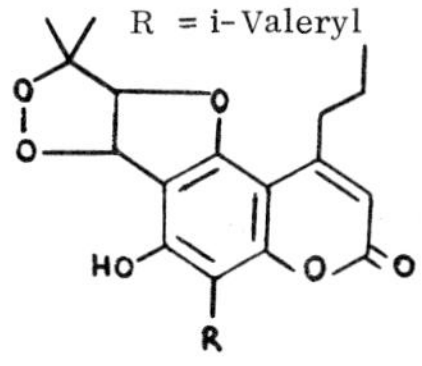

Tet. Lett., 1970, 3975.

0903-017

R = n-Butyryl

$C_{21}H_{26}O_6$ 374

Tet. Lett., 1970, 3975.

0903-013

R = α-Me-Butyryl

$C_{22}H_{26}O_7$ 402

Tet. Lett., 1970, 3975.

0903-018

R = i-Valeryl

$C_{22}H_{28}O_6$ 388

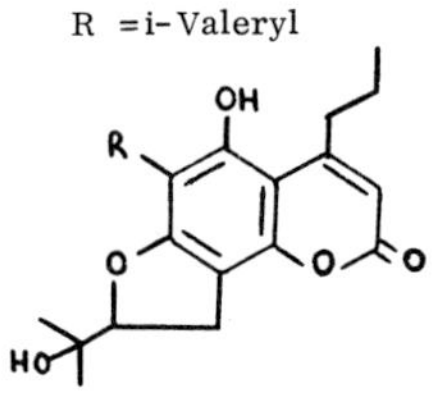

Tet. Lett., 1970, 3980.

0903-019

MAMMEA C/BB $C_{24}H_{32}O_5$ 400

100°C

R = n-pentyl

JCS, 1967, 2425.

0903-024

MAB-4 $C_{22}H_{28}O_6$ 388

84°C

R = α - Me- Butyryl

Tet. Lett., 1970, 3983.

0903-020

DALBERGICHROMENE $C_{16}H_{14}O_3$ 254
OI
100°C

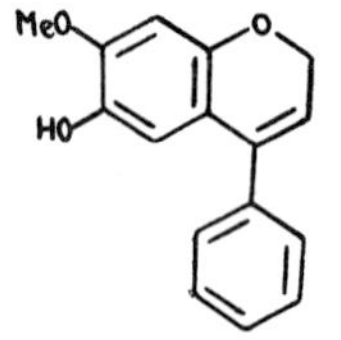

Tet., 1971, 799.

0999-001

MAB-6 $C_{22}H_{26}O_5$ 370

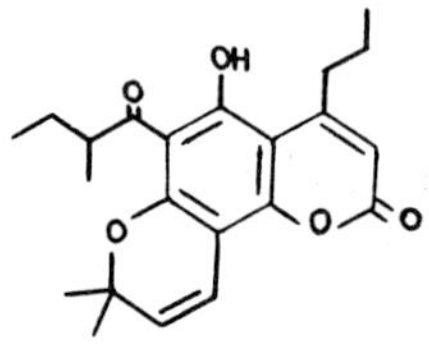

Tet. Lett., 1970, 3983.

0903-021

KUHLMANNENE $C_{17}H_{16}O_4$ 284

139°C

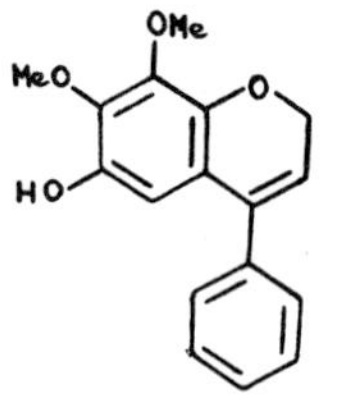

Chem. Comm., 1968,
1390.

0999-002

TOMENTOLIDE-B $C_{22}H_{24}O_5$ 368
OI
159°C

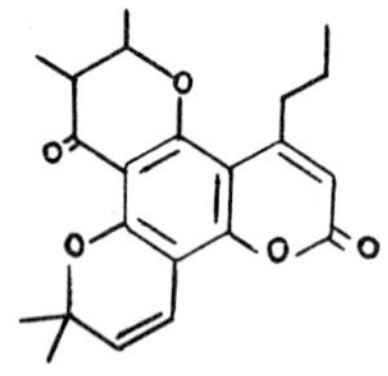

Tet. Lett., 1967, 2633.

0903-022

OBTUSAFURAN $C_{16}H_{16}O_3$ 256
+47°
111°C

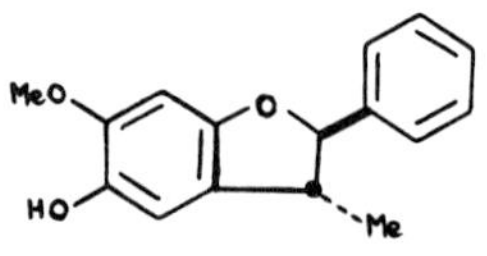

Chem. Comm., 1968,
1394.

0999-003

FERRUOL-A $C_{23}H_{30}O_5$ 386

130°C

R = .CH(Me) Et

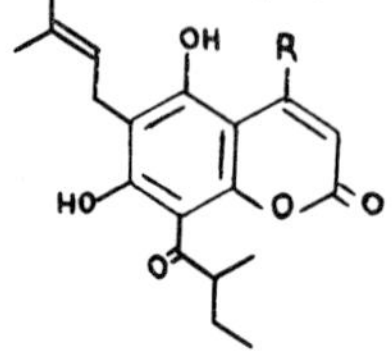

Tet., 1967, 4161.

0903-023

MELANOXIN $C_{17}H_{18}O_5$ 302
-49°a
108°C

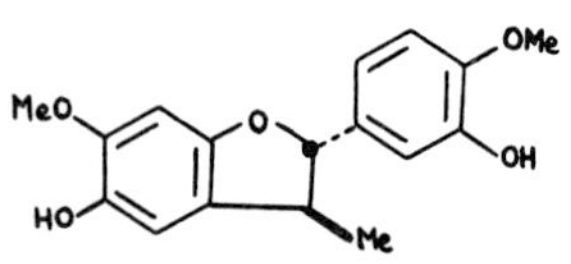

Tet., 1969, 4409.

0999-004

104

AUTUMNALIN	$C_{17}H_{14}O_6$	314 OI 245^0C

Tet. Lett., 1970, 475.

999-005

EUCOMIN, 4'-de-Me-5-O-Me- -dihydro-	$C_{17}H_{16}O_5$	300 -38^0d 196^0C

Experientia. 1970 (26)
472.

0999-010

AUTUMNALIN, 3,9-dihydro-	$C_{17}H_{16}O_6$	316 -10^0d 208^0C

*dihydro

Tet. Lett., 1970, 475.

0999-006

PUNCTATIN	$C_{17}H_{14}O_6$	314 OI 190^0C

R = H

Experientia, 1970 (26)
472.

0999-011

EUCOMIN, 4'-des-Me-	$C_{16}H_{12}O_5$	284 OI 211^0C

R = H

Experientia, 1970 (26)
472.

0999-007

PUNCTATIN, 3,9-dihydro-	$C_{17}H_{16}O_6$	316 -37^0d 205^0C

R = H

* dihydro

Experientia. 1970 (26)
472.

0999-012

EUCOMIN	$C_{17}H_{14}O_5$	334 OI 195^0C

R = Me

Tet. Lett., 1967. 3479.

0999-008

PUNCTATIN, 4'-O-methyl-	$C_{18}H_{16}O_6$	328 OI 214^0C

R = Me

Tet. Lett., 1970. 475.

0999-013

EUCOMOL	$C_{17}H_{16}O_6$	352 -32^0c 135^0C

Tet. Lett., 1967, 3479.

0999-009

PUNCTATIN, 4'-O-Me-3,9-dihydro-	$C_{18}H_{18}O_6$	330 amorph.

R = Me

*dihydro

Experientia, 1970 (26)
472.

0999-014

EUCOMNALIN	$C_{17}H_{14}O_6$	314
		OI
		245^0C

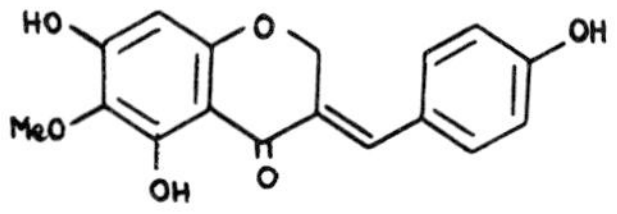

Tet. Lett., 1970, 475.

0999-015

ANGOLENSIN	$C_{16}H_{16}O_4$	272
		-123^0
		119^0C

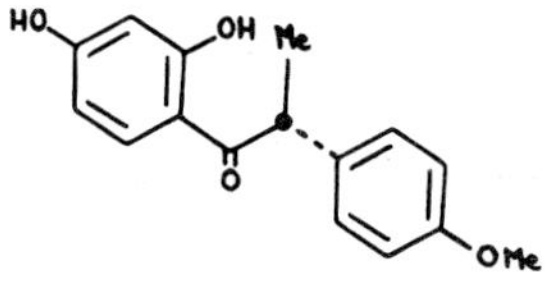

Aust. J. C., 1965, 1787.

0999-017

MARGINALIN	$C_{15}H_{10}O_4$	254
		OI

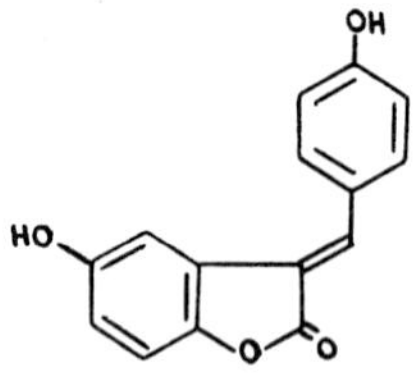

Ann. Chem., 1970 (734) 116.

0999-016

This class includes the chalcones and benzylstyrenes, and is subclassified according to the oxygenation pattern of the shikimate-derived ring. The miscellaneous section (1099) includes less common oxygenation patterns and also some other biogenetically interesting C$_6$-C$_3$-C$_6$ variations (e.g., the "interrupted-aromatic" ring of grandiflorone, 1099-015).

1001	
1002	
1003	
1004	
1099	MISC.

OBTUSASTYRENE	$C_{15}H_{14}O$	210
		OI

Chem. Comm. , 1968, 1390.

1001-001

VIOLASTYRENE, iso-	$C_{17}H_{18}O_3$	270
		OI
		87^0C

Chem. Comm. , 1968, 1390.

1001-006

OBTUSTYRENE	$C_{16}H_{16}O_2$	240
		OI

Chem. Comm. , 1968, 1390.

1001-002

FLEMICHAPPARIN	$C_{16}H_{14}O_4$	270
		OI
		159^0C

Tet. , 1971, 2111.

1001-007

LONCHOCARPIN	$C_{20}H_{18}O_3$	306
		OI
		108^0C

Chem. Abstr. , 1954, 10744.

1001-003

FLEMICHAPPARIN-A	$C_{20}H_{18}O_4$	322
		OI
		192^0C

Tet. Lett. , 1970, 4367.

1001-008

MUCRONUSTYRENE	$C_{17}H_{18}O_3$	270
		OI

Chem. Comm. , 1968, 1390.

1001-004

CHALCONE, 2',6'-di-OH-4'-OMe-	$C_{16}H_{14}O_4$	270
		OI
		161^0C

Chem. Ind. , 1969, 1779.

1001-009

VIOLASTYRENE	$C_{17}H_{18}O_3$	270
		OI
		85^0C

Chem. Comm. , 1968, 1390.

1001-005

deleted

1001-010

CARDAMONIN $\quad$ $C_{16}H_{14}O_4$ $\quad$ 270 $\quad$ 207^0C Chem. Abstr., 1968 (69) 35863. 1001-011	PEDICININ, O-Me- $\quad$ $C_{17}H_{14}O_6$ $\quad$ 314 OI 111^0C $\quad$ R = Me $\quad$ Rev. Pure & App. Chem., 1951 (1) 186. 1001-016
MALLOTUS CHALCONE $\quad$ $C_{21}H_{20}O_4$ $\quad$ 336 $\quad$ 121^0C JCS, 1968, 2627. 1001-012	$C_{21}H_{22}O_{10}$ $\quad$ 434 Chem. Ind., 1967, 1526. 1001-017
PEDICIN $\quad$ $C_{18}H_{18}O_6$ $\quad$ 330 J. Ind. Chem. Soc., 1939 (16) 1. 1001-013	PONGAMOL $\quad$ $C_{18}H_{14}O_4$ $\quad$ 294 OI 128^0C JCS, 1955, 2048. 1001-018
PEDICELLIN $\quad$ $C_{20}H_{22}O_6$ $\quad$ 358 OI 93^0C JCS, 1941, 662. 1001-014	$C_{18}H_{20}O_4$ $\quad$ 300 OI 116^0C Acta Chem. Scand., 1971, 1929. 1001-019
PEDICININ $\quad$ $C_{16}H_{12}O_6$ $\quad$ 300 OI 204^0C $\quad$ R = H J. Ind. Chem. Soc., 1937 (14) 703. 1001-015	deleted 1001-020

deleted

1001-021

LIQUIRITIN, iso- $C_{21}H_{22}O_9$ 418

R = Glucosyl

Ber., 1881, 2463.

1002-002

CHALCONE, 2,4-di-OH-6-OMe- $C_{16}H_{14}O_4$ 270
OI

R = H

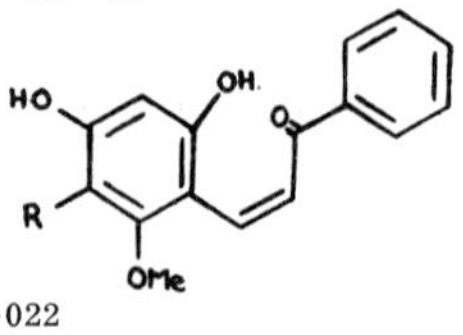

JCS, 1971, 3967.

1001-022

LIQUIRITIGENIN, 4'-O-Me-iso- $C_{16}H_{14}O_4$ 270
OI
174^0C

R = Me

JCS, 1954, 2562.

1002-003

CHALCONE, 2,4-di-OH-5-Me-6-OMe- $C_{17}H_{16}O_4$ 284
OI

R = Me

JCS, 1971, 3967.

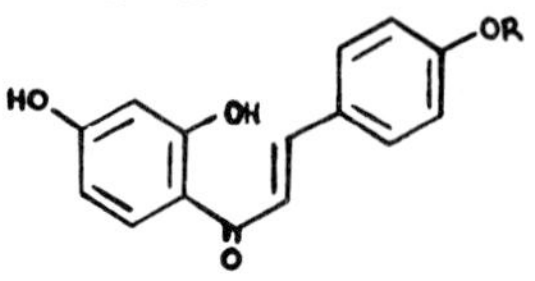

1001-023

LIQUIRITIGENIN, 4'-Glucosyl-iso- $C_{21}H_{22}O_9$ 418

R = Glucosyl

Chem. Abstr., 1971
(74) 50573.

1002-004

CHALCONE, 2-OH-4,6-diOMe- $C_{17}H_{16}O_4$ 284
OI
114^0C

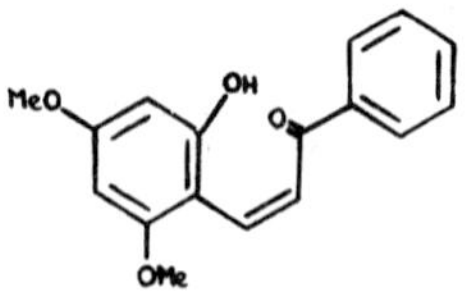

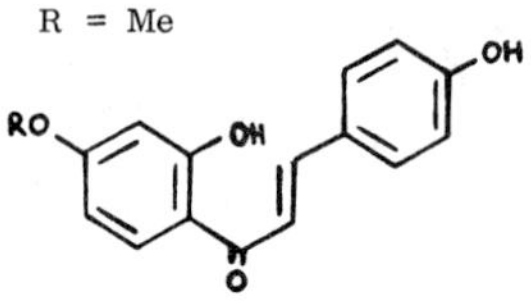

Chem. Abstr., 1955
(49) 11273.

1001-024

LIQUIRITIGENIN, 4'-diglucosyl-iso- $C_{27}H_{32}O_{11}$ 580

R = Glu-Glu

Arch. Biochem. Biophys.,
1956 (60) 329.

1002-005

LIQUIRITIGENIN, iso- $C_{15}H_{12}O_4$ 256
OI
202^0C

R = H

JCS, 1953, 2185.

1002-001

SALIPURPOL, iso- $C_{15}H_{12}O_5$ 272
OI

R = H

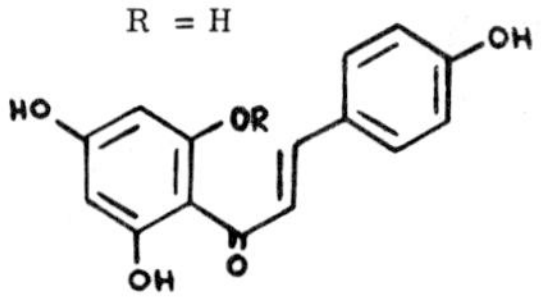

Ber., 1943 (76) 386.

1002-006

SALIPURPOSIDE, iso-	$C_{21}H_{22}O_{10}$	434

R = Glucosyl

Ber., 1943 (76) 386.

1002-007

SAKURANIN, neo-	$C_{22}H_{26}O_{10}$	450

R = Glucosyl

Chem. Abstr., 1955 (49) 4942.

1002-012

PHLORETIN	$C_{15}H_{14}O_5$	274 OI 257^0C

R = H

Ber., 1942 (75) 645.

1002-008

ASEBOGENIN	$C_{16}H_{16}O_5$	288 OI 168^0C

R = H

J. Chem. Soc. Jap., 1936 (57) 1141.

1002-013

PHLORIZIN	$C_{21}H_{24}O_{10}$	436

R = Glucosyl

J. Pharm. Soc. Jap., 1955 (75) 603.

1002-009

ASEBOTIN	$C_{22}H_{26}O_{10}$	450

R = Glucosyl

J. Chem. Soc. Jap., 1936 (57) 1141.

1002-014

GLYCYPHYLLIN	$C_{21}H_{26}O_9$	422

R = Rhamnosyl

JCS, 1886 (49) 857.

1002-010

CHALCONE, 2'-OH-4, 4', 6', - -tri-OMe-	$C_{18}H_{18}O_5$	314 OI 115^0C

Z. Nat., 1963 (18 B) 370.

1002-015

CHALCONOSAKURANETIN	$C_{16}H_{16}O_5$	288 OI

R = H

Chem. Abstr., 1955 (49) 4942.

1002-011

XANTHOHUMOL	$C_{21}H_{22}O_5$	354 172^0C

Ann., 1963 (663) 74.

1002-016

BAVACHALCONE, iso- $C_{20}H_{20}O_4$ 324
OI
155^0C

R = H

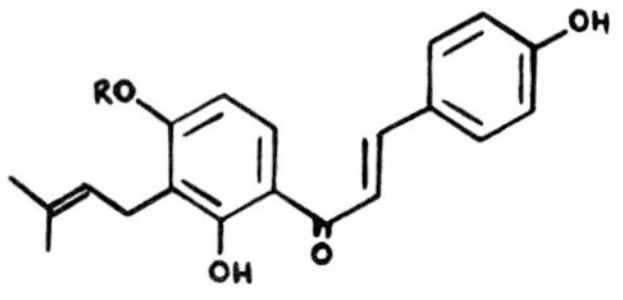

Tet. Lett., 1968, 2401.

1002-017

CARTHAMIN, iso- $C_{21}H_{22}O_{11}$ 450

228^0C

JCS, 1930, 752.

1002-022

BAVACHALCONE $C_{21}H_{22}O_4$ 338
OI
161^0C

R = Me

Tet. Lett., 1968, 2401.

1002-018

CARTHAMONE $C_{21}H_{20}O_{11}$ 448

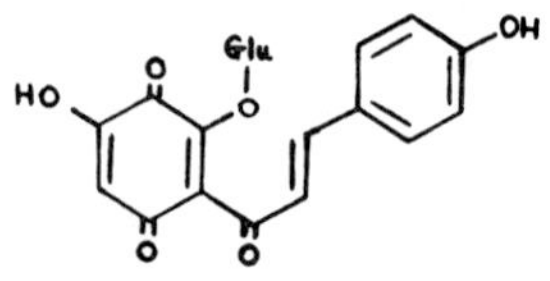

Curr. Sci., 1960 (29) 57.

1002-023

SOPHORADIN $C_{30}H_{36}O_4$ 460
OI
161^0C

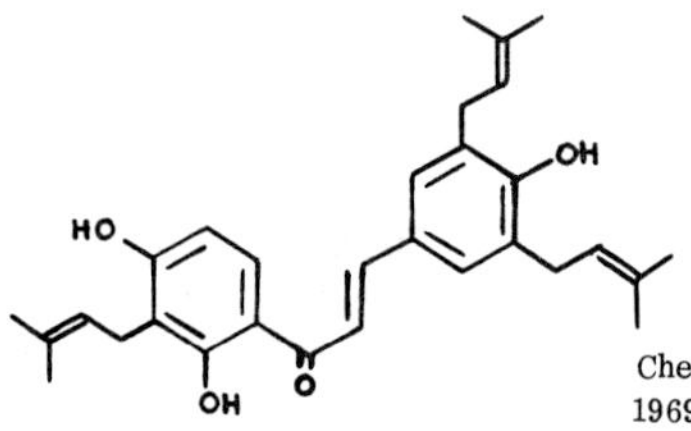

Chem. Pharm. Bull.,
1969, 1299.

1002-019

BUTEIN $C_{15}H_{12}O_5$ 272
OI
214^0C

R = H

JACS, 1956, 825.

1003-001

SOPHORADOCHROMENE $C_{30}H_{34}O_4$ 458
OI
154^0C

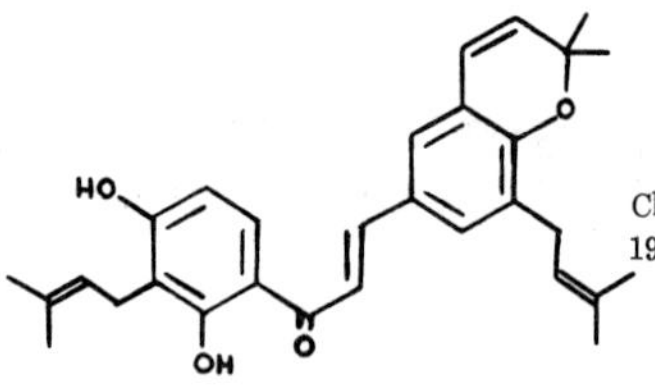

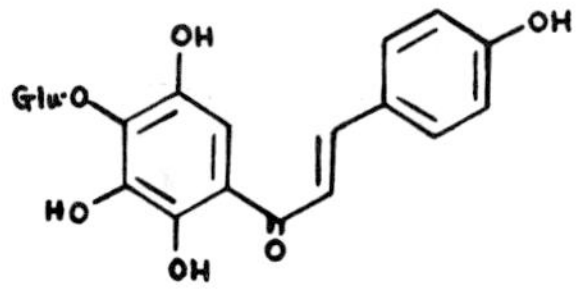

Chem. Pharm. Bull.,
1970, 742.

1002-020

COREOPSIN $C_{21}H_{22}O_{10}$ 434

193^0C

R = Glucosyl

JACS, 1953, 1900.

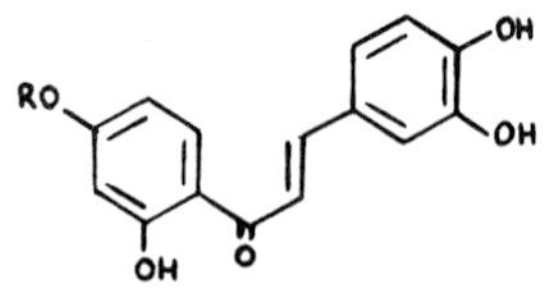

1003-002

CHALCONOCARTHAMIDIN $C_{15}H_{12}O_6$ 288

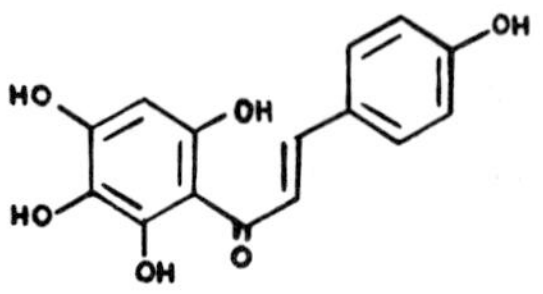

J. Chem. Soc. Jap.,
1930 (51) 237.

1002-021

MONOSPERMOSIDE $C_{21}H_{22}O_{10}$ 434

194^0C

R = H

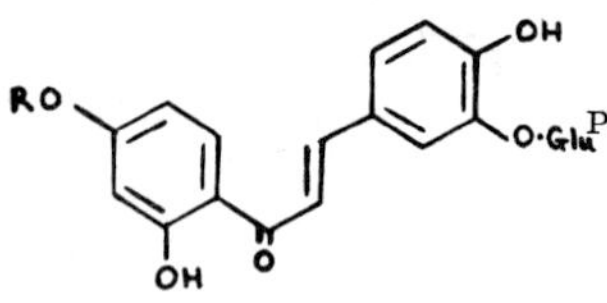

Phytochem., 1970, 2231.

1003-003

BUTRIN, iso- $C_{27}H_{32}O_{15}$ 596 190^{0}C R = Glucosyl JCS, 1955, 1589. 1003-004	LANCEOLETIN $C_{16}H_{14}O_6$ 302 OI R = H JACS, 1953, 1900. 1003-009
STILLOPSIDIN $C_{15}H_{12}O_6$ 288 OI 232^{0}C JACS, 1956, 1196. 003-005	LANCEOLIN $C_{22}H_{24}O_{11}$ 464 R = Glucosyl JACS, 1953, 1900. 1003-010
OKANIN $C_{15}H_{12}O_6$ 288 OI 238^{0}C JACS, 1956, 825. 1003-006	ASPALATHIN $C_{21}H_{24}O_{11}$ 452 Tet. Lett., 1965, 3497. 1003-011
deleted 1003-007	ROBTEIN $C_{15}H_{12}O_6$ 288 OI Biochem, J., 1962 (84) 416. 1004-001
MAREIN $C_{21}H_{22}O_{11}$ 450 JACS, 1956, 825; 929. 1003-008	ECHINATIN $C_{16}H_{14}O_4$ 270 210^{0}C Tet. Lett., 1971, 2567. 1099-001

MUCRONULASTYRENE $C_{17}H_{18}O_4$ 286

R = H

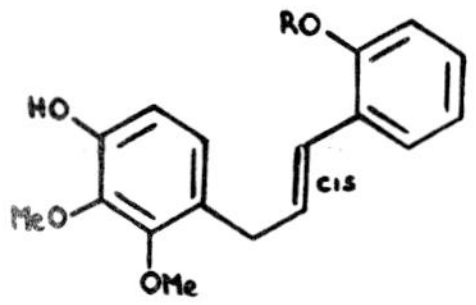

Chem. Comm., 1968, 1390.

1099-002

FLEMINGIN-A $C_{25}H_{26}O_5$ 406
-4^0c
149^0C

R = H

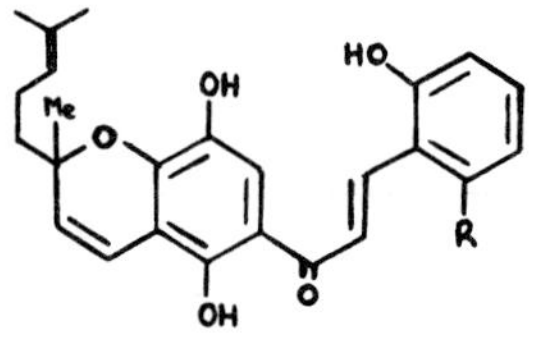

Tet., 1968, 500.

1099-007

VILLOSTYRENE $C_{18}H_{20}O_4$ 300

R = Me

Chem. Comm., 1968, 1390.

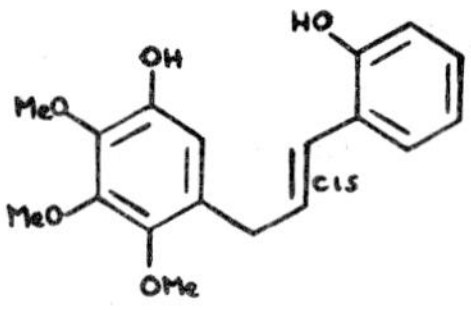

1099-003

FLEMINGIN-B $C_{25}H_{26}O_6$ 422
$+7^0$e
177^0C

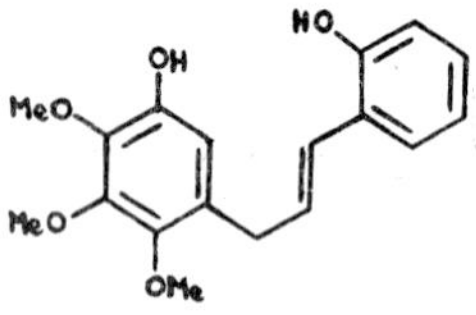

R = OH

Tet., 1968, 500.

1099-008

KUHLMANNISTYRENE $C_{18}H_{20}O_5$ 316
OI

Chem. Comm., 1968, 1390.

1099-004

FLEMINGIN, homo- $C_{26}H_{30}O_5$ 438
OI
161^0C

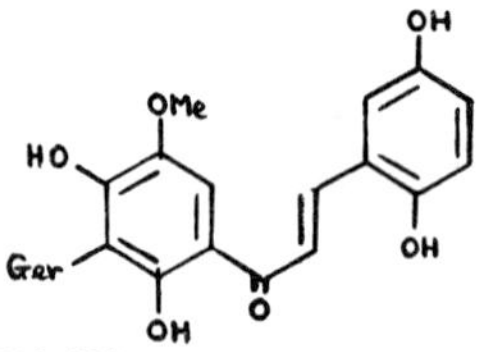

Tet., 1968, 500.

1099-009

PETROSTYRENE $C_{18}H_{20}O_5$ 316
OI

Chem. Comm., 1968, 1390.

1099-005

FLEMINGIN-C $C_{25}H_{26}O_5$ 406
$+2^0$e
181^0C

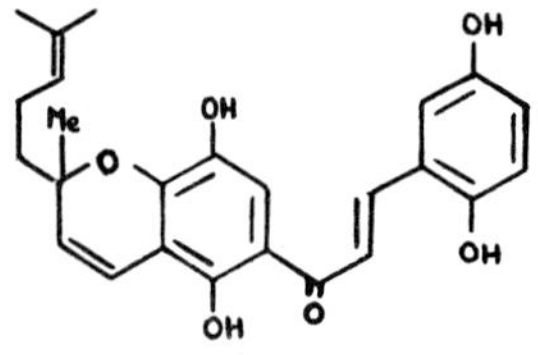

Tet., 1968, 500.

1099-010

FLEMINGIN, deoxy-homo- $C_{25}H_{28}O_5$ 408
OI

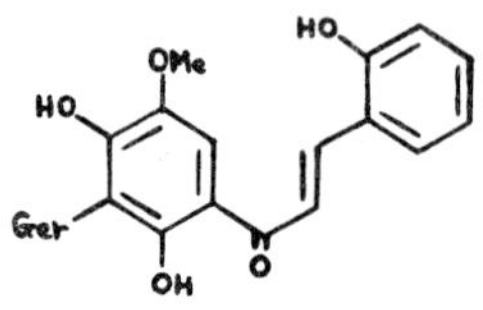

Chem. Abstr., 1968 (68) 114755.

1099-006

HYSSOPIN $C_{16}H_{12}O_6$ 300
OI
266^0C

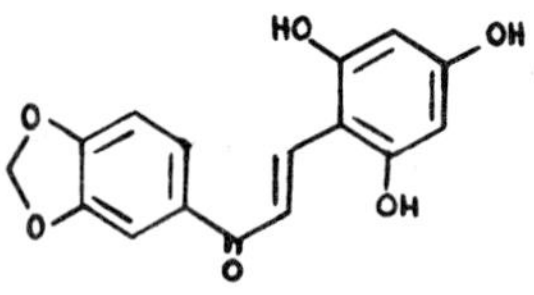

JACS, 1929, 1267.

1099-011

KURARIDIN $C_{26}H_{30}O_6$ 438
+7⁰e
115⁰C

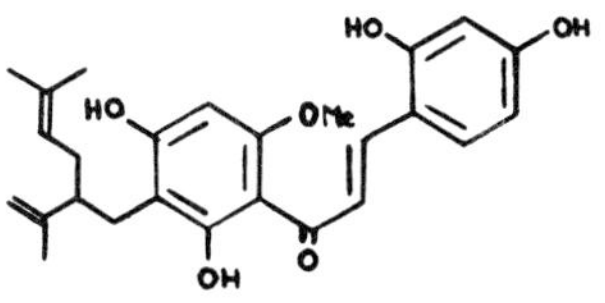

Chem. Pharm. Bull.,
1971, 2126.

1099-012

$C_{18}H_{20}O_4$ 300
138⁰C

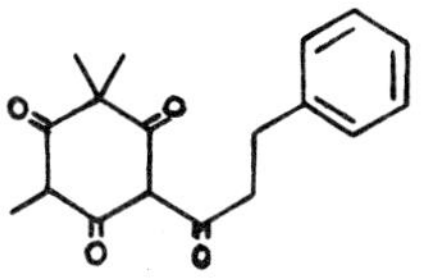

Acta Chem. Scand.,
1971, 1929.

1099-014

OBTUSAQUINONE $C_{16}H_{14}O_3$ 254
OI
178⁰C

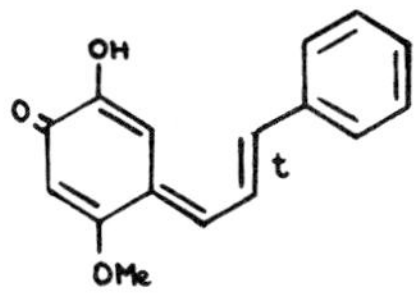

Chem. Comm., 1968,1396.
Tet. Lett., 1972, 2149.

1099-013

GRANDIFLORONE $C_{19}H_{22}O_4$ 314
OI
32⁰C

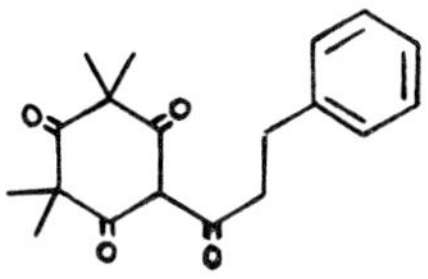

JCS, 1966, 1496.

1099-015

The flavones are a very large class of naturally occurring compounds (approximately 400 included here) that can be suitably subclassified according to the oxygenation patterns of the shikimate-derived aromatic ring (C). In addition the presence or absence of an oxy substitutent at position 3 is a noteworthy feature for subclassification purposes:

1101	No oxy substituent	1102	as 1101 + 3-oxy
1103	4'-oxy	1104	as 1103 + 3-oxy
1105	3',4'-dioxy	1106	as 1105 + 3-oxy
1107	3',4',5'-trioxy	1108	as 1107 + 3-oxy
1199	Misc.		

FLAVONE	$C_{15}H_{10}O_2$	222 OI 100^0C
		JACS, 1945, 491.
1101-001		

TECTOCHRYSIN	$C_{16}H_{12}O_4$	268 OI 165^0C
R = Me		
		Chem. Abstr., 1955 (49) 11273.
1101-006		

FLAVONE, 5-OH-	$C_{15}H_{10}O_3$	238 OI 159^0C
		JCS, 1934, 1483.
1101-002		

AEQUINOCTIN	$C_{21}H_{20}O_9$	430 245^0C
R = Glucosyl		
		Can. J. C., 1971. 49.
1101-007		

LANCEOLATIN-B	$C_{17}H_{10}O_3$	262 OI 138^0C
		Chem. Abstr., 1955 (49) 10278.
1101-003		

CHRYSIN, Glucuronosyl-	$C_{21}H_{18}O_{10}$	444 -112^0 225^0C
R = Glucuronosyl		
		Biochem. J., 1955. 58.
1101-008		

CHRYSIN	$C_{15}H_{10}O_4$	254 OI 290^0C
R = H		
		Ber., 1877 (10) 176.
1101-004		

STROBOCHRYSIN	$C_{16}H_{12}O_4$	268 OI 285^0C
		Chem. Abstr., 1955 (49) 8739.
1101-009		

TORINGIN	$C_{21}H_{20}O_9$	416 240^0C
R = Glucosyl		
		J. Pharm. Soc. Jap., 1909, 1.
1101-005		

GAMATIN	$C_{18}H_{12}O_4$	292 OI 181^0C
		Chem. Abstr., 1956, 13008.
1101-010		

BAICALEIN

$C_{15}H_{10}O_5$ 270
OI
265^0C

R = H

JACS, 1955, 5390.

1101-011

PRIMETIN

$C_{15}H_{10}O_4$ 254
OI
231^0C

JCS, 1939, 1922.

1101-016

BAICALIN

$C_{21}H_{18}O_{11}$ 460
-145^0
223^0C

R = Glucuronosyl

JCS, 1936, 592.

1101-012

WOGONIN

$C_{16}H_{12}O_5$ 284
OI
203^0C

JCS, 1938, 1555.

1101-017

OROXYLIN-A

$C_{16}H_{12}O_5$ 284
OI
220^0C

JCS, 1936, 591.

1101-013

ALNETIN

$C_{18}H_{16}O_6$ 328
OI
100^0C

Bull. Chem. Soc. Jap.,
1971, 2761.

1101-018

PONGAMIA GLABRA FURANO-
FLAVONE

$C_{18}H_{12}O_4$ 292
OI
190^0C

JCS, 1963, 163.

1101-014

TACHROSIN

$C_{23}H_{20}O_6$ 392
OI
226^0C

J. S. Afr. Chem. Inst.,
1971 (24) 1.

1101-019

FLAVONE, 5, 6-diOMe-

$C_{17}H_{14}O_4$ 282
OI
199^0C

JCS, 1956, 4170.

1101-015

GALANGIN

$C_{15}H_{10}O_5$ 270
OI
220^0C

R = R' = H

JCS, 1933, 368.

1102-001

IZALPININ	$C_{16}H_{12}O_5$ 284 OI 195^{0}C	
R = H ; R' = Me		
	J. Pharm. Soc. Jap., 1935, 229.	
1102-002		

ALNUSTIN	$C_{18}H_{16}O_6$ 328 OI 176^{0}C	
R = Me		
	Bull. Chem. Soc. Jap., 1971, 2761.	
1102-007		

GALANGIN 3-ME ETHER	$C_{16}H_{12}O_5$ 284 OI 300^{0}C
R = Me ; R' = H	
	JCS, 1925, 181.
1102-003	

GNAPHALIIN $C_{17}H_{14}O_6$ 314 OI 174^{0}C

R = H

Tet. Lett., 1969, 431.
Ber., 1971 (104) 2381.

1102-008

SERICETIN $C_{25}H_{24}O_5$ 404 OI 158^{0}C

Proc. C. S., 1960, 177.

1102-004

GNAPHALIIN 3-ME ETHER	$C_{18}H_{16}O_6$ 328 OI 177^{0}C
R = Me	
	Ber., 1971 (104) 2381.
1102-009	

KARANJIN $C_{18}H_{12}O_4$ 292 OI 159^{0}C

Chem. Abstr., 1968 (69) 41688.

1102-005

GNAPHALIIN, iso- $C_{17}H_{14}O_6$ 314 OI 217^{0}C

Ber., 1971 (104) 2381.

1102-010

ALNUSIN $C_{16}H_{12}O_6$ 356 OI 240^{0}C

R = H

Bull. Chem. Soc. Jap., 1971, 2761.

1102-006

FLAVONE, 3, 5, 7-**tri**-OH-6, 8- -di-OMe- $C_{17}H_{14}O_7$ 330 OI

Arch. Pharm., 1971 (304) 213.

1102-011

PRATOL	$C_{16}H_{12}O_4$	268 OI 264^0C

JCS, 1926, 2344.

1103-001

	$C_{21}H_{18}O_{11}$	446 -103^0 173^0C

R = Glucuronosyl

Ber., 1971 (104) 2681.

1103-006

APIGENIN $C_{15}H_{10}O_5$ 270   OI 349^0C

R = H

Chem. Pharm. Bull., 1955 (3) 469.

1103-002

$C_{27}H_{30}O_{14}$ 578

R = neohesperidosyl

Pharm. Weekbl., 1971 (106) 337.

1103-007

THALICTIIN $C_{21}H_{20}O_{10}$ 432 -116^0 238^0C

R = Galactosyl

Ber., 1969, 792.

1103-003

RHOIFOLIN $C_{27}H_{30}O_{14}$ 578 -160^0m 250^0C

R = .Rha-Glu

Khim. Prir. Soedin., 1970 (6) 365.

1103-008

$C_{22}H_{20}O_{11}$ 460 246^0C

R = 7-Me-Galacturonosyl

Phytochem., 1970, 1595.

1103-004

APIIN $C_{26}H_{28}O_{14}$ 564 -130^0 236^0C

R = .Apiosyl-Glu

J. Pharm. Soc. Jap., 1935, 977.

1103-009

COSMOSIIN $C_{21}H_{20}O_{10}$ 432 226^0C

R = Glucosyl

J. Pharm. Soc. Jap., 1940, 502.

1103-005

$C_{26}H_{28}O_{14}$ 564

R = .Xylosyl-Glu

Khim. Prir. Soedin., 1969, 595.

1103-010

APIGENIN-5-ME ETHER	$C_{16}H_{12}O_5$	284
		OI

R = Me

Arch. Pharm., 1970 (303) 792.

1103-011

VITEXIN, 2''-O-xylosyl-	$C_{26}H_{28}O_{14}$	564

R = .Glucosyl-Xylosyl

JOC, 1968, 1571.

1103-016

APIGENIN-5-GLUCOSIDE	$C_{21}H_{20}O_{10}$	432
		295^0C

R = Glucosyl

Ber., 1943, 776.

1103-012

VITEXIN, 2''-para-OH-benzoyl-	$C_{28}H_{24}O_{12}$	552
		204^0C

R = 2''-(p-OH-benzoyl)-Glucosyl

Chem. Ind., 1966, 625.

1103-017

VITEXIN, iso-	$C_{21}H_{20}O_{10}$	432
		$+16^0$e
		239^0C

R = H

Tet. Lett., 1966, 3657.

1103-013

VICENIN-2	$C_{27}H_{30}O_{15}$	594
		235^0C

JOC, 1968, 1571.

1103-018

SAPONARIN	$C_{27}H_{30}O_{15}$	604
		-8^0w
		228^0C

R = Glucosyl

J. Pharm. Soc. Jap., 1944, 304.

1103-014

VIOLANTHIN	$C_{27}H_{30}O_{15}$	594
		229^0C

Isomer of Vicenin-2?

Tet. Lett., 1965, 1707.

1103-019

VITEXIN	$C_{21}H_{20}O_{10}$	432
		-14^0p
		260^0C

R = Glucosyl

Tet., 1958 (3) 269.

1103-015

GENKWANIN	$C_{16}H_{12}O_5$	284
		OI
		286^0C

R = H

Tet. Lett., 1968, 3447.

1103-020

GENKWANIN, gluco-	$C_{22}H_{22}O_{10}$	446
		263^0C
R = Glucosyl		
	JACS, 1954, 5559.	
1103-021		

ACACIIN	$C_{28}H_{32}O_{14}$	592
		-85^0p
		263^0C
R = .Glu-Rha		
	Chem. Abstr., 1951 (45) 7977.	
1103-026		

SWERTISIN	$C_{22}H_{22}O_{10}$	446
		-10^0p
		243^0C

Tet. Lett., 1966, 1611.

1103-022

FORTUNELLIN	$C_{28}H_{32}O_{14}$	592
		215^0C
R =. Glu-Rha		
	J. Pharm. Soc. Jap., 1958, 1311.	
1103-027		

ACACETIN	$C_{16}H_{12}O_5$	284
		OI
		262^0C
R = H		

JCS, 1951, 691.

1103-023

	$C_{28}H_{28}O_{17}$	636
		-48^0p
		191^0C
R = (Glucuronosyl)$_2$		
	Chem. Pharm. Bull., 1971, 148. Tet. Lett., 1970, 2935.	
1103-028		

ACACETIN-7-ME ETHER	$C_{17}H_{14}O_5$	298
		OI
		174^0C
R = Me		
	Phytochem., 1971, 1942. Tet. Lett., 1968, 3447.	
1103-024		

VITEXIN, 4'-O-Me-	$C_{22}H_{22}O_{10}$	446
		-22^0p
		236^0C

Tet. Lett., 1966, 3657.

1103-029

LINARIN	$C_{28}H_{32}O_{14}$	592
		-100^0
		265^0C
R = Rutinosyl		
	JCS, 1951, 691.	
1103-025		

FLAVONE, 5-OH-4',7-diOMe-6-Me-	$C_{18}H_{16}O_5$	312
		OI
		188^0C

Aust. J. C., 1964 (17) 692.

1103-030

EUCALYPTIN $C_{19}H_{18}O_5$ 326
OI
185°C

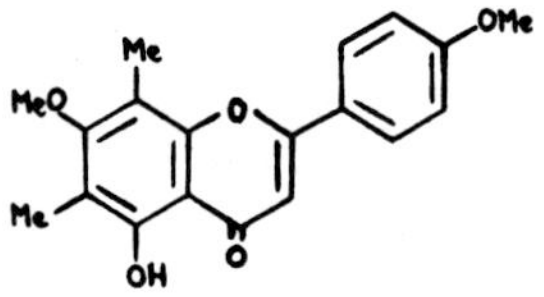

Ind. J. C. , 1966 (4) 481.

1103-031

SORBARIN $C_{21}H_{20}O_{10}$ 432
>300°C

R = Rhamnosyl

Chem. Pharm. Bull. ,
1970, 916.

1103-036

SCUTELLAREIN $C_{15}H_{10}O_6$ 286
OI
>340°C

R = H

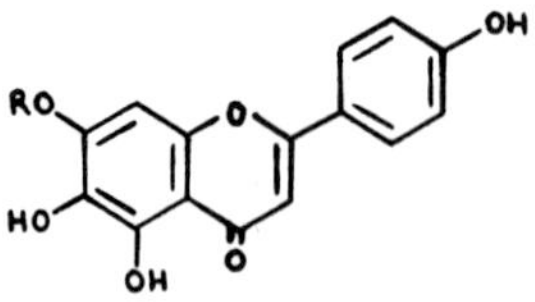

JOC, 1966, 3228.

1103-032

$C_{26}H_{30}O_{16}$ 610

R = Glucobiosyl

Khim. Prir. Soedin. .
1970 (6) 534.

1103-037

SORBIFOLIN $C_{16}H_{12}O_6$ 300
OI
291°C

R = Me

Chem. Pharm. Bull. .
1970, 916.

1103-033

SCUTELLARIN $C_{21}H_{18}O_{12}$ 462
-140°
>310°C

R = Glucuronosyl

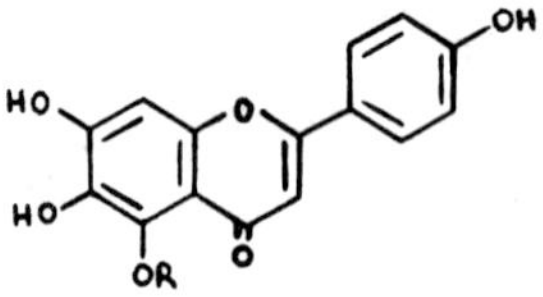

Biochem. J. , 1955 (59) 58.

1103-038

$C_{21}H_{20}O_{11}$ 448

R = Glucosyl

Khim. Prir. Soedin. ,
1970 (6) 534.

1103-034

SCUTELLAREIN, 4'-O-Me- $C_{16}H_{12}O_6$ 300
OI
251°C

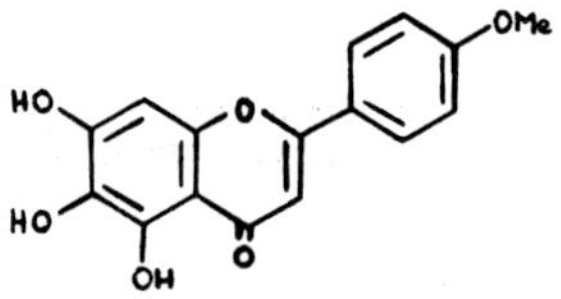

Khim. Prir. Soedin. ,
1971 (7) 373.

1103-039

$C_{21}H_{18}O_{12}$ 462

R = Glucuronosyl

Khim. Prir. Soedin. ,
1969 (5) 596.

1103-035

HISPIDULIN $C_{16}H_{12}O_6$ 300
OI
291°C

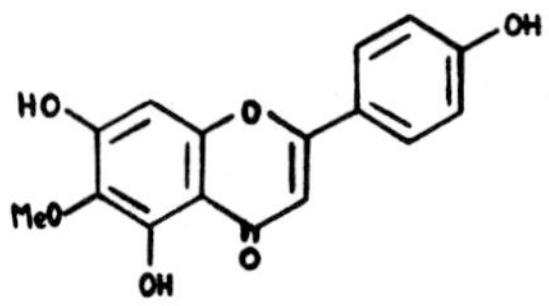

JOC, 1966, 3228.
Ind. J. C. , 1966, 173.

1103-040

CIRSIMARITIN	$C_{17}H_{14}O_6$	314 OI 263°C

R = H

Arch. Pharm., 1971 (304) 557.
Phytochem., 1970, 227.

1103-041

LADANIN	$C_{23}H_{24}O_{11}$	476 -97° 218°C

R = Glu

Chem. Abstr., 1970 (72) 63168.

1103-046

CIRSIMARITIN-4'-O-RUTINOSIDE	$C_{29}H_{34}O_{15}$	622 279°C

R = Rutinosyl

Phytochem., 1971 (10) 452.

1103-042

FLAVONE, 4',6,7-tri-OMe-5-OH-	$C_{18}H_{16}O_6$	328 OI

J. Pharm. Sci., 1968, 1037.

1103-047

PECTOLINARIGENIN	$C_{17}H_{14}O_6$	**314** OI 219°C

R = H

Monatsh. Chem., 1932 (60) 8.

1103-043

SUDACHITIN, des-OMe-	$C_{17}H_{14}O_7$	298

R = H

Bull. Chem. Soc. Jap., 1961, 1547.

1103-048

PECTOLINARIN	$C_{29}H_{34}O_{15}$	622 -98° 256°C

R = .Rha-Glu

Ber., 1941 (74) 1818.

1103-044

NEVADENSIN	$C_{18}H_{16}O_7$	312 OI 194°C

R = Me

JOC, 1966, 3228.

1103-049

LADANEIN	$C_{17}H_{14}O_6$	314 OI 311°C

R = H

Chem. Abstr., 1970 (73) 63168.

1103-045

XANTHOMICROL	$C_{18}H_{16}O_7$	312 OI 229°C

Tet., 1961 (14) 297.

1103-050

TANGERETIN, 5-O-des-Me- $C_{19}H_{18}O_7$ 326
OI
176^0C

R = H

Tet., 1965, 1441.

1103-051

FLAVONE, 3, 4', 7-tri-OH- $C_{15}H_{10}O_5$ 270
OI
310^0C

Biochem. J., 1955, 582.

1104-001

TANGERETIN $C_{20}H_{20}O_7$ 368
OI
154^0C

R = Me

Tet., 1965, 1441.

1103-052

PRATOLETIN $C_{15}H_{10}O_6$ 286
OI
285^0C

Acta Phytochim.,
1943 (13) 99.

1104-002

BAYIN $C_{21}H_{20}O_9$ 416
-1^0e
220^0C

Chem. Ind., 1962, 1720.

1103-053

KAEMPFEROL $C_{15}H_{10}O_6$ 286
OI
280^0C

R = H

JCS, 1955, 2948.

1104-003

LIGNOSIDE $C_{34}H_{34}O_{17}$ 714
-92^0
246^0C

R = 6-Ac-Glucosyl
R' = 2, 3-di-OH-dihydro-coniferyl

Chem. Abstr., 1970
(73) 120453.

1103-054

KAEMPFEROL 3-ME-ETHER $C_{16}H_{12}O_6$ 300
OI

R = Me

Tet. Lett., 1970, 1601.

1104-004

TILIANIN $C_{22}H_{22}O_{10}$ 446
-64^0
245^0C

Ber., 1967 (100) 2783.
Tet. Lett., 1967, 1453.

1103-055

JUGLANIN $C_{20}H_{18}O_{10}$ 418
-169^0
225^0C

R = Arabinosyl

Natwiss., 1955 (42) 181.

1104-005

TRIFOLIN	$C_{21}H_{20}O_{11}$	448	NICOTOFLORIN	$C_{27}H_{30}O_{15}$	594 -29° (180°C)
R = Galactosyl			R = Rutinosyl		
	Acta Phytochim., 1943, 99.			Natwiss., 1955(42) 607.	
1104-006			1104-011		
ASTRAGALIN, iso-	$C_{21}H_{20}O_{11}$	448		$C_{26}H_{28}O_{15}$	580
R = α-D-Glucosyl			R = .Apiosyl-Glu		
	Khim. Prir. Soedin., 1970 (6) 628.			Phytochem., 1970, 2053.	
1104-007			1104-012		
ASTRAGALIN	$C_{21}H_{20}O_{11}$	448 178°C	PANASENOSIDE	$C_{27}H_{30}O_{16}$	610
R = β-D-Glucosyl			R = .Gal-Glu		
	Arch. Pharm., 1958, 113.			Chem. Abstr., 1970 (73) 127785.	
1104-008			1104-013		
KAEMPFEROL-3-GLUCURONIDE	$C_{21}H_{18}O_{12}$	462	SOPHORAFLAVANOLOSIDE	$C_{27}H_{30}O_{16}$	610 -61°e 208°C
R = Glucuronosyl			R = Sophorosyl		
	Ber., 1970, 3678.			Bull. Soc. Chim. Biol., 1938 (20) 459.	
1104-009			1104-014		
AFZELIN	$C_{21}H_{20}O_{10}$	432 230°C	RUSTOSIDE	$C_{26}H_{28}O_{15}$	580
R = Rhamnosyl			R = .Xylosyl-Glu		
	Natwiss., 1959 (46) 358.			Khim. Prir. Soedin., 1970 (6) 636.	
1104-010			1104-015		

$C_{27}H_{30}O_{15}$ 594 R = . Gal-Rha Curr. Sci. , 1971 (40) 106. 1104-016	KAEMPFEROL-3,7-DIGLUCOSIDE $C_{27}H_{30}O_{16}$ 610 R = R' = Glu Compt. Rendu, 1970 (271D) 1128. 1104-021

| $C_{33}H_{40}O_{20}$ 756
 345^0C

 R = . Rhamnosyl(3,4-di-Glu)

 Phytochem. , 1970, 441.

 1104-017 | KAEMPFERITRIN $C_{27}H_{30}O_{14}$ 578
 -250^0
 202^0C

 R = R' = Rha

 Nature, 1951 (168) 788.

 1104-022 |

| POPULNIN $C_{21}H_{20}O_{11}$ 448

 R = Glucosyl

 Compt. Rendu, 1971 (271) 1128.

 1104-018 | LEPIDOSIDE $C_{26}H_{28}O_{14}$ 564
 -65^0m
 260^0C

 R = Xylosyl ; R' = Rha

 Khim. Prir. Soedin. . 1970 (6) 127.

 1104-023 |

| $C_{21}H_{20}O_{10}$ 432
 -165^0m
 232^0C

 R = Rhamnosyl

 Khim. Prir. Soedin. , 1969 (5) 441.

 1104-019 | KAEMPFEROL, 3-diglucosyl-7--glucosyl- $C_{33}H_{40}O_{21}$ 772
 -16^0w

 R = Glu-Glu ; R' = Glu

 Compt. Rend. . 1970 (271C) 769.

 1104-024 |

| EQUISETRIN $C_{27}H_{30}O_{16}$ 610
 196^0C

 R = Glu-Glu

 J. Pharm. Soc. Jap. , 1940 (8) 179.

 1104-020 | ROBININ $C_{33}H_{40}O_{19}$ 740
 -122^0
 196^0C

 R = . Gal-Rha ; R' = Rha

 Ber. , 1941, 1783.

 1104-025 |

TILIROSIDE $C_{30}H_{26}O_{13}$ 594

                                   250^0C

R = Glu ; R' = Coumaroyl

Natwiss., 1959, 358.

1104-026

AMURENSIN $C_{26}H_{30}O_{12}$ 598

                                   290^0C

R = H

JACS, 1953, 5507.

1104-031

$C_{26}H_{28}O_{14}$ 564
-25^0c
216^0C

Arch. Pharm., 1957
(290) 342.

1104-027

PHELLOSIDE $C_{32}H_{40}O_{17}$ 696
-90^0
282^0C

R = Glucosyl

Chem. Abstr., 1968
(69) 41709.

1104-032

ICARITIN, nor-anhydro- $C_{20}H_{22}O_8$ 390
                                     OI
                                   226^0C

R = H

Chem. Abstr., 1970
(73) 42404.

1104-028

ICARIIN, nor- $C_{32}H_{40}O_{16}$ 680
-93^0p
236^0C

J. Pharm. Soc. Jap.,
1955, 719.

1104-033

ICARITIN, iso-anhydro- $C_{21}H_{24}O_8$ 404
                                     OI
                                     275^0C

R = Me

Chem. Abstr., 1970
(73) 42404.

1104-029

PHELLATIN $C_{26}H_{30}O_{12}$ 534

Khim. Prir. Soedin.,
1970 (6) 762.

1104-034

ICARITIN, nor- $C_{20}H_{20}O_7$ 372
                                     OI
                                     307^0C

JACS, 1953, 5507.

1104-030

KUMATAKENIN $C_{17}H_{14}O_6$ 314
                                     OI
                                     246^0C

Aust. J. C., 1965, 1441.

1104-035

KAEMFERIDE $C_{16}H_{12}O_6$ 300 OI 228^{0}C	KEYAKININ $C_{22}H_{22}O_{11}$ 462 234^{0}C	
R = H	R = Glucosyl	
	Aust. J. C., 1966, 705.	
JCS, 1926, 2336.		
1104-036	1104-041	
MUMENIN $C_{22}H_{22}O_{11}$ 462 278^{0}C	ERMANIN $C_{17}H_{14}O_6$ 314 OI 231^{0}C	
R = Glucosyl	Tet. Lett., 1971, 1767.	
JOC, 1959, 408.		
1104-037	1104-042	
ICARITIN $C_{21}H_{22}O_7$ 386 OI 239^{0}C	FLAVONE, 7, 4'-di-OMe-3, 5-di-OH- $C_{17}H_{14}O_6$ 314 OI 180^{0}C	
R = R' = H	R = H	
J. Pharm. Soc. Jap., 1955, 719.	Tet. Supp. No. 8 (I), 71.	
1104-038	1104-043	
ICARIIN $C_{33}H_{42}O_{16}$ 694 -87^{0}p 231^{0}C	FLAVONE, 3, 7, 4'-tri-OMe-5-OH- $C_{18}H_{16}O_6$ 328 OI 145^{0}C	
R = Rha ; R' = Glu	R = Me	
J. Pharm. Soc. Jap., 1955, 719.	Tet. Supp. No. 8 (I), 71.	
1104-039	1104-044	
RHAMNOCITRIN $C_{16}H_{12}O_6$ 300 OI 222^{0}C		
R = H	deleted	
JCS, 1947, 122.		
1104-040	1104-045	

KAEMPFEROL, 6-OMe-	$C_{16}H_{12}O_7$	316 OI 270^0C

Compt. Rend., 1971
(272C) 1529.

1104-046

PENDULETIN	$C_{18}H_{16}O_7$	344 OI 217^0C

R = H

Aust. J. C., 1967, 1049.

1104-051

FLAVONE, 4',5,7-tri-OH-3, -6-di-OMe-	$C_{17}H_{14}O_7$	330 OI

R = H

Phytochem., 1971 (10)
450.

1104-047

PENDULIN	$C_{24}H_{26}O_{12}$	506 -34^0p 178^0C

R = Glucosyl

Tet., 1960 (4) 132.

1104-052

	$C_{23}H_{24}O_{12}$	492

R = Glucosyl

Phytochem., 1971 (10)
450.

1104-048

MIKANIN	$C_{18}H_{16}O_7$	344 OI

JCS, 1965, 6371.

1104-053

EUPALITIN	$C_{17}H_{14}O_7$	330 OI 290^0C

R = H

Tet., 1970, 2851.

1104-049

VOGELETIN	$C_{16}H_{12}O_7$	316 OI 283^0C

R = H

Tet. Lett., 1965, 3849.

1104-054

EUPALIN	$C_{23}H_{24}O_{11}$	476 -129^0p 208^0C

R = Rhamnosyl

Tet., 1970, 2851.

1104-050

VOGELOSIDE	$C_{22}H_{22}O_{12}$	478

R = Glucosyl

Chem. Abstr., 1967
(66) 10815.

1104-055

FLAVONE, 5-OH-3, 4', 6, 7--tetra-OMe- $C_{19}H_{18}O_7$ 358 OI 173^{0}C

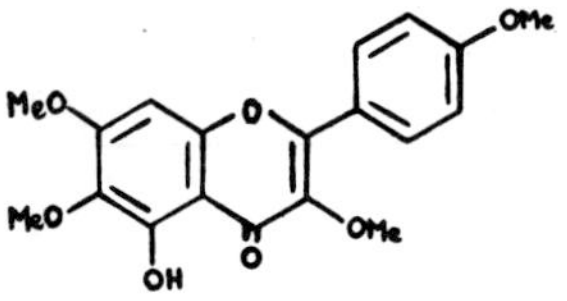

Aust. J. C., 1966 (19) 2133.

1104-056

PRUDOMESTIN-3-ME ETHER $C_{18}H_{16}O_7$ 344 OI 174^{0}C

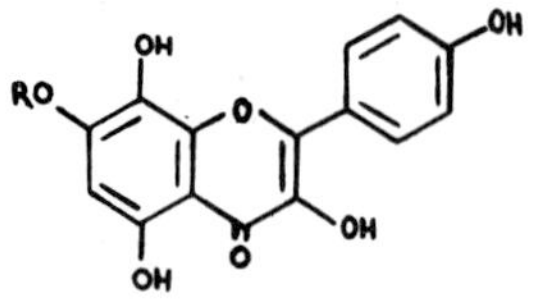

R = Me

Aust. J. C., 1965, 1871.

1104-061

HERBACETIN $C_{15}H_{10}O_7$ 302 OI 281^{0}C

R = H

JCS, 1938, 56.

1104-057

TAMBULIN $C_{18}H_{16}O_7$ 344 OI 205^{0}C

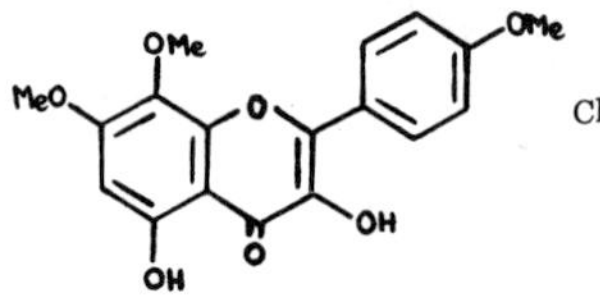

Chem. Abstr., 1949, 226.

1104-062

HERBACITRIN $C_{21}H_{20}O_{12}$ 464 248^{0}C

R = Glucosyl

Proc. Ind. Acad. Sci., 1939 (9A) 365.

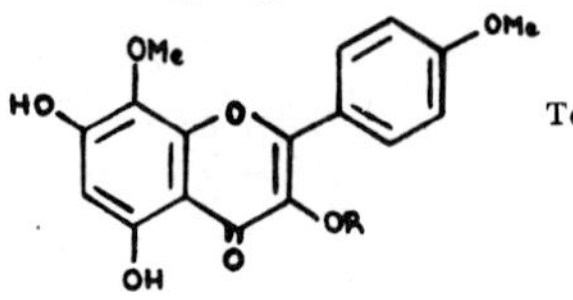

1104-058

FLAVONE, 5, 4'-diOH-3, 7, 8--triOMe- $C_{18}H_{16}O_7$ 344 OI 266^{0}C

R = H

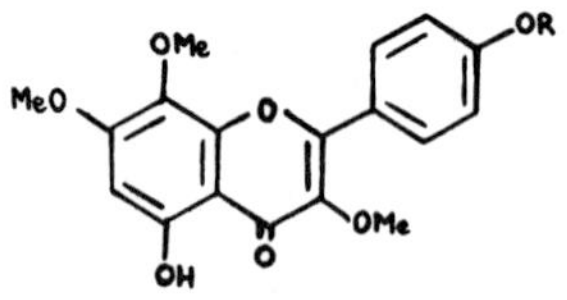

Aust. J. C., 1968, 2085.

1104-063

FLAVONE, 3, 8-diOMe-4', 5, 7--triOH- $C_{17}H_{14}O_7$ 330 OI 243^{0}C

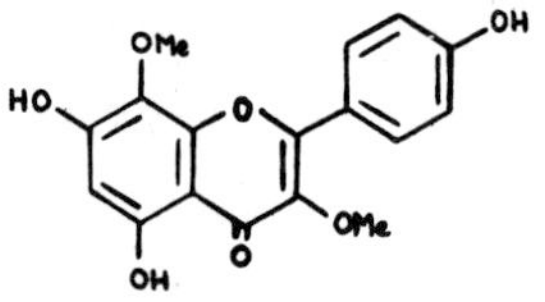

Aust. J. C., 1967, 1049.

1104-059

FLINDULATIN $C_{19}H_{18}O_7$ 358 OI 161^{0}C

R = Me

Aust. J. C., 1954 (7) 181.

1104-064

PRUDOMESTIN $C_{17}H_{14}O_7$ 330 OI 210^{0}C

R = H

Tet., 1966, 941.

1104-060

AURANETIN $C_{20}H_{20}O_7$ 368 OI 141^{0}C

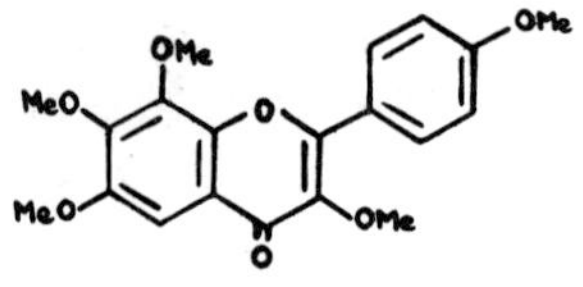

Ber., 1942, 2083.

1104-065

CALYCOPTERIN	$C_{19}H_{18}O_8$ 370 OI 227°C	
		$C_{27}H_{30}O_{16}$ 610
		R = Laminaribiosyl
	Helv., 1934, 1560.	Khim. Prir. Soedin., 1969, 595.
1104-066		1105-004
AURANETIN, 5-OH-	$C_{20}H_{20}O_8$ 384 OI 126°C	CAESIOSIDE $C_{26}H_{28}O_{15}$ 580 226°C
		R = Primeverosyl
	Tet., 1960 (8) 64.	Tet. Lett., 1968, 2781.
1104-067		1105-005
LUTEOLIN	$C_{15}H_{10}O_6$ 286 OI 330°C	GRAVEOBIOSIDE-A $C_{26}H_{28}O_{15}$ 580
R = H		R = . Glu-Apiosyl
	JACS, 1942, 1704.	Chem. Abstr., 1954, 3483. Chem. Ind., 1953, 85.
1105-001		1105-006
LUTEOLIN 7-ME-ETHER	$C_{16}H_{12}O_6$ 300	$C_{27}H_{30}O_{15}$ 594
R = Me		R = Rutinosyl
	Arch. Pharm., 1971 (304) 557.	Khim. Prir. Soedin., 1970 (6) 470.
1105-002		1105-007
LUTEOLIN, gluco-	$C_{21}H_{20}O_{11}$ 448 240°C	$C_{27}H_{30}O_{15}$ 594
R = Glucosyl		R = . Rha-Glu (Cf 007 ?)
	Phytochem., 1971, 490. Natwiss., 1959, 558.	Chem. Abstr., 1971 (74) 95429.
1105-003		1105-008

$C_{27}H_{30}O_{16}$ 610 R = . Gal-Glu Khim. Prir. Soedin., 1970 (6) 470. 1105-009	GALUTEOLIN $C_{21}H_{20}O_{11}$ 448 280^0C R = Glu ; R' = H J. Pharm. Soc. Jap., 1940, 449. 1105-014
ORIENTIN, iso- $C_{21}H_{20}O_{11}$ 448 $+31^0$ 235^0C Phytochem., 1971 (10) 490. 1105-010	LUTEOLIN-5,7-DIGLUCOSIDE $C_{27}H_{30}O_{16}$ 610 R = R' = Glu Lloydia, 1971 (34) 258. 1105-015
ORIENTIN $C_{21}H_{20}O_{11}$ 448 $+18^0$p 266^0C Chem. Ind., 1964, 499. 1105-011	$C_{21}H_{20}O_{11}$ 448 177^0C Experientia, 1970 (26) 1192. 1105-016
LUCENIN-1 $C_{27}H_{30}O_{16}$ 610 Tet. Lett., 1965, 1105. 1105-012	LUTEOLIN-3'-GLUCOSIDE-7- BISULFATE $C_{21}H_{20}O_{14}S$ 528 Z. Nat., 1971 (26B) 490. 1105-017
deleted 1105-013	NEPITRIN $C_{22}H_{22}O_{11}$ 462 Ind. J. C., 1970 (8) 1074. 1105-018

134

SPINOSIDE $C_{28}H_{32}O_{15}$ 608 Khim. Prir. Soedin., 1970 (6) 626. 1105-019	$C_{22}H_{20}O_{12}$ 476 R = Glucuronosyl Phytochem. 1971 (10) 490. 1105-024
SWERTIAJAPONIN $C_{22}H_{22}O_{11}$ 462 $-3^0 p$ $265^0 C$ Tet. Lett., 1966, 1611. 1105-020	$C_{28}H_{32}O_{15}$ 608 R = Rutinosyl Phytochem. 1971 (10) 490. 1105-025
CHRYSOERIOL $C_{16}H_{12}O_6$ 300 OI $330^0 C$ R = H Arch. Pharm., 1917 (255) 308. 1105-021	GRAVEOBIOSIDE-B $C_{27}H_{30}O_{15}$ 594 R = . Glu-Apiosyl Chem. Abstr. 1954 (48) 3483. 1105-026
VELUTIN $C_{17}H_{14}O_6$ 314 OI $226^0 C$ R = Me JOC, 1970, 3989. 1105-022	SALICAPRENE $C_{27}H_{30}O_{15}$ 594 $246^0 C$ R = . Gal-Ara Chem. Abstr. 1970 (73) 22133. 1105-027
$C_{22}H_{22}O_{11}$ 462 R = Glucosyl Phytochem., 1970 (10) 490. 1105-023	ORTANTHOSIDE $C_{28}H_{32}O_{16}$ 624 R = Galactobiosyl Khim. Prir. Soedin., 1971 (7) 117. 1105-028

135

DIOSMETIN $C_{16}H_{12}O_6$ 300
OI
258^0C

R = H

JCS, 1930, 817.

1105-029

DIOSMETIN, 8-C-Glucosyl- $C_{22}H_{22}O_{11}$ 462

268^0C

R = H ; R' = Glu

JOC, 1968, 1571.

1105-034

PILLOIN $C_{17}H_{14}O_6$ 314
OI
237^0C

R = Me

JOC, 1971, 3829.

1105-030

DIOSMETIN, 6,8-di-C-Glucosyl- $C_{28}H_{32}O_{16}$ 624

R = R' = Glu

JOC, 1968, 1571.

1105-035

$C_{22}H_{22}O_{11}$ 462

R = Glucosyl

Khim. Prir. Soedin. .
1970 (6) 626.

1105-031

FLAVOYADORININ - B $C_{23}H_{24}O_{11}$ 476

R = Glu

Agr. Biol. Chem. , 1970
(34) 900.

1105-036

DIOSMIN $C_{28}H_{32}O_{15}$ 608

279^0C

R = . Glu-Rha

J. Pharm. Soc. Jap. ,
1938 (58) 639.

1105-032

FLAVOYADORININ-B, homo- $C_{28}H_{32}O_{15}$ 608

R = . Glu-Apiosyl

Agr. Biol. Chem. , 1970
(34) 900.

1105-037

DIOSMETIN, 6-C-Glucosyl- $C_{22}H_{22}O_{11}$ 462

244^0C

R = Glu ; R' = H

JOC, 1968, 1571.

1105-033

FLAVONE, 5-OH-3', 4', 7-tri-OMe- $C_{18}H_{16}O_6$ 328
OI
163^0C

Planta Med. , 1970
(18) 332.

1105-038

ARTHRAXIN	$C_{21}H_{16}O_9$	412
		-29^0e
		336^0C

JCS, 1971, 1982.

1105-039

PEDALIIN	$C_{22}H_{22}O_{12}$	478
		$+28^0$
		254^0C

R = Glucosyl

Ind. J. C., 1970, 1074.

1105-044

PONGAGLABRONE	$C_{18}H_{10}O_5$	306
		OI
		233^0C

Tet., 1963, 221.

1105-040

GENKWANIN, 3'-OH-6-OMe-	$C_{17}H_{14}O_7$	330

R = Me

Arch. Pharm., 1971
(304) 557.

1105-045

PINNATIN	$C_{19}H_{12}O_6$	336
		OI
		233^0C

Chem. Abstr., 1956,
13008.

1105-041

BATATIFOLIN	$C_{16}H_{12}O_7$	316
		OI
		251^0C

Ber., 1970 (103) 1822.

1105-046

EUPAFOLIN	$C_{16}H_{12}O_7$	316
		OI
		272^0C

Arch. Pharm., 1971
(304) 557.
Tet., 1969, 1603.

1105-042

FLAVONE, 4',5,7-triOH-3',6-diOMe-	$C_{17}H_{14}O_7$	330

R = H

Chem. Abstr., 1968
(69) 27184.

1105-047

PEDALITIN	$C_{16}H_{12}O_7$	316
		OI

R = H

Tet., 1969, 1603.

1105-043

JACEOSIDE	$C_{23}H_{24}O_{12}$	492
		-70^0m
		225^0C

R = Glucosyl

Tet. Lett., 1969, 3411.

1105-048

137

EUPATORIN	$C_{18}H_{16}O_7$	344
		OI
		197^0C

Tet., 1969, 1603.

1105-049

FLAVONE, 5-OH-7, 8, 3', 4'- -tetra-OMe-	$C_{19}H_{18}O_7$	358

Tet., 1968, 2121.

1105-054

EUPATILIN	$C_{18}H_{16}O_7$	344
		OI
		235^0C

Tet., 1969, 1603.

1105-050

SUDACHITIN	$C_{18}H_{16}O_8$	360
		OI

Bull. Chem. Soc. Jap., 1961, 1547.

1105-055

FLAVONE, 5, 6, 7, 3', 4'-penta-OMe-	$C_{20}H_{20}O_7$	372
		OI
		177^0C

Chem. Ind., 1960, 264.

1105-051

ACEROSIN	$C_{18}H_{16}O_8$	360
		OI
		241^0C

Tet., 1967, 3557.

1105-056

deleted

1105-052

GARDENIN-D	$C_{19}H_{18}O_8$	374
		OI
		191^0

Ind. J. C., 1970 (8) 398.

1105-057

HYPOLAETIN-7-GLUCOSIDE	$C_{21}H_{20}O_{12}$	464
		244^0C

Phytochem., 1971 (10) 434.

1105-053

HYMENOXIN	$C_{19}H_{18}O_8$	374
		OI
		212^0C

JOC, 1967, 3254.

1105-058

NOBILETIN	$C_{21}H_{22}O_8$	402
		OI
		135^0C

JCS, 1938, 1003.

1105-059

FISETIN 3-ME-ETHER	$C_{16}H_{12}O_6$	300
		269^0C

R = Me

Chem. Comm., 1968, 1246.

1106-002

LUCIDIN	$C_{18}H_{14}O_8$	358
		OI
		256^0C

R = H

JCS, 1965, 2743.

1105-060

FLAVONE, 3,7,3'-triOH-4'-OMe-	$C_{16}H_{12}O_6$	300
		OI
		288^0C

Biochem. J.,
1956 (60) 582.

1106-003

LUCIDIN DI-ME ETHER	$C_{20}H_{18}O_8$	386
		OI
		172^0C

R = Me

JCS, 1965, 2743.

1105-061

KANUGIN, des-OMe-	$C_{18}H_{14}O_6$	326
		OI
		147^0C

JCS, 1956, 2176.

1106-004

LUTEOLIN-6-C-XYLOSIDE, -7-O-rhamnosyl-	$C_{26}H_{28}O_{15}$	580
		198^0C

Phytochem., 1971 (10)
677.

1105-062

PONGACHROMENE	$C_{22}H_{18}O_6$	378
		OI
		195^0C

Tet., 1969, 1063.

1106-005

FISETIN	$C_{15}H_{10}O_6$	286
		OI
		348^0C

R = H

JCS, 1926, 2334.

1106-001

PONGAPIN	$C_{19}H_{12}O_6$	336
		OI
		190^0C

Tet., 1958 (2) 207.

1106-006

FLAVONE, 3,5,8,3',4'-penta-OH- $C_{15}H_{10}O_7$ 298
OI
270^0C

Proc. Ind. Acad. Sci.,
1954 (39A) 296.

1106-007

POLYSTACHOSIDE $C_{20}H_{18}O_{11}$ 434
-26^0m
246^0C

R = β-L-arabinosyl

Arch. Pharm., 1955
(288) 419.

1106-012

FLAVONE, 3,7,8,3',4'-penta-OH- $C_{15}H_{10}O_7$ 298
OI
315^0C

JCS, 1954, 1399.

1106-008

FOENICULIN $C_{20}H_{18}O_{11}$ 434
256^0C

R = arabinosyl

J. Pharm. Soc. Jap.,
1959, 986.

1106-013

QUERCETIN $C_{15}H_{10}O_7$ 302
OI
317^0C

R = H

JCS, 1927, 239.

1106-009

HYPERIN $C_{21}H_{20}O_{12}$ 464
-59^0
238^0C

R = Galactosyl

Arch. Pharm., 1955
(58) 362.

1106-014

AVICULARIN $C_{20}H_{18}O_{11}$ 434
-116^0m
216^0C

R = α-L-arabofuranosyl

J. Pharm. Soc. Jap.,
1960 (80) 102.

1106-010

QUERCITRIN, iso- $C_{21}H_{20}O_{12}$ 464
-22^0
242^0C

R = Glucosyl

JACS, 1949, 2658.

1106-015

GUAIJAVERIN $C_{20}H_{18}O_{11}$ 434
256^0C

R = α-L-arabopyranosyl

JCS, 1958, 3320.

1106-011

$C_{23}H_{22}O_{13}$ 506
167^0C

R = Ac-Glucosyl

Phytochem., 1971 (10)
2547.

1106-016

QUERCITURONE	$C_{21}H_{18}O_{13}$	478
		-23^0e
		182^0C
R = Glucuronosyl		
Nature, 1955, 176.		
1106-017		

QUERCETIN-3-GENTIOBIOSIDE	$C_{27}H_{30}O_{17}$	626
R = Gentiobiosyl		
Khim. Prir. Soedin., 1970, 761.		
1106-022		

QUERCITRIN	$C_{21}H_{20}O_{11}$	448
		183^0C
R = Rhamnosyl		
J. Pharm. Soc. Jap., 1958, 1302.		
1106-018		

	$C_{27}H_{30}O_{16}$	610
R = . Gal^6-Rha		
Khim. Prir. Soedin., 1970 (6) 629.		
1106-023		

RAYNOUTRIN	$C_{20}H_{18}O_{11}$	434
		-175^0e
		204^0C
R = Xylosyl		
J. Pharm. Soc. Jap., 1956, 323.		
1106-019		

RUTIN	$C_{27}H_{30}O_{16}$	610
		-36^0p
		191^0C
R = Rutinosyl		
JCS, 1966, 1140.		
1106-024		

MERATIN	$C_{27}H_{30}O_{17}$	626
		-52^0m
		180^0C
R = Glu-Glu		
Helv., 1945, 1157.		
1106-020		

	$C_{26}H_{28}O_{15}$	580
R = . Ara-Rha		
Khim. Prir. Soedin., 1969 (5) 597.		
1106-025		

	$C_{27}H_{30}O_{17}$	626
R = α-Glu^2-Glu		
Chem. Abstr., 1971 (74) 126531.		
1106-021		

QUERCIMERITRIN	$C_{21}H_{20}O_{12}$	464
		250^0C
R = Glu		
Compt. Rend., 1970 (271) 1128.		
1106-026		

$C_{21}H_{20}O_{11}$ 448

R = Rha

Chem. Abstr., 1971
(74) 20349.

1106-027

$C_{33}H_{40}O_{21}$ 772

R = Glu ; R' = . Glu-Rha

Compt. Rend., 1970
(271D) 2408.

1106-032

QUERCETIN-3,7-DIGLUCOSIDE $C_{27}H_{30}O_{17}$ 626

R = R' = Glu

Chem. Abstr., 1971
(74) 20345.

1106-028

$C_{26}H_{28}O_{16}$ 596

R = Glu ; R' = Xyl

Chem. Abstr., 1971
(74) 20345.

1106-033

$C_{26}H_{28}O_{16}$ 596

R = Glu ; R' = Ara

Khim. Prir. Soedin.,
1970 (6) 630.

1106-029

PINOQUERCETIN $C_{16}H_{12}O_7$ 316
OI
290^0C

Chem. Abstr., 1956
(50) 15750.

1106-034

$C_{27}H_{30}O_{16}$ 610

R = Glu ; R' = Rha

Chem. Abstr., 1971
(74) 20345.

1106-030

FLAVONE, 5,7,3',4'-tetra OH-
3-OMe- $C_{16}H_{12}O_7$ 316
276^0C

R = H

Aust. J. C., 1968, 2349.

1106-035

PETIOLAROSIDE $C_{27}H_{30}O_{16}$ 610
-110^0e
191^0C

R = Rha ; R' =Glu

Chem. Abstr., 1971
(74) 85192.
Compt. Rend., 1970
(270D) 2710.

1106-031

TRANSILIN $C_{22}H_{22}O_{12}$ 478
-73^0m
244^0C

R = Glucosyl

Khim. Prir. Soedin.,
1969, 439.

1106-036

AZALEATIN	$C_{16}H_{12}O_7$	316
		OI
		322^0C

R = H

JACS, 1956, 4725.

1106-037

CARYATIN	$C_{17}H_{14}O_7$	330
		OI
		300^0C

Nature, 1966 (212) 1065.

1106-042

AZALEIN	$C_{22}H_{22}O_{11}$	462
		183^0C

R = Rhamnosyl

JACS, 1956, 4725.

1106-038

ISORHAMNETIN	$C_{16}H_{12}O_7$	316
		OI
		306^0C

R = H

JCS, 1926, 2336.

1106-043

RHAMNETIN	$C_{16}H_{12}O_7$	316
		OI
		295^0C

R = H

Ber., 1944 (77) 211.

1106-039

PERSICARIN	$C_{16}H_{11}O_{10}SK$	434
		OI
		280^0C

R = SO$_3$K

J. Pharm. Soc. Jap.,
1940 (60) 174.

1106-044

XANTHORHAMNIN	$C_{34}H_{42}O_{20}$	770

R = Rhamninosyl

JCS, 1927, 234.

1106-040

DISTICHIN	$C_{21}H_{20}O_{11}$	448
		-178^0
		262^0C

R = Arabinosyl

J. Pharm. Soc. Jap.,
1960, 102.

1106-045

PEUMOSIDE	$C_{27}H_{30}O_{15}$	594
		196^0C

Natwiss., 1965 (52) 161.

1106-041

CACTICIN	$C_{22}H_{22}O_{12}$	478
		-54^0p
		268^0C

R = Glucosyl

Tet. Lett., 1966, 567.
Ber., 1966 (99) 1384.

1106-046

ALLIOSIDINE $C_{27}H_{30}O_{15}$ 594
-74^0w
200^0C

R = . Rha^2-Ara

Khim. Prir. Soedin.,
1970 (6) 201.

1106-047

BOLDOSIDE $C_{28}H_{32}O_{16}$ 624

215^0C

R = Glu ; R' = Rha

Ber., 1967, 2301.
Natwiss., 1965 (52) 161.

1106-052

$C_{28}H_{32}O_{17}$ 640

R = . Glu^6-Gal

Khim. Prir. Soedin.,
1971 (7) 117.

1106-048

ISORHAMNETIN-3-GALACTOSIDE-
-7-GLUCOSIDE $C_{28}H_{32}O_{17}$ 640

R = Glu ; R' = Gal

Compt. Rend. . 1971
(272D) 2616.

1106-053

NARCISSIN $C_{28}H_{32}O_{16}$ 624

181^0C

R = Rutinosyl

Bull. Chem. Soc. Jap.,
1957, 862.

1106-049

BRASSICOSIDE $C_{34}H_{42}O_{22}$ 802

210^0C

R = Glu-Glu ; R' = Glu

Ber., 1967 (100) 2301.
Tet. Lett. . 1966. 567.

1106-054

$C_{34}H_{42}O_{21}$ 786

R = . Gal-Rha-Glu

Chem. Abstr. . 1955
(49) 14922.

1106-050

DACTYLIN $C_{28}H_{32}O_{17}$ 640

188^0C

JOC, 1957, 189.
JACS, 1931, 2744.

1106-055

LUTEOSIDE $C_{28}H_{32}O_{16}$ 624

R = Rha ; R' = Glu

Compt. Rend., 1970
(270D)2710.

1106-051

QUERCETIN-3, 3'-DIME ETHER $C_{17}H_{14}O_7$ 330
OI
258^0C

Tet. Lett., 1970, 1601.

1106-056

RHAMNAZIN	$C_{17}H_{14}O_7$	330
		OI
		317^0C

R = H

MeO

OH

OH

OMe

JCS, 1946, 771.

1106-057

TAMARIXIN	$C_{16}H_{11}O_{10}SK$	434
		OI
		316^0C

R = SO₃K

$R = SO_3K$

Chem. Abstr. 1966
(65) 15309.

1106-062

PERSICARIN-7-ME ETHER	$C_{17}H_{13}O_{10}SK$	448
		OI
		212^0C

$R = SO_3K$

Chem. Abstr., 1955,
9634.

1106-058

TAMARIXIN	$C_{22}H_{22}O_{12}$	478
		316^0C

R = Glucosyl

JCS, 1954, 3063.

1106-063

ALBOSIDE	$C_{23}H_{24}O_{12}$	492
		179^0C

R = Glucosyl

Tet. Lett., 1968, 2301.

1106-059

OMBUIN	$C_{17}H_{14}O_7$	330
		OI
		230^0C

R = H

MeO

OH

OMe

OH

OR

O

J. Sci. Ind. Res., 1956
(15B) 263.

1106-064

TAMARIXETIN	$C_{16}H_{12}O_7$	316
		OI
		259^0C

R = H

HO

OH

OH

OMe

OH

OR

O

JCS, 1954, 3063.

1106-060

OMBUOSIDE	$C_{29}H_{34}O_{16}$	638
		-43^0p
		195^0C

R = Rutinosyl

Chem. Abstr., 1953
(46) 12376.

1106-065

TAMARIXETIN 3-ME ETHER	$C_{17}H_{14}O_7$	330
		OI
		235^0C

R = Me

Phytochem., 1971 (10)
664.

1106-061

AYANIN	$C_{18}H_{16}O_7$	344
		OI
		172^0C

MeO

OMe

OH

OH

OMe

O

JCS, 1952, 92.

1106-066

QUERCETIN 3', 4', 7- TRIME ETHER	$C_{18}H_{16}O_7$	344 OI 190^{0}C

R = H

Tet. Lett., 1970, 1601.

1106-067

PATULETIN	$C_{16}H_{12}O_8$	332 OI 263^{0}C

R = H

Proc. Ind. Acad. Sci., 1941, 643.

1106-072

	$C_{24}H_{26}O_{11}$	490 190^{0}C

R = Rhamnosyl

Phytochem., 1971 (10) 2256.

1106-068

PATULITRIN	$C_{22}H_{22}O_{13}$	494

R = Glucosyl

Chem. Abstr., 1957 (51) 8083.

1106-073

QUERCETAGETIN	$C_{15}H_{10}O_8$	314 OI 324^{0}C

R = R' = H

JCS, 1929, 74.

1106-069

AXILLARIN	$C_{17}H_{14}O_8$	346 OI 208^{0}C

Experientia, 1968 (24) 769.

1106-074

TAGETIIN	$C_{21}H_{20}O_{13}$	480 -110^0

R = Glu ; R' = H

J. Pharm. Soc. Jap., 1957, 31.

1106-070

EUPATOLITIN	$C_{17}H_{14}O_8$	346 OI 286^{0}C

R = H

Tet., 1970, 2851.

1106-075

QUERCETAGITRIN	$C_{21}H_{20}O_{13}$	480 237^{0}C

R = H ; R' = Glu

Proc. Ind. Acad. Sci., 1948 (28A) 94.

1106-071

EUPATOLIN	$C_{23}H_{24}O_{12}$	492 -146^{0}p 200^{0}C

R = Rhamnosyl

Tet., 1970, 2851.

1106-076

CHRYSOSPLENOL-D $C_{18}H_{16}O_8$ 260
OI
237°C

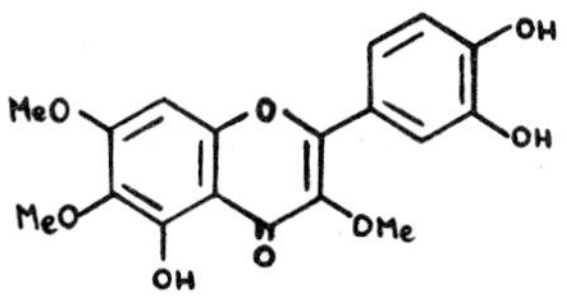

Chem. Pharm. Bull.,
1968, 2310.

1106-077

CHRYSOSPLENOSIDE-B $C_{25}H_{28}O_{13}$ 536

227°C

R = Glucosyl

Chem. Pharm. Bull.,
1968, 2310.

1106-082

SPINACETIN $C_{17}H_{14}O_8$ 346
OI
235°C

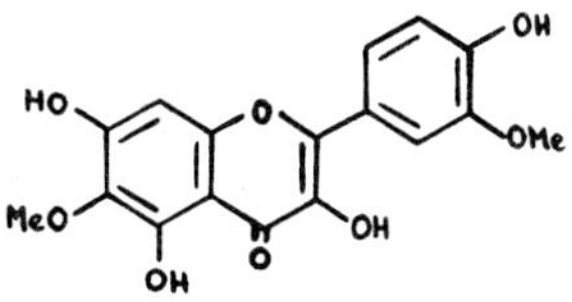

JOC, 1961, 4718.

1106-078

CENTAUREIDIN $C_{18}H_{16}O_8$ 360
OI
196°C

R = H

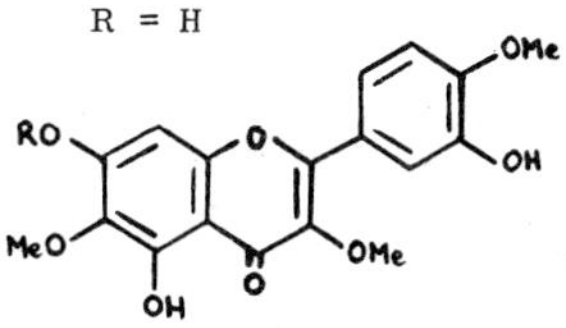

Experientia, 1968
(24) 880.
Ber., 1964, 1666.

1106-083

JACEIDIN $C_{18}H_{16}O_8$ 360
OI
127°C

R = H

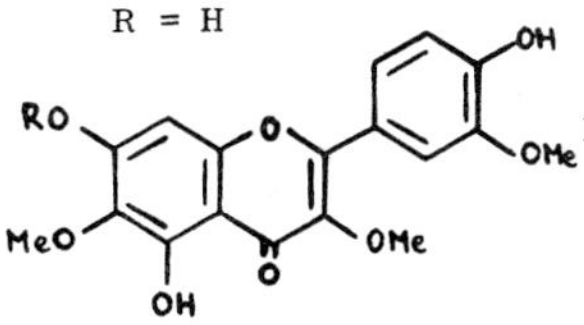

Ber., 1964, 610.

1106-079

CENTAUREIN $C_{24}H_{36}O_{13}$ 522

R = Glucosyl

Ber., 1964, 1666.

1106-084

JACEIN $C_{24}H_{26}O_{13}$ 522
-73°m
206°C

R = Glucosyl

Ber., 1964, 610.

1106-080

AYANIN B, oxy- $C_{18}H_{16}O_8$ 360
OI
208°C

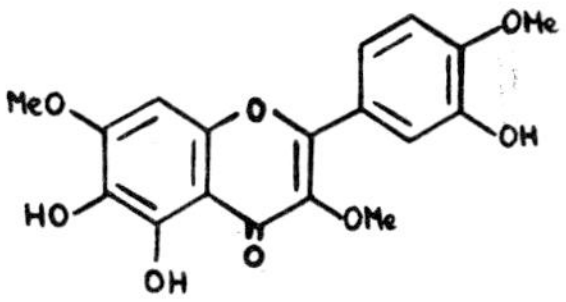

JCS, 1956, 1369.

1106-085

POLYCLADIN $C_{19}H_{18}O_8$ 374
OI
203°C

R = H

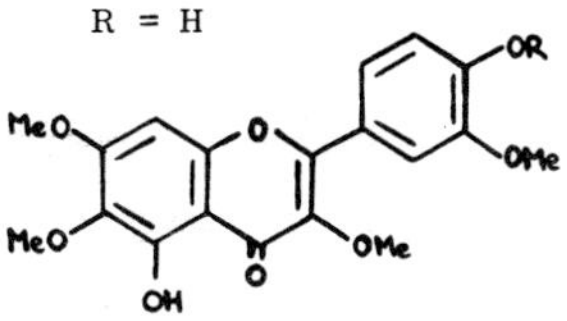

Gazz., 1957 (87) 1185.

1106-081

CASTICIN $C_{19}H_{18}O_8$ 374
OI
187°C

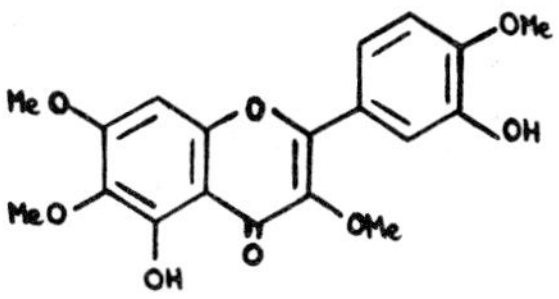

Tet. Lett., 1964, 323.

1106-086

ARTEMETIN	$C_{20}H_{20}O_8$	388
		OI
		163^0C

JOC, 1961, 3014.

1106-087

GOSSYPIN	$C_{21}H_{20}O_{13}$	480
		229^0C

R = H ; R' = Glu

Proc. Ind. Acad. Sci.,
1936 (4A) 54.

1106-092

MELISIMPLIN	$C_{19}H_{16}O_8$	372
		OI
		235^0C

R = H

JCS, 1955, 3908.

1106-088

GOSSYPITRIN	$C_{21}H_{20}O_{13}$	480
		241^0C

R = Glu; R' = H

JCS, 1916, 145.
Phytochem., 1971, 3287.

1106-093

MELISIMPLEXIN	$C_{20}H_{18}O_8$	386
		OI
		185^0C

R = Me

JCS, 1955, 3908.

1106-089

FLAVONE, 7-OMe-3, 5, 8, 3', 4'-penta-OH-	$C_{16}H_{12}O_8$	332
		OI
		261^0C

Chem. Abstr., 1968
(69) 10328.

1106-094

MELITERNATIN	$C_{19}H_{14}O_8$	370
		OI
		198^0C

JCS, 1951, 3131.

1106-090

CORNICULÀTUSIN	$C_{16}H_{12}O_8$	332
		OI
		276^0C

R = H

Tet. Lett., 1970, 803.

1106-095

GOSSYPETIN	$C_{15}H_{10}O_8$	318
		OI
		314^0C

R = R' = H

JCS, 1929, 74.

1106-091

CORNICULATUSIN 3-GALACTOSIDE	$C_{22}H_{22}O_{13}$	494
		216^0C

R = Galactosyl

Tet. Lett., 1970, 803.

1106-096

FLAVONE, 5,7,3',4'-tetra-OH-3,8-di-OMe- C$_{17}$H$_{14}$O$_8$ 346 OI 302^{0}C

R = Me

Ber., 1968, 3987.

1106-097

TAMBULETIN C$_{23}$H$_{24}$O$_{13}$ 508 OI 270^{0}C

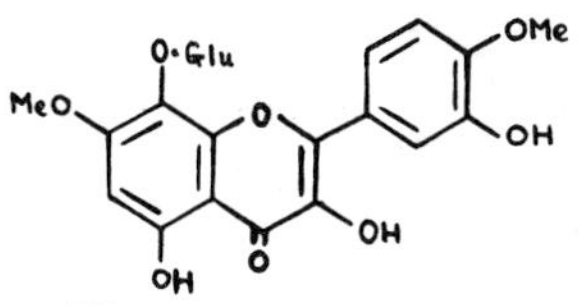

Phytochem., 1971 (10) 883.

1106-102

FLAVONE, 5,3',4'-triOH-3,7,8-triOMe- C$_{18}$H$_{16}$O$_8$ 360 OI 250^{0}C

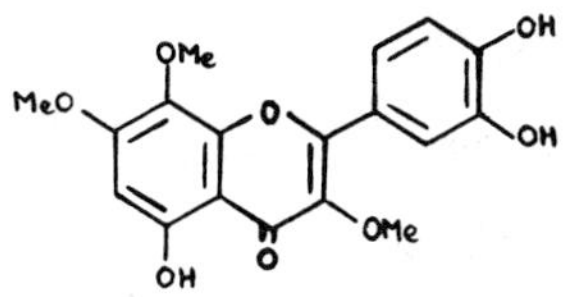

Tet., 1965, 3219.
Ber., 1968, 3987.

1106-098

FLAVONE, 4',5,7-triOH-3,3',8-triOMe- C$_{17}$H$_{14}$O$_8$ 346 OI 216^{0}C

R = H

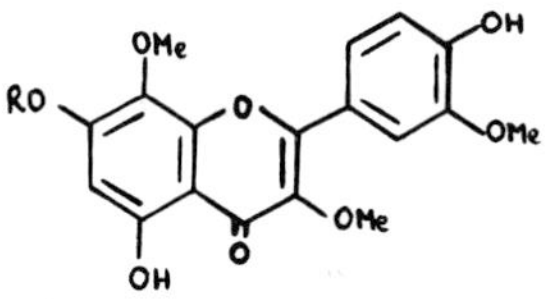

Aust. J. C., 1967, 1049.

1106-103

WHARANGIN C$_{17}$H$_{12}$O$_8$ 344 OI 277^{0}C

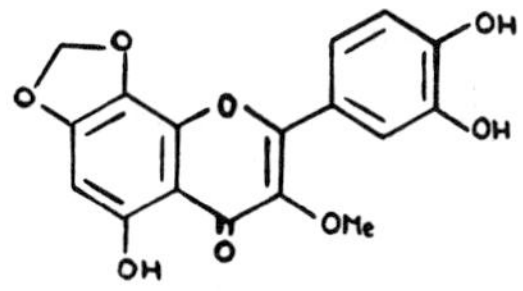

JCS, 1951, 3131.

1106-099

TERNATIN C$_{19}$H$_{18}$O$_8$ 374 OI 211^{0}C

R = Me

JCS, 1950, 2376.

1106-104

LIMOCITRIN C$_{17}$H$_{14}$O$_8$ 346 OI 275^{0}C

R = H

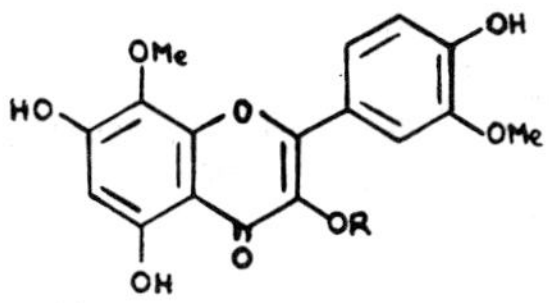

JACS, 1957, 6567.

1106-100

GOSSYPETIN PENTA ME-ETHER C$_{20}$H$_{20}$O$_8$ 388 OI 162^{0}C

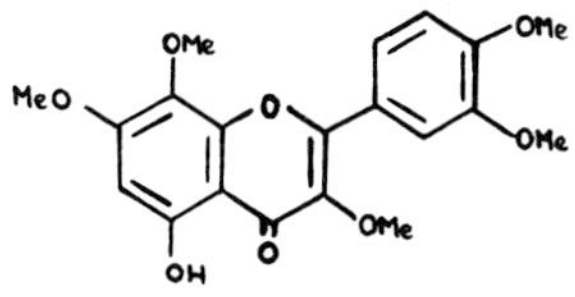

Ber., 1967 (100) 2296.
Aust. J. C., 1964 (17)934.

1106-105

LIMOCITRIN, 3-Glucosyl- C$_{23}$H$_{24}$O$_{13}$ 508 245^{0}C

R = Glucosyl

Tet., 1964, 2313.

1106-101

MELITERNIN C$_{20}$H$_{18}$O$_8$ 386 OI 186^{0}C

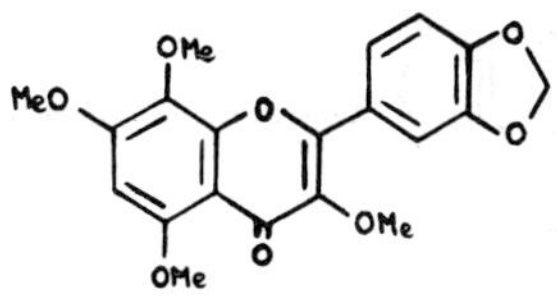

JCS, 1950, 2376.

1106-106

LIMOCITROL	$C_{18}H_{16}O_9$	376 OI 221°C

R = H

Tet., 1964, 2313.

1106-107

TRANSILITIN	$C_{16}H_{12}O_7$	316 OI 255°C

Khim. Prir. Soed., 1969, 439.

1106-112

LIMOCITROL, 3-glucosyl-	$C_{24}H_{26}O_{14}$	538 204°C

R = Glucosyl

Tet., 1964, 2313.

1106-108

QUERCETIN, 3-Rhamno-gluco- -glucosyl-	$C_{33}H_{40}O_{21}$	772

J. Agr. Chem. Soc. Jap., 1954 (28) 190.

1106-113

LIMOCITROL, iso-	$C_{18}H_{16}O_9$	376 OI 237°C

R = H

Tet., 1964, 2313.

1106-109

SPIRAEOSIDE	$C_{21}H_{20}O_{12}$	464 212°C

Natwiss., 1959, 427.

1106-114

LIMOCITROL-3-GLUCOSIDE. iso-	$C_{21}H_{26}O_{14}$	538 225°C

R = Glucosyl

Tet., 1964, 2313.

1106-110

GOSSPETIN TETRA-ME ETHER	$C_{19}H_{18}O_8$	374 OI 184°C

Aust. J. C., 1964 (17) 934.

1106-115

MELIBENTIN	$C_{21}H_{20}O_9$	416 OI 134°C

Aust. J. C., 1965, 2021.

1106-111

FLAVONE-3'-GLUCOSIDE, -5, 7, 3', 4', 5'-PENTA-OH-	$C_{21}H_{20}O_{12}$	464 283°C

Chem. Abstr., 1970 (73) 32284.

1107-001

TRICIN	$C_{17}H_{14}O_7$	330
		OI
		289^0C
R = R' = H		
		JCS, 1950, 116.
1107-002		

FLAVONE, 5, 6, 7, 3', 4', 5'-HEXA- -OMe-	$C_{21}H_{22}O_8$	402
		OI
		J. Ind. Chem. Soc., 1971 (48) 80.
1107-007		

TRICIN-5-GLUCOSIDE	$C_{23}H_{24}O_{12}$	492
R = Glu ; R' = H		
		Phytochem., 1971 (10) 490.
1107-003		

ACRAMMERIN	$C_{16}H_{12}O_8$	332
		OI
		338^0C
R = H		
		Chem. Abstr., 1956, 2558.
1107-008		

TRICIN-7-GLUCOSIDE	$C_{23}H_{24}O_{12}$	492
R = H ; R' = Glu		
		Phytochem., 1971 (10) 490.
1107-004		

FLAVONE, 5, 7-DI-OH-8, 3', 4', 5'- -TETRA-OMe-	$C_{19}H_{18}O_8$	358
		OI
		226^0C
R = Me		
		Ind. J. C., 1971 (9) 189.
1107-009		

TRICIN-7-GLUCURONIDE	$C_{23}H_{22}O_{13}$	506
R = H ; R' = Glucuronosyl		
		Phytochem., 1971 (10) 490.
1107-005		

GARDENIN - E	$C_{19}H_{18}O_9$	390
		OI
		232^0C
R = R' = H		
		Ind. J. C., 1970 (8) 398.
1107-010		

CORYMBOSIN	$C_{19}H_{18}O_7$	358
		OI
		188^0C
		Tet. Lett., 1967, 4580.
1107-006		

GARDENIN-C	$C_{20}H_{20}O_9$	404
		OI
		180^0C
R = H ; R' = Me		
		Ind. J. C., 1970 (8) 398.
1107-011		

GARDENIN-A	$C_{21}H_{22}O_9$	418
		OI
		164^0C
R = R' = Me		
		Ind. J. C., 1970 (8) 573.
1107-012		

	$C_{21}H_{20}O_{13}$	480
R = Galactosyl		
		Pharmazie., 1969 (24) 283.
1108-004		

SCAPOSIN $\qquad$ $C_{19}H_{18}O_9$ $\qquad$ 374

OI

211^0C

Tet., 1968, 3675.

1107-013

	$C_{27}H_{30}O_{18}$	642
		195^0C
R = Gal-Gal		
		Arch. Pharm., 1957 (290) 338.
1108-005		

MYRICETIN $\qquad$ $C_{15}H_{10}O_8$ $\qquad$ 318

OI

358^0C

R = H

JOC, 1956, 534.

1108-001

CANNABISCITRIN $\qquad$ $C_{21}H_{20}O_{13}$ $\qquad$ 480

220

R = H ; R' = Glu

Proc. Ind. Acad. Sci., 1946 (23A) 296.

1108-006

MYRICITRIN, Iso-	$C_{21}H_{20}O_{13}$	480
		-36^0
		276^0C
R = Glucosyl		
		Khim. Prir. Soedin., 1968 (4) 50.
1108-002		

MYRICETIN 3, 3'-DI-GALACTOSIDE	$C_{27}H_{30}O_{18}$	642
R = R' = Gal		
		Chem. Abstr., 1970 (73) 22117.
1108-007		

MYRICITRIN $\qquad$ $C_{21}H_{20}O_{12}$ $\qquad$ 464

200^0C

R = Rhamnosyl

JACS, 1919, 208.
JCS Jap., 1931, 193.

1108-003

PINOMYRICETIN $\qquad$ $C_{16}H_{12}O_8$ $\qquad$ 332

OI

$>336^0$C

Chem. Abstr., 1956, 15750.

1108-008

ANNULATIN $C_{16}H_{12}O_8$ 332
OI

R = H

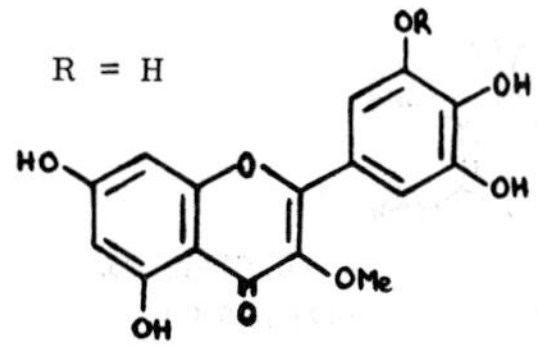

Phytochem., 1967, 1111.

1108-009

MYRICETIN-3',4',7-TRIME-
ETHER $C_{18}H_{16}O_8$ 360
OI
233⁰C

R = H

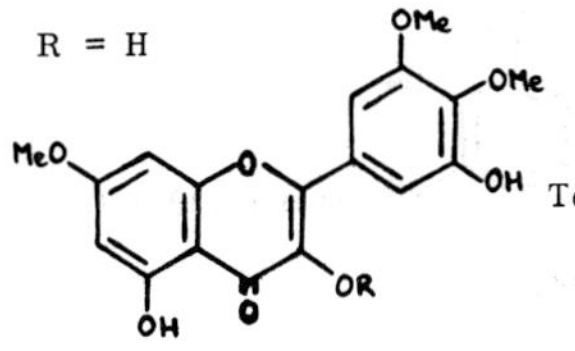

Tet. Lett., 1970, 1601.

1108-014

MYRICETIN, 3-O-Me-3'-O-Glu- $C_{22}H_{22}O_{13}$ 494

R = Glucosyl

Phytochem., 1970, 2413.

1108-010

MYRICETIN-3,7,3',4'-TETRA-
ME ETHER $C_{19}H_{18}O_8$ 374
OI
150⁰C

R = Me

Tet., 1967, 2295.

1108-015

EUROPETIN $C_{16}H_{12}O_8$ 332
OI

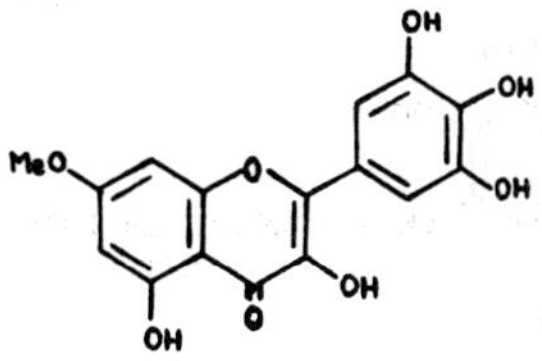

Phytochem., 1967, 1111.

1108-011

COMBRETOL $C_{20}H_{20}O_8$ 388
OI
144⁰C

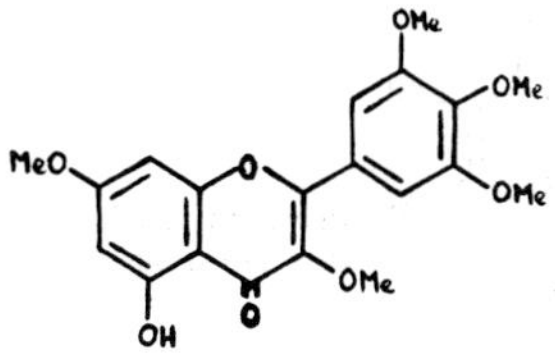

JCS, 1966, 125.

1108-016

MEARNSITRIN $C_{16}H_{12}O_8$ 332
OI

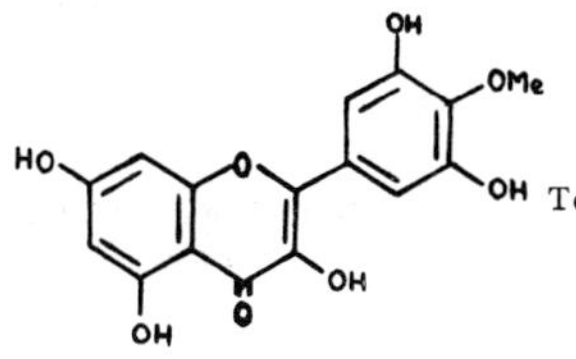

Tet. Lett., 1967, 2519.

1108-012

ROBINETIN $C_{15}H_{10}O_7$ 302
OI
326⁰C

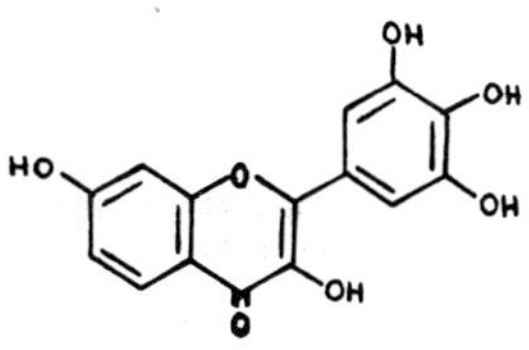

JCS, 1932, 1107.

1108-017

SYRINGETIN $C_{17}H_{14}O_8$ 346
OI

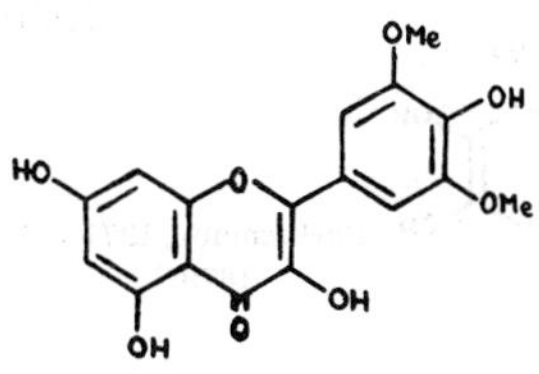

Phytochem., 1967, 1111.

1108-013

KANUGIN $C_{19}H_{16}O_7$ 356
OI
204⁰C

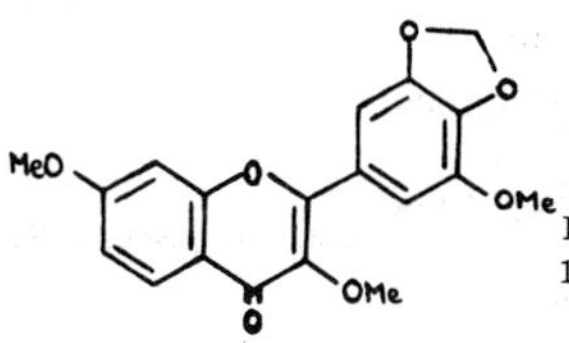

Proc. Ind. Acad. Sci.,
1946 (23A) 147.

1108-018

FLAVONE, 3, 5', 6, 7-tetra-OMe- 3', 4'-diOH- $C_{19}H_{18}O_8$ 374

Natwiss., 1966, 19.

1108-019

HIBISCETIN HEPTA-ME-ETHER $C_{22}H_{24}O_9$ 432 OI

J. Ind. Chem. Soc., 1971 (48) 80.

1108-024

FLAVONE, 3, 3', 4', 5, 5', 6, 7- -hepta-OMe- $C_{22}H_{24}O_9$ 432 OI 156^{0}C

JOC, 1968, 3574.

1108-020

DIGICITRIN $C_{21}H_{22}O_{10}$ 434 OI 178^{0}C

Ber., 1966, 3218.

1108-025

HIBISCETIN $C_{15}H_{10}O_9$ 334 OI 350^{0}C

R = H

Proc. Ind. Acad. Sci., 1944 (19A) 88.

1108-021

APULEITRIN $C_{19}H_{18}O_9$ 390 OI 176^{0}C

R = H

Phytochem., 1971 (10) 2433.

1108-026

HIBISCITRIN $C_{21}H_{30}O_{14}$ 496 239^{0}C

R = Glucosyl

Chem. Abstr., 1949 (43) 1408.

1108-022

APULEIRIN $C_{20}H_{20}O_9$ 404 OI 219^{0}C

R = Me

Phytochem., 1971 (10) 2433.

1108-027

FLAVONE, 5, 7, 3'-triOH-3, 8, 4', 5'- - tetraOMe- $C_{19}H_{18}O_9$ 390 OI 215^{0}C

Aust. J. C., 1968, 2529.

1108-023

MYRICETIN-7-GLUCOSIDE $C_{21}H_{20}O_{13}$ 480 239^{0}C

Phytochem., 1971 (10) 1679.

1108-028

DATISCETIN	$C_{15}H_{10}O_6$	286 OI 277°C

JCS, 1955, 166.

1199-001

FLAVONE, 3',5,6-tri-OMe-	$C_{18}H_{16}O_5$	312 OI

JOC, 1968, 3577.

1199-006

PTAEROXYLOL	$C_{16}H_{12}O_6$	300 OI

Chem. Abstr., 1954, 8489.

1199-002

WIGHTIN	$C_{18}H_{16}O_7$	344 OI 188°C

Tet., 1965 (21) 3237.

1199-007

ECHIOIDININ	$C_{16}H_{12}O_5$	284 OI 265°C

R = H

Tet., 1965 (21) 2633.

1199-003

ARTOCARPETIN, nor-	$C_{15}H_{10}O_6$	286 OI 330°C

R = R' = H

Tet. Lett., 1965, 663.

1199-008

ECHIOIDIN	$C_{22}H_{22}O_{10}$	446 -24° 277°C

R = Glucosyl

Tet., 1965 (21) 3715.

1199-004

ARTOCARPESIN	$C_{20}H_{18}O_5$	338 OI 250°C

R = 3,3-DMA; R' = H

Ind. J. C., 1966, 406.
Tet. Lett., 1965, 663.

1199-009

FLAVONE, 2',5,6-tri-OMe-	$C_{18}H_{16}O_5$	312 OI 125°C

JOC, 1968, 3580.

1199-005

MULBERRIN	$C_{25}H_{26}O_6$	422 OI 154°C

R = R' = 3,3-DMA

Tet. Lett., 1968, 1715.

1199-010

ARTOCARPETIN $C_{16}H_{12}O_6$ 300
OI
310^0C

R = H

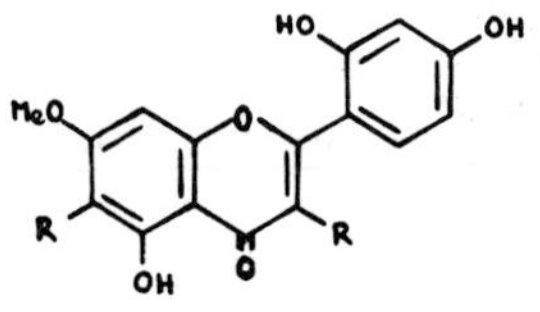

J. Sci. Ind. Res.,
1960 (19 B) 470.

1199-011

MORIN $C_{15}H_{10}O_7$ 302
OI
289^0C

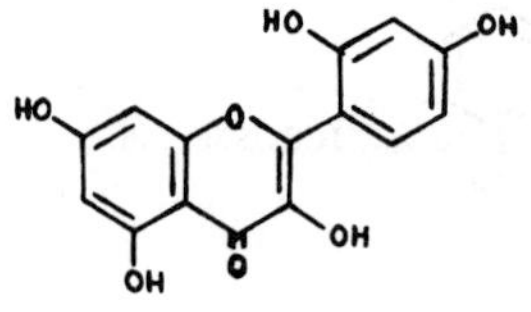

JACS, 1951, 3340.

1199-016

ARTOCARPIN $C_{26}H_{28}O_6$ 436
OI
175^0C

R = 3, 3-DMA

J. Sci. Ind. Res.,
1961 (20 B) 112.

1199-012

CHRYSOSPLIN $C_{19}H_{18}O_8$ 374
OI

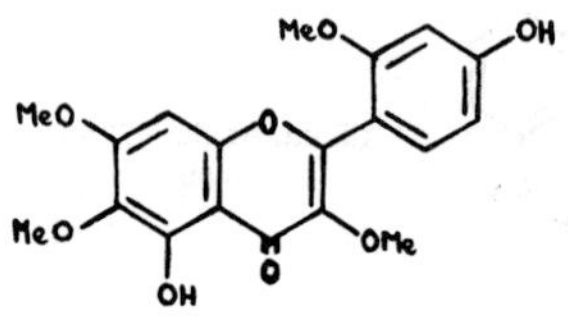

Khim. Prir. Soedin.,
1970 (6) 268.

1199-017

MULBERROCHROMENE $C_{25}H_{24}O_6$ 420
OI
234^0C

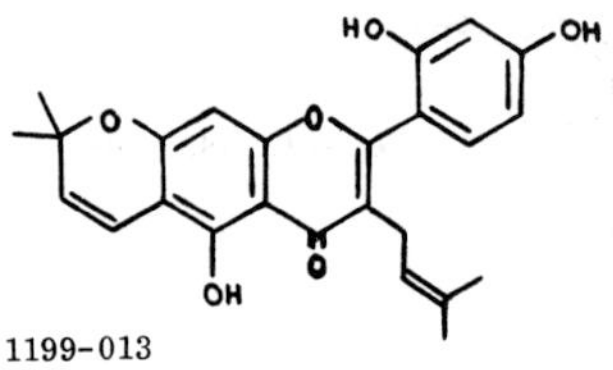

Tet. Lett., 1968, 1715.

1199-013

ZAPOTININ $C_{18}H_{16}O_6$ 328
OI
224^0C

R = H

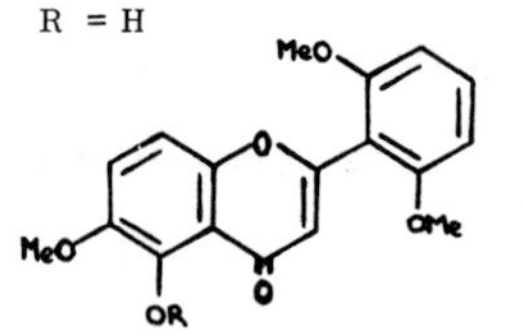

Tet., 1967, 2413.

1199-018

CYCLOMULBERRIN $C_{25}H_{24}O_6$ 420
231^0C

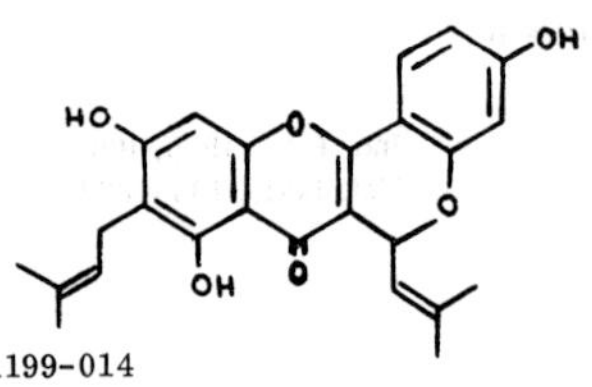

Tet. Lett., 1968, 1715.

1199-014

ZAPOTIN $C_{19}H_{18}O_6$ 342
OI
150^0C

R = Me

Tet. Lett.. 1968, 3993.

1199-019

CYCLOMULBERROCHROMENE $C_{25}H_{22}O_6$ 418
233^0C

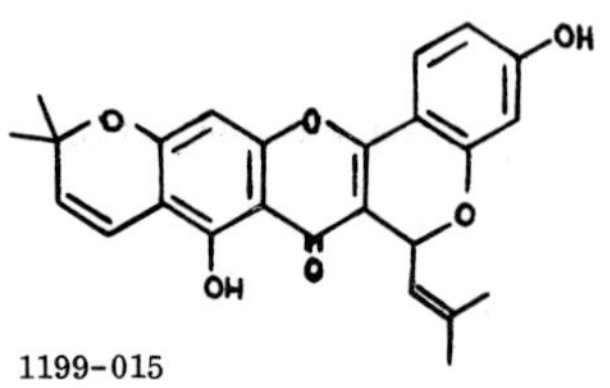

Tet. Lett., 1968, 1715.

1199-015

FLAVONE, 3', 5, 5', 6-tetra-OMe- $C_{19}H_{18}O_6$ 342
OI

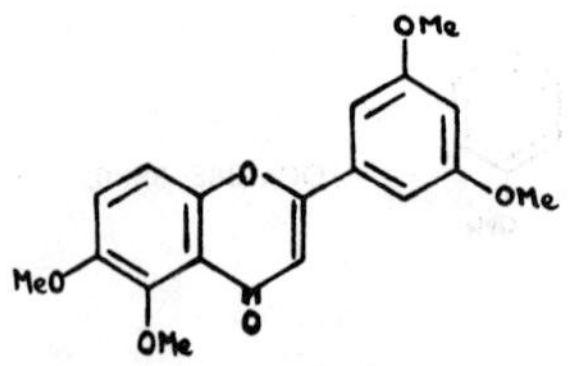

JOC, 1968, 3577.

1199-020

SERPYLLIN	$C_{20}H_{20}O_8$	388
		OI
		170^0C

Tet., 1968, 7027.

1199-021

APULEIN	$C_{20}H_{20}O_9$	404
		212^0C

R = Me

Phytochem., 1971, 2433.

1199-026

AYANIN-A, oxy-	$C_{18}H_{16}O_8$	360
		OI
		230^0C

R = H

JCS, 1954, 4587.

1199-022

CHLORFLAVONIN	$C_{18}H_{15}ClO_7$	378
		OI
		212^0C

JCS, 1969, 2418.

1199-027

CHRYSOPLENOSIDE	$C_{24}H_{26}O_{13}$	522
		274^0C

R = Glucosyl

Chem. Abstr., 1969 (70) 44853.

1199-023

FICININE	$C_{20}H_{19}NO_4$	337
		235^0C

Tet. Lett., 1965, 1987.

1199-028

AYANIN-A, 5-O-ME-OXY-	$C_{19}H_{18}O_8$	374
		OI
		259^0C

Phytochem., 1971 (10) 2433.

1199-024

FICININE, Iso-	$C_{20}H_{19}NO_4$	337
		168^0C

Tet. Lett., 1965, 1987.

1199-029

APULEIN, 5-O-desMe-	$C_{19}H_{18}O_9$	390
		OI
		229^0C

R = H

Phytochem., 1971, 2433.

1199-025

FLAVONE, 2',5,8-TRI-OH-	$C_{15}H_{10}O_5$	270
		OI

Compt. Rend., 1971 (272D) 2961.

1199-030

APULEIDIN $C_{18}H_{16}O_8$ 360

155^{0}C

R = H

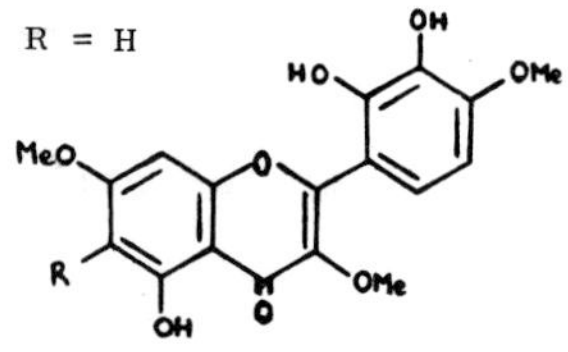

Phytochem., 1971 (10)
2433.

1199-031

ARTOCARPIN DIME ETHER, Iso- $C_{28}H_{30}O_6$ 462

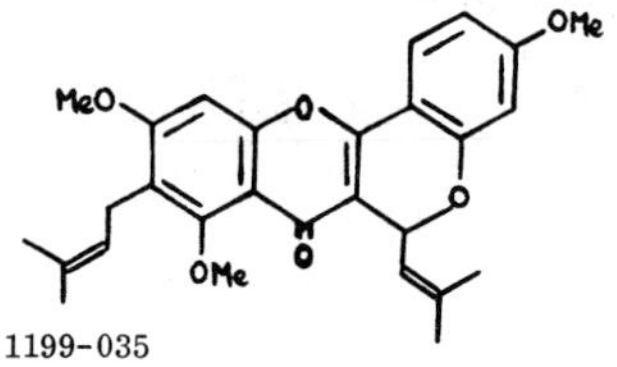

Chem. Abstr., 1966
(64) 17563.

1199-035

APULEISIN $C_{18}H_{16}O_9$ 376
OI
194^{0}C

R = OH

Phytochem., 1971 (10)
2433.

1199-032

GARCININ $C_{22}H_{16}O_8$ 408
OI

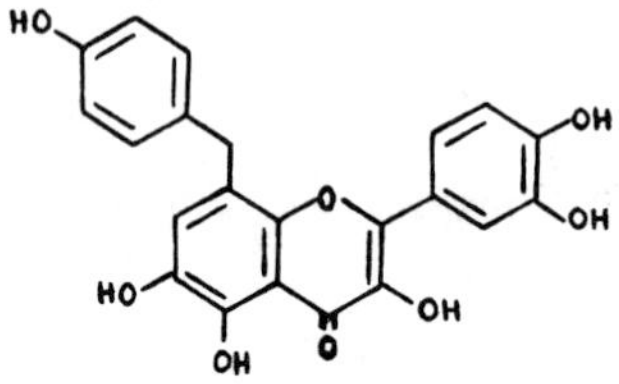

JCS, 1933, 610.

1199-036

ARTOCARPIN, cyclo- $C_{26}H_{26}O_6$ 434

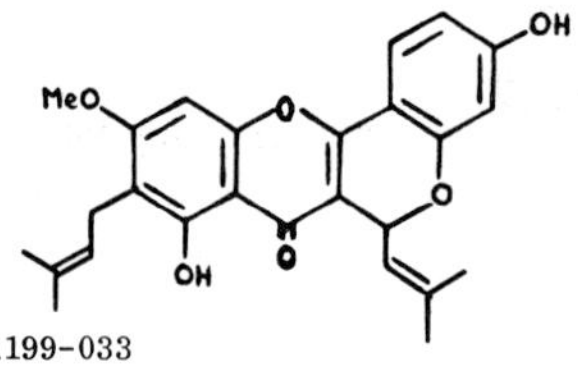

Tet. Lett., 1964, 125.

1199-033

CYCLOHETEROPHYLLIN $C_{30}H_{30}O_7$ 502

205^{0}C

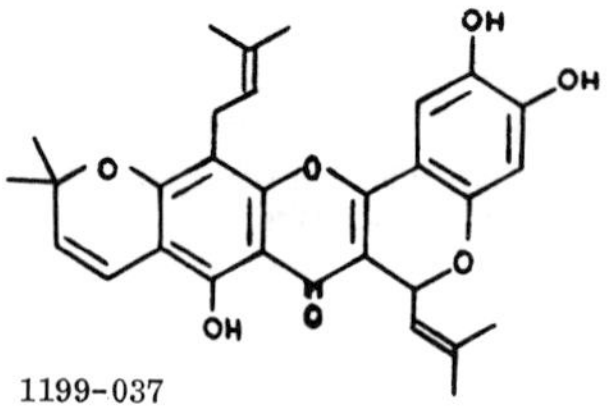

Ind. J. C., 1971 (9) 7.

1199-037

ARTOCARPIN DIME ETHER $C_{28}H_{32}O_6$ 464
OI
151^{0}C

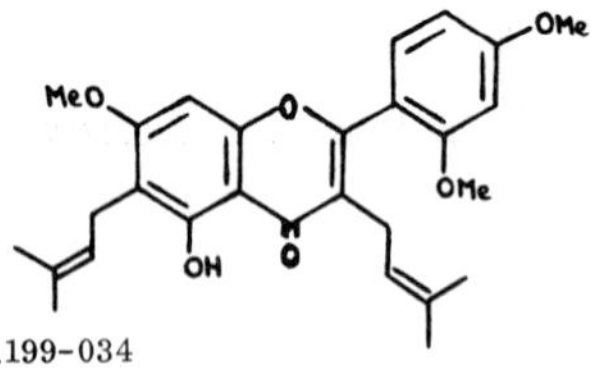

Chem. Abstr., 1966
(64) 17563.

1199-034

12 | FLAVANONES

The flavanones have been subclassified in a similar manner to the flavones.

1201	No oxy substituent	1202	as 1201 + 3-oxy
1203	4'-oxy	1204	as 1203 + 3-oxy
1205	3', 4'-dioxy	1206	as 1205 + 3-oxy
1207	3', 4', 5'-trioxy	1208	as 1207 + 3-oxy
1299	Misc.		

PINOCEMBRIN $C_{15}H_{12}O_4$ 256
-146^0m

R = H

Tet. Lett., 1967, 481.

1201-001

STROBOPININ $C_{16}H_{14}O_4$ 270
-60^0m
226^0C

Chem. Pharm. Bull.,
1957 (5) 195.

1201-006

VERECUNDIN $C_{21}H_{22}O_9$ 418

R = Glucosyl

JACS, 1957, 450 & 1738.

1202-002

CRYPTOSTROBIN $C_{16}H_{14}O_4$ 270
-33^0m
202^0C

Chem. Pharm. Bull.,
1957 (5) 195.

1201-007

PINOCEMBRIN-7-NEOHESPERIDOSIDE $C_{27}H_{32}O_{13}$ 564
278^0C

R = Neohesperidosyl

Phytochem., 1970, 1877.

1201-003

MATTEUCINOL, des-OMe- $C_{17}H_{16}O_4$ 284
-50^0a
202^0C

Ber., 1936 (69) 1893.

1201-008

ALPINETIN $C_{16}H_{14}O_4$ 270
225^0C

JCS, 1950, 3117.

1201-004

WOGONIN, dihydro- $C_{16}H_{14}O_5$ 286

Bull. Soc. Chim. Fr.,
1957, 192.

1201-009

PINOSTROBIN $C_{16}H_{14}O_4$ 270
-56^0c
112^0C

Acta Chem. Scand.,
1950, 1042.

1201-005

PEDICIN, Iso- $C_{18}H_{18}O_6$ 330

Rev. Pure App. Chem.,
1951 (1) 186.

1201-010

MALLOTUS FLAVANONE	$C_{21}H_{20}O_4$	336
		145^0C

JCS, 1968, 2627.

1201-011

ALNUSTINOL	$C_{16}H_{14}O_6$	302
		175^0C

Bull. Chem. Soc. Jap.,
1971, 2761.

1202-005

PINOBANKSIN	$C_{15}H_{12}O_5$	272
		$+14^0$
		177^0C

Acta Chem. Scand.,
1949, 1375.

1202-001

FLAVANONE, 3, 7-di-OH-6-OMe-	$C_{16}H_{14}O_5$	286

Chem. Abstr., 1971
(74) 72459.

1202-006

ALPINONE	$C_{16}H_{14}O_5$	286
		-20^0c
		186^0C

R = H

Chem. Ind., 1955, 443.

1202-002

NARINGENIN	$C_{15}H_{12}O_5$	272
		0^0
		259^0C

R = R' = H

JACS, 1957, 450; 1738.

1203-001

ALPINONE, 3-O-Ac-	$C_{18}H_{16}O_6$	328
		$+17^0p$
		135^0C

R = Ac

Acta Chem. Scand.,
1956, 393.

1202-003

SALIPURPOSIDE	$C_{21}H_{22}O_{10}$	434
		-96^0e
		225^0C

R = Glu ; R' = H

Compt. Rendu, 1954
(238) 2112.
Arch. Pharm., 1959
(292) 398.

1203-002

deleted

1202-004

PRUNIN	$C_{21}H_{22}O_{10}$	434
		-42^0
		225^0C

R = H; R' = Glu

JACS, 1957, 1738.

1203-003

NARINGIN $C_{27}H_{32}O_{14}$ 580
-90^0e
171^0C

R = H; R' = . Glu-Rha

JACS, 1952, 3614.

1203-004

SAKURANIN $C_{22}H_{24}O_{10}$ 448
-107^0a
214^0C

R = Glucosyl

JACS, 1955, 3557.

1203-009

NARIRUTIN $C_{27}H_{32}O_{14}$ 580
162^0C

R = H ; R' = Rutinosyl

Tet. Lett., 1968, 1635.

1203-005

SAKURANETIN, L-iso- $C_{16}H_{14}O_5$ 286
-20^0
194^0C

R = H

JACS, 1957, 450.

1203-010

NARINGENIN, 5, 7-di-Glucosyl- $C_{27}H_{32}O_{15}$ 596

R = R' = Glu

Planta Med., 1971
(19) 311.

1203-006

SAKURANIN, iso- $C_{22}H_{24}O_{10}$ 448
-41^0
190^0C

R = Glucosyl

JACS, 1957, 450.

1203-011

SELINONE $C_{20}H_{20}O_5$ 340
-50^0e
152^0C

Chem. Ind., 1971, 355.
Tet. Lett., 1959 (9) 853.

1203-007

PONCIRIN $C_{28}H_{34}O_{14}$ 594
-105^0m
210^0C

R = Rutinosyl

Tet. Lett., 1968, 1635.
Compt. Rendu, 1946
(223) 45.

1203-012

SAKURANETIN $C_{16}H_{14}O_5$ 286
$+11^0$
150^0C

R = H

JACS, 1957, 1738.

1203-008

CARTHAMIDIN $C_{15}H_{12}O_6$ 288
218^0C

R = H

Chem. Abstr., 1950
(44) 3491.

1203-013

CARTHAMIN, neo-	$C_{21}H_{22}O_{11}$	450
		228^0C
R = Glucosyl		
		Curr. Sci., 1960 (29)54.
1203-014		

FARREROL $C_{17}H_{16}O_5$ 300

-20^0e

214^0C

R = H

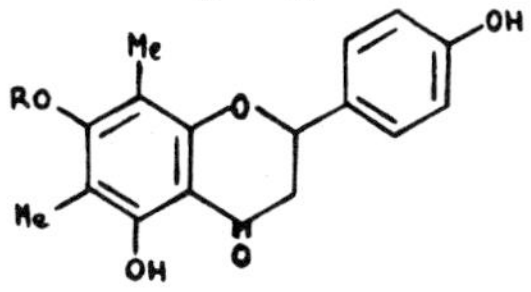

Chem. Ind., 1956, 738.

JCS, 1955, 3740.

1203-019

HEMIPHLOIN $C_{21}H_{22}O_{10}$ 434

$+40^0$

210^0C

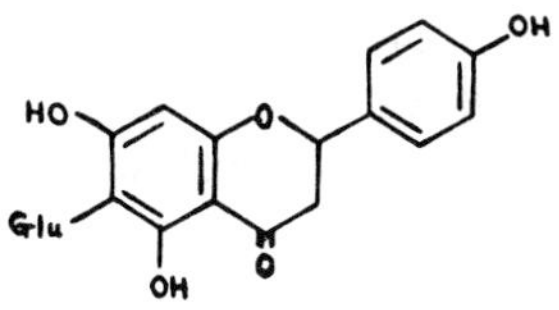

Aust. J. C., 1965, 531.

1203-015

ANGOPHOROL $C_{18}H_{18}O_6$ 314

Rac

167^0C

R = Me

JCS, 1960, 2063.

1203-020

HEMIPHLOIN, Iso- $C_{21}H_{22}O_{10}$ 434

-12^0

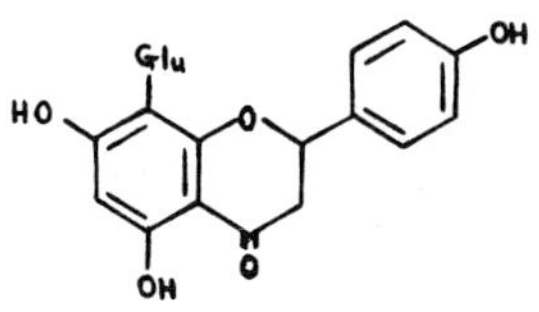

Aust. J. C., 1965, 147;

531.

1203-016

MATTEUCINOL $C_{18}H_8O_5$ 314

-40^0a

118^0C

R = H

JCS, 1954, 2782.

1203-021

PORIOL $C_{16}H_{14}O_5$ 284

R = H

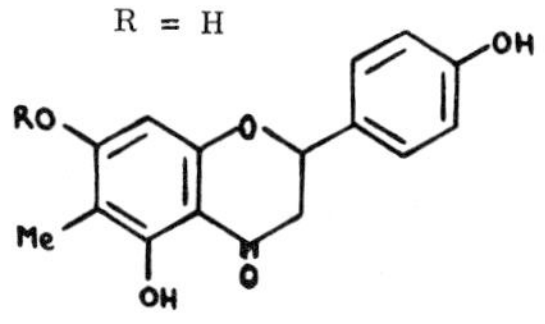

Can. J. C., 1967, 1020.

1203-017

MATTEUCININ $C_{30}H_{38}O_{15}$ 638

-29^0a

140^0C

R = Glu-Glu

JCS, 1960, 3197.

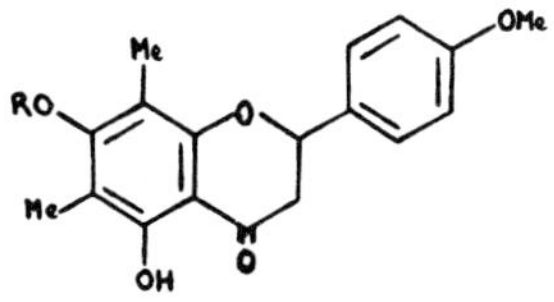

1203-022

PORIOLIN $C_{22}H_{24}O_{10}$ 446

-20^0a

175^0C

R = Glucosyl

Aust. J. C., 1969, 483.

1203-018

XANTHOHUMOL, iso- $C_{21}H_{22}O_5$ 354

198^0C

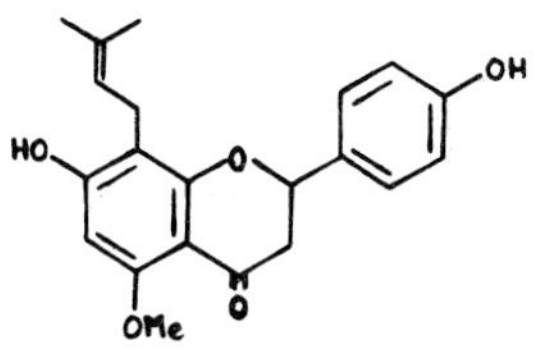

Bull. Soc. Chim. Belges,

1957 (66) 452.

1203-023

LIQUIRITIGENIN	$C_{15}H_{12}O_4$	256 -92^0m 208^0C

R = R' = H

JCS, 1938, 1320.
Arch. Biochem. Biophys.,
1956 (60) 329.

1203-024

BAVACHININ	$C_{21}H_{22}O_4$	338 -10^0c 155^0C

R = Me

Tet. Lett., 1968, 2401.

1203-029

LIQUIRITIN	$C_{21}H_{22}O_9$	418 -68^0 212^0C

R = Glu; R' = H

JCS, 1938, 1320.
Arch. Biochem. Biophys.,
1956 (60) 329.

1203-025

BAVACHIN, iso-	$C_{20}H_{20}O_4$	324 -4^0e 188^0C

Tet. Lett., 1968, 2401.

1203-030

LIQUIRITIGENIN, 7-Glucosyl-	$C_{21}H_{22}O_9$	418

R = H; R' = Glu

Arch. Biochem. Biophys.,
1956 (60) 329.

1203-026

SOPHORANONE	$C_{30}H_{36}O_4$	460 -13^0e 108^0C

Chem. Pharm. Bull.,
1969, 1302.

1203-031

LIQUIRITIGENIN, 7-diglucosyl-	$C_{27}H_{32}O_{14}$	580

R = H; R' = Glu-Glu

Arch. Biochem. Biophys.,
1956 (60) 329.

1203-027

SOPHORANOCHROMENE	$C_{30}H_{34}O_4$	458 -64^0e 152^0C

Chem. Pharm. Bull.,
1969, 1302.

1203-032

BAVACHIN	$C_{20}H_{20}O_4$	324 -29^0e 192^0C

R = H

Tet. Lett., 1968, 2401.

1203-028

FLAVANONE, 4',7,8-tri-OH-	$C_{15}H_{12}O_5$	272 104^0C

Biochem. J., 1966 (98)
493.

1203-033

AROMADENDRIN	$C_{15}H_{12}O_6$	288 $+26^0$ 248^0C

R = H

JACS, 1957, 1738.

1204-001

PHELLAVIN	$C_{26}H_{32}O_{12}$	536

Khim. Prir. Soedin., 1970 (6) 762.

1204-006

AROMADENDRIN, 7-O-Me-	$C_{16}H_{14}O_6$	302 $+30^0$m 188^0C

R = Me

Phytochem., 1971 (10) 1972.

1204-002

PHELLAMURETIN	$C_{20}H_{20}O_6$	356 220^0C

JACS, 1953, 5507.

1204-007

SINENSIN	$C_{21}H_{22}O_{11}$	450

R = Glucosyl

Biochem. J., 1961 (78) 298.

1204-003

PHELLAMURIN	$C_{26}H_{32}O_{12}$	536 205^0C

R = H

JACS, 1953, 5507.

1204-008

AROMADENDRIN 7-RHAMNOSIDE	$C_{24}H_{22}O_{10}$	434 -99^0e 161^0C

R = Rhamnosyl

Aust. J. C., 1960 (13) 150.

1204-004

PHELLOSIDE, dihydro-	$C_{32}H_{42}O_{17}$	698 151^0C

R = Glucosyl

Chem. Abstr., 1968 (69) 41709.

1204-009

ENGELITIN	$C_{21}H_{22}O_{10}$	434 -16^0p 177^0C

J. Pharm. Soc. Jap., 1959, 555.

1204-005

KEYAKINOL	$C_{21}H_{22}O_{11}$	450 214^0C

Chem. Abstr., 1956 (50) 14729.

1204-010

ERIODICTYOL R = H 1205-001	$C_{15}H_{12}O_6$ 288 265^0C JACS, 1940, 3258.	FLAVANONE, 7-glucosyloxy-6-Me- -3', 4', 5-tri-OH- 1205-006	$C_{22}H_{24}O_{11}$ 464 255^0C Can. J. C., 1969, 869.
ERIODICTYOL, 7-glucosyl- R = Glucosyl 1205-002	$C_{21}H_{22}O_{11}$ 450 176^0C Tet. Lett., 1966, 5133.	ERIODICTYOL, homo- 1205-007	$C_{16}H_{14}O_6$ 302 -28^0e 224^0C J. Pharm. Soc. Jap., 1929, 64;71.
ERIODICTIN R = Rhamnosyl 1205-003	$C_{21}H_{22}O_{10}$ 434 -51^0 186^0C Nature, 1936 (138) 1057.	HESPERETIN R = H 1205-008	$C_{16}H_{14}O_6$ 302 -38^0e 224^0C J. Pharm. Soc. Jap., 1928, 207;940.
ERIOCITRIN R = Rutinosyl JACS, 1960, 2803. 1205-004	$C_{27}H_{32}O_{15}$ 596 160^0C	HESPERETIN 7-ME ETHER R = Me Tet. Lett., 1966, 1293. 1205-009	$C_{17}H_{16}O_6$ 316 164^0C
ERIODICTYOL-5, 3'-DIGLUCOSIDE $C_{27}H_{32}O_{16}$ 612 1205-005	Planta Med., 1971 (19) 311.	HESPERIDIN R = Rutinosyl Ber., 1938, 2511. JCS, 1931, 1704. 1205-010	$C_{28}H_{34}O_{15}$ 510 -88^0p 260^0C

HESPERIDIN, neo-	$C_{28}H_{34}O_{15}$	510
		-105^0e
		244^0C
R = Neohesperidosyl		
	Helv., 1949, 714.	
1205-011		

COREOPSIN, iso-	$C_{21}H_{22}O_{10}$	434
R = Glucosyl		
	Chem. Abstr., 1955 (49) 8856.	
1205-016		

	$C_{22}H_{24}O_6$	384
		-34^0c
		166^0C
R = 3,3-DMA		

Aust. J. C., 1958, 376.

1205-012

MONOSPERMOSIDE, iso-	$C_{21}H_{22}O_{10}$	434
		163^0C
R = H		

Phytochem., 1970, 2231.

1205-017

	$C_{26}H_{32}O_6$	440

R =

Aust. J. C., 1965 (18) 1649.

1205-013

BUTRIN	$C_{27}H_{32}O_{15}$	596
R = Glucosyl		
	Chem. Abstr., 1950 (44) 3097.	
1205-018		

CYRTOMINETIN	$C_{17}H_{16}O_6$	316

Chem. Pharm. Bull., 1956 (4) 24.

1205-014

OKANIN, iso-	$C_{15}H_{12}O_6$	288
		140^0C
R = H		

JCS, 1951, 569.
JACS, 1957, 214.

1205-019

BUTIN	$C_{15}H_{12}O_5$	272
		205^0C
R = H		

JCS, 1904, 1459.

1205-015

FLAVANOMAREIN	$C_{21}H_{32}O_{11}$	450
R = Glucosyl		
	JACS, 1957, 214.	
1205-020		

BUTIN, 8-OMe-	$C_{16}H_{14}O_6$	302
		196^0C
		JACS, 1943, 677.
1205-021		

GLUCODISTYLIN	$C_{21}H_{22}O_{12}$	466
R = Glucosyl		
		J. Pharm. Soc. Jap., 1956, 343.
1206-003		

FLAVANONE, 3',4',7,8-tetra-OMe-	$C_{19}H_{20}O_6$	344
		143^0C
		JACS, 1944, 486.
1205-022		

ASTILBIN	$C_{21}H_{22}O_{11}$	450
		180^0C
R = Rhamnosyl		
		J. Pharm. Soc. Jap., 1952, 469;578.
1206-004		

PLATHYMENIN	$C_{15}H_{12}O_6$	288
		229^0C
		JCS, 1953, 1055.
1205-023		

PADMATIN	$C_{16}H_{14}O_7$	318
		170^0C
R = H		
		Tet., 1959 (5) 91.
1206-005		

TAXIFOLIN	$C_{15}H_{12}O_7$	304
		$+42^0$a
		222^0C
R = H		
		Acta Chem. Scand., 1955, 1728.
		JCS, 1955, 2948.
1206-001		

PADMATIN, glucosyl-	$C_{22}H_{24}O_{12}$	470
R = Glucosyl		
		Phytochem., 1971 (10) 2256.
1206-006		

HULTENIN	$C_{16}H_{14}O_7$	318
		$+7^0$
		208^0C
R = Me		
		Chem. Abstr., 1966 (65) 5435.
1206-002		

QUERCETIN, 6-Me-dihydro-	$C_{16}H_{14}O_7$	318
		$+7^0$
		194^0C
		Can. J. Chem., 1971, 49.
1206-007		

DEODARIN $C_{16}H_{14}O_7$ 318
+28°a
250°C

Curr. Sci., 1971 (40) 464.

1206-008

AMPELOPSIN $C_{15}H_{12}O_8$ 320

245°C

Ann., 1940 (544) 253.

1208-002

FUSTIN $C_{15}H_{12}O_6$ 288

217°C

J. Chem. Soc. Jap.,
1951 (72) 223.

1206-009

CITRONETIN $C_{16}H_{14}O_5$ 286

204°C

R = H

J. Pharm. Soc. Jap.,
1931, 578.

1299-001

MELANOXETIN PENTA-ME
ETHER, dihydro- $C_{20}H_{22}O_7$ 374
+88°e
147°C

Tet., 1963, 1371.

1206-010

CITRONIN $C_{28}H_{34}O_{14}$ 594

R = .Rha-Glu

J. Pharm. Soc. Jap.,
1931, 578.

1299-002

ROBTIN $C_{15}H_{12}O_6$ 288
-3°m
276°C

Biochem. J., 1962 (82)
324; 1964 (84) 416.

1207-001

STEPPOGENIN $C_{15}H_{12}O_6$ 288
+9°m
256°C

R = H

Chem. Abstr., 1968
(69) 41743.

1299-003

ROBINETIN, dihydro- $C_{15}H_{12}O_7$ 304
+1°
226°C

Ann., 1954 (587) 207.

1208-001

ARTOCARPANONE $C_{16}H_{14}O_6$ 302
OI
210°C

R = Me

J. Sci. Ind. Res., 1960
(19B) 470.

1299-004

STEPPOSIDE $C_{21}H_{22}O_{11}$ 450
-52°m
149°C

R = Glucosyl

Chem. Abstr., 1968
(69) 41743.

1299-005

SILYBIN $C_{25}H_{22}O_{10}$ 482
158°C

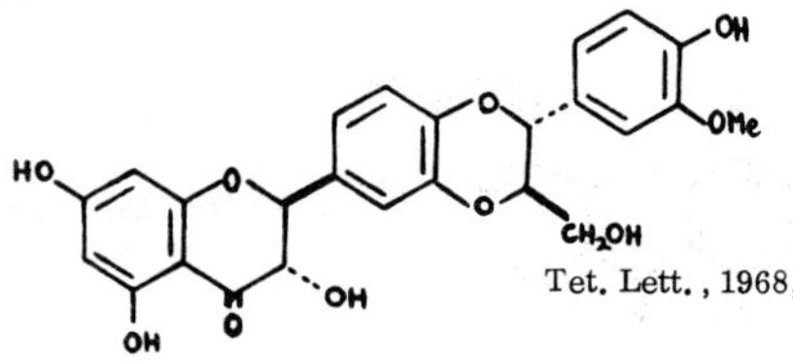

Tet. Lett., 1968, 2911.

1299-010

KURARINONE, nor- $C_{25}H_{28}O_6$ 424
+8°e
133°C

Chem. Pharm. Bull.,
1971, 2126.

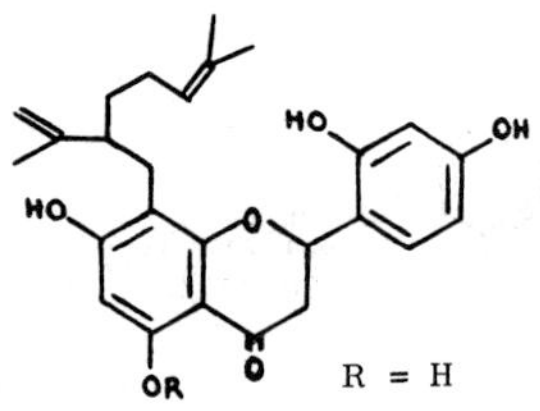

R = H

1299-006

SILYCHRISTIN $C_{25}H_{22}O_{10}$ 482
+81°p
175°C

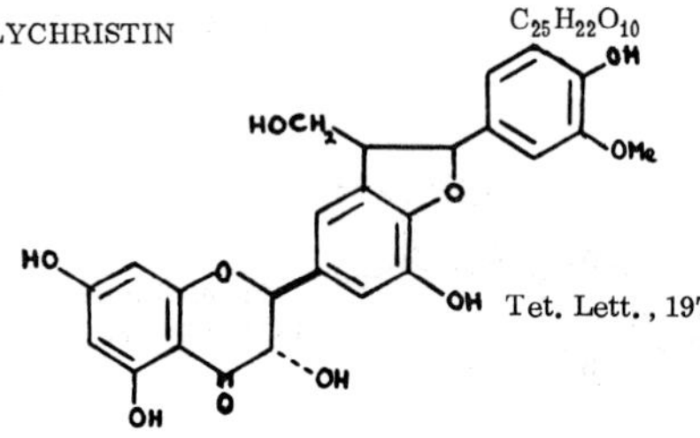

Tet. Lett., 1971, 1895.

1299-011

KURARINONE $C_{26}H_{30}O_6$ 438
+25°w
121°C

R = Me

Chem. Pharm. Bull.,
1971, 2126.

1299-007

SILYDIANIN $C_{25}H_{22}O_{10}$ 482
+175°a
191°C

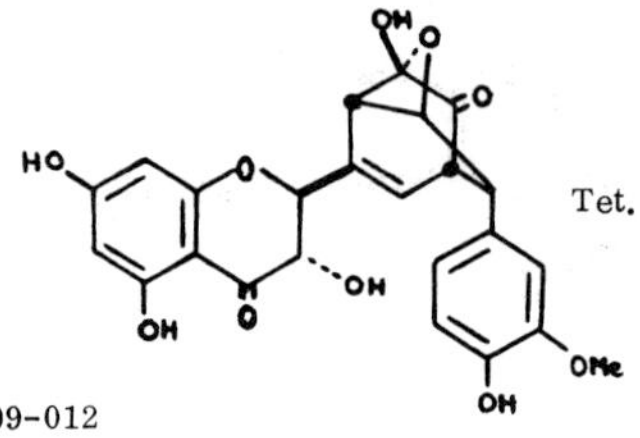

Tet. Lett., 1970, 2675.

1299-012

ARTOCARPIN, dihydro-cyclo- $C_{26}H_{28}O_6$ 436

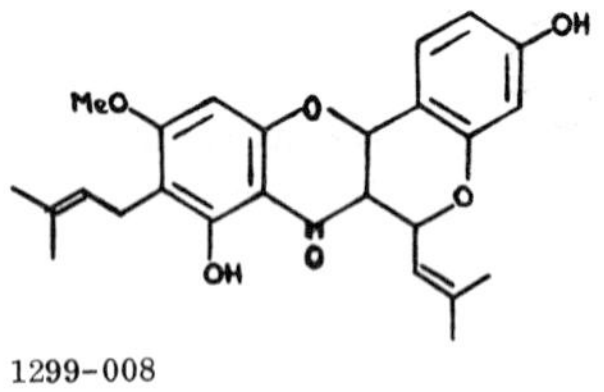

Chem. Abstr., 1966
(64) 17563.

1299-008

ROTTLERIN, iso- $C_{30}H_{28}O_8$ 516
OI
182°C

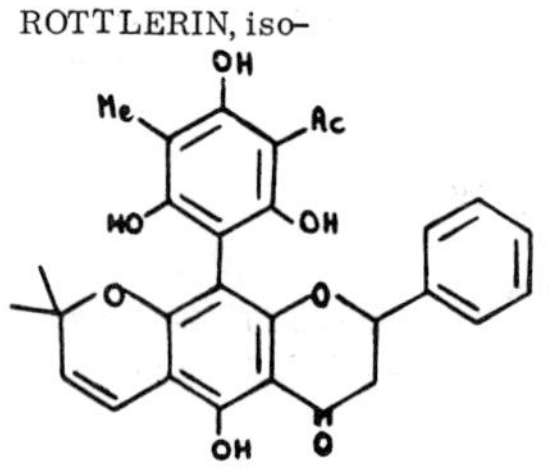

JCS, 1939, 1587.

1299-013

FARREROL, proto- $C_{17}H_{18}O_6$ 318
210°C

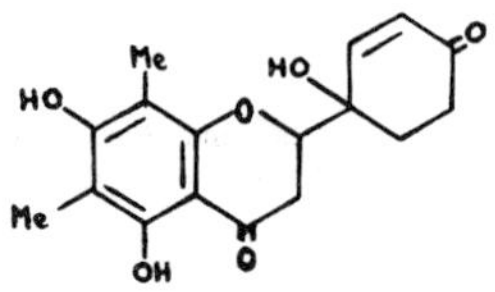

Chem. Pharm. Bull.,
1970, 596.

1299-009

OBTUSIFOLINE $C_{24}H_{22}O_7$ 422
203°C

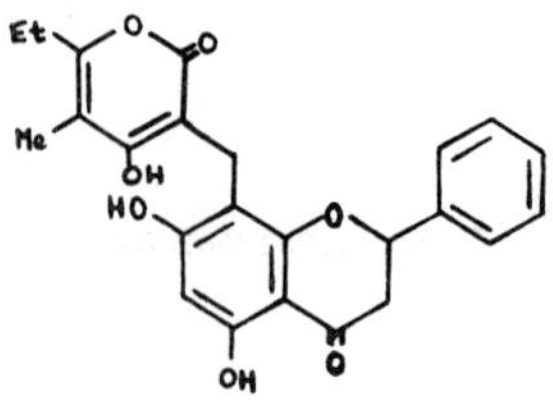

Z. Nat., 1970 (25B) 989.

1299-014

13 | FLAVANS

Included in this flavan class are the flavanols and flavandiols. The subclassification summarized below follows that employed for the other flavonoid classes.

1301	No oxy substituent on ring C
1302	4'-oxy
1303	3', 4'-dioxy
1304	3', 4', 5'-trioxy
1399	Misc.

KOABURANIN $C_{21}H_{24}O_8$ 404

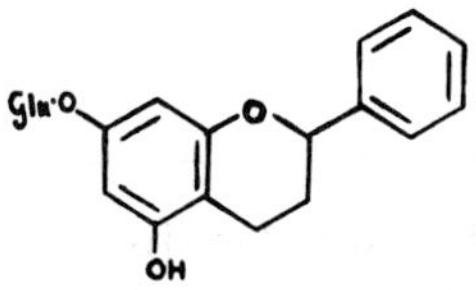

Chem. Abstr., 1971
(74) 1107.

1301-001

FLAVAN, (-)-4'-OH-7-OMe-8-Me- $C_{17}H_{18}O_3$ 270
-22^0c
126^0C

R = Me

Aust. J. C., 1971, 1257.
Tet. Lett., 1970, 1037.

1302-003

FLAVAN-7-OL, 5-OMe- $C_{16}H_{16}O_3$ 256
-6^0c
86^0C

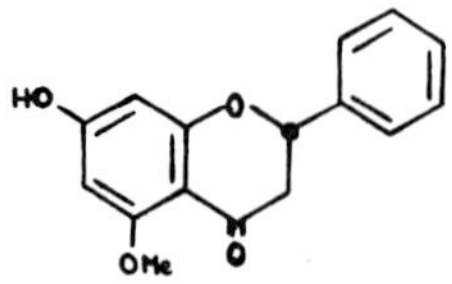

JCS, 1971, 3967.

1301-002

FLAVAN, 4', 5, 7-tri-OMe- $C_{18}H_{20}O_4$ 300
Rac
110^0C

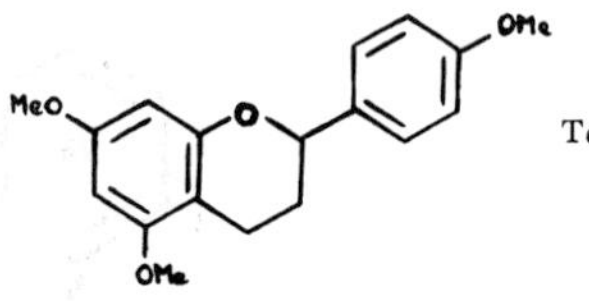

Tet. Lett., 1964, 2211.

1302-004

FLAVAN-7-OL, 5-OMe-6-Me- $C_{17}H_{18}O_3$ 270
-9^0c
123^0C

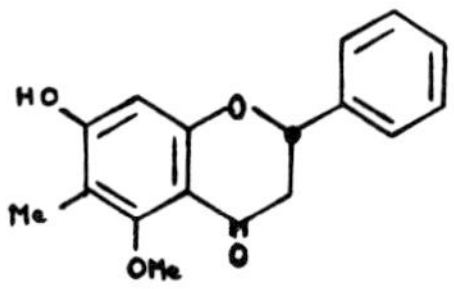

JCS, 1971, 3967.

1301-003

AFZELECHIN, (+)- $C_{15}H_{14}O_5$ 274

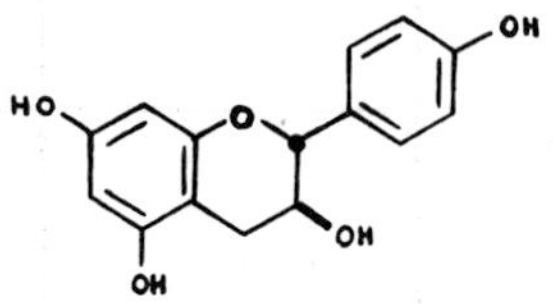

Aust. J. C., 1960, 390.

1302-005

FLAVAN, 4'-OMe- $C_{16}H_{16}O_2$ 240

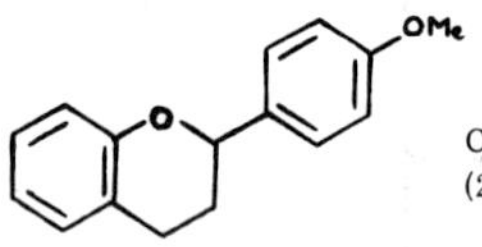

Compt. Rend., 1964
(259) 4167.

1302-001

TERACACIDIN, (-)- $C_{15}H_{14}O_6$ 290
-71^0e
226^0C

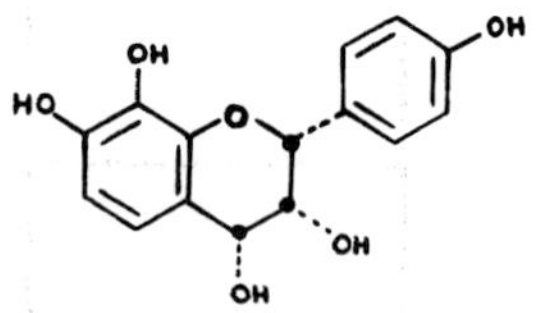

Aust. J. C., 1967, 2191.

1302-006

FLAVAN, (-)-4'-OH-7-OMe- $C_{16}H_{16}O_3$ 256
-16^0e
149^0C

R = H

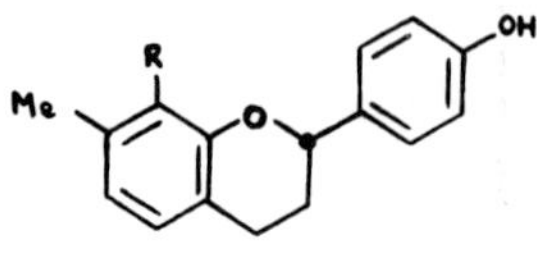

Aust. J. C., 1971, 1257.
Tet. Lett., 1970, 1037.

1302-002

TERACACIDIN, iso- $C_{15}H_{14}O_6$ 290

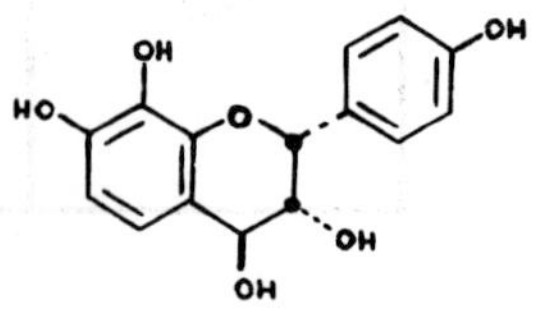

Aust. J. C., 1967, 2191.

1302-007

$C_{15}H_{14}O_6$ 290

Biochem. J., 1966
(98) 493.

1302-008

GUIBOURTACACIDIN, 4-epi- $C_{15}H_{14}O_5$ 274

4-epimer

Biochem. J., 1965 (96) 36.

1302-013

FISTUCACIDIN $C_{15}H_{14}O_6$ 290
Rac
246^0C

Tet., 1967, 515.

Stereo (?)

1302-009

AFZELECHIN, (-) - $C_{15}H_{14}O_5$ 274
-20^0
$>300^0$C

Chem. Abstr., 1968
(69) 77066.

1302-014

FLAVAN-3, 4, 4'7- TETRAOL,
- 8- OMe- $C_{16}H_{16}O_6$ 304

JCS, 1970, 1800.

1302-010

AFZELECHIN, epi-(-) - $C_{15}H_{14}O_5$ 274
-59^0e
242^0C

2-epimer

JCS, 1955, 2948.

1302-015

GUIBOURTACACIDIN $C_{15}H_{14}O_5$ 274
$+12^0$a
170^0C

JCS, 1970, 1800.
Biochem. J., 1963 (87) 439.

1302-011

LEUCOPELARGONIDIN $C_{15}H_{14}O_6$ 290
$+9^0$e
220^0C

Tet. Lett., 1967, 2443.

1302-016

GUIBOURTACACIDIN, 3-epi- $C_{15}H_{14}O_5$ 274

3-epimer

JCS, 1970, 1800.
Biochem. J., 1965 (96) 36.

1302-012

$C_{22}H_{26}O_{11}$ 466

Chem. Abstr., 1970
(73) 63217.

1302-017

$C_{20}H_{22}O_6$ 258 -93^0a $>320^0$C

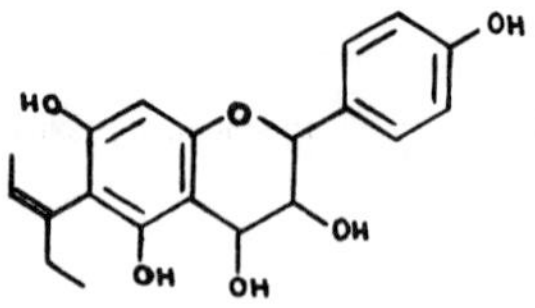

Ind. J. C., 1966, 73.

1302-018

CATECHIN-3-GALLATE, (+)- $C_{22}H_{18}O_{10}$ 442 $+56^0$e

R = Galloyl

JCS, 1969, 1824.

1303-004

FLAVAN-3,4-DIOL, 5,4'-di-OH- $C_{15}H_{14}O_5$ 274

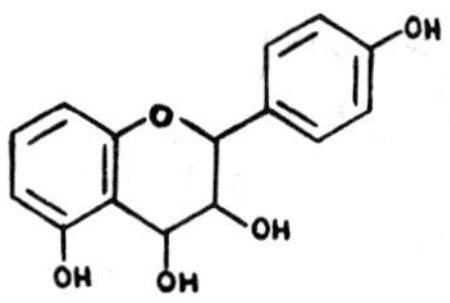

Chem. Abstr., 1969 (70) 47233.

1302-019

POLYDINE $C_{20}H_{22}O_{10}$ 422 -122^0m 192^0C

Ara·O-

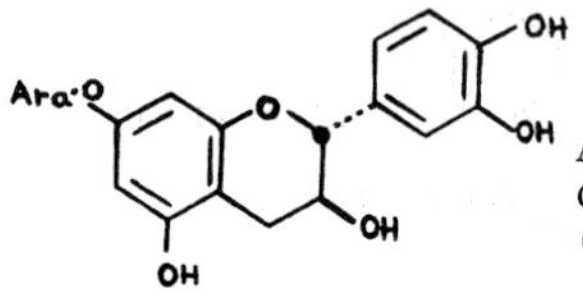

Ann., 1970 (734) 46. Coll. Czech. Chem. Comm., 1967, 3075.

1303-005

FISETINIDOL, (-)- $C_{15}H_{14}O_5$ 274 -9^0 214^0C

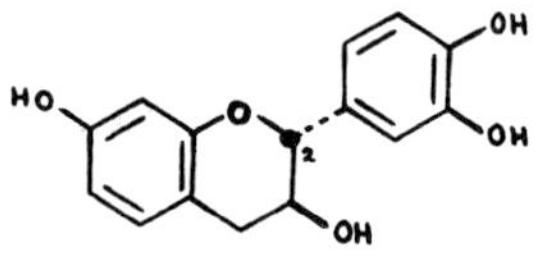

Chem. Abstr., 1968 (69) 77066.

1303-001

CATECHIN, (-)- $C_{15}H_{14}O_6$ 290 -17^0w 216^0C

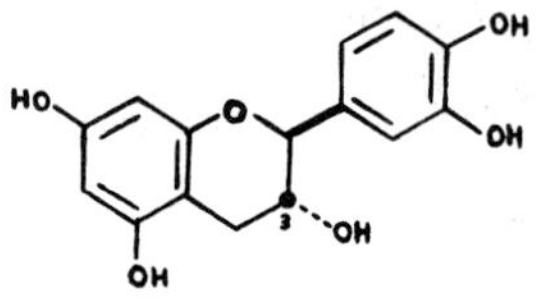

Ann., 1924, 276.

1303-006

FISETINIDOL, epi- $C_{15}H_{14}O_5$ 274 $+82^0$ 120^0C

2-epimer

JCS, 1966, 1644.

1303-002

CATECHIN, (+)-epi- $C_{15}H_{14}O_6$ 290 $+70^0$e 236^0C

3-epimer

JACS, 1931, 1500.

1303-007

CATECHIN, (+)- $C_{15}H_{14}O_6$ 290 $+18^0$w 176^0C

R = H

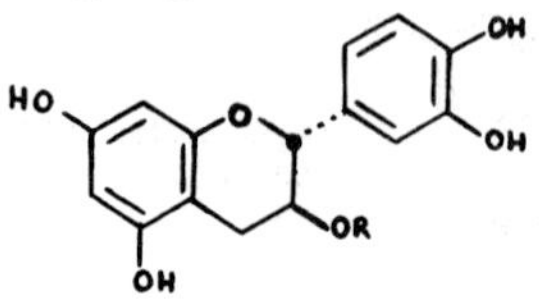

JOC, 1953, 521.

1303-003

CATECHIN, (-)-epi- $C_{15}H_{14}O_6$ 290 -69^0e 242^0C

R = H

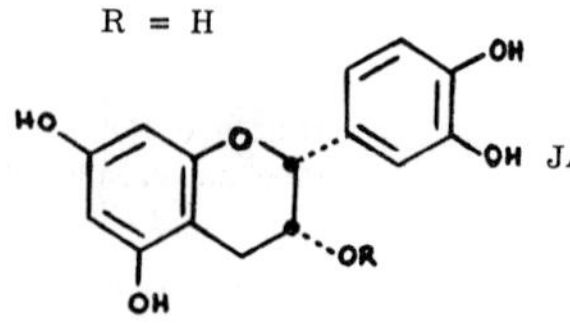

JACS, 1931, 1500.

1303-008

CATECHIN, 3-O-glucosyl-epi- $C_{21}H_{24}O_{11}$ 452

R = Glucosyl

J. App. Chem. USSR, 1946 (19) 1197.

1303-009

$C_{15}H_{14}O_6$ 290
-29°
148°C

Chem. Ind., 1964, 1799.

1303-014

CATECHIN, 3-O-Galloyl-epi-(-)- $C_{22}H_{18}O_{10}$ 342
-177°e
253°C

R = Galloyl

JCS'1948, 2249.

1303-010

$C_{15}H_{14}O_6$ 290

amorph.

4-epimer

Chem. Ind., 1964, 1799.

1303-015

MOLLISACACIDIN $C_{15}H_{14}O_6$ 290
+13°m
240°C

JCS, 1962, 4502.

1303-011

$C_{15}H_{14}O_6$ 290

amorph.

2-epimer

Chem. Ind., 1964, 1799.

1303-016

MOLLISACACIDIN, 4-epi- $C_{15}H_{14}O_6$ 290

106°C

4-epimer

JCS, 1970, 1800.

1303-012

LEUCOFISETINIDIN, (+)- $C_{15}H_{14}O_6$ 290
+4°a
amorph.

(Stereo ?)

Aust. J. C., 1968, 2353.

1303-017

LEUCOFISETINIDIN, (-)- $C_{15}H_{14}O_6$ 290
-11°m
135°C

JCS, 1959, 1402.
Biochem. J., 1964 (90) 343.

1303-013

LEBBECACIDIN $C_{15}H_{14}O_6$ 290

200°C

Chem. Abstr., 1966 (64) 2045.

1303-018

LEUCOCYANIDIN, (+)- $C_{15}H_{14}O_7$ 306
(+)
350^{0}C

(Stereo ?)

R = H

JCS, 1961, 2787.

1303-019

MELACACIDIN, iso- $C_{15}H_{14}O_7$ 306

4-epimer

JCS, 1962, 4502.

1303-024

LEUCOCYANIDIN, (+)- $C_{15}H_{14}O_7$ 306
(+)
−

R = H
Stereoisomer

Tet., 1959 (6) 21.

1303-020

FLAVAN-3, 3'4, 4', 7- PENTAOL- $C_{16}H_{16}O_7$ 320
-8- OMe-

JCS, 1970, 1800.

1303-025

LEUCOCYANIDIN GALLATE $C_{22}H_{18}O_{11}$ 458
+23^{0}a
−

R = Galloyl

Aust. J. C., 1964, 803.

1303-021

ROBINETINIDOL $C_{15}H_{14}O_6$ 290
−10^0
204^{0}C

R = H (Stereo ?)

Biochem. J., 1962 (84)
416.

1304-001

$C_{22}H_{26}O_{12}$ 482

240^{0}C

Phytochem., 1971 (10)
2256.

1303-022

RHOBIDANOL, (+)- $C_{15}H_{14}O_6$ 290

206^{0}C

R = H   (Stereo ?)

Chem. Abstr., 1968
(68) 59546.

1304-002

MELACACIDIN, (-)- $C_{15}H_{14}O_7$ 306
−75^{0}e
229^{0}C

JCS, 1962, 4502.

1303-023

RHOBIDANOL-3-GALLATE $C_{22}H_{18}O_{10}$ 442

214^{0}C

R = Galloyl

Chem. Abstr., 1968
(68) 59546.

1304-003

GALLOCATECHIN, (+)- $C_{15}H_{14}O_7$ 306
+15[0]
188[0]C

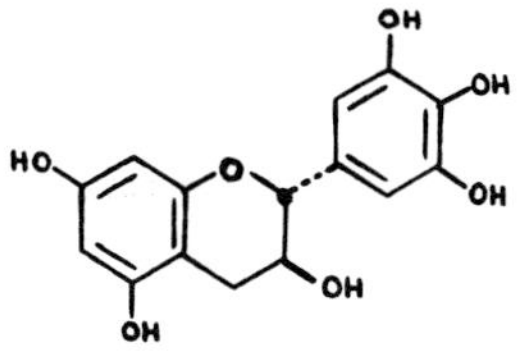

JCS, 1957, 3586.

1304-004

OURATEA-CATECHIN $C_{16}H_{16}O_7$ 320
-62[0]e
142[0]C

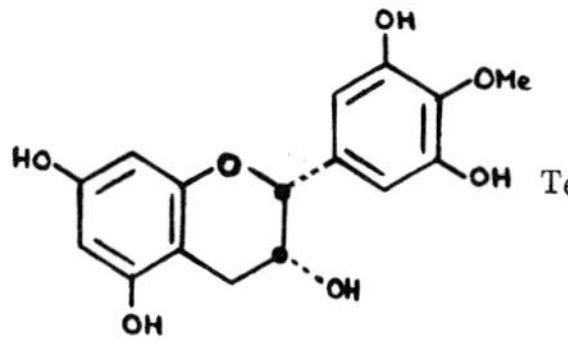

Tet. Lett., 1967, 4211.

1304-009

GALLOCATECHIN GALLATE, (-)- $C_{22}H_{18}O_{11}$ 358
-179[0]e
216[0]C

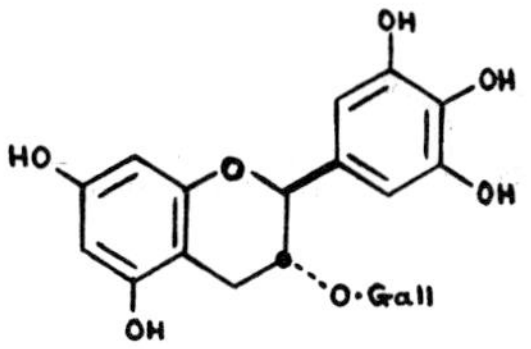

JCS, 1948, 2249.

1304-005

LEUCOROBINETINIDIN $C_{15}H_{14}O_7$ 306
+35[0]
174[0]C

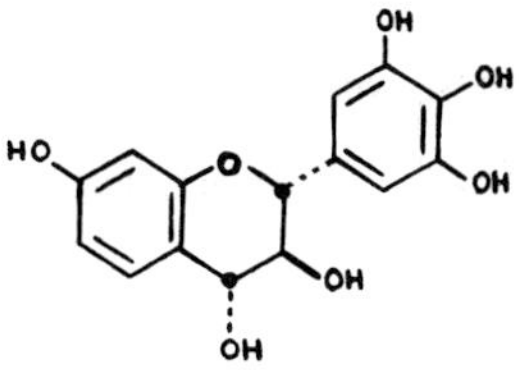

Biochem. J., 1964
(90) 343.

1304-010

GALLOCATECHIN, epi-(-)- $C_{15}H_{14}O_7$ 306
-50[0]a
218[0]C

R = H

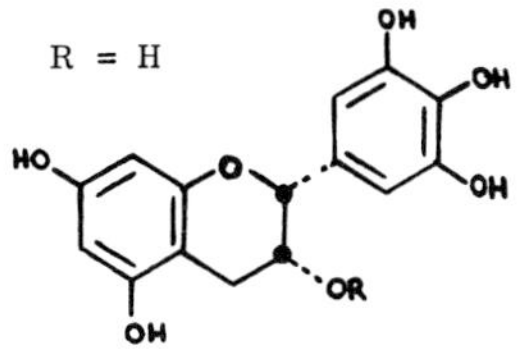

JCS, 1957, 3586.

1304-006

FLAVAN, 3, 4, 5, 7, 3', 4', 5'-hepta-OH- $C_{15}H_{14}O_8$ 322
(-)
300[0]C

Stereoisomer

Tet., 1965 (21) 1445.

1304-011

GALLOCATECHIN GALLATE,
-epi-(-)- $C_{22}H_{18}O_{11}$ 358
-190[0]e
254[0]C

R = Galloyl

JCS, 1957, 3587.

1304-007

AURICULACACIDIN, (-)- $C_{15}H_{14}O_6$ 290
212[0]C

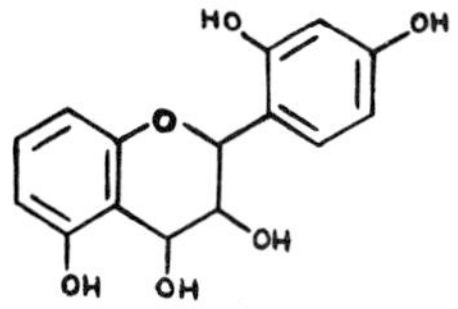

Aust. J. C., 1968, 1635.

1399-001

GALLOCATECHIN, DL- $C_{15}H_{14}O_7$ 306
OI

Racemate

Bull. Agr. Chem. Soc.
Jap., 1939 (15) 636.

1304-008

DRACORHODIN, nor- $C_{16}H_{12}O_3$ 252
OI
120[0]C

R = H

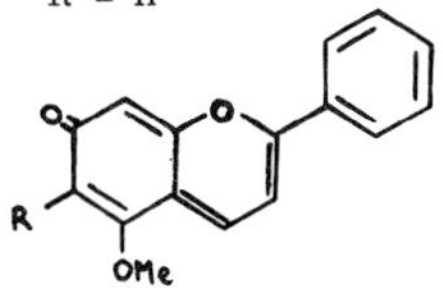

JCS, 1971, 3967.

1399-002

DRACORHODIN	$C_{17}H_{14}O_3$	266
		OI
		168^0C
	R = Me	
		JCS, 1950, 1882.
1399-003		

CARAJURIN	$C_{17}H_{14}O_5$	298
		OI
		206^0C
	R = Me	
		JCS, 1927, 3015.
1399-005		

CARAJURONE $\qquad$ $C_{16}H_{12}O_5$ $\qquad$ 274
OI
184^0C

R = H

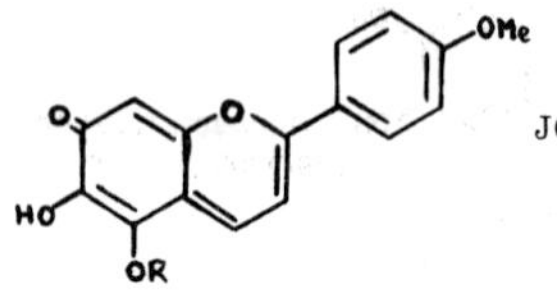

JCS, 1927, 3015.

1399-004

CYANOMACLURIN $\qquad$ $C_{15}H_{12}O_6$ $\qquad$ 288
$+192^0$w
250^0C

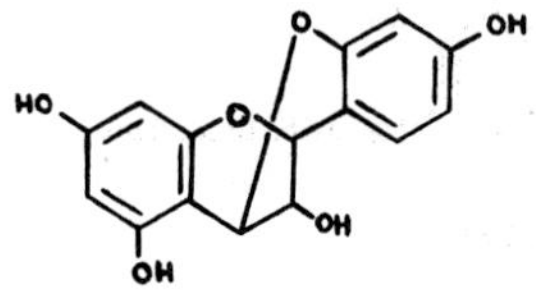

Tet. Lett., 1966, 5357.

1399-006

The isoflavonoid class embraces the isoflavones, isoflavanones, isoflavans, and 3-phenylcoumarins. The subclassification employed here is similar to the other flavonoid classes in that it is based on the oxygenation patterns of the C ring; in addition the 3-phenylcoumarin and isoflavans have been placed into their own separate subclasses. Classes 1401 to 1403 include both iso-flavanones and isoflavones. Classes 1404 and 1405 include any oxygenation patterns.

1401	4'-oxy
1402	3',4'-dioxy
1403	3',4',5'-trioxy
1404	
1405	
1499	Misc.

GENISTEIN	$C_{15}H_{10}O_5$	270 OI 292^{0}C

R = H

Chem. Comm., 1969, 830.

1401-001

ALPINUMISOFLAVONE	$C_{20}H_{16}O_5$	336 OI 213^{0}C

JCS, 1971, 3389.

1401-006

GENISTIN	$C_{21}H_{20}O_{10}$	432 -21^{0}p 255^{0}C

R = Glucosyl

Ber., 1943 (76) 1110.

1401-002

CHANDALONE	$C_{25}H_{24}O_5$	404 OI 64^{0}C

JCS, 1969, 374.

1401-007

SPHAEROBIOSIDE	$C_{27}H_{30}O_{14}$	578 -73^{0}p 204^{0}C

R = .Glu-Rha

Ber., 1967 (100) 101.

1401-003

WARANGALONE	$C_{25}H_{24}O_5$	404 OI 164^{0}C

JCS, 1966, 701; 1969, 374.

1401-008

SOPHORICOSIDE	$C_{21}H_{20}O_{10}$	432 -47^0 297^{0}C

R = Glucosyl

Chem. Ind., 1954, 518.

1401-004

OSAJIN, iso-	$C_{25}H_{24}O_5$	404

JACS, 1946, 406.

1401-009

SOPHORICOBIOSIDE	$C_{27}H_{30}O_{14}$	574 -72^{0}p 247^{0}C

R = .Glu-Rha

Ber., 1942 (75) 482.

1401-005

OSAJIN	$C_{25}H_{24}O_5$	404 OI 191^{0}C

JCS, 1969, 374.

1401-010

PRUNETIN	$C_{16}H_{12}O_5$	288 OI 240^0C	SCANDINONE	$C_{26}H_{26}O_5$	418 OI 208

PRUNETIN $C_{16}H_{12}O_5$ 288
                                        OI
                                        240^0C

Chem. Comm., 1969, 830.

R = H

1401-011

SCANDINONE $C_{26}H_{26}O_5$ 418
                                          OI
                                          208

JCS, 1969, 374.

1401-016

PRUNETRIN $C_{22}H_{22}O_{10}$ 450
                                            -15^0p
                                            236^0C

R = Glucosyl

Ber., 1957 (90) 836.

1401-012

TECTORIGENIN $C_{16}H_{12}O_6$ 300
                                            OI
                                            230^0C

Chem. Z., 1929 (I) 912.

R = H

1401-017

BIOCHANIN-A $C_{16}H_{12}O_5$ 284
                                           OI
                                           214^0C

Aust. J. C., 1966, 1755.

R = H

1401-013

TECTORIGENIN, 7-O-Me- $C_{17}H_{14}O_6$ 314
                                                    OI
                                                    228^0C

R = Me

Ind. J. C., 1965, 474.

1401-018

SISSOTRIN $C_{22}H_{22}O_{10}$ 446
                                            -35^0f
                                            220^0C

R = Glucosyl

Phytochem., 1965, 89.
Tet. Lett., 1965, 3191.

1401-014

TECTORIDIN $C_{22}H_{22}O_{11}$ 462
                                             -29^0p
                                             258^0C

R = Glucosyl

Ber., 1960, 1269.

1401-019

LANCEOLARIN $C_{27}H_{30}O_{14}$ 578
                                              -97^0m
                                              168^0C

R = .Glu-Apiose

Tet., 1967, 405.

1401-015

MUNINGIN $C_{17}H_{14}O_6$ 314
                                        OI
                                        285^0C

JCS, 1952, 96.

1401-020

IRISOLIDONE	$C_{17}H_{14}O_6$	314

JOC, 1965, 3561.

1401-021

FORMONONETIN	$C_{16}H_{12}O_4$	268 OI 259^0C

R = H

Aust. J. C., 1966, 1755.

1401-026

TECTORIGENIN, 7, 4'-diMe-	$C_{18}H_{16}O_6$	328 184^0C

Curr. Sci., 1966 (34) 431.

1401-022

ONONIN	$C_{22}H_{22}O_9$	430 -24^0 245^0C

R = Glucosyl

JCS. 1933. 274.

1401-027

DAIDZEIN	$C_{15}H_{10}O_4$	254 OI 320^0C

R = H

Ann., 1931 (489) 118.

1401-023

ISOFLAVONE, 4'-OH-7-OMe-	$C_{16}H_{12}O_4$	268

R = H

Chem. Abstr., 1969
(70) 44797.

1401-028

DAIDZIN	$C_{21}H_{20}O_9$	416 -36^0 235^0C

R = Glucosyl

Ann., 1931 (489) 118 : 129.

1401-024

DURLETTONE	$C_{21}H_{20}O_4$	336 OI 137^0C

R = 3, 3-DMA

Tet., 1967. 4741.

1401-029

PUERARIN	$C_{21}H_{20}O_9$	416

Chem. Pharm. Bull.,
1959, 134.

1401-025

IRISOLONE	$C_{17}H_{12}O_6$	312 OI 269^0C

Tet., 1961 (16) 201.

1401-030

182

AFROMOSIN	$C_{17}H_{14}O_5$	298
		229^0C

JCS, 1960, 1491.

1401-031

PRATENSEIN	$C_{16}H_{12}O_6$	300
		OI
		272^0C

JOC, 1963, 2336.

1402-005

OROBOL	$C_{15}H_{10}O_6$	286
		OI
		270^0C

R = H

JCS, 1949, 1571.

1402-001

DERRUSTONE	$C_{18}H_{14}O_6$	326
		OI
		153^0c

JCS, 1969, 365

1402-006

SANTAL	$C_{16}H_{12}O_6$	300
		OI
		223^0C

R = Me

JCS, 1949, 1571.

1402-002

DERRUBONE	$C_{21}H_{18}O_6$	366
		OI
		211^0C

JCS, 1969, 365.

1402-007

POMIFERIN	$C_{25}H_{24}O_6$	420
		OI
		200^0C

JCS, 1951, 3104.

1402-003

ROBUSTONE	$C_{21}H_{16}O_6$	364
		OI
		172^0C

R = H

JCS, 1969, 365.

1402-008

POMIFERIN, iso-	$C_{25}H_{24}O_6$	420
		OI
		265^0C

JACS, 1946, 406.

1402-004

ROBUSTONE ME ETHER	$C_{22}H_{18}O_6$	378
		OI
		170^0C

R = Me

JCS, 1969, 365.

1402-009

ISOFLAVONE, 3', 4', 7-tri-OH- $C_{15}H_{10}O_5$ 270
OI

R = H

Chem. Abstr., 1969
(70) 44797.

1402-010

MAXIMIN $C_{21}H_{18}O_5$ 350
OI
133^0C

R = 3, 3-DMA

Tet., 1962, 1445.

1402-015

ISOFLAVONE, 4', 7-di-OH-3-OMe- $C_{16}H_{12}O_5$ 284
OI

R = Me

Chem. Abstr., 1969
(70) 44797.

1402-011

CLADRIN $C_{17}H_{14}O_5$ 298
OI
258^0C

R = H

Tet., 1969, 3887.

1402-016

SAYANEDINE $C_{17}H_{14}O_5$ 298
OI
166^0C

Chem. Pharm. Bull.,
1970, 1872.

1402-012

CABREUVIN $C_{18}H_{16}O_5$ 312
OI
165^0C

R = Me

Ber., 1958 (91) 2858.

1402-017

BAPTIGENIN, pseudo- $C_{16}H_{10}O_5$ 282
OI
297^0C

R = H

JCS, 1953, 1852.

1402-013

ISOFLAVANONE, 6, 7-diOMe-3', $C_{18}H_{16}O_6$ 328
-4'-methylenedioxy-
202^0C

JCS, 1969, 1787.

1402-018

BAPTISIN, pseudo- $C_{28}H_{30}O_{14}$ 590
-98^0
250^0C

R = .Glu-Rha

Monatsh., 1929 (53/4)
454.

1402-014

DURMILLONE $C_{22}H_{18}O_6$ 378
OI
184^0C

Tet. Lett., 1967, 4741.
JOC, 1967, 1055.

1402-019

MAXIMA ISOFLAVONE-A	$C_{17}H_{10}O_6$	310 OI 227°C

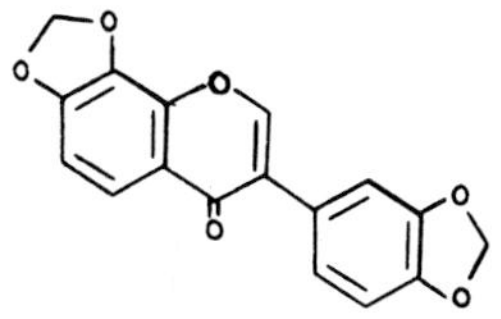

Tet., 1962, 1445.

1402-020

ISOFLAVONE, 3', 6, 7-triOMe-4', -5'-methylenedioxy-	$C_{19}H_{16}O_7$	356 OI 212°C

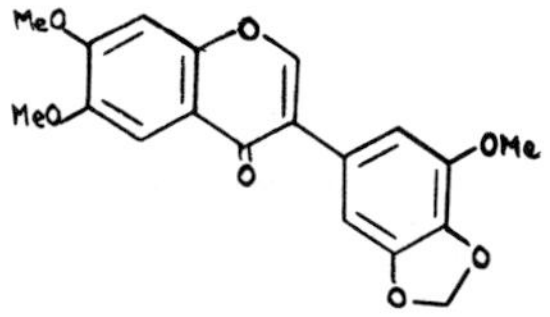

JCS, 1969, 1787.

1403-003

CLADRASTIN	$C_{18}H_{16}O_6$	328 OI 206°C

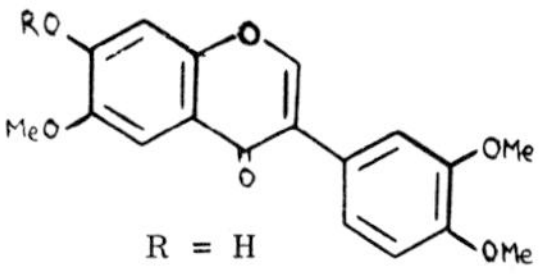

R = H

Tet., 1969, 3887.

1402-021

DERRUSNIN	$C_{19}H_{16}O_7$	356 OI 187°C

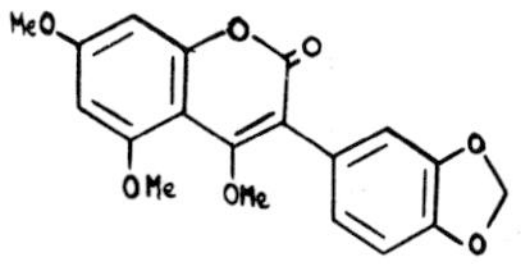

JCS, 1969, 365.

1404-001

ISOFLAVONE, 3'.4'.6.7-tetraOMe-	$C_{19}H_{18}O_6$	342 OI 188°C

R = Me

JCS. 1969. 1787.

1402-022

ROBUSTIC ACID	$C_{22}H_{20}O_6$	380 OI 208°C

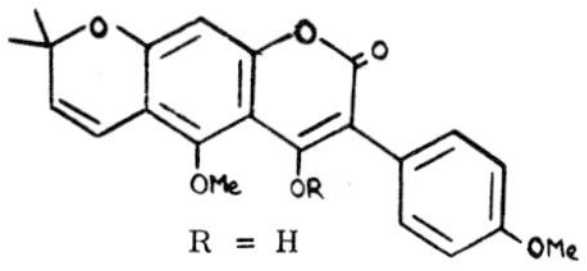

R = H

JOC, 1969, 365.

1404-002

IRIGENIN	$C_{18}H_{16}O_8$	360 OI 185°C

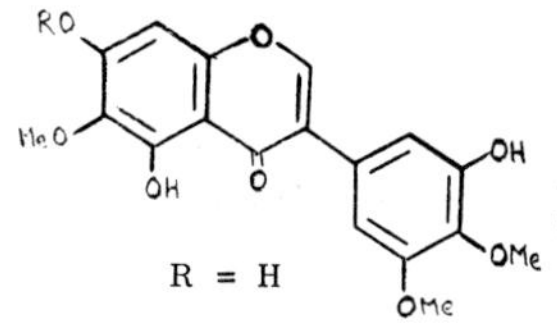

R = H

Tet. Lett., 1960, 23.

1403-001

ROBUSTIC ACID, ME ETHER	$C_{23}H_{22}O_6$	394 OI 196°C

R = Me

JCS, 1969, 365.

1404-003

IRIDIN	$C_{24}H_{26}O_{13}$	522 208°C

R = Glucosyl

JCS, 1928, 1022.

1403-002

ROBUSTIN	$C_{22}H_{18}O_7$	394 OI 206°C

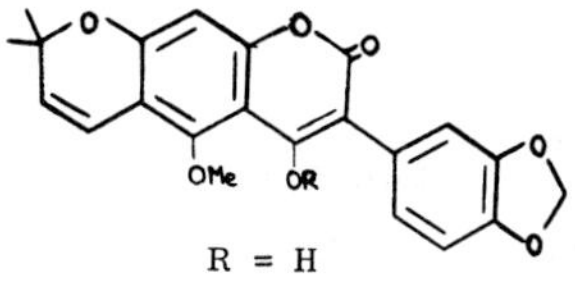

R = H

JCS, 1969, 365.

1404-004

ROBUSTIN, ME ETHER	$C_{23}H_{20}O_7$	408 OI 210^{0}C
R = Me		
	JCS. 1969, 365.	
1404-005		

NEOFOLIN	$C_{20}H_{14}O_7$	366 OI
R = OMe		
	J. S. Afr. Chem. Inst., 1966(19) 24.	
1404-010		

LONCHOCARPIC ACID	$C_{26}H_{26}O_6$	434 OI 212^{0}C

R = H

JCS, 1969, 374.

1404-006

EQUOL	$C_{15}H_{14}O_3$	242 190^{0}C

Chem. Comm., 1968, 1265.

1405-001

LONCHOCARPENIN	$C_{27}H_{28}O_6$	448 OI 182^{0}C
R = Me		
	JCS. 1969, 374.	
1404-007		

VESTITOL, (+)-	$C_{16}H_{16}O_4$	272 156^{0}C

Chem. Comm., 1968, 1263.

1405-002

SCANDENIN	$C_{26}H_{26}O_6$	434 OI 232^{0}C

JCS, 1969, 374.

1404-008

LAXIFLORAN	$C_{17}H_{18}O_5$	302

JCS, 1969, 888.

1405-003

PACHYRRIZIN	$C_{19}H_{12}O_6$	336 OI 208^{0}C

R = H

Monatsh., 1957, 541.

1404-009

MUCRONULATOL, (-)-	$C_{17}H_{18}O_5$	302 145^{0}C

Chem. Comm., 1968, 1263.

1405-004

DUARTIN, (-)- $C_{18}H_{20}O_6$ 332

149°C

Chem. Comm., 1968, 1263.

1405-005

TLATLANCUAYIN $C_{18}H_{14}O_6$ 326
OI
148°C

Tet., 1962, 559.

1499-001

MUCROQUINONE, (-)- $C_{17}H_{16}O_6$ 316

192°C

Chem. Comm., 1968, 1263.

1405-006

PODOSPICATIN $C_{17}H_{14}O_7$ 330
OI
215°C

Tet., 1959 (6) 143.

1499-002

LONCHOCARPAN $C_{18}H_{20}O_6$ 332

156°C

JCS, 1969, 888.

1405-007

SOPHOROL $C_{16}H_{10}O_6$ 298
OI

Bull. Chem. Soc. Jap., 1966, 1525.

1499-003

LICORICIDIN $C_{25}H_{32}O_5$ 412
+20°m
161°C

Chem. Pharm. Bull., 1968, 1932.

1405-008

FERREIRIN $C_{16}H_{14}O_6$ 302
OI
211°C

JCS, 1952, 4752.

R = H

1499-004

ISOFLAVAN, 2'-OH-4,7-DI-OMe- $C_{17}H_{18}O_4$ 286

Chem. Abstr., 1971 (74) 72459.

1405-009

FERREIRIN, homo- $C_{17}H_{16}O_6$ 316

168°C

R = Me

JCS, 1953, 3454.

1499-005

AURICULATIN $C_{25}H_{24}O_6$ 420 OI 136^{0}C		**NEPSEUDIN** $C_{20}H_{18}O_6$ 338 OI 115^{0}C
R = H JCS, 1968, 1899. 1499-006		JCS, 1963, 1569. 1499-011
AURICULIN $C_{26}H_{26}O_6$ 434 OI 125^{0}C R = Me Tet., 1970, 5041. 1499-007		**CAVIUNIN** $C_{19}H_{18}O_8$ 374 OI 192^{0}C JOC, 1961 (26) 2453. 1499-012
AURICULATIN, iso- $C_{25}H_{24}O_6$ 420 OI 133^{0}C Tet., 1970, 5041. 1499-008		**TOXICAROL ISOFLAVONE** $C_{23}H_{22}O_7$ 410 OI 219^{0}C Experientia, 1971 (27) 875. 1499-013
MUNETONE $_0H_{24}O_5$ 416 OI 192^{0}C Proc. Chem. Soc., 1963, 179. 1499-009		**ISOFLAVONE, 2', 4', 5'-6, 7-penta- -OMe-** $C_{20}H_{20}O_7$ 372 OI 171^{0}C JCS, 1969, 1787. 1499-014
MUNDULONE $C_{26}H_{26}O_6$ 434 -12^0c 180^{0}C J. Ind. C. S., 1962 (39) 475. 1499-010		**MILLDURONE** $C_{19}H_{16}O_7$ 356 OI 235^{0}C JCS, 1969, 1787. Tet., 1967, 4741. 1499-015

MAXIMA ISOFLAVONE-C $C_{22}H_{20}O_6$ 380
OI
144^0C

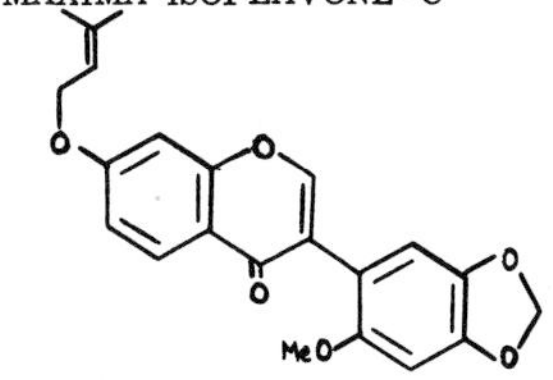

Curr. Sci., 1955 (24) 337.

1499-016

RAUTENONE, neo- $C_{19}H_{14}O_6$ 338

180^0C

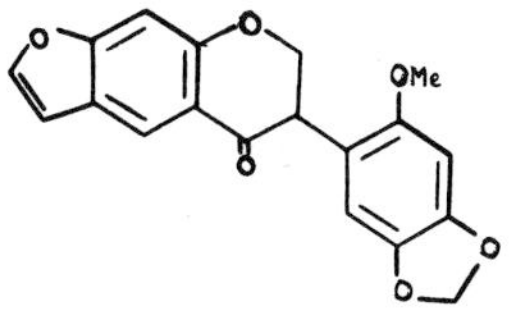

JCS, 1963, 1569.

1499-019

JAMAICIN $C_{22}H_{18}O_6$ 378
OI
194^0C

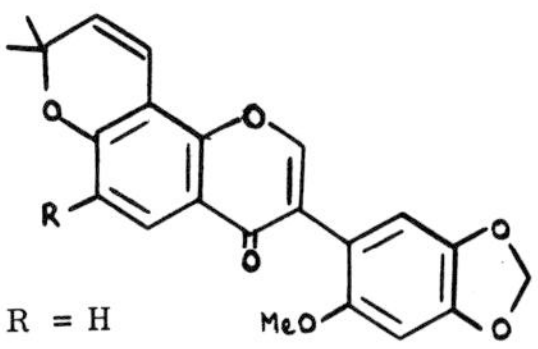

R = H

Tet. Supp. No. 7, 1966,
333.

1499-017

PISCERYTHRONE $C_{21}H_{20}O_7$ 384
OI
183^0C

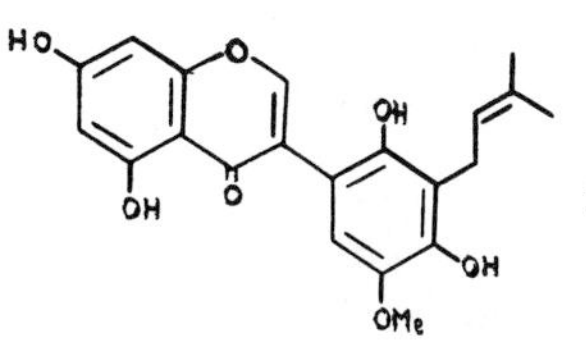

Tet. Supp. No. 7, 1966,
333.

1499-020

ICHTHYNONE $C_{23}H_{20}O_7$ 408
OI
202^0C

R = OMe

Tet., 1964, 1317.

1499-018

FERRUGONE $C_{23}H_{20}O_7$ 408
OI
168^0C

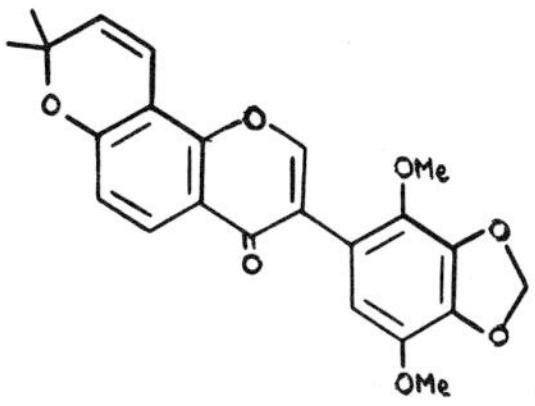

JOC, 1967, 1055.

1499-021

The dimeric flavonoids are found linked through a variety of positions and at various oxidation levels. There unfortunately exists at present some ambiguity in the actual linkage in a few of the 8-4/6-4 biflavonoids. Trimeric and tetrameric flavonoids are included here under class 1599.

1501	8-3 linked	1502	8-4 linked
1503	8-6 linked	1504	8-8 linked
1505	8-3' linked	1506	6-4 linked
1599	Misc.		

GB-1A $C_{30}H_{22}O_{10}$ 542

Tet. Lett., 1967, 1767.

1501-001

GB-1 $C_{30}H_{22}O_{11}$ 558

Tet. Lett., 1967, 1767.

1501-006

TALBOTAFLAVONE $C_{30}H_{20}O_{10}$ 540

300^{0}C

R = H

Phytochem.,
1970, 881.

1501-002

FUKUGETIN, (+)- $C_{30}H_{20}O_{11}$ 558
+170^{0}m
244^{0}C

R = H

Tet. Lett., 1969, 121.

1501-007

VOLKENSIFLAVONE $C_{30}H_{20}O_{10}$ 540

amorph.

R = H
stereoisomer

Phytochem., 1970, 225.

1501-003

FUKIGISIDE $C_{36}H_{30}O_{16}$ 720

243^{0}C

R = Glucosyl

Tet. Lett., 1970, 1717.

1501-008

VOLKENSIFLAVONE, (+-)- $C_{30}H_{20}O_{10}$ 540
OI
292^{0}C

R = H
Racemate

Phytochem., 1970, 221.
Tet. Lett., 1970, 4203.

1501-004

MORELLOFLAVONE $C_{30}H_{20}O_{11}$ 558
OI
298^{0}C

R = H
Racemate

Tet. Lett., 1969, 121.

1501-009

SPICATASIDE $C_{36}H_{30}O_{15}$ 702

233^{0}C

R = Glucosyl

Tet. Lett., 1970, 4203.

1501-005

FUKUGETIN, 3'-OMe- $C_{31}H_{22}O_{11}$ 572
Rac
290^{0}C

Tet. Lett., 1969, 121.

1501-010

XANTHOCHYMUSSIDE $C_{36}H_{32}O_{16}$ 722

219°C

Tet. Lett., 1970, 4203.

1501-011

LEUCOFISETINIDIN-(+)-CATECHIN $C_{30}H_{26}O_{11}$ 562

Chem. Comm.,
1966, 370.

1502-004

GB-2 $C_{30}H_{22}O_{12}$ 574

Tet. Lett., 1967, 1767.

1501-012

PROCYANIDIN-B1 $C_{30}H_{26}O_{12}$ 578

Ann., 1971 (748) 218.
Gazz., 1971 (101) 245.

1502-005

XANTHORRHONE $C_{32}H_{28}O_7$ 524

195°C

Aust. J. C., 1969, 423.

R = H

1502-001

PROCYANIDIN-B2 $C_{30}H_{26}O_{12}$ 578

*epimer

JCS, 1972. 1387.

1502-006

XANTHORRHONE, 14-OH- $C_{32}H_{28}O_8$ 540

192°C

R = OH

Tet. Lett., 1967, 481.

1502-002

PROCYANIDIN-B3 $C_{30}H_{26}O_{12}$ 578
+17°a

JCS, 1972, 1387.

1502-007

LEUCOPELARGONIDIN DIMER $C_{30}H_{26}O_{11}$ 562

185°C

Tet. Lett., 1967, 2443.

1502-003

PROCYANIDIN-B4 $C_{30}H_{26}O_{12}$ 578

*epimer

JCS, 1972, 1387.

1502-008

$C_{31}H_{28}O_{12}$ 592 +54° Chem. Abstr., 1968, 114366. 1502-009	CUPRESSUFLAVONE-7,7''-DIME ETHER R = Me; R' = H 1504-002	$C_{32}H_{22}O_{10}$ 566 >300°C Tet. Lett., 1970, 2937.

$C_{30}H_{26}O_{13}$ 594 -88° 200°C Tet., 1966, 2367. 1502-010	W11 $C_{32}H_{22}O_{10}$ 566 +65° 302°C R = H; R' = Me Tet. Lett., 1968, 5515. 1504-003

AGATHISFLAVONE 7-ME ETHER $C_{31}H_{20}O_{10}$ 552 R = H Tet. Lett., 1970, 2937. Experientia, 1969 (25) 351. 1503-001	WB1 $C_{34}H_{26}O_{10}$ 594 (-) 151°C R = R' = Me Tet. Lett., 1968, 5515. 1504-004

AGATHIS FLAVONE-4''',7-diMe-ether- $C_{32}H_{22}O_{10}$ 566 R = Me Tet. Lett., 1970, 2937. Experientia, 1969 (25) 351. 1503-002	CUPRESSUFLAVONE-7-ME-ETHER $C_{31}H_{20}O_{10}$ 552 188°C R = H Tet. Lett., 1970, 2937. 1504-005

CUPRESSUFLAVONE $C_{30}H_{18}O_{10}$ 538 >360°C Tet., 1967, 397. R = R' = H 1504-001	CUPRESSUFLAVONE, 4',7,7''-tri-O-Me- $C_{33}H_{24}O_{10}$ 580 R = Me Chem. Pharm. Bull., 1971, 1500. 1504-006

AMENTOFLAVONE $C_{30}H_{18}O_{10}$ 538
OI
300^0C

Natwiss., 1965 (52) 161.

R = R' = H

1505-001

SOTETSUFLAVONE $C_{31}H_{20}O_{10}$ 552
OI
264^0C

R = Me; R' = H

JACS, 1960, 1505.

1505-002

SEQUOIAFLAVONE $C_{31}H_{20}O_{10}$ 552
OI

R = H; R' = Me

Phytochem.,
1971 (10) 436.

1505-003

BILOBETIN $C_{31}H_{20}O_{10}$ 552
OI
248^0C

JCS, 1963, 1477.

R = H; R' = Me

1505-004

PODOCARPUSFLAVONE-A $C_{31}H_{20}O_{10}$ 552
OI

R = Me; R' = H

Tet. Lett., 1968, 2339.

1505-005

GINKGETIN, iso- $C_{32}H_{22}O_{10}$ 566
OI

R = R' = Me

JCS, 1963, 1477.

1505-006

GINKGETIN $C_{32}H_{22}O_{10}$ 566
OI

Phytochem.,
1971 (10) 2541.

1505-007

AMENTOFLAVONE-7'', 4''-DIME-
-ETHER, 2, 3-DIHYDRO- $C_{32}H_{24}O_{10}$ 568

296^0C

Phytochem.,
1971 (10) 2465.

1505-008

AMENTOFLAVONE, 4', 7''-di-O-Me- $C_{32}H_{22}O_{10}$ 566
OI

Chem. Pharm. Bull.,
1971. 1500.

1505-009

PODOCARPUSFLAVONE-B $C_{32}H_{22}O_{10}$ 566
OI
281^0C

Phytochem., 1971, 2787.
Tet. Lett., 1968, 2339.

R = H

1505-010

HEVEAFLAVONE	$C_{33}H_{24}O_{10}$ 580	
	OI	
	>305°C	
R = Me		
	Tet. Lett., 1969, 2017.	
1505-011		

GLOBIFLORIN-3B1 $C_{30}H_{26}O_{10}$ 546

JCS, 1969, 2572.

1506-001

SCIADOPITYSIN $C_{33}H_{24}O_{10}$ 580
OI

Chem. Pharm. Bull.,
1959, 821.

1505-012

GLOBIFLORIN-3B2 $C_{30}H_{26}O_{10}$ 546

*epimer

JCS, 1969, 2572.

1506-002

SCIADOPITYSIN, 2,3-dihydro- $C_{33}H_{26}O_{10}$ 582

151°C

*dihydro

Phytochem.,
1971 (10) 2465.

1505-013

PROANTHOCYANIDIN COOH $C_{31}H_{26}O_{12}$ 590

Chem. Comm.,
1970, 492.

1506-003

KAYAFLAVONE $C_{33}H_{24}O_{10}$ 580
OI

Chem. Pharm. Bull.,
1961 (9) 358.

R = H

1505-014

BILEUCOFISETINIDIN $C_{30}H_{26}O_{11}$ 562

JCS, 1967, 1217.

1506-004

AMENTOFLAVONE, 4,4''',7,7''-
-tetra-OMe- $C_{34}H_{26}O_{10}$ 594
+41°
273°C

R = Me

Experientia, 1969, 350.

1505-015

BILEUCOFISETINIDIN $C_{30}H_{26}O_{11}$ 562

*epimer

JCS, 1967, 1217.

1506-005

BILEUCOFISETINIDIN	$C_{30}H_{26}O_{11}$	562

**epimer

JCS, 1967, 1217.

1506-006

HINOKIFLAVONE	$C_{30}H_{18}O_{10}$	538
		OI

R = R' = H

Tet. Lett. , 1967, 397.

1599-003

LEUCOROBINETINIDIN-(+)-CATECHIN	$C_{30}H_{26}O_{12}$	578

R = H

Chem. Comm. , 1966, 370.

1506-007

HINOKIFLAVONE, 2, 3-dihydro-	$C_{30}H_{20}O_{10}$	540
		$314^{0}C$

R = R' = H

*dihydro

Phytochem. , 1971 (10) 2465.

1599-004

LEUCOROBINETINIDIN-(+)- -GALLOCATECHIN	$C_{30}H_{26}O_{13}$	594

R = OH

Chem. Comm. , 1966, 370.

1506-008

CRYPTOMERIN, neo-	$C_{31}H_{20}O_{10}$	552

R = Me; R' = H

Tet. Lett. , 1968. 2339.

1599-005

CRYPTOMERIN-A	$C_{31}H_{20}O_{10}$	552
		OI
		$310^{0}C$

Chem. Pharm. Bull. , 1966 (14) 1408.

R = H

1599-001

CRYPTOMERIN, iso-	$C_{31}H_{20}O_{10}$	552
		OI

R = H; R' = Me

Tet. Lett. , 1968, 2339.

1599-006

CRYPTOMERIN-B	$C_{32}H_{22}O_{10}$	566
		OI
		$302^{0}C$

R = Me

Chem. Pharm. Bull. , 1966 (14) 1408.

1599-002

CHAMAECYPARIN	$C_{32}H_{22}O_{10}$	566
		OI

R = R' = Me

Tet. Lett. , 1968, 2340.

1599-007

PROPELARGONIDIN $C_{30}H_{26}O_{12}$ 578

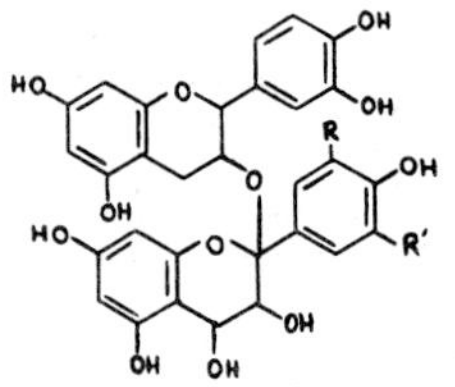

Chem. Abstr.,
1968 (69) 27182.

R = R' = H

1599-008

HEMLOCK TANNIN $C_{30}H_{26}O_{12}$ 578

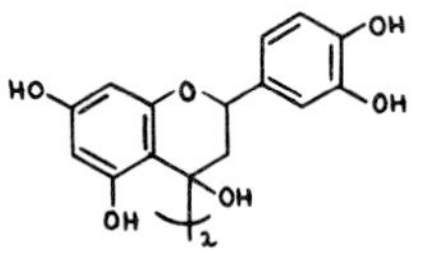

JCS, 1934, 1506.

1599-013

PROANTHOCYANIDIN (Cacao) $C_{30}H_{26}O_{13}$ 594
$+40^0$a
200^0C

R = OH
R' = H

Tet. Lett., 1962, 1073.

1599-009

ZEYHERIN $C_{30}H_{22}O_{11}$ 558

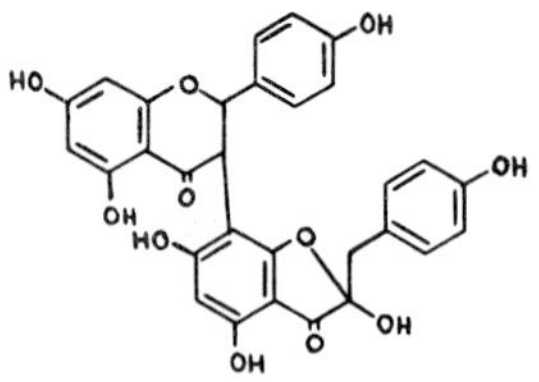

Tet. Lett., 1971, 1647.

1599-014

PROANTHOCYANIDIN (Barley) $C_{30}H_{26}O_{13}$ 594

R = OH; R' = H
(Cf 009 ?)

Chem. Abstr.,
1968 (69) 27182.

1599-010

DRACORUBIN, nor- $C_{31}H_{22}O_5$ 474
-77^0m
256^0C

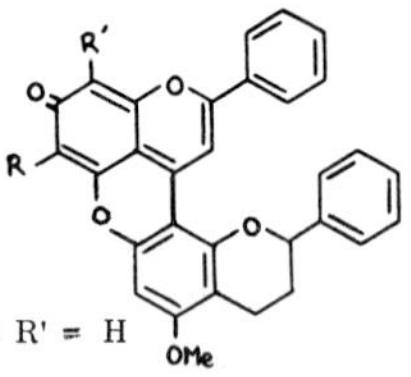

JCS, 1971. 3967.

R = R' = H   OMe

1599-015

PRODELPHINIDIN $C_{30}H_{26}O_{14}$ 610

R = R' = OH

Chem. Abstr.,
1963 (69) 27182.

1599-011

DRACORUBIN $C_{32}H_{24}O_5$ 488
-35^0c
315^0C

R, R' = H, Me (?)

JCS, 1950, 3117.

1599-016

LEUCOPELARGONIDIN DIMER $C_{30}H_{26}O_{11}$ 562

180^0C

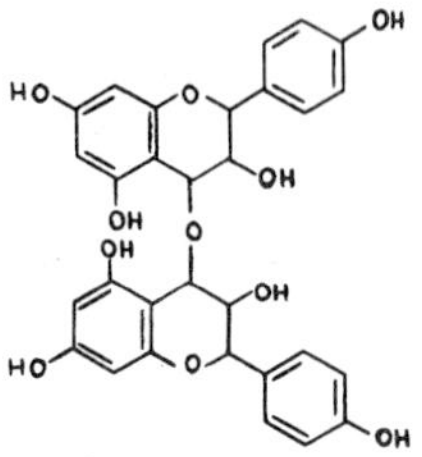

Tet. Lett., 1967, 2443.

1599-012

LEUCOPELARGONIDIN DIOXAN
-DIMER $C_{30}H_{24}O_{10}$ 544

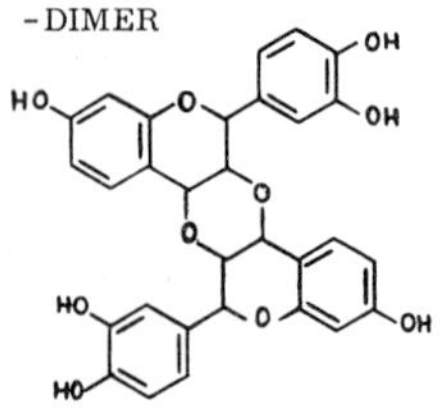

JCS, 1969, 897.

1599-017

THEAFLAVIN R = R' = H 1599-018	$C_{29}H_{24}O_{12}$	564 240^0C Tet. Lett., 1970, 5237.	THEAFLAVIC ACID 1599-023	$C_{21}H_{16}O_{10}$	428 Tet. Lett., 1970, 5247.
THEAFLAVIN-2A R = Galloyl; R' = H 1599-019	$C_{36}H_{28}O_{16}$	716 Tet. Lett., 1970, 5237.	THEAFLAVIC ACID, epi- *epimer 1599-024	$C_{21}H_{16}O_{10}$	428 Tet. Lett., 1970, 5247.
THEAFLAVIN-2B R = H; R' = Galloyl 1599-020	$C_{36}H_{28}O_{16}$	716 Tet. Lett., 1970, 5237.	PROCYANIDINO-(-)-EPICATECHIN 1599-025	$C_{30}H_{24}O_{12}$	576 $+64^0a$ Tet. Lett., 1966, 429.
THEAFLAVIN-3 R = R' = Galloyl 1599-021	$C_{43}H_{32}O_{30}$	868 Tet. Lett., 1970, 5237.	1599-026	$C_{60}H_{50}O_{21}$	3106 Chem. Comm., 1971, 1257.
THEAFLAVIN, iso- R = R' = H *epimer 1599-022	$C_{29}H_{24}O_{12}$	564 Tet. Lett., 1970, 5241.			

LEUCOPELARGONIDIN TETRAMER $C_{60}H_{50}O_{20}$ 3090

>330^0C

Tet. Lett., 1967, 2443.

1599-027

16 | PTEROCARPANS

The heterocyclization of a 4-oxy isoflavan provides the basic pterocarpan nucleus; where the isoflavan precursor is a 3-phenylcoumarin, the coumestan nucleus is supplied instead.

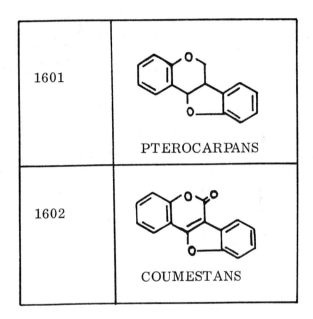

1601	PTEROCARPANS
1602	COUMESTANS

PTEROCARPAN, (-)-3-OH-9-OMe-	$C_{16}H_{14}O_4$	270 -226^0C 128^0C

R = H

JCS, 1969, 1109.

1601-001

PTEROCARPAN, 3, 4, 9-tri-OMe-	$C_{18}H_{18}O_5$	314 -158^0 119^0C

R = Me

JCS, 1969, 1109.

1601-006

PTEROCARPAN, (+)-3-OH-9-OMe-	$C_{16}H_{14}O_4$	270

R = H (antipode)

Chem. Abstr., 1971
(74) 72459.

1601-002

MAACKIAIN, (-)-	$C_{16}H_{12}O_5$	284 -252^0c 178^0C

R = H

Chem. Pharm. Bull.,
1968, 771.

1601-007

PTEROCARPAN, 3, 9-di-OMe-(+)-	$C_{17}H_{16}O_4$	284 -225^0c 88^0C

R = Me

Tet., 1967 (23) 1817.
JCS, 1969, 1109.

1601-003

TRIFOLIRHIZIN	$C_{22}H_{22}O_{10}$	446 -183^0e 143^0C

R = Glucosyl

Acta Chem. Scand.,
1961 (15) 936.

1601-008

PTEROCARPEN, 3, 9-di-OMe-	$C_{17}H_{14}O_4$	282 OI 196^0C

JCS, 1969, 1109.
Chem. Comm., 1965, 309.

1601-004

MAACKIAIN, D-	$C_{16}H_{12}O_5$	284

R = H (antipode)

Chem. Pharm. Bull.,
1963, 172.

1601-009

PTEROCARPAN, 3-OH-4, 9-di-OMe-	$C_{19}H_{18}O_6$	342 -116^0c 111^0C

R = H

JCS, 1969, 1109.

1601-005

SOPHOJAPONICIN	$C_{22}H_{22}O_{10}$	446 -104^0 203^0C

R = Glucosyl (antipode)

Chem. Pharm. Bull.,
1963, 172.

1601-010

MAACKIAIN, DL-	C$_{16}$H$_{12}$O$_5$	284 Rac 200^{0}C		

R = H (Racemate)

Experientia, 1966 (22) 777.

1601-011

PTEROCARPAN, 3,4-diOMe-8,9- -methylenedioxy-   C$_{18}$H$_{16}$O$_6$ 328 -202^0 245^{0}C

R = Me

JCS, 1969, 1109.

1601-016

PTEROCARPIN   C$_{17}$H$_{14}$O$_5$ 298 -208^{0}c 164^{0}C

Chem. Pharm. Bull., 1968, 771.

1601-012

PHILENOPTERAN   C$_{17}$H$_{16}$O$_6$ 316   186^{0}C

R = H

JCS, 1969, 887.

1601-017

FLEMICHAPPARIN-B   C$_{17}$H$_{12}$O$_5$ 296 OI 179^{0}C

Chem. Ind., 1970, 745.

1601-013

PHILENOPTERAN, 9-O-Me-   C$_{18}$H$_{18}$O$_6$ 330   184^{0}C

R = Me

JCS, 1969, 887.

1601-018

PTEROCARPIN, (-)-2-OH-   C$_{17}$H$_{14}$O$_6$ 314 -228^{0}C 239^{0}C

Tet., 1970, 5007.

1601-014

PTEROCARPAN, (-)-3-OH-2-(3,3- DMA)-8,9-methylenedioxy-   C$_{21}$H$_{20}$O$_5$ 352   146^{0}C

Chem. Abstr., 1971 (74) 45669.

1601-019

PTEROCARPAN, 3-OH-4-OMe-8, -9-methylenedioxy-   C$_{17}$H$_{14}$O$_6$ 314 -197^{0}c 160^{0}C

R = H

JCS, 1969, 1109.

1601-015

LEIOCARPIN   C$_{21}$H$_{18}$O$_5$ 350   99^{0}C

Chem. Abstr., 1971 (74) 72455.

1601-020

PHASEOLLIDIN	$C_{20}H_{20}O_4$ 324	GANGETIN	$C_{26}H_{28}O_5$ 420 -205^0c 99^0C
Tet. Lett., 1972, 1673.		JCS, 1971, 2420.	
1601-021		1601-026	

PHASEOLLIN	$C_{20}H_{18}O_4$ 322 178^0C	FICININ	$C_{19}H_{14}O_6$ 338 -200^0c 180^0C
Tet. Lett., 1964, 438.		Chem. Abstr., 1966 (65) 13677.	
1601-022		1601-027	

FICIFOLINOL	$C_{25}H_{28}O_4$ 392 -218^0c 149^0C	PISATIN	$C_{17}H_{14}O_6$ 314 $+280^0e$ 61^0C
J. S. Afr. Chem. Inst., 1970 (23) 24.		Tet. Lett., 1962, 673	
1601-023		1601-028	

FOLITENOL	$C_{25}H_{26}O_4$ 390 -233^0c 128^0C	COUMESTROL	$C_{15}H_8O_5$ 268 OI 385^0C
J. S. Afr. Chem. Inst., 1970 (23) 24.		R = H	JACS, 1958, 3969.
1601-024		1602-001	

FOLININ	$C_{25}H_{26}O_4$ 390 -230^0c 181^0C	COUMESTROL DIME ETHER	$C_{17}H_{12}O_5$ 296 OI 111^0C
J. S. Afr. Chem. Inst., 1970 (23) 24.		R = Me	JCS, 1969, 1109.
1601-025		1602-002	

204

LUCERNOL	$C_{15}H_8O_6$	284 OI >350^{0}C

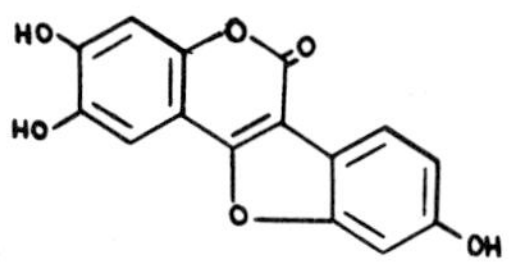

Tet. Lett., 1967, 2153.

1602-003

REPENSOL	$C_{15}H_8O_6$	284 OI

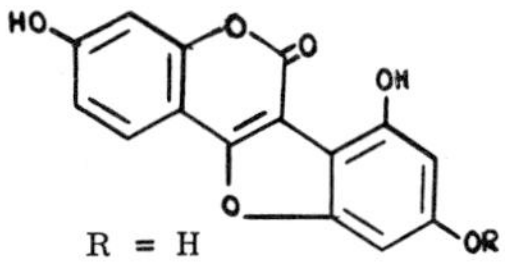

R = H

Phytochem., 1971 (10) 466.

1602-008

COUMESTAN, 7-OH-11, 12-diOMe-	$C_{17}H_{12}O_6$	312 OI 306^{0}C

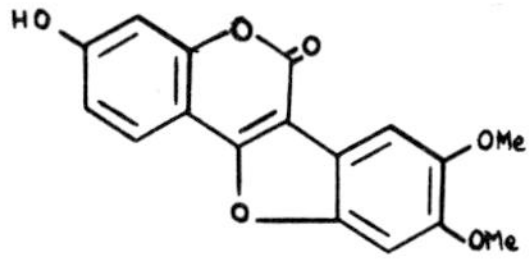

JOC, 1966 (31) 988.

1602-004

TRIFOLIOL	$C_{16}H_{10}O_6$	298 OI 332^{0}C

R = Me

Tet., 1964 (20) 1963.

1602-009

MEDICAGOL	$C_{16}H_8O_6$	296 OI 324^{0}C

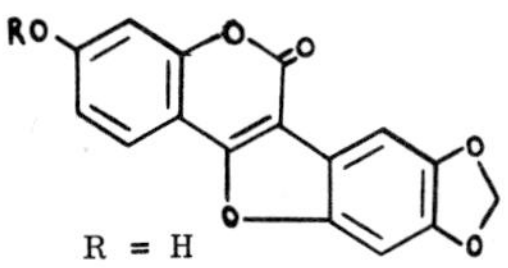

R = H

JOC, 1965, 2353.

1602-005

LISETIN	$C_{21}H_{18}O_7$	382 OI 284^{0}C

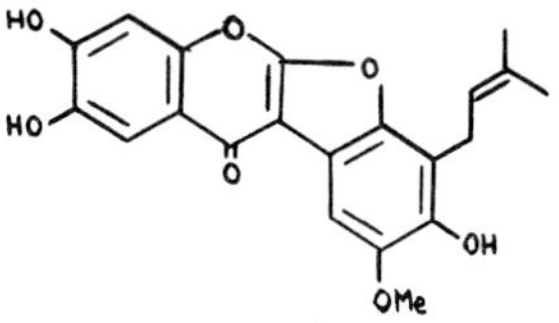

Chem. Comm., 1966, 305.

1602-010

FLEMICHAPPARIN-C	$C_{17}H_{10}O_6$	310 OI 272^{0}C

R = Me

Chem. Ind., 1970, 1113.

1602-006

WEDELOLACTONE	$C_{16}H_{10}O_7$	314 OI 280^{0}C

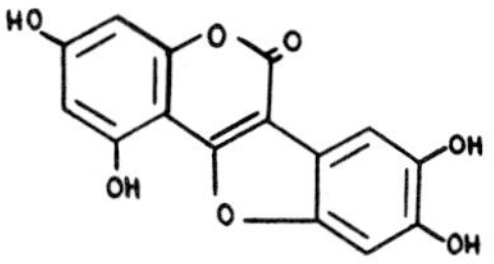

JCS, 1950, 629.

1602-011

EROSNIN	$C_{18}H_8O_6$	320 OI

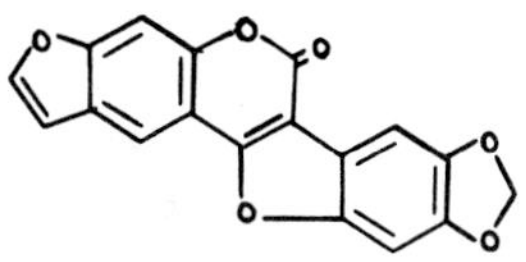

Helv., 1959, 61.

1602-007

1701	4'-oxy
1702	3', 4'-dioxy
1703	3', 4', 5'-trioxy

APIGENIDIN $C_{15}H_{11}O_4+$ 255
            OI
          $Cl^-:300^0C$

R = H

JCS, 1925, 1128.
Nature, 1932, 21.

1701-001

$C_{26}H_{29}O_{14}+$ 565

R = .Glu-Xyl

Experientia, 1971 (27)
1006.

1701-006

GESNERIN $C_{21}H_{21}O_9+$ 417

R = Glucosyl

Nature, 1932 (130) 21.

1701-002

PELARGONIN $C_{27}H_{31}O_{15}+$ 595

Natwiss., 1970 (57) 394.

1701-007

PELARGONIDIN $C_{15}H_{11}O_5+$ 271
            OI

R = H

JCS, 1922, 1577.

1701-003

AURANTINIDIN $C_{15}H_{11}O_6+$ 287
            OI

Phytochem., 1967, 1111.

1701-008

CALLISTEPHIN $C_{21}H_{21}O_{10}+$ 433

R = Glucosyl

JCS, 1928, 1460.

1701-004

LUTEOLINIDIN $C_{15}H_{11}O_5+$ 271
            OI

Chem. Ind., 1960, 229.

1702-001

PELARGONIDIN 3-RHAMNO-
GLUCOSIDE $C_{27}H_{31}O_{14}+$ 579

R = .Glu-Rha

Experientia, 1971 (27)
1006.

1701-005

CYANIDIN $C_{15}H_{11}O_6+$ 287
            OI
        $Cl^-: 200^0C$

R = H

JCS, 1935, 426.

1702-002

CYANIDIN 3-ARABINOSIDE	$C_{20}H_{19}O_{10}+$	419
R = Arabinosyl		
	Biochem. J., 1957 (65) 177.	
1702-003		

LYCORICYANIN	$C_{26}H_{29}O_{15}+$	581
R = Primeverosyl		
	Acta Phytochim., 1942 (13) 19.	
1702-008		

IDAEIN	$C_{21}H_{21}O_{11}+$	449
R = Galactosyl		
	JCS, 1931, 2722. Nature, 1955 (175) 634.	
1702-004		

KERACYANIN	$C_{27}H_{31}O_{15}+$	595
	$Cl^-: 175^0C$	
R = . Glu-Rha		
	Chem. Pharm. Bull., 1954 (2) 41.	
1702-009		

CHRYSANTHEMIN	$C_{21}H_{21}O_{11}+$	449
R = Glucosyl		
	J. Food Sci., 1970 (35) 41.	
1702-005		

CYANIDIN-3-RUTINOSIDE	$C_{27}H_{31}O_{15}+$	595
R = Rutinosyl		
	J. Food Sci., 1970 (35) 237.	
1702-010		

GOSSYPICYANIN	$C_{26}H_{29}O_{15}+$	581
R = .Xyl-Glu		
	Chem. Abstr., 1971 (74) 1063.	
1702-006		

CYANIDIN-3-SOPHOROSIDE	$C_{27}H_{31}O_{16}+$	611
R = Sophorosyl		
	J. Food Sci., 1970 (35) 237.	
1702-011		

FRITILLARICYANIN	$C_{26}H_{29}O_{14}+$	565
		182^0C
R = .Xyl-Rha		
	Experientia, 1958 (14) 274.	
1702-007		

CYANIDIN-3-DIGLUCOSIDE	$C_{27}H_{31}O_{16}+$	611
R = Glu-Glu (Cf 012 ?)		
	J. Food Sci., 1971, 101.	
1702-012		

MECOCYANIN $C_{27}H_{31}O_{16}+$ 611		PEONIDIN-3-ARABINOSIDE $C_{21}H_{21}O_{10}+$ 433
R = Gentiobiosyl		R = Arabinosyl
JCS, 1934, 1608.		J. Am. Hort. Sci., 1970 (95) 283.
1702-013		1702-018

$C_{33}H_{41}O_{20}+$ 757		PEONIDIN-3-GALACTOSIDE $C_{22}H_{23}O_{11}+$ 463
R = .Glu-Rha-Glu		R = Galactosyl
J. Food Sci., 1970 (35) 237.		J. Am. Hort. Sci., 1970 (95) 283.
1702-014		1702-019

CYANIN $C_{27}H_{31}O_{16}+$ 611		OXYCOCCICYANIN $C_{22}H_{23}O_{11}+$ 463
R = R' = Glu		R = Glucosyl

Natwiss., 1970 (57) 394.		JCS, 1931, 2722.
1702-015		1702-020

SHISONIN $C_{36}H_{37}O_{18}+$ 697		PEONIDIN-3-XYLOSYGLUCOSIDE $C_{27}H_{31}O_{15}+$ 595
R = 6-Coumaroyl-Glu		R = .Glu-Xyl
R' = Glu		
Agr. Biol. Chem., 1966 (30) 420.		Phytochem., 1970, 2411.
1702-016		1702-021

PEONIDIN $C_{16}H_{13}O_6+$ 301 OI		PEONIN $C_{28}H_{33}O_{16}+$ 625
R = H		

JCS, 1926, 1968.		Ber., 1949 (82) 481.
1702-017		1702-022

ROSINIDIN $C_{17}H_{15}O_6+$ 315
OI

Nature, 1958 (181) 26.

1702-023

DELPHINIDIN-3-RHAMNOSIDE $C_{21}H_{21}O_{11}+$ 449

R = Rhamnosyl

Helv., 1927 (10) 67.

1703-005

DELPHINIDIN $C_{15}H_{11}O_7+$ 303
OI
Cl^-: 350°C

R = H

JCS, 1925, 166.

1703-001

TULIPANIN $C_{27}H_{31}O_{16}+$ 611

R = .Glu-Rha

Botan. Mag. (Tokyo),
1956 (69) 462.

1703-006

DELPHINIDIN-3-ARABINOSIDE $C_{20}H_{19}O_{11}+$ 435

R = Arabinosyl

Phytochem., 1971, 2235.

1703-002

HYACIN $C_{27}H_{31}O_{17}+$ 627

R = Glu-Glu

J. Chem. Soc. Jap.,
1935 (56) 1043.

1703-007

EMPETRIN $C_{21}H_{21}O_{12}+$ 465
Cl: 177°C

R = Galactosyl

J. Am. Hort. Sci.,
1970 (95) 283.

1703-003

DELPHIN $C_{27}H_{31}O_{17}+$ 627
Cl^-: 202°C

R = R' = Glu

JACS, 1959, 4110.

1703-008

DELPHINIDIN-3-GLUCOSIDE $C_{21}H_{21}O_{12}+$ 465

R = Glucosyl

JCS, 1934, 1039.

1703-004

CAYRATININ $C_{42}H_{47}O_{23}+$ 919

R = .Glu-Glu-Cinn
R' = Glu

Chem. Abstr., 1971
(74) 10390.

1703-009

NASUNIN	$C_{42}H_{47}O_{23}+$	919

Cl^-: 180^0C

R = .Glu-Rha-p-Coum
R' = Glu

Agr. Biol. Chem.,
1966 (30) 420.

1703-010

PETUNIDIN-3-XYLOSYL-GLUCOSIDE	$C_{27}H_{31}O_{16}+$	611

R = .Glu-Xyl

Phytochem., 1970, 2411.

1703-015

PETUNIDIN	$C_{16}H_{13}O_7+$	317
		OI

R = H

JCS, 1930, 793.

1703-011

PETUNIDIN-5-XYLOSIDE	$C_{21}H_{21}O_{11}+$	449

Chem. Abstr., 1952
(46) 581.

1703-016

PETUNIDIN-3-ARABINOSIDE	$C_{21}H_{21}O_{11}+$	449

R = Arabinosyl

J. Am. Hort. Sci.,
1970 (95) 283.

1703-012

PETUNIN	$C_{28}H_{33}O_{17}+$	641

JACS, 1940, 2808.

1703-017

PETUNIDIN-3-GALACTOSIDE	$C_{22}H_{23}O_{12}+$	479

R = Galactosyl

J. Am. Hort. Sci.,
1970 (95) 283.

1703-013

MALVIDIN	$C_{17}H_{15}O_7+$	331
		OI
	Cl^-:	300^0C

R = H

JCS, 1928, 1541.

1703-018

PETUNIDIN-3-MONOGLUCOSIDE	$C_{22}H_{23}O_{12}+$	479

R = Glucosyl

J. Food Sci., 1970
(35) 41.

1703-014

MALVIDIN-3-ARABINOSIDE	$C_{22}H_{23}O_{11}+$	463

R = Arabinosyl

J. Am. Hort. Sci.,
1970 (95) 283.

1703-019

OENIN	$C_{23}H_{25}O_{12}+$	493

R = Glucosyl

JCS, 1931, 2701.

1703-021

HIRSUTIDIN	$C_{18}H_{17}O_7+$	345
		OI

R = H

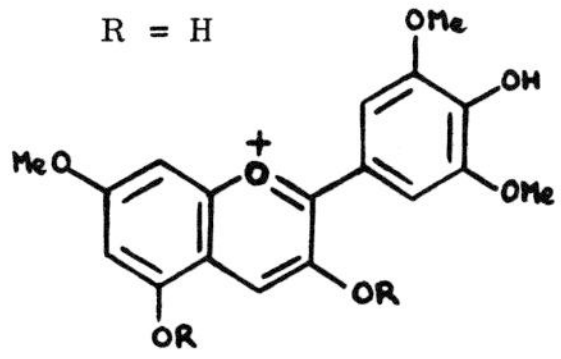

JCS, 1930, 793.

1703-024

ENSATIN	$C_{38}H_{41}O_{17}+$	769
	Cl^-:	$176^0 C$

R = .Glu-Glu-p-Coum

Acta Phytochim. (Jap.),
1939 (11) 65.

1703-022

HIRSUTIN	$C_{30}H_{37}O_{17}+$	669

R = Glucosyl

JCS, 1932, 2293.

1703-025

MALVIDIN-3, 5-DIGLUCOSIDE	$C_{29}H_{35}O_{17}+$	655

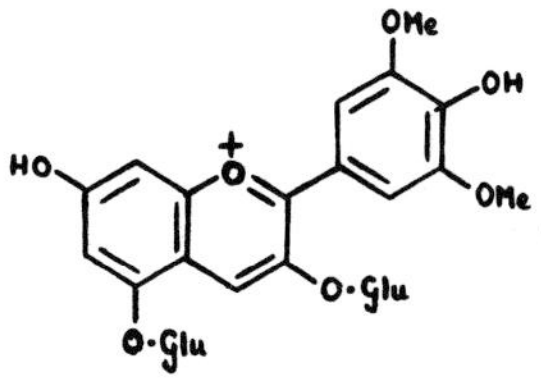

Chem. Abstr., 1971
(74) 61581.

1703-023

1801	ROTENOIDS
1802	PELTOGYNANS
1803	AURONES

MUNDUSERONE	$C_{19}H_{18}O_6$	342
		$+103^0 c$
		$162^0 C$

Tet., 1967, 2505.

1801-001

TOXICAROL, dehydro-	$C_{23}H_{20}O_7$	408
		OI
		$234^0 C$

* dehydro

JACS, 1933, 422.

1801-006

DEGUELIN	$C_{23}H_{22}O_6$	394
		$-101^0 b$
		$168^0 C$

JACS, 1940, 2520.

1801-002

MILLETTONE, (-)-	$C_{22}H_{18}O_6$	378

Tet., 1967, 4747.

1801-007

DEGUELIN, dehydro-	$C_{23}H_{20}O_6$	392
		OI
		$232^0 C$

* dehydro

JACS, 1933, 422.

1801-003

MILLETTONE, dehydro-	$C_{22}H_{16}O_6$	376
		OI
		$358^0 C$

* dehydro

Tet. Supp., No. 7, 333.

1801-008

TEPHROSIN, iso-	$C_{23}H_{22}O_7$	410
		Rac
		$252^0 C$

R = H; R' = OH

artifact ?

JACS, 1932, 4454

1801-004

MILLETTONE, iso-	$C_{22}H_{18}O_6$	378

Tet. Supp., No. 7, 333.

1801-009

TOXICAROL	$C_{23}H_{22}O_7$	410
		$-67^0 b$
		$125^0 C$

JCS, 1940, 1178.

1801-005

MILLETTOSIN	$C_{22}H_{18}O_7$	394
		Oil

Tet., 1967, 4741.

1801-010

ROTENONE	$C_{23}H_{22}O_6$	394
		-225^0b
		164^0C

Chem. Comm., 1968, 234.

1801-011

DALPANOL	$C_{23}H_{24}O_7$	410
		-136^0c
		196^0C

JCS, 1971, 29.

1801-016

ROTENONE, dehydro-	$C_{23}H_{20}O_6$	392
		-41^0c
		218^0C

*dehydro

Science, 1933 (77) 311.

1801-012

SUMATROL	$C_{23}H_{22}O_7$	410
		-184^0b
		195^0C

Tet. Supp. No. 7, 333.

1801-017

deleted

1801-013

ELLIPTONE	$C_{20}H_{16}O_6$	352
		-18^0b
		179^0C

JCS, 1942, 587.

1801-018

AMORPHIGENIN	$C_{23}H_{22}O_7$	410
		-126^0c
		196^0C

CH_2OR

JCS, 1964, 6023.

R = H

1801-014

DOLINEONE	$C_{19}H_{12}O_6$	336
		$+135^0$c
		234^0C

JCS, 1963, 1569.

R = H

1801-019

AMORPHIN	$C_{34}H_{40}O_{16}$	704
		-124^0 m
		155^0C

R = Vicianosyl

JCS, 1967, 1796.

1801-015

DOLINEONE, 12a-OH-	$C_{19}H_{12}O_7$	352
		$+137^0$c
		180^0C

R = OH

Tet., 1970, 3023.

1801-020

MALACCOL $C_{20}H_{16}O_7$ 368
-190^0c
225^0C

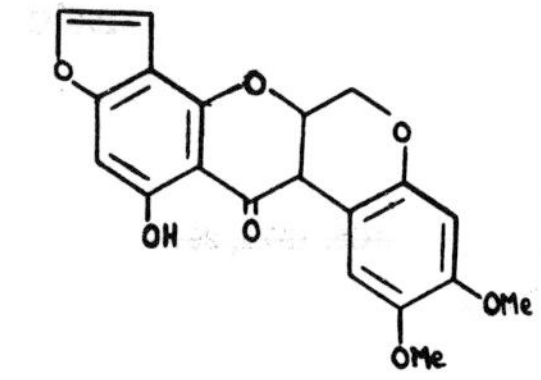

JCS, 1941, 878.

1801-021

PELTOGYNAN-14-ONE, 2,3,9- $C_{17}H_{12}O_7$ 328
-tri-OH-8-OMe-6, 13-dehydro-

Chem. Abstr., 1971 (74)
72457.

1802-001

PACHYRRHIZONE $C_{20}H_{14}O_7$ 366
$+100^0c$
235^0C

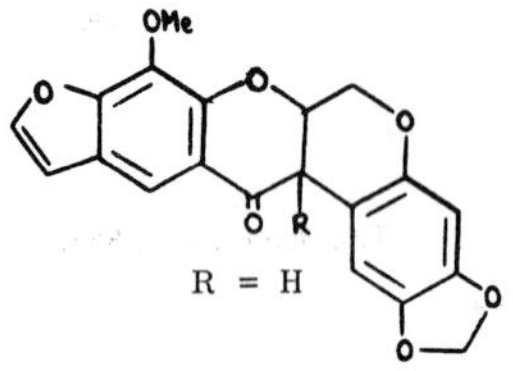

R = H

Helv., 1953, 664.

1801-022

CROMBEONE $C_{16}H_{10}O_7$ 314

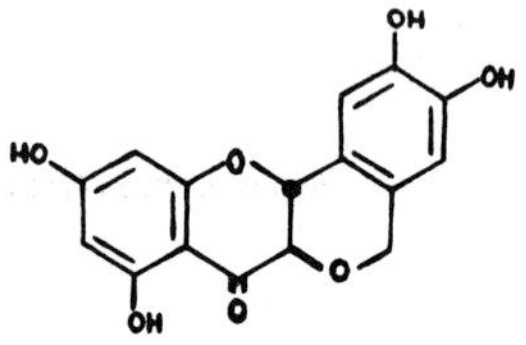

Chem. Comm., 1971,
116.

1802-002

PACHYRRIZONE, 12a-OH- $C_{20}H_{14}O_8$ 382
$+126^0c$
214^0C

R = OH

Tet., 1970, 3023.

1801-023

MOPANIN $C_{16}H_{10}O_6$ 298
OI

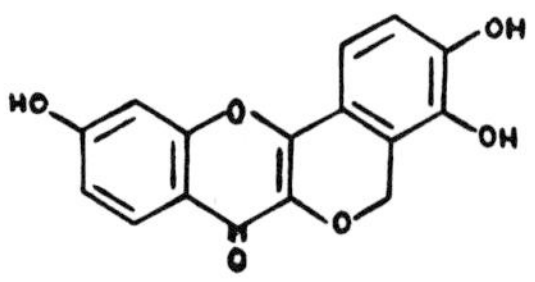

JCS, 1967, 1407.

1802-003

MYRICONOL $C_{23}H_{22}O_6$ 394
-60^0e
114^0C

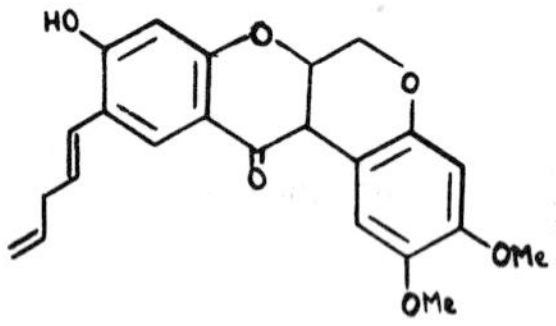

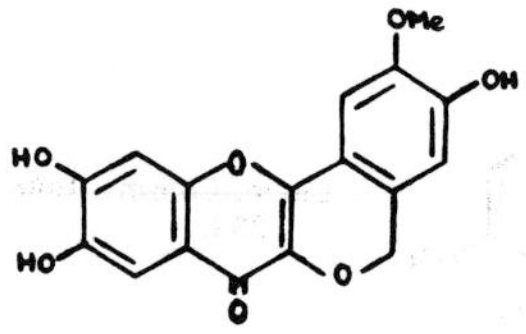

Curr. Sci., 1963 (32) 16.

1801-024

GUARABIN $C_{17}H_{14}O_7$ 330

280^0C

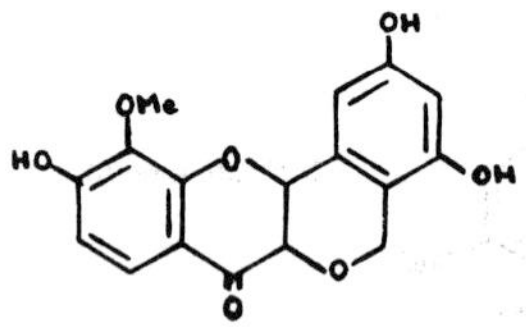

Chem. Abstr., 1971
(74) 72457.

1802-004

TEPHROSIN $C_{23}H_{22}O_7$ 410
Rac
198^0C

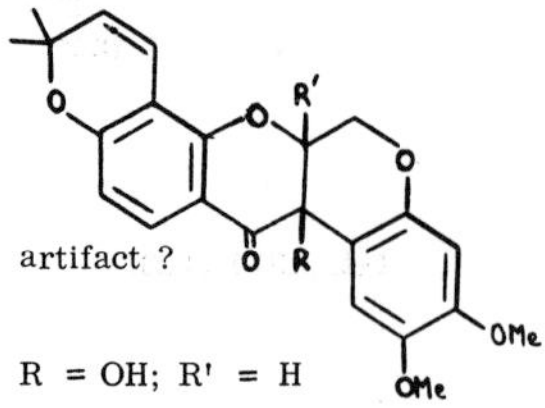

artifact ?

JACS, 1943, 27.

R = OH; R' = H

1801-025

GUARABIN, Iso- $C_{17}H_{14}O_7$ 330

259^0C

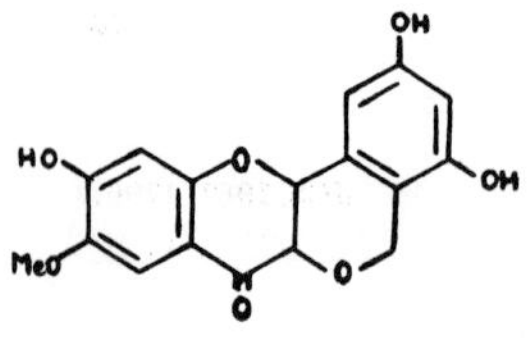

Chem. Abstr., 1971
(74) 72457.

1803-005

DISTEMONANTHIN

$C_{17}H_{10}O_9$ 346
OI
351^0C

JCS, 1954, 4594.

1802-006

MOPANOL, (+)-

$C_{16}H_{14}O_6$ 302
$+209^0$

JCS, 1966, 1644.

1802-007

MOPANOL-B

$C_{16}H_{14}O_6$ 302

*epimer

JCS, 1966, 1644.

1802-008

PELTOGYNOL, (+)-

$C_{16}H_{14}O_6$ 302
$+288^0$

JCS, 1966, 1644.

1802-009

PELTOGYNOL-B

$C_{16}H_{14}O_6$ 302

*epimer

JCS, 1966, 1644.

1802-010

AUREUSIDIN

$C_{15}H_{10}O_6$ 286
OI
272^0C

R = R' = H

JACS, 1955, 4622.

1803-001

AUREUSIN

$C_{21}H_{20}O_{11}$ 448

R = Glu; R' = H

JCS, 1955, 4622.

1803-002

CERNUOSIDE

$C_{21}H_{20}O_{11}$ 448
254^0C

R = H; R' = Glu

JACS, 1955, 4622.

1803-003

RENGASIN

$C_{16}H_{12}O_6$ 300
OI
315^0C

R = H; R' = Me

Tet. Lett., 1963, 1999.

1803-004

MARITIMETIN

$C_{15}H_{10}O_6$ 286
OI

R = H

JACS, 1956, 825.

1803-005

MARITIMEIN	$C_{21}H_{20}O_{11}$	448

R = Glucosyl

JACS, 1957, 214.

1803-006

SULPHURETIN, 6-diglucosyl-	$C_{27}H_{30}O_{15}$	594

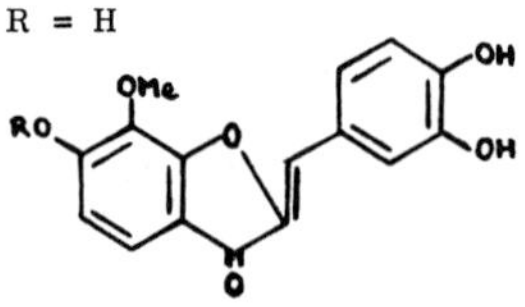

R = Glu-Glu

Arch. Biochem. Biophys.,
1956 (60) 329.

1803-011

LEPTOSIDIN	$C_{16}H_{12}O_6$	300 OI 253°C

R = H

JACS, 1951, 5765.

1803-007

PALASITRIN	$C_{27}H_{30}O_{15}$	594 200°C

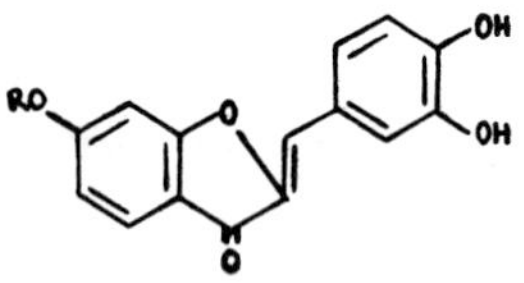

JCS, 1955, 1589.

1803-012

LEPTOSIN	$C_{22}H_{22}O_{11}$	462 230°C

R = Glucosyl

JACS, 1944, 486.

1803-008

BRACTEIN	$C_{21}H_{20}O_{12}$	464 245°C

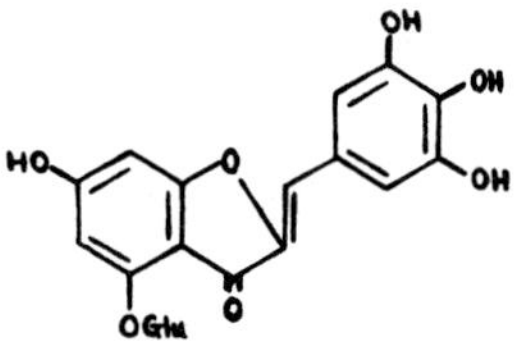

Tet. Lett., 1962, 599.

1803-013

SULPHURETIN	$C_{15}H_{10}O_5$	270 OI 282°C

R = H

JACS, 1953, 1900.

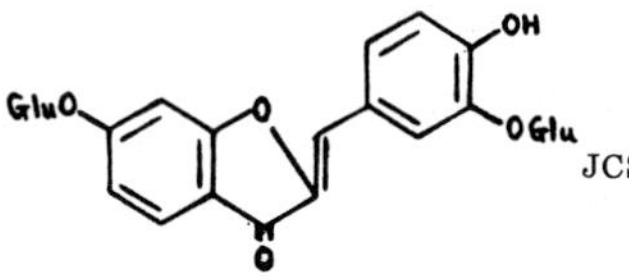

1803-009

MAESOPSIN	$C_{15}H_{12}O_6$	288 219°C

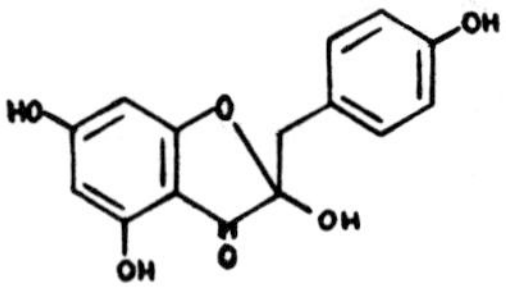

Aust. J. C., 1962 (15) 314.

1803-014

SULPHUREIN	$C_{21}H_{20}O_{10}$	432 185°C

R = Glucosyl

JACS, 1956, 825.

1803-010

Two biogenetically quite distinct groups of compounds have been collected together in class 19. Both groups are nonaromatic, one (1902) because it includes the shikimic acid prearomatics and the other (1901) because the aromatic ring has been further oxidized (often to an epoxide). This latter class (1901) is biogenetically related to the orcinol (21) and benzoquinone (20) classes. The origin of the shikimic acid prearomatics lay in the condensation of appropriate carbohydrate precursors (see Chart 1, page 4).

PHYLLOSTINE $C_7H_6O_4$ 154
   -106^0e
   56^0C

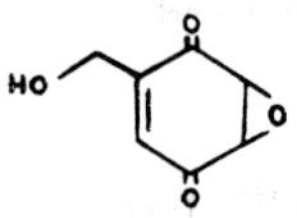

Agr. Biol. Chem.,
1971 (35) 105.

1901-001

SPINULOSIN–HYDRATE $C_8H_{10}O_6$ 202

182^0C

Chem. Pharm. Bull.,
1967 (15) 427.

1901-006

TERREMUTIN $C_7H_8O_4$ 154

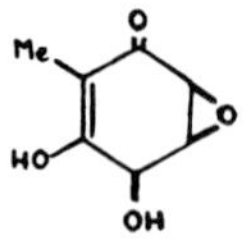

Tet., 1968, 4839.

1901-002

SPINULOSIN QUINOL–HYDRATE $C_{18}H_{12}O_6$ 204

191^0C

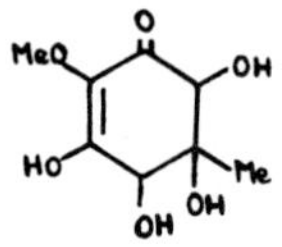

Chem. Pharm. Bull.,
1970, 561.

1901-007

TERREIC ACID $C_7H_6O_4$ 154
   -29^0
   127^0C

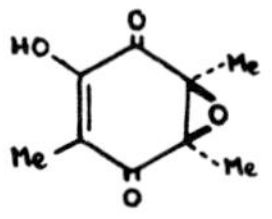

JACS, 1958, 5536.

1901-003

EPOXYDONE $C_7H_8O_4$ 156
   $+93^0$m
   42^0C

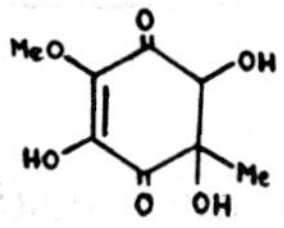

Helv., 1965, 204.

1901-008

FUMIGATIN OXIDE $C_8H_8O_5$ 184

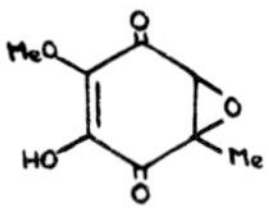

Chem. Pharm. Bull.,
1967 (15) 427.

1901-004

PANEPOXYDONE $C_{11}H_{14}O_4$ 210
   -61^0
   Oil

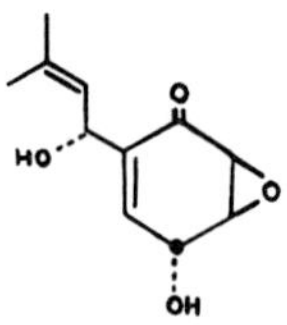

Helv., 1970 (53) 1577.

1901-009

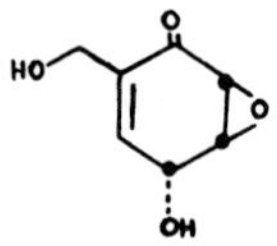

SPINULOSIN QUINOL, dihydro- $C_8H_{12}O_5$ 188

166^0C

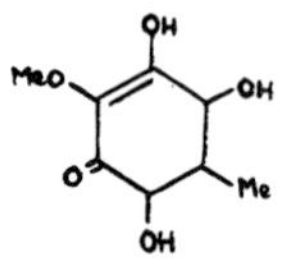

Chem. Pharm. Bull.,
1967 (15) 427.

1901-005

PANEPOXYDONE, iso- $C_{11}H_{14}O_4$ 210
   $+163^0$m
   124^0C

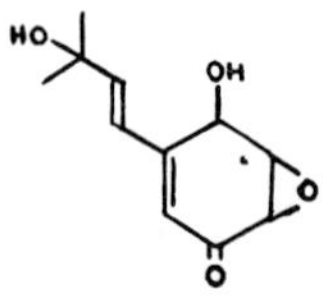

Helv., 1970 (53) 1577.

1901-010

PANEPOXYDONE, neo-

$C_{11}H_{14}O_4$
210
-64^0e
156^0C

Helv., 1970 (53) 1577.

1901-011

SENEPOXIDE

$C_{18}H_{18}O_7$
346
-197^0c
83^0C

Tet., 1968, 1633.

1901-016

PANEPOXYDIONE

$C_{11}H_{12}O_4$
208
$+223^0$c
Oil

Helv., 1970 (53) 1577.

1901-012

CROTEPOXIDE

$C_{18}H_{18}O_8$
362
$+74^0$c
150^0C

JACS, 1968, 2982.

1901-017

PANEPOXYDOL, neo-

$C_{11}H_{16}O_4$
212
$+130^0$m
164^0C

Helv., 1970 (53) 577.

1901-013

PIPOXIDE

$C_{21}H_{18}O_6$
366
$+24^0$c
153^0C

Tet., 1970, 4403.

1901-018

PANEPOXYDOL, 7-Deoxy-

$C_{11}H_{16}O_3$
196
-42^0c
104^0C

Helv., 1970 (53) 1577.

1901-014

GLIOROSEIN

$C_{10}H_{14}O_4$
198
OI
48^0C

Proc. C. S., 1958, 343.

1901-019

SENEOL

$C_{19}H_{22}O_8$
378
-128^0c
114^0C

Tet., 1968, 1633.

1901-015

SHIKIMIC ACID

$C_7H_{10}O_5$
174
-246^0w
184^0C

R = H

J. Biol. Chem., 1956 (220) 177.

1902-001

223

DACTYLIFRIC ACID	$C_{16}H_{16}O_8$	336 -124^0e 225^0C

R = Caffeoyl

Biochem. Biophys. Res.
Comm. , 1964 (14) 124.

1902-002

QUINIC ACID, 3-O-p-coumaroyl-	$C_{16}H_{18}O_8$	338 -54^0m 248^0C

JCS, 1961, 5153.

1902-007

SHIKIMIC ACID, dihydro-	$C_7H_{12}O_5$	176 -63^0w 135^0C

R = H

*dihydro

Biochem. J. , 1957 (66)
283.

1902-003

QUINIC ACID, 3-O-Feruoyl-	$C_{17}H_{20}O_9$	368 196^0C

Tet. , 1962, 1207.

1902-008

SHIKIMIC ACID, 5-dehydro-	$C_7H_8O_5$	172 -57^0e 151^0C

JACS, 1953, 5567.

1902-004

QUINIC ACID, 4-O-Caffeoyl-	$C_{16}H_{18}O_9$	354 -69^0

Chem. Ind. , 1964, 1984.

1902-009

QUINIC ACID	$C_7H_{12}O_6$	192 -43^0w 162^0C

Ber. , 1954 (87) 793.

1902-005

QUINIC ACID, 5-Caffeoyl-	$C_{16}H_{18}O_9$	354 -230^0e

Planta Med. Phytother. ,
1970 (4) 56.

1902-010

CHLOROGENIC ACID	$C_{16}H_{18}O_9$	354 -34^0w 208^0C

3-O-Caffeoyl-Quinic Acid

JACS, 1956, 825.

1902-006

QUINIC ACID, 1, 3-Dicaffeoyl-	$C_{25}H_{24}O_{12}$	516

Planta Med. Phytother. ,
1970 (4) 56.

1902-011

CYNARINE	$C_{25}H_{24}O_{12}$	516
		-59^0m
		228^0C
1,4-Di-Caffeoyl-Quinic Acid		
Nature, 1954 (174) 1062.		
1902-012		

QUINIC ACID, dehydro- $C_7H_{10}O_6$ 190

141^0C

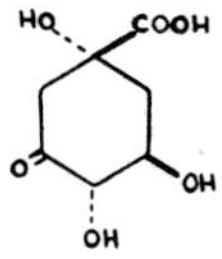

JACS, 1953, 5572.

1902-017

QUINIC ACID, 3,4-dicaffeoyl- $C_{25}H_{24}O_{12}$ 516

-225^0m

Tet. Lett., 1964, 2851.

1902-013

CHORISMIC ACID $C_{10}H_8O_6$ 224

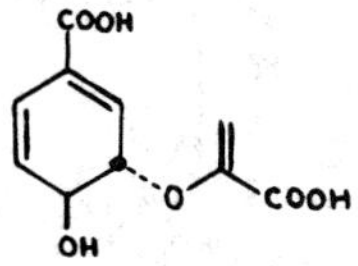

JACS, 1969, 5893.

1902-018

QUINIC ACID, 3,5-dicaffeoyl- $C_{25}H_{24}O_{12}$ 516

-198^0m

JCS, 1964, 2137.

1902-014

PREPHENIC ACID $C_{10}H_{10}O_6$ 226

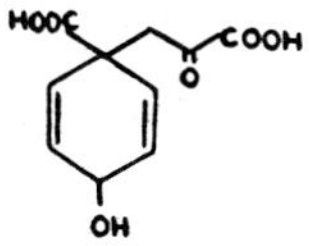

Science, 1954 (119) 774.

1902-019

QUINIC ACID, 4,5-dicaffeoyl- $C_{25}H_{24}O_{12}$ 516

-170^0m

Tet. Lett., 1964, 2851.

1902-015

THEOGALLIN $C_{14}H_{16}O_{10}$ 344

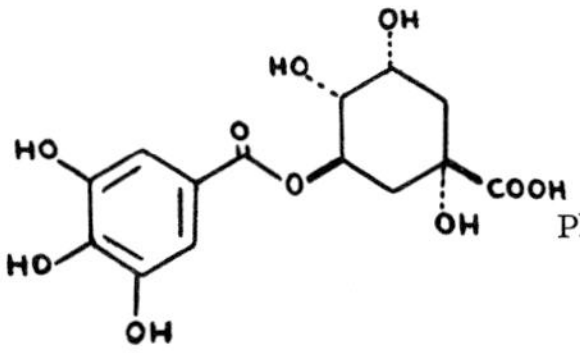

Phytochem., 1971 (10) 1671.

1902-020

---

deleted

1902-016

Meaningful biogenetic analysis of the benzoquinones in terms of their structural features is a dubious enterprise owing to the wide spectrum of possible precursors and modifying reactions that could lead to the benzoquinones. Several of the mold benzoquinones have been examined biosynthetically, and an acetate/malonate origin established via an appropriate orcinol-type phenolic precursor. The ubiquinones, on the other hand, do not incorporate acetate into the quinone ring, and, from the evidence thus far at hand, shikimic acid would appear to provide the appropriate phenolic precursor.

The biogenetic details being ambiguous, the subclassification of the benzoquinones has been carried out according to the number of oxy functions on the quinone ring. Class 2005 includes the various types of dimeric benzoquinone (i.e., including those that possess a carbon chain connecting the two rings).

2001	No oxy substituent
2002	One oxy substituent
2003	Two oxy substituents
2004	Three oxy substituents
2005	Dimeric Compounds

BENZOQUINONE, p-	$C_6H_4O_2$ 108 OI 116^{0}C	XYLOQUINONE, para- $C_8H_8O_2$ 136 OI 125^{0}C
JCS, 1949, 2115.		JACS, 1955, 4942.
2001-001		2001-006

TOLUQUINONE, p- R = Me	$C_7H_6O_2$ 122 OI 69^{0}C	CUMOQUINONE, pseudo- $C_9H_{10}O_2$ 152 OI 32^{0}C
Biochimia, 1957 (22) 236.		JACS, 1955, 4942.
2001-002		2001-007

GENTISYLQUINONE R = CH$_2$OH	$C_7H_6O_3$ 138 OI 76^{0}C	BENZOQUINONE, 3,3-DMA- $C_{11}H_{12}O_2$ 176 OI 30^{0}C
Helv., 1947, 1.		Ber., 1966 (99) 885.
2001-003		2001-008

BENZOQUINONE, 2-Et- R = Ethyl	$C_8H_8O_2$ 136 OI 40^{0}C	BENZOQUINONE, 5-(3,3-DMA)-2-Me- R = 3,3-DMA	$C_{12}H_{14}O_2$ 190 OI Oil
Biochem. J., 1948 (43) 474.		JCS, 1968, 859.	
2001-004		2001-009	

XYLOQUINONE, ortho-	$C_8H_8O_2$ 136 OI 55^{0}C	BENZOQUINONE, 5-geranyl-2-Me- R = Geranyl	$C_{17}H_{22}O_2$ 258 OI Oil
JACS, 1956, 774.		JCS, 1968, 857.	
2001-005		2001-010	

PLASTOQUINONE-3 n = 3 2001-011	$C_{23}H_{32}O_2$	340 OI Oil JACS, 1965, 1402.
TOCOPHEROLQUINONE, α- 2001-016	$C_{29}H_{50}O_3$	446 Oil Biochem. J. , 1964 (90) 41.
PLASTOQUINONE-4 n = 4 2001-012	$C_{28}H_{40}O_2$	408 OI Oil Z. Nat. , 1963 (18 B) 446.
BENZOQUINONE, 2-OMe- 2002-001	$C_7H_6O_3$	138 OI 144^0C JCS, 1952, 4821.
PLASTOQUINONE-8 n = 8 2001-013	$C_{48}H_{72}O_2$	680 Phytochem. , 1970, 737.
COPRININ 2002-002	$C_8H_8O_3$	152 OI 175^0C Biochem. J. , 1969 (114) 369.
PLASTOQUINONE-9 n = 9 2001-014	$C_{53}H_{80}O_2$	748 OI 48^0C JACS, 1959, 2026.
PRIMIN 2002-003	$C_{12}H_{16}O_3$	208 OI 62^0C Z. Nat. , 1967 (22 B) 287.
PLASTOQUINONE, phytyl- 2001-015	$C_{28}H_{46}O_2$	414 Oil Phytochem. , 1970, 213.
deleted 2002-004		

UBIQUINONE-4, 5-des-OMe- n = 4 2002-005	$C_{28}H_{40}O_3$ Biochim. Biophys. Acta, 1969 (176) 895.	424 OI
UBIQUINONE-9, 5-des-OMe- n = 9 2002-010	$C_{53}H_{80}O_3$ Tet. Lett., 1967, 1237.	764 OI 52^0C
UBIQUINONE-5, 5-des-OMe- n = 5 2002-006	$C_{33}H_{48}O_3$ Biochim. Biophys. Acta, 1969 (176) 895.	492 OI
UBIQUINONE-10, 5-des-OMe- n = 10 2002-011	$C_{58}H_{88}O_3$ JACS, 1966, 4754.	832 OI
UBIQUINONE-6, 5-des-OMe- n = 6 2002-007	$C_{38}H_{56}O_3$ Biochim. Biophys. Acta, 1969 (176) 895.	560 OI
UBIQUINONE-10, 5-des-OMe- -dihydro- 2002-012	$C_{58}H_{90}O_3$ Biochem. J., 1970 (117) 799.	834 OI
UBIQUINONE-7, 5-des-OMe- n = 7 2002-008	$C_{43}H_{64}O_3$ Biochim. Biophys. Acta, 1969 (176) 895.	628 OI
UBIQUINONE-9, 5-des-OMe-3-des- -Me- n = 9 2002-013	$C_{52}H_{78}O_3$ Biochim. Biophys. Acta, 1969 (176) 895.	750 OI
UBIQUINONE-8, 5-des-OMe- n = 8 2002-009	$C_{48}H_{72}O_3$ Ann., 1969 (729) 158.	696 OI 42^0C
UBIQUINONE-10, 5-des-OMe-3- -des-Me- n = 10 2002-014	$C_{57}H_{86}O_3$ JACS, 1966, 4754. Biochim. Biophys. Acta, 1969 (176) 895.	818 OI

UBIQUINONE-10, 5-des-OMe-3- -des-Me-dihydro- $C_{57}H_{88}O_3$ 820
OI

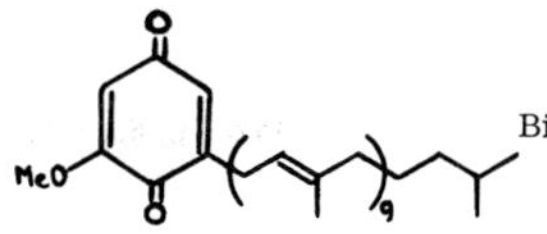

Biochem. J., 1970 (117) 799.

2002-015

SHANORELLIN $C_9H_{10}O_4$ 182
OI
121^0C

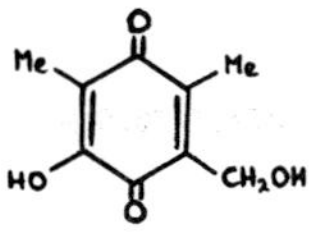

Phytochem., 1968, 2177.

2002-020

RHODOQUINONE-9 $C_{53}H_{81}NO_3$ 779
OI

n = 9

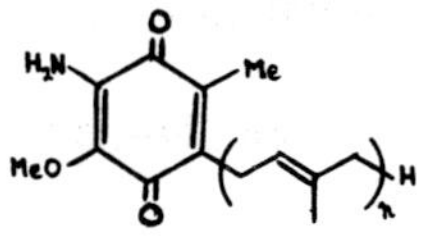

Tet. Lett., 1969, 1969.

2002-016

deleted

2002-021

RHODOQUINONE-10 $C_{58}H_{89}NO_3$ 847
OI
70^0C

n = 10

JACS, 1968, 5587.

2002-017

SARCODONTIC ACID $C_{22}H_{32}O_5$ 376
OI
123^0C

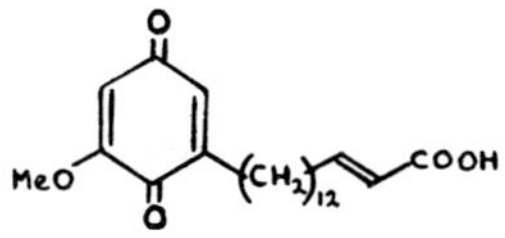

Coll. Czech. Chem. Comm., 1965 (30) 2626.

2002-022

deleted

2002-018

BENZOQUINONE, 5,6-di-OH-2,3- -diMe- $C_8H_8O_4$ 168
OI
182^0C

R = H

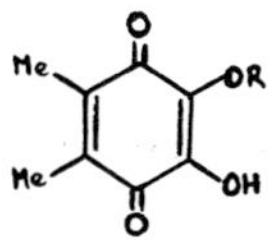

Acta Chem. Scand., 1964 (18) 2303.

2002-023

BENZOQUINONE, 2-OMe-3-Me- $C_8H_8O_3$ 152
OI
19^0C

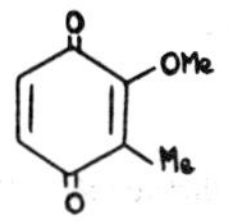

Ann. Entomol. Soc. Am., 1962 (55) 261.

2002-019

BENZOQUINONE, 5-OH-6-OMe- -2,3-diMe- $C_9H_{10}O_4$ 182
OI
70^0C

R = Me

Acta Chem. Scand., 1964 (18) 2303.

2002-024

BENZOQUINONE, 2, 5-di-OMe-	$C_8H_8O_4$	168 OI 250°C

JCS, 1955, 575.

2003-001

AURANTIOGLIOCLADIN	$C_{10}H_{12}O_4$	196 OI 63°C

Proc. C. S. , 1958, 343.

2003-006

BENZOQUINONE, 2, 6-di-OMe-	$C_8H_8O_4$	168 OI 254°C

JCS, 1952, 4821.

2003-002

EMBELIN	$C_{17}H_{26}O_4$	294 OI 142°C

$R = n - C_{11}H_{23}$

Helv. , 1948, 2237.

2003-007

BENZOQUINONE, 2, 5-di-OH-3-Et-	$C_8H_8O_4$	168 OI 130°C

JOC, 1966, 3645.

2003-003

RAPANONE	$C_{19}H_{30}O_4$	322 OI 141°C

$R = n - C_{13}H_{27}$

JACS, 1948, 71.

2003-008

FUMIGATIN	$C_8H_8O_4$	168 OI 116°C

Chem. Ind. , 1954, 1327.

2003-004

BENZOQUINONE. 2-OH-5-OMe- -3-tridecyl-	$C_{20}H_{32}O_4$	336 OI

$R = n- C_{13}H_{27}$

Phytochem. , 1968 (7) 773.

2003-009

TOLUQUINONE, 6-OH-4-OMe-	$C_8H_8O_4$	168 OI 202°C

Acta Chem. Scand. , 1966, 45.

2003-005

BENZOQUINONE, 2-OH-5-OMe- -3-tridecenyl-	$C_{20}H_{30}O_4$	334 OI

$R = C_{13}H_{25}$

Phytochem. , 1968 (7) 773.

2003-010

BENZOQUINONE, 2-OH-5-OMe- -3-pentadecenyl- R = $C_{15}H_{29}$ Phytochem., 1968 (7) 773. 2003-011	$C_{22}H_{34}O_4$	362 OI 67^0C
BOVINONE n = 4 JCS, 1969, 2398. 2003-016	$C_{23}H_{36}O_4$	412 OI 85^0C

BHOGATIN $C_{16}H_{24}O_4$ 280 OI 156^0C

R = $n-C_9H_{19}$

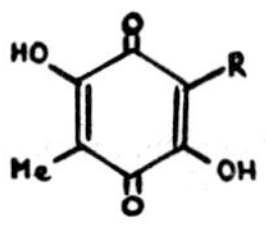

Phytochem., 1970, 415.

2003-012

UBIQUINONE-1 $C_{14}H_{18}O_4$ 250

n = 1

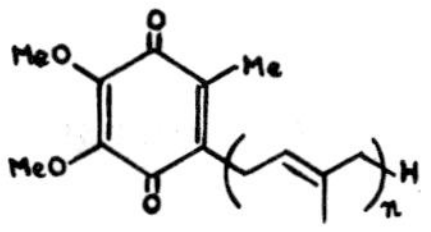

Biochemistry, 1967 (6) 2861.

2003-017

MAESAQUINONE $C_{26}H_{42}O_4$ 418 OI 122^0C

R = .$CH_2(CH_2)_{12}CH:CH(CH_2)_3Me$

JACS, 1948, 71.

2003-013

UBIQUINONE-2 $C_{19}H_{26}O_4$ 318

n = 2

Biochemistry, 1967 (6) 2861.

2003-018

POLYGONAQUINONE $C_{28}H_{48}O_4$ 448 OI 133^0C

R = $n-C_{21}H_{43}$

Tet., 1964, 2319.

2003-014

UBIQUINONE-3 $C_{24}H_{34}O_4$ 386

n = 3

Biochemistry, 1967 (6) 2861.

2003-019

BOVOQUINONE-3 $C_{21}H_{28}O_4$ 344 OI 91^0C

n = 3

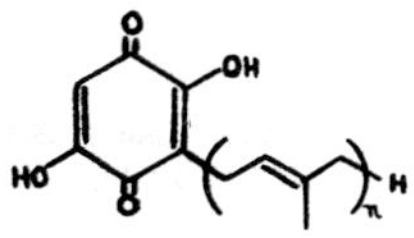

JCS, 1971, 2582.

2003-015

UBIQUINONE-4 $C_{29}H_{42}O_4$ 454

n = 4

Biochemistry, 1967 (6) 2861.

2003-020

UBIQUINONE-5	$C_{34}H_{50}O_4$	522 OI
n = 5		
	Biochem. Biophys. Res. Comm., 1966 (24) 252.	
2003-021		

UBIQUINONE-10	$C_{59}H_{90}O_4$	862 OI 50^0C
n = 10		
	Biochem. J., 1958 (68) 16 P & 29 P.	
2003-026		

UBIQUINONE-6	$C_{39}H_{58}O_4$	590 OI 16^0C
n = 6		
	JACS, 1958, 4751.	
2003-022		

UBIQUINONE-11	$C_{64}H_{98}O_4$	950
n = 11		
	Phytochem., 1970 (9) 355.	
2003-027		

UBIQUINONE-7	$C_{44}H_{66}O_4$	658 OI 31^0C
n = 7		
	JACS, 1958, 4751.	
2003-023		

UBIQUINONE-12	$C_{69}H_{106}O_4$	1018
n = 12		
	Phytochem., 1970 (9) 355.	
2003-028		

UBIQUINONE-8	$C_{49}H_{74}O_4$	726 OI 37^0C
n = 8		
	JACS, 1958, 4751.	
2003-024		

UBIQUINONE-10, X-dihydro-	$C_{59}H_{92}O_4$	864 OI 29^0C
terminally reduced		
	Biochemistry, 1963 (2) 196.	
2003-029		

UBIQUINONE-9	$C_{54}H_{82}O_4$	794 OI 45^0C
n = 9		
	JACS, 1958, 4751.	
2003-025		

UBIQUINONE-10, tetrahydro-	$C_{59}H_{94}O_4$	866 29^0C
Not terminally reduced		
	J. Biol. Chem., 1962 (237) PC2715.	
2003-030		

UBIQUINONE-10, 5-O-des-Me- $C_{58}H_{88}O_4$ 848
OI

Biochem. Biophys. Res.
Comm., 1967 (28) 324.

2003-031

RUBROGLIOCLADIN $C_{20}H_{26}O_8$ 394
OI
74^0C

JCS, 1953, 815.

2005-004

SPINULOSIN $C_8H_8O_5$ 184
OI
203^0C

Biochem. J., 1947 (41) 569.

2004-001

VILANGIN $C_{35}H_{52}O_8$ 600
OI
264^0C

JOC, 1961, 4529.

2005-005

PHOENICIN $C_{14}H_{10}O_6$ 274
OI
230^0C

Helv., 1943, 2031

2005-001

DIBOVOQUINONE-3, 3, methylene- $C_{43}H_{56}O_8$ 700
OI
198^0C

JCS, 1971. 2582.

2005-006

OOSPOREIN $C_{14}H_{10}O_8$ 306
OI
260^0C

JCS, 1955, 2163.

2005-002

DIBOVOQUINONE-3, 4 $C_{47}H_{62}O_8$ 754
OI
138^0C

JCS, 1971, 2582.

2005-007

deleted

2005-003

DIBOVOQUINONE-4, 4 $C_{52}H_{70}O_8$ 822
OI
133^0C

JCS, 1971, 2582.

2005-008

AMITENONE	$C_{53}H_{72}O_8$	836
		OI
		188^0C

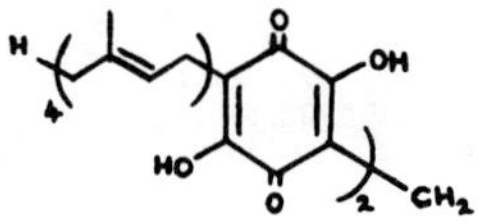

Tet. Lett., 1968, 5067.

2005-009

ARDISIAQUINONE-B	$C_{30}H_{40}O_8$	528
		OI
		119^0C

R = H ; R' = Me

Tet. Lett., 1968, 1387.

2005-011

ARDISIAQUINONE-A	$C_{30}H_{40}O_8$	528
		OI
		154^0C

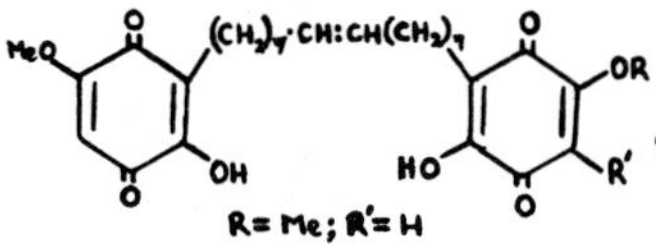

R= Me; R'= H

Tet. Lett., 1968, 1387.

2005-010

Acetate Derived Aromatics:
ACETATE/MALONATE BIOGENETIC PATHWAY

Known as the "polyketide" or "acetogenic" pathway, this route to a wide variety of acyclic, alicyclic, heterocyclic, and aromatic compounds can be visualized in terms of two basic building blocks and the normal repertoire of organic chemistry. The fundamental building blocks consist of a starter acyl unit and a condensing acyl unit. Typical starters are acetate, propionate, iso-butyrate, benzoate, and cinnamate; the condensing unit is most generally malonate although an increasing number of compounds are being found the biogenesis of which suggests a propionate condensing unit. In the discussion acetate will be used as the typical starter and malonate the condensing unit. It may be mentioned here that this pathway is frequently referred to as the "acetate" pathway, whereas malonate is the usual building block; in that malonate is derived from acetate by carboxylation the acetate label is not a serious misnomer (also bearing in mind that acyl intermediates are actually involved). The combination of one acetyl unit with several malonyl units provides a linear polyketide chain that can undergo a variety of chemical modifications, three of which are illustrated below:

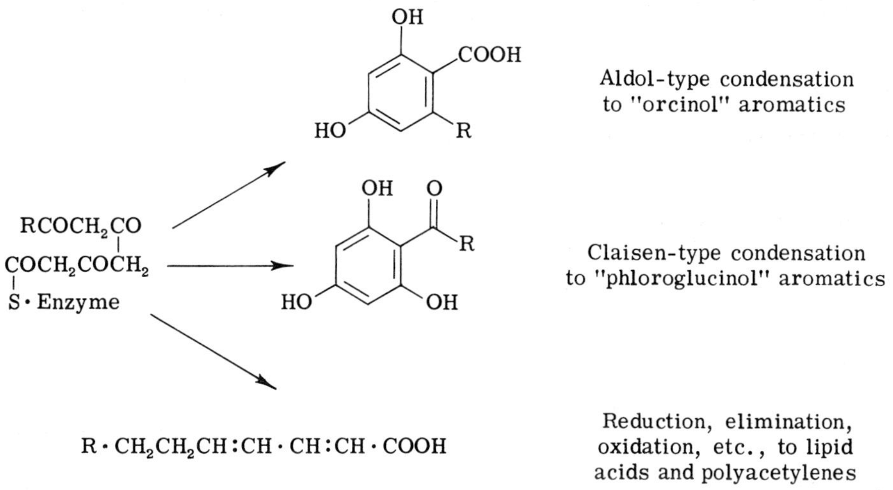

Aldol-type condensation to "orcinol" aromatics

Claisen-type condensation to "phloroglucinol" aromatics

Reduction, elimination, oxidation, etc., to lipid acids and polyacetylenes

Larger polyketide chains can be folded and condensed to provide the more complex aromatics such as the fungal anthraquinones and the tetracyclines. Chart 3 summarizes a series of such condensations for precursor chains of from 4 to 10 acetate units. The linear acetogenins will be discussed further in the introduction to Class 31.

Number of acetate units	Condensation modes illustrated by examples		
4	Xanthoxylin (2201)	Orsellinic Acid (2101)	
5	Eugenin (2203)	Reticulol (2104)	Curvulic Acid (2103)
	Flaviolin (2505)		
6	Sorbicillin (2205)	Diaporthin (2106)	α-Sorigenin ∗
	2505-004	∗ The majority of β-methyl-naphthyls most likely derive from shikimic acid and mevalonic acid	

Number of acetate units	Condensation modes illustrated by examples		
7	Monocerone (2107)	Nor-rubrofusarin(2504)	Griseophenone-C(2802)
	Alternariol (2801)	Vioxanthin (2510)	Xanthomegnin (2510)
	Fusarubin (2505)	Atrovenetin (2701)	
8	Lasiodiplodin (2108)	Erythrostominone	Actinorhodin (2510)
	Endrocrocin (2604)	Laccaic Acid (2604)	Curvularin (2108)

Number of acetate units	Condensation modes illustrated by examples	
9	Zearalenone (2109)	Frenolicin (2504)
	Ptilometric Acid (2604)	Rhodocomatulin Me Ether (2605)
10	Avermutin (2605)	Eta-pyrromycinone (2703)
	Daunomycinone (2703)	

21	$n = 1,3,5$ etc Orcinols	22	$n = 1,3,5$ etc Phloroglucinols
23	Depsidones	24	Xanthones
25	Naphthoquinones etc.	26	Anthraquinones etc.
27	e.g. Polycyclics		
2801	Biphenyls	2803	Grisans

241

The "orcinol" aromatics are characterized by functionalization consistent with an aldol-type condensation of a polyketide precursor:

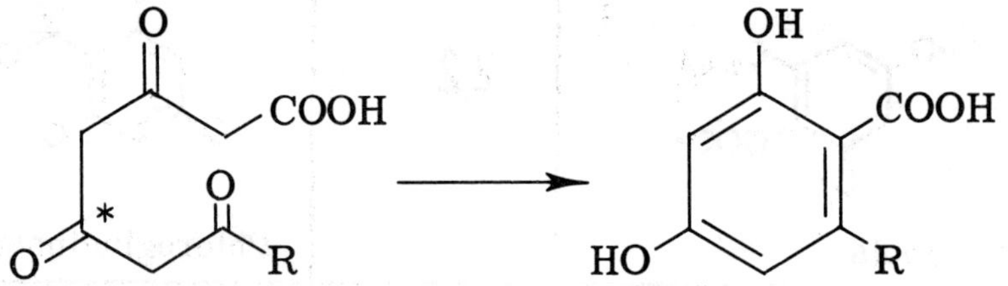

It should be appreciated that the oxo function (*) is not mechanistically demanded for such a condensation, and as such could be present in the linear precursor in a reduced form:

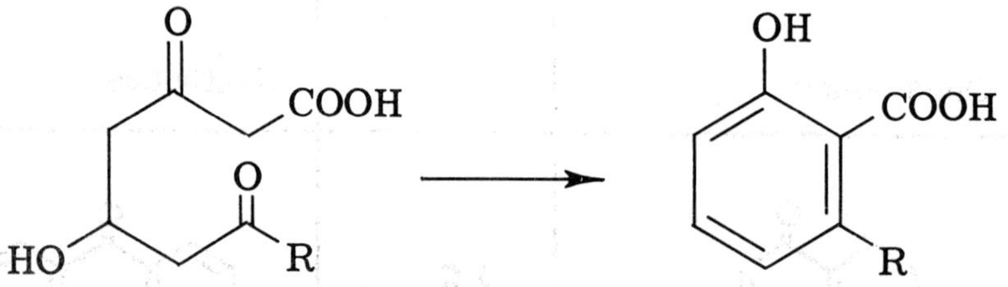

Similarly decarboxylation, alkylation, halogenation, and oxidation are processes that may substantially modify the orcinol nucleus. In general these secondary processes (illustrated on Chart 4) do not completely mask the orcinol origin of the products, but some ambiguities are evident (e.g., gentisyl alcohol has been placed in the C_6C_1 class although its biogenesis could equally well be via m-cresol).

The classification employed here for the orcinols basically uses the concept of the number of "acetate" units needed to derive a compound. In the case of those compounds derived from 4 "acetate" units further subclassification was deemed useful so that the phthalic acid derivatives could be brought together (2102). Similarly the compounds derived from 5 "acetate" units have been broken down into three more useful categories.

2101-015   α-Orcinol   Sparassol   2101-007

Dihydro-fumigation   Orsellinic Acid   2101-011

Everninocin   2102-007   Rhizonic Acid

2101-018   m-cresol   6-Me-Salicylic   Gladiolic
                                      Acid   Acid

243

2101	4 "acetate" units	2102	also: aldehydo analogs, phthalic anhydrides, phthalides
2103	5 "acetate" units	2104	5 "acetate" units
2105	5 "acetate" units	2106	6 "acetate" units
2107	7 "acetate" units	2108	8 "acetate" units
2109	9 "acetate" units	2110	10 "acetate" units
2111	11 "acetate" units	2112	12 "acetate" units
2199	Miscellaneous		

ORSELLINIC ACID	$C_8H_8O_4$	168 OI 176°C
R = H	Z. Nat., 1959 (14B) 69.	
2101-001		

EVERNINIC ACID	$C_9H_{10}O_4$	182 OI 165°C
	JCS, 1932, 1392.	
2101-006		

ORSELLINIC ACID ET ESTER	$C_{10}H_{12}O_4$	196 OI 130°C
R = Et	JCS, 1969, 704.	
2101-002		

ORSELLINIC ACID, 5-Me-	$C_9H_{10}O_4$	182 OI
	Biochem. J., 1964, 43.	
2101-007		

MONTAGNETOL	$C_{12}H_{16}O_7$	272 +16°w 135°C
R = .CH₂CHOH.CHOH.CH₂OH	Proc. Ind. Acad. Sci., 1942 (15A) 429.	
2101-003		

ORSELLINIC ACID, 3-Me-	$C_9H_{10}O_4$	182 OI 184°C
R = H	Chem. Comm., 1972, 391.	
2101-008		

ORSELLINIC ACID ME ETHER	$C_9H_{10}O_4$	182 OI 144°C
R = H	Sci. Proc. Roy. Dublin Soc., 1943 (23) 143.	
2101-004		

RHIZONIC ACID	$C_{10}H_{12}O_4$	196 OI 234°C
R = Me	Ber., 1934 (67) 1793.	
2101-009		

SPARASSOL	$C_{10}H_{12}O_4$	196 OI 67°C
R = Me	Ber., 1924 (57A) 471.	
2101-005		

ORSELLINIC ACID, 3-formyl-	$C_9H_8O_5$	196 OI 173°C
	Helv., 1933 (16) 282.	
2101-010		

BENZOIC ACID, 2,4-diOH-3,5,6--tri-Me-	$C_{10}H_{12}O_4$	196 OI 192^0C

Experientia, 1967, 703.

2101-011

SYLVOPINOL	$C_8H_{10}O_3$	154 OI 85^0C

Chem. Abstr., 1970
(73) 63192.

2101-016

SALICYLIC ACID, 6-Me-	$C_8H_8O_3$	152 OI 178^0C

Experientia, 1958 (14)
38.

2101-012

CRESOL, meta-	C_7H_8O	108 OI Oil

Chem. Abstr., 1956
(50) 1684.

2101-017

ORCINOL, α-	$C_7H_8O_2$	122 OI 107^0C

Ann., 1848 (68) 99.

2101-013

BENZYL ALCOHOL, m-OH-	$C_7H_8O_2$	124 OI 68^0C

Arch. Biochem. Biophys.,
1964, 156.

2101-018

ORCINOL, β-	$C_8H_{10}O_2$	138 OI 162^0C

Curr. Sci., 1966 (35) 147.

2101-014

FUMIGATIN, dihydro-	$C_8H_{12}O_4$	170 OI 100^0C

JCS, 1938, 439.

2101-019

XYLENE, 3,5-di-OH-ortho-	$C_8H_{10}O_2$	138 OI 116^0C

Acta Chem. Scand.,
1965 (19) 414.
Tet. Lett., 1969, 4675.

2101-015

SICCANOCHROMENIC ACID, Pre-	$C_{23}H_{32}O_4$	372

Tet. Lett., 1969, 2457.

2101-020

SICCANOCHROMENIC ACID	$C_{23}H_{30}O_4$	370
	Tet. Lett., 1969, 2457.	
2101-021		

LL-Z1272-β	$C_{23}H_{32}O_3$	356 OI 97°C
R = H	Tet., 1969, 1323.	
2101-026		

SICCANOCHROMENE, Pre-	$C_{22}H_{32}O_2$	328
	Tet. Lett., 1969, 2457.	
2101-022		

LL-Z1272-α	$C_{23}H_{31}O_3Cl$	390 OI 73°C
R = Cl	Tet., 1969, 1323.	
2101-027		

SICCANOCHROMENE-E	$C_{22}H_{30}O_3$	342 -86°c 192°C
	Tet. Lett., 1969, 2457.	
2101-023		

LL-Z1272-ε	$C_{23}H_{32}O_4$	372 +6°m 172°C
R = H	Tet., 1969, 1323.	
2101-028		

SICCANIN	$C_{22}H_{30}O_3$	342 -150°c 138°C
	Tet. Lett., 1967, 2177.	
2101-024		

LL-Z1272-δ	$C_{23}H_{31}O_4Cl$	406 +6°m 130°C
R = Cl	Tet., 1969, 1323.	
2101-029		

GRIFOLIN	$C_{22}H_{32}O_2$	252 OI 43°C
	Tet., 1963, 2079.	
2101-025		

LL-Z1272-γ	$C_{23}H_{29}O_4Cl$	404 -31° 173°C
R = H	Tet., 1969, 1323.	
2101-030		

LL-Z1272-Zeta	$C_{25}H_{31}O_6Cl$	462
		-15^0m
		157^0C
R = OAc		
	Tet., 1969, 1323.	
2101-031		

QUADRILINEATIN	$C_{10}H_{10}O_4$	194
		OI
		172^0C
	Biochem. J., 1957 (67) 155.	
2102-002		

EVERNINOCIN	$C_{15}H_{18}O_7Cl_2$	380
R = H	Chem. Comm., 1971, 746.	
2101-032		

SALICYLIC ACID, 6-CHO-	$C_8H_6O_4$	166
		OI
	Biochim. Biophys. Acta, 1958 (28) 21.	
2102-003		

EVERNINONITROSE	$C_{23}H_{30}NO_{11}Cl_2$	566
		-64^0c
		amorph.
R = Evernitrose		
	Chem. Comm., 1971, 746.	
2101-033		

PHTHALIC ACID, 3-OH-	$C_8H_6O_5$	182
		OI
		166^0C
	Experientia, 1958 (14) 38.	
2102-004		

ZINNIOL	$C_{15}H_{22}O_4$	266
		OI
	Can. J. C., 1968, 767.	
2101-034		

HEMIPIC ACID	$C_{10}H_{10}O_6$	226
		OI
		160^0C
	Helv., 1945 (28) 722.	
2102-005		

FLAVIPIN	$C_9H_8O_5$	196
		OI
		233^0C
	Biochem. J., 1956 (63) 395.	
2102-001		

HEMIPIC ACID, meta-	$C_{10}H_{10}O_6$	226
		OI
		175^0C
	Helv., 1945 (28) 722.	
2102-006		

PHTHALIC ACID, 3,5-di-OH- $C_8H_6O_6$ 198
OI
188^0C

HO-COOH / COOH / OH

JCS, 1942, 368.

2102-007

PHTHALIDE, 5,7-DI-OMe- $C_{10}H_{10}O_4$ 194
OI

R = Me

Arch. Pharm., 1971
(304) 228.

2102-012

PHTHALIC ACID, 4,5-di-OH-3-OMe- $C_9H_8O_7$ 228
OI

HO-COOH / HO- / COOH / OMe

Chem. Abstr., 1955,
14888.

2102-008

PHTHALIDE, 5-OMe-7-glucosyloxy- $C_{15}H_{18}O_9$ 342
-49^0p
194^0C

R = Glucosyl

Coll. Czech. Chem.
Comm., 1959 (24) 3938.

2102-013

PHTHALIC ACID, 4,6-DI-OH-3-OMe- $C_9H_8O_7$ 228
OI
193^0C

OH / COOH / HO- / COOH / OMe

J. Sci. Ind. Res.,
1954 (13B) 842.

2102-009

NIDULOL $C_{10}H_{10}O_4$ 194
OI
234^0C

HO- / Me- / OMe

Chem. Abstr., 1968
(69) 106158.

2102-014

MECONIN $C_{10}H_{10}O_4$ 194
OI
102^0C

MeO- / OMe

Helv., 1945 (28) 722.

2102-010

PHTHALIDE, 7-OH-4,6-di-Me- $C_{10}H_{10}O_3$ 178
OI
157^0C

Me / Me- / OH

Biochem. J., 1952, 648.

2102-015

PHTHALIDE, 7-OH-5-OMe- $C_9H_8O_4$ 180
OI
187^0C

R = H

MeO- / OH

Arch. Pharm., 1971
(304) 228.

2102-011

PHTHALIDE, 6-Me-4,5,7-tri-OH- $C_9H_8O_5$ 196
OI
180^0C

Me- / OH / HO- / Me

Tet. Lett., 1969, 4675.

2102-016

BENZOFURAN, 1, 3-dihydro-5-Me- $C_9H_{10}O_4$ 182
-4, 6, 7-tri-OH-iso-

OI

123^{0}C

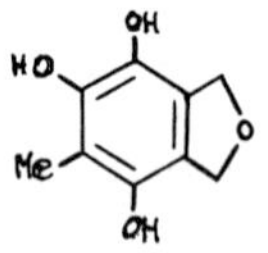

Tet. Lett., 1969, 4675.

2102-017

GLADIOLIC ACID, dihydro- $C_{11}H_{12}O_5$ 224

OI

135^{0}C

R = H

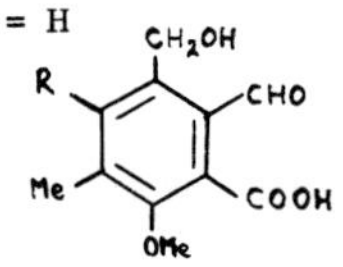

Biochem. J., 1952
(50) 635.

2102-022

MYCOPHENOLIC LACTONE $C_{17}H_{20}O_7$ 336

219^{0}C

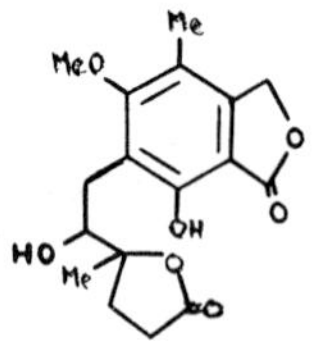

Tet. Lett., 1966, 5107.

2102-018

CYCLOPOLIC ACID $C_{11}H_{12}O_6$ 240

OI

147^{0}C

R = OH

Biochem. J., 1952
(50) 610.

2102-023

MYCOCHROMENIC ACID $C_{17}H_{18}O_6$ 318

164^{0}C

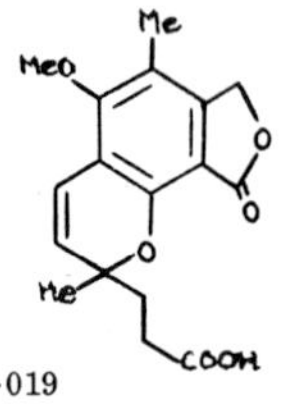

Tet. Lett., 1966, 5107.

2102-019

GLADIOLIC ACID $C_{11}H_{10}O_5$ 222

OI

159^{0}C

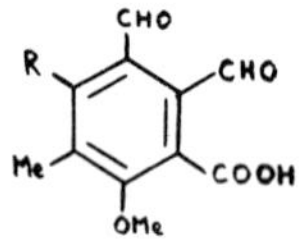

Biochem. J., 1952
(50) 648.

R = H

2102-024

MYCOPHENOLIC ACID $C_{17}H_{20}O_6$ 320

OI

141^{0}C

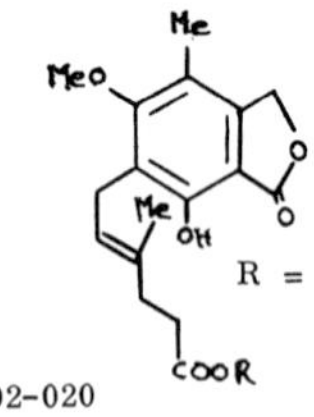

Biochem. J., 1952 (50)
630.

R = H

2102-020

CYCLOPALDIC ACID $C_{11}H_{10}O_6$ 238

OI

224^{0}C

R = OH

JCS, 1960, 866.

2102-025

MYCOPHENOLIC ACID
ET ESTER $C_{19}H_{24}O_6$ 348

OI

90^{0}C

R = Et

Tet. Lett., 1966, 5107.

2102-021

PHTHALIDE, 4-OH- $C_8H_6O_3$ 150

OI

255^{0}C

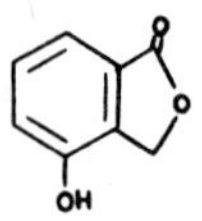

Nature, 1966 (210) 1261.

2102-026

PHTHALIC ACID	$C_8H_6O_4$	166
		OI
		231^0C
R = H		
		Phytochem., 1969, 1309.
2102-027		

ASPERUGIN	$C_{24}H_{32}O_5$	400
		OI
R = Me		
		JCS, 1965, 4672.
2102-032		

PHTHALIC ACID ME ESTER	$C_9H_8O_4$	180
		OI
		85^0C
R = Me		
		Phytochem., 1969, 1309.
2102-028		

CURVULINIC ACID	$C_{10}H_{10}O_5$	210
		OI
		218^0C
		Tet., 1967, 3801.
R = H		
2103-001		

FOMECIN-A	$C_8H_8O_5$	184
		OI
		160^0C
		Can. J. C., 1964 (42) 1595.
2102-029		

CURVULIN	$C_{12}H_{14}O_5$	238
		OI
		145^0C
R = Et		
		Tet., 1967, 3801.
2103-002		

FOMECIN-B	$C_8H_6O_5$	182
		OI
		230^0C
		Can. J. C., 1964 (42) 1595.
2102-030		

CURVULIC ACID	$C_{11}H_{12}O_6$	240
		OI
		154^0C
		JCS, 1967, 117.
R = H		
2103-003		

ASPERUGIN-B	$C_{23}H_{30}O_5$	386
		OI
		Phytochem., 1967, 1157.
R = H		
2102-031		

CURVIN	$C_{13}H_{16}O_6$	268
		OI
		123^0C
R = Et		
		Tet., 1965, 1411.
2103-004		

FUSCIN, dihydro- $C_{15}H_{18}O_5$ 278

206^{0}C

JCS, 1956, 1028.

2103-005

CITRINONE, des-COOH-dihydro- $C_{12}H_{14}O_4$ 222

JCS, 1968, 85.

2104-004

FUSCIN $C_{15}H_{16}O_6$ 280
OI
230^{0}C

Biochem. J., 1948
(43) 528.

2103-006

$C_{13}H_{14}O_4$ 234
+33^0
191^{0}C

JCS, 1968, 85.

2104-005

RAMULOSIN $C_{10}H_{14}O_3$ 182
+18^{0}e
120^{0}C

Biochem. J., 1964
(93) 92.

2104-001

MELLEIN, OMe- $C_{11}H_{12}O_4$ 208
-51^{0}m
76^{0}C

JOC, 1968, 1577.

2104-006

RAMULOSIN, 6-OH- $C_{10}H_{14}O_4$ 198
+92^{0}m
133^{0}C

Tet. Lett., 1970, 2377.

2104-002

$C_{11}H_{11}O_4Cl$ 242
-71^{0}m
170^{0}C

JOC, 1968, 1577.

2104-007

CANESCIN $C_{15}H_{14}O_7$ 306
+18^0
201^{0}C

Tet. Lett., 1969, 1518.

2104-003

$C_{11}H_{11}O_4Cl$ 242
-68^{0}m
124^{0}C

JCS, 1969, 2187.

2104-008

	$C_{11}H_{10}O_4Cl_2$	276
		-142^0m
		225^0C
		JOC, 1968, 1577.
2104-009		

MELLEIN, (-)-	$C_{10}H_{10}O_3$	178
		-128^0c
		58^0C
		JCS, 1955, 2871.
2104-014		

ISOCOUMARIN, 8-OH-6-OMe-3-Me-	$C_{11}H_{10}O_4$	206
		OI
		129^0C
		Tet. Lett., 1971, 1337.
2104-010		

ISOCOUMARIN, 8-OH-3-Me-	$C_{10}H_8O_3$	176
		OI
		99^0C
		Arkiv. Kemi, 1960 (15) 131.
2104-015		

ISOCOUMARIN, 3,8-di-OH-3-Me- -dihydro-	$C_{10}H_{10}O_4$	194
		-40^0
		121^0C
		Agr. Biol. Chem., 1970 (34) 1296.
2104-011		

deleted

2104-016

ISOCOUMARIN, 4,8-di-OH-3-Me- -dihydro-	$C_{10}H_{10}O_4$	194
(stereochemistry ?) Cf 013		0^0
		109^0C
		Agr. Biol. Chem., 1970 (34) 1296.
2104-012		

MELLEIN, 5-Me-	$C_{11}H_{12}O_3$	192
		-105^0c
		126^0C
		Tet. Lett., 1966, 3723.
2104-017		

MELLEIN, cis-4-OH-	$C_{10}H_{10}O_4$	194
		112^0C
		JCS, 1971, 1624.
2104-013		

OOSPOLACTONE	$C_{11}H_{10}O_3$	190
		OI
		129^0C
		Agr. Biol. Chem., 1966 (30) 193.
2104-018		

KIGELIN, 6-de-Me-

$C_{11}H_{12}O_5$ 224

R = H

Phytochem., 1971 (10)
1603.

2104-019

CITRININ

$C_{13}H_{14}O_5$ 250
-35^0e
179^0C

JOC, 1964, 766.

2104-024

KIGELIN

$C_{12}H_{14}O_5$ 238
-80^0
144^0C

R = Me

Phytochem., 1971 (10)
1603.

2104-020

CITRINONE, dihydro-

$C_{13}H_{14}O_6$ 266
$+104^0$c
140^0C

JCS, 1968, 85.

2104-025

ISOCOUMARIN, 6,7,8-triOH-3-Me- $C_{10}H_8O_5$ 208
OI
234^0C

R = R' = H

Tet. Lett., 1971, 1337.

2104-021

ISOCOUMARIN, 7-COOH-3, 4-
-dihydro-8-OH-3-Me-

$C_{11}H_{10}O_5$ 222

R = OH; R' = H

Tet. Lett., 1971, 4033.

2104-026

RETICULOL

$C_{11}H_{10}O_5$ 222
OI
193^0C

R = Me; R' = H

Tet. Lett., 1971, 1337.
Experimentia, 1964 (20)
258.

2104-022

ACHRATOXIN-B

$C_{20}H_{19}NO_6$ 369
-35^0e
221^0C

R = (N) Phenylalanine
R' = H

JCS, 1965, 7083.

2104-027

ISOCOUMARIN, 8-OH-6, 7-di-OMe-
-3-Me-

$C_{12}H_{12}O_5$ 236
OI
199^0C

R = R' = Me

Agr. Biol. Chem., 1971
(35) 363.
Tet. Lett., 1971, 1337.

2104-023

OCHRATOXIN-A

$C_{20}H_{18}NO_6Cl$ 403
-118^0c
169^0C

R = (N) Phenylalanine
R' = Cl

JCS, 1965, 7083.

2104-028

OCHRATOXIN- C	$C_{21}H_{20}NO_6Cl$	417

R = (N) Phenylalanine Ethyl Ester
R' = Cl

JCS, 1965, 7083.

2104-029

SCLEROTININ-A	$C_{12}H_{14}O_5$	238

R = Me

Tet. Lett., 1969, 631.

2104-034

OCHRATOXIN-A, 4-OH-	$C_{20}H_{18}NO_7Cl$	419
		217^0C

R = (N) phenylalanine
R' = Cl
4α - H

Tet. Lett., 1971, 4033.

2104-030

SCLEROTININ-B	$C_{11}H_{12}O_5$	224

R = H

Tet. Lett., 1969, 631.

2104-035

OOSPOGLYCOL	$C_{11}H_{10}O_5$	222
		$+62^0p$
		116^0C

Agr. Biol. Chem., 1963
(27) 822.

2104-031

ORSELLINIC ACID, α-Ac-	$C_{10}H_{10}O_5$	210
		OI
		155^0C

Biochem. J., 1933 (27)
634.

2105-001

OOSPONOL	$C_{11}H_8O_5$	220
		OI
		176^0C

Agr. Biol. Chem., 1963
(27) 817.

2104-032

ORSELLINIC ACID, α-Ac-α-OH-	$C_{10}H_{10}O_6$	226
		203^0C

Biochem. J., 1932 (26)
1441.

2105-002

ISOCOUMARIN, 4-Ac-6, 8-di-OH- -5-Me-	$C_{12}H_{10}O_5$	234
		179^0C

JCS, 1966, 126.

2104-033

ORSELLINIC ACID, α-Ac-α-oxo-	$C_{10}H_8O_6$	224
		OI
		130^0C

Biochem. J., 1933 (27)
634.

2105-003

USTIC ACID 2105-004	$C_{11}H_{12}O_7$ 256 OI 169^0C Biochem. J., 1951 (48) 53.	CANNABIGEROLIC ACID 2106-002 $C_{22}H_{28}O_4$ 356 OI Tet., 1965, 1223.

USTIC ACID $C_{11}H_{12}O_7$ 256
OI
169^0C

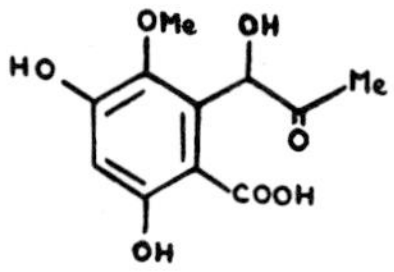

Biochem. J., 1951 (48) 53.

2105-004

CANNABIGEROLIC ACID $C_{22}H_{28}O_4$ 356
OI

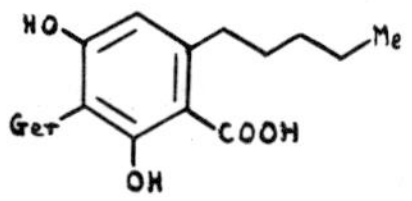

Tet., 1965, 1223.

2106-002

CANNABIDIVARIN $C_{19}H_{26}O_2$ 286
-139^0c

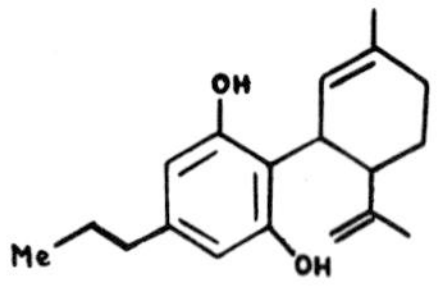

Tet. Lett., 1969, 145.

2105-005

CANNABICHROMENIC ACID $C_{22}H_{30}O_4$ 358
$+5^0c$

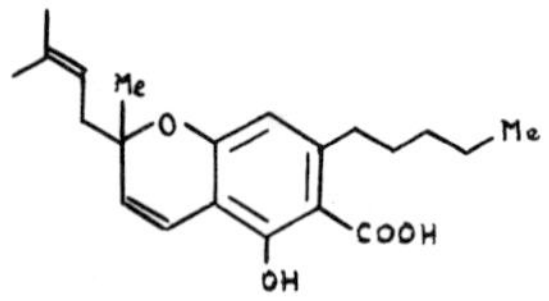

Chem. Pharm. Bull., 1968 (6) 1157.

2106-003

PHENOL-A $C_{11}H_{16}O_3$ 196
178^0C

R = H

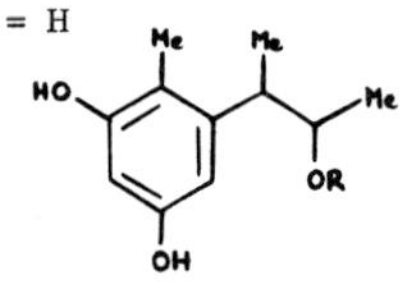

JCS, 1968, 85.

2105-006

CANNABIDIOLIC ACID $C_{22}H_{30}O_4$ 358

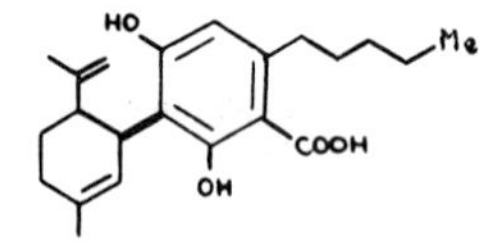

Tet., 1965, 1223.

2106-004

PHENOL-A ALDEHYDE $C_{12}H_{16}O_4$ 224
132^0C

R = CHO

JCS, 1968, 85.

2105-007

CANNABIELSOIC ACID A $C_{22}H_{30}O_6$ 374
Oil

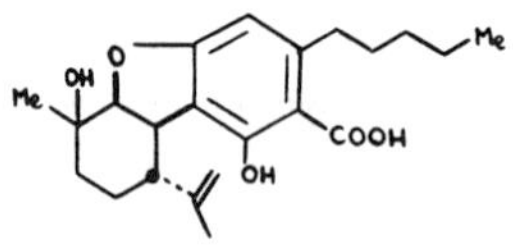

Chem. Comm., 1970, 273.

2106-005

OLIVETOL-COOH $C_{12}H_{16}O_4$ 224
OI
142^0C

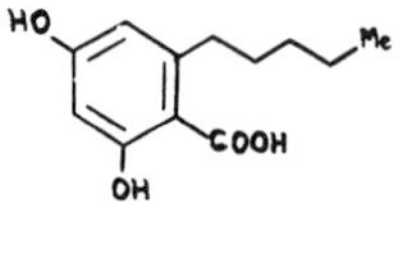

Ber., 1937 (70) 206.

2106-001

CANNABINOLIC ACID,
-tetrahydro- $C_{22}H_{30}O_4$ 358
-222^0c

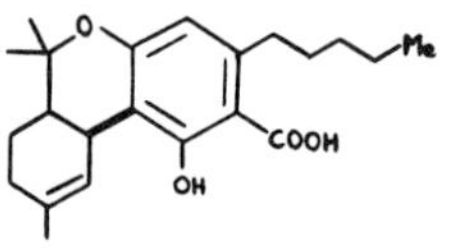

Chem. Pharm. Bull., 1967, 1075.

2106-006

CANNABINOLIC ACID	$C_{22}H_{26}O_4$ 354 OI m. e. 86^{0}C	CANNABINOL, tetrahydro-Δ-1-
	Tet. , 1965, 1223.	
2106-007		2106-012

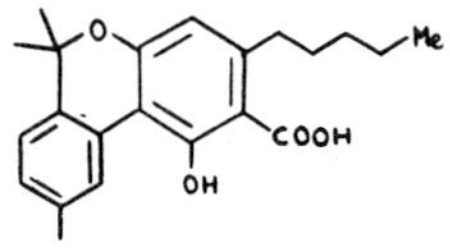

CANNABINOL, tetrahydro-Δ-1- $C_{21}H_{30}O_2$ 314

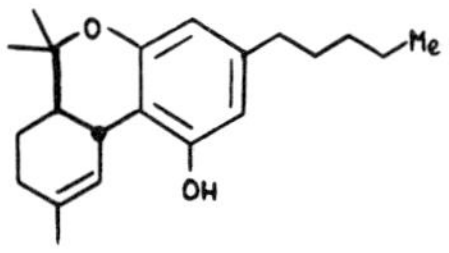

Tet. Lett. , 1967, 119.

CANNABINOLIC ACID B, -tetrahydro- $C_{22}H_{30}O_4$ 358 -202^0 185^{0}C

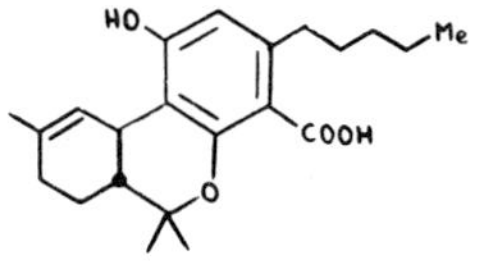

Tet. Lett. , 1969, 2339.

2106-008

CANNABINOL, tetrahydro-Δ-6- $C_{21}H_{30}O_2$ 314

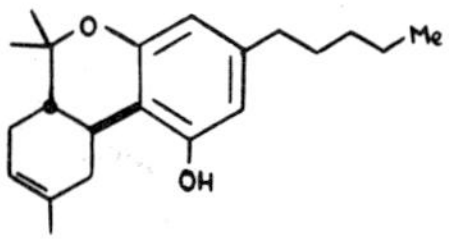

JACS, 1966, 1832.

2106-013

CANNABIGEROL ME ETHER $C_{22}H_{34}O_2$ 330

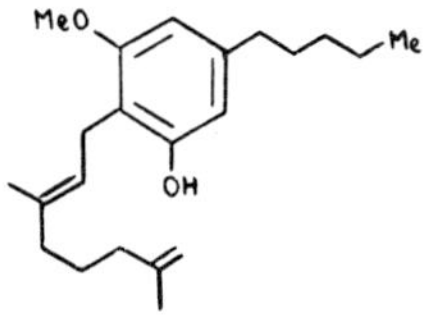

Chem. Pharm. Bull. , 1968, 1164.

2106-009

CANNABINOL $C_{21}H_{26}O_2$ 310 OI 77^{0}C

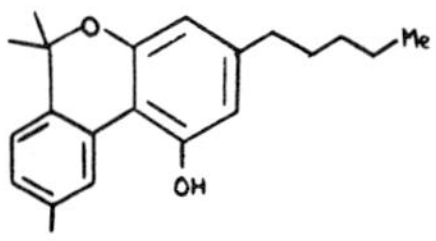

JACS, 1944, 1868.

2106-014

CANNABIDOL $C_{21}H_{20}O_2$ 314 -125^0 66^{0}C

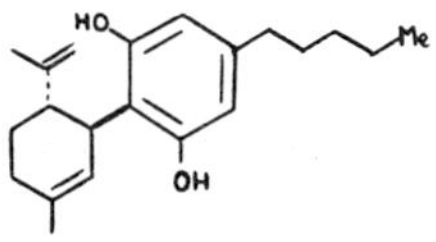

JACS, 1940, 2566.

2106-010

CANNABICYCLOL $C_{21}H_{30}O_2$ 314 146^{0}C

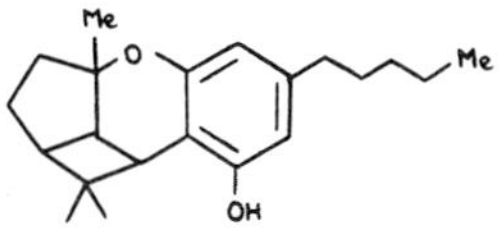

Chem. Comm. , 1968, 894.

2106-015

CANNABICHROMENE $C_{21}H_{30}O_2$ 314 -9^{0}c 145^{0}C

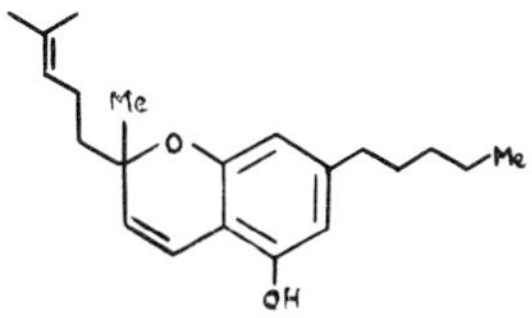

Chem. Comm. , 1966, 20. Tet. Lett. , 1966, 1477.

2106-011

CANNABIDIOL-COOH TETRA- HYDRO-CANNABITRIOL ESTER $C_{43}H_{56}O_7$ 684

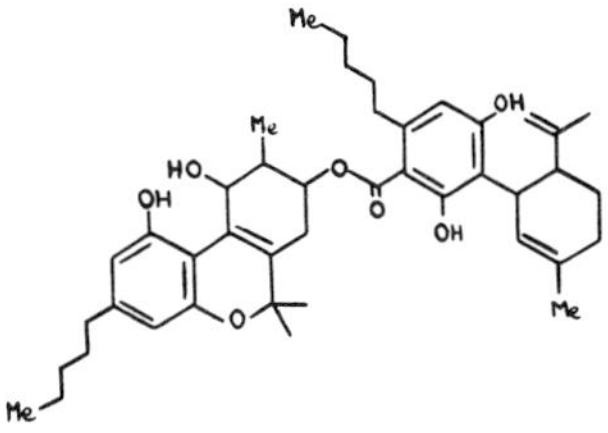

Tet. , 1968, 5379.

2106-016

DIAPORTHIN	$C_{13}H_{14}O_5$ 250 +54°c 84°C	PHTHALIDE, n-Bu-	$C_{12}H_{14}O_2$ 190 Oil

DIAPORTHIN $C_{13}H_{14}O_5$ 250 +54°c 84°C

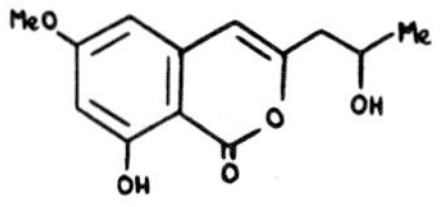

Helv., 1966, 1283.

2106-017

PHTHALIDE, n-Bu- $C_{12}H_{14}O_2$ 190 Oil

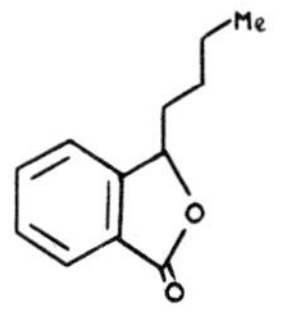

Helv., 1943 (26) 1281.

2106-022

ASCOCHITINE $C_{15}H_{16}O_5$ 276

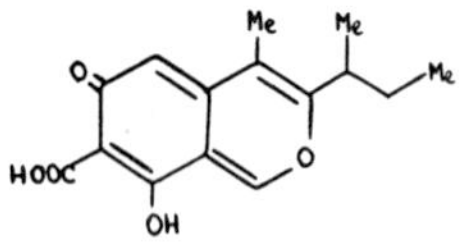

Chem. Comm., 1966, 620.

2106-018

LIGUSTICUM LACTONE $C_{12}H_{12}O_2$ 188 OI Oil

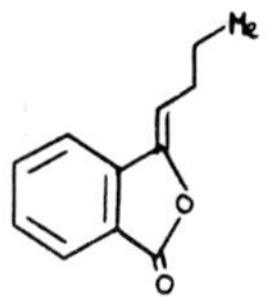

Chem. Pharm. Bull., 1960, 243.

2106-023

CNILILIDE $C_{12}H_{18}O_2$ 194

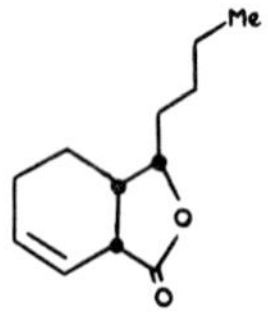

Tet., 1965, 1432.

2106-019

LIGUSTILIDE $C_{12}H_{14}O_2$ 190 OI

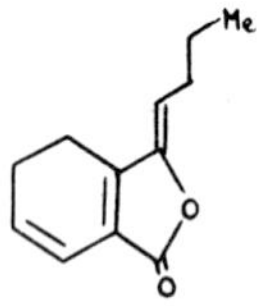

Tet., 1964 (20) 1971.

2106-024

CNILILIDE, neo- $C_{12}H_{18}O_2$ 194

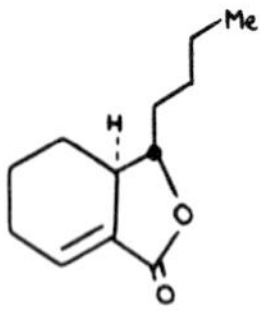

Tet., 1965, 1432.

2106-020

CNIDIUM LACTONE $C_{12}H_{18}O_2$ 194 -72°c Oil

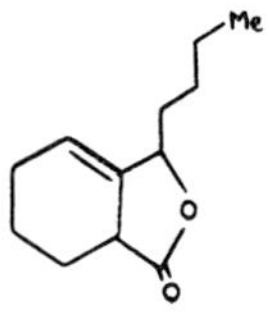

Tet., 1964 (20) 1976.

2106-025

SEDANONIC ACID ANHYDRIDE $C_{12}H_{16}O_2$ 192

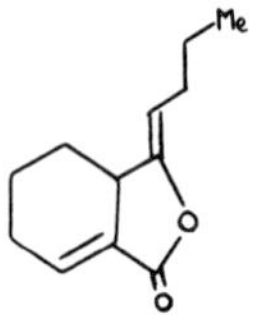

Helv., 1943 (26) 1281.

2106-021

SEDANONIC ACID $C_{12}H_{18}O_3$ 210

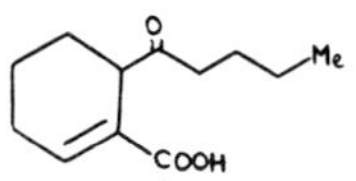

Tet., 1964 (20) 1976.

2106-026

MICONIDIN	$C_{12}H_{18}O_3$	210 OI
		Gazz., 1971, 41.
2106-027		

MONOCERIN	$C_{16}H_{20}O_6$	308 $+53^0$m 65^0C
		JCS, 1970, 2598.
	R = H	
2107-002		

MITORUBRIN	$C_{21}H_{18}O_7$	382 -405^0d 218^0C
	R = Me	Chem. Comm., 1970. 101.
2106-028		

MONOCERIN, OH-	$C_{16}H_{20}O_7$	324 104^0C
	R = OH	JCS, 1970, 2598.
2107-003		

MITORUBRINOL	$C_{21}H_{18}O_8$	398 -375^0d 220^0C
	R = CH_2OH	JACS, 1965. 3484.
2106-029		

AUROGLAUCIN	$C_{19}H_{22}O_3$	298 OI 153^0C
		JCS, 1965, 1231.
2107-004		

MITORUBRINIC ACID	$C_{21}H_{16}O_9$	412
	R = COOH	Gen. Microbiol. 1967 (15) 93.
2106-030		

FLAVOGLAUCIN	$C_{19}H_{28}O_3$	196 OI 105^0C
		Chem. Pharm. Bull., 1953 (1) 162.
2107-005		

MONOCERONE	$C_{16}H_{18}O_7$	322 113^0C
		JCS, 1970, 2598.
2107-001		

PYRICULOL	$C_{14}H_{16}O_4$	248 -54^0c 96^0C
		Tet. Lett., 1969, 3977.
2107-006		

FREQUENTIN $C_{14}H_{20}O_4$ 252 +68[0]c 134[0]C

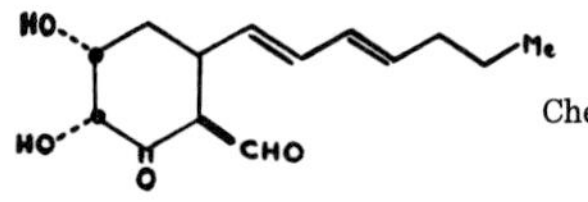

Chem. Ind., 1963, 412.

2107-007

RUBROROTIORIN $C_{23}H_{23}O_5Cl$ 414 -368[0]c 172[0]C

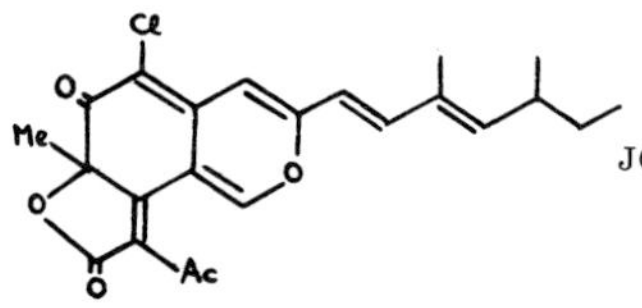

JCS, 1971, 3575.

2108-003

PALITANTIN $C_{14}H_{22}O_4$ 254 +4[0]c 163[0]C

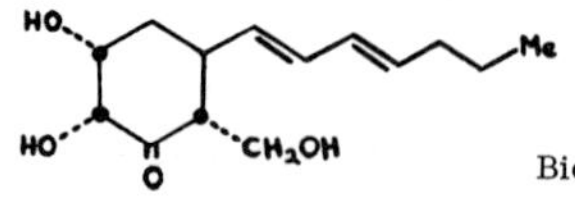

Biochem. J., 1960, 91.

2107-008

PERSOONOL $C_{17}H_{26}O_2$ 262 OI Oil

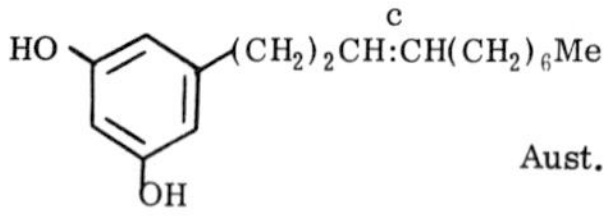

HO-〇-(CH$_2$)$_2$CH:CH(CH$_2$)$_6$Me

Aust. J. C., 1971, 1925.

2109-001

PULVILLORIC ACID $C_{15}H_{18}O_5$ 278 +72[0] 74[0]C

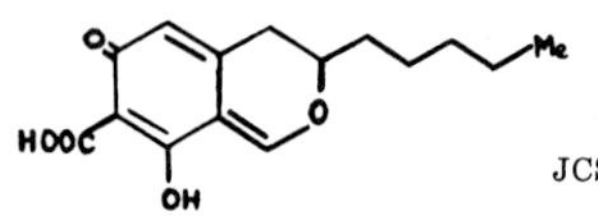

JCS, 1966, 1608.

2107-009

MONORDEN $C_{18}H_{17}O_6Cl$ 364 +203[0] 203[0]C

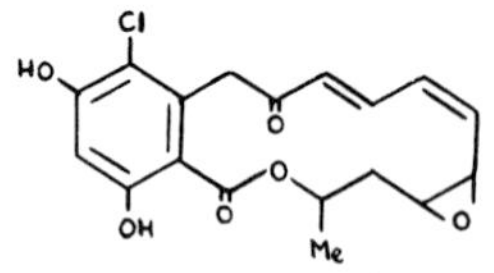

Tet. Lett., 1964, 869.

2109-002

SCLEROTIORIN $C_{21}H_{23}O_5Cl$ 390 +500[0] 206[0]C

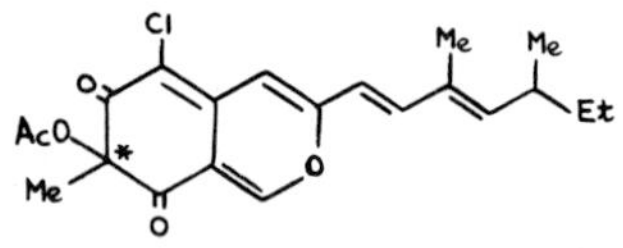

JCS, 1959, 3004.
JCS, 1964. 16.

2108-001

GREVILLOL $C_{19}H_{32}O_2$ 292 OI 82[0]C

HO-〇-$C_{13}H_{27}$

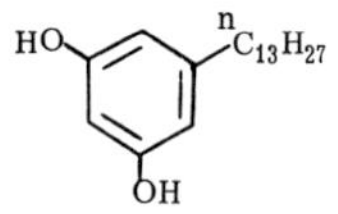

Aust. J. C., 1965, 2015.

2110-001

SCLEROTIORIN, 7-epi- $C_{21}H_{23}O_5Cl$ 390 -389[0] 206[0]C

*epimer

Chem. Ind., 1963, 1625.

2108-002

GINKGOLIC ACID, hydro- $C_{22}H_{36}O_3$ 348 OI 93[0]C

HO-〇-$C_{15}H_{31}$ COOH

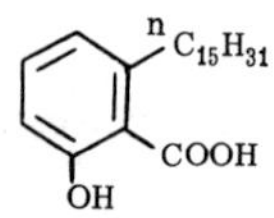

JOC, 1949, 1039.
Nature, 1954 (174) 604.

2111-001

260

GINKGOLIC ACID $\quad\quad$ $C_{22}H_{34}O_3$ $\quad$ 346
OI

structure: benzene ring with $(CH_2)_7CH{:}CH(CH_2)_5Me$, COOH, OH

Nature, 1954 (174) 604.

2111-002

RESORCINOL, 5-pentadec-10-
-enyl- $\quad\quad$ $C_{21}H_{34}O_2$ $\quad$ 318
OI

structure: HO-benzene ring with $(CH_2)_9CH{:}CH(CH_2)_3Me$, OH

Aust. J. C., 1968, 2979.

2111-007

ANACARDIC ACID-III $\quad$ $C_{22}H_{32}O_3$ $\quad$ 344
OI
26^0C

structure: benzene ring with $(CH_2)_7CH{:}CHCH_2CH{:}CH(CH_2)_2Me$, COOH, OH

Nature, 1954 (174) 604.

2111-003

CARDOL-I $\quad\quad$ $C_{21}H_{32}O_2$ $\quad$ 316
OI
Oil

structure: HO-benzene ring with $(CH_2)_7(CH_2CH{:}CH)_2Et$, OH

Nature, 1953 (171) 841.

2111-008

ANACARDIC ACID-IV $\quad$ $C_{22}H_{30}O_3$ $\quad$ 342
OI

structure: benzene ring with $(CH_2)_7CH{:}CHCH_2CH{:}CHCH_2CH{:}CH_2$, COOH, OH

Nature, 1954 (174) 604.

2111-004

CARDOL-II $\quad\quad$ $C_{21}H_{30}O_2$ $\quad$ 314
OI
Oil

structure: HO-benzene ring with $(CH_2)_7(CH{:}CHCH_2)_2CH{:}CH_2$, OH

Nature, 1953 (171) 841.

2111-009

RESORCINOL, 5-n-pentadecyl- $\quad$ $C_{21}H_{36}O_2$ $\quad$ 320
OI

structure: HO-benzene ring with $^nC_{15}H_{31}$, OH

Aust. J. C., 1968, 2979.

2111-005

LEPROSOL-I, β- $\quad\quad$ $C_{23}H_{40}O_2$ $\quad$ 348
OI

structure: HO-benzene ring with Me, $C_{15}H_{31}$, Me, OMe

JCS, 1969, 61.

2111-010

BILOBOL $\quad\quad$ $C_{21}H_{34}O_2$ $\quad$ 318
OI
36^0C

structure: HO-benzene ring with $(CH_2)_7CH{:}CH(CH_2)_5Me$, OH

Aust. J. C., 1968, 2979.

2111-006

GINKGOL, hydro- $\quad\quad$ $C_{21}H_{36}O$ $\quad$ 304
OI
54^0C

structure: benzene ring with $^nC_{15}H_{31}$, OH

Helv., 1947 (30) 675.

2111-011

GINKGOL $C_{21}H_{34}O$ 302 OI Oil

(benzene ring with $(CH_2)_7CH{:}CH(CH_2)_5Me$ and OH)

JOC, 1949, 849.

2111-012

URUSHIOL-II $C_{21}H_{34}O_2$ 318 OI Oil

(benzene ring with $(CH_2)_7CH{:}CH(CH_2)_5Me$ and OH, OH)

JACS, 1954, 5070.

2111-017

ANACARDOL-III $C_{21}H_{32}O$ 300 OI

(benzene ring with $(CH_2)_7(CH{:}CHCH_2)_2Et$ and OH)

Nature, 1953 (171) 841.

2111-013

URUSHIOL-III $C_{21}H_{32}O_2$ 316 OI Oil

(benzene ring with $(CH_2)_7(CH{:}CHCH_2)_2Et$ and OH, OH)

JACS, 1954, 5070.

2111-018

ANACARDOL-IV $C_{21}H_{30}O$ 298 OI

(benzene ring with $(CH_2)_7(CH{:}CHCH_2)_2CH{:}CH_2$ and OH)

Nature, 1953 (171) 841.

2111-014

URUSHIOL-IV $C_{21}H_{30}O_2$ 314 OI Oil

(benzene ring with $(CH_2)_7CH{:}CHCH_2(CH{:}CH)_2Me$ and OH, OH)

JACS, 1954, 5070.

2111-019

URUSHIOL-I $C_{21}H_{36}O_2$ 320 OI 59°C

(benzene ring with $n\text{-}C_{15}H_{31}$ and OH, OH)

JACS, 1954, 5070.

2111-015

URUSHIOL-V $C_{21}H_{30}O_2$ 314 OI Oil

(benzene ring with $(CH_2)_7(CH{:}CHCH_2)_2CH{:}CH_2$ and OH, OH)

JACS, 1954, 5070.

2111-020

RENGHOL $C_{21}H_{34}O_2$ 318 OI 15°C

(benzene ring with $(CH_2)_9CH{:}CH(CH_2)_3Me$ and OH, OH)

Rec. Trav. Chim., 1938 (57) 225.

2111-016

PELANDJUAIC ACID $C_{24}H_{36}O_3$ 372 OI 25°C

(benzene ring with OH, COOH, and $(CH_2)_7(CH{:}CH.CH_2)_2Bu$)

Rec. Trav. Chim., 1941 (60) 678.

2112-001

LACCOL	$C_{23}H_{36}O_2$	344
		OI
		23^0C

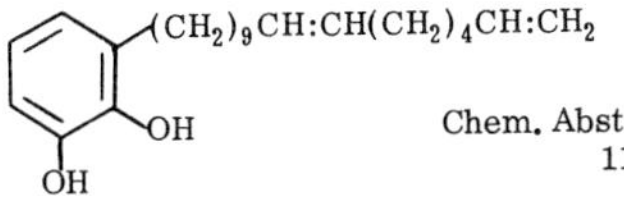

Chem. Abstr., 1953, 11363.

2112-002

ACETOPHENONE, 2-OH-6-Me-	$C_9H_{10}O_2$	150
		OI
		Oil

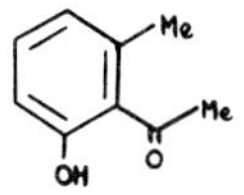

Aust. J. C., 1969, 775.

2199-002

GLUTARENGOL	$C_{23}H_{38}O_2$	346

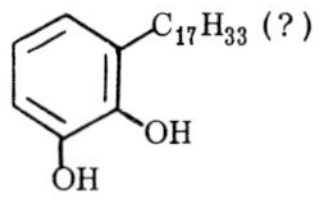

Rec. Trav. Chim., 1941 (60) 656.

2112-003

LEPROSOL-II, β-	$C_{26}H_{46}O_2$	390
		OI

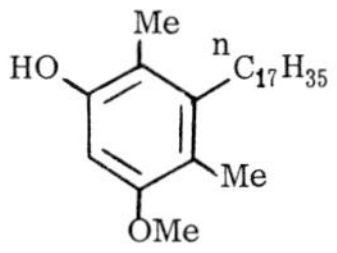

JCS, 1969, 61.

2112-004

PIROLAGENIN	$C_{17}H_{24}O_2$	260
		OI
		Oil

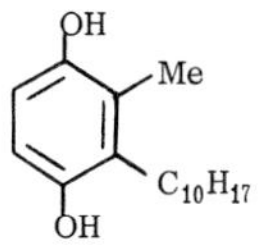

Chem. Abstr., 1953, 1638.

2199-004

RESORCINOL, 5-heneicosyl-	$C_{27}H_{48}O_2$	404
		OI
		100^0C

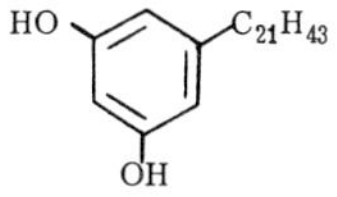

JOC, 1964, 435.

2114-001

FUSAMARIN	$C_{18}H_{24}O_4$	304

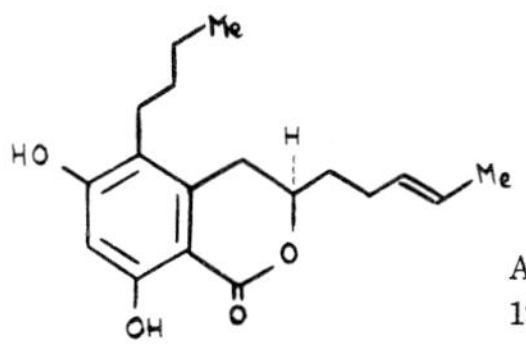

Agr. Biol. Chem., 1970 (34) 760.

2199-005

CURVULOL	$C_{10}H_{12}O_4$	196
		$+19^0e$
		204^0C

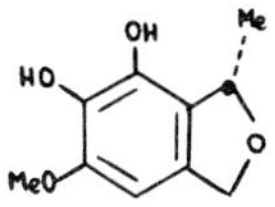

Tet., 1967, 3801.

2199-001

MONOCEROLIDE	$C_{13}H_{12}O_7$	280
		199^0C

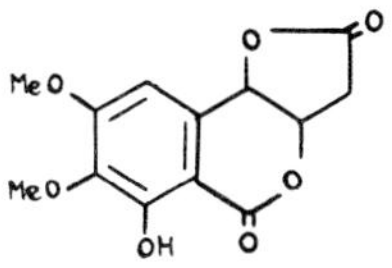

JCS, 1970, 2598.

2199-006

ALOESIN $C_{19}H_{22}O_9$ 394
+58^{0}e
144^{0}C

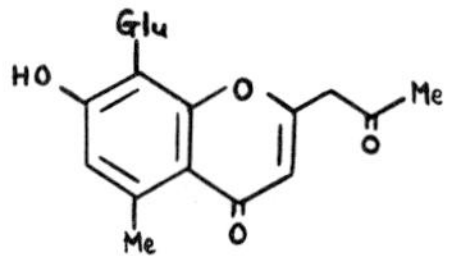

JCS, 1970, 2581.
Planta Med., 1971 (19)
322.

2199-007

PANOSIALIN-III $C_{22}H_{38}O_8S_2$ 494

R = iso-hexadecyl

J. Ant., 1971 (24) 870.

2199-012

CITROMYCETIN $C_{14}H_{10}O_7$ 322
OI
284^{0}C

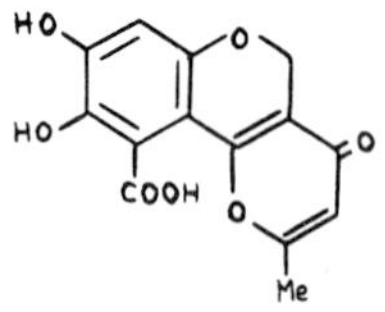

JCS, 1951, 2013.

2199-008

PHTHALIDE, 3-isobutylidenyl- $C_{12}H_{12}O_2$ 188
OI
Oil

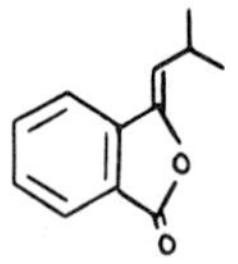

Ann., 1908 (359) 213.

2199-013

FULVIC ACID $C_{14}H_{12}O_8$ 308
246^{0}C

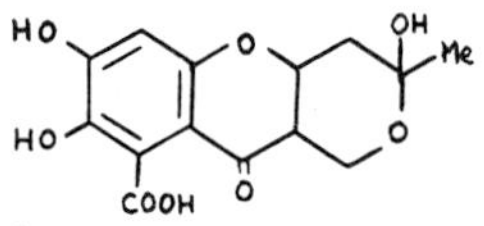

Nature, 1957 (179) 366.

2199-009

PHTHALIDE, 3-isovalylidenyl- $C_{13}H_{14}O_2$ 202
OI
Oil

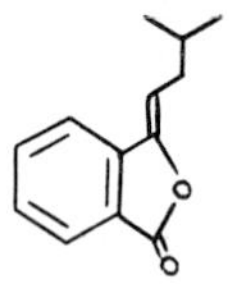

JOC, 1963, 987.

2199-014

PANOSIALIN-I $C_{21}H_{36}O_8S_2$ 480
R=isopentadecyl

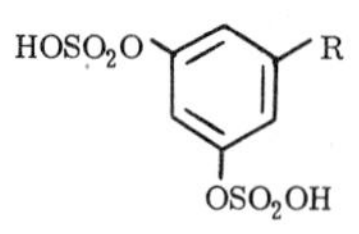

J. Ant., 1971 (24) 870.

2199-010

PHTHALIDE, 3-isovalylidenyl-
-3α, 4-dihydro- $C_{13}H_{16}O_2$ 204
Oil

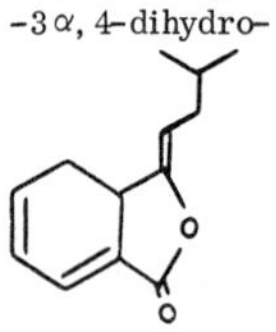

JOC, 1963, 985.

2199-015

PANOSIALIN-II $C_{21}H_{36}O_8S_2$ 480

R = n-pentadecyl

J. Ant., 1971 (24) 870.

2199-011

22 | PHLOROGLUCINOLS

The Claisen-type condensation of a polyketide precursor provides the structural features of the phloroglucinol nucleus:

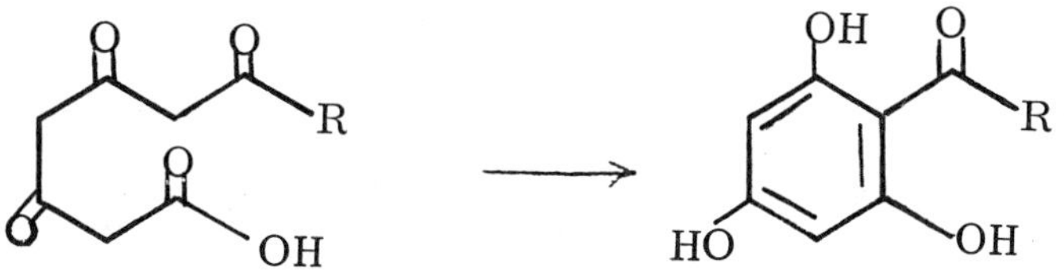

C-alkylation (methylation and isoprenylation) is prevalent among the phloroglucinols owing to the presence of the beta-diketo moiety; in some cases geminal bisalkylation destroys the aromaticity, but for simplicity these are retained in the phloroglucinol class.

Another feature of the phloroglucinol class is the relative abundance of non-acetyl starter units; compounds the biogenesis of which suggests a starter unit other than acetyl have been brought together as one class (2206).

2201	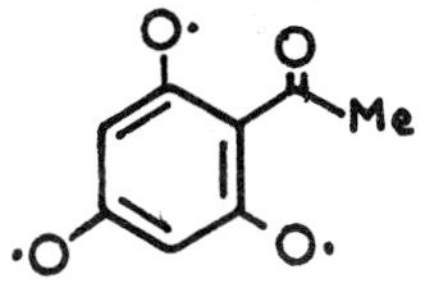 4 "acetate" units	2202	as 2201 C-isoprenylated
2203	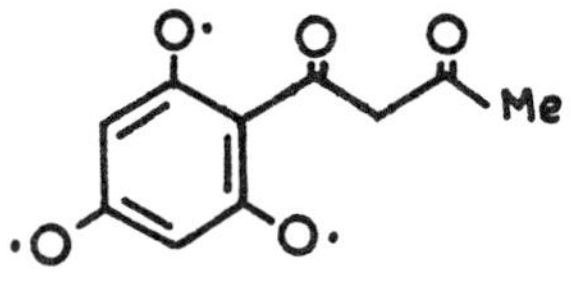 5 "acetate" units	2204	as 2203 C-isoprenylated
2205	6 "acetate" units	2206	Nonacetyl Starters
2207	Dimers, Trimers, etc.	2299	Miscellaneous

ACETOPHENONE, 2-OH-	$C_8H_8O_2$	136 OI Oil
	JCS, 1899, 66.	
2201-001		

CLAVATOL	$C_{10}H_{12}O_3$	180 OI 183^0C
	JCS, 1947, 611.	
2201-006		

PAEONOL	$C_9H_{10}O_3$	166 OI 48^0C
R = Me		
	Chem. Pharm. Bull., 1966, 779.	
2201-002		

ACETOPHENONE, 2,6-di-OH-	$C_8H_8O_3$	152 OI 156^0C
	JCS, 1960, 654.	
2201-007		

POPULNEOL	$C_{15}H_{14}O_3$	242 OI
R = .CH₂Ph		
	Ind. J. C., 1971 (9) 286.	
2201-003		

ACETOPHENONE, 2-OH-5-OMe-	$C_9H_{10}O_3$	166 OI 50^0C
R = H		
	Compt. Rend., 1935 (200) 1990.	
2201-008		

PEONOL, gluco-	$C_{15}H_{20}O_8$	328
R = Glucosyl		
	J. Prakt. Chem., 1942 (160) 33.	
2201-004		

ACETOPHENONE, 2-Glucosyloxy- -5-OMe-	$C_{15}H_{20}O_8$	328 140^0C
R = Glucosyl		
	Bull. Soc. Chim. Biol., 1936 (18) 1405.	
2201-009		

PEONOL PRIMEVEROSIDE	$C_{20}H_{28}O_{12}$	460 142^0C
R = Primeverosyl		
	Compt. Rend., 1970 (270D) 2872.	
2201-005		

	deleted	
2201-010		

XANTHOXYLIN R = H	$C_{10}H_{12}O_4$	196 OI 82^0C J. Am. Pharm. Ass., 1954 (43) 43. 2201-011	ANGUSTIONE	$C_{11}H_{16}O_3$	196 Oil Aust. J. C., 1968, 2825. 2201-016

XANTHOXYLIN

R = H

2201-011

ANGUSTIONE

$C_{11}H_{16}O_3$ 196

Oil

Aust. J. C., 1968, 2825.

2201-016

XANTHOXYLIN, Me-

R = Me

$C_{11}H_{14}O_4$ 210
OI
143^0C

Tet., 1961 (16) 206.

2201-012

ANGUSTIONE, dehydro-

$C_{11}H_{14}O_3$ 194

Oil

Aust. J. C., 1968, 2825.

2201-017

PHLORACETOPHENONE
TRI-ME ETHER

$C_{11}H_{14}O_4$ 210
OI
102^0C

Chem. Zentr., 1933
(II) 3132.

2201-013

REMIROL, pre-

$C_{14}H_{18}O_4$ 250
OI
174^0C

Tet. Lett., 1971, 3179.

2202-001

ACETOPHENONE, 2-OH-3, 4, 6-
-tri-OMe-

$C_{11}H_{14}O_5$ 226
OI
111^0C

Chem. Abstr., 1969
(70) 26401.

2201-014

ACRONYLIN

$C_{14}H_{18}O_4$ 250
OI

Tet. Lett., 1971, 3179.
Chem. Ind., 1971, 654.

2202-002

PHENYLGLYOXYLIC ACID,
-2, 4, 5-tri-OH-

$C_8H_6O_6$ 198
OI
193^0C

JCS, 1950, 3380.

2201-015

EVODIONOL

R = H

$C_{14}H_{16}O_4$ 248
OI
86^0C

JCS, 1950, 2376.

2202-003

EVODIONOL ME ETHER R = Me 2202-004	$C_{15}H_{18}O_4$	262 OI 79^0C Chem. Abstr., 1950, 1230.
ENCECALIN, des-O-Me- R = H 2202-009	$C_{13}H_{14}O_3$	218 OI 77^0C Ber., 1970, 90.

EVODIONOL, allo- R = H 2202-005	$C_{14}H_{16}O_4$	248 OI 71^0C Chem. Abstr., 1950, 1230.
ENCECALIN R = Me 2202-010	$C_{14}H_{16}O_3$	232 OI Phytochem., 1969, 1293.

EVODIONOL ME ETHER, allo- R = Me 2202-006	$C_{15}H_{18}O_4$	262 OI 106^0C JCS, 1950, 2376.
RIPARIOCHROMENE-A R = H 2202-011	$C_{14}H_{16}O_4$	248 OI 88^0C Acta Chem. Scand., 1969, 3605.

EVODIONOL, iso- 2202-007	$C_{14}H_{16}O_4$	248 OI 128^0C Tet. Lett., 1971, 3179.
RIPARIOCHROMENE-A ME ETHER R = Me 2202-012	$C_{15}H_{18}O_4$	262 OI Oil Acta Chem. Scand., 1969, 3605.

EVODIONE 2202-008	$C_{16}H_{20}O_5$	292 OI 57^0C JCS, 1948, 2005.
RIPARIOCHROMENE-B R = Ac 2202-013	$C_{15}H_{16}O_5$	276 OI 145^0C Acta Chem. Scand., 1969, 3605.

RIPARIOCHROMENE-C	$C_{17}H_{20}O_5$	304 OI 110°C
R = i-butyryl		
	Acta Chem. Scand., 1969, 3605.	
2202-014		

EUPARIN, dihydro-	$C_{13}H_{14}O_3$	218 69°C
*dihydro		
	Ber., 1970, 90.	
2202-019		

CHROMENE, 6-Ac-2, 2-di-Me-3-	$C_{13}H_{14}O_2$	202 OI Oil
	Ber., 1970, 90.	
2202-015		

TREMETONE, OH-	$C_{13}H_{14}O_3$	218 (-)
	Tet., 1964, 1419.	
2202-020		

FRANKLINONE	$C_{19}H_{22}O_4$	314 OI 128°C
	Tet., 1961 (16) 206.	
2202-016		

AGERATONE	$C_{15}H_{14}O_4$	274 OI 123°C
	Acta Chem. Scand., 1970, 721.	
2202-021		

REMIROL	$C_{14}H_{16}O_4$	248 +66° 77°C
	Tet. Lett., 1969, 4674.	
2202-017		

AGERATONE, dihydro-	$C_{15}H_{16}O_5$	276 Oil
*dihydro		
	Acta Chem. Scand., 1970, 721.	
2202-022		

EUPARIN	$C_{13}H_{12}O_3$	216 OI 120°C
	JCS, 1939, 925.	
2202-018		

EUPARIN, OMe-	$C_{14}H_{14}O_4$	246 OI 95°C
	Ber., 1970, 90.	
2202-023		

EUPARIN, dihydro–OMe–	$C_{14}H_{16}O_4$	248 Oil	BUTYROPHENONE, 2,6-diOH–	$C_{10}H_{12}O_3$	180 OI 117^0C

*dihydro

Ber., 1970, 90.

JCS, 1960, 654.

2202-024

2203-002

REMIRIDIOL	$C_{14}H_{16}O_5$	264 $+63^0$ 97^0C

Tet. Lett., 1970, 3945.

2202-025

ASPIDINOL	$C_{12}H_{16}O_4$	224 OI 156^0C

Helv., 1920 (3) 392.

2203-003

	$C_{14}H_{16}O_3$	232 68^0C

Ber., 1970, 90.

2202-026

EUGENITOL	$C_{11}H_{10}O_4$	206 OI 291^0C

R = H; R' = Me

Chem. Ind., 1955, 1009.

2203-004

EUPARONE ME ETHER	$C_{13}H_{14}O_2$	202 OI 141^0C

Phytochem., 1969, 1295.

2202-027

EUGENITOL, iso–	$C_{11}H_{10}O_4$	206 OI 148^0C

R = Me; R' = H

Helv., 1950 (33) 1770.

2203-005

EUGENONE	$C_{13}H_{16}O_5$	252 OI 98^0C

Helv., 1948, 748.

2203-001

EUGENIN	$C_{11}H_{10}O_4$	206 OI 120^0C

R = H

Helv., 1948 (31) 1603.

2203-006

270

EUGENITIN	$C_{12}H_{12}O_4$	220 OI 163^0C

R = Me

Helv., 1949 (32) 813.

2203-007

CHROMANONE, 5-OH-2-Me- $C_{10}H_{10}O_3$ 178

31^0C

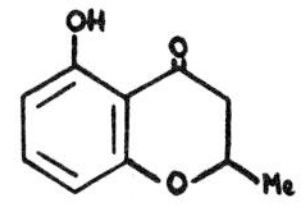

JCS, 1960, 654.

2203-012

	$C_{14}H_{16}O_5$	264 OI 130^0C

R = CH_2OEt

JCS, 1969, 704.

2203-008

CHROMONE, 5-OH-2-Me- $C_{10}H_8O_3$ 176

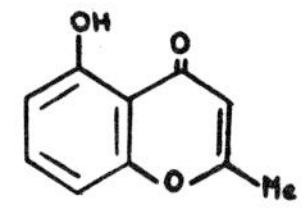

JCS, 1960, 654.

2203-013

LEPRARIC ACID $C_{18}H_{18}O_8$ 382
OI
155^0C

R = $CH_2OCOCH:C(Me)CH_2COOH$

JCS, 1969, 704.

2203-009

LEPTORUMOL $C_{12}H_{14}O_4$ 222
OI
218^0C

R = H

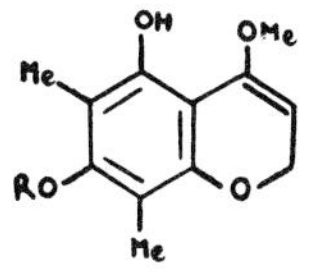

Chem. Abstr., 1969
(70) 35079.

2203-014

ANGUSTIFOLIONOL $C_{13}H_{14}O_4$ 234
OI
119^0C

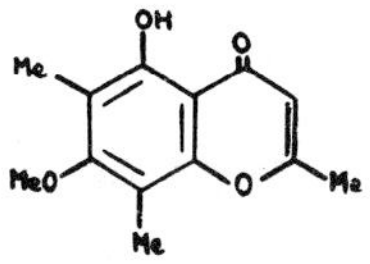

Aust. J. C., 1954, 168.

2203-010

LEPTORUMOLIN $C_{18}H_{24}O_9$ 388

254^0C

R = Glucosyl

Chem. Abstr., 1969
(70) 35079.

2203-015

SORDIDONE $C_{11}H_9O_4Cl$ 240
OI
261^0C

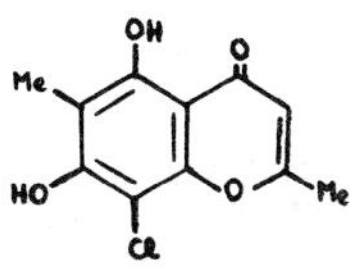

Phytochem., 1969, 1301.
Chem. Comm., 1968, 154.

2203-011

PEUCENIN $C_{15}H_{16}O_4$ 260
OI
211^0C

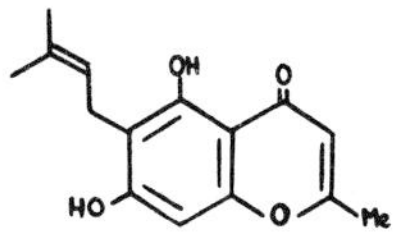

JCS, 1967, 604.

2204-001

PEUCENIN, hetero-	$C_{15}H_{16}O_4$	260
		OI
		193^0C
	Tet. Lett., 1967, 2737.	
2204-002		

INOPHYLLOIDIC ACID	$C_{32}H_{46}O_6$	526
	Tet. Lett., 1968, 3285.	
2204-007		

PEUCENIN 7-ME ETHER, hetero-	$C_{16}H_{18}O_4$	274
		OI
R = Me		
R' = H		
	Tet. Lett., 1967, 2737.	
2204-003		

BRASILIENSIC ACID	$C_{32}H_{46}O_6$	526
	Tet. Lett., 1968, 3285.	
2204-008		

PEUCENIN DIME ETHER, hetero-	$C_{17}H_{20}O_4$	286
		OI
		154^0C
R = R' = Me		
	Tet. Lett., 1967, 2737.	
2204-004		

BRASILIENSIC ACID, iso-	$C_{32}H_{46}O_6$	526
	cis isomer	
	Tet. Lett., 1968, 3285.	
2204-009		

PAPUANIC ACID	$C_{25}H_{36}O_6$	432
		75^0C
	JOC, 1968, 4191.	
2204-005		

KARENIN, desoxy-	$C_{15}H_{14}O_4$	258
		OI
		134^0C
	JCS, 1967, 145.	
2204-010		

PAPUANIC ACID, iso-	$C_{25}H_{36}O_6$	432
	*epimer	
	JOC, 1968, 4191.	
2204-006		

KARENIN	$C_{15}H_{14}O_5$	274
		OI
		204^0C
	R = CH_2OH	
	JCS, 1967, 145.	
	Tet. Lett., 1967, 3459.	
2204-011		

PTAEROXYLIN, dehydro- $C_{15}H_{12}O_4$ 256
OI
161^{0}C

Tet. Lett., 1967, 3459.

2204-012

SPATHELIACHROMENE, Me-iso- $C_{16}H_{16}O_4$ 272
OI

Chem. Abstr., 1971
(75) 129961.

2204-017

PTAEROXYLINOL $C_{15}H_{14}O_5$ 274
OI
135^{0}C

Tet. Lett., 1967, 3459.

2204-013

SPATHELIACHROMENE $C_{15}H_{14}O_4$ 258
OI

R = H

Chem. Abstr., 1971
(75) 129961.

2204-018

PTAEROGLYCOL $C_{15}H_{14}O_6$ 290

234^{0}C

Tet. Lett., 1967, 3461.

2204-014

$C_{20}H_{22}O_4$ 326
OI

R = 3,3-DMA

Chem. Abstr., 1971
(75) 129961.

2204-019

PTAEROCYCLIN $C_{15}H_{14}O_7$ 306
OI
266^{0}C

Tet. Lett., 1967, 2737.

2204-015

$C_{20}H_{20}O_4$ 324
OI

R = Senecioyl

Chem. Abstr., 1971
(75) 129961.

2204-020

PTAEROXYLONE $C_{14}H_{10}O_5$ 258
OI
215^{0}C

Tet. Lett., 1967, 3459.

2204-016

GAMMANDOL $C_{15}H_{16}O_5$ 276

R = H

Khim. Prir. Soedin.,
1970, 759.

2204-021

HAMAUDOL	$C_{15}H_{16}O_5$	276

R = H (Cf 021 ?)

Khim. Prir. Soedin.,
1970 (6) 468.

2204-022

PTAEROXYLIN, allo- $\quad C_{15}H_{14}O_4 \quad$ 258
OI
151^0C

R = CH_3

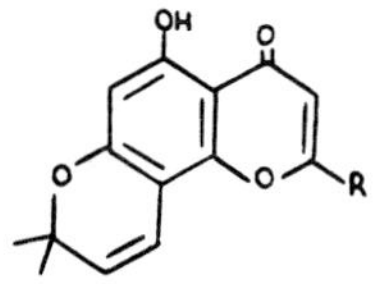

JCS, 1966, 111.

2204-027

HAMAUDOL ACETATE $\quad C_{17}H_{18}O_6 \quad$ 318

R = Ac

Khim. Prir. Soedin.,
1970 (6) 468.

2204-023

PTAEROCHROMENOL $\quad C_{15}H_{14}O_5 \quad$ 274
OI
176^0C

R = CH_2OH

Tet. Lett., 1967, 2737.

2204-028

SESELIRIN $\quad C_{19}H_{20}O_6S \quad$ 376

R = -CO-CH=CH·SMe

Khim. Prir. Soedin.,
1970 (6) 412.

2204-024

SORBIFOLIN $\quad C_{20}H_{22}O_5 \quad$ 342
OI
144^0C

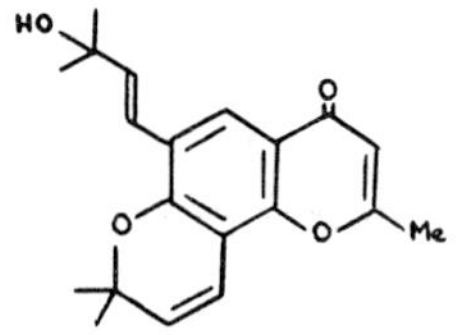

JCS, 1967, 2540.

2204-029

BLANCOIC ACID $\quad C_{24}H_{32}O_6 \quad$ 416

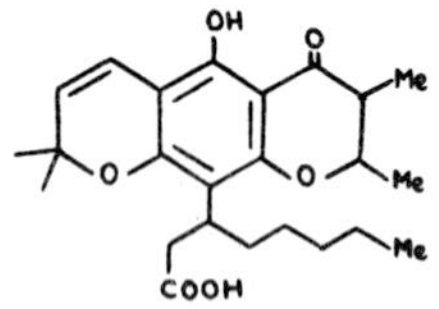

JOC, 1968, 4187.

2204-025

VISNAMMINOL $\quad C_{15}H_{16}O_5 \quad$ 276
$+96^0c$
160^0C

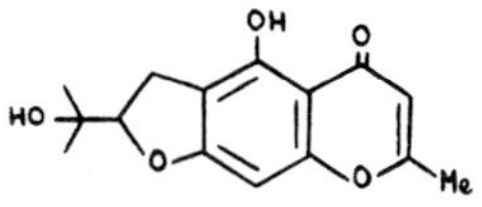

Helv., 1956 (39) 923.

2204-030

APETALIC ACID $\quad C_{22}H_{28}O_6 \quad$ 388
$+28^0c$
117^0C

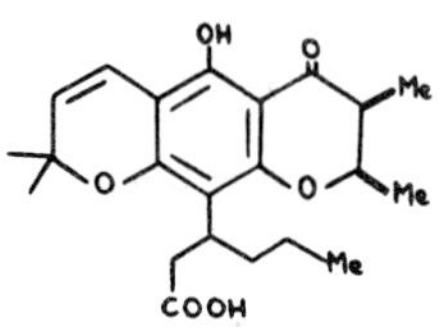

Tet. Lett., 1967, 4177.

2204-026

UMTATIN $\quad C_{15}H_{12}O_5 \quad$ 272
-56^0c
178^0C

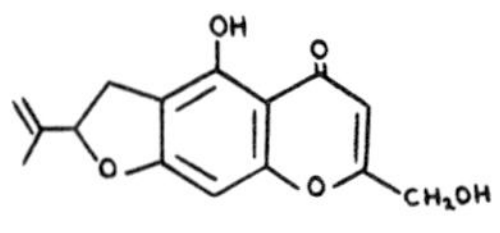

Tet. Lett., 1967, 2737.

2204-031

274

VISNAGIN	$C_{13}H_{10}O_4$	230
		OI
R = Me		111^0C
	JACS, 1951, 782.	
2204-032		

AMMIOL	$C_{14}H_{12}O_6$	276
		OI
		211^0C
R = CH$_2$OH		
	JOC, 1961, 886.	
2204-037		

KHELLOL	$C_{13}H_{10}O_5$	246
		OI
R = CH$_2$OH		
	JACS, 1951, 782.	
2204-033		

ERIOSTEMOIC ACID	$C_{20}H_{24}O_5$	344
		OI
		101^0C
	Aust. J. C., 1963 (16) 123.	
2204-038		

KHELLININ	$C_{19}H_{20}O_{10}$	408
		143^0C
	Ber., 1941 (74) 1549.	
2204-034		

ERIOSTOIC ACID	$C_{20}H_{24}O_5$	344
		OI
		174^0C
	Aust. J. C., 1962 (15) 812.	
2204-039		

KHELLINOL	$C_{13}H_{10}O_5$	246
		OI
		203^0C
	JACS, 1953, 3265.	
2204-035		

LL-D253-GAMMA	$C_{13}H_{14}O_4$	234
		$+26^0$m
		158^0C
	JOC, 1972, 1636.	
2204-040		

KHELLIN	$C_{14}H_{12}O_5$	260
		OI
R = Me		154^0C
	JACS, 1953, 3265.	
2204-036		

LL-D253-ALPHA	$C_{13}H_{16}O_5$	252
		$+25^0$m
		188^0C
R = H	JOC, 1972, 1636.	
2204-041		

LL-D253-BETA	$C_{15}H_{18}O_6$	294
		180^0C
R = Ac		
	JOC, 1972, 1636.	
2204-042		

AGGLOMERONE	$C_{13}H_{18}O_4$	238
		OI
		75^0C
	Aust. J. C., 1967, 1419.	
	JCS, 1967, 1839.	
2206-002		

URSONIC ACID	$C_{15}H_{16}O_5$	276
		OI
	Khim. Prir. Soedin., 1970 (6) 421.	
2204-043		

TASMANONE	$C_{14}H_{20}O_4$	252
		OI
	Aust. J. C., 1968, 2825.	
2206-003		

URSININ	$C_{16}H_{18}O_6$	306
		OI
	Khim. Prir. Soedin., 1970 (6) 421.	
2204-044		

FLAVESONE	$C_{14}H_{20}O_4$	252
		OI
	Aust. J. C., 1968, 2825.	
2206-004		

SORBICILLIN	$C_{14}H_{16}O_3$	232
		OI
		113^0C
	JACS, 1948, 4240.	
2205-001		

BAECKEOL	$C_{13}H_{18}O_4$	238
		OI
		104^0C
R = H		
	JCS, 1940, 1208.	
2206-005		

deleted		
2206-001		

CONGLOMERONE	$C_{13}H_{18}O_4$	238
		OI
		63^0C
R = Me		
	Chem. Abstr., 1940, 2346.	
2206-006		

276

XANTHOSTEMONE	$C_{12}H_{16}O_3$	208 OI Oil

Aust. J. C., 1968, 2825.

2206-007

TORQUATONE	$C_{16}H_{24}O_4$	280 OI 40^0C

Aust. J. C., 1962 (15) 144.

2206-012

COHUMULONE	$C_{20}H_{28}O_5$	348 -195^0m

R = OH

JCS, 1954, 2400.

2206-008

HUMULONE	$C_{21}H_{30}O_5$	362 -212^0e 64^0C

R = OH

Tet., 1970, 385.

2206-013

COHUMULONE, 4-deoxy-	$C_{20}H_{28}O_4$	332 OI 88^0C

R = H

Chem. Comm., 1967, 1212.

2206-009

HUMULONE, 4-deoxy-	$C_{21}H_{30}O_4$	346 OI 83^0C

R = H

Angew. Chem., 1958 (70) 343.

2206-014

COLUPULONE	$C_{25}H_{36}O_4$	400 OI 92^0C

R = 3,3-DMA

JCS, 1955, 174.

2206-010

LUPULONE	$C_{26}H_{38}O_4$	414 OI 92^0C

R = 3,3-DMA

Chem. Ind., 1955, 1595.
JCS, 1955, 174.

2206-015

LEPTOSPERMONE	$C_{15}H_{22}O_4$	266 OI

Aust. J. C., 1968, 2825.

2206-011

ADLUPULONE	$C_{26}H_{38}O_4$	414 82^0C

JCS, 1955, 174.

2206-016

G1 $C_{15}H_{22}O_5$ 282 0^0 98^0C

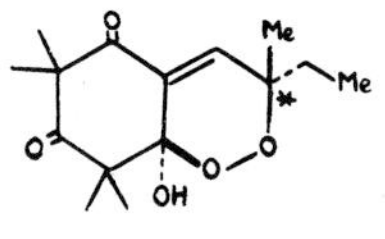

Tet. Lett., 1971, 1353.

2206-017

USNIC ACID, (-)- $C_{18}H_{16}O_7$ 344 -495^0C 203^0C

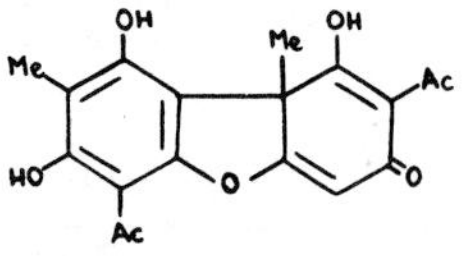

Tet. Lett., 1966, 5211.

2207-001

G2 $C_{15}H_{22}O_5$ 282 0^0 128^0C

*epimer

Tet. Lett., 1971, 1353.

2206-018

USNIC ACID, (+)- $C_{18}H_{16}O_7$ 344 $+495^0$c 203^0C

Enantiomer

Chem. Comm., 1966, 656.

2207-002

G3 $C_{14}H_{20}O_5$ 268 0^0 170^0C

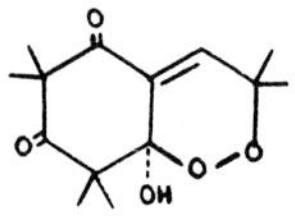

Tet. Lett., 1971, 1353.

2206-019

USNIC ACID, (+)-iso- $C_{18}H_{16}O_7$ 344 $+500^0$d 151^0C

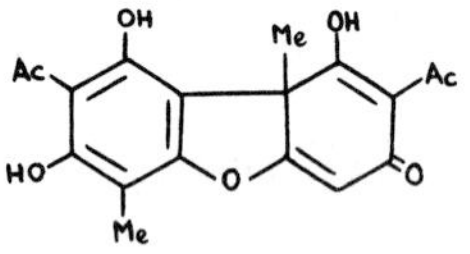

Tet. Lett., 1967, 4867.

2207-003

ADHUMULONE $C_{21}H_{30}O_5$ 362

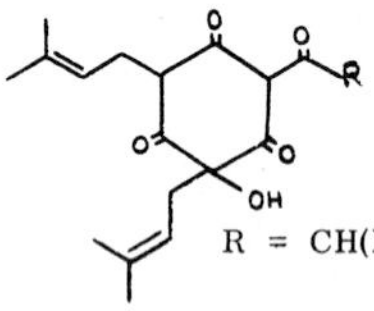

R = CH(Me) Et

Chem. Ind., 1955, 992.

2206-020

RHODOMYRTOXIN, pseudo- $C_{24}H_{28}O_7$ 428 201^0C

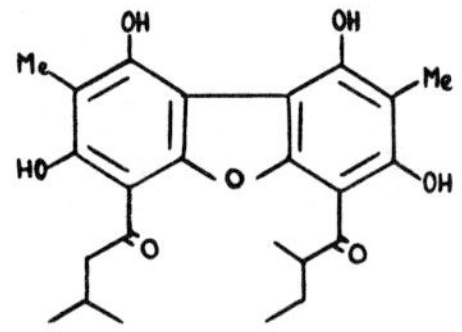

JCS, 1969, 2403.

2207-004

HUMULONE, pre- $C_{22}H_{32}O_5$ 376 -172^0 Oil

R = $CH_2CH_2CHMe_2$

Bull. Soc. Chim. Belg., 1962 (71) 438.

2206-021

PHLORASPIN $C_{23}H_{28}O_8$ 432   OI 212^0C

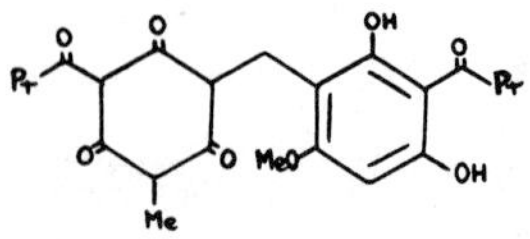

Acta Chem. Scand., 1961 (15) 1777.

2207-005

PROTOKOSIN	$C_{25}H_{32}O_8$	460 +8°c 182°C
		JCS, 1937, 562.
2207-006		

FLAVASPIDIC ACID AB	$C_{22}H_{26}O_8$	418 OI 211°C
R = Me		Acta Chem. Scand., 1964, 344.
2207-011		

KOSIN, α-	$C_{25}H_{32}O_8$	460 OI
R = H; R' = Me		Ber., 1956 (89) 2600. JCS, 1952, 3102.
2207-007		

FLAVASPIDIC ACID PB	$C_{23}H_{28}O_8$	432 OI 170°C
R = Et		Acta Chem. Scand., 1964, 344.
2207-012		

KOSIN, β-	$C_{25}H_{32}O_8$	460 OI
R = Me; R' = H		JCS, 1952, 3102.
2207-008		

FLAVASPIDIC ACID BB	$C_{24}H_{30}O_8$	446 OI 92°C
R = Pr		JCS, 1953, 1828.
2207-013		

DESASPIDIN-AB	$C_{22}H_{26}O_8$	418 OI 146°C
R = Me		Acta Chem. Scand., 1964, 344.
2207-009		

ASPIDIN	$C_{25}H_{32}O_8$	460 OI 124°C
R = H; R' = Me		Helv., 1956, 153.
2207-014		

DESASPIDIN-PB	$C_{23}H_{28}O_8$	432 OI 142°C
R = Et		Acta Chem. Scand., 1964, 344.
2207-010		

PARAASPIDIN	$C_{25}H_{32}O_8$	460 OI 124°C
R = Me; R' = H		Acta Chem. Scand., 1962 (16) 1251.
2207-015		

ASBASPIDIN-AA $C_{21}H_{24}O_8$ 404

OI

170^0C

R = Me

Acta Chem. Scand.,
1964 (18) 344.

2207-016

FILIXIC ACID PBP $C_{34}H_{40}O_{12}$ 640

OI

193^0C

R = R' = Et

Acta Chem. Scand.,
1963 (17) 191.

2207-021

ALBASPIDIN-PP $C_{23}H_{28}O_8$ 432

OI

136^0C

R = Et

Acta Chem. Scand.,
1964 (18) 344.

2207-017

FILIXIC ACID PBB $C_{35}H_{42}O_{12}$ 654

OI

185^0C

R = Pr; R' = Et

Acta Chem. Scand.,
1963 (17) 191.

2207-022

ALBASPIDIN-BB $C_{25}H_{32}O_8$ 460

OI

148^0C

R = Pr

Acta Chem. Scand.,
1964 (18) 344.

2207-018

FILIXIC ACID BBB $C_{36}H_{44}O_{12}$ 668

OI

173^0C

R = R' = Pr

Acta Chem. Scand.,
1963 (17) 191.

2207-023

ULIGINOSIN-A $C_{28}H_{36}O_8$ 512

OI

161^0C

JACS, 1968, 4721.

2207-019

TRIDESASPIDIN $C_{35}H_{42}O_{12}$ 654

OI

150^0C

Acta Chem. Scand.,
1963, 2361.

2207-024

ULIGINOSIN-B $C_{28}H_{34}O_8$ 498

OI

142^0C

JACS, 1968, 4721.

2207-020

TRIFLAVASPIDIC ACID $C_{35}H_{42}O_{12}$ 654

OI

170^0C

R = R' = H

Acta Chem. Scand.,
1963, 2361.

2207-025

TRIASPIDIN	$C_{36}H_{44}O_{12}$	668 OI 158°C

R = H; R' = Me

Acta Chem. Scand., 1963, 2361.

2207-026

| PHLOROGLUCINOL | $C_6H_6O_3$ | 126 OI 216°C |

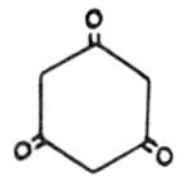

JOC, 1955, 1402.

2299-001

TRISPARAASPIDIN $C_{36}H_{44}O_{12}$ 668 OI

R = Me; R' = H

Helv., 1970 (53) 2176.

2207-027

FILICINIC ACID $C_8H_{10}O_3$ 154 OI 214°C

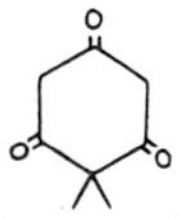

Chem. Ind., 1954, 546.

2299-002

FLAVASPIDIC ACID, methylene--bis-nor- $C_{47}H_{56}O_{16}$ 876 OI 160°C

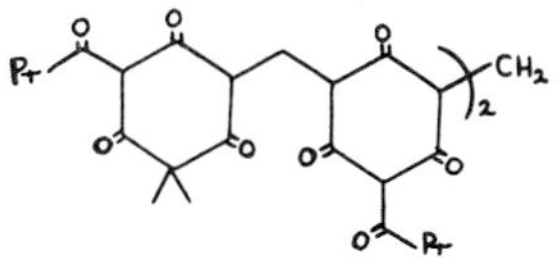

Acta Chem. Scand., 1963, 2370.

2207-028

SYCARPIC ACID $C_{10}H_{14}O_3$ 182 OI 186°C

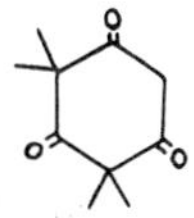

Aust. J. C., 1960 (13) 385.

2299-003

ACROVESTONE $C_{32}H_{42}O_8$ 554 0° 142°C

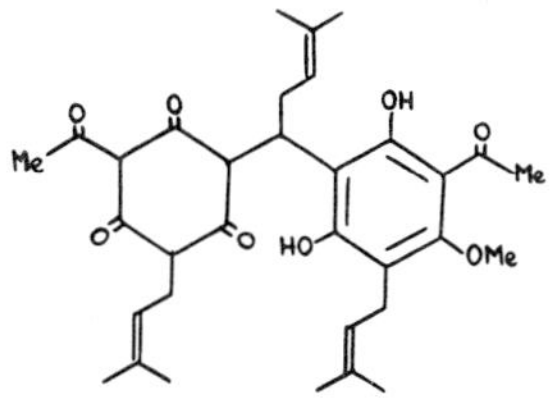

Chem. Abstr., 1964 (61) 14573.

2207-029

MELICOPOL $C_{19}H_{26}O_6$ 350 OI 133°C

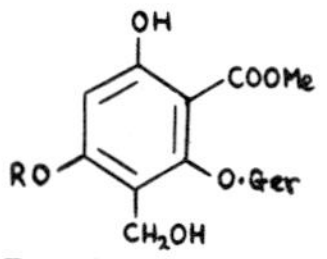

R = H

Aust. J. C., 1965, 2021.

2299-004

RHODOMYRTOXIN $C_{24}H_{28}O_7$ 428 OI 198°C

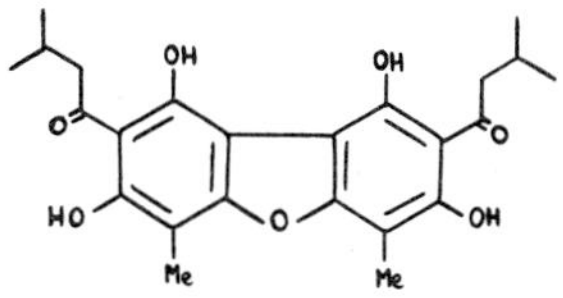

JCS, 1969, 2403.

(?)

2207-030

MELICOPOL, Me- $C_{20}H_{28}O_6$ 364 OI 103°C

R = Me

Aust. J. C., 1965, 2021.

2299-005

PHLOROPYRONE $C_{21}H_{26}O_7$ 390
OI
111^0C

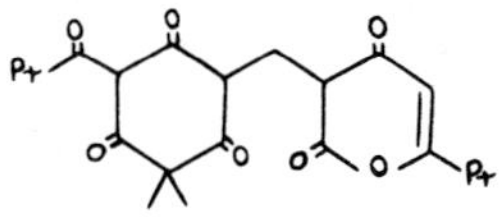

Acta Chem. Scand.,
1961 (15) 839.

2299-006

$C_{41}H_{70}O_4$ 626
OI

$R = {}^{n}C_{31}H_{63}$

Tet. Lett., 1970, 1039.

2299-009

$C_{37}H_{62}O_4$ 570
OI

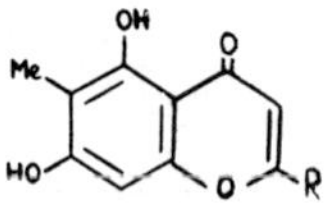

$R = {}^{n}C_{27}H_{55}$

Tet. Lett., 1970, 1039.

2299-007

$C_{13}H_{12}O_4$ 132
OI
210^0C

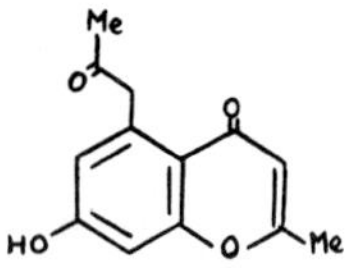

Tet., 1971, 981.

2299-010

$C_{39}H_{66}O_4$ 598
OI

$R = {}^{n}C_{29}H_{59}$

Aust. J. C., 1971, 1257.
Tet. Lett., 1970, 1039.

2299-008

23 | DEPSIDES

The depsides are basically orcinol esters of orcinol acids, and the depsidones are heterocyclic derivatives of the depsides. There are now several compounds known where more than two orcinol units are combined, and these have been placed together in Class 2303.

2301	Depsides
2302	Depsidones
2303	$n \geqslant 1$
2399	Miscellaneous

LECANORIC ACID $C_{16}H_{14}O_7$ 318
OI
175^0C

R = H

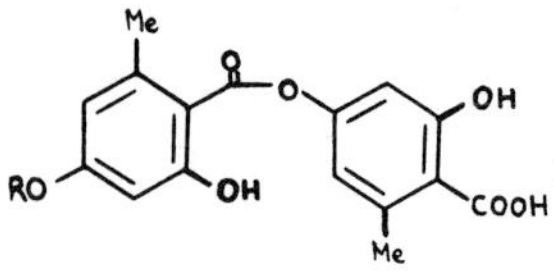

Ber., 1932, 983.

2301-001

EVERNIC ACID $C_{17}H_{16}O_7$ 332
OI
169^0C

R = Me

JCS, 1932, 1388.

2301-002

DIPLOSCHISTESIC ACID $C_{16}H_{14}O_8$ 334
OI
174^0C

Ber., 1936, 2327.

2301-003

OBTUSATIC ACID $C_{18}H_{18}O_7$ 346
OI
208^0C

J. Pharm. Soc. Jap.,
1936 (56) 237.

2301-004

TUMIDULIN $C_{17}H_{14}O_7Cl_2$ 400
OI
175^0C

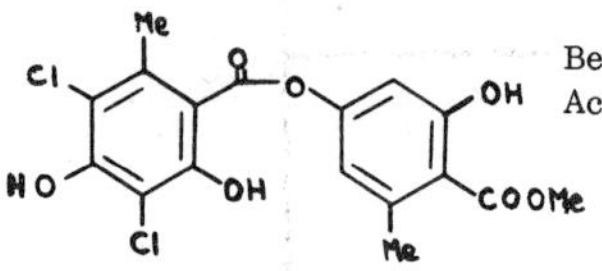

Ber., 1966, 1106.
Acta Chem. Scand.,
1965, 1188.

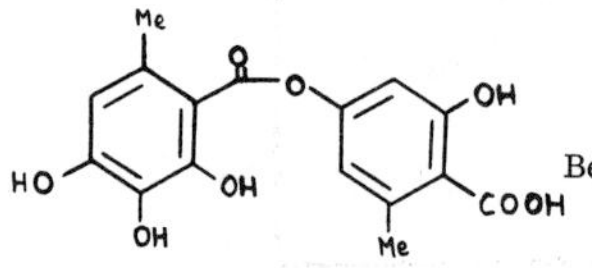

2301-005

BARBATIC ACID, 4-O-des-Me- $C_{18}H_{18}O_7$ 346
OI
176^0C

R = H

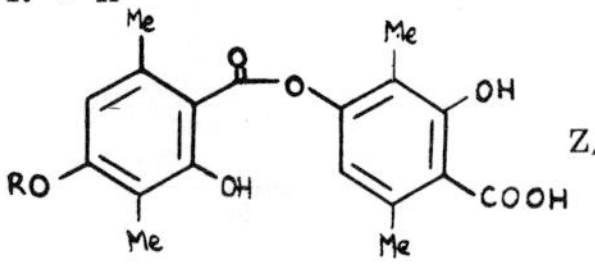

Z. Nat., 1968, 856.

2301-006

BARBATIC ACID $C_{19}H_{20}O_7$ 360
OI
187^0C

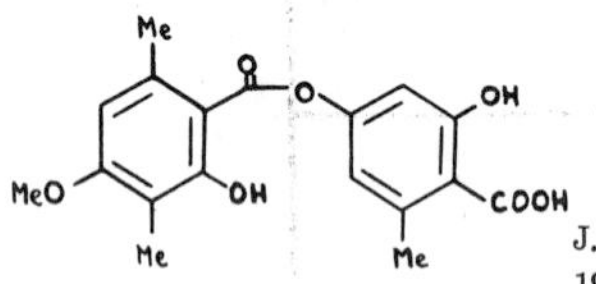

R = Me

J. Pharm. Soc. Jap.,
1936 (56) 237.

2301-007

BAEOMYCESIC ACID $C_{19}H_{18}O_8$ 374
OI
233^0C

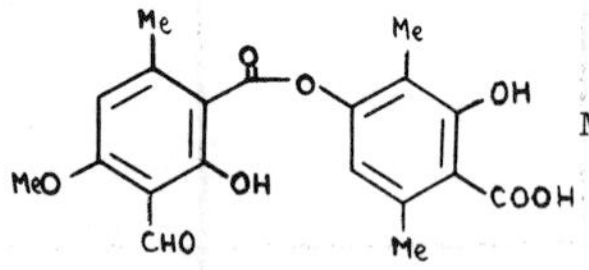

Monatsh., 1935 (66) 57.

2301-008

ATRANORIN $C_{19}H_{18}O_8$ 374
OI
196^0C

R = H

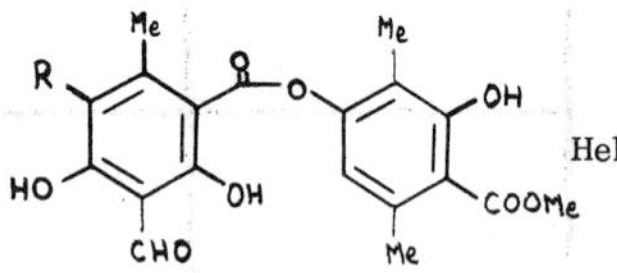

Helv., 1926, 650.

2301-009

ATRANORIN, Cl- $C_{19}H_{17}O_8Cl$ 408
OI
208^0C

R = Cl

Monatsh., 1934 (64) 126.

2301-010

DIFFRACTAIC ACID $C_{20}H_{22}O_7$ 374
OI
190^0C

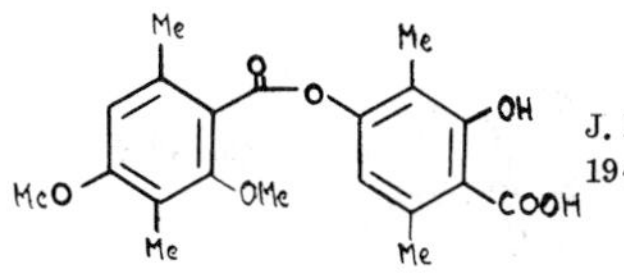

J. Pharm. Soc. Jap.,
1940 (60) 177.

2301-011

NEPHROARCTIN $C_{20}H_{20}O_7$ 372
0^0
192^0C

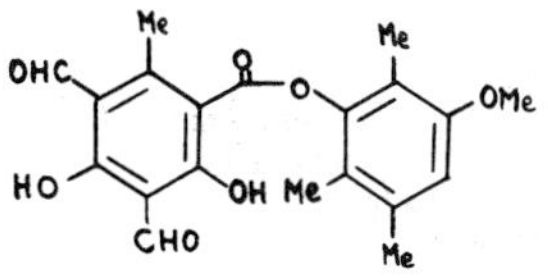

Chem. Comm., 1969, 78.

2301-016

SQUAMATIC ACID $C_{19}H_{18}O_9$ 390
OI
219^0C

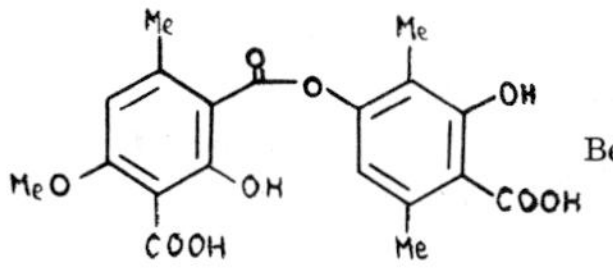

Ber., 1937, 62.

2301-012

SPHAEROPHORIN $C_{23}H_{28}O_7$ 416
OI
140^0C

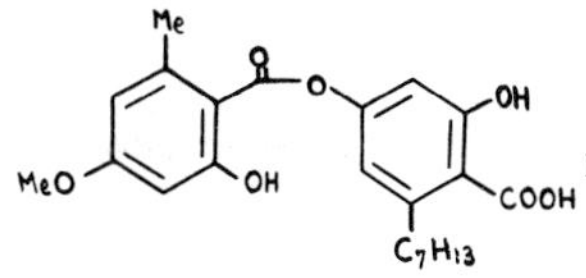

Ber., 1934, 416.

2301-017

THAMNOLIC ACID, hypo- $C_{19}H_{18}O_{10}$ 406
OI
218^0C

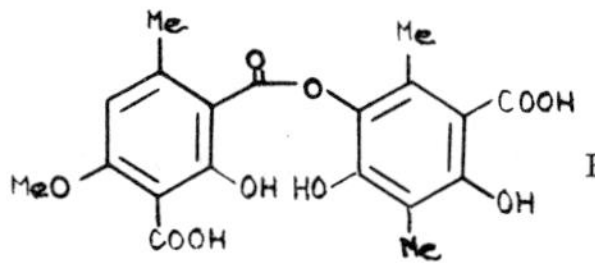

Ber., 1941, 824.

2301-013

DIVARICATIC ACID $C_{21}H_{24}O_7$ 388
OI
137^0C

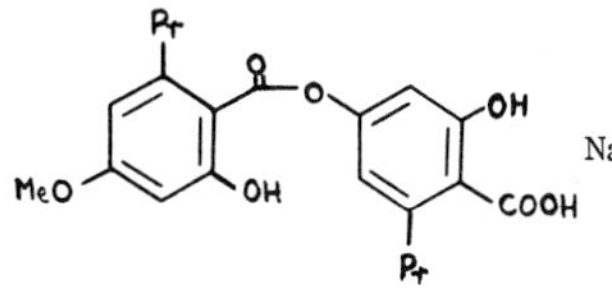

Natwiss., 1963 (50) 645.

2301-018

THAMNOLIC ACID $C_{19}H_{16}O_{11}$ 420
OI
223^0C

R = COOH

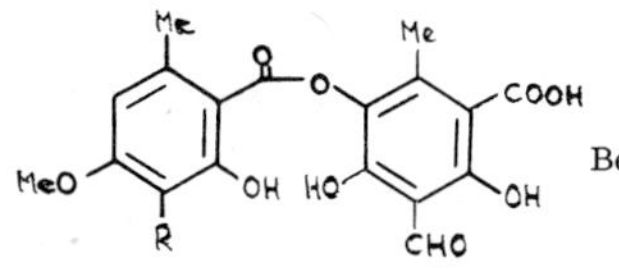

Ber., 1939, 1402.

2301-014

SEKIKAIC ACID $C_{22}H_{26}O_8$ 418
OI
150^0C

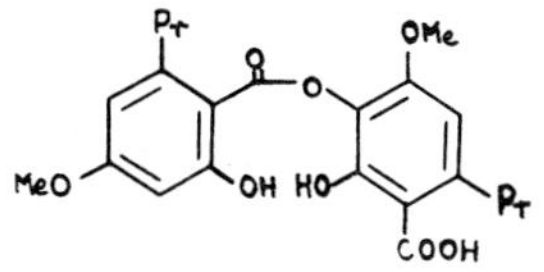

Z. Nat., 1966, 90.

2301-019

THAMNOLIC ACID, haema- $C_{19}H_{16}O_{10}$ 404
OI
203^0C

R = CHO

JCS, 1967, 1603.

2301-015

STERNOSPORIC ACID $C_{23}H_{28}O_7$ 416
OI
112^0C

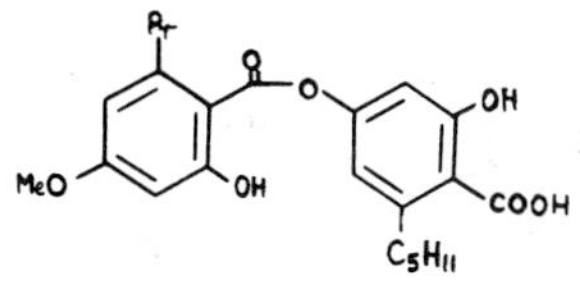

Phytochem., 1970, 841.

2301-020

285

PALUDOSIC ACID	$C_{23}H_{28}O_8$	432 OI

R = H

Chem. Abstr., 1968 (69) 8863.

2301-021

ANZIAIC ACID	$C_{24}H_{30}O_7$	430 OI 124^0C

Ber., 1937, 1826.

2301-026

MEROCHLOROPHEIC ACID	$C_{24}H_{30}O_8$	446 OI 165^0C

R = Me

Phytochem., 1965, 133.

2301-022

PERLATOLIC ACID	$C_{25}H_{32}O_7$	444 OI 108^0C

R = H

Ber., 1935, 634.

2301-027

SEKIKAIC ACID, homo-	$C_{24}H_{30}O_8$	446 OI 133^0C

R = H

Ber., 1937, 1822.

2301-023

PLANA ACID	$C_{27}H_{36}O_7$	472 OI 110^0C

R = Me

Z. Nat., 1965, 1119.

2301-028

BONINIC ACID	$C_{25}H_{32}O_8$	460 OI 135^0C

R = Me

Ber., 1937, 1817.

2301-024

GLOMELLIFERIC ACID	$C_{25}H_{30}O_8$	458 OI

Chem. Abstr., 1951 (45) 2939.

2301-029

IMBRICARIC ACID	$C_{23}H_{28}O_7$	416 OI 125^0C

Ber., 1937, 1823.

2301-025

OLIVETORIC ACID	$C_{26}H_{32}O_8$	472 OI 151^0C

Science, 1964 (143) 255.

R = H

2301-030

CONFLUENTIC ACID $C_{28}H_{36}O_8$ 500
                                                     OI
                                                     157[0]C

R = Me

Ber., 1962, 328.

2301-031

PHENARCTIN $C_{21}H_{20}O_9$ 416
                                                     OI
                                                     167[0]C

Acta Chem. Scand.,
1969, 3601.

2301-036

MICROPHYLLIC ACID $C_{29}H_{36}O_9$ 528
                                                     OI
                                                     116[0]C

Ber., 1935, 2022.

2301-032

VIRENSIC ACID $C_{18}H_{14}O_8$ 358
                                                     OI
                                                     246[0]C

Tet., 1961 (12) 173.

2302-001

ARTHONIOIC ACID $C_{29}H_{36}O_9$ 528
                                                     OI
                                                     168[0]C

Z. Nat., 1970 (25B) 49.

2301-033

PSOROMIC ACID $C_{18}H_{14}O_8$ 358
                                                     OI
                                                     265[0]C

J. Pharm. Soc. Jap.,
1937 (70) 811.

2302-002

ERYTHRIN $C_{20}H_{22}O_{11}$ 438
                                                     +11[0]
                                                     148[0]C

Cf 2301-001
R = Erythritol

J. Pharm. Soc. Jap.,
1941 (61) 108.

2301-034

PROTOCETRARIC ACID, hypo- $C_{18}H_{12}O_{11}$ 404
                                                     OI
                                                     242[0]C

Phytochem., 1965, 951.

2302-003

CRYPTOCHLOROPHEIC ACID $C_{25}H_{32}O_8$ 460
                                                     OI
                                                     182[0]C

Phytochem., 1965, 133.

2301-035

PROTOCETRARIC ACID $C_{18}H_{14}O_9$ 374
                                                     OI
                                                     250[0]C
R = H

Ber., 1934 (67) 311.

2302-004

PROTOCETRARIC ACID, fumaryl- $C_{22}H_{16}O_{12}$ 472
OI
255^0C

$R = .CO.CH \overset{t}{:} CH.COOH$

Ber., 1934, 311.

2302-005

HAIDERIN $C_{19}H_{19}O_6Cl$ 388

170^0C

R = H

Pak. J. Sci. Ind. Res.,
1970 (13) 364.

2302-010

VICANICIN $C_{17}H_{14}O_5Cl_2$ 368
OI
250^0C

R = H

Tet. Lett., 1959 (9) 1.

2302-006

RUBININ $C_{20}H_{21}O_6Cl$ 402

162^0C

R = Me

Pak. J. Sci. Ind. Res.,
1970 (13) 364.

2302-011

CALOPLOICIN $C_{17}H_{13}O_5Cl_3$ 402
OI
260^0C

R = Cl

Chem. Pharm. Bull.,
1971, 1070.

2302-007

SHIRIN $C_{19}H_{17}O_6Cl_3$ 456

118^0C

Pak. J. Sci. Ind. Res.,
1970 (13) 364.

2302-012

DIPLOICIN $C_{16}H_{10}O_5Cl_4$ 422
OI
232^0C

Sci. Proc. Roy. Dubli
Soc., 1948 (24) 319.

2302-008

YASIMIN $C_{19}H_{18}O_5$ 326
OI
200^0C

R = H

Pak. J. Sci. Ind. Res.,
1970 (13) 244.

2302-013

GANGALEOIDIN $C_{18}H_{14}O_7Cl_2$ 412
OI
214^0C

Sci. Proc. Roy. Dublin
Soc., 1943 (23) 143.

2302-009

NIDULIN, des-Cl-nor- $C_{19}H_{16}O_5Cl_2$ 394
OI
214^0C

R = Cl

JCS, 1956, 3545.

2302-014

NIDULIN, nor-	$C_{19}H_{15}O_5Cl_3$	428
		OI
		186°C

R = H

JCS, 1956, 3545.

2302-015

CONSTICTIC ACID	$C_{19}H_{14}O_{10}$	402
		198°C

R = Me

Chem. Pharm. Bull.,
1970, 2364.

2302-020

NIDULIN	$C_{20}H_{17}O_5Cl_3$	442
		OI
		180°C

R = Me

JCS, 1956, 3545.

2302-016

PANNARIN	$C_{18}H_{15}O_6Cl$	362
		OI
		217°C

J. Pharm. Soc. Jap.,
1941 (61) 332.

2302-021

STICTIC ACID, nor-	$C_{18}H_{12}O_9$	372
		283°C

R = H

Natwiss., 1963 (50) 646.

2302-017

VARIOLARIC ACID	$C_{16}H_{10}O_7$	314
		OI
		296°C

Sci. Proc. Roy. Dublin
Soc., 1943 (23) 71.

2302-022

STICTIC ACID	$C_{19}H_{14}O_9$	386
		OI
		269°C

R = Me

Ber., 1936 (69) 1370.

2302-018

LOBARIC ACID	$C_{25}H_{28}O_8$	456
		OI
		192°C

R = Me
R' = H

Ber., 1937 (70) 206.

2302-023

SALAZINIC ACID	$C_{18}H_{12}O_{10}$	388
		260°C

R = H

Ber., 1933, 1215.

2302-019

LOXODIN	$C_{25}H_{28}O_8$	456
		OI
		132°C

R = H; R' = Me

Phytochem., 1970, 1139.

2302-024

LOBARIDONE, nor- $C_{23}H_{26}O_6$ 514
OI
188^0C

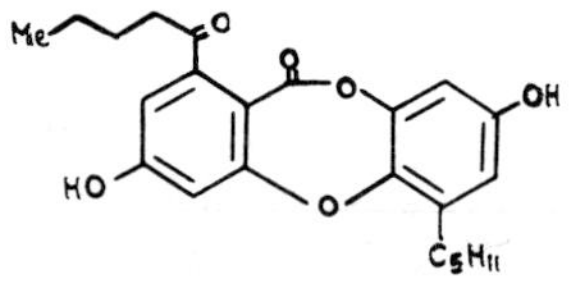

Aust. J. C., 1960, 314.

2302-025

GYROPHORIC ACID $C_{24}H_{20}O_{10}$ 468
OI
220^0C

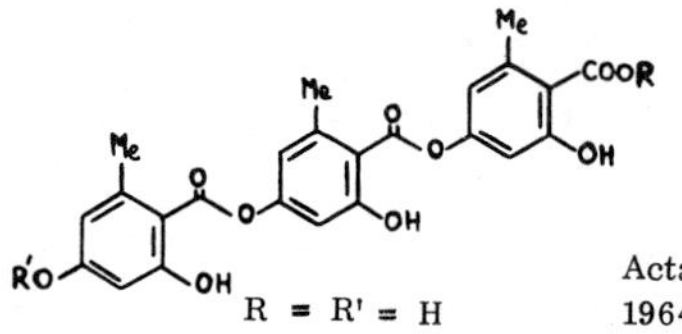

R = R' = H

Acta Chem. Scand.,
1964 (18) 329.

2303-001

COLENSOIC ACID $C_{25}H_{30}O_7$ 442
OI
174^0C

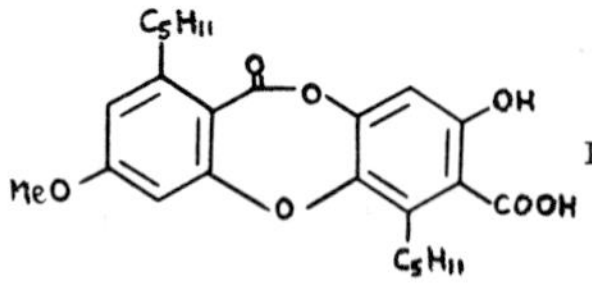

Phytochem., 1970, 2569.

2302-026

GYROPHORIC ACID ME ESTER $C_{25}H_{22}O_{10}$ 482
OI
288^0C

R = Me; R' = H

Phytochem., 1970, 2061.

2303-002

PHYSODIC ACID $C_{26}H_{30}O_8$ 470
OI
205^0C

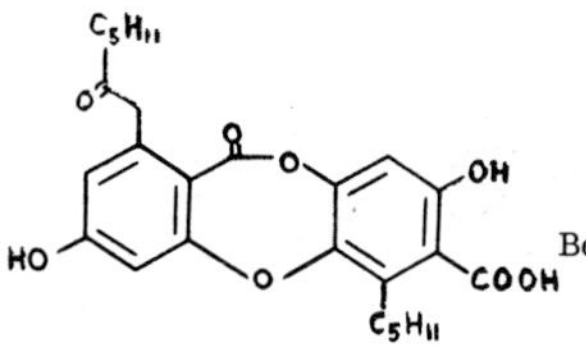

Ber., 1935 (68) 1500.

2302-027

TENUIORIN $C_{26}H_{24}O_{10}$ 496
OI
238^0C

R = R' = H

Phytochem., 1969, 2345.

2303-003

ALECTORONIC ACID $C_{28}H_{32}O_9$ 512
OI
193^0C

R = H

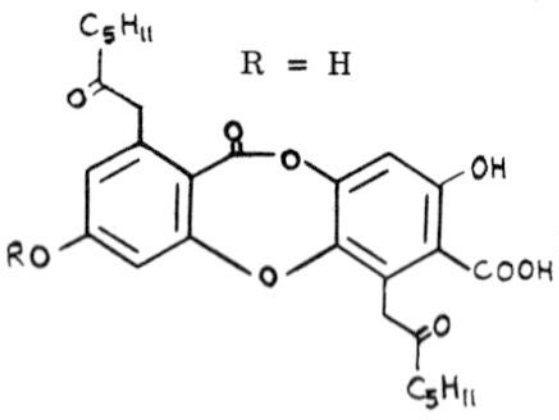

Ber., 1937 (70) 206.

2302-028

UMBILICARIC ACID $C_{25}H_{22}O_{10}$ 482
OI
185^0C

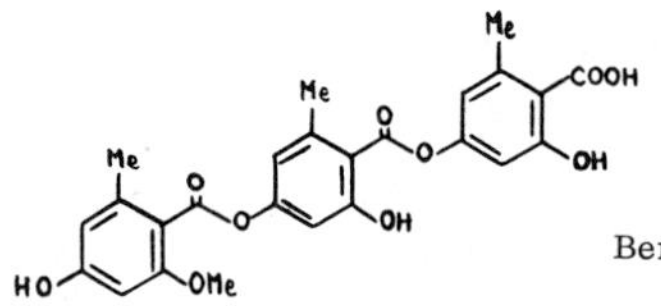

Ber., 1937 (70) 204.

2303-004

COLLATOLIC ACID, α- $C_{29}H_{34}O_9$ 526
OI
125^0C

R = Me

Ber., 1933 (66) 649.

2302-029

HIASIC ACID $C_{24}H_{20}O_{11}$ 484
OI
190^0C

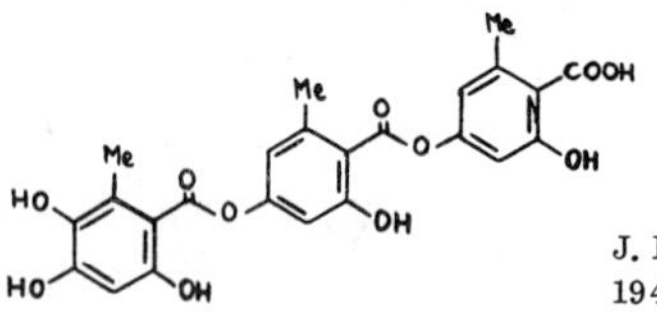

J. Pharm. Soc. Jap.,
1942 (62) 339.

2303-005

APHTHOSIN	$C_{34}H_{30}O_{13}$	646
		OI
		300^0C

Phytochem., 1970, 2587.

2303-006

SCHIZOPELTIC ACID	$C_{19}H_{18}O_7$	358
		OI
		230^0C

R = Me

Acta Chem. Scand.,
1967, 1111.
Z. Nat., 1970 (25B) 265.

2399-005

ALECTORIALIC ACID	$C_{18}H_{16}O_9$	376
		OI
		177^0C

R = Me

Acta Chem. Scand.,
1970, 345.

2399-001

DIDYMIC ACID	$C_{22}H_{26}O_5$	370
		OI
		172^0C

J. Pharm. Soc. Jap.,
1944 (68) 41.

2399-006

BARBATOLIC ACID	$C_{18}H_{14}O_{10}$	390
		OI
		206^0C

R = CHO

Z. Nat., 1966, 91.

2399-002

PORPHYRILIC ACID	$C_{16}H_{10}O_7$	314
		OI
		282^0C

Acta Chem. Scand.,
1956 (10) 1404.

2399-007

ASTERRIC ACID	$C_{17}H_{16}O_8$	348
		OI

Biochem. J., 1964, 43.

2399-003

STREPSILIN	$C_{15}H_{10}O_5$	270
		OI
		324^0C

J. Pharm. Soc. Jap.,
1944 (64) 20.

2399-008

PANNARIC ACID	$C_{16}H_{12}O_7$	316
		OI
R = H		244^0C

Acta Chem. Scand.,
1959 (13) 1855.

2399-004

PICROLICHENIC ACID	$C_{25}H_{30}O_7$	442
		178^0C

Acta Chem. Scand.,
1958 (12) 147.

2399-009

GEODOXIN

$C_{17}H_{12}O_8Cl_2$ 415
Rac
217^0C

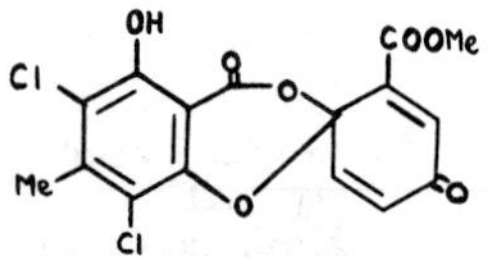

JCS, 1959, 2831.

2399-010

While the xanthones are an easily identified heterocyclic class of aromatics, their biogenetic origins are far more complex. It would seem on inspection that the heterocyclization of a benzophenone precursor would suitably rationalize the biogenesis of these compounds. However, the isolation of the ergochrome pigments (2406) suggests the possible intervention of an anthraquinone precursor (and the anthraquinones themselves are by no means a homogeneous biogenetic class).

The classification employed here is thus simply based on the degree of oxygenation of the xanthone nucleus—a special class being provided for dimeric compounds (2406). The miscellaneous Class (2499) includes complex xanthonoids, especially those in which the aromaticity has been destroyed by substitution or reduction.

2401 Mono-oxy	2402 Di-oxy
2403 Tri-oxy	2404 Tetra-oxy
2405 Penta-oxy	2406 Bi-Xanthyls
2499 Misc.	

XANTHONE, 2-OH- $C_{13}H_8O_3$ 212 OI 241°C

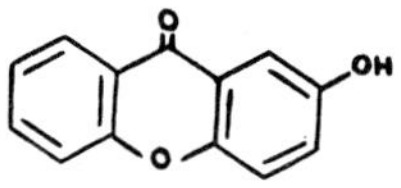

Chem. Abstr., 1967 (67) 108528.

2401-001

XANTHONE, 1,5-di-OH- $C_{13}H_8O_4$ 228 OI 269°C

R = H

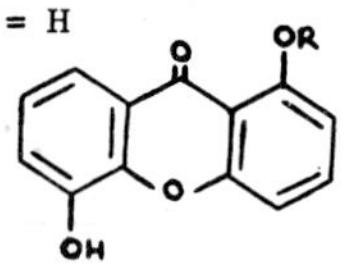

JCS, 1968, 2581.

2402-003

XANTHONE, 4-OH- $C_{13}H_8O_3$ 212 OI 246°C

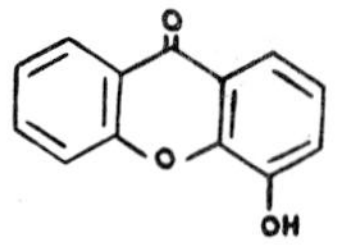

Tet. Lett., 1966, 6089.

2401-002

XANTHONE, 5-OH-1-OMe- $C_{14}H_{10}O_4$ 242 OI 251°C

R = Me

JCS, 1969, 2421.

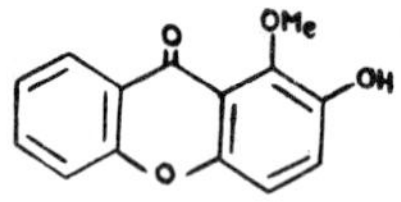

2402-004

CASSIAXANTHONE $C_{15}H_8O_7$ 300 OI 330°C

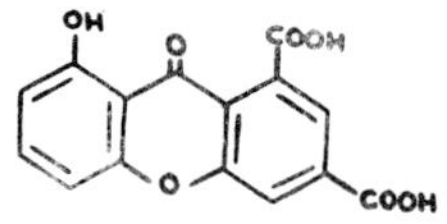

Phytochem., 1970, 1153.

2401-003

GUANANDIN $C_{18}H_{16}O_4$ 296 OI 213°C

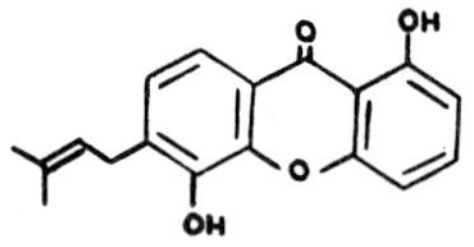

Tet., 1968, 3059.

2402-005

XANTHONE, 2-OH-1-OMe- $C_{14}H_{10}O_4$ 242 OI 170°C

Chem. Abstr., 1967 (67) 108528.

2402-001

GUANANDIN, oxy- $C_{18}H_{16}O_5$ 312 OI

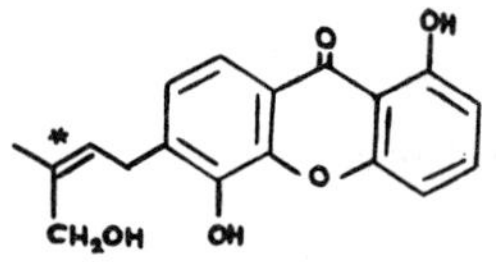

Tet., 1968, 3059.

2402-006

XANTHONE, 1,2-methylenedioxy- $C_{14}H_8O_4$ 240 OI 218°C

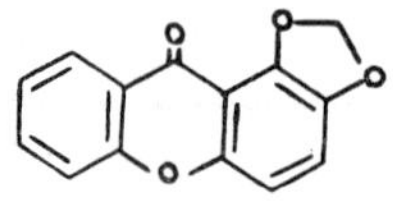

Phytochem., 1970, 453.

2402-002

GUANANDIN, dihydro-oxy- $C_{18}H_{18}O_5$ 314

*dihydro

Tet., 1968, 3059.

2402-007

GUANANDIN, dehydro-cyclo-   294
OI
168°C

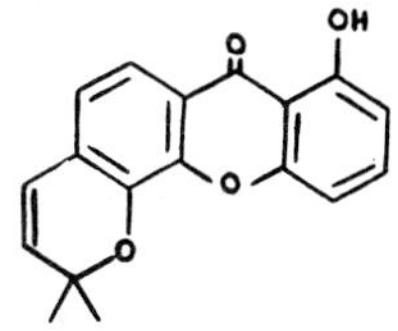

Tet., 1968, 1601.

2402-008

GUANANDIN, iso- $C_{18}H_{16}O_4$ 296
OI
175°C

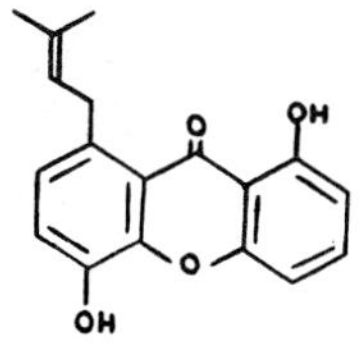

Tet., 1968, 1601.

2402-009

SCRIBLITIFOLIC ACID $C_{19}H_{18}O_6$ 342
0°
165°C

JCS, 1967, 871.

2402-010

EUXANTHONE $C_{13}H_8O_4$ 228
OI
240°C

JCS, 1931, 1709.

2402-011

deleted

2402-012

PINSELIN $C_{16}H_{12}O_6$ 300
OI
225°C

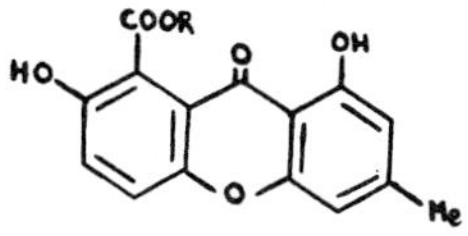

R = Me

Chem. Comm., 1971, 423.

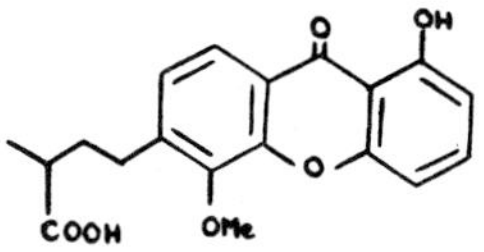

2402-013

PINSELIC ACID $C_{15}H_{10}O_6$ 286
OI
200°C

R = H

J. Biochem., 1953 (40)
451.

2402-014

XANTHONE, 5-OH-3-OMe- $C_{14}H_{10}O_4$ 242
OI

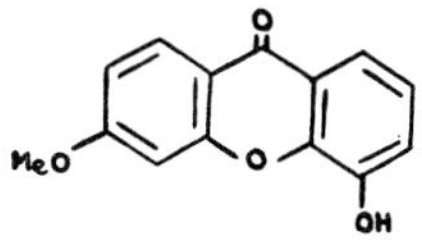

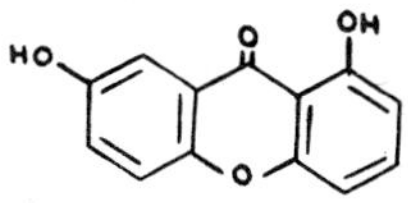

Chem. Abstr., 1967
(66) 102437.

2402-015

XANTHONE, 2,3-diOH-1-OMe- $C_{14}H_{10}O_5$ 258
OI
173°C

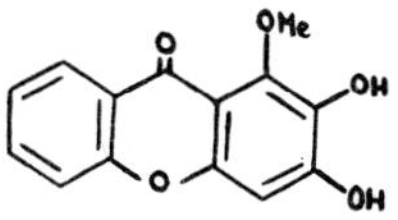

Chem. Abstr., 1969
(70) 823.

2403-001

XANTHONE, 3-OH-1,2-diOMe- $C_{15}H_{12}O_5$ 272
OI
237°C

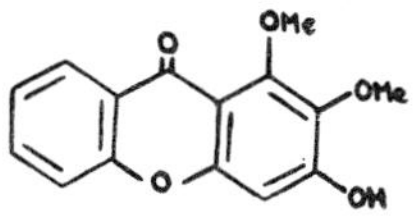

Chem. Abstr., 1969
(70) 823.

2403-002

GLOBUXANTHONE	$C_{18}H_{16}O_5$	312 OI 162^0C

JCS, 1966, 2186.

2403-003

XANTHONE, 5-OH-1, 3-diOMe-	$C_{15}H_{12}O_5$	272 OI 264^0C

R = Me; R' = H

Tet., 1966, 1777.

2403-008

XANTHONE, 2, 8-diOH-1-OMe-	$C_{14}H_{10}O_5$	258 OI 198^0C

R = H

Tet., 1966, 1785.

2403-004

XANTHONE, 1, 3, 5-tri-OMe-	$C_{16}H_{14}O_5$	286 OI

R = R' = Me

Tet., 1969 (25) 1961.

2403-009

XANTHONE, 8-OH-1, 2-diOMe-	$C_{15}H_{12}O_5$	272 OI 174^0C

R = Me

Tet., 1966, 1785.

2403-005

XANTHONE, 2-(3, 3-DMA)-1, 3, 5- -tri-OH-	$C_{18}H_{16}O_5$	312 OI

JCS, 1967, 2500.

2403-010

MESUAXANTHONE-A	$C_{14}H_{10}O_5$	258 OI 271^0C

R = R' = H

Tet., 1967, 243.

2403-006

JACAREUBIN, 6-des-OH-	$C_{18}H_{14}O_5$	310 OI 213^0C

JCS, 1967, 2500.
Tet., 1968, 1601.

2403-011

XANTHONE, 1-OH-3, 5-di-OMe-	$C_{15}H_{12}O_5$	272 OI 173^0C

R = H; R' = Me

Phytochem., 1971 (10)
2425.

2403-007

GARTANIN, 8-desoxy-	$C_{23}H_{24}O_5$	380 OI 165^0C

Tet., 1971, 3919.

2403-012

LICHEXANTHONE, nor-	$C_{14}H_{10}O_5$	258 OI 273°C

R = R' = H

Acta Chem. Scand., 1968, 1698.

2403-013

GENTISEIN	$C_{13}H_8O_5$	244 OI 317°C

R = R' = R'' = H

Tet., 1969, 1507.

2403-018

GRISEOXANTHONE-C	$C_{15}H_{12}O_5$	272 OI 255°C

R = Me; R' = H

JCS, 1960, 4628.

2403-014

GENTISIN	$C_{14}H_{10}O_5$	258 OI 270°C

R = R'' = H; R' = Me

JCS, 1927, 1983.

2403-019

LICHEXANTHONE	$C_{16}H_{14}O_5$	286 OI 190°C

R = R' = Me

Bull. Chem. Soc. Jap., 1942 (17) 202.

2403-015

GENTISIN, iso-	$C_{14}H_{10}O_5$	258 OI 241°C

R = R' = H; R'' = Me

JCS, 1927, 1983.

2403-020

ARTHOTHELIN	$C_{14}H_7O_5Cl_3$	361 OI 276°C

X = H

Acta Chem. Scand., 1968, 1698.

2403-016

GENTIOSIDE	$C_{25}H_{28}O_{14}$	552 -107°e 266°C

R = H; R' = Glu; R'' = Me

Gazz. Chim. Ital., 1955 (85) 1007.

2403-021

THIOPHANIC ACID	$C_{14}H_6O_5Cl_4$	395 243°C

X = Cl

Chem. Comm., 1968, 155.

2403-017

XANTHONE, 1-OH-3,7-diOMe-	$C_{15}H_{12}O_5$	272 OI 167°C

R = H; R' = R'' = Me

Tet., 1969, 1507.

2403-022

XANTHONE, 1, 3, 7-triOMe-	$C_{16}H_{14}O_5$	286 OI 174°C

R = R' = R'' = Me

Tet., 1969 (25) 1961.

2403-023

MESUAXANTHONE-B	$C_{13}H_8O_5$	244 OI 290°C

R = H

Tet., 1967, 243.

2403-028

MBARRAXANTHONE, iso-	$C_{18}H_{16}O_5$	312 OI

JCS, 1969, 486.

2403-024

BUCHANAXANTHONE	$C_{14}H_{10}O_5$	258 OI 244°C

R = Me

JCS, 1968, 2579.

2403-029

MBARRAXANTHONE	$C_{18}H_{16}O_5$	312 OI

JCS, 1966, 2265.

2403-025

XANTHONE, 1, 6, 7-triOH-	$C_{16}H_{14}O_5$	286 OI 187°C

JCS, 1969, 2201.

2403-030

OSAJAXANTHONE	$C_{18}H_{14}O_5$	310 OI 248°C

JCS, 1969, 486.

2403-026

TOVOXANTHONE	$C_{18}H_{14}O_5$	310 OI 226°C

Chem. Abstr., 1971 (74) 72454.

2403-031

RAVENELIN	$C_{14}H_{10}O_5$	258 OI 268°C

Biochem. J., 1936 (30) 1303.

2403-027

XANTHONE, 3, 4-diOH-2-OMe-	$C_{14}H_{10}O_5$	258 OI 244°C

R = H

Tet., 1966, 1777.

2403-032

XANTHONE, 4-OH-2, 3-diOMe-	$C_{15}H_{12}O_5$	272
		OI
		218^0C
R = Me		
		Tet., 1966, 1777.
2403-033		

STERIGMATOCYSTIN, desMe-	$C_{17}H_{10}O_6$	310
		-483^0c
		254^0C
R = R' = H		Chem. Comm., 1970, 1069.
2403-038		

XANTHONE, 3-OH-2, 4-diOMe-	$C_{15}H_{12}O_5$	272
		OI
		225^0C
		Chem. Abstr., 1969 (70) 823.
2403-034		

STERIGMATOCYSTIN	$C_{18}H_{12}O_6$	324
		-387^0c
		246^0C
R = Me; R' = H		
		Tet., 1968, 717.
2403-039		

XANTHONE, 4-OH-2, 3-methylene- -dioxy-	$C_{14}H_8O_5$	256
		OI
R = H		299^0C
		Phytochem., 1970, 2537.
2403-035		

STERIGMATOCYSTIN ME ETHER	$C_{19}H_{14}O_6$	338
		265^0C
R = R' = Me		
		Tet., 1968, 717.
2403-040		

XANTHONE, 4-OMe-2, 3-methylene- -dioxy-	$C_{15}H_{10}O_5$	270
		OI
		238^0C
R = Me		
		Tet., 1966, 1777.
2403-036		

STERIGMATOCYSTIN, dihydro- -O-Me-	$C_{19}H_{16}O_6$	340
		282^0C
		Tet. Lett., 1970, 3109.
2403-041		

MACULATOXANTHONE	$C_{28}H_{30}O_6$	462
		$+19^0m$
		175^0C
		Tet. Lett., 1969, 4893.
2403-037		

ASPERTOXIN	$C_{19}H_{14}O_7$	354
		Tet. Lett., 1968, 3207.
2403-042		

XANTHONE, 1, 3, 5-tri-OH-2-OMe- $C_{14}H_{10}O_6$ 274
OI

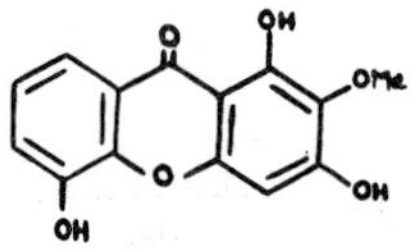

Ind. J. C., 1971 (9) 772.

2404-001

SYMPHOXANTHONE $C_{18}H_{16}O_6$ 328
OI
223^0C

JCS, 1966, 2186.

2404-006

XANTHONE, 1-OH-2, 3, 5-triOMe- $C_{16}H_{14}O_6$ 302
OI
188^0C

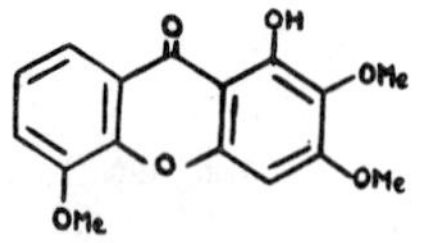

Tet., 1969 (25) 1961.

2404-002

SWERTIANIN, nor- $C_{13}H_8O_6$ 260
OI
335^0C

Chem. Pharm. Bull.,
1969, 155.

2404-007

XANTHONE, 1-OH-2, 3, 7-triOMe- $C_{16}H_{14}O_6$ 302
OI

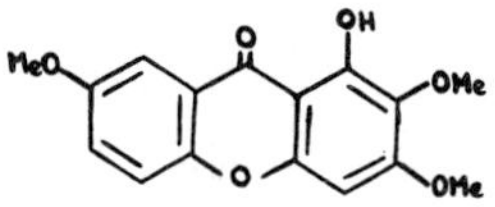

Tet., 1969 (25) 1961.

2404-003

XANTHONE, 1, 3, 8-tri-OH-7-OMe- $C_{14}H_{10}O_6$ 274
OI
299^0C

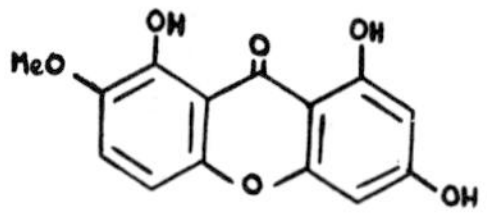

Phytochem., 1970, 2537.

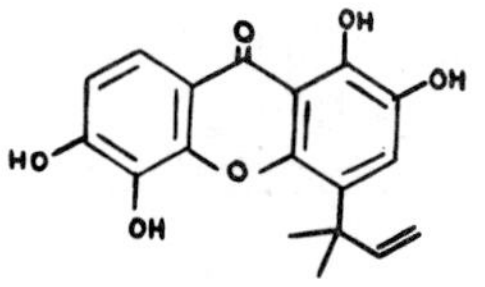

2404-008

XANTHONE, 2-OH-1, 3, 7-triOMe- $C_{16}H_{14}O_6$ 302
OI
192^0C

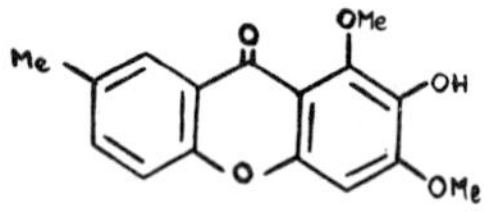

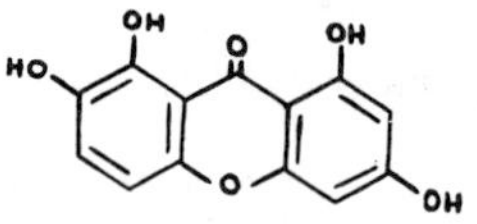

Tet., 1969 (25)1961.

2404-004

SWERTIANIN $C_{14}H_{10}O_6$ 274
OI
220^0C

Chem. Pharm. Bull.,
1969, 155.

2404-009

XANTHONE, 3, 8-di-OH-1, 2-di-
-OMe- $C_{15}H_{12}O_6$ 288
OI
169^0C

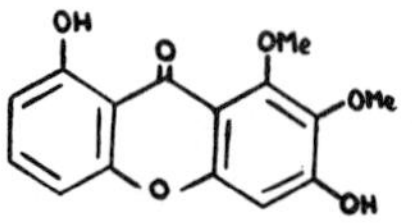

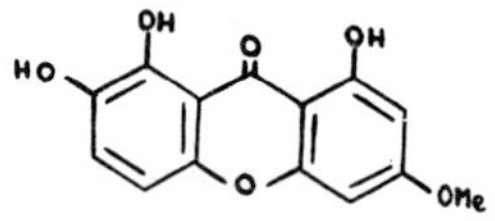

Ind. J. C., 1971 (9) 772.

2404-005

GENTIACAULEIN $C_{15}H_{12}O_6$ 288
OI

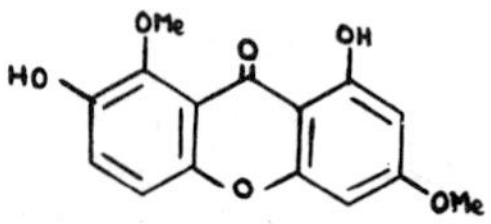

Chem. Abstr., 1967
(67) 64182.

2404-010

SWERTIANIN, Me-	$C_{15}H_{12}O_6$	288 OI 190°C

Chem. Pharm. Bull., 1969, 155.

2404-011

XANTHONE, 1-OH-3,4,5-triOMe-	$C_{16}H_{14}O_6$	302 OI

R = H; R' = Me

Tet., 1969 (25) 1961.

2404-016

SWERTININ	$C_{15}H_{12}O_6$	288 OI 217°C

J. Sci. Ind. Res., 1954 (13B) 175.

2404-012

XANTHONE, 1,3,4,5-tetra-OMe-	$C_{17}H_{16}O_6$	316 OI

R = R' = Me

Tet., 1969 (25) 1961.

2404-017

XANTHONE, 1-OH-3,7,8-triOMe-	$C_{16}H_{14}O_6$	302 OI

Chem. Abstr., 1969 (70) 3743.

2404-013

XANTHONE, 1,4,7-triOH-3-OMe-	$C_{14}H_{10}O_6$	274 OI 275°C

JCS, 1969, 2201.

2404-018

DECUSSATIN	$C_{16}H_{14}O_6$	302 OI 150°C

J. Sci. Ind. Res., 1954 (13B)175.

2404-014

XANTHONE, 1,3-diOH-4,7-diOMe-	$C_{15}H_{12}O_6$	288 OI

R = R' = H

Tet., 1969 (25) 1961.

2404-019

XANTHONE, 1,3-diOH-4,5-diOMe-	$C_{15}H_{12}O_6$	288 OI

R = R' = H

Tet., 1969 (25) 1947.

2404-015

XANTHONE, 1-OH-3,4,7-triOMe-	$C_{16}H_{14}O_6$	302 OI

R = H; R' = Me

Tet., 1969 (25) 1961.

2404-020

XANTHONE, 1,3,4,7-tetra-OMe- $C_{17}H_{16}O_6$ 316 OI R = R' = Me Tet., 1969 (25) 1961. 2404-021	UGAXANTHONE $C_{18}H_{16}O_6$ 328 OI JCS, 1966, 2265. 2404-026	
XANTHONE, 1,3,5,6-tetra-OH- $C_{13}H_8O_6$ 260 OI R = H JCS, 1966, 430. 2404-022	JACAREUBIN $C_{18}H_{14}O_6$ 326 OI 257°C JCS, 1953, 3932. R = H 2404-027	
XANTHONE, 1,3,5-tri-OH-6-OMe- $C_{14}H_{10}O_6$ 274 OI 286°C R = Me Phytochem., 1971 (10) 2425. 2404-023	MACLURAXANTHONE $C_{23}H_{22}O_6$ 394 OI 181°C R = 1,1-DMA JOC, 1964, 692. 2404-028	
XANTHONE, 1,3,5,6-tetra-OMe- $C_{17}H_{16}O_6$ 316 OI 143°C Phytochem., 1970, 450. 2404-024	BELLIDIFOLIN, des-Me- $C_{13}H_8O_6$ 260 OI 316°C Tet., 1965, 1449. R = R' = H 2404-029	
UGAXANTHONE, iso- $C_{18}H_{16}O_6$ 328 OI JCS, 1969, 486. 2404-025	BELLIDIFOLIN $C_{14}H_{10}O_6$ 274 OI 270°C R = Me; R' = H Tet., 1965, 1449. 2404-030	

SWERCHIRIN	$C_{15}H_{12}O_6$	288 OI 186°C

R = R' = Me

Tet., 1965, 1449.

2404-031

MANGIFERIN, homo- — $C_{20}H_{20}O_{11}$ — 436 +9° 250°C

R = Me

Chem. Pharm. Bull., 1968, 760.

2404-036

GARTANIN — $C_{23}H_{24}O_6$ — 396 OI 167°C

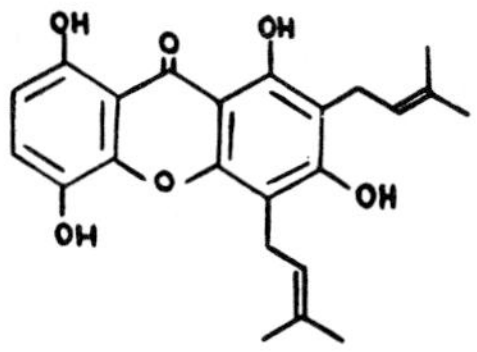

Tet., 1971, 3919.

2404-032

MANGIFERIN, iso- — $C_{19}H_{18}O_{11}$ — 422 +6°p 260°C

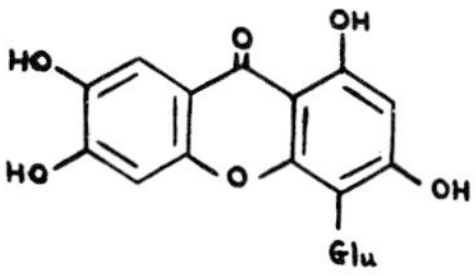

Tet. Lett., 1969, 941.

2404-037

STERIGMATOCYSTIN, 5-OMe- — $C_{19}H_{14}O_7$ — 354 -360°c 223°C

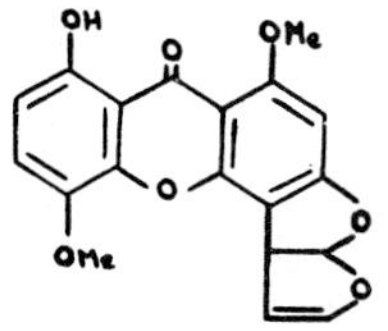

Chem. Comm., 1968, 1574.

2404-033

$C_{18}H_{16}O_6$ — 328 OI

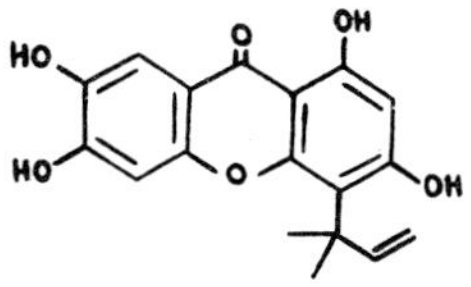

JCS, 1971, 1332.

2404-038

XANTHONE, 1,3,6,7-tetra-OH- — $C_{13}H_8O_6$ — 260 OI

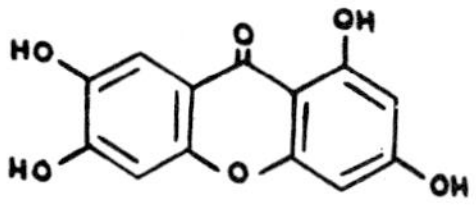

JCS, 1966, 430.

2404-034

MANGOSTIN, γ- — $C_{23}H_{24}O_6$ — 396 OI 207°C

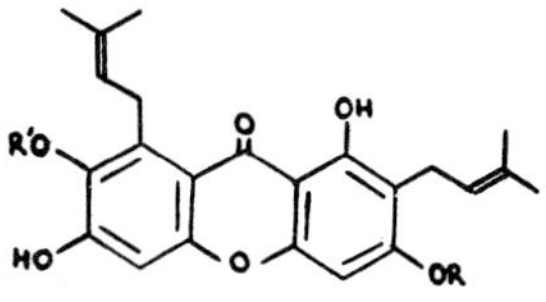

R = R' = H

Ind. J. C., 1971, 505.
Aust. J. C., 1970, 2541.

2404-039

MANGIFERIN — $C_{19}H_{18}O_{11}$ — 422 270°C

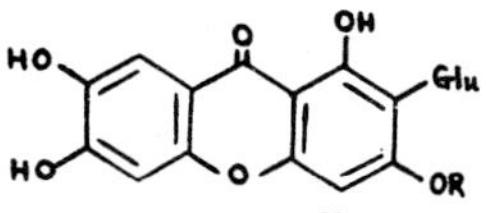

R = H

Bull. Chem. Soc. Jap., 1957 (30) 625.

2404-035

MANGOSTIN — $C_{24}H_{26}O_6$ — 410 OI 181°C

R = H; R' = Me

Chem. Comm., 1968, 211.

2404-040

MANGOSTIN, β-	$C_{25}H_{28}O_6$	424 OI 175^{0}C

R = R' = Me

Can. J. C., 1968, 3770.

2404-041

XANTHONE, 2-OH-1, 3, 4, 7-tetra- -OMe	$C_{17}H_{16}O_7$	332 OI 145^{0}C

R = H

Tet., 1969 (25) 1961.

2405-003

BELLIDIFOLIN, iso-	$C_{14}H_{10}O_6$	274 OI 264^{0}C

Tet., 1965, 1449.

2404-042

POLYGALAXANTHONE-B	$C_{18}H_{18}O_7$	346 OI 121^{0}C

R = Me

Bull. Soc. Chim. Fr., 1967, 130.

2405-004

CELIBIXANTHONE	$C_{19}H_{16}O_6$	342 OI 220^{0}C

Tet., 1963 (19) 667.

2404-043

POLYGALAXANTHONE-A	$C_{17}H_{14}O_7$	330 OI 170^{0}C

Tet., 1969, 5295.

2405-005

XANTHONE, 1-OH-2, 3, 4, 5-tetra- OMe-	$C_{17}H_{16}O_7$	332 OI

Tet., 1969 (25) 1961.

2405-001

XANTHONE, 1-OMe-2, 3:6, 7-bis- -CH$_2$OH-	$C_{16}H_{10}O_7$	314 OI 251^{0}C

Tet., 1969, 4415.

2405-006

XANTHONE, 1-OH-2, 3, 4, 7-tetra- -OMe-	$C_{17}H_{16}O_7$	332 OI 119^{0}C

Tet., 1969 (25) 1961.

2405-002

CORYMBIFERIN	$C_{15}H_{12}O_7$	304 OI 268^{0}C

Tet., 1965, 3687.

2405-007

BELLIDIN, 4,7-di-OMe- $C_{15}H_{12}O_7$ 304
OI
220^0C

Tet., 1965, 3687.

2405-008

SECALONIC ACID A $C_{32}H_{30}O_{14}$ 638
-73^0c
260^0C

A

Chem. Comm., 1971, 111.

2406-001

XANTHONE, 1,3,6,7,8-penta-OH- $C_{13}H_8O_7$ 276
OI

Phytochem., 1971 (10)
2425.

2405-009

SECALONIC ACID D $C_{32}H_{30}O_{14}$ 638
$+82^0$c
254^0C

A – A (Enantiomer)

Chem. Comm., 1971. 111.

2406-002

XANTHONE, 1,3,8-tri-OH-6,7-
-di-OMe- $C_{15}H_{12}O_7$ 304
OI
290^0C

R = H

Phytochem., 1971 (10)
2425.

2405-010

SECALONIC ACID B $C_{32}H_{30}O_{14}$ 638
$+196^0$
255^0C

B

Chem. Comm., 1971, 111.

2406-003

XANTHONE, 1,8-di-OH-3,6,7-tri-
-OMe- $C_{16}H_{14}O_7$ 318
OI
241^0C

R = Me

Phytochem., 1971 (10)
2425.

2405-011

SECALONIC ACID C $C_{32}H_{30}O_{14}$ 638
$+23^0$a
160^0C

A – B

Chem. Comm., 1971, 111.

2406-004

XANTHONE, 1-OH-3,6,7,8-tetra-
-OMe- $C_{17}H_{16}O_7$ 332
OI
171^0C

Phytochem., 1971 (10)
2425.

2405-012

ERGOFLAVIN $C_{30}H_{26}O_{14}$ 610
$+37^0$a
$>360^0$C

C

Tet., 1965, 1417.

2406-005

ERGOCHRYSIN-A	$C_{31}H_{28}O_{14}$ 624 -37^0a 198^0C	
A - C		
	Chem. Comm., 1971, 111.	
2406-006		

ERGOCHROME-CD	$C_{31}H_{30}O_{15}$ 642	
C - D		
	Tet. Lett., 1965, 2031.	
2406-011		

ERGOCHRYSIN-B	$C_{31}H_{28}O_{14}$ 624 $+82^0a$ 197^0C	
B - C		
	Tet., 1965, 1417.	
2406-007		

ERGOXANTHIN	$C_{31}H_{28}O_{14}$ 624 $+124^0c$ 186^0C	

Chem. Comm., 1971, 111.	
2406-012	

ERGOCHROME-DD	$C_{32}H_{34}O_{16}$ 674	

Tet. Lett., 1965, 2031.	
2406-008	

MORELLIN, desoxy-	$C_{33}H_{38}O_6$ 530	
R = R' = CH_3		

Ind. J. C., 1964 (2) 405.	
2499-001	

ERGOCHROME-AD	$C_{32}H_{32}O_{15}$ 656	
A - D		
	Tet. Lett., 1965, 2035.	
2406-009		

MORELLIN	$C_{33}H_{36}O_7$ 544	
R = CHO; R' = CH_3		
	Tet. Lett., 1963, 459.	
2499-002		

ERGOCHROME-BD	$C_{32}H_{34}O_{16}$ 674	
B - D		
	Tet. Lett., 1965, 2031.	
2406-010		

MORELLIN, iso-	$C_{33}H_{36}O_7$ 544	
R = CH_3; R' = CHO		
	Tet. Lett., 1963, 459.	
2499-003		

MORELLIN, dihydro-iso-	$C_{33}H_{38}O_7$	546	GUTTILACTONE, β-	$C_{29}H_{40}O_5$	468

MORELLIN, dihydro-iso- $C_{33}H_{38}O_7$ 546

R = CH_3 ; R' = CHO

(✻ dihydro)   Ind. J. C. , 1964 (2) 405.

2499-004

GUTTILACTONE, β- $C_{29}H_{40}O_5$ 468

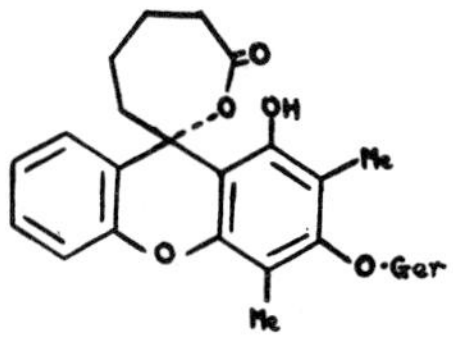

Angew Chem. , 1962 (1) 455.

2499-008

MORELLIC ACID $C_{33}H_{36}O_8$ 560
240^{0}C

R = COOH; R' = CH_3

Tet. Lett. , 1966, 687.

2499-005

ARUGOSIN-A $C_{25}H_{28}O_6$ 424

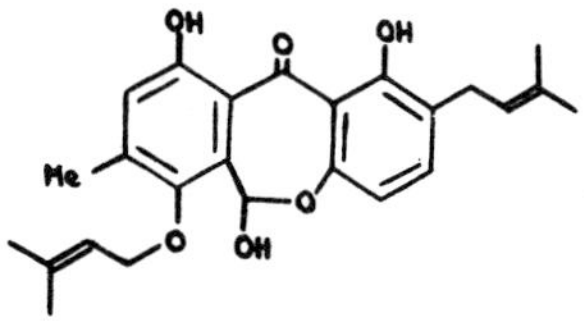

JCS, 1970, 1178.

2499-009

MORELLIC ACID, iso- $C_{33}H_{36}O_8$ 560
272^{0}C

R = CH_3 ; R' = COOH

Tet. Lett. , 1966, 687.

2499-006

ARUGOSIN-B $C_{25}H_{28}O_6$ 424

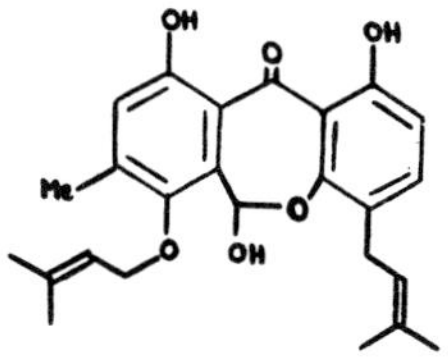

JCS, 1970, 1178.

2499-010

GAMBOGIC ACID $C_{38}H_{44}O_8$ 619

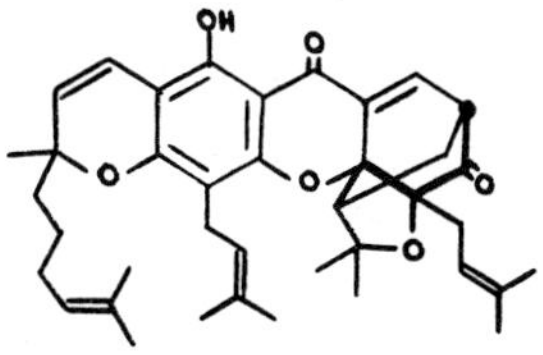

JCS, 1966, 772.

2499-007

APHLOIOL $C_{14}H_{14}O_8$ 310
304^{0}C

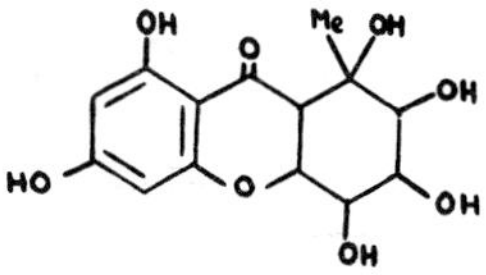

Bull. Soc. Chim. Fr. , 1964, 376.

2499-011

The naphthyl aromatics include the naphthalene phenolics and the naptho-quinones. While this group has been placed in the section of aromatic classes generally derived from acetate, there is now sufficient experimental evidence to indicate that shikimate too can be a precursor of some of the naphthyls (especially the 2-methylnaphthoquinones).

The classification employed here is based upon the degree of oxygenation of the naphthyl nucleus (2501-2508) with some further classes reserved for the binaphthyls (2509-2511).

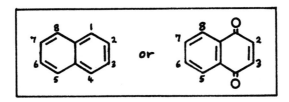

2501 Mono-oxy	2502 Di-oxy
2503 Tri-oxy	2504 Tetra-oxy
2505 Penta-oxy	2506 Hexa-oxy
2507 Hepta-oxy	2508 Octa-oxy
2509 1, 1'-binaphthyls	2510 2, 2'-binaphthyls
2511 1, 2'-binaphthyls	2599 Misc.

NAPHTHALENE, 1-OMe- R = Me JCS, 1968, 850. 2501-001	$C_{11}H_{10}O$ 158 OI Oil	NAPHTHOL, 2-Ac-1- Curr. Sci., 1970 (39) 209. 2501-006 $C_{12}H_{10}O_2$ 186 OI

NAPHTHALENE, 1-OMe-

R = Me

$C_{11}H_{10}O$ 158
OI
Oil

JCS, 1968, 850.

2501-001

NAPHTHOL, 2-Ac-1-

$C_{12}H_{10}O_2$ 186
OI

Curr. Sci., 1970 (39) 209.

2501-006

NAPHTHOL, O-DMA-1-

R = 3,3-DMA

$C_{15}H_{16}O$ 212
OI
Oil

JCS, 1968, 854.

2501-002

NAPHTHOL, 4-OMe-1-

$C_{11}H_{10}O_2$ 174
OI
125^0C

JCS, 1968, 854.

2502-001

NAPHTHOL, O-i-Val-1-

R = isopentyl

$C_{15}H_{18}O$ 214
OI
Oil

JCS, 1968, 854.

2501-003

RENIFOLIN

R, R' = (H, Glu) ?

$C_{18}H_{24}O_7$ 352

JCS, 1968, 857.

2502-002

$C_{15}H_{16}O$ 212
OI
pic: 136^0C

Natwiss., 1967 (54) 118.
JCS, 1968, 854.

2501-004

NAPHTHOQUINONE, 2-Me-

$C_{11}H_8O_2$ 172
OI
106^0C

J. Biol. Chem., 1940
(133) 391.

2502-003

$C_{15}H_{14}O$ 210
OI
44^0C

JCS, 1968, 854.

2501-005

NAPHTHOQUINONE, 6-Me-

$C_{11}H_8O_2$ 172
OI
90^0C

Acta Chem. Scand.,
1948 (2) 192.

2502-004

CHIMAPHILIN

$C_{12}H_{10}O_2$ 186
OI
114^0C

Natwiss., 1968, 445.

2502-005

MENAQUINONE-1

$C_{16}H_{16}O_2$ 240
OI
Oil

n = 1

JCS, 1967, 2100.

2502-010

LAPACHOL, deoxy-

$C_{15}H_{14}O_2$ 226
OI
60^0C

JCS, 1967, 2100.

2502-006

MENAQUINONE-6

$C_{41}H_{56}O_2$ 580
OI
50^0C

n = 6

Helv., 1958 (41) 786.

2502-011

LAPACHENOLE

$C_{16}H_{16}O_2$ 240
OI
62^0C

JCS, 1968, 850.

2502-007

MENAQUINONE-7

$C_{46}H_{64}O_2$ 648
OI
54^0C

n = 7

J. Biol. Chem., 1939
(131) 327.

2502-012

LAPACHENOLE, nor-dihydro-

$C_{15}H_{16}O_2$ 228
OI
127^0C

R = H

JCS, 1968, 850.

2502-008

MENAQUINONE-8

$C_{51}H_{72}O_2$ 716
OI

n = 8

Nature, 1964 (204) 80.

2502-013

LAPACHENOLE, dihydro-

$C_{16}H_{18}O_2$ 242
OI
77^0C

R = Me

JCS, 1968, 850.

2502-009

MENAQUINONE-9

$C_{56}H_{80}O_2$ 784
OI
59^0C

n = 9

Helv., 1960 (43) 433.

2502-014

MENAQUINONE-8, dihydro- $C_{51}H_{74}O_2$ 718

m = 6

Biochem. J., 1965 (97)
766.

2502-015

MENAQUINONE-6, 2-desMe- $C_{40}H_{54}O_2$ 566
OI

n = 6

Biochemistry, 1964 (3)
949.

2502-020

MENQUINONE-9, dihydro- $C_{56}H_{82}O_2$ 786

m = 7

Chem. Abstr., 1967
(67) 53905.

2502-016

MENAQUINONE-7, 2-desMe- $C_{45}H_{62}O_2$ 634
OI

n = 7

Biochemistry, 1964 (3)
949.

2502-021

PHYLLOQUINONE $C_{31}H_{46}O_2$ 450

Oil

Helv., 1964 (47) 221.

2502-017

MENAQUINONE-8, 2-des-Me- $C_{50}H_{70}O_2$ 702
OI

n = 8

Biochemistry, 1969 (8)
4651.

2502-022

CHLOROBIUMQUINONE $C_{46}H_{62}O_3$ 662
OI
50^0C

JACS, 1969, 6889.

2502-018

MENAQUINONE-9, 2-des-Me- $C_{55}H_{78}O_2$ 770
OI

n = 9

J. Biol. Chem., 1965
(240) 3425.

2502-023

MENAQUINONE-5, 2-desMe- $C_{35}H_{46}O_2$ 498
OI

n = 5

Biochemistry, 1964 (3)
949.

2502-019

NAPHTHA-1.8-DIOL ME ETHER $C_{11}H_{10}O_2$ 174

JCS 1960. 654.

2502-024

NAPHTHA-1,8-DIOL DIME ETHER $C_{12}H_{12}O_2$ 188
OI
159^0C

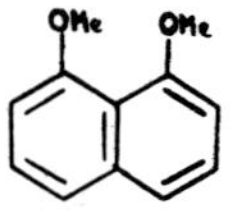

JCS, 1960, 654.

2502-025

MUSIZIN $C_{13}H_{12}O_3$ 216
OI

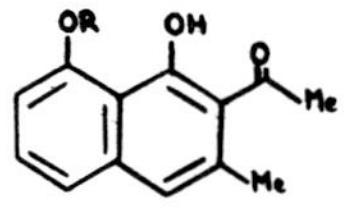

R = H

JCS, 1963, 1340.

2502-030

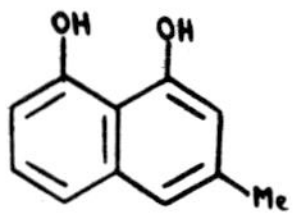

NAPHTHALENE-1,8-DIOL, 3-Me- $C_{11}H_{10}O_2$ 174
OI
140^0C

JCS, 1965, 1533.

2502-026

MUSIZIN, 8-glucosyl- $C_{19}H_{22}O_8$ 378
-119^0e
203^0C

R = Glucosyl

JCS, 1963, 1340.

2502-031

NAPHTHALDEHYDE, 4-OH-5-
-OMe-2-
R = H; R' = Me $C_{12}H_{10}O_3$ 202
OI
89^0C

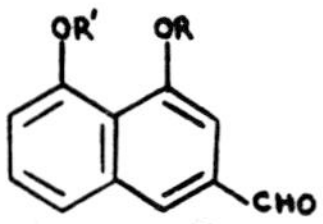

Phytochem., 1971 (10)
458.

2502-027

DIANELLIN $C_{25}H_{32}O_{12}$ 524
-119^0e
167^0C

R = Rutinosyl

Aust. J. C., 1961 (14) 637.

2502-032

NAPHTHALDEHYDE, 5-OH-4-OMe- $C_{12}H_{10}O_3$ 202
OI
104^0C

R = Me; R' = H

JCS, 1970, 626.

2502-028

SORIGENIN, β- $C_{12}H_8O_4$ 216
OI
238^0C

R = R' = H

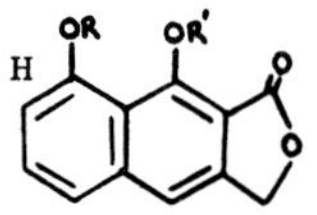

Chem. Ind., 1959, 1576.

2502-033

NAPHTHALDEHYDE, 4,5-diOMe-2- $C_{13}H_{12}O_3$ 216
OI
109^0C

R = R' = Me

JCS, 1970, 626.

2502-029

SORININ, β- $C_{23}H_{26}O_{13}$ 510

amorph.

R, R' = (H, Primeverosyl) ?

J. Agr. Chem. Soc. Jap.,
1944, 283.

2502-034

ELEUTHEROL 2502-035	$C_{14}H_{12}O_4$ 244 +90°c 203°C Helv., 1950 (33) 595.	

NAPHTHOIC ACID, 6, OH-5, 7-di- -OMe-2- 2503-001	$C_{13}H_{12}O_5$ 248 OI 216°C Tet., 1969, 2334.

XANTHORRHOEIN 2502-036	$C_{14}H_{14}O_2$ 214 +56°b 68°C Tet. Lett., 1964, 1623.

NAPHTHOIC ACID LACTONE, -6-OH-3-CH₂OH-5, 7-diOMe-2- 2503-002	$C_{14}H_{12}O_5$ 260 OI 189°C Tet., 1969, 3226.

XANTHORRHOEOL R = H 2502-037	$C_{15}H_{14}O_3$ 242 +143°c 121°C Aust. J. C., 1965, 575.

LAWSONE R = H 2503-003	$C_{10}H_6O_3$ 174 OI 193°C Tet. Lett., 1969, 4777.

XANTHORRHOEIN, Ac- R = Me 2502-038	$C_{16}H_{16}O_3$ 256 +127°c 125°C Aust. J. C., 1965, 575. Tet. Lett., 1964, 1623.

LAWSONE ME ETHER R = Me 2503-004	$C_{11}H_8O_3$ 188 OI 183°C J. Biol. Chem., 1948 (174) 335.

NAPHTHALENE, 2, 3-methylene- -dioxy- 2502-039	$C_{11}H_8O_2$ 172 OI 96°C Bull. Chem. Soc. Jap., 1969 (42) 3591.

PHTHIOCOL 2503-005	$C_{11}H_8O_3$ 188 OI 174°C Nature, 1949 (163) 365.

LAPACHOL	$C_{15}H_{14}O_3$	242 OI 140^{0}C

R = H

JCS, 1968, 850.

2503-006

LAPACHONE, β-	$C_{15}H_{14}O_3$	242 OI 155^{0}C

JCS, 1968, 850.

2503-011

LAPACHOL ME ETHER	$C_{16}H_{16}O_3$	256 OI 55^{0}C

R = Me

JCS, 1967, 2100.

2503-007

LAPACHONE, dehydro-iso-α-	$C_{15}H_{12}O_3$	240 -32^{0}b 105^{0}C

Natwiss., 1968 (55)38.

2503-012

LOMATIOL	$C_{15}H_{14}O_4$	258 OI 128^{0}C

JACS, 1948, 617.

2503-008

NAPHTHOQUINONE, α-Et-furano-	$C_{14}H_{10}O_3$	226 OI 143^{0}C

Natwiss., 1968 (55) 38.

2503-013

LAPACHONE, α-	$C_{15}H_{14}O_3$	242 OI 119^{0}C

JCS, 1967, 2100.

2503-009

NAPHTHOPYRAN-5, 10-DIONE, -3-Me-	$C_{14}H_{10}O_3$	226 OI 193^{0}C

Natwiss., 1968 (55)38.

2503-014

LAPACHONE, dehydro-α-	$C_{15}H_{12}O_3$	240 OI 143^{0}C

JCS, 1967, 2100.

2503-010

NAPHTHOPYRAN-5, 10-DIONE, -3-OH-2-Me-	$C_{14}H_{10}O_4$	242 +45^{0}p 185^{0}C

Natwiss., 1968 (55) 38.

2503-015

DUNNIONE	$C_{15}H_{14}O_3$	242 +310°c 98°C	DIOSQUINONE	$C_{10}H_6O_3$	174 OI 199°C
	JCS, 1940, 1493.			JOC, 1964, 3617.	
2503-016			2503-021		

MACASSAR-II	$C_{13}H_{14}O_3$	218 OI 107°C	NAPHTHOQUINONE, 8-OMe-3- -Me-1, 2-	$C_{12}H_{10}O_3$	202 OI 144°C
R = H	JCS, 1965, 2355.			Ind. J. C. , 1968 (6) 681.	
2503-017			2503-022		

MACASSAR-III	$C_{14}H_{16}O_3$	232 OI 70°C	MUSIZIN, 6-OH-	$C_{13}H_{12}O_4$	232 OI
R = Me	JCS, 1965, 2355.		R = H	Tet. Lett. , 1969, 471.	
2503-018			2503-023		

NAPHTHALDEHYDE, 6-OH-4, 5- -diOMe-2-	$C_{13}H_{12}O_4$	232 OI 164°C	MUSIZIN-8-GLUCOSIDE, 6-OH-	$C_{19}H_{22}O_9$	394 214°C
R = H	JCS, 1970, 626.		R = Glucosyl	Tet. Lett. , 1969, 471.	
2503-019			2503-024		

NAPHTHALDEHYDE, 4, 5, 6-tri- -OMe-2-	$C_{14}H_{14}O_4$	246 OI 123°C	TORACHRYSONE	$C_{14}H_{14}O_4$	246 OI 214°C
R = Me	JCS, 1970, 626.			Chem. Pharm. Bull. , 1969, 454.	
2503-020			2503-025		

SORIGENIN, α- $C_{13}H_{10}O_5$ 246
OI
228^0C

R = R' = H

Agr. Biol. Chem.,
1963 (27) 40.

2503-026

FLUORAPHIN, Ac- $C_{23}H_{26}O_{11}$ 478

R = 6-Ac-glucosyl

Tet. Lett., 1969, 471.

2503-031

SORININ, α- $C_{24}H_{28}O_{14}$ 540
159^0C

R, R' = (H, Primeverosyl) ?

Chem. Abstr., 1939
(33) 6301.

2503-027

$C_{19}H_{18}O_6$ 342
120^0C

Tet. Lett., 1971, 3501.

2503-032

NERIAPHIN, dihydro- $C_{21}H_{26}O_{10}$ 438

Tet. Lett., 1969, 471.

2503-028

$C_{17}H_{16}O_5$ 300
183^0C

Tet. Lett., 1971, 3501.

2503-033

NERIAPHIN $C_{21}H_{24}O_{10}$ 436
214^0C

* oxo

Tet. Lett., 1969, 471.

2503-029

FLAVOSPERONE $C_{16}H_{14}O_4$ 270
OI

JCS, 1963, 4868.

2503-034

FLUORAPHIN $C_{21}H_{24}O_{10}$ 436

R = Glucosyl

Tet. Lett., 1969, 471.

2503-030

ELEUTHERINOL $C_{15}H_{12}O_4$ 256
OI
300^0C

Aust. J. C., 1953 (6) 373.

2503-035

NAPHTHALDEHYDE, 4, 5, 8- -triOMe-2-	$C_{14}H_{14}O_4$	246 OI 112^{0}C

JCS, 1970, 626.

2503-036

JUGLONE, 7-Me-dihydro-	$C_{11}H_{10}O_3$	190 OI

*dihydro

Nature, 1952 (169) 974.

2503-041

JUGLONE, α-hydro-	$C_{10}H_8O_3$	176 OI 169^{0}C

R = H

JCS, 1955, 904.

2503-037

STYPANDRONE	$C_{13}H_{10}O_4$	230 OI 136^{0}C

Aust. J. C. , 1965 (18) 218.

2503-042

JUGLONE 4-GLUCOSIDE, -α-hydro-	$C_{16}H_{18}O_8$	338 -97^{0}m 216^{0}C

R = Glucosyl

JCS, 1955, 904.

2503-038

SHINANOLONE	$C_{11}H_{12}O_3$	192 -23^{0}c 110^{0}C

Chem. Pharm. Bull. , 1971, 2314.

2503-043

JUGLONE	$C_{10}H_6O_3$	174 OI 154^{0}C

JCS, 1955, 904.

2503-039

SHINANOLONE, iso-	$C_{11}H_{12}O_3$	192

*epimer

Chem. Pharm. Bull. , 1971, 2314.

2503-044

JUGLONE, 7-Me-	$C_{11}H_8O_3$	188 OI 126^{0}C

Nature, 1952 (169) 974.

2503-040

SCLERONE	$C_{10}H_{10}O_3$	178

Agr. Biol. Chem. Jap. , 1968 (32) 1471.

2503-045

PLUMBAGIN

$C_{11}H_8O_3$ 188   OI 77^0C

R = H

Nature, 1952 (169) 974.

2503-046

MOLLISIN

$C_{14}H_{10}O_4Cl_2$ 313   OI 202^0C

Rec. Trav. Chim.,
1964 (83) 1023.

2503-051

PLUMBAGIN ME ETHER

$C_{12}H_{10}O_3$ 202   OI 94^0C

R = Me

Ind. J. C., 1968 (6) 681.

2503-047

NAPHTHOQUINONE, 2, 3-diOMe-

$C_{12}H_{10}O_4$ 218   OI 115^0C

2504-001

PLUMBAGIN, 3-Cl-

$C_{11}H_7O_3Cl$ 222   OI 125^0C

Acta Chem. Scand.,
1968, 2722.

2503-048

NAPHTHALENONE, 3, 4-dihydro-
-3, 4, 8-tri-OH-

$C_{10}H_{10}O_4$ 194 -36^0m 191^0C

Tet. Lett., 1972, 13.

2504-002

ELEUTHERIN

$C_{16}H_{16}O_4$ 272 $+346^0$c 175^0C

JCS, 1951, 612.

2503-049

JUGLONE, 2-OH-6-Et-

$C_{12}H_{10}O_4$ 218   OI 220^0C

JOC, 1966, 3645.

2504-003

ELEUTHERIN, iso-

$C_{16}H_{16}O_4$ 272 -46^0c 177^0C

* epimer

JCS, 1951, 612.

2503-050

VENTILAGONE

$C_{16}H_{16}O_5$ 288 $+430^0$c 205^0C

Aust. J. C., 1963 (16)
695.

2504-004

deleted 2504-005	DROSERONE-5-ME-ETHER $\quad$ C$_{12}$H$_{10}$O$_4$ $\quad$ 218 $\quad$ OI $\quad$ 173^{0}C R = Me Ind. J. C., 1968 (6) 681. 2504-010
NAPHTHALDEHYDE, 5-OH-4, 6, 8- -triOMe-2- $\quad$ C$_{14}$H$_{14}$O$_5$ $\quad$ 262 $\quad$ OI $\quad$ 174^{0}C JCS, 1970, 626. R = H 2504-006	deleted 2504-011
NAPHTHALDEHYDE, 4, 5, 6, 8- -tetra-OMe-2- $\quad$ C$_{15}$H$_{16}$O$_5$ $\quad$ 276 $\quad$ OI $\quad$ 120^{0}C R = Me JCS, 1970, 626. 2504-007	DIOMELQUINONE-A $\quad$ C$_{12}$H$_{10}$O$_4$ $\quad$ 218 $\quad$ OI $\quad$ 152^{0}C R = H Ann., 1966 (691) 172. 2504-012
JUGLONE, 3-OMe-7-Me- $\quad$ C$_{12}$H$_{10}$O$_4$ $\quad$ 218 $\quad$ OI $\quad$ 210^{0}C Chem. Pharm. Bull., 1971, 851. 2504-008	DIOMELQUINONE-A ME ETHER $\quad$ C$_{13}$H$_{12}$O$_4$ $\quad$ 232 $\quad$ OI $\quad$ 184^{0}C R = Me Ind. J. C., 1968 (6) 681. 2504-013
DROSERONE $\quad$ C$_{11}$H$_8$O$_4$ $\quad$ 204 $\quad$ OI $\quad$ 181^{0}C R = H JCS, 1937, 1597. 2504-009	CARYOPTERONE, α- $\quad$ C$_{15}$H$_{12}$O$_4$ $\quad$ 256 $\quad$ OI $\quad$ 144^{0}C Helv., 1969, 808. 2504-014

LAMBERTELLIN $C_{14}H_8O_5$ 256
OI
253^0C

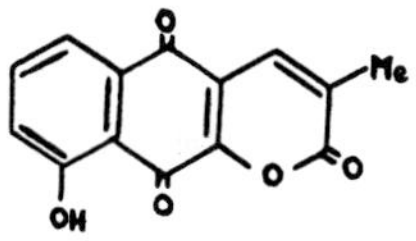

JCS, 1970, 109.

2504-015

RUBROFUSARIN GENTIOBIOSIDE $C_{27}H_{32}O_{15}$ 596
-62^0c
187^0C

R = Me; R' = Gentiobiosyl

Chem. Pharm. Bull.,
1969, 458.

2504-020

SAPTARANGI QUINONE A $C_{33}H_{52}O_5$ 528
178^0C

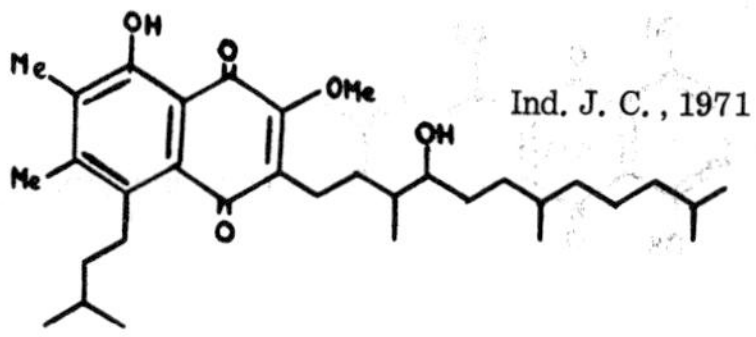

Ind. J. C., 1971 (9) 117.

2504-016

COMANTHERIN SULPHATE
(SODIUM SALT) $C_{16}H_{13}O_8SNa$ 388
OI

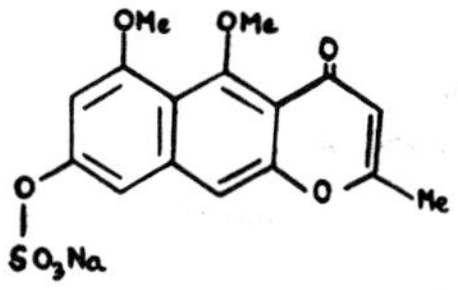

Aust. J. C., 1970, 2325.

2504-021

RUBROFUSARIN, nor- $C_{14}H_{10}O_5$ 258
OI

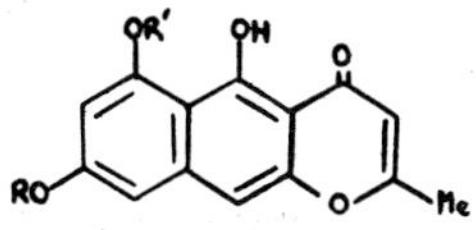

R = R' = H

Chem. Abstr., 1967
(66) 94864.

2504-017

COMAPARVIN SULPHATE $C_{17}H_{16}O_6S$ 380
OI

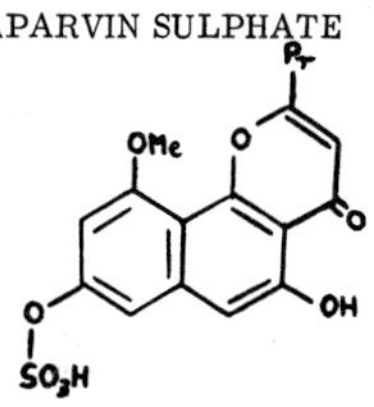

Aust. J. C., 1971, 1487.

2504-022

FONSECIN $C_{15}H_{12}O_5$ 272
OI
198^0C

R = H; R' = Me

Nature, 1962 (195) 502.

2504-018

NAPHTHAZARIN, 2-Me- $C_{11}H_8O_4$ 204
OI
174^0C

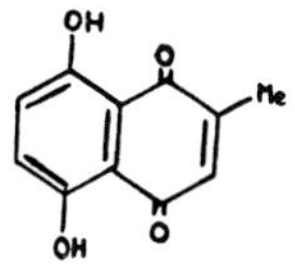

Acta Chem. Scand.,
1968, 2722.

2504-023

RUBROFUSARIN $C_{15}H_{12}O_5$ 272
OI
211^0C

R = Me; R' = H

Chem. Ind., 1961, 289.

2504-019

NAPHTHAZARIN, 2, 7-di-Me- $C_{12}H_{10}O_4$ 218
OI
125^0C

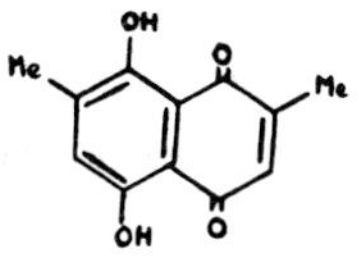

Biochemistry, 1965 (4)
1150.

2504-024

ALKANNAN	$C_{16}H_{18}O_4$	274
		OI
		99^0C

Ann., 1936 (521) 1.

2504-025

ALKANNIN, O-tetracryl-	$C_{22}H_{24}O_6$	384
		-169^0e

Tet. Lett., 1966, 3677.

2504-030

ALKANNIN	$C_{16}H_{16}O_5$	288
		(-)
		148^0C

R = H

Natwiss., 1935, 246.

2504-026

SHIKONIN	$C_{16}H_{16}O_5$	288
		145^0C

R = H

Chem. Ind., 1961, 947.

2504-031

ALKANNIN, O-Ac-	$C_{18}H_{18}O_6$	330
		104^0C

R = Ac

Phytochem., 1971 (10)
1909.

2504-027

SKIKONIN, O-Ac-	$C_{18}H_{18}O_6$	330
		$+26^0$c
		85^0C

R = Ac

Phytochem., 1969 (8)
1587.

2504-032

ALKANNIN, O-senecioyl-	$C_{21}H_{22}O_6$	370
		116^0C

R = Senecioyl

Phytochem., 1971 (10)
1909.

2504-028

SHIKONIN, isobutyryl-	$C_{20}H_{22}O_6$	358
		90^0C

R = Isobutyl

Tet. Lett., 1965, 4737.

2504-033

ALKANNIN, O-Angeloyl-	$C_{21}H_{22}O_6$	330
		Oil

R = Angeloyl

Ber., 1935 (68) 1487.

2504-029

SHIKONIN, senecioyl-	$C_{21}H_{22}O_6$	370
		114^0C

R = Senecioyl

Tet. Lett., 1965, 4737.

2504-034

R = $C_{21}H_{24}O_7$ 388

91^{0}C

Tet. Lett., 1966, 3677.

2504-035

FLAVIOLIN $C_{10}H_6O_5$ 206
OI
250^{0}C

R = H

Chem. Ind., 1954, 1110.

2505-001

ARNEBIN-5 $C_{16}H_{18}O_5$ 290
OI
111^{0}C

Phytochem., 1971 (10)
1909.

2504-036

FLAVIOLIN DI-ME ETHER $C_{12}H_{10}O_5$ 234
OI
267^{0}C

R = Me

Chem. Comm., 1968, 71.
JOC, 1966, 1496.

2505-002

ARNEBIN-6 $C_{18}H_{20}O_7$ 348

89^{0}C

Phytochem., 1971 (10)
1909.

2504-037

NAPHTHOQUINONE, 6-Et-5-OH-2, $C_{14}H_{14}O_5$ 262
-7-di-OMe-   OI
187^{0}C

Experientia, 1969 (25)
474.

2505-003

ARNEBIN-2 $C_{21}H_{24}O_6$ 372

Phytochem., 1971 (10)
1909.

2504-038

JUGLONE, 2, 7-diOH-6-Ac- $C_{12}H_8O_6$ 248
OI
215^{0}C

JOC, 1966, 3645.

2505-004

FRENOLICIN $C_{18}H_{18}O_7$ 346
-38^{0}m
160^{0}C

JACS, 1968, 1325.

2504-039

COMAPARVIN SULPHATE, 6-OMe- $C_{18}H_{18}O_9S$ 410
OI

Aust. J. C., 1971, 1487.

R = H

2505-005

COMAPARVIN-5-ME-ETHER SULPHATE, 6-OMe-	$C_{19}H_{20}O_9S$	424 OI
R = Me		
	Aust. J. C., 1971, 1487.	
2505-006		

NAPHTHAZARIN, 6-Et-2-OH-	$C_{12}H_{10}O_5$	266 OI 204°C
R = Et		

JOC, 1966, 3645.

2505-011

NAPHTHOPURPURIN	$C_{10}H_6O_5$	206 OI 200°C
R = H		

JOC, 1966, 3645.

2505-007

NAPHTHAZARIN, 6-Ac-2-OH-	$C_{12}H_8O_6$	248 OI 180°C
R = Ac		
	JOC, 1966, 3645.	
2505-012		

DROSERONE, OH-	$C_{11}H_8O_5$	220 OI 192°C
R = Me		
	JCS, 1935, 334.	
2505-008		

JAVANICIN, nor-	$C_{14}H_{12}O_6$	276 OI 202°C
R = H		

JOC, 1968, 4299.

2505-013

NAPHTHOQUINONE, 2-Et-3,5,8--triOH-	$C_{12}H_{10}O_5$	234 OI 187°C
R = Et		
	Monatsh., 1947 (77) 251.	
2505-009		

JAVANICIN	$C_{15}H_{14}O_6$	290 OI 208°C
R = Me		
	JCS, 1947, 1021.	
2505-014		

NAPHTHAZARIN, 2-OH-3-Ac-	$C_{12}H_8O_6$	248 OI 164°C
R = Ac		
	JOC, 1966, 3645.	
2505-010		

SOLANIOL	$C_{15}H_{16}O_6$	292 +122°m 192°C

Tet., 1968, 4745.

2505-015

FUSARUBIN $C_{15}H_{14}O_7$ 306

218^{0}C

Can. J. C. , 1965 (43) 2423.

2505-016

NAPHTHAZARIN, 2- Me- 8- OH- 2- -H-pyrano- $C_{14}H_{10}O_6$ 274
OI
170^{0}C

Tet. Lett. , 1968, 4581.

2505-021

FUSARUBIN, O-deMe-anhydro- $C_{14}H_{10}O_6$ 274
OI
203^{0}C

R = H

JCS, 1970, 930.

2505-017

SPINOCHROME- B $C_{10}H_6O_6$ 222
OI
> 260^{0}C

R = H

JOC, 1966, 3645.

2506-001

FUSARUBIN, anhydro- $C_{15}H_{12}O_6$ 288
OI
193^{0}C

R = Me

JCS, 1947, 1021.

2505-018

JUGLONE, 2, 3, 7-triOH-6-Et- $C_{12}H_{10}O_6$ 250
OI
266^{0}C

R = Et

JOC, 1966, 3645.

2506-002

HAEMOVENTUSIN $C_{15}H_{22}O_7$ 314

Acta Chem. Scand. , 1971 (25) 483.

2505-019

JUGLONE, 2, 3, 7-tri- OH-6-Ac- $C_{12}H_8O_7$ 264
OI
250^{0}C

R = Ac

JOC, 1966, 3645.

2506-003

NAPHTHOQUINONE, 2-acetonyl- -3, 5-di- OH-7- OMe- $C_{14}H_{12}O_6$ 276
OI
208^{0}C

Aust. J. C. , 1972, 875.

2505-020

CORDEAUXIONE $C_{14}H_{12}O_7$ 292
OI
194^{0}C

Helv. , 1955, 215.

2506-004

JUGLONE, 3-Et-2,6,7-tri-OH- $C_{12}H_{10}O_6$ 250
OI
220^0C

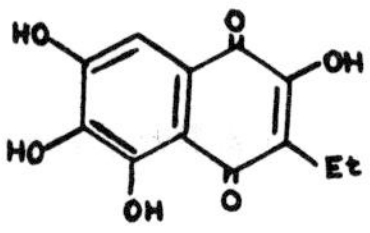

Experientia, 1967 (23)
624.

2506-005

SPINOCHROME-A $C_{12}H_8O_7$ 264
OI
192^0C

R = H

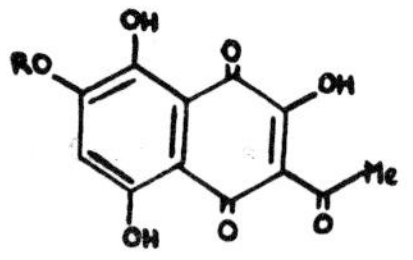

JOC, 1966, 3645.

2506-010

SPINOCHROME-S $C_{12}H_8O_7$ 264
OI
278^0C

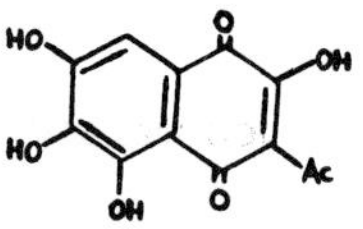

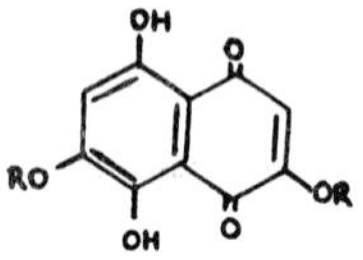

Aust. J. C., 1967, 1693.

2506-006

NAPHTHAZARIN, 3-Ac-2-OH-7-OMe- $C_{13}H_{10}O_7$ 278
OI
247^0C

R = Me

Experientia, 1967 (23)
624.

2506-011

MOMPAIN $C_{10}H_6O_6$ 222
OI
265^0C

Chem. Pharm. Bull.,
1967, 380.

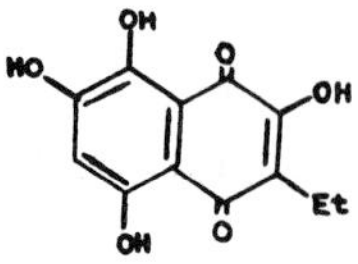

R = H

2506-007

$C_{14}H_{10}O_6$ 274
OI
166^0C

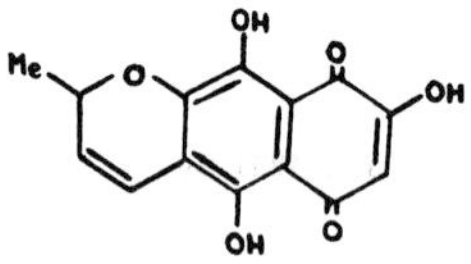

Tet. Lett., 1968, 4581.

2506-012

NAPHTHOQUINONE, 5,8-diOH-
-2,7-diOMe- $C_{12}H_{10}O_6$ 250
OI
276^0C

R = Me

JCS, 1966, 1496.

2506-008

ERYTHROSTOMINOL, desoxy- $C_{17}H_{18}O_7$ 334

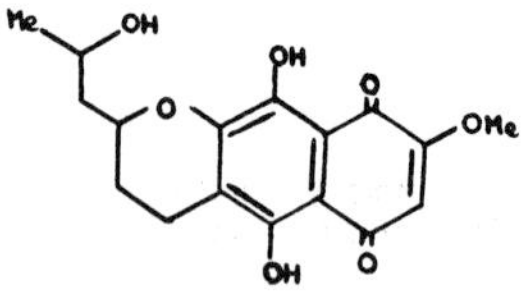

Chem. Comm., 1970, 209.

2506-013

NAPHTHAZARIN, 2,7-di-OH-3-Et- $C_{12}H_{10}O_6$ 250
OI
191^0C

JOC, 1966, 3645.

2506-009

ERYTHROSTOMINONE, desoxy- $C_{17}H_{16}O_7$ 332
R = H

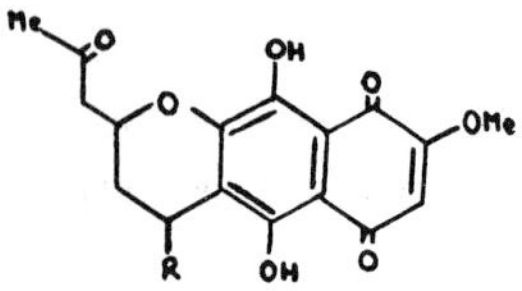

Chem. Comm., 1970, 209.

2506-014

ERYTHROSTOMINONE	$C_{17}H_{16}O_8$	348 +231° 185°C
R = OH		
	Chem. Comm., 1970, 209.	
2506-015		

NAPHTHAZARIN, 6-Et-3,7-di-OH- -2-OMe-	$C_{13}H_{12}O_7$	280 OI 203°C
	Tet., 1967 (23) 3271.	
2507-004		

LOMANDRONE	$C_{15}H_{16}O_6$	292 OI 118°C
	Aust. J. C., 1970, 1695.	
2506-016		

NAPHTHAZARIN, 6-Et-2,7-di-OH- -3-OMe-	$C_{13}H_{12}O_7$	280 OI 180°C
	Tet., 1967 (23) 3271.	
2507-005		

SPINOCHROME-D	$C_{10}H_6O_7$	238 OI sub. 285°C
R = H		
	JOC, 1966, 3645.	
2507-001		

LOMASTILONE	$C_{15}H_{14}O_8$	322 172°C
	Aust. J. C., 1970, 1695.	
2507-006		

ECHINOCHROME-A	$C_{12}H_{10}O_7$	266 OI 223°C
R = Et		
	JOC, 1966, 3645.	
2507-002		

LOMAZARIN	$C_{15}H_{16}O_8$	324 100°C
	Aust. J. C., 1970, 1695.	
2507-007		

SPINOCHROME-C	$C_{12}H_8O_8$	280 OI 247°C
R = Ac		
	JOC, 1966, 3645.	
2507-003		

SPINOCHROME-E	$C_{10}H_6O_8$	254 OI 300°C
	JCS, 1966, 426.	
2508-001		

NAMAKOCHROME

$C_{11}H_8O_8$

268
OI
218^0C

Bull. Chem. Soc. Jap.,
1960 (33) 1234.

2508-002

MYCOCHRYSONE

$C_{20}H_{12}O_7$

364
-331^0
180^0C

JCS, 1969, 2059.

2509-003

NAPHTHAZARIN, 2, 6-diOH-3, 7-
- di-OMe-

$C_{12}H_{10}O_8$

282
OI
253^0C

Experientia, 1967 (23)
624.

2508-003

APHININ-SM

$C_{36}H_{38}O_{16}$

726

Nature, 1966 (210) 395.

2509-004

NAPHTHAZARIN, 2, 7-di-OH-3,
- 6-di-OMe-

$C_{12}H_{10}O_8$

282
OI
218^0C

Experientia, 1967 (23)
624.

2508-004

CEPHALOCHROMIN

$C_{28}H_{22}O_{10}$

518
$+510^0c$
300^0C

Proc. C. S., 1964, 195.

2509-005

DINAPHTHYL, 4, 5, 4', 5'-tetra-OH-
-1, 1'-

$C_{20}H_{14}O_4$

318
OI

JCS, 1960, 654.

2509-001

USTILAGINOIDIN-A

$C_{28}H_{18}O_{10}$

514
-380^0d
300^0C

Tet. Lett., 1963, 1777.

2509-006

MARITINONE

$C_{22}H_{14}O_6$

374
OI
195^0C

Chem. Pharm. Bull.,
1971, 851.

2509-002

USTILAGINOIDIN-B

$C_{28}H_{18}O_{11}$

530
-360^0d
300^0C

Tet. Lett., 1963, 1777.

2509-007

328

USTILAGINOIDIN-C $C_{28}H_{18}O_{12}$ 546 −260°d 300°C

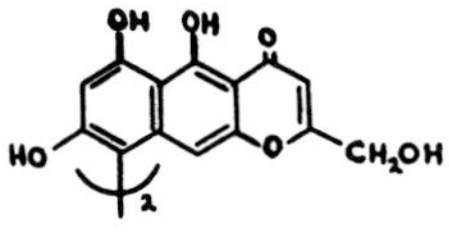

Tet. Lett., 1963, 1777.

2509-008

DIOSPYRIN $C_{22}H_{14}O_6$ 374 OI 287°C

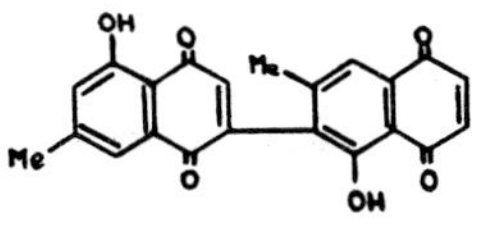

Tet. Lett., 1967, 1313.
Ind. J. C., 1970 (8) 569.

2510-003

XYLINDEIN $C_{32}H_{24}O_{10}$ 568 300°C

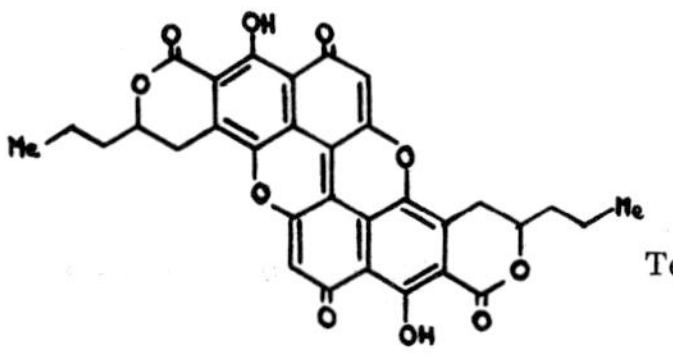

Tet., 1965, 2095.

2509-009

ELLIPTINONE $C_{22}H_{14}O_6$ 374 0° 310°C

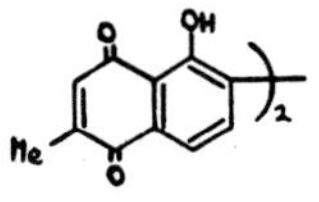

Chem. Pharm. Bull., 1971, 2271.
JCS, 1968, 2279.

2510-004

PROTOAPHIN-FB $C_{36}H_{44}O_{19}$ 780 210°C

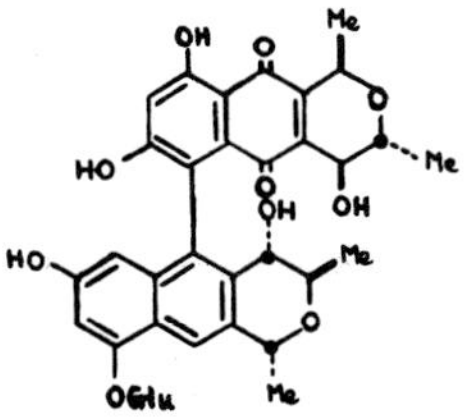

JCS, 1964, 51.

2509-010

MAMEGAKINONE $C_{22}H_{14}O_6$ 374 OI 256°C

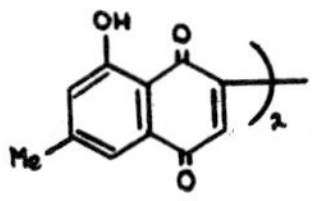

Chem. Pharm. Bull., 1971, 2271.

2510-005

DIOSPYROL $C_{22}H_{18}O_4$ 346 OI

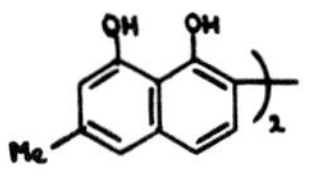

Tet. Lett., 1967, 4859.

2510-001

BIRAMENTACEONE $C_{22}H_{14}O_6$ 374 OI 235°C

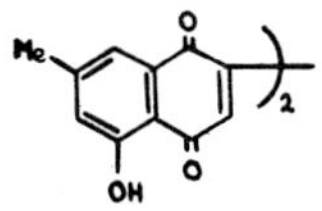

Phytochem., 1969 (8) 1951.

2510-006

DIOSPYROS $C_{22}H_{18}O_4$ 346 OI

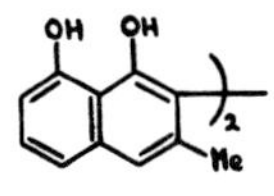

JCS, 1957, 2233.

2510-002

DIANELLINONE $C_{26}H_{18}O_8$ 458 OI 286°C

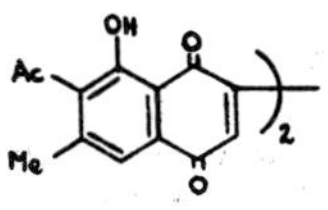

Aust. J. C., 1965 (18) 218.

2510-007

VIOXANTHIN

$C_{30}H_{26}O_{10}$ 546
OI
192^{0}C

R=H

Can. J. C., 1969, 1223.

2510-008

AUROFUSARIN

$C_{30}H_{18}O_{12}$ 570
OI
330^{0}C

JCS, 1966, 2234.

2510-013

VIRIDITOXIN

$C_{34}H_{30}O_{14}$ 662
-202^{0}c
243^{0}C

R = COOMe

Tet. Lett., 1971, 4705.

2510-009

ACTINORHODIN

$C_{32}H_{26}O_{14}$ 634

335^{0}C

Angew. Chem., 1964 (76)
863.

2510-014

XANTHOMEGNIN

$C_{30}H_{22}O_{12}$ 574
-156^{0}c
340^{0}C

Can. J. C., 1969, 1223.

2510-010

TECTOL

$C_{30}H_{26}O_4$ 450
OI
216^{0}C

R=H

Tet. Lett., 1963, 1269.

2510-015

FUSCOFUSARIN

$C_{30}H_{20}O_{11}$ 550
OI
>300^{0}C

Chem. Pharm. Bull.,
1968, 2213.

2510-011

TECTOL, tetrahydro-

$C_{30}H_{30}O_4$ 454
OI
249^{0}C

R = H
(* dihydro)

2510-016

$C_{24}H_{20}O_6$ 404
OI
317^{0}C

Chem. Comm., 1970,
1461.

2510-012

TECTOL DIME ETHER,
—tetrahydro-

$C_{32}H_{34}O_4$ 482
OI
186^{0}C

R = Me
(* dihydro)

2510-017

TECTOL, dehydro-	$C_{30}H_{24}O_4$	448 OI 194°C

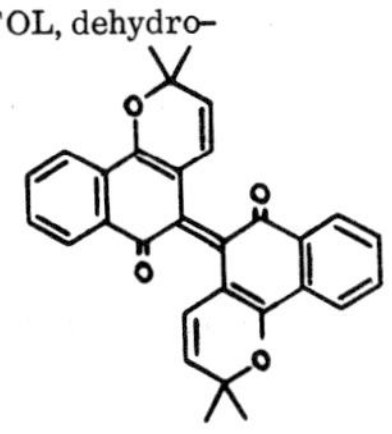

TECTOL, dehydro-　$C_{30}H_{24}O_4$　448 OI 194°C

JCS, 1968, 850.

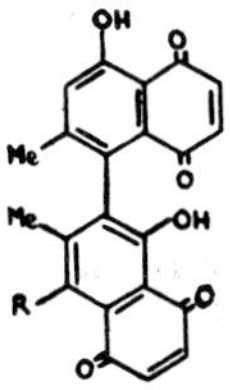

2510-018

deleted

2599-002

DIOSPYRIN, iso-　$C_{22}H_{14}O_6$　374 -150°c 227°C

R = H

JCS, 1968, 2279.

2511-001

XANTHODACTYNAPHIN-JC/2　$C_{30}H_{28}O_{11}$　564

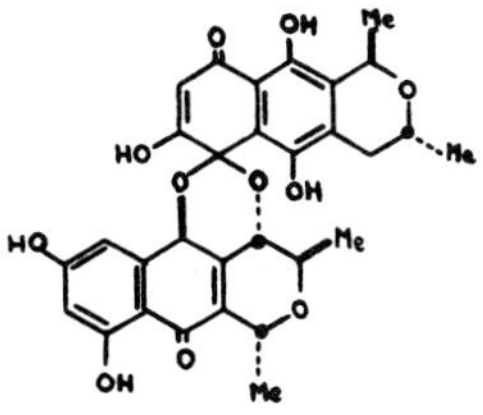

JCS, 1967, 712.

2599-003

DIOSPYRIN, OH-iso-　$C_{22}H_{14}O_7$　390 -720°d 276°C

R = OH

Chem. Pharm. Bull., 1971, 851.

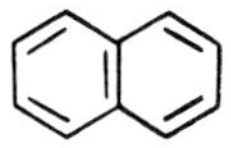

2511-002

XANTHODACTYNAPHIN-JC/1　$C_{30}H_{28}O_{12}$　580

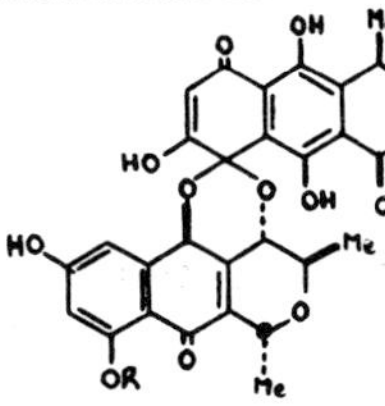

R = H　JCS, 1967, 712.

2599-004

AURASPERONE-A　$C_{32}H_{26}O_{10}$　570 OI

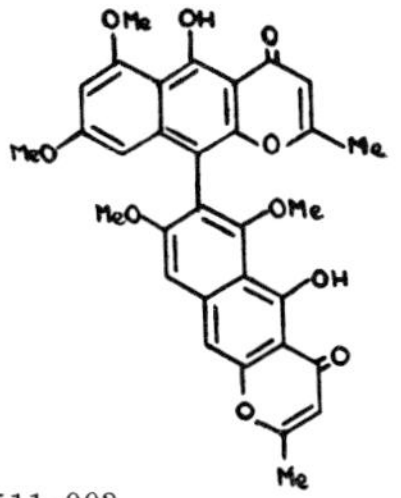

Chem. Abstr., 1967 (67) 53986.

2511-003

PROTODACTYNAPHIN-JC/1　$C_{36}H_{38}O_{17}$　742

R = Glucosyl

JCS, 1967, 720.

2599-005

NAPHTHALENE　$C_{10}H_8$　128 OI 80°C

Chem. Abstr., 1947 (41) 3510.

2599-001

RHODODACTYNAPHIN-JC/2　$C_{30}H_{28}O_{11}$　564

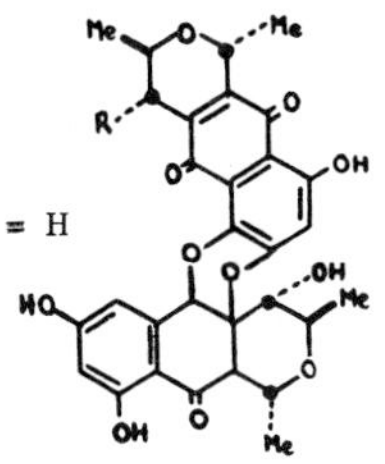

R = H　JCS, 1967, 712.

2599-006

RHODODACTYNAPHIN-JC/1 $C_{30}H_{28}O_{12}$ 580

R = OH

JCS, 1967, 712.

2599-007

TRIANELLINONE $C_{39}H_{24}O_{12}$ 684
OI
300^0C

R = Ac

Tet. Lett., 1970, 583.

2599-010

DIOSPYRIN, bis-iso- $C_{44}H_{26}O_{12}$ 746
-678^0c
320^0C

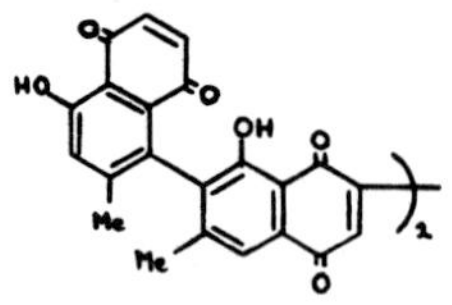

Tet. Lett., 1970, 7.

2599-008

DIOSPYRIN, neo- $C_{22}H_{14}O_6$ 374
OI

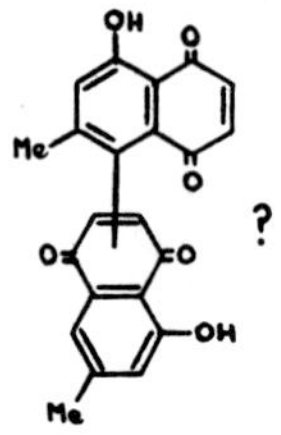

Chem. Pharm. Bull., 1971, 851.

2599-011

XYLOSPYRIN $C_{33}H_{18}O_9$ 558

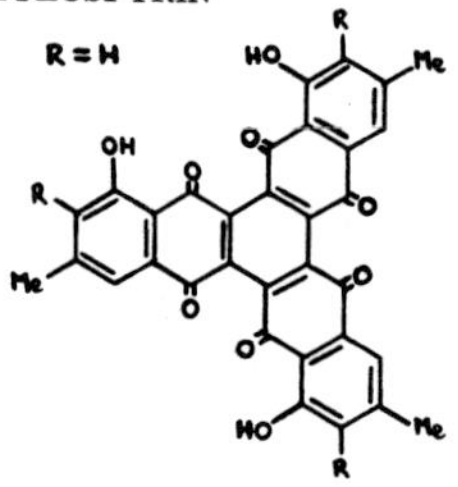

Tet. Lett., 1970, 1739.

2599-009

26 | ANTHRAQUINONES

This class of tricyclic aromatics includes the anthraquinones, anthrones, and phenolic anthracenes. The classification is according to the degree of oxygenation of the anthraquinonoid nucleus (2601-2606), with additional classes reserved for the dimeric anthraquinones (2607/8) and bisanthones (2609). The anthrones will be found along with other miscellaneous anthraquinones in Class 2699. It has been established that shikimic acid is a precursor of some of the plant anthraquinones, but for obvious convenience all anthraquinones have been placed in this structural category.

2601 No oxy substituents	2602 Mono-oxy
2603 Di-oxy	2604 Tri-oxy
2605 Tetra-oxy	2606 Penta-oxy
2607 2, 2'-Bisanthraquinones	2608 1, 1'-Bisanthraquinones
2609 10, 10'-Bisanthrones	2699 Misc.

ANTHRAQUINONE (artefact ?)	$C_{14}H_8O_2$	208 OI 286^{0}C
		Biochem. J., 1955 (60) 583.
2601-001		

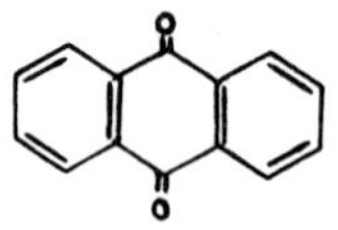

ANTHRAQUINONE, 2-COOH- $C_{15}H_8O_4$ 252 OI 291^{0}C

R = COOH

JCS, 1967, 2100.

2601-006

TECTOQUINONE   R = Me $C_{15}H_{10}O_2$ 222 OI 176^{0}C

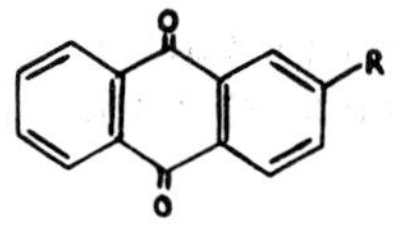

JCS, 1967, 2100.

2601-002

ANTHRAQUINONE, 1-OH-   R = H $C_{14}H_8O_3$ 224 OI 194^{0}C

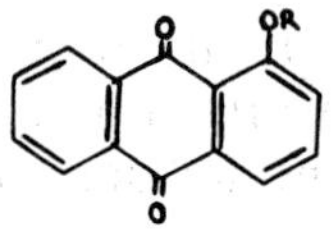

JCS, 1967, 2100.

2602-001

ANTHRAQUINONE, 2-CH_2OH- $C_{15}H_{10}O_3$ 238 OI 192^{0}C

R = CH_2OH

JCS, 1967, 2100.

2601-003

ANTHRAQUINONE, 1-OMe- $C_{15}H_{10}O_4$ 238 OI 170^{0}C

R = Me

JCS, 1967, 2100.

2602-002

ANTHRAQUINONE, 2-CH_2OAc- $C_{17}H_{12}O_4$ 280 OI 151^{0}C

R = CH_2OAc

JCS, 1967, 2100.

2601-004

ANTHRAQUINONE, 2-OH-   R = H $C_{14}H_8O_3$ 224 OI 304^{0}C

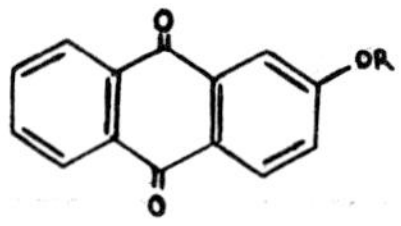

JCS, 1968, 854.

2602-003

ANTHRAQUINONE, 2-CHO- $C_{15}H_8O_3$ 236 OI 189^{0}C

R = CHO

JCS, 1967, 2100.

2601-005

ANTHRAQUINONE, 2-OMe- $C_{15}H_{10}O_3$ 238 OI 196^{0}C

R = Me

JCS, 1968, 854.

2602-004

ANTHRAQUINONE, 1-OH-2-Me-	$C_{15}H_{10}O_3$	238
		OI
R = H		185°C

JCS, 1968, 854.

2602-005

PACHYBASIN	$C_{15}H_{10}O_3$	238
		OI
R = H		178°C

Chem. Pharm. Bull.,
1955, 156.

2602-010

ANTHRAQUINONE, 1-OMe-2-Me-	$C_{16}H_{12}O_3$	252
		OI
R = Me		155°C

JCS, 1968, 854.

2602-006

PACHYBASIN ME ETHER	$C_{16}H_{12}O_3$	252
		OI
R = Me		190°C

Helv., 1963 (46) 1377.

2602-011

FERRUGINOL	$C_{15}H_{10}O_4$	254
		OI
		203°C

Tet. Lett., 1971, 4681.

2602-007

ALIZARIN	$C_{14}H_8O_4$	240
		OI
R = H		290°C

JCS, 1968, 2437.

2603-001

ANTHRAQUINONE, 3-OH-2-Me-	$C_{15}H_{10}O_3$	238
		OI
R = H		302°C

JCS, 1949, 1241.

2602-008

ALIZARIN 1-ME ETHER	$C_{15}H_{10}O_4$	254
		OI
R = Me		183°C

JCS, 1968, 2437.

2603-002

ANTHRAQUINONE, 3-OMe-2-Me-	$C_{16}H_{12}O_3$	252
		OI
R = Me		195°C

Phytochem., 1968 (7)
1423.

2602-009

ALIZARIN 2-ME ETHER	$C_{15}H_{10}O_4$	254
		OI
R = Me		234°C

JCS, 1968, 854.

2603-003

RUBERYTHIC ACID	$C_{25}H_{26}O_{13}$	534
		259^0C
R = Primeverosyl		
	Ber., 1939 (72) 913.	
2603-004		

XANTHOPURPURIN 3-ME ETHER	$C_{15}H_{10}O_4$	254
		OI
		193^0C
R = H; R' = Me		
	JCS, 1968, 2437.	
2603-009		

ALIZARIN, 3-Me-	$C_{15}H_{10}O_4$	254
		OI
		250^0C
R = H		
	Phytochem., 1968 (7) 1421.	
2603-005		

XANTHOPURPURIN DI-ME ETHER	$C_{16}H_{12}O_4$	268
		OI
		155^0C
R = R' = Me		
	JCS, 1968, 2437.	
2603-010		

DIGITOLUTEIN	$C_{16}H_{12}O_4$	268
		OI
		222^0C
R = Me		
	Bull. Soc. Chim. Fr., 1955, 108.	
2603-006		

RUBIADIN	$C_{15}H_{10}O_4$	254
		OI
		302^0C
R = R' = H		
	JCS, 1968, 854.	
2603-011		

XANTHOPURPURIN	$C_{14}H_8O_4$	240
		OI
		270^0C
R = R' = H		
	JCS, 1968, 884.	
2603-007		

RUBIADIN 1-ME ETHER	$C_{16}H_{12}O_4$	268
		OI
		283^0C
R = Me; R' = H		
	Aust. J. C., 1962 (15) 332.	
2603-012		

XANTHOPURPURIN 1-ME ETHER	$C_{15}H_{10}O_4$	254
		OI
		312^0C
R = Me; R' = H		
	Rev. Fac. Qium. Ind. Agr., 1945 (14) 163.	
2603-008		

RUBIADIN 3-GLUCOSIDE	$C_{21}H_{20}O_9$	416
		270^0C
R = H; R' = Glu		
(artefact ?)		
	JCS, 1936, 1714.	
2603-013		

RUBIADIN PRIMVEROSIDE	$C_{26}H_{28}O_{13}$	548
		249^0C

R = H; R' = Primeverosyl

JCS, 1936, 1714.

2603-014

DAMNACANTHAL	$C_{16}H_{10}O_5$	284
		OI
		211^0C

R = Me

J. Pharm. Soc. Jap.,
1955 (75) 219.

2603-019

LUCIDIN $C_{15}H_{10}O_5$ 270 OI $>330^0C$

R = H

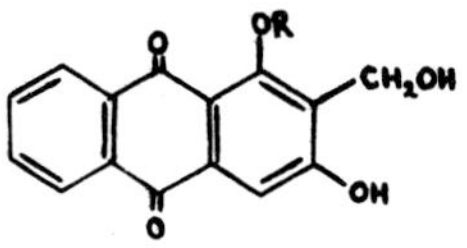

JCS, 1953, 306.

2603-015

MUNJISTIN $C_{15}H_8O_6$ 284 OI 232^0C

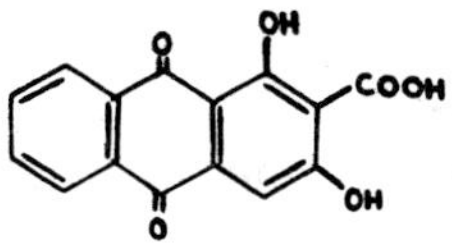

Chem. Pharm. Bull.,
1962 (10) 985.

2603-020

DAMNACANTHOL $C_{16}H_{12}O_5$ 286 OI 288^0C

R = Me

Aust. J. C., 1962 (15) 336.

2603-016

XANTHOPURPURIN, 6-Me- $C_{15}H_{10}O_4$ 254 OI 269^0C

R = H

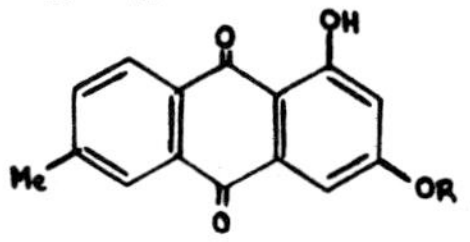

JCS, 1894, 851.

2603-021

IBERICIN $C_{17}H_{14}O_5$ 300 OI 182^0C

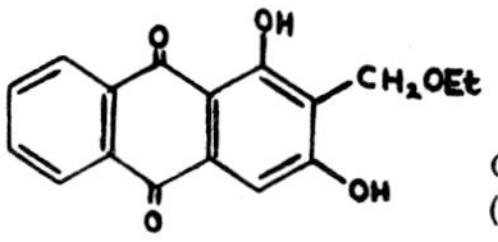

Chem. Abstr., 1968
(68) 29464.

(artefact ?)

2603-017

XANTHOPURPURIN 3-ME ETHER, -6-Me- $C_{16}H_{12}O_4$ 268 OI 184^0C

R = Me

Can. J. C., 1969 (47) 767.

2603-022

DAMNACANTHAL, nor- $C_{15}H_8O_5$ 270 OI 219^0C

R = H

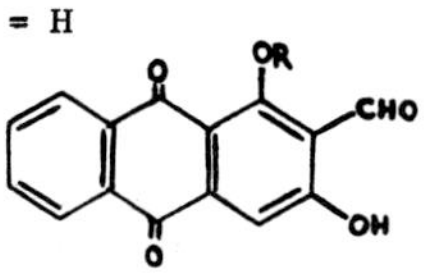

Aust. J. C., 1962 (15) 332.

2603-018

QUINIZARIN, 2-Me- $C_{15}H_{10}O_3$ 238 OI 176^0C

R = Me

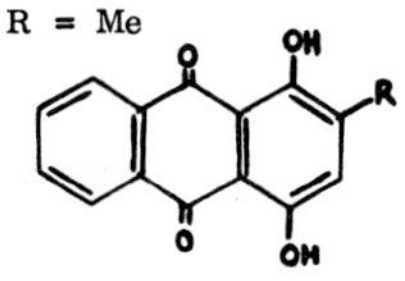

Natwiss., 1965 (52) 262.

2603-023

DIGIFERROL R = CH$_2$OH 2603-024	$C_{15}H_{10}O_4$ 254 OI 201^{0}C Tet. Lett., 1971, 4681.	ALOE-EMODIN R = H 2603-029	$C_{15}H_{10}O_5$ 270 OI 224^{0}C JCS, 1913, 2006.

SORANJIDIOL $C_{15}H_{10}O_4$ 254   OI   288^{0}C

JCS, 1952, 1718.

2603-025

ALOE-EMODIN 8-GLUCOSIDE $C_{21}H_{20}O_{10}$ 432   222^{0}C

R = Glucosyl

Natwiss., 1964 (51) 310.

2603-030

PHOMARIN $C_{15}H_{10}O_4$ 254   OI   254^{0}C

Biochem. J., 1966, 112.

2603-026

RHEIN $C_{15}H_8O_6$ 284   OI   310^{0}C

R = R' = H

Tet., 1967, 515.

2603-031

CHRYSOPHANOL $C_{15}H_{10}O_4$ 254   OI   196^{0}C

R = H

Rec. Trav. Chim.,
1960 (79) 1289.

2603-027

RHEIN 8-GLUCOSIDE $C_{21}H_{18}O_{11}$ 446   266^{0}C

R = Glucosyl ; R' = H

Chem. Abstr., 1969
(70) 9366.

2603-032

CHRYSOPHANEIN $C_{21}H_{20}O_9$ 416   -81^0 246^{0}C

R = Glucosyl

Natwiss., 1964 (51) 310.

2603-028

RHEIN-1, 8-DIGLUCOSIDE $C_{27}H_{28}O_{16}$ 608

R = R' = Glucosyl

J. Pharm. Pharm.,
1961 (13) 639.

2603-033

HYSTAZARIN ME ETHER	$C_{15}H_{10}O_4$ 254 OI 232^{0}C	

JCS, 1907, 2066.

2603-034

ANTHRAGALLOL 1, 2-DIME ETHER	$C_{16}H_{12}O_5$ 284 OI 231^{0}C	

JCS, 1955, 3298.

2604-004

PROTETRONE	$C_{19}H_{13}NO_8$ 383 OI 188^{0}C	

JACS, 1968, 7126.

2603-035

ANTHRAGALLOL 1, 3-DIME ETHER	$C_{16}H_{12}O_5$ 284 OI 212^{0}C	

JCS, 1929, 1399.

2604-005

ANTHRAGALLOL	$C_{14}H_8O_5$ 256 OI 312^{0}C	

JCS, 1949, 1241.

2604-001

ANTHRAGALLOL 2, 3-DIME ETHER	$C_{16}H_{12}O_5$ 284 OI 167^{0}C	

JCS, 1968, 2437.

2604-006

ANTHRAGALLOL 2-ME ETHER	$C_{15}H_{10}O_5$ 270 OI 220^{0}C	

JCS, 1949, 1246.

2604-002

ANTHRAGALLOL TRI-ME ETHER	$C_{17}H_{14}O_5$ 298 OI 165^{0}C	

Leather Sci., 1968
(15) 49.

2604-007

ANTHRAGALLOL 3-ME ETHER	$C_{15}H_{10}O_5$ 270 OI 242^{0}C	

JCS, 1968, 2437.

2604-003

PURPURIN	$C_{14}H_8O_5$ 256 OI 263^{0}C	

JCS, 1968, 2437.

2604-008

PURPURIN, pseudo- $C_{15}H_8O_7$ 300
OI
229^0C

R = H

COOH

JCS, 1936, 1714.

2604-009

MORINDONIN $C_{27}H_{30}O_{15}$ 594

255^0C

R = Gentianosyl

J. Sci. Ind. Res., 1950
(19B) 433.

2604-014

GALIOSIN $C_{26}H_{26}O_{16}$ 594

100^0C

R = Primeverosyl

JCS, 1936, 1714.

2604-010

OBTUSIFOLIN $C_{16}H_{12}O_5$ 284
OI
238^0C

Chem. Pharm. Bull.,
1958 (65) 315.

2604-015

MORINDONE $C_{15}H_{10}O_5$ 270
OI
284^0C

R = H

JACS, 1925, 283.

2604-011

NATALOE- EMODIN 8-ME $C_{16}H_{12}O_5$ 284
ETHER   OI

Phytochem., 1964 (3)
383.

2604-016

MORINDIN $C_{26}H_{28}O_{14}$ 564

264^0C

R = Primeverosyl

Ann. Pharm. Fr., 1954
(12) 794.

2604-012

JUZUNOL $C_{16}H_{12}O_6$ 300
OI
$>300^0C$

R = CH_2OH

Chem. Pharm. Bull.,
1960 (8) 417.

2604-017

MORINDIN $C_{27}H_{30}O_{14}$ 578
-91^0d
264^0C

R = Rutinosyl

JCS, 1963, 3471.

2604-013

JUZUNAL $C_{16}H_{10}O_6$ 298
OI
248^0C

R = CHO

J. Pharm. Soc. Jap.,
1955, 219.

2604-018

MACROSPORIN	$C_{16}H_{12}O_5$ 284 OI 310^0C Bull. Agr. Soc. Jap., 1959 (23) 547.	DERMOLUTEIN, 5-Cl- $C_{17}H_{11}O_7Cl$ 362 OI $>280^0C$ Ber., 1969 (102) 4104.
2604-019		2604-024
ERYTHROLACCIN, deoxy- R = H	$C_{15}H_{10}O_5$ 270 OI 300^0C Tet. Lett., 1968, 2231.	EMODIN, 3-Ac- $C_{17}H_{12}O_6$ 312 OI Chem. Comm., 1970, 1577.
2604-020		2604-025
LACCAIC ACID D R = COOH	$C_{16}H_{10}O_7$ 314 OI. 300^0C Tet. Lett., 1968, 2231.	COELULATIN, desoxy- $C_{15}H_{10}O_5$ 270 OI R = Me Planta Med., 1971 (19) 299.
2604-021		2604-026
ENDOCROCIN R = H	$C_{16}H_{10}O_7$ 314 OI 340^0C Chem. Pharm. Bull., 1953 (1) 160.	COELULATIN $C_{15}H_{10}O_6$ 286 OI decomp. $>250^0C$ R = CH_2OH Aust. J. C., 1962 (15) 336.
2604-022		2604-027
DERMOLUTEIN R = Me	$C_{17}H_{12}O_7$ 328 OI 270^0C Ber., 1969 (102) 4104.	EMODIN $C_{15}H_{10}O_5$ 270 R = H OI 255^0C JACS, 1934, 1312.
2604-023		2604-028

EMODIN, 8-O-glucosyl-	$C_{21}H_{20}O_{10}$	432
		190^0C
R = Glucosyl		
	Chem. Pharm. Bull., 1968 (16) 2299.	
2604-029		

MADAGASCIN	$C_{20}H_{18}O_5$	338
		OI
		156^0C
R = 3,3-DMA		
	Tet. Lett.. 1964, 1431.	
2604-034		

FRANGULIN-A	$C_{21}H_{20}O_9$	416
		-134^0
		228^0C
R = β-L-Rhamnosyl		
	Natwiss., 1964 (51) 310.	
2604-030		

QUESTIN $C_{16}H_{12}O_5$ 284 OI 302^0C

Acta Chem. Scand., 1969 (23) 3493.

2604-035

FRANGULIN-B	$C_{21}H_{20}O_9$	416
		196^0C
R = α-L-Rhamnosyl		
	Z. Nat., 1969 (24B) 1408.	
2604-031		

PHYSCION $C_{16}H_{12}O_5$ 284 OI 210^0C

R = H

JCS, 1970, 307.

2604-036

FRANGULIN-A, gluco-	$C_{24}H_{30}O_{14}$	578
		215^0C
R = ·β-Rha^4—Glu		
	Z. Nat., 1969 (24B) 1408.	
2604-032		

PHYSCIONIN	$C_{22}H_{22}O_{10}$	446
		235^0C
R = Glucosyl		
	Z. Nat.. 1963 (18B) 89.	
2604-037		

FRANGULIN-B, gluco-	$C_{27}H_{30}O_{14}$	578
R = ·α-Rha^4—Glu		
	Z. Nat., 1969 (24B) 1408.	
2604-033		

PHYSCION, glucorhamnosyl-	$C_{28}H_{32}O_{14}$	592
		220^0C
R = ·Rha^4—Glu		
	Phytochem., 1971 (10) 1921.	
2604-038		

EMODIN, 7-CL- $C_{15}H_9O_5Cl$ 304 OI 286°C

R = H

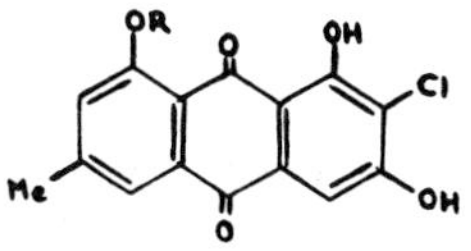

Tet. Lett., 1969, 919.

2604-039

EMODIN-6, 8-DIME ETHER $C_{17}H_{14}O_5$ 298 OI 212°C

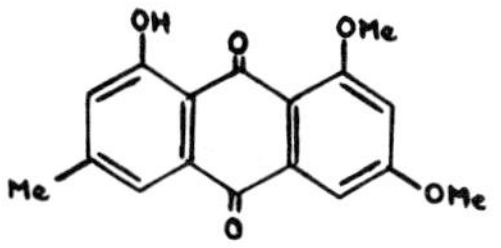

Phytochem., 1971 (10) 1379.

2604-044

EMODIN-1-ME ETHER, 7-Cl- $C_{16}H_{11}O_5Cl$ 318 OI 280°C

R = Me

Arkiv Kemi, 1969 (30) 217.

2604-040

CITREOROSEIN $C_{15}H_{10}O_6$ 286 OI 288°C

R = H

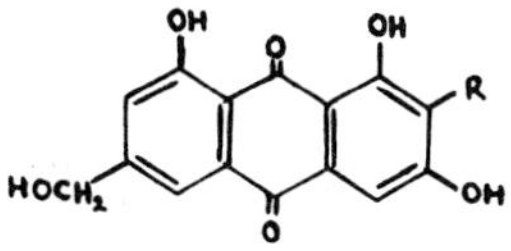

Biochem. J., 1940 (34) 159.

2604-045

EMODIN, 5, 7-di-Cl- $C_{15}H_8O_5Cl_2$ 338 OI 268°C

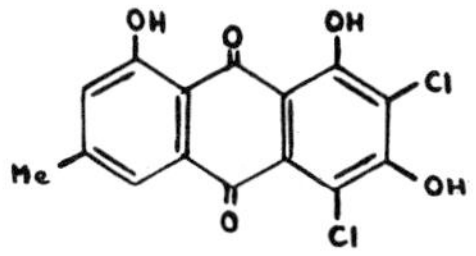

Tet. Lett., 1968, 1150.

2604-041

CITREOROSEIN, 7-Cl- $C_{15}H_9O_6Cl$ 320 OI 297°C

R = Cl

Chem. Pharm. Bull., 1968, 305.

2604-046

FRAGILIN $C_{16}H_{11}O_5Cl$ 318 OI 268°C

R = H

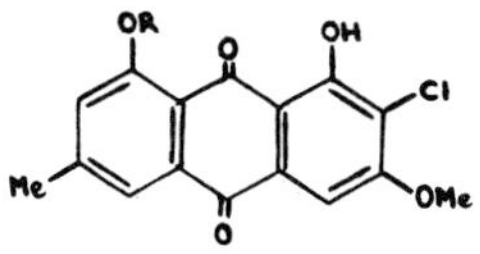

Acta Chem. Scand., 1967 (21) 2889.

2604-042

ROSEOPURPURIN $C_{16}H_{12}O_6$ 300 OI 286°C

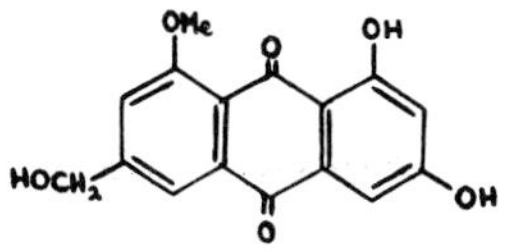

Biochem. J., 1940, 577.

2604-047

FRAGILIN ME ETHER $C_{17}H_{13}O_5Cl$ 332 OI 254°C

R = Me

Acta Chem. Scand., 1967 (21) 2889.

2604-043

QUESTINOL $C_{16}H_{12}O_6$ 300 OI 281°C

R = Me; R' = H

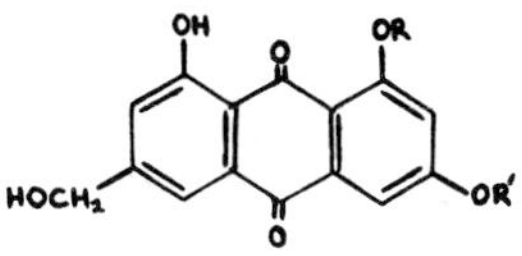

Biochem. J., 1964 (92) 369.

2604-048

TELOSCHISTIN	$C_{16}H_{12}O_6$	300
		OI
		246^0C
R = H; R' = Me		
	Chem. Pharm. Bull., 1956, 298.	
2604-049		

ISLANDICIN	$C_{15}H_{10}O_5$	270
		OI
		218^0C
	Aust. J. C., 1963 (16) 695.	
2604-054		

CITREOROSEIN-6, 8-DIME ETHER	$C_{17}H_{14}O_6$	314
		OI
		238^0C
R = R' = Me		
	Phytochem., 1971 (10) 1379.	
2604-050		

HELMINTHOSPORIN	$C_{15}H_{10}O_5$	270
		OI
		227^0C
	JCS, 1933, 488.	
2604-055		

FALLACINAL	$C_{16}H_{10}O_6$	298
		OI
		251^0C
	Chem. Pharm. Bull., 1956 (4) 298.	
2604-051		

RHODOPTILOMETRIN	$C_{17}H_{14}O_6$	314
		218^0C
	Aust. J. C., 1967, 541.	
2604-056		

EMODIC ACID	$C_{15}H_8O_7$	300
R = H		OI
		364^0C
	Phytochem., 1968 (7) 1183.	
2604-052		

RHODOPTILOMETRIN, iso-	$C_{17}H_{14}O_6$	314
		276^0C
	Aust. J. C., 1967, 541.	
2604-057		

PARIETINIC ACID	$C_{16}H_{10}O_7$	314
		OI
		300^0C
R = Me		
	Phytochem., 1968 (7) 1183.	
2604-053		

NALGIOVENSIN	$C_{18}H_{16}O_6$	327
		40^0c
		200^0C
R = H		
	JCS, 1961, 4691.	
2604-058		

NALGIOLAXIN R = Cl 2604-059	$C_{18}H_{15}O_6Cl$ 362 $+40^0c$ 248^0C JCS, 1967, 2570.	

CEREOALBOLINIC ACID $C_{16}H_{10}O_8$ 330 OI 290^0C

R = COOH

Tet. Lett., 1968, 2229.

2605-003

PTILOMETRIC ACID $C_{18}H_{14}O_7$ 342 OI 299^0C

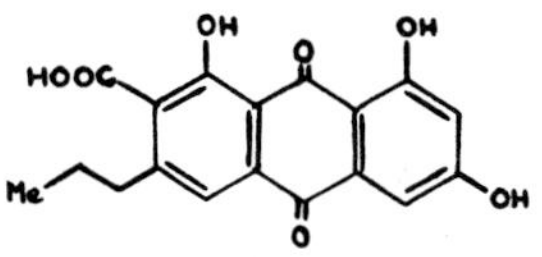

Aust. J. C., 1967, 541.

2604-060

EMODIN, 7-OH- $C_{15}H_{10}O_6$ 286 OI 297^0C

R = H

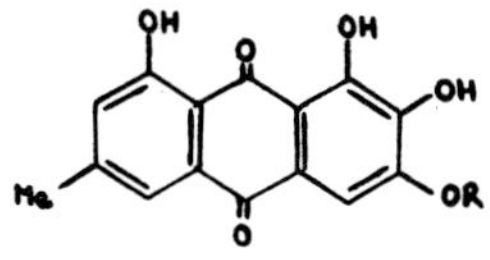

Chem. Comm., 1970, 1577.

2605-004

VERSICOLORIN-A, 6-deoxy- $C_{18}H_{10}O_6$ 322 -439^0d 245^0C

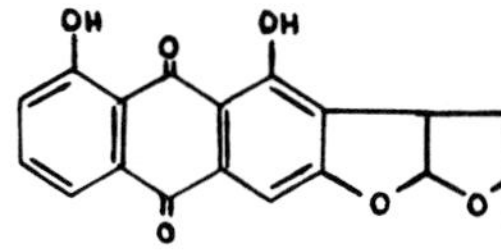

Chem. Comm., 1970, 1069.

2604-061

DERMOGLAUCIN $C_{16}H_{12}O_6$ 300 OI 233^0C

R = Me

Tet. Lett., 1966, 3077.

2605-005

COPAREOLATIN $C_{15}H_{10}O_6$ 286 OI 298^0C

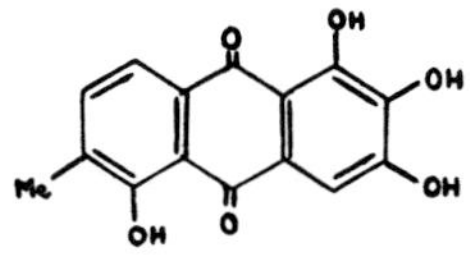

JCS, 1952, 1718.

2605-001

ANTHRAQUINONE, 1,7-DI-OH-
-6,8-DI-OMe-3-Me- $C_{17}H_{14}O_6$ 314 OI 239^0C

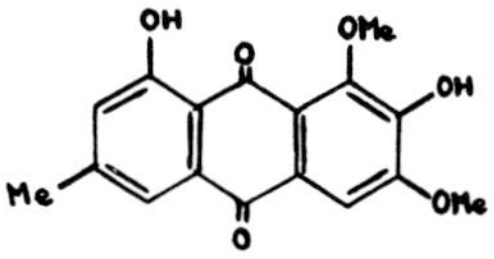

Phytochem., 1971 (10) 1379.

2605-006

ERYTHROLACCIN, iso- $C_{15}H_{10}O_6$ 286 OI 320^0C

R = H

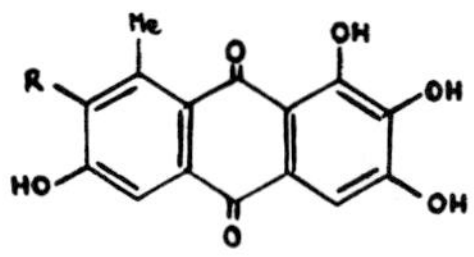

Ind. J. C., 1970 (8) 109.

2605-002

CLAVORUBIN $C_{16}H_{10}O_8$ 330 OI 232^0C

R = COOH

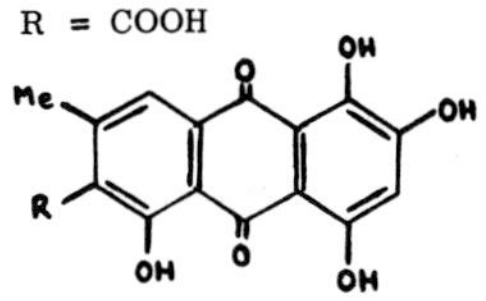

Angew. Chem., 1959 (71) 407.

2605-007

EMODIN, 2-Ac-5-OH- $\quad$ $C_{17}H_{12}O_7$ $\quad$ 328 OI R = Acetyl Chem. Comm., 1970, 1577. 2605-008	KERMESIC ACID $\quad$ $C_{16}H_{10}O_8$ $\quad$ 330 OI 250°C R = H Tet. Lett., 1968, 2223. 2605-013
EMODIN, 5-OH- $\quad$ $C_{15}H_{10}O_6$ $\quad$ 286 OI R = H Chem. Comm., 1970, 1577. 2605-009	CARMINIC ACID $\quad$ $C_{22}H_{20}O_{13}$ $\quad$ 492 +52°w – R = Glucosyl Ind. J. C., 1965 (3) 92. 2605-014
PAPULOSIN $\quad$ $C_{15}H_9O_6Cl$ $\quad$ 320 OI 273°C R = Cl Tet. Lett., 1969, 919. 2605-010	ERYTHROLACCIN $\quad$ $C_{15}H_{10}O_6$ $\quad$ 286 OI 314°C Tet., 1967, Supp. 8/2) 229. 2605-015
CITREOROSEIN, 8-OH- $\quad$ $C_{15}H_{10}O_7$ $\quad$ 302 OI Chem. Comm., 1970, 1577. 2605-011	ALATERNIN $\quad$ $C_{15}H_{10}O_6$ $\quad$ 286 OI 310°C R = H Tet. Lett., 1959, 22. 2605-016
XANTHORIN $\quad$ $C_{16}H_{12}O_6$ $\quad$ 300 OI 253°C Tet. Lett., 1967, 4719. 2605-012	ALTERNIN 6-ME ETHER $\quad$ $C_{16}H_{12}O_6$ $\quad$ 300 OI 267°C R = Me Can. J. C., 1969, 767. 2605-017

CATENARIN $C_{15}H_{10}O_6$ 286
OI
246^0C

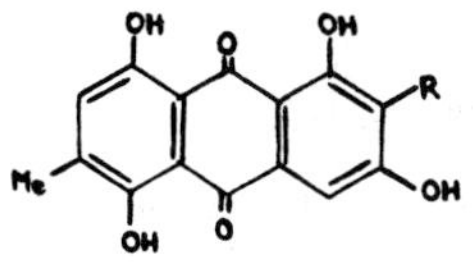

Biochem. J., 1940 (34)
1124.

2605-018

DERMORUBIN $C_{17}H_{11}O_8Cl$ 378
OI
$>300^0C$

R = Cl

Ber., 1969, 4104.

2605-023

ANTHRAQUINONE, 2-Cl-1,3,5,8-
-tetra-OH-6-Me- $C_{15}H_9O_6Cl$ 320
OI

R = Cl

Acta Chem. Scand.,
1969, 2241.

2605-019

ANTHRAQUINONE, 1,3,5-tri-OH-
-8-OMe-2-Me- $C_{22}H_{22}O_{11}$ 462

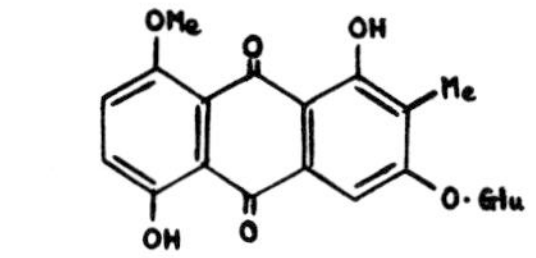

Planta Med., 1971 (19)
299.

2605-024

ERYTHROGLAUCIN $C_{16}H_{12}O_6$ 300
OI
205^0C

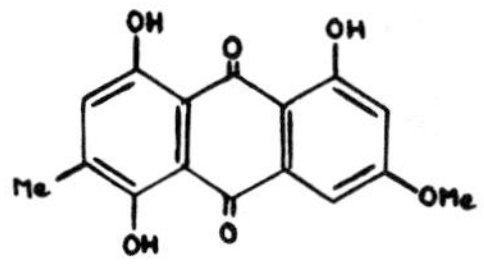

Biochem. J., 1939 (33)
1291.

2605-020

NODOSOSIDE $C_{20}H_{20}N_2O_{11}$ 464

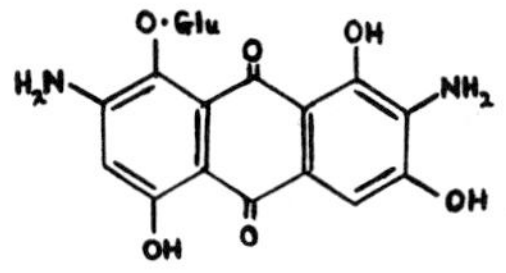

Planta Med., 1971 (19)
222.

2605-025

TRITISPORIN $C_{15}H_{10}O_7$ 302
OI
261^0C

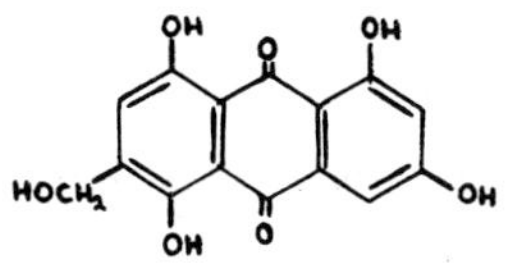

Biochem. J., 1956 (64)
464.

2605-021

CYNODONTIN $C_{15}H_{10}O_6$ 286
OI
260^0C

R = Me

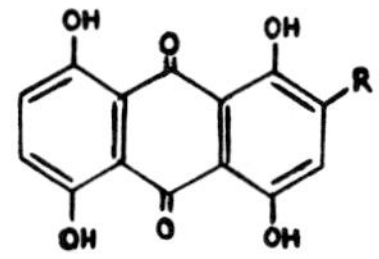

Nature, 1961 (190) 441.

2605-026

DERMORUBIN $C_{17}H_{12}O_8$ 344
OI
$>300^0C$

R = H

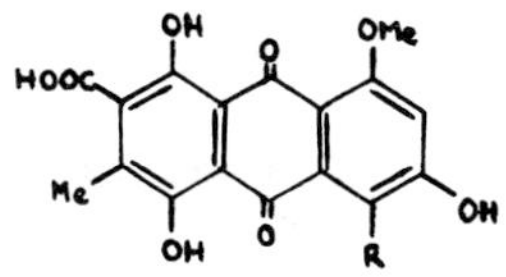

Ber., 1969, 4104.

2605-022

TECTOLEAFQUINONE $C_{19}H_{14}O_6$ 338
OI

R =

Tet. Lett., 1965, 2623.

2605-027

RHODOCOMATULIN 6-ME ETHER R = H	$C_{19}H_{16}O_7$	356 OI 251^{0}C

Aust. J. C., 1967, 515.

2605-028

AVERYTHRIN R = H	$C_{20}H_{18}O_6$	354 OI 230^{0}C

Chem. Comm., 1966, 913.

2605-033

RHODOCOMATULIN 6,8--DIME ETHER R = Me	$C_{20}H_{18}O_7$	370 OI 209^{0}C

Aust. J. C., 1967, 515.

2605-029

AVERYTHRIN 6-ME ETHER R = Me	$C_{21}H_{20}O_6$	368 OI 215^{0}C

Phytochem., 1970, 2061.

2605-034

SOLORINIC ACID, nor- R = H	$C_{20}H_{18}O_7$	370 OI 269^{0}C

JCS, 1966, 1727.

2605-030

AVERMUTIN	$C_{20}H_{18}O_7$	370 +226^{0}d 271^{0}C

Tet., 1969, 1497.

2605-035

SOLORINIC ACID R = Me	$C_{21}H_{20}O_7$	384 OI 204^{0}C

JCS, 1966, 1727.

2605-031

AVERUFIN R = R' = H	$C_{20}H_{16}O_7$	368 1^{0}e 281^{0}C

Chem. Comm., 1971, 1060.

2605-036

AVERANTIN	$C_{20}H_{20}O_7$	372 (-) 233^{0}C

JCS, 1966, 855.

2605-032

AVERUFIN 5-ME ETHER R = Me; R' = H	$C_{21}H_{18}O_7$	382

Chem. Comm., 1966, 912.

2605-037

AVERUFIN 5,7-DIME ETHER	$C_{22}H_{20}O_7$	396

R = R' = Me

Chem. Comm., 1966, 912.

2605-038

VERSICOLORIN-C $C_{18}H_{12}O_7$ 340
0^0
350^0C

Racemic Versicolorin-B

Tet., 1969, 1497.

2605-043

VERSICONOL $C_{18}H_{16}O_8$ 352
-36^0d
265^0C

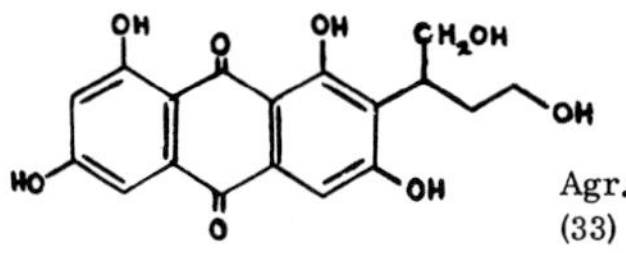

Agr. Biol. Chem., 1969 (33) 131.

2605-039

AVERSIN $C_{20}H_{16}O_7$ 368
-222^0c
217^0C

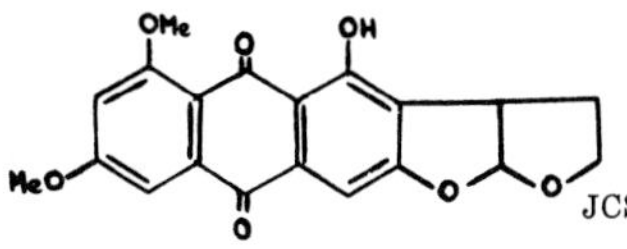

JCS, 1971, 3899.

2605-044

VERSICOLORIN-A $C_{18}H_{10}O_7$ 338
-354^0d
289^0C

R = H

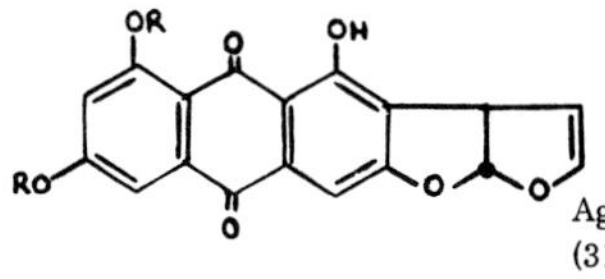

Agr. Biol. Chem., 1967 (31) 11.

2605-040

DOTHISTROMIN $C_{18}H_{12}O_9$ 372

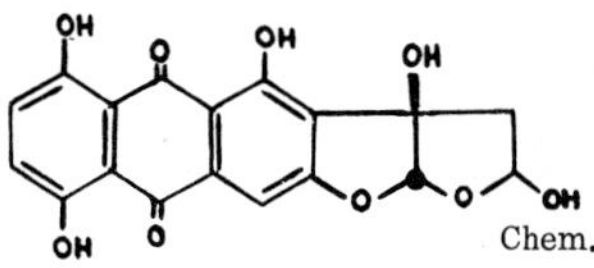

Chem. Comm., 1970, 1705.

2605-045

VERSICOLORIN-A 6,8-DIME -ETHER $C_{20}H_{14}O_7$ 366
-320^0d
285^0C

R = Me

Agr. Biol. Chem., 1971 (35) 444.

2605-041

LACCAIC ACID C $C_{25}H_{17}N_1O_{13}$ 539

R = COOH; R' = NH_2

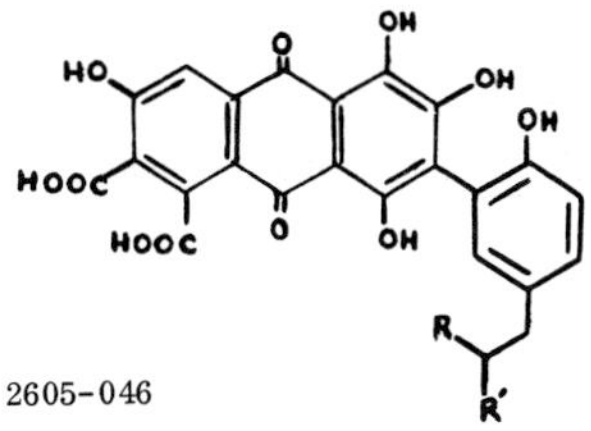

Ind. J. C., 1969 (7) 188.

2605-046

VERSICOLORIN-B $C_{18}H_{12}O_7$ 340
-223^0c
298^0C

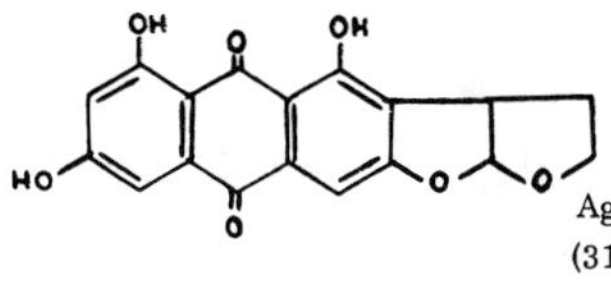

Agr. Biol. Chem., 1967 (31) 11.

2605-042

LACCAIC ACID E $C_{24}H_{17}NO_{11}$ 495
OI

R = H; R' = NH_2

Ind. J. C., 1969 (7) 188.

2605-047

LACCAIC ACID A	$C_{26}H_{19}NO_{12}$	537 OI 230°C

R = H; R' = NHAc

JCS, 1967, 842.

2605-048

AURANTIO-OBTUSIN	$C_{17}H_{14}O_7$	430 OI 265°C

R = R' = H

Chem. Pharm. Bull., 1960 (8) 246.

2606-003

LACCAIC ACID B	$C_{24}H_{16}O_{12}$	496 OI

R = H; R' = OH

Tet. Lett., 1967, 2437.

2605-049

OBTUSIN	$C_{18}H_{16}O_7$	444 OI 242°C

R = Me; R' = H

Chem. Pharm. Bull., 1960 (8) 246.

2606-004

ANTHRAQUINONE, 1,4,5,8--tetra-OH-2,6-di-Me-	$C_{16}H_{12}O_6$	300 OI 263°C

Aust. J. C., 1968 (21) 783.

2605-050

CHRYSO-OBTUSIN	$C_{19}H_{18}O_7$	458 OI 220°C

R = R' = Me

Chem. Pharm. Bull., 1969 (17) 454.

2606-005

DERMOCYBIN	$C_{16}H_{12}O_7$	316 OI 228°C

Biochem. J., 1961 (80) 387.

2606-001

VENTIMALIN	$C_{15}H_{10}O_7$	302 OI 278°C

Aust. J. C., 1963 (16) 695.

2606-006

ANTHRAQUINONE, 8-OMe-3-Me--1,2,6,7-tetra-OH-	$C_{16}H_{12}O_7$	316 OI 330°C

Phytochem., 1971 (10) 1379.

2606-002

VIMINALIN	$C_{17}H_{14}O_7$	430 OI 279°C

Aust. J. C., 1963 (16) 695.

2606-007

RUBROCOMATULIN 6-ME ETHER C$_{19}$H$_{16}$O$_8$ 472 OI 298^{0}C

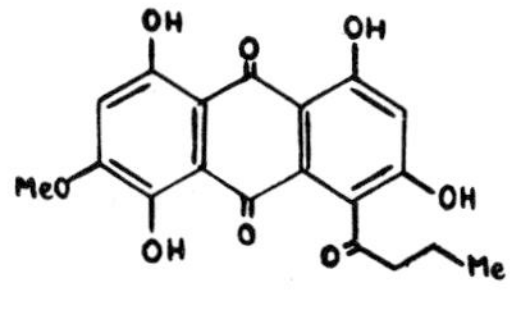

Aust. J. C., 1967 (20) 515.

2606-008

CASSIAMIN-B C$_{30}$H$_{18}$O$_{10}$ 538 OI > 350^{0}C

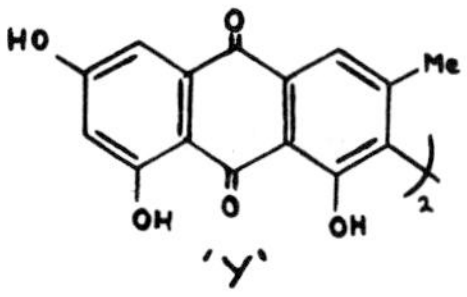

Ind. J. C., 1970, 109.

'Y'

2607-003

ASPERTHECIN C$_{15}$H$_{10}$O$_8$ 318 OI 370^{0}C

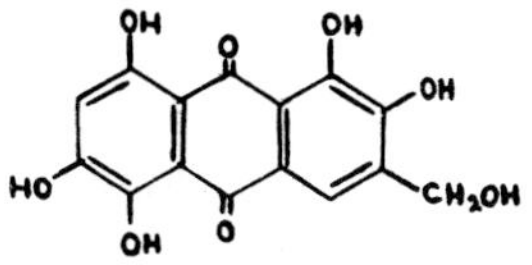

Biochem. J., 1955 (59) 475.

2606-009

CASSIAMINE-A C$_{30}$H$_{18}$O$_9$ 522 - 356^{0}C

X—Y

Tet. Lett., 1964, 3023.

2607-004

EMODIN, 2, 5, 7-tri-OH- C$_{15}$H$_{10}$O$_8$ 318 OI 365^{0}C

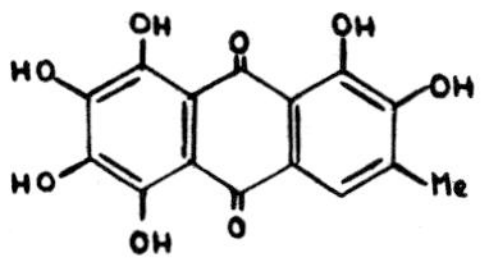

Tet. Lett., 1970, 445.

2606-010

MADAGASCARIN C$_{30}$H$_{18}$O$_{10}$ 538

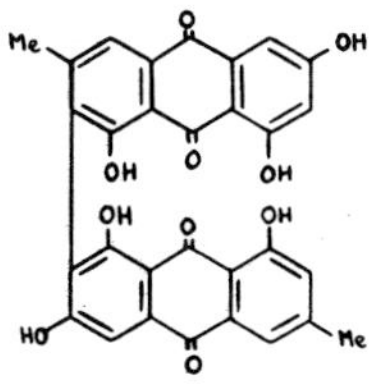

Aust. J. C., 1972, 843.

2607-005

CHRYSOTALUNIN C$_{30}$H$_{18}$O$_8$ 506 OI 360^{0}C

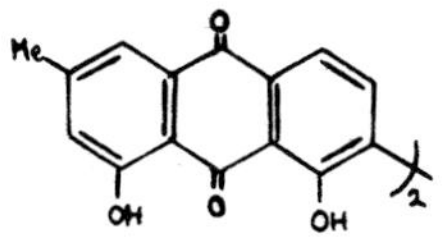

Chem. Ind., 1970, 1353.

2607-001

JULICHROME Q1.1 C$_{38}$H$_{34}$O$_{14}$ 714 -83^{0}m 215^{0}C

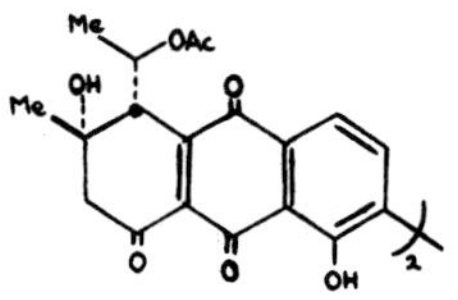

Tet., 1968, 4233.

Q1

2607-006

CASSIAMINE-C C$_{30}$H$_{18}$O$_8$ 506 OI 350^{0}C

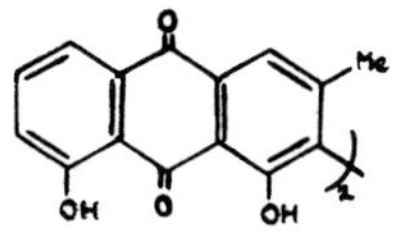

Ind. J. C., 1970, 109.

'X'

2607-002

JULICHROME Q2.2 C$_{38}$H$_{30}$O$_{12}$ 678 250^{0}C

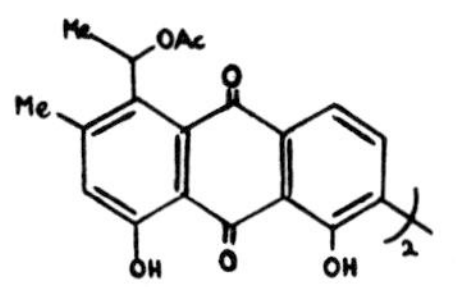

Tet., 1971, 1557.

Q2

2607-007

JULICHROME Q3.3 $C_{38}H_{38}O_{16}$ 750

$>300^0C$

Tet., 1970, 5201.

Q3

2607-008

JULICHROME Q8.8 $C_{38}H_{38}O_{14}$ 718

Tet., 1970, 5201.

2607-011

Q4

JULICHROME Q1.2 $C_{38}H_{32}O_{13}$ 696

168^0C

Q1- Q2

Tet., 1970, 5719.

2607-012

JULICHROME Q5.5 $C_{34}H_{22}O_{10}$ 590

OI

$>300^0C$

Tet., 1970, 5719.

Q5

2607-009

JULICHROME Q1.3 $C_{38}H_{36}O_{15}$ 732

200^0C

Q1- Q3

Tet., 1969, 3007.

2607-013

JULICHROME Q6.6 $C_{38}H_{38}O_{12}$ 686

241^0C

Tet., 1971, 1557.

Q6

2607-010

JULICHROME Q1.4 $C_{38}H_{34}O_{15}$ 730

200^0C

Q1 - Q4

Tet., 1969, 3007.

2607-014

Q7

JULICHROME Q1.5 $C_{36}H_{28}O_{12}$ 652

$>300^0C$

Q1 - Q5

Tet.. 1971, 1557.

2607-015

JULICHROME Q1.6	$C_{38}H_{36}O_{13}$	700
		193^0C
Q1 - Q6		
	Tet., 1971, 1557.	
2607-016		

JULICHROME Q3.5	$C_{36}H_{30}O_{13}$	670
		228^0C
Q3 - Q5		
	Tet., 1971, 1557.	
2607-021		

JULICHROME Q1.7	$C_{38}H_{36}O_{15}$	732
		290^0C
Q1 - Q7		
	Tet., 1970, 5201.	
2607-017		

JULICHROME Q3.8	$C_{38}H_{38}O_{15}$	734
		$+47^0m$
		179^0C
Q3 - Q8		
	Tet., 1970, 5201.	
2607-022		

JULICHROME Q2.3	$C_{38}H_{34}O_{14}$	714
		165^0C
Q2 - Q3		
	Tet., 1969, 3007.	
2607-018		

JULICHROME Q4.5	$C_{36}H_{28}O_{13}$	668
		243^0C
Q4 - Q5		
	Tet., 1971, 1557.	
2607-023		

JULICHROME Q2.5	$C_{36}H_{26}O_{11}$	634
		$> 300^0C$
Q2 - Q5		
	Tet., 1971, 1557.	
2607-019		

JULICHROME Q5.6	$C_{36}H_{30}O_{11}$	638
		194^0C
Q5 - Q6		
	Tet., 1971, 1557.	
2607-024		

JULICHROME Q3.4	$C_{38}H_{36}O_{16}$	748
		$> 280^0C$
Q3 - Q4		
	Tet., 1969, 3007.	
2607-020		

FLAVOMANNIN	$C_{30}H_{26}O_{10}$	546
		-1400^0e
		225^0C
	JCS, 1968, 2560.	
2607-025		

353

JULICHROME Q1.9 $C_{38}H_{34}O_{16}$ 746

175°C

Tet., 1970, 5201.

2607-026

AUROSKYRIN $C_{30}H_{18}O_9$ 522
(+)
300°C

Chrysophanol- Emodin

Tet. Lett., 1968, 1881.

2608-005

RUGULOSIN, di-anhydro- $C_{30}H_{18}O_8$ 506
(+)
321°C

Tet. Lett., 1968, 1881.

(Chrysophanol)
2608-001

RHODOISLANDIN-B $C_{30}H_{18}O_{10}$ 538
(+)
300°C

Emodin- Islandicin

Tet. Lett., 1968, 1881.

2608-006

IRIDOSKYRIN $C_{30}H_{18}O_{10}$ 538
(+)
358°C

Biochem. J., 1954 (57) 212.

(Islandicin)
2608-002

RHODOISLANDIN-A $C_{30}H_{18}O_{10}$ 538
(+)
300°C

(Chrysophanol)

(Catenarin)
Tet. Lett., 1968, 1881.

2608-007

SKYRIN $C_{30}H_{18}O_{10}$ 538
(+)
300°C

Chem. Pharm. Bull., 1965 (13) 385.

(Emodin)
2608-003

AURANTIOSKYRIN $C_{30}H_{18}O_9$ 522
(+)
300°C

Emodin- Catenarin

Tet. Lett., 1968, 1881.

2608-008

ROSEOSKYRIN $C_{30}H_{18}O_9$ 522
(+)
300°C

Chrysophanol- Islandicin

Tet. Lett., 1968, 1881.

2608-004

PUNICOSKYRIN $C_{30}H_{18}O_{11}$ 554
(+)
300°C

Islandicin-Catenarin

Tet. Lett., 1968, 1881.

2608-009

SKYRINOL	$C_{30}H_{18}O_{12}$ 570 (+) 360°C	Chem. Pharm. Bull., 1955 (3) 286.

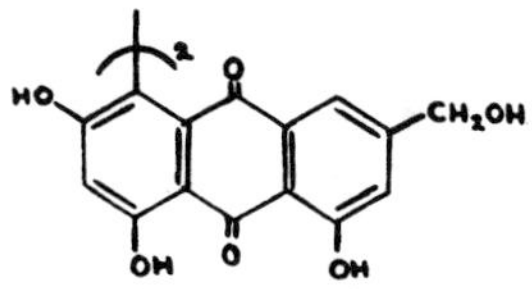

(Citreorosein)

2608-010

ALOE-EMODIN-10,10'-BIANTHRONE	$C_{30}H_{22}O_8$ 510	Lloydia, 1965 (28) 63.

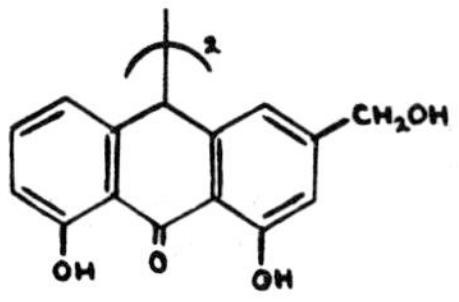

2609-002

SKYRIN, oxy- $C_{30}H_{18}O_{11}$ 554 (+) 360°C

Emodin-Citreorosein

Chem. Pharm. Bull., 1957 (5) 573.

2608-011

EMODIN-10,10'-BIANTHRONE $C_{30}H_{22}O_8$ 510

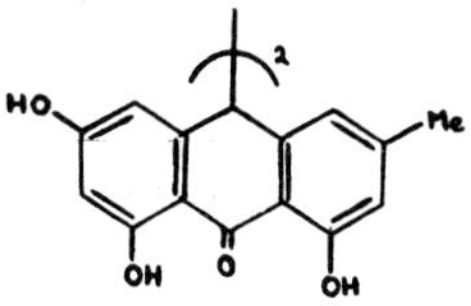

Planta Med., 1967 (15) 233.

2609-003

FUSAROSKYRIN $C_{32}H_{22}O_{12}$ 598   300°C

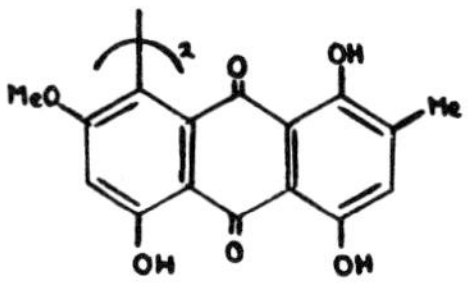

Chem. Ind., 1961, 1754.

2608-012

PALMIDIN-B $C_{30}H_{22}O_7$ 494

Aloe-Emodin-Chrysophanol

Planta Med., 1967 (15) 233.

2609-004

PENICILLIOPSIN $C_{30}H_{22}O_8$ 510   OI   330°C

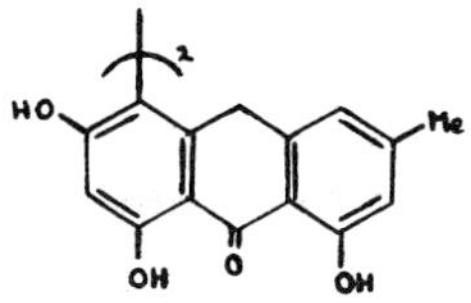

Angew., 1955 (67) 706.

2608-013

PALMIDIN-C $C_{30}H_{22}O_7$ 494

Emodin-Chrysophanol

Planta Med., 1967 (15) 233.

2609-005

ARAROBINOL $C_{30}H_{22}O_6$ 478   224°C

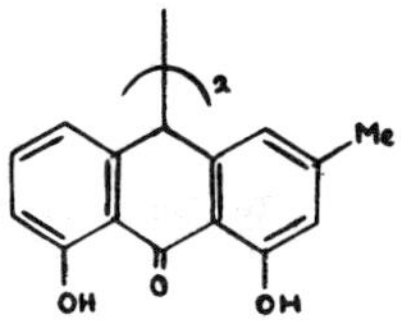

JCS, 1925 (948) 1160.

(Chrysophanol)

2609-001

PALMIDIN-A $C_{30}H_{22}O_8$ 510

Aloe-emodin - Emodin

Planta Med., 1967 (15) 233.

2609-006

PALMIDIN-D $C_{31}H_{24}O_8$ 524

Emodin

(Physcion)

Ann. Pharm. Fr., 1968 (26) 673.

2609-007

SENNOSIDE-A $C_{42}H_{38}O_{20}$ 862 -164^0a $> 200^0$C

Helv., 1950, 313.

2609-012

SENNIDINE-A $C_{30}H_{18}O_{10}$ 538 $+180^0$

(Rhein)

Helv., 1950, 313.

2609-008

SENNOSIDE-B $C_{42}H_{38}O_{20}$ 862 -100^0a 180^0C

Stereoisomer of Sennoside-A

Forschr. Chem. Org. Nat., 1950 (7) 248.

2609-013

RHEIDIN-B $C_{30}H_{20}O_8$ 508

Rhein - Chrysophanol

Pharm. Weekbl., 1964 (99) 613.

2609-009

SENNOSIDE-C $C_{42}H_{40}O_{19}$ 848 -21^0w 197^0C

Helv., 1965 (48) 1911.

2609-014

RHEIDIN-A $C_{30}H_{20}O_9$ 524

Rhein-Emodin

Pharm. Weekbl., 1964 (99) 613.

2609-010

SENNOSIDE-D $C_{42}H_{40}O_{19}$ 848 $+3^0$w 210^0C

Stereoisomer of Sennoside-C

Helv., 1965 (48) 1911.

2609-015

RHEIDIN-C $C_{31}H_{22}O_9$ 538

Rhein- Physcion

Pharm. Weekbl., 1964 (99) 613.

2609-011

FLAVOOBSCURIN-A $C_{30}H_{19}O_8Cl_3$ 612 $+30^0$a $> 360^0$C

R = H

Tet. Lett., 1968, 3749.

2609-016

FLAVOOBSCURIN-B$_1$	C$_{30}$H$_{18}$O$_8$Cl$_4$	646
		+35^{0}a
		360^{0}C
R = Cl		
		Tet. Lett., 1968, 3749.
2609-017		

ALOINOSIDE-B	C$_{27}$H$_{32}$O$_{13}$	564
		-45^{0}d
		233^{0}C
R = Rhamnosyl		
		Z. Nat., 1964 (19B) 222.
2699-004		

FLAVOOBSCURIN-B$_2$	C$_{30}$H$_{18}$O$_8$Cl$_4$	646
		+7^{0}a
		360^{0}C
R = Cl		
(Stereoisomer)		
		Tet. Lett., 1968, 3749.
2609-018		

RHEIN-9-ANTHRONE-8--GLUCOSIDE   C$_{21}$H$_{20}$O$_{10}$ 432

J. Pharm. Pharm., 1961 (13) 639.

2699-005

CHRYSAROBIN   C$_{15}$H$_{12}$O$_3$ 240   OI   207^{0}C

Can. J. C., 1963, 1622.

2699-001

EMODIN-ANTHRONE   C$_{15}$H$_{12}$O$_4$ 256   OI   250^{0}C

R = H

Natwiss., 1953, 509.

2699-006

BARBALOIN   C$_{21}$H$_{22}$O$_9$ 418   +21^{0}w   148^{0}C

R = H

Aust. J. C., 1955, 523.

2699-002

MADAGASCIN ANTHRONE   C$_{20}$H$_{20}$O$_4$ 324   OI   168^{0}C

R = 3,3-DMA

Tet. Lett., 1964, 1431.

2699-007

BARBALOIN, iso-   C$_{21}$H$_{22}$O$_9$ 418   -19^0

R = H
(Stereoisomer)

Helv., 1952, 17.

2699-003

PHYSCION-10-ANTHRONE   C$_{16}$H$_{14}$O$_4$ 270   OI   260^{0}C

Biochem. J., 1939 (33) 1291.

2699-008

HARONGIN ANTHRONE	$C_{30}H_{36}O$	460 OI 208^0C

Tet. Lett., 1964, 1431.

2699-009

FLAVOSKYRIN	$C_{15}H_{12}O_5$	272 -295^0d 208^0C

Chem. Pharm. Bull.,
1956 (4) 303.

2699-014

HARUNGANIN	$C_{30}H_{36}O_4$	460 OI -

JACS, 1962, 2653.

2699-010

BOSTRYCIN	$C_{16}H_{16}O_8$	336 223^0C

Tet., 1970, 1339.

2699-015

EMODINANTHRONE, -10-glucosyloxy-	$C_{21}H_{22}O_{10}$	434 220^0C

Helv., 1946, 189.

2699-011

ALTERSOLANOL-B	$C_{16}H_{16}O_6$	304 229^0C

R = H

Can. J. C., 1969, 767.

2699-016

ANTHRONE, 2-Cl-1,3,8-triOH- -6-Me-	$C_{15}H_{11}O_4Cl$	290 OI 228^0C

Chem. Pharm. Bull.,
1968, 304.

2699-012

ALTERSOLANOL-A	$C_{16}H_{16}O_8$	336 -290^0p 218^0C

R = OH

Can. J. C., 1969, 767.

2699-017

HOMONATALOIN	$C_{22}H_{24}O_{10}$	448 -111^0e 203^0C

Phytochem., 1964 (3)
383.

2699-013

CATENARIN, 5,6-dihydro-	$C_{15}H_{12}O_6$	288 95^0C

JCS, 1968, 1941.

2699-018

CATENARIN, 5,6,7,8-tetrahydro-	$C_{15}H_{14}O_6$	290
		~130°C
*dihydro		JCS, 1968, 1941.
2699-019		

LUTEOSKYRIN	$C_{30}H_{22}O_{12}$	574
		-880°a
		273°C
R = H		
		Tet. Lett., 1969, 767.
2699-022		

ORUWAL	$C_{17}H_{14}O_3$	266
		OI
		157°C
		Chem. Comm., 1972, 405.
2699-020		

OMe / CHO / OMe

RUBROSKYRIN	$C_{30}H_{22}O_{12}$	574
		289°C
		Tet. Lett., 1968, 5558.
2699-023		

RUGULOSIN	$C_{30}H_{22}O_{10}$	542
		(+)
		293°C
R = H		Tet. Lett., 1968, 6135.
2699-021		

PHYSCION-9-ANTHRONE	$C_{16}H_{14}O_4$	270
		OI
		182°C
R = Me		Biochem. J., 1939 (33) 1291.
2699-024		

This class has been provided as a "catch-all" for the variety of polycyclic aromatics not already covered by the main classes. Thus, the olivomycins and chromomycins are sufficiently distinctive and unlike the anthraquinones as to warrant separate classification (2701). Recently several phenanthrene derivatives have been isolated, and these are included here (2701) along with the herqueinones. The tetracyclines (2702) are also a small but distinctive group of polycyclics with the unique malonamidyl starter unit. The tetracyclic rhodomycins will be found here (2703) as also will the angular tetracyclics such as tetrangomycin and aquayamycin. The more complex polycyclics, as, for example, the chrysoaphins and duclauxins, have been posted to that final catch-all, Class 2799. With the nature of this polycyclics class it will be apparent that the biogenetic origins of its constituent members are likely to be rather diverse. In many cases, however, the familiar beta-diketo oxygenation pattern will be evident, and a polyketide chain can be traced through the molecule. Notable exceptions include some of the phenanthrene compounds that probably derive from phenolic coupling of a stilbene precursor and hopeaphenol, which, on inspection, reveals four stilbene units condensed together.

2701	Tricyclics (excl. Anthraquinones and Anthrones)	2702	Tetracyclines (incl. tetracyclic precursors)
2703	Tetracyclics (other than 2702)	2704	Misc. Polycyclics

CERVICARCIN $C_{19}H_{20}O_9$ 392
-60^0e
205^0C

Chem. Comm., 1970, 1693.
Agr. Biol. Chem., 1968
(32) 209.

2701-001

OLIVOMYCIN-D $C_{47}H_{66}O_{22}$ 982
-25^0e

R = .Olivose^3—Olivose

R' = .Oliose^4—Olivomose

Tet. Lett.. 1966, 1643.

2701-006

OLIVIN $C_{20}H_{22}O_9$ 406
$+60^0$e
190^0C

R = R' = H

Chem. Comm., 1967, 9.

2701-002

CHROMOMYCINONE $C_{21}H_{24}O_9$ 420
176^0C

R = R' = H

Tet. Lett., 1970, 1329.

2701-007

OLIVOMYCIN-A $C_{58}H_{84}O_{26}$ 1196
-35^0e
160^0C

R = .Olivose^3—Olivose^3—Olivomycose
                           4|
                    i-Butyryl

R' = .(3-Ac-Oliose)^4—Olivomose   Tet. Lett., 1966, 1643.

2701-003

CHROMOMYCIN-A4 $C_{48}H_{68}O_{22}$ 996
-47^0e

R = .Olivose^3—Olivose

R' = .Chromose-D^4—Olivomose

Nature, 1968 (218) 193.

2701-008

OLIVOMYCIN-B $C_{56}H_{80}O_{26}$ 1168
-28^0e

R = .Olivose^3—Olivose^3—(4-Ac-Olivomycose)

R' = .(3-Ac-Oliose)^4—Olivomose

Tet. Lett., 1966, 1463.

2701-004

AUREOLIC ACID $C_{52}H_{76}O_{24}$ 1084
-51^0e
181^0C

R = . Mycarose^4—Oliose^3—Olivose

R' = .Olivose^3—Olivose

Tet. Lett., 1970. 1329.

2701-009

OLIVOMYCIN-C $C_{56}H_{82}O_{25}$ 1154
-17^0e

R = .Olivose^3—Olivose^3—Olivomycose
                          4|
                   i-Butyryl

R' = .Oliose^4—Olivomose   Tet. Lett., 1966, 1643.

2701-005

ABURAMYCIN-C $C_{57}H_{84}O_{25}$ 1168
-17^0e
amorph.

R = .Olivose^3—Olivose^3—Olivomycose
                          4|
                i-Butyryl

R' = .Oliose^4—Olivomose   Nature, 1968 (218) 193.

2701-010

CHROMOMYCIN-A3	$C_{57}H_{82}O_{26}$	1182 -26^0e 183^0C

R = .Olivose$^{\underline{3}}$—Olivose$^{\underline{3}}$—Chromose-B

R' = .Chromose-D$^{\underline{4}}$—Olivomose

Nature, 1968 (218) 193.
Tet. Lett. , 1966. 1643.

2701-011

PHENANTHRENE, 2, 6-di-OH-3, 4, -7-tri-OMe- $C_{17}H_{16}O_5$ 300   OI 298^0C

JCS, 1971, 3070.

2701-016

CHROMOMYCIN-A2	$C_{59}H_{86}O_{26}$	1210 -61^0e

R = .Olivose$^{\underline{3}}$—Olivose$^{\underline{3}}$—Olivomycose
                                              $\overset{4}{|}$
                                         i-Butyryl

R' = .Chromose-D$^{\underline{4}}$—Olivomose

Nature. 1968 (218) 193.
Tet. Lett. , 1966. 1643.

2701-012

PHENANTHRENE, 2, 6-di-OH-3, 4.- $C_{17}H_{18}O_5$ 302   OI 208^0C
-7-tri-OMe-dihydro-

*dihydro

JCS, 1971, 3070.

2701-017

ORCHINOL $C_{16}H_{16}O_3$ 256   OI 127^0C

Helv. , 1963, 1354.

2701-013

PHENANTHRENE, 4, 6, 7-tri-OH- $C_{16}H_{14}O_5$ 286   OI 232^0C
-2, 3-di-OMe-

JCS, 1971, 3070.

2701-018

PHENANTHRENE, 2. 5-di-OH-3. $C_{16}H_{16}O_4$ 272   OI 129^0C
-7-di-OMe-dihydro-

JCS, 1971, 3070.

2701-014

PHENANTHRENE, 4. 6, 7-tri-OH-2, $C_{16}H_{16}O_5$ 288   OI 190^0C
-3-di-OMe-dihydro-

*dihydro

JCS, 1972, 206.

2701-019

PHENANTHRENE, 2-OH-3, 5, 7- $C_{17}H_{18}O_4$ 286   OI
-tri-OMe-

Chem. Abstr. , 1971
(75) 85214.
Tet. Lett. , 1972, 4869.

2701-015

PHENANTHRENE, 4, 7-di-OH-2, $C_{17}H_{16}O_5$ 300   OI 195^0C
-3, 6-tri-OMe-

JCS, 1972, 207.

2701-020

PHENANTHRENE, 4,7-di-OH-2,3,- -6-tri-OMe-dihydro- *dihydro JCS, 1971, 3070. 2701-021	$C_{17}H_{18}O_5$	302 OI 148^0C

DEUTICULATOL $C_{15}H_{10}O_4$ 254 OI 162^0C — J. Chin. Chem. Soc., 1947 (15) 21. — 2701-026

PHENANTHRENE, 1-OH-2,5,6,7--tetra-OMe- $C_{18}H_{18}O_5$ 314 OI — Chem. Abstr., 1971 (75) 85214. — 2701-022

PILOQUINONE $C_{21}H_{20}O_5$ 352 OI 177^0C R = H — Bull. Soc. Chim. Fr., 1963, 1918. — 2701-027

PHENANTHRENE, 2-OH-1,7-di--OMe-5,6-methylenedioxy- $C_{17}H_{14}O_5$ 298 OI — Chem. Abstr., 1971 (75) 85214. — 2701-023

PILOQUINONE, 4-OH- $C_{21}H_{20}O_6$ 368 OI 175^0C R = OH — Bull. Soc. Chim. Fr., 1969, 3100. — 2701-028

PHENANTHRENE, 1,2,7-tri-OMe--5,6-methylenedioxy- $C_{18}H_{16}O_5$ 312 OI 151^0C — Tet. Lett., 1969, 67. — 2701-024

ATROVENETIN $C_{19}H_{18}O_6$ 342 $+154^0d$ 295^0C — JCS, 1965, 1097. — 2701-029

PHENANTHRENE QUINONE, 1,3,4--tri-OMe-2,7- $C_{17}H_{14}O_5$ 298 OI — Chem. Abstr., 1971 (75) 85214. — 2701-025

HERQUEINONE, nor- $C_{19}H_{18}O_7$ 358 1080^0p 279^0C R = H — Tet., 1962 (18) 839. — 2701-030

HERQUEINONE	$C_{20}H_{20}O_7$	372
		$+440^0$e
		226^0C
R = Me		
	JOC, 1970, 179.	
2701-031		

TETRACYCLINE, Br-	$C_{22}H_{23}N_2O_8Br$	523
		-196^0
		171^0C
R = Br		
	Il Farmaco Ed. Sci., 1955 (10) 337.	
2702-003		

HERQUEINONE, iso-	$C_{20}H_{20}O_7$	372
R = Me		
(* epimer)		
	Tet. Lett., 1970, 963.	
2701-032		

TETRACYCLINE, 6-des-Me-	$C_{21}H_{22}N_2O_8$	430
		-36^0
		242^0C
R = H		

	JACS, 1957, 4564.	
2702-004		

SCHIZANDRIN	$C_{24}H_{32}O_7$	432
		$+78^0$c
		129^0C

	Tet. Lett., 1961, 730.	
2701-033		

TETRACYCLINE, 7-Cl-6-Des-Me-	$C_{21}H_{22}N_2O_8Cl$	465
		-258^0
		176^0C
R = Cl		
	Ant. Chemoth., 1959 (9) 13.	
2702-005		

TETRACYCLINE	$C_{22}H_{24}N_2O_8$	444
		-239^0m
		172^0C
R = H		

	Chem. Comm., 1967, 77.	
2702-001		

TETRACYCLINE, oxy-	$C_{22}H_{24}N_2O_9$	460
		$+27^0$m
		185^0C
R = H		

	JACS, 1965, 134.	
2702-006		

AUREOMYCIN	$C_{22}H_{23}N_2O_8Cl$	479
		-275^0m
		168^0C
R = Cl		
	JACS, 1963, 851.	
2702-002		

TETRACYCLINE, 7-Cl-5-OH-	$C_{22}H_{23}N_2O_9Cl$	494
		HCl: -237^0
R = Cl		
	JACS, 1966, 3647.	
2702-007		

TETRACYCLINE, 7-Cl-dehydro- $C_{22}H_{21}N_2O_8Cl$ 476
HCl:$+16^0$

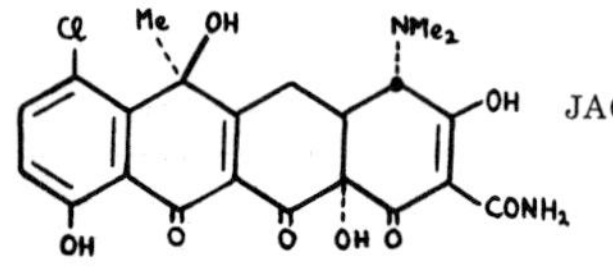

JACS, 1958, 5572.

2702-008

CHELOCARDIN $C_{22}H_{21}NO_7$ 411

215^0C

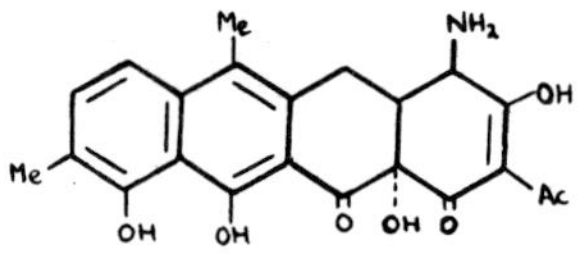

JACS, 1970, 6070.

2702-014

TETRACYCLINE, anhydro-des-
-Me-Cl- $C_{21}H_{20}N_2O_7Cl$ 447

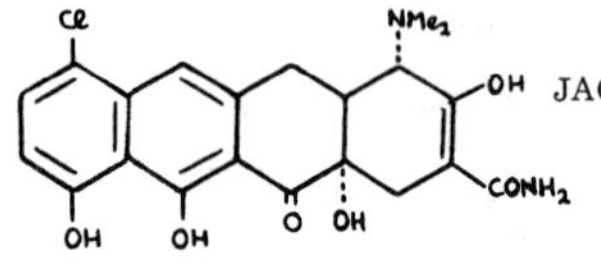

JACS, 1969, 206.

2702-010

PRETETRAMID, 4-OH-6-Me- $C_{20}H_{15}NO_7$ 381
OI
260^0C

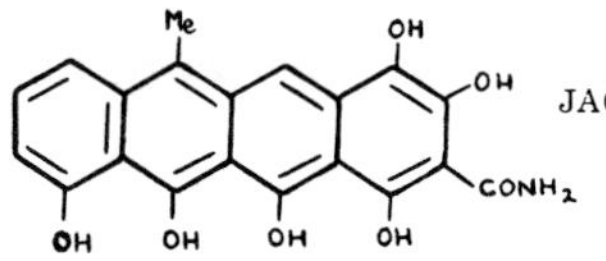

JACS, 1965 (87) 1973.

2702-015

TETRACYCLINE, di-des-N-Me-
-des-Me-anhydro-Cl- $C_{19}H_{16}N_2O_7Cl$ 419

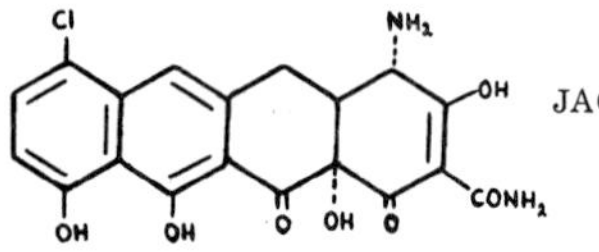

JACS, 1968, 2201.

2702-011

TETRACYCLINE, 2-Ac-2-
-decarboxamido- $C_{23}H_{25}NO_8$ 443
-125^0

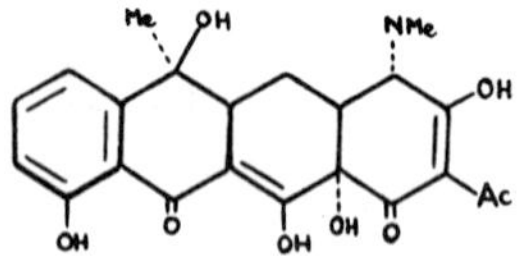

Antibiotiki, 1971, 22.

2702-012

PROTETRONE, Me- $C_{20}H_{17}NO_8$ 399

130^0C

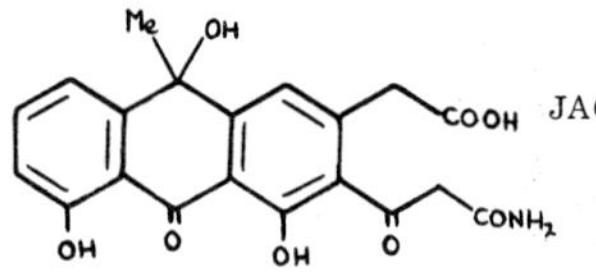

JACS, 1968, 7127.

2702-017

TERRAMYCIN-X $C_{23}H_{25}NO_9$ 459
HCl: -47^0
HCl: 202^0C

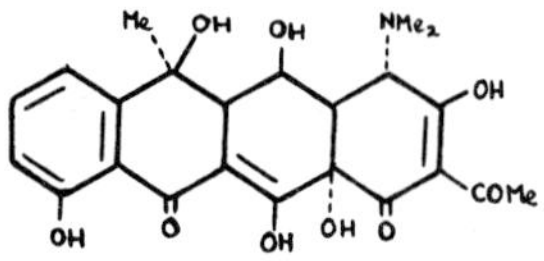

JACS, 1960, 5934.

2702-013

GALIRUBINONE-B1 $C_{22}H_{16}O_6$ 376
OI
236^0C

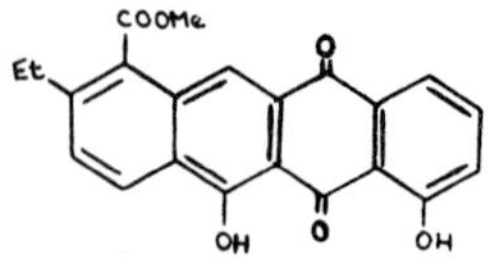

Ber., 1967 (100) 2561.

2703-001

RHODOMYCINONE, 10-desoxy-γ- $C_{20}H_{18}O_6$ 354

232^{0}C

Ber., 1965 (98) 3785.

2703-002

RHODOMYCINONE, β-1- $C_{19}H_{16}O_7$ 356

~260^{0}C

Ber., 1965 (98) 3785.

2703-007

CITROMYCINONE, γ- $C_{20}H_{18}O_6$ 354

R = H   207^{0}C

Ber., 1968 (101) 1341.

2703-003

RHODOMYCINONE, γ- $C_{20}H_{18}O_7$ 370

R = H   235^{0}C

Ber., 1967 (99) 3578.

2703-008

CITROMYCINONE. α- $C_{20}H_{18}O_7$ 370

136^{0}C

R = OH

Ber.. 1968 (101) 1341.

2703-004

RHODOMYCIN-I, γ- $C_{28}H_{33}NO_9$ 527

R = Rhodosamine

Natwiss., 1961 (48) 717.

2703-009

PYRROMYCINONE, eta(1)- $C_{21}H_{14}O_7$ 378

R = Me   291^{0}C

COOMe

Tet. Lett., 1968, 1587.

2703-005

RHODOMYCIN-II, γ- $C_{36}H_{48}N_2O_{11}$ 684

R = .(Rhodosamine)$_2$

Natwiss., 1961 (48) 717.

2703-010

PYRROMYCINONE. eta- $C_{22}H_{16}O_7$ 392

236^{0}C

R = Et

Tet. Lett., 1968. 471.

2703-006

RHODOMYCIN-III, γ- $C_{42}H_{58}N_2O_{14}$ 810

R = .(Rhodosamine)$_2$-(2-deoxy-fucose)

Natwiss., 1961 (48) 717.

2703-011

RHODOMYCIN-IV, γ-	$C_{48}H_{68}N_2O_{16}$	924

R = .(Rhodosamine)$_2$-(2-deoxy-fucose)-

— Rhodinose

Natwiss., 1961 (48) 717.

2703-012

RHODOMYCINONES, α- $C_{20}H_{18}O_8$ 386

218^{0}C

Ber., 1967 (100) 3578.

2703-017

AKLAVINONE, 7-deoxy- $C_{22}H_{20}O_7$ 396

R = H 224^{0}C

Tet. Lett., 1968, 4719.

2703-013

RHODOMYCINONE, α-2- $C_{20}H_{18}O_8$ 386

208^{0}C

Ber., 1968 (101) 1341.

2703-018

AKLAVINONE $C_{22}H_{20}O_8$ 412
+142^{0}c
172^{0}C

R = OH

Tet. Lett., 1960 (8) 28.
Chem. Pharm. Bull.,
1968 (16) 1251.

2703-014

RHODOMYCINONE, β- $C_{20}H_{18}O_8$ 386

R = R' = H 225^{0}C

Ber., 1967, 3578.

2703-019

PYRROMYCINONE, zeta- $C_{22}H_{20}O_8$ 412
+74^{0}c
216^{0}C

Ber., 1967 (100) 2561.
Natwiss., 1960 (47) 135.

2703-015

RHODOMYCIN-B $C_{28}H_{33}NO_{10}$ 543
+174^0

R = Rhodosamine

R' = H

Tet. Lett., 1969, 415.

2703-020

RHODOMYCINONE, zeta- $C_{22}H_{20}O_8$ 412

274^{0}C

Ann., 1966 (696) 145.

2703-016

RHODOMYCIN-A $C_{36}H_{48}N_2O_{12}$ 700
+178^0
189^{0}C

R = R' = Rhodosamine

Tet. Lett., 1969, 415.

2703-021

RHODOMYCIN-IV, ẟ-	$C_{48}H_{68}N_2O_{17}$	940

R = .(Deoxy-fucose)-Rhodinose

R' = (Rhodosamine)$_2$

Tet. Lett., 1969, 415.

2703-022

RHODOMYCINONE, ẟ-	$C_{22}H_{20}O_9$	428

196⁰C

Tet., 1963, 395.

2703-027

DAUNOMYCINONE	$C_{21}H_{18}O_8$	398

+193⁰d
213⁰C

R = H

Tet. Lett., 1968, 3891.

2703-023

RHODOMYCINONE, ε-	$C_{22}H_{20}O_9$	428

210⁰C

Ann., 1966 (696) 145.

2703-028

DAUNOMYCIN	$C_{27}H_{29}NO_{10}$	527

HCl: +253⁰m
HCl: 189⁰C

R = Daunosamine

Tet. Lett., 1968, 3891.

2703-024

PYRROMYCINONE, ε-	$C_{22}H_{20}O_9$	428

+143⁰c
213⁰C

R = H

Tet. Lett., 1960 (8) 25.

2703-029

DAUNOMYCINONE, dihydro-	$C_{21}H_{20}O_8$	400

R = H

Fr. Pat. 2008463.

2703-025

PYRROMYCIN	$C_{30}H_{35}NO_{11}$	585

HCl: +132⁰m
HCl: 162⁰C

R = Rhodosamine

Ber., 1959 (92) 1904.

2703-030

DAUNOMYCIN, dihydro-	$C_{27}H_{31}NO_{10}$	529

R = Daunosamine

Fr. Pat. 2008463.

2703-026

CINERUBINE-A	$C_{42}H_{41}NO_{16}$	815

R = Rhodosamine
(2-deoxy-fucose)
Cinerulose-A

Ant. Ag. Chemother., 1970, 68.

2703-031

RHODOMYCINONE, θ-	$C_{22}H_{20}O_9$	428 +192° 220°C	ADRIAMYCIN	$C_{27}H_{29}NO_{11}$	543

RHODOMYCINONE, θ- $C_{22}H_{20}O_9$ 428 +192° 220°C

JCS, 1964, 3927.

2703-032

ADRIAMYCIN $C_{27}H_{29}NO_{11}$ 543

R = Daunosamine

Tet. Lett., 1969. 1007.

2703-037

RHODOMYCINONE, β-iso- $C_{20}H_{18}O_9$ 402 amorph.

R = H

Ber., 1967 (100) 3578.

2703-033

RHODOMYCINONE, ε-iso- $C_{22}H_{20}O_{10}$ 444 228°C

Ber., 1961 (94) 2174.

2703-038

RHODOMYCIN-II, β-iso- $C_{36}H_{48}N_2O_{13}$ 716

R = Rhodosamine

Tet. Lett., 1969, 415.

2703-034

CHROMOCYCLOMYCINONE $C_{24}H_{24}O_9$ 456

R = R' = H

Chem. Comm., 1968, 762.

2703-039

RHODOMYCINONE, zeta-iso- $C_{22}H_{20}O_9$ 428 259°C

Ber., 1961 (94) 2174.

2703-035

CHROMOCYCLOMYCIN $C_{48}H_{64}O_{21}$ 976 -180°e 197°C

R = Mycarose
R' = .Olivose-Oliose-Mycarose

Chem. Comm., 1968. 762.

2703-040

ADRIAMYCINONE $C_{21}H_{18}O_9$ 414 +188°d 224°C

R = H

Tet. Lett., 1969, 1007.

2703-036

PILLAROMYCINONE $C_{20}H_{18}O_7$ 370 +550°m 202°C

R = H

Chem. Pharm. Bull., 1970, 1706.

2703-041

PILLAROMYCIN-A $\quad$ C$_{28}$H$_{28}$O$_{11}$ $\quad$ 540

$\qquad$ -37^{0}m

$\qquad$ 208^{0}C

R = pillarose

Chem. Pharm. Bull.,
1970, 1720.

2703-042

RABELOMYCIN $\quad$ C$_{19}$H$_{14}$O$_6$ $\quad$ 338

J. Ant., 1970 (23) 437.

2703-047

GRANATICIN $\quad$ C$_{22}$H$_{20}$O$_{10}$ $\quad$ 444

$\qquad$ 205^{0}C

Helv., 1968, 1269.

2703-043

AQUAYAMYCIN $\quad$ C$_{25}$H$_{26}$O$_{10}$ $\quad$ 486

$\qquad$ +100^{0}d

$\qquad$ 190^{0}C

Tet., 1970, 5171.

2703-048

TETRANGULOL $\quad$ C$_{19}$H$_{12}$O$_4$ $\quad$ 304

$\qquad$ OI

$\qquad$ 200^{0}C

JOC, 1966, 2920.

2703-044

RHODOMYCINONE, γ-iso- $\quad$ C$_{20}$H$_{18}$O$_8$ $\quad$ 386

$\qquad$ amorph.

Tet. Lett., 1968, 4719.

2703-049

TETRANGOMYCIN $\quad$ C$_{19}$H$_{14}$O$_5$ $\quad$ 322

$\qquad$ +42^{0}c

$\qquad$ 183^{0}C

JOC, 1966, 2920.

2703-045

LACHNANTHOCARPONE $\quad$ C$_{19}$H$_{12}$O$_3$ $\quad$ 288

$\qquad$ OI

Tet. Lett., 1969, 4325.

2704-001

OCHROMYCINONE $\quad$ C$_{19}$H$_{14}$O$_4$ $\quad$ 306

$\qquad$ +204^{0}c

$\qquad$ 152^{0}C

Tet. Lett., 1967, 1449.

2703-046

LACHNANTHOFLUORONE $\quad$ C$_{19}$H$_{10}$O$_3$ $\quad$ 286

$\qquad$ OI

Tet. Lett., 1969, 4325.

2704-002

HAEMOCORIN AGLYCONE $C_{20}H_{14}O_4$ 318

227^{0}C

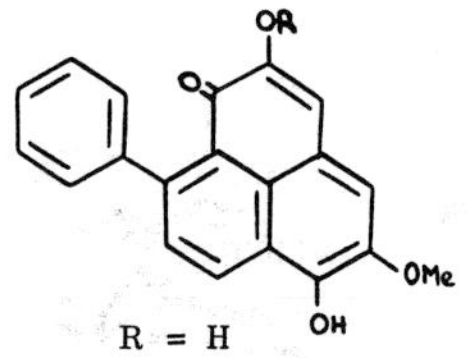

R = H

Chem. Comm., 1968, 1557.

2704-003

PURPUROGENONE $C_{29}H_{20}O_{11}$ 544
+254^{0}d
~300^{0}C

R = OH

JCS, 1971, 3488.

2704-008

HAEMOCORIN $C_{32}H_{34}O_{14}$ 642

263^{0}C

R = Cellobiosyl

Aust. J. C., 1958, 230.

2704-004

PURPUROGENONE, deoxy- $C_{29}H_{20}O_{10}$ 528
+242^{0}d
>300^{0}C

R = H

JCS, 1971, 3493.

2704-009

RESISTOMYCIN $C_{22}H_{16}O_6$ 376
-32^{0}m
315^{0}C

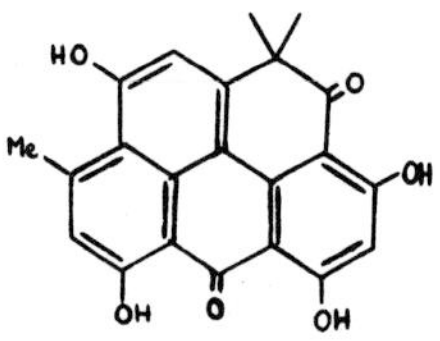

Chem. Comm., 1968, 374.

2704-005

QUINOCYCLINE-A, iso- $C_{33}H_{32}N_2O_{10}$ 616

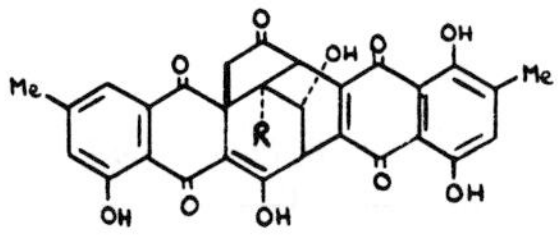

HCl:+12^{0}C

JACS, 1964, 5368.

2704-010

RESISTOFLAVIN $C_{22}H_{16}O_7$ 392
-96^{0}p
239^{0}C

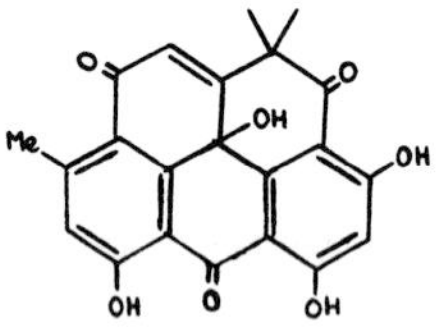

Tet., 1970, 5875.

2704-006

ARENICOCHROMINE $C_{21}H_{12}O_6$ 360

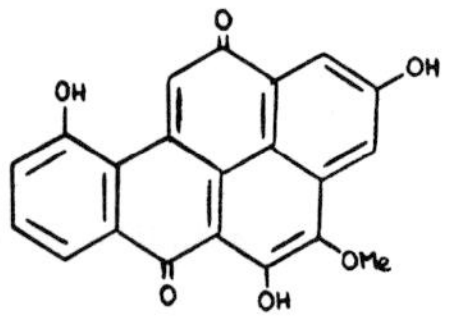

Chem. Comm., 1970, 550.

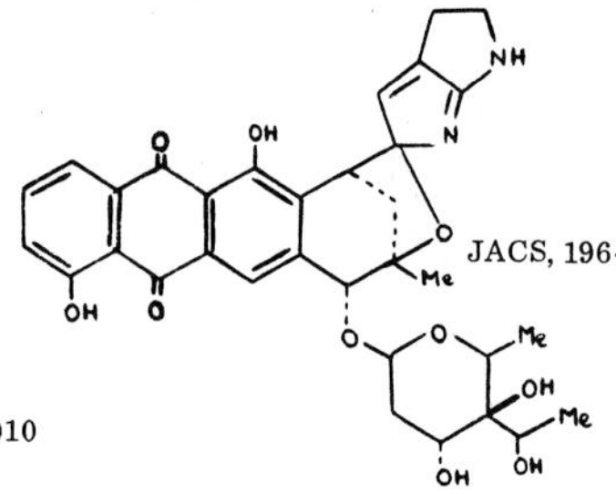

2704-011

PHENOCYCLINONE $C_{35}H_{24}O_{14}$ 668

amorph.

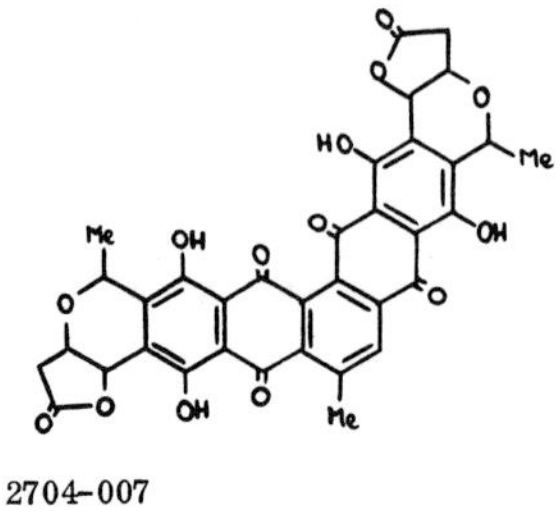

Ber., 1970 (103) 708.

2704-007

PERYLENE-3, 10-QUINONE, $C_{20}H_{10}O_4$ 314
-4, 9-diOH- OI

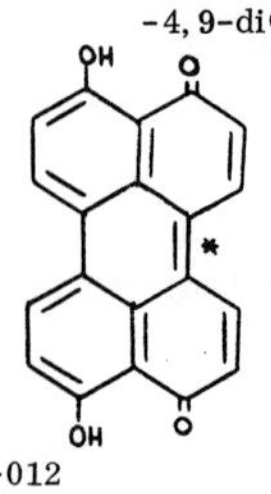

Chem. Ind., 1956, 376.

2704-012

ASPERGILLINE C20H12O4 316 *dihydro Chem. Abstr., 1968 (69) 27099. 2704-013	CERCOSPORIN C29H26O10 534 271°C Chem. Comm., 1971, 1463. 2704-018
ELSINOCHROME-A C30H24O10 544 248°C R = R' = Ac JCS, 1969, 1219. 2704-014	PROTOHYPERICIN C30H18O8 OI Angew. Chem., 1955 (67) 706. 2704-019
ELSINOCHROME-B C30H26O10 546 208°C R = Ac; R' = .CHOH.Me JCS, 1969, 1219. 2704-015	HYPERICIN C30H16O8 504 OI 300°C R = Me Natwiss., 1951, 47. 2704-020
ELSINOCHROME-C C30H28O10 548 293°C R = R' = CHOH.Me JCS, 1969, 1219. 2704-016	HYPERICIN, pseudo- C32H20O10 564 R = CHOH.Me Natwiss., 1954, 86. 2704-021
ELSINOCHROME C30H26O10 546 160°C JCS, 1970, 2159. 2704-017	HOPEAPHENOL C56H42O12 906 -407°e 351°C Chem. Comm., 1966, 439. JCS, 1965, 406. 2704-022

NEOCLAUXIN

$C_{26}H_{16}O_9$ 472

~250^{0}C

Tet. Lett., 1966, 2867.

2704-023

CHRYSOAPHIN-SL1

$C_{30}H_{24}O_9$ 528

238^{0}C

JCS, 1969, 629.

2704-028

DUCLAUXIN, des-Ac-

$C_{27}H_{20}O_{10}$ 504

253^{0}C

R = H

Tet. Lett., 1966, 2867.

2704-024

CHRYSOAPHIN-SL3

$C_{30}H_{24}O_9$ 528

* epimer

JCS, 1969, 629.

2704-029

DUCLAUXIN

$C_{29}H_{22}O_{11}$ 546
+272^{0}c
230^{0}C

R = Ac

Tet. Lett., 1966, 2883.

2704-025

CHRYSOAPHIN-SL2

$C_{30}H_{24}O_9$ 528

214^{0}C

* epimer

** epimer

JCS, 1969, 627.

2704-030

XENOCLAUXIN

$C_{28}H_{18}O_{11}$ 530
+310^0
300^{0}C

Tet. Lett., 1966, 2867.

2704-026

XANTHOAPHIN-SL1

$C_{30}H_{26}O_{10}$ 546

204^{0}C

JCS, 1969, 627.

2704-031

CRYPTOCLAUXIN

$C_{29}H_{22}O_{12}$ 562

300^{0}C

Tet. Lett., 1966, 2867.

2704-027

XANTHOAPHIN-SL2

$C_{30}H_{26}O_{10}$ 546

* epimer

JCS, 1969, 629.

2704-032

RHODOAPHIN-BE $C_{30}H_{22}O_{10}$ 542

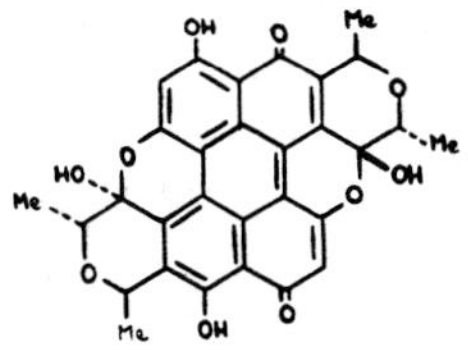

–

JCS, 1967, 704.

2704-033

CHARTREUSIN $C_{32}H_{32}O_{14}$ 640

$+132^{0}p$

$185^{0}C$

Digitalose
|
Fucose

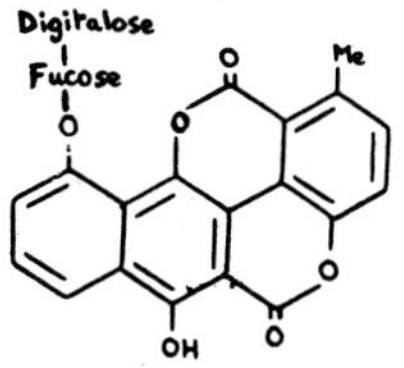

JACS, 1962, 4011.

2704-034

| 28 | $C_6 C_n C_6$ |

Several groups of compounds of the generic type $C_6 C_n C_6$ have been assembled in this class—the notable exception being the $C_6 C_3 C_6$ group, which has been posted elsewhere (Class 10). Two other groups of compounds have been included in this class: the grisans (because of their relationship to some of the benzophenones) and the turrianes (oxidatively coupled $C_6 C_n C_6$ macrocycles). The biogenetic origins of these various groups differ considerably both from group to group and within each group. Thus, the benzophenones can derive from the condensation of a heptaketide chain (e.g., the griseophenones) or from the condensation of an acetate-extended benzoyl moiety of shikimate origin (e.g., the cotoins). The biphenyls are to some extent a mystery class in that there would appear to be no clear-cut representatives of the biogenetic shikimate/acetate route known to date (although aucuparin may prove to be the first). Altenusin and alternariol are clearly derived from a heptaketide precursor. The aucuparins, sappanin, and urolithin-B remain somewhat uncommitted biogenetically.

2801	(cf: 0105 Ellagic Acids) Biphenyls
2802	Benzophenones
2803	Grisans
2804	Stilbenes
2805	$C_6-C_n-C_6$ (n $\geqslant$ 4)
2806	Turrianes

AUCUPARIN	$C_{14}H_{14}O_3$	230 OI 101^0C

Acta Chem. Scand., 1963 (17) 1151.

2801-001

ALTENUSIN	$C_{15}H_{14}O_6$	290 OI 196^0C

Aust. J. C., 1970, 2343.

2801-006

AUCUPARIN, OMe-	$C_{15}H_{16}O_4$	260 OI 121^0C

Acta Chem. Scand., 1963 (17) 1157.

2801-002

ALTENUSIN, dehydro-	$C_{15}H_{12}O_6$	288 190^0C

Chem. Comm., 1971, 393.
Aust. J. C., 1970, 2343.

2801-007

SAPPANIN	$C_{12}H_{10}O_4$	218 210^0C

Monatsh., 1930 (55) 342.

2801-003

BOTRALLIN	$C_{16}H_{14}O_7$	318 OI

Rec. Trav. Chim., 1968 (87) 940.

2801-008

UROLITHIN-B	$C_{13}H_8O_3$	212 OI 224^0C

Biochem. J., 1964 (93) 474.

2801-004

BENZOPHENONE, p-OH-	$C_{13}H_{10}O_2$	198 OI 134^0C

Chem. Abstr., 1948 (42) 2730.

2802-001

ALTERNARIOL	$C_{14}H_{10}O_5$	258 OI 350^0C

Biochem. J., 1953 (55) 421.

2801-005

COTOIN	$C_{14}H_{12}O_4$	244 OI 131^0C

R = R' = H

J. Am. Pharm. Ass., 1922 (11) 904.

2802-002

HYDROCOTOIN $C_{15}H_{14}O_4$ 258
OI
98^0C

R = Me; R' = H

Monatsh., 1897 (18) 736.

2802-003

PROTOCOTOIN $C_{16}H_{14}O_6$ 302

R = H

Monatsh., 1928 (49) 429.

2802-008

HYDROCOTOIN, Me- $C_{16}H_{16}O_4$ 272
OI
114^0C

R = R' = Me

J. Am. Pharm. Ass.,
1938 (27) 95.

2802-004

PROTOCOTOIN, Me- $C_{17}H_{16}O_6$ 316
OI
134^0C

R = Me

Ber., 1892 (25) 1119.

2802-009

BENZOPHENONE, 6-OH-2, 4-di-
-OMe-3-Me- $C_{16}H_{16}O_4$ 272

Aust. J. C., 1963 (16) 282.

2802-005

BENZOPHENONE, 2, 3', 4, 6-
-tetra-OH- $C_{13}H_{10}O_5$ 246
OI
237^0C

Tet., 1969, 1507.

2802-010

SCLEROIN $C_{15}H_{14}O_5$ 274
OI
144^0C

Tet., 1965, 2697.

2802-006

SULOCHRIN $C_{17}H_{16}O_7$ 332
OI
262^0C

JCS, 1972, 240.

2802-011

MACLURIN $C_{13}H_{10}O_6$ 262
OI
221^0C

Chem. Abstr., 1943
(37) 5195.

2802-007

GUTTIFERIN, α- $C_{33}H_{38}O_8$ 552
-475^0c
114^0C

Experientia, 1961 (17) 213.

2802-012

GRISEOPHENONE-C	$C_{16}H_{16}O_6$	304 OI 183^0C

R = R' = H

Biochem. J., 1963 (88) 349.

2802-013

GEODIN, (+)-bis-de-Cl-	$C_{17}H_{14}O_7$	330 $+186^0$e 167^0C

R = H

Biochem. J., 1964, 43.

2803-001

GRISEOPHENONE-B	$C_{16}H_{15}O_6Cl$	338 OI 205^0C

R = H; R' = Cl

Biochem. J., 1961 (81) 28.

2802-014

GEODIN, (-)-bis-de-Cl-	$C_{17}H_{14}O_7$	330 -103^0e 144^0C

R = H

Biochem. J., 1964, 43.

2803-002

GRISEOPHENONE-A	$C_{17}H_{17}O_6Cl$	352 OI 214^0C

R = Me; R' = Cl

Biochem. J., 1961 (81) 28.

2802-015

TRYPACIDIN	$C_{18}H_{16}O_7$	344 -103^0 229^0C

R = Me

JCS, 1965, 6658.

2803-003

GRISEOPHENONE-Y	$C_{17}H_{17}O_5Cl$	336 OI 181^0C

JCS, 1960, 4628.

2802-016

ERDIN	$C_{16}H_{10}O_7Cl_2$	384 Rac 211^0C

R = H

JCS, 1958, 1767.

2803-004

GEODIN, dihydro-	$C_{17}H_{14}O_7Cl_2$	400

Biochem. J., 1964 (90) 43.

2802-017

GEODIN	$C_{17}H_{12}O_7Cl_2$	398 (+) 235^0C

R = Me

JCS, 1958, 1767.

2803-005

GRISEOFULVIN, de-Cl-	$C_{17}H_{18}O_6$	318 +390°a 180°C
R = H	JCS, 1953, 1697.	
2803-006		

STILBENE, trans- $C_{14}H_{12}$ 180 OI
Tet. Lett., 1969, 3235.
2804-001

GRISEOFULVIN $C_{17}H_{17}O_6Cl$ 352 +417°a 220°C
R = Cl
Chem. Ind., 1961, 792.
2803-007

STILBENE, 4-OH- $C_{14}H_{12}O$ 196 OI 186°C
R = H
J. Sci. Ind. Res., 1954 (13B) 835.
2804-002

GRISEOFULVIN, Br- $C_{17}H_{17}O_6Br$ 397 204°C
R = Br
JCS, 1963, 1050.
2803-008

STILBENE, 4-OMe- $C_{15}H_{14}O$ 210 OI 136°C
R = Me
J. Sci. Ind. Res., 1954 (13B) 835.
2804-003

GRISEOFULVIN, dehydro- $C_{17}H_{15}O_6Cl$ 350
JCS, 1960, 4628.
2803-009

PINOSYLVIN $C_{14}H_{12}O_2$ 212 OI 156°C
R = R' = H
J. Sci. Ind. Res., 1954 (13B) 835.
2804-004

GRISEOFULVIN, de-Cl-6'-O--des-Me- $C_{16}H_{16}O_6$ 304
JCS, 1960, 4628.
2803-010

PINOSYLVIN ME ETHER $C_{15}H_{14}O_2$ 226 OI 122°C
R = Me; R' = H
Chem. Abstr., 1956 (50) 1990.
2804-005

382

PINOSYLVIN DIME ETHER $C_{16}H_{16}O_2$ 240
OI
56^0C

R = R' = Me

J. Sci. Ind. Res., 1954
(13B) 835.

2804-006

PELLEPIPHYLLIN $C_{16}H_{18}O_3$ 258
OI
71^0C

Coll. Czech. Chem.
Comm., 1970, 1926.

2804-011

PINOSYLVIN ME ETHER,
-dihydro- $C_{15}H_{16}O_2$ 228
OI
51^0C

Chem. Abstr., 1954
(48) 12922.

2804-007

RESVERATROL $C_{14}H_{12}O_3$ 228
OI
261^0C

R = H

Ber., 1941 (74) 867.

2804-012

HYDRANGEIC ACID $C_{15}H_{12}O_4$ 256
OI
180^0C

Ber., 1930 (63) 429.

2804-008

PICEID $C_{20}H_{22}O_8$ 390
-76^0
225^0C

R = Glucosyl

Phytochem., 1971 (10) 607.

2804-013

HYDRANGEIC ACID, dihydro- $C_{15}H_{14}O_4$ 258
OI
192^0C

Z. Nat., 1971 (26B) 738.
Nature, 1969 (223) 1176.

2804-009

PICEID, 4'-O-Me- $C_{21}H_{24}O_8$ 404
-50^0
227^0C

Aust. J. C., 1971, 2427.

2804-014

HYDRANGENOL $C_{15}H_{12}O_4$ 256
182^0C

Can. J. Biochem. Physiol.,
1962 (40) 449.

2804-010

PTEROSTILBENE $C_{16}H_{16}O_3$ 256
OI
88^0C

JCS, 1953, 3693.

2804-015

PHYLLODULCINOL $C_{16}H_{14}O_5$ 286
+75°a
120°C

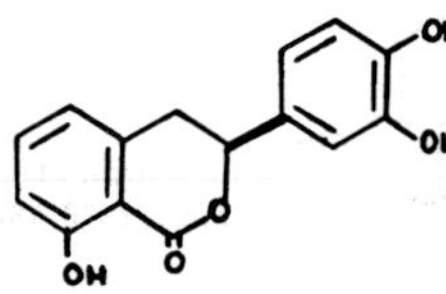

Chem. Ind., 1959, 671.

2804-016

STILBENE, 3, 4, 5, 3', 5'- Penta- OH- $C_{14}H_{12}O_5$ 260
OI
245°C

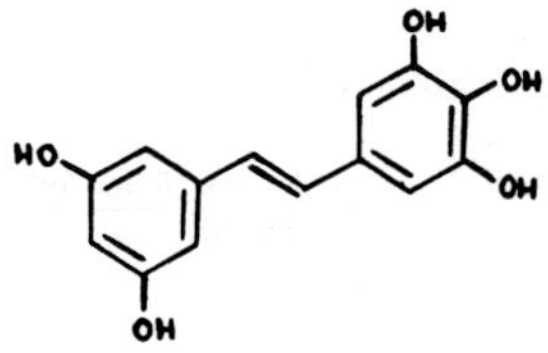

JCS, 1956, 4477.

2804-021

STILBENE 5-GLUCOSIDE,
-3, 5, 4'-triOH-3'-OMe- $C_{21}H_{24}O_9$ 420
-54°a
amorph.

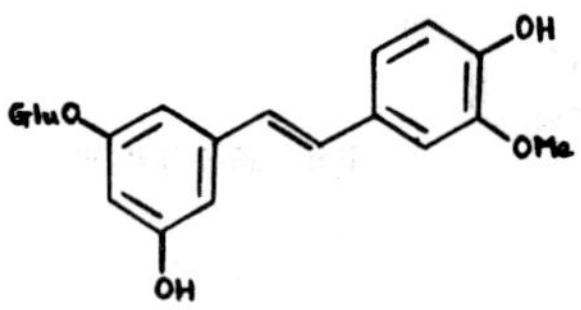

Can. J. C., 1968, 2525.

2804-017

RESVERATROL, oxy- $C_{14}H_{12}O_4$ 244
OI
199°C

R = H

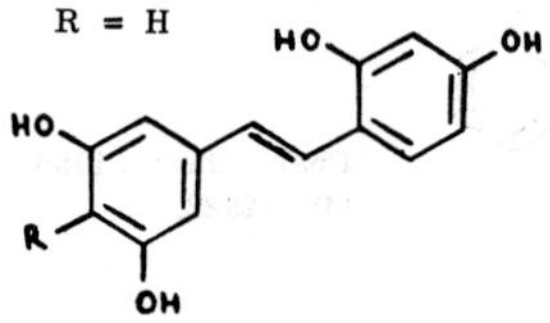

Tet. Lett., 1970, 4051.

2804-022

PICEATANNOL $C_{14}H_{12}O_4$ 244
OI
229°C

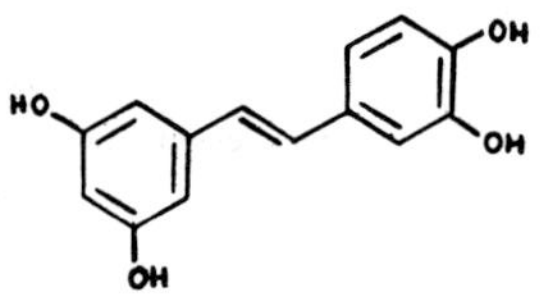

Tet. Lett., 1970, 4051.

2804-018

CHLOROPHORIN $C_{24}H_{28}O_4$ 380
OI
158°C

R = Geranyl

JCS, 1950, 3547.

2804-023

RHAPONTIGENIN $C_{15}H_{14}O_4$ 258
OI
185°C

R = H

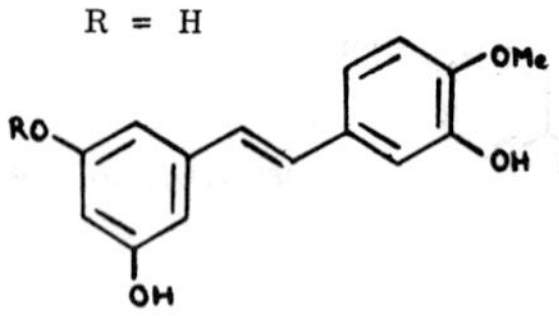

Tet. Lett., 1970, 4051.

2804-019

PTEROFURAN $C_{16}H_{14}O_5$ 286
OI
208°C

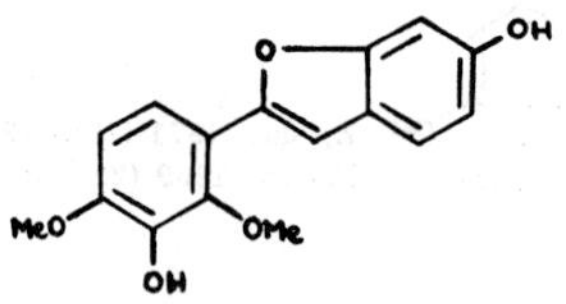

Aust. J. C., 1964, 379.

2804-024

RHAPONTIN $C_{21}H_{24}O_9$ 420
-60°a
237°C

R = Glucosyl

J. Pharm. Soc. Jap.,
1938 (58) 405.

2804-020

$C_{16}H_{12}O_5$ 284
OI
118°C

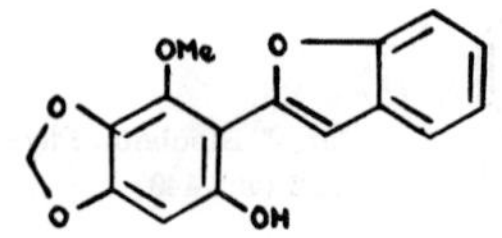

JACS, 1959, 4979.

2804-025

ONONETIN $C_{15}H_{14}O_4$ 258
     OI
     159^0C

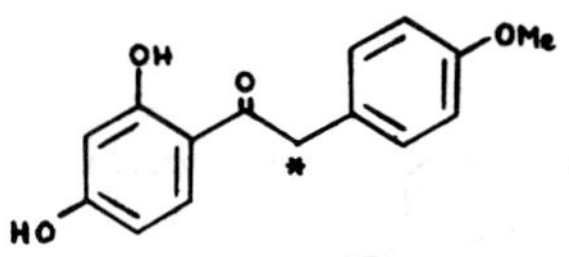

Aust. J. C., 1966, 1755.

2804-026

PENTANOL, 1, 5-di-Ph-3- $C_{17}H_{20}O$ 240
     OI
     48^0C

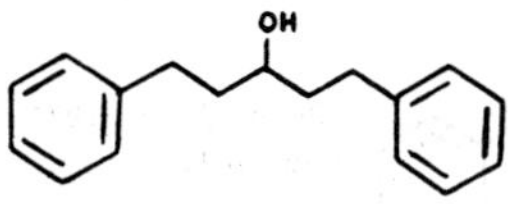

Aust. J. C., 1962, 819.

2805-003

ONONETIN, oxo- $C_{15}H_{12}O_5$ 272
     OI
     172^0C

* oxo

Aust. J. C., 1966, 1755.

2804-027

PENTANE-1, 3-DIOL, 1, 5-di-Ph- $C_{17}H_{20}O_2$ 256
     -19^0e
     89^0C

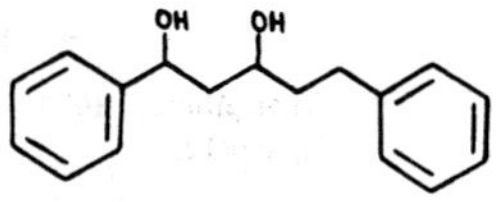

Aust. J. C., 1962, 819.

2805-004

DIBENZYL, 3, 4'-di-OH-4, 5- -di-OMe- $C_{16}H_{18}O_4$ 274
     OI
     129^0C

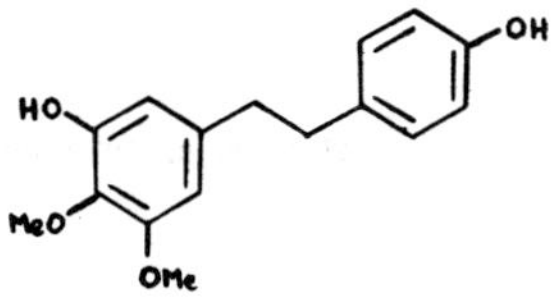

JCS, 1972, 206.

2804-028

HEPTAN-3, 5-DIOL, 1, 7-di-Ph- $C_{19}H_{24}O_2$ 284

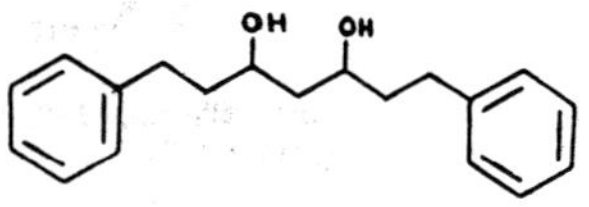

Khim. Prir. Soedin., 1970 (6) 463.

2805-005

BUTANE-2, 3-DIOL, 1, 4-di-Ph- $C_{10}H_{18}O_2$ 242
     $+5^0c$
     146^0C

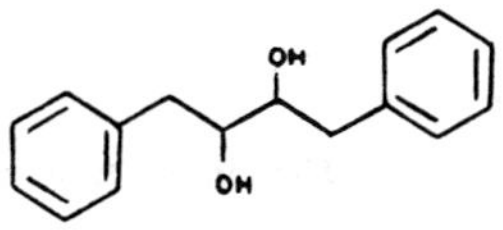

Helv., 1963 (46) 1083.

2805-001

YASHABUSHI-KETOL, dihydro- $C_{19}H_{22}O_2$ 282
     36^0C

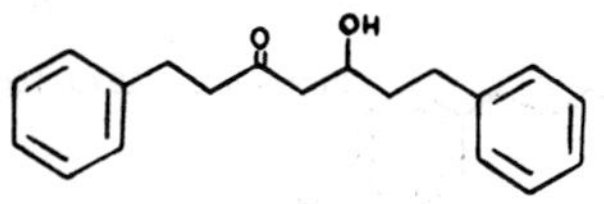

Bull. Chem. Soc. Jap., 1970, 575.

2805-006

AGRIMOLIDE $C_{18}H_{18}O_5$ 314
     $+8^0a$
     176^0C

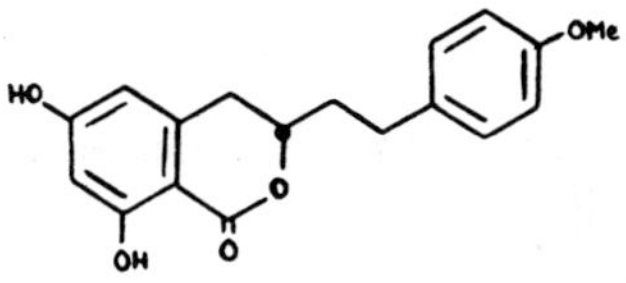

Tet. Lett., 1968, 4115.

2805-002

YASHABUSHI-KETOL $C_{19}H_{20}O_2$ 280
     $+29^0c$
     60^0C

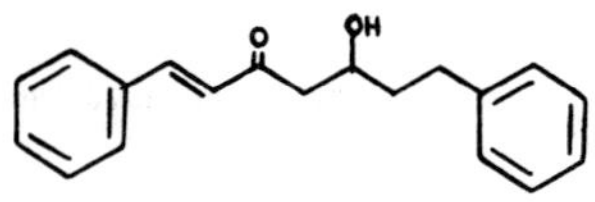

Bull. Chem. Soc. Jap., 1970, 575.

2805-007

DICOUMAROYL-METHANE	$C_{19}H_{16}O_4$	308
		OI
		224^0C

R = R' = H

Chem. Abstr. , 1971
(74) 61612.

2805-008

CENTROLOBIN, (+)-des-Me-	$C_{19}H_{22}O_3$	298
		247^0C

R = H

Phytochem. , 1970, 1869.

2805-013

COUMAROYL-FERULOYL- -METHANE	$C_{20}H_{18}O_5$	338
		OI
		168^0C

R = Me; R' = H

Chem. Abstr. , 1971
(74) 61612.

2805-009

CENTROLOBIN, (-)-des-Me-	$C_{19}H_{22}O_3$	298
		246^0C

R = H
(Enantiomer)

Phytochem. , 1970. 1869.

2805-014

CURCURMIN	$C_{21}H_{20}O_6$	368
		OI
		183^0C

R = R' = Me

Arch. Pharm. , 1932
(270) 413.

2805-010

CENTROLOBIN, (+)-	$C_{20}H_{24}O_3$	312
		$+97^0m$
		85^0C

R = Me

Phytochem. , 1970, 1869.

2805-015

CENTROLOBOL, (-)-	$C_{19}H_{24}O_3$	300
		-9^0m
		129^0C

Phytochem. , 1970, 1869.

2805-011

CENTROLOBIN, (-)-	$C_{20}H_{24}O_3$	312
		-98^0m
		85^0C

R = Me
(Enantiomer)

Phytochem. , 1970, 1869.

2805-016

CENTROLOBOL, (+)-	$C_{19}H_{24}O_3$	300
		$+8^0m$
		129^0C

*epimer

Phytochem. , 1970, 1869.

2805-012

STRIATOL	$C_{28}H_{42}O_4$	442
		OI
		132^0C

Aust. J. C. , 1968, 2993.

2805-017

ROBUSTOL	$C_{26}H_{36}O_4$	412 OI 144^0C	ASADANOL, iso-	$C_{19}H_{18}O_6$	342

Tet. Lett., 1970, 325.

Chem. Abstr., 1967 (67) 2840.

2805-018

2806-005

MYRICANOL	$C_{21}H_{26}O_5$	358 -66^0c 105^0C	TURRIANE, 17, 19, 24-triOH-22-OMe-	$C_{27}H_{38}O_4$	426 OI

JCS, 1971, 3634.

R = H

Aust. J. C., 1970 (23) 147.

2806-001

2806-006

MYRICANONE	$C_{21}H_{24}O_5$	356 OI 195^0C	TURRIANE, 17, 24-diOH-19, 22--diOMe-	$C_{28}H_{40}O_4$	440 OI

*oxo

R = Me

JCS, 1971, 3634.

Aust. J. C., 1970 (23) 147.

2806-002

2806-007

ASADANIN	$C_{19}H_{20}O_6$	344 $+84^0c$ 238^0C	TURRI-4-ENE, 17, 19, 24-triOH--22-OMe-	$C_{27}H_{36}O_4$	436 OI

Chem. Abstr., 1967 (67) 2839.

R = H

Aust. J. C., 1970. 147.

2806-003

2806-008

ASADANOL, epi-	$C_{19}H_{18}O_6$	342	TURRI-4-ENE, 17, 24-diOH-19, 22--diOMe-	$C_{28}H_{38}O_4$	438 OI

Chem. Abstr., 1967 (67) 2839.

R = Me

Aust. J. C., 1970. 147.

2806-004

2806-009

TURRI-6-ENE, 17. 19, 24-triOH-
-22-OMe-
$C_{27}H_{36}O_4$ 436
OI

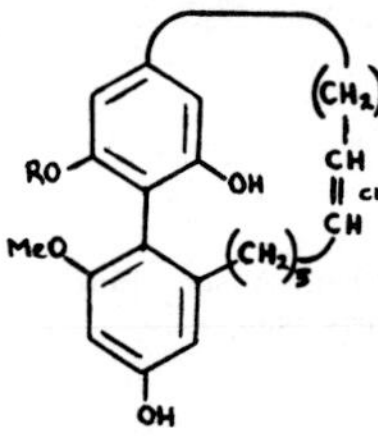

R = H   Aust. J. C. , 1970, 147.

2806-010

TURRI-6-ENE, 17, 24-diOH-19,
-22-diOMe-
$C_{28}H_{38}O_4$ 438
OI

R = Me

Aust. J. C. , 1970, 147.

2806-011

This cyclopentanoids class is purely structurally based, and its members are derived from several biogenetic origins. Many of the simplest cyclopentanoids are in fact somewhat ambiguous as to their biogenesis. The prostaglandins appear to derive from a C_{20} linear precursor by appropriate cyclization, whereas several highly substituted cyclopentanones can be regarded as the products of ring contraction of phenolic precursors.

2901	Simple Cyclopentanoids
2902.	Prostaglandins
2903	Cyclopentanoids derived by ring contraction of a phenolic precursor
2999	Misc.

CYCLOPENTANONE C_5H_8O 84
OI
Oil

Ber., 1898 (31) 1885.

2901-001

CALDARIOMYCIN $C_5H_8O_2Cl_2$ 171
121^0C

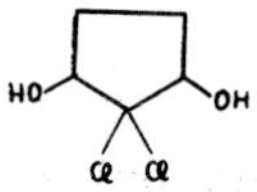

J. Biol. Chem., 1959
(234) 2560.

2901-006

deleted

2901-002

ALEPROLIC ACID $C_6H_8O_2$ 112

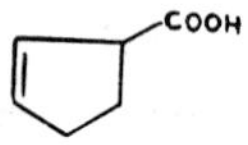

JACS, 1937, 2349.

2901-007

CYCLOPENT-2-ENONE, 2-Me- C_6H_8O 96
OI
Oil

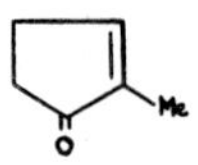

Compt. Rend., 1914
(158) 506.

2901-003

TETRAPHYLLIN-A $C_{12}H_{17}NO_6$ 271
-14^0w
117^0C

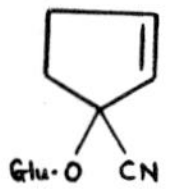

Phytochem., 1971 (10)
1373.

2901-008

CYCLOPENTANONE, 2, 4, 4-tri-Me- $C_8H_{14}O$ 126
Oil

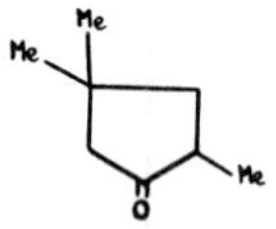

Helv., 1944, 51.

2901-004

TETRAPHYLLIN-B $C_{12}H_{17}NO_7$ 287
-36^0w
170^0C

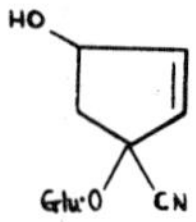

Phytochem., 1971 (10)
1373.

2901-009

SARKOMYCIN $C_7H_8O_3$ 140
-32^0m
Oil

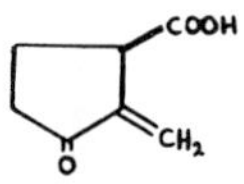

JACS, 1967, 2330.

2901-005

ALEPRESTIC ACID $C_{10}H_{16}O_2$ 168
impure

n = 4

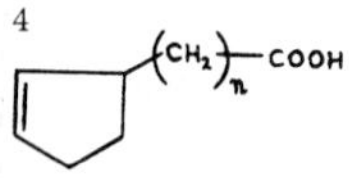

JACS, 1939, 2349.

2901-010

ALEPRYLIC ACID	$C_{12}H_{20}O_2$	196 $+91^0$ 32^0C

n = 6

JACS, 1939, 2349.

2901-011

HYDNOCARPIC ACID, keto-	$C_{16}H_{26}O_3$	266 OI 108^0C

n = 10

JCS, 1937, 955.

2901-016

ALEPRIC ACID	$C_{14}H_{24}O_2$	224 $+77^0$ 48^0C

n = 8

JACS, 1939, 2349.

2901-012

CHAULMOOGRIC ACID, keto-	$C_{18}H_{30}O_3$	294 OI 116^0C

n = 12

JCS, 1937, 955.

2901-017

HYDNOCARPIC ACID	$C_{16}H_{28}O_2$	252 $+70^0c$ 60^0C

n = 10

Biochem. J., 1948
(42) 581.

2901-013

OUDENONE	$C_{12}H_{16}O_3$	208 -11^0e 78^0C

JACS, 1971, 1285.

2901-018

CHAULMOOGRIC ACID	$C_{18}H_{32}O_2$	280 $+62^0$ 68^0C

n = 12

JACS, 1955, 3807.

2901-014

JASMONE	$C_{11}H_{16}O$	164 OI Oil

JCS, 1955, 1512.

2901-019

GORLIC ACID	$C_{18}H_{30}O_2$	278 $+61^0$ Oil

JACS, 1939, 3442.

2901-015

CINEROLONE	$C_{10}H_{14}O_2$	166 Oil

JCS, 1950, 1152.

2901-020

PYRETHROLONE $C_{11}H_{14}O_2$ 178 $+18^0$ Oil JCS, 1951, 2906. 2901-021	**PROSTAGLANDIN-A2, 19-OH-** $C_{20}H_{30}O_5$ 350 R = OH J. Biol. Chem., 1966 (241) 257. 2902-004
REDUCTIC ACID, Me- $C_6H_8O_3$ 128 84^0C Ann., 1949 (563) 37. 2901-022	**PROSTAGLANDIN-A2, 15-epi-** $C_{20}H_{30}O_4$ 334 R = R' = H Tet. Lett., 1969, 5185. 2902-005
PROSTAGLANDIN-A1 $C_{20}H_{32}O_4$ 336 R = H J. Biol. Chem., 1966 (241) 257. 2902-001	**PROSTAGLANDIN A2 METHYL -ESTER, Ac-15-epi-** $C_{23}H_{34}O_5$ 390 R = Ac R' = Me Tet. Lett., 1969, 5185. 2902-006
PROSTAGLANDIN-A1, 19-OH- $C_{20}H_{32}O_5$ 352 R = OH J. Biol. Chem., 1966 (241) 257. 2902-002	**PROSTAGLANDIN-B1** $C_{20}H_{32}O_4$ 336 R = H J. Biol. Chem., 1966 (241) 257. 2902-007
PROSTAGLANDIN-A2 $C_{20}H_{30}O_4$ 334 R = H J. Biol. Chem., 1966 (241) 257. 2902-003	**PROSTAGLANDIN-B1, 19-OH-** $C_{20}H_{32}O_6$ 352 R = OH J. Biol. Chem., 1966 (241) 257. 2902-008

PROSTAGLANDIN-B2 $C_{20}H_{30}O_4$ 334

R = H

J. Biol. Chem.,
1966 (241) 257.

2902-009

PROSTAGLANDIN-F1α $C_{20}H_{36}O_5$ 356

102^0C

JACS. 1968. 3247.

2902-014

PROSTAGLANDIN-B2, 19-OH- $C_{20}H_{30}O_5$ 350

R = OH

J. Biol. Chem.,
1966 (241) 257.

2902-010

PROSTAGLANDIN-F2α $C_{20}H_{34}O_5$ 354

Oil

JOC, 1969, 3552.

2902-015

PROSTAGLANDIN-E1 $C_{20}H_{34}O_5$ 354

116^0C

Nature, 1966 (212) 38.

2902-011

PROSTAGLANDIN-F3α $C_{20}H_{32}O_5$ 352

Biochim. Biophys. Acta,
1964 (84) 707.

2902-016

PROSTAGLANDIN-E2 $C_{20}H_{32}O_5$ 352

Oil

JACS, 1969, 5675.

2902-012

HUMULINONE $C_{21}H_{30}O_6$ 378

74^0C

Tet., 1960 (9) 271.

2903-001

PROSTAGLANDIN-E3 $C_{20}H_{30}O_5$ 350

JACS, 1963, 1879.

2902-013

COHULUPONE $C_{19}H_{26}O_4$ 318
OI
Oil

R = .CMe2

JCS, 1963, 1769.

2903-002

HULUPONE $C_{20}H_{28}O_4$ 332
OI
Oil

R = .CH$_2$CHMe$_2$

JCS, 1963, 1764.

2903-003

CRYPTOSPORIOPSIN $C_{10}H_{10}O_4Cl_2$ 264
+129^{0}C
133^{0}C

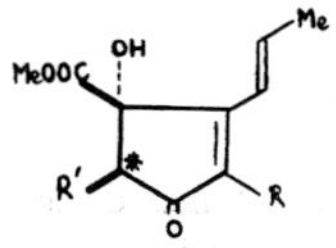

R = R' = Cl

Can. J. C., 1969, 3700.
JACS, 1969, 157.

2903-008

ADHULUPONE $C_{20}H_{28}O_4$ 332
Oil

R = .CH(Me) Et

JCS, 1964, 4774.

2903-004

CRYPTOSPORIOPSIN. des-Cl- $C_{10}H_{11}O_4Cl$ 230
92^{0}C

R = H; R' = Cl

JACS, 1969, 160.

2903-009

HUMULINIC ACID, pre- $C_{16}H_{24}O_4$ 280
96^{0}C

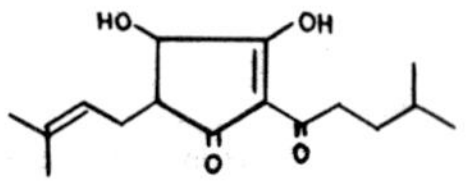

Bull. Soc. Chim. Belg.,
1962 (71) 438.

2903-005

CRYPTOSPORIOPSIN, des-Cl-epi- $C_{10}H_{11}O_4Cl$ 230
84^{0}C

R = H; R' = Cl (* epimer)

JACS. 1969, 160.

2903-010

HULUPINIC ACID $C_{15}H_{20}O_4$ 264
OI
168^{0}C

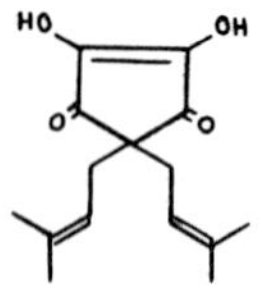

JCS, 1964, 952.

2903-006

CRYPTOSPORIOPSIN, dihydro- $C_{10}H_{12}O_4Cl_2$ 266
-90^{0}m
121^{0}C

JCS. 1969. 2187.

2903-011

CALYTHRONE $C_{12}H_{16}O_3$ 208
Oil

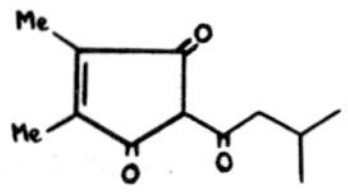

Aust. J. C., 1968, 2825.

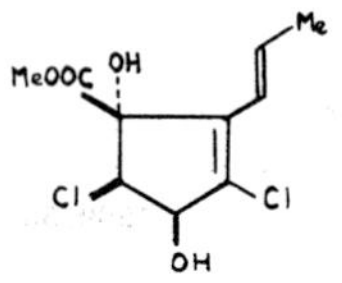

2903-007

$C_{21}H_{32}O_8$ 412
66^{0}C

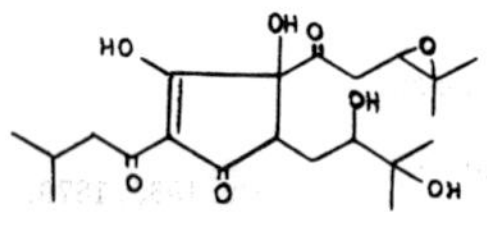

JCS. 1968. 1193.

2903-012

LUCIDONE	$C_{15}H_{12}O_4$	256 OI 165^0C

R = H

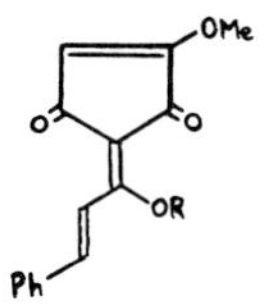

Tet. Lett., 1968, 4243.

2903-013

BONGKREKIC ACID	$C_{28}H_{38}O_7$	486 Na: $+165^0c$

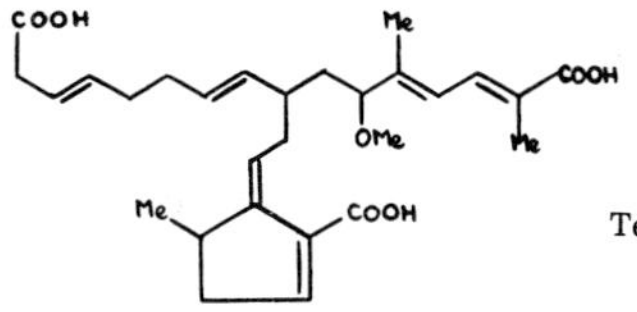

Tet., 1971, 1839.

2999-001

LUCIDONE, Me-	$C_{16}H_{14}O_4$	270 OI 116^0C

R = Me

Tet. Lett., 1968, 4243.

2903-014

BREFELDIN A	$C_{16}H_{24}O_4$	280 $+96^0m$ 204^0C

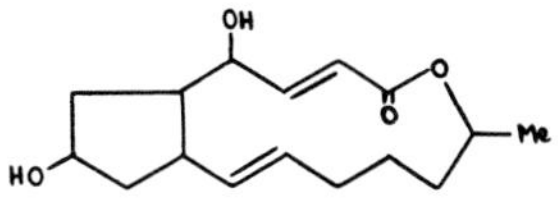

Agr. Biol. Chem., 1970 (34) 395.

2999-002

LINDERONE	$C_{16}H_{14}O_5$	286 92^0C

R = H

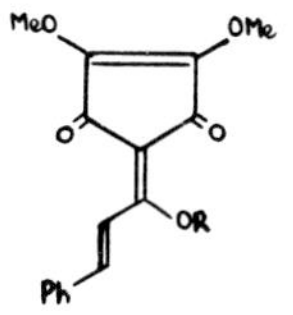

JCS, 1962, 4338.

2903-015

JASMONIC ACID LACTONE, 5'-OH-	$C_{12}H_{16}O_3$	208 -260e 104^0C

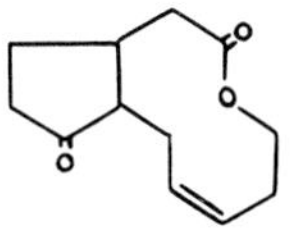

Helv., 1964 (47) 1152.

2999-003

LINDERONE, Me-	$C_{17}H_{16}O_5$	300 84^0C

R = Me

JCS, 1962, 4338.

2903-016

JASMONIC ACID ISOLEUCINYL AMIDE	$C_{18}H_{29}NO_4$	323

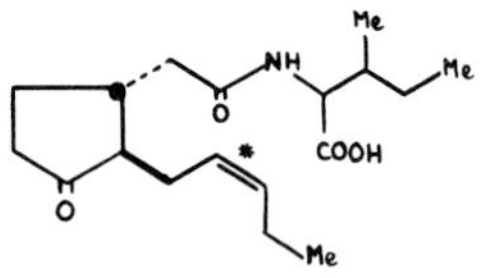

JCS, 1970, 1839.

2999-004

	$C_{17}H_{18}O_3$	270 OI Oil

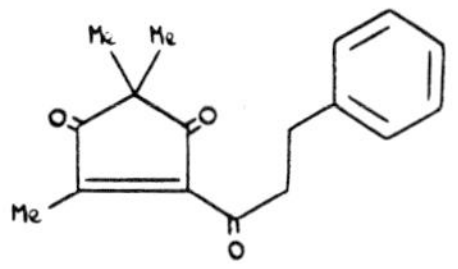

Acta Chem. Scand., 1971, 1929.

2903-017

JASMONIC ACID ISOLEUCINYL AMIDE, dihydro-	$C_{18}H_{31}NO_4$	325 148^0C

*dihydro

JCS, 1970, 1839.

2999-005

TERREIN $C_{10}H_{14}O_3$ 182
(+)
127^0C

JCS, 1955, 1028.

2999-006

There are a number of small heterocyclics known which do not conveniently belong in any of the structural/biogenetic class thus far provided. As these groups of compounds are structurally distinctive, they have been assembled in this class. It should be a point of caution to observe that some of these heterocycles (e.g., furans) can be part of a linear polyketide chain, and in such cases Class 32 should be consulted. Class 3005 includes the butenolides, tetronic acids, and bisbutenolides.

3001	Cycloheptanoids	3002	γ-Pyrones
3003	α-Pyrones	3004	Furans
3005	Butenolides		

STIPITATIC ACID R = H 3001-001	$C_8H_6O_5$ JACS, 1971, 3534.	182 OI 282^0C
SEPEDONIN 3001-006	$C_{11}H_{12}O_5$ Chem. Comm., 1968, 1669.	224
COMPOUND – T R = Et 3001-002	$C_{10}H_{10}O_5$ Arch. Biochem. Biophys., 1959 (81) 169.	210 OI 150^0C
SEPEDONIN, anhydro- 3001-007	$C_{11}H_{10}O_4$ Can. J. C., 1965 (43) 1835.	206 OI 205^0C
STIPITATONIC ACID 3001-003	$C_9H_4O_6$ JACS, 1971, 3534.	208 237^0C
PURPUROGALLIN (Artifact ?) 3001-008	$C_{11}H_8O_5$ JCS, 1951, 1318.	220 OI 274^0C
PUBERULIC ACID 3001-004	$C_8H_6O_6$ Chem. Comm., 1972, 655.	198 OI 319^0C
DICTYOPTERENE-C' 3001-009	$C_{11}H_{18}$ JACS, 1971, 3087.	150
PUBERULONIC ACID 3001-005	$C_9H_4O_7$ Chem. Comm., 1972, 655.	224 OI 298^0C
DICTYOPTERENE-D' 3001-010	$C_{11}H_{16}$ Science, 1971 (171) 815. JACS, 1971, 3087.	148

MECONIC ACID, pyro-	$C_5H_4O_3$	112 OI 118^0C

J. Pharm. Soc. Jap.,
1954 (74) 109.

3002-001

MALTOL, 6-OH-2, 3-dihydro-	$C_6H_8O_4$	144 $+200^0$m 82^0C

JCS, 1971, 3069.

3002-006

RUBIGINOL	$C_5H_4O_4$	128 OI 203^0C

Proc. Acad. Jap., 1956
(32) 595.

3002-002

KOJIC ACID	$C_6H_6O_4$	142 OI 150^0C

JCS, 1924, 575.

3002-007

MALTOL	$C_6H_6O_3$	126 OI 164^0C

R = H

Aust. J. C., 1968, 1927.

3002-003

KOJIC ACID, iso-	$C_6H_6O_4$	142 OI 183^0C

Proc. Acad. Jap., 1956
(32) 600.

3002-008

INNOVANOSIDE	$C_{21}H_{22}O_{10}$	434 -132^0m 124^0C

R = 6-coumaroyl-Glucosyl

Chem. Pharm. Bull.,
1970, 856.

3002-004

KOJIC ACID, 3-OH-	$C_6H_6O_5$	158 187^0C

Agr. Biol. Chem., 1961
(25) 802.

3002-009

MALTOL, 5-OH-	$C_6H_6O_4$	142 184^0C

Agr. Biol. Chem., 1961
(25) 939.

3002-005

VERSICOLIN	$C_7H_8O_3$	140 OI

Tet. Lett., 1969, 4871.

3002-010

PYRONE, 3-CHO-2, 3-dihydro-γ- $C_6H_6O_3$ 126
OI

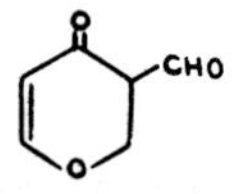

Biochemistry, 1969
(8) 4172.

3002-011

---

deleted

3002-016

COMENIC ACID $C_6H_4O_5$ 180
OI
118^0C

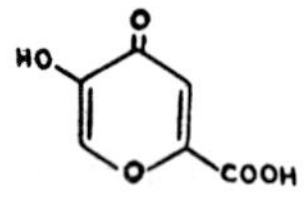

Proc. Acad. Jap., 1956
(32) 595.

3002-012

DAUCIC ACID $C_7H_8O_7$ 204

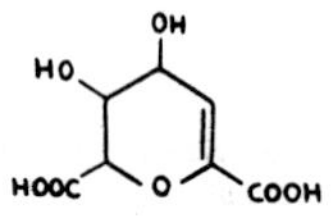

Nature, 1971 (232) 423.

3002-017

RUBIGINIC ACID $C_6H_4O_6$ 196
OI
230^0C

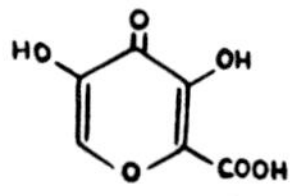

Bull. Agr. Chem. Soc.,
1955, 97.

3002-013

PARASORBIC ACID $C_6H_8O_2$ 112
$+209^0$e

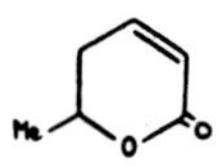

JACS, 1957, 2267.

3003-001

CHELIDONIC ACID $C_7H_4O_6$ 184
OI
262^0C

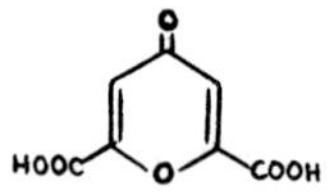

Can. J. Biochem.,
1971 (49) 412.

3002-014

HEXANOIC ACID LACTONE,
-4-Me-5-OH- $C_7H_{12}O_2$ 128

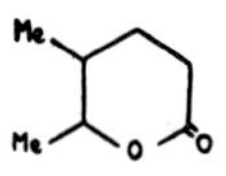

Acta Chem. Scand.,
1968, 2041.

3003-002

MECONIC ACID $C_7H_4O_7$ 200
OI

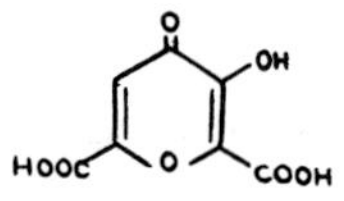

Ber., 1954 (87) 1440.

3002-015

DIVALONIC ACID $C_6H_{10}O_3$ 130

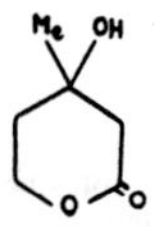

JACS, 1956, 5273.

3003-003

OPUNTIOL	$C_7H_8O_4$	156 OI 180°C

Tet., 1965, 93.

3003-004

TETRACETIC ACID LACTONE	$C_8H_8O_4$	168 OI 118°C

Chem. Comm., 1968. 1127.

3003-009

PYRONE-6-COOH, 3, 4-di-Me-α-	$C_8H_8O_4$	168 OI 184°C

JACS, 1958, 2541.

3003-005

ASPERLINE, 4-OH-6, 7-desoxy-	$C_8H_{10}O_3$	154 +175°c 51°C

Tet. Lett., 1969, 1791.

3003-010

PESTALOTIN	$C_{11}H_{18}O_4$	214 OI 88°C

Tet. Lett., 1971, 3137.

3003-006

ANTIBIOTIC U-13, 933	$C_{10}H_{12}O_5$	212 +345°e 72°C

Tet. Lett., 1966. 1969.

3003-011

CLIVONECIC ACID	$C_{10}H_{12}O_4$	196 -208°e 143°C

Coll. Czech. Chem. Comm., 1969, 1465.

3003-007

ASPERGILLUS LACTONE	$C_9H_{12}O_4$	184 -16°c 111°C

JCS. 1967, 2242.
Tet. Lett., 1970. 1867.

3003-012

TRIACETIC LACTONE, Me-	$C_7H_8O_3$	140 OI 213°C

JACS, 1964, 1264.

3003-008

HELIPYRONE	$C_{17}H_{20}O_6$	320 OI 217°C

Tet. Lett., 1970, 3369.

3003-013

FURAN	C_4H_4O	68 OI Oil

Ber., 1880 (13) 879.

3004-001

FURAN-3-AL, 4,5-di-Me-	$C_7H_8O_2$	124 OI Oil

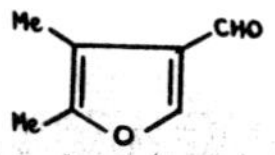

Compt. Rend., 1909 (149) 795.

3004-006

SYLVAN	C_5H_6O	82 OI Oil

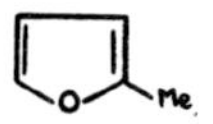

Ber., 1898 (31) 37.

3004-002

FUROIC ACID, β-	$C_5H_4O_4$	112 OI 122°C

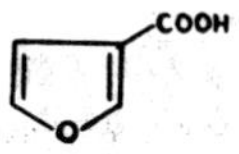

JACS, 1933, 2903.

3004-007

FURFURYL ALCOHOL, α-	$C_5H_6O_2$	98 OI Oil

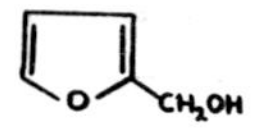

Compt. Rend., 1909 (149) 630.

3004-003

FUROIC ACID, 5-CH$_2$OH-2-	$C_6H_6O_4$	142 OI 167°C

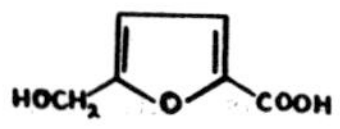

Bull. Agr. Chem. Soc. Jap., 1955 (19) 84.

3004-008

FURFURAL	$C_5H_4O_2$	96 OI Oil

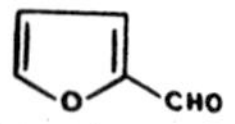

J. Agr. Food Chem., 1970 (18) 538.

3004-004

FURFURAL, 5-CH$_2$OH-	$C_6H_6O_3$	126 OI

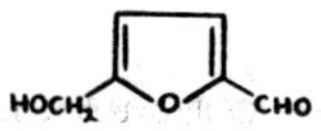

Bull. Agr. Chem. Soc. Jap., 1931 (7) 819.

3004-009

FURFURAL, 5-Me-	$C_6H_6O_2$	110 OI Oil

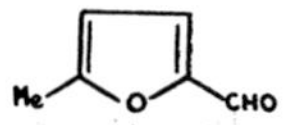

Chem. Abstr., 1956 (50) 15028.

3004-005

FURAN-3-ONE, 2-Me-tetrahydro-	$C_5H_8O_2$	100

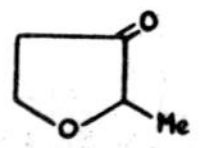

Tet., 1964, 1763.

3004-010

FURAN-3-ONE, 4-OH-2,5-di-Me--2,3-dihydro-	$C_6H_8O_3$	128 OI 70^{0}C
		JOC, 1966, 2391.
3004-011		

FURAN-3-ONE, 2,3-dihydro-2-OH--2,4-diMe-5-t-propenyl-	$C_9H_{12}O_3$	168 99^{0}C
		JCS, 1971, 2261.
3004-016		

OOSPOLIDE	$C_8H_{10}O_5$	186 OI 111^{0}C
		Tet. Lett., 1968, 4231.
3004-012		

JACONECIC ACID	$C_{10}H_{16}O_6$	232 +28^{0}e 184^{0}C
		JACS, 1959, 5201.
3004-017		

FURAN, 2-furylmethyl-	$C_9H_8O_2$	148 OI Oil
		Tet., 1964 (20) 2951.
3004-013		

BUT-3-ENOLIDE, 3-OMe-	$C_5H_6O_3$	114 OI 57^{0}C
		Acta Chem. Scand., 1954 (8) 525.
3005-001		

FURAN, (-)-trans-2,5-di-Et-	$C_8H_{16}O$	128 -10^{0}e Oil
		Bull. Chem. Soc. Jap., 1970, 3947.
3004-014		

NARTHESIDE	$C_{11}H_{16}O_9$	292 +27^{0}w 184^{0}C
		Chem. Abstr., 1961 (55) 14599.
3005-002		

BOTRYODIPLODIN	$C_7H_{12}O_3$	144 42^{0}C
		Chem. Comm., 1969, 1414.
3004-015		

BUTYROLACTONE, α-methylene-	$C_5H_6O_2$	98 OI Oil
		JACS, 1946, 2332.
3005-003		

PROTOANEMONIN	$C_5H_4O_2$	96 OI Oil

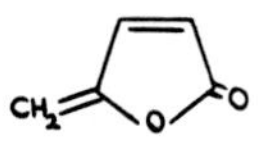

JACS, 1946, 2510.

3005-004

HIBISCUSIC ACID	$C_6H_6O_7$	190 $+122^0$ 182^0C

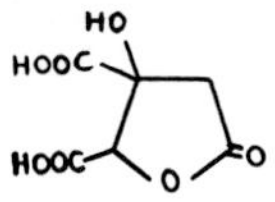

Chem. Abstr., 1949, (43) 7644.

3005-009

TETRONIC ACID, γ-Me-	$C_5H_6O_3$	114 (-) 115^0C

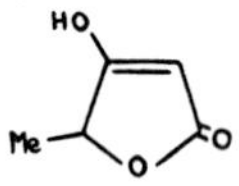

Biochem. J., 1935 (29) 1300.

3005-005

deleted

3005-010

FURAN-2-ONE, 5-OH-3-Vinyl-5H-	$C_6H_6O_3$	126 OI Oil

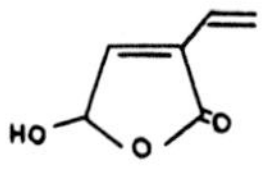

Proc. C. S., 1963, 183.
J. Antibiotics, 1962 (15A) 130.

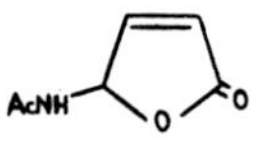

3005-006

BIGLANDULINIC ACID	$C_9H_{10}O_6$	214 OI 171^0C

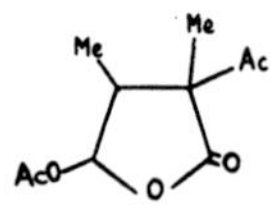

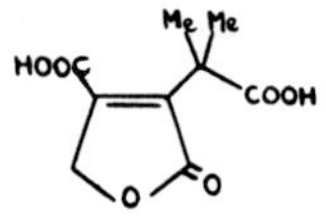

Chem. Abstr., 1939 (33) 1297.

3005-011

BUTENOLIDE, 4-NHAc-2-	$C_6H_7NO_3$	141 0^0 115^0C

Tet., 1968, 1225.

3005-007

PENICILLIC ACID	$C_8H_{10}O_4$	170 85^0C

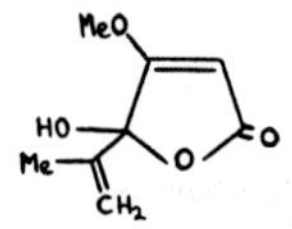

Arch. Biochem., 1944 (5) 279.

3005-012

ACETOMYCIN	$C_{10}H_{14}O_5$	214 -167^0e 115^0C

Helv., 1963, 605.

3005-008

ASCADIOL	$C_7H_8O_4$	OI 66^0C

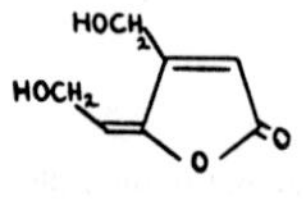

Chem. Pharm. Bull., 1971, 1786.

3005-013

NIGROSPORA LACTONE $C_8H_{12}O_3$ 156
+49⁰c
Oil

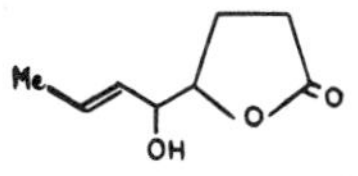

Tet. Lett., 1969, 1791.

3005-014

CARLOSIC ACID $C_{10}H_{12}O_6$ 228

181⁰C

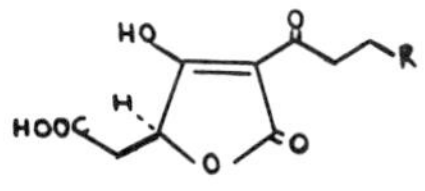

R = Me

J. Biol. Chem., 1962, 859.
Chem. Comm., 1972, 1047.

3005-019

ASPERTETRONIN-A $C_{16}H_{20}O_4$ 276
+133⁰c
72⁰C

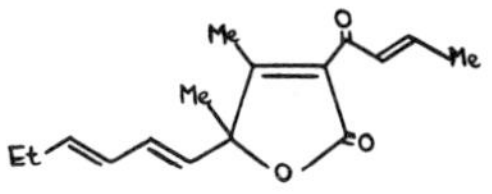

JCS, 1968, 58.

3005-015

CARLIC ACID $C_{10}H_{12}O_6$ 228
-160⁰w
176⁰C

R = CH_2OH

J. Biol. Chem., 1962,
859.

3005-020

ASPERTETRONIN-B $C_{16}H_{22}O_5$ 294
-70⁰c
Oil

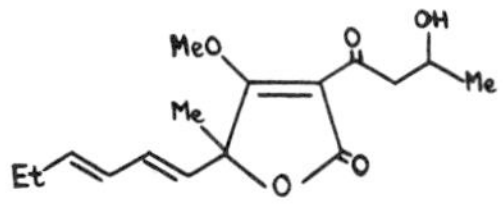

JCS, 1969, 58.

3005-016

CAROLIC ACID $C_9H_{10}O_4$ 182

123⁰C

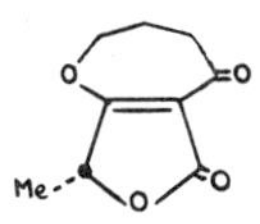

J. Biol. Chem., 1962, 859.
Acta Chem. Scand.,
1968 (22) 3251.

3005-021

BOVOLIDE $C_{11}H_{16}O_2$ 180
OI

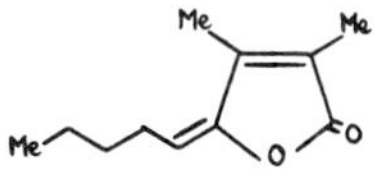

Rec. Trav. Chim.,
1966, 43.

3005-017

deleted

3005-022

ITACONITIN $C_{14}H_{14}O_5$ 264
OI

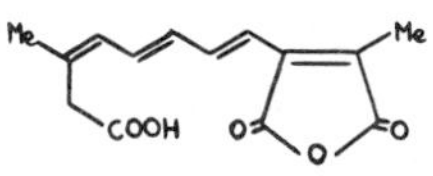

Chem. Ind., 1954, 805.

3005-018

MINEOLUTEIC ACID $C_{16}H_{26}O_7$ 330
+108⁰a
171⁰C

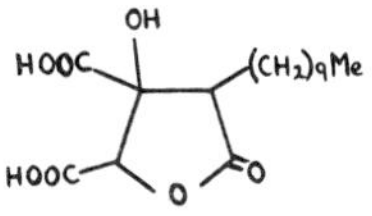

Biochem. J., 1934 (28)
828.

3005-023

NEPHROSTERANIC ACID	$C_{17}H_{30}O_4$	298	LICHESTERINIC ACID, L-Allo-proto-	$C_{19}H_{32}O_4$	324	

NEPHROSTERANIC ACID $C_{17}H_{30}O_4$ 298

95°C

Ber., 1937 (70) 227.

n = 10

3005-024

LICHESTERINIC ACID, L-Allo-proto- $C_{19}H_{32}O_4$ 324 -102°c 107°C

n = 12 (stereoisomer)

Ber., 1937 (70) 1053.

3005-029

NEPHROMOPSIC ACID $C_{19}H_{34}O_4$ 326

137°C

n = 12

Ber., 1935 (68) 995.

3005-025

ROCCELLARIC ACID $C_{19}H_{34}O_4$ 326 +35°c 110°C

Z. Nat., 1967, 666.

3005-030

NEPHROSTERINIC ACID $C_{17}H_{28}O_4$ 296 +11°c 96°C

n = 10

Ber., 1937 (70) 227.

3005-026

LICHESTERINIC ACID, (-)- $C_{19}H_{32}O_4$ 324 -32°c 124°C

J. Pharm. Soc. Jap., 1941 (61) 266.

3005-031

LICHESTERINIC ACID, D-Proto- $C_{19}H_{32}O_4$ 324 +13°c 107°C

n = 12

Ber., 1937 (70) 1053.

3005-027

SPICULISPORIC ACID $C_{17}H_{28}O_6$ 328 145°C

Chem. Abstr., 1950, (44) 3899.

3005-032

LICHESTERINIC ACID, L-Proto- $C_{19}H_{32}O_4$ 324 -13°c 108°C

n = 12 (enantiomer)

J. Pharm. Soc. Jap., 1927 (539) 1.

3005-028

TERRESTRIC ACID $C_{11}H_{14}O_4$ 210 89°C

Biochem., 1936 (30) 2194.

3005-033

VIRIDICATIC ACID $C_{12}H_{16}O_6$ 256

175^{0}C

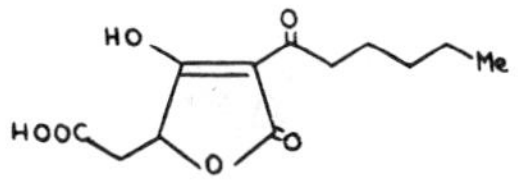

Biochem. J., 1960 (74) 369.

3005-034

TETRENOLIN $C_{11}H_{12}O_4$ 208
OI
127^{0}C

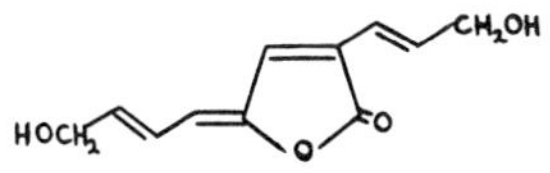

Tet., 1969, 5677.

3005-039

RANUNCULIN $C_{11}H_{16}O_8$ 276
-81^{0}w
141^{0}C

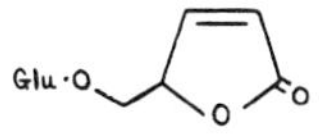

Biochem. J., 1951 (49) 332.

3005-035

deleted

3005-040

ASCORBIC ACID $C_6H_8O_6$ 176
+48^{0}m
191^{0}C

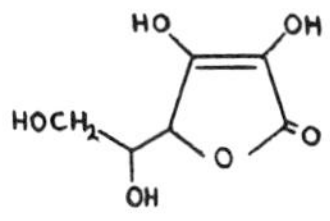

Helv., 1945 (28) 248.

3005-036

$C_{26}H_{42}O_8$ 482

82^{0}C

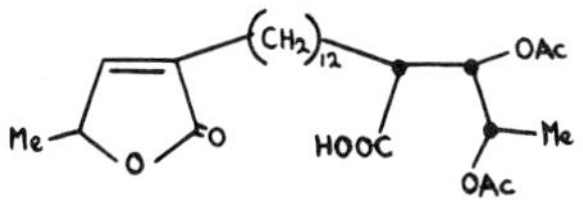

JOC, 1971, 719.

3005-041

ACARENOIC ACID $C_{17}H_{28}O_4$ 296
-39^{0}c
145^{0}C

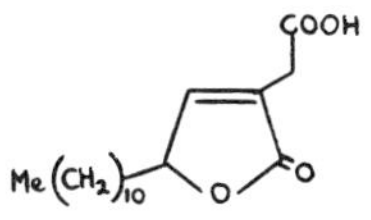

Acta Chem. Scand., 1967(21) 1993.

3005-037

ANCEPSENOLIDE $C_{22}H_{34}O_4$ 362

92^{0}C

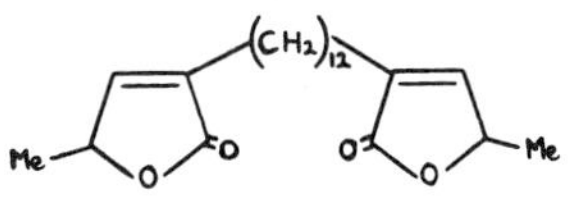

Tet. Lett., 1966, 97.

3005-042

CAROLINIC ACID $C_9H_{10}O_6$ 214

129^{0}C

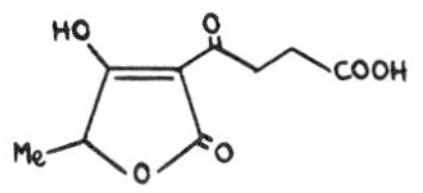

Biochem. J., 1935 (29) 1881.

3005-038

ANCEPSENOLIDE, OH- $C_{22}H_{36}O_5$ 380
+3^0
123^{0}C

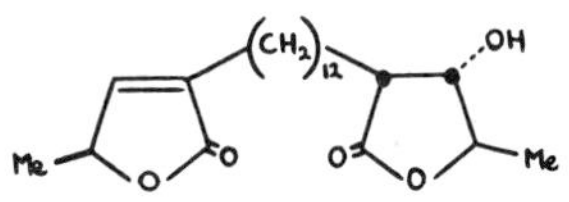

JOC, 1969, 1989.

3005-043

ANEMONIN	$C_{10}H_8O_4$	192 OI 158^0C

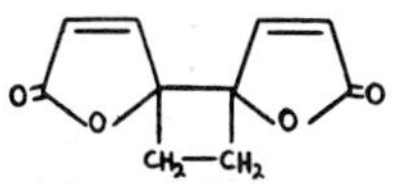

JACS, 1965, 3251.

3005-044

CANADENSOLIDE	$C_{11}H_{14}O_4$	210 -141^0 46^0C

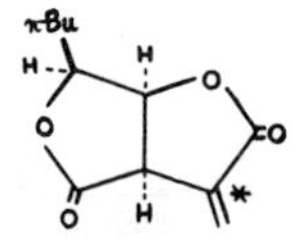

Chem. Comm.,
1971, 1561.

3005-048

ETHISOLIDE	$C_9H_{10}O_4$	182 -214^0e 122^0C

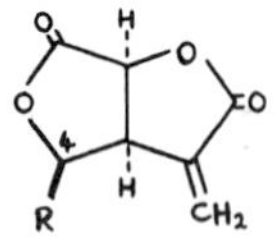

R = Et

JCS, 1971, 2431.

3005-045

CANADENSOLIDE, dihydro-	$C_{11}H_{16}O_4$	212 -31^0 94^0C

dihydro (α-Me)

Tet. Lett., 1968. 727.

3005-049

AVENACIOLIDE, 4-iso-	$C_{15}H_{22}O_4$	266 -154^0e 130^0C

R = $(CH_2)_7Me$

JCS, 1971 2431.

3005-046

CITRIC ACID LACTONE, -allo-iso-	$C_6H_6O_6$	174 $+42^0w$ 140^0C

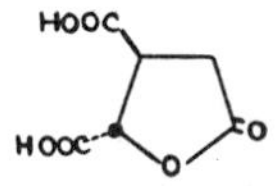

Bull. Agr. Chem. Soc.
Jap., 1957 (21) 263.

3005-050

AVENACIOLIDE	$C_{15}H_{22}O_4$	266 -42^0 49^0C

R = $(CH_2)_7Me$
4-epimer

JCS, 1963. 5385.

3005-047

MONOCROTALIC ACID	$C_8H_{12}O_5$	188 -5^0w 182^0C

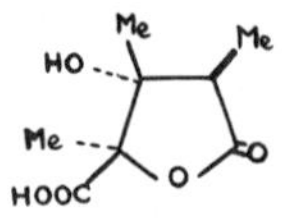

JCS, 1969, 1386.

3005-051

The linear acetogenins include the fatty acids, alkanes, alkenes and alkynes, the polyacetylenes, long-chain alcohols, ketones, and aldehydes. Related to these biogenetically are the various oxygen and sulphur heterocycles found in Classes 32 and 33, and the macrocyclic lactones of Class 34. Covered by Class 31 also are the branched-chain compounds (3194-3197), including the special case where the branching is in the form of a cyclopropane ring (e.g., the dictyopterenes). The fatty acid esters of glycerol (the glycerides) have been placed in an individual class (3198).

The classification of subclasses is determined by the length of the carbon chain—thus derivatives of pentane will be found in Class 3105 (except iso-pentanes, which are regarded as branched butanes and as such are located in Class 3194).

Derivatives found in Class 31:

carboxylic acids, esters, and amides
aldehydes and ketones
primary, secondary, and tertiary alcohols,
    diols, triols, etc.
methylthiols, sulphones, sulphoxides
epoxides and glycosides
saturated hydrocarbons, alkenes, alkynes,
    polyenes and polyacetylenes
halogenated derivatives

NB: sulphides will be found in Class 3301.

$31nn$	Where nn is the number of carbon atoms in the linear chain
3194	C_4 branched chain
3195	C_5 branched chain
3196	C_6 branched chain
3197	C_n branched chain $(n \geqslant 7)$
3198	Glycerides

METHYL ALCOHOL	CH_4O	32 OI	ETHANOL	C_2H_6O	46 OI
MeOH		JACS, 1925, 1751.	MeCH$_2$OH		Ann., 1875(177) 344.
3101-001			3102-002		
METHYLMERCAPTAN	CH_4S	48 OI Oil	ACETALDEHYDE	C_2H_4O	44 OI
MeSH		Biochem. J., 1935 (29) 1297.	Me.CHO		Arch. Pharm., 1894 (232) 642.
3101-002			3102-003		
FORMALDEHYDE	CH_2O	30 OI Gas	GLYCOLALDEHYDE	$C_2H_4O_2$	60 OI 76^{0}C
H.CHO		J. Am. Pharm. Ass., 1933 (22) 214.	HOCH$_2$CHO		Nature, 1951, 180.
3101-003			3102-004		
FORMIC ACID	CH_2O_2	46 OI 9^{0}C	ACETIC ACID	$C_2H_4O_2$	60 OI 17^{0}C
H.COOH		Chem. Abstr., 1956 (50) 1684.	MeCOOH		Ann., 1864 (130) 364.
3101-004			3102-005		
ETHYLENE	C_2H_4	28 OI Gas	ETHYL ACETATE	$C_4H_8O_2$	88 OI
CH$_2$:CH$_2$		Plant Physiol., 1951 (26) 304.	MeCOOEt		Trans. Roy. Soc., 1931 (B. 220) 331.
3102-001			3102-006		

ACETIC ACID, fluoro-	$C_2H_3O_2F$	78 OI 35^0C
FCH$_2$COOH		Nature, 1964 (201) 827.
3102-007		

ALLYL ALCOHOL	C_3H_6O	58 OI Oil
CH$_2$:CH.CH$_2$OH		Arch. Biochem. Biophys., 1964 (107) 137.
3103-002		

GLYCOLIC ACID	$C_2H_4O_3$	76 OI 78^0C
HOCH$_2$COOH		Ber. , 1932 (65) 642.
3102-008		

GLYCEROL	$C_3H_8O_3$	92 OI 18^0C
HOCH$_2$CH(OR)CH$_2$OH R = H		Acta Chem. Scand. , 1956 (10) 1096.
3103-003		

GLYCXYLIC ACID	$C_2H_2O_3$	74 OI 98^0C
OHC.COOH		Acta Chem. Scand. , 1957 (11) 1431.
3102-009		

FLUORIDOSIDE	$C_9H_{18}O_8$	254 $+165^0w$ 128^0C
R = Galactopyranosyl		JACS, 1954, 2221.
3103-004 X-ref:3504-005		

OXALIC ACID	$C_2H_2O_4$	90 OI 101^0C
HOOC.COOH		Science, 1943 (97) 262.
3102-010		

PROPIONALDEHYDE	C_3H_6O	58 OI Oil
CH$_3$CH$_2$CHO		Chem. Abstr. , 1956(50) 13184.
3103-005		

PROPANOL	C_3H_8O	60 OI
Et.CH$_2$OH		JACS, 1913, 90.
3103-001		

METHYLGLYOXAL	$C_3H_4O_2$	72 OI Oil
CH$_3$COCHO		JACS, 1939, 725.
3103-006		

GLYCERALDEHYDE	$C_3H_6O_3$	90 +14° Syrup
HOCH$_2$CHOH.CHO		
	JOC, 1943, 111.	
3103-007		

LACTIC ACID, D-(-)-	$C_3H_6O_3$	90 -2.6°w 53°C
enantiomer		
	Can. J. Microbiol., 1965 (11) 319.	
3103-012		

ACETONE	C_3H_6O	58 OI
CH$_3$COCH$_3$		
	Chem. Abstr., 1953 (47) 172.	
3103-008		

PYRUVIC ACID	$C_3H_4O_3$	88 OI
CH$_3$COCOOH		
	J. Biol. Chem., 1952 (196) 853.	
3103-013		

ACETONE, di-OH-	$C_3H_6O_3$	90 OI 70°C
HOCH$_2$COCH$_2$OH		
	Experientia, 1952, 445.	
3103-009		

PYRUVIC ACID, phospho-enol-	$C_3H_5O_6P$	168
CH$_2$:C(OPO$_3$H$_2$)COOH		
	Biochem. Z., 1934 (273) 60.	
3103-014		

PROPIONIC ACID	$C_3H_6O_2$	74 OI
CH$_3$CH$_2$COOH		
	Monatsh., 1905 (26) 727.	
3103-010		

GLYCERIC ACID, phospho-	$C_3H_7O_7P$	186
HOCH$_2$CH(OPO$_3$H$_2$)COOH		
	Biochem. Z., 1935 (273) 239.	
3103-015		

LACTIC ACID, L-(+)-	$C_3H_6O_3$	90 +3°w 53°C
COOH / HO—H / CH$_3$		
	Can. J. Microbiol., 1965 (11) 319	
3103-011		

PYRUVIC ACID, OH-	$C_3H_4O_4$	104 OI
HOCH$_2$COCOOH		
	Acta Chem. Scand., 1957 (11) 1431.	
3103-016		

TARTRONIC ACID	$C_3H_4O_5$	120 OI 163^0C
HOOC.CHOH.COOH		Nature, 1951 (167) 905.
3103-017		

PROPYL-MERCAPTAN	C_3H_8S	76 OI
$CH_3CH_2CH_2SH$		Biochem. J., 1949 (44) 87.
3103-022		

MESOXALIC ACID	$C_3H_2O_5$	118 OI 120^0C
HOOCCOCOOH		Ber., 1913 (46) 3862.
3103-018		

PROPYLPHOSPHONIC ACID, (-)-cis-1, 2-epoxy-	$C_3H_7O_3P$	122 94^0C

		Chem. Abstr., 1970 (73) 2663.
3103-023		

MALONIC ACID	$C_3H_4O_4$	104 OI 135^0C
HOOCCH_2COOH		J. Pharm. Soc. Jap., 1955 (75) 761.
3103-019		

GLYCEROL, 1, 1-O-β-D- -Galactofuranosyl-	$C_9H_{18}O_8$	254 -78^0w Syrup

		Acta Chem. Scand., 1967 (21) 2083.
3103-024		

GLYCERIC ACID, L-(-)-	$C_3H_6O_4$	106 $+13^0$ 138^0C
HOCH_2CHOH.COOH		Chem. Abstr., 1965 (63) 18650g.
3103-020		

PROPIONIC ACID ME ESTER, -β-SMe-	$C_5H_{10}O_2S$	134 OI Oil
MeSCH_2CH_2COOMe		JACS, 1945, 1646.
3103-025		

ACRYLIC ACID	$C_3H_4O_2$	72 OI 13^0C
CH_2:CH.COOH		JACS, 1945, 1646.
3103-021		

PROPIONIC ACID, β-(diMe- -sulphonium hydroxide)-	$C_5H_{12}O_3S$	152 OI 134^0C
HO^- $Me_2\overset{+}{S}OCH_2CH_2COOH$		Chem. Ind., 1954, 729.
3103-026		

BUTYLENE	C_4H_8	56
		OI
MeCH:CHMe		
	Chem. Zentr., 1922 (II) 1195.	
3104-001		

BUTAN-2-ONE, 3-OH-	$C_4H_8O_2$	88
		OI
		15^0C
MeCOCHOH.Me		
	JACS, 1945, 494.	
3104-006		

BUTANOL	$C_4H_{10}O$	74
		OI
MeCH$_2$CH$_2$CH$_2$OH		
	Chem. Abstr., 1949 (43) 4427.	
3104-002		

DIACETYL	$C_4H_6O_2$	86
		OI
MeCO.COMe		
	Nature. 1953 (172) 412.	
3104-007		

BUT-2-EN-1-OL	C_4H_8O	72
		OI
MeCH:CH.CH$_2$OH		
	Compt. Rend., 1950 (231) 872.	
3104-003		

deleted	
3104-008	

BUT-3-EN-1-OL	C_4H_8O	72
		OI
HOCH$_2$CH$_2$CH:CH$_2$		
	Compt. Rend., 1950 (231) 872.	
3104-004		

BUTYRALDEHYDE, 3-OH-	$C_4H_8O_2$	88
		Oil
MeCHOH.CH$_2$CHO		
	Chem. Zentr., 1940 (I) 1683.	
3104-009		

BUTAN-2,3-DIOL	$C_4H_{10}O_2$	90
MeCHOH.CHOH.Me		
	Nature, 1946 (157) 336.	
3104-005		

BUTYRIC ACID	$C_4H_8O_2$	88
		OI
MeCH$_2$CH$_2$COOH		
	Ann., 1872 (162) 193.	
3104-010		

CROTONIC ACID	$C_4H_6O_2$	86
		OI
		72^0C
MeCH:CHCOOH		
	Chem. Abstr., 1956 (50) 15028.	
3104-011		

FUMARIC ACID	$C_4H_4O_4$	116
		OI
		301^0C
HOOC.CH:CH.COOH		
	Natwiss., 1955 (42) 441.	
3104-016		

CROTONIC ACID, iso-	$C_4H_6O_2$	86
		OI
		15^0C
cis isomer		
	Chem. Abstr., 1951 (45) 10507.	
3104-012		

FUMARIC ACID, epoxy-	$C_4H_4O_5$	132
		-117^0w
		185^0C

$$HOOC-\overset{O}{\triangle}-COOH$$

J. Biol. Chem., 1963 (238) 843.	
3104-017	

BUTYRIC ACID, α-keto-β-OH-	$C_4H_6O_4$	114
MeCHOH.COCOOH		
	Acta Chem. Scand., 1955, 188.	
3104-013		

MALIC ACID, D(+)-	$C_4H_6O_5$	134
		$+5^0a$
		100^0C
HOOC.CH_2CHOH.COOH		
	Chem. Zentr., 1913 (I) 645.	
3104-018		

BUTYRIC ACID, α-keto-γ-OH-	$C_4H_6O_4$	114
HOCH_2CH_2COCOOH		
	Acta Chem. Scand., 1955, 188.	
3104-014		

MALIC ACID, L-	$C_4H_6O_5$	134
		-1^0w
		99^0C
enantiomer		
	Chem. Ind., 1936 (55) 155.	
3104-019		

SUCCINIC ACID	$C_4H_6O_4$	118
		OI
		186^0C
HOOC.CH_2CH_2COOH		
	Biochem. Z., 1930 (219) 103.	
3104-015		

OXALACETIC ACID	$C_4H_4O_5$	132
		OI
HOOC.CH_2COCOOH		
	Acta Chem. Scand., 1957 (11) 1431.	
3104-020		

TARTARIC ACID, D- $C_4H_6O_6$ 150 $+12^0$w 172^{0}C

COOH
―OH
HO―
COOH

Biochem. Z., 1923(136)291

3104-021

BUTAN-2-ONE C_4H_8O 72 OI Oil

$CH_3CH_2COCH_3$

Chem. Abstr., 1938 (32) 5155.

3104-026

TARTARIC ACID, L- $C_4H_6O_6$ 150 -12^0w 169^{0}C

enantiomer

Nature, 1951 (168) 271.

3104-022

PENTANAL $C_5H_{10}O$ 86 OI

$Me(CH_2)_3CHO$

JOC, 1948, 443.

3105-001

MALEIC ACID, di-OH- $C_4H_4O_6$ 148 OI 155^{0}C

HO   COOH
  ＼ ／
   ‖
  ／ ＼
HO   COOH

JACS, 1953, 6244.

3104-023

PENTAN-2-ONE $C_5H_{10}O$ 86 OI

$MeCH_2CH_2COMe$

JACS, 1945, 1646.

3105-002

ERYTHRITOL, meso- $C_4H_{10}O_4$ 122 OI 120^{0}C

CH$_2$O R
―OH   R = H
―OH
CH$_2$OH

Z. Physiol. Chem., 1936 (243) 103.

3104-024

VALERIC ACID, N- $C_5H_{10}O_2$ 102 OI

$Me(CH_2)_3COOH$

JACS, 1945, 1646.

3105-003

ERYTHRITOL, 1-β-mannopyranosyl-meso- $C_{10}H_{20}O_9$ 284 -37^0 160^{0}C

R = Mannosyl

Can. J. Biochem. Physiol.. 1956 (34) 10.

3104-025

GLUTARIC ACID $C_5H_8O_4$ 136 OI 97^{0}C

$HOOC.CH_2CH_2CH_2COOH$

Bull. Agr. Chem. Soc., 1942 (18) 93.

3105-004

GLUTARIC ACID, α-OH-	$C_5H_8O_5$	148 $+2^0$w 72^0C	HEXANOL	$C_6H_{14}O$	102 OI
HOOC.CH_2CH_2CHOH.COOH	Ber., 1891 (24) 3299.		Me$(CH_2)_4CH_2$OH	Chem. Abstr., 1953 (47) 11667.	
3105-005			3106-003		
GLUTARIC ACID, α-keto-	$C_5H_6O_5$	146 OI 115^0C	HEX-3-EN-1-OL	$C_6H_{12}O$	100 OI Oil
HOOC.CH_2CH_2COCOOH	JACS, 1954, 2392.		MeCH_2CH:CHCH_2CH_2OH	JCS, 1950, 873.	
3105-006			3106-004		
GLUTACONIC ACID, trans-	$C_5H_6O_4$	130 OI 138^0C	HEXANAL	$C_6H_{12}O$	100 OI Oil
HOOC.CH_2CH:CH.COOH	Bull. Agr. Chem. Soc. Jap., 1942 (18) 93.		Me$(CH_2)_4$CHO	J. Agr. Food Chem., 1970 (18) 538.	
3105-007			3106-005		
DIALLYL	C_6H_{10}	82 OI	HEX-2-EN-1-AL	$C_6H_{10}O$	98 OI Oil
CH_2:CHCH$_2CH_2$CH:CH_2	Chem. Abstr., 1952 (46) 7712.		MeCH_2CH_2CH:CHCHO	Compt. Rend., 1931 (192) 1467.	
3106-001			3106-006		
HEXA-1,3,5-TRIYNE	C_6H_2	74 OI	SOYANAL	$C_6H_{10}O_2$	114 OI Oil
H(C ≡ C)$_3$H	Tet. Lett., 1966, 4223.		MeCO$(CH_2)_3$CHO	Chem. Abstr., 1951 (45) 7015.	
3106-002			3106-007		

CAPROIC ACID	$C_6H_{12}O_2$	116
		OI
$Me(CH_2)_4COOH$		
	Chem. Abstr.,	
	1946 (40) 6758.	
3106-008		

HEXA-2, 4-DIENOIC ACID	$C_8H_{12}O_2S$	172
ME ESTER, 5-SMe-		OI
$MeC(SMe):CH.CH:CH.COOMe$		
	Ber., 1966, 2096.	
3106-013		

HEX-2-ENOIC ACID	$C_6H_{10}O_2$	114
		OI
		34^0C
$MeCH_2CH_2CH:CHCOOH$		
	Chem. Zentr.,	
	1930 (I) 447.	
3106-009		

HEXANOIC ACID, 3-OH-	$C_6H_{12}O_3$	132
$MeCH_2CH_2CHOH.CH_2COOH$		
	Angew Chem.,	
	1971 (10) 124.	
3106-014		

SORBIC ACID	$C_6H_8O_2$	112
		OI
		133^0C
$Me(CH:CH)_2COOH$		
	JCS, 1965, 5651.	
3106-010		

CAPROIC ACID, α-keto-	$C_6H_{10}O_3$	130
		OI
$Me(CH_2)_3COCOOH$		
	Acta Chem. Scand.,	
	1957 (11) 1431.	
3106-015		

HEX-2-ENE-4-YNOIC ACID	$C_7H_8O_2$	124
ME ESTER, cis-		OI
$MeC\overset{c}{\vdots}C.CH:CH.COOMe$		
	Ber., 1966, 2096.	
3106-011		

ADIPIC ACID	$C_6H_{10}O_4$	146
		OI
		150^0C
$HOOC(CH_2)_4COOH$		
	Ber., 1891 (24) 3299.	
3106-016		

HEX-2-ENE-4-YNOIC ACID	$C_7H_8O_2$	124
ME ESTER, trans-		OI
trans isomer		
	Ber., 1966, 2096.	
3106-012		

ADIPIC ACID, α-keto-	$C_6H_8O_5$	160
		OI
		124^0C
$HOOC(CH_2)_3COCOOH$		
	Acta Chem. Scand.,	
	1954, 1720.	
3106-017		

HEPTANE	C_7H_{16}	100 OI Oil
$CH_3.(CH_2)_5.CH_3$		JACS, 1948, 2014.
3107-001		

HEPT-4-ENAL, C-	$C_7H_{12}O$	112 OI Oil
$Me.CH_2.CH:CH.CH_2.CH_2.CHO$		Nature, 1964 (202) 552.
3107-006		

HEPTANOL	$C_7H_{16}O$	116 OI Oil
$CH_3.(CH_2)_5.CH_2OH$		Chem. Abstr., 1953 (47) 11667.
3107-002		

HEPT-4-ENAL, t-	$C_7H_{12}O$	112 OI Oil
(trans)		Tet. Lett., 1966, 2479.
3107-007		

HEPTAN-2-OL	$C_7H_{16}O$	116 Oil
$CH_3.(CH_2)_4.CHOH.CH_3$		Compt. Rend., 1909 (149) 630.
3107-003		

HEPTAN-2-ONE	$C_7H_{14}O$	114 OI Oil
$Me.CO.(CH_2)_4.Me$		Nature, 1965 (206) 530.
3107-008		

HEPTANAL	$C_7H_{14}O$	114 OI Oil
$CH_3.(CH_2)_5.CHO$		Rec. Trav. Chim., 1931 (50) 708.
3107-004		

HEPTANOIC ACID	$C_7H_{14}O_2$	130 OI Oil
$CH_3(CH_2)_5.COOH$		Arch. Pharm., 1939, 65.
3107-009		

HEPT-2-EN-1-AL	$C_7H_{12}O$	112 OI Oil
$CH_3.(CH_2)_3.CH:CH.CHO$		Chem. Abstr., 1948 (42) 3975.
3107-005		

PIMELIC ACID, α-keto-	$C_7H_{10}O_5$	174 OI 94^0C
$HOOC.(CH_2)_4.CO.COOH$		Acta Chem. Scand., 1954, 1720.
3107-010		

PIMELIC ACID, γ-OH-α-keto-	$C_7H_{10}O_6$	190	OCT-1-EN-3-OL, ℓ-	$C_8H_{16}O$	128
					Oil
HOOC.CH$_2$.CH$_2$.CHOH.CH$_2$.CO.COOH	Acta Chem. Scand., 1954, 1720.		CH$_3$.(CH$_2$)$_4$.CHOR.CH:CH$_2$ R = H	Chem. Abstr., 1954 (48) 3932.	
3107-011			3108-005		

OCTAN-1-OL	$C_8H_{18}O$	130 OI Oil	OCT-1-EN-3-OL (-), Ac-	$C_{10}H_{18}O_2$	170 +4° Oil
CH$_3$.(CH$_2$)$_6$.CH$_2$OH	Ann., 1873 (166) 80.		R = Ac	Acta Chem. Scand., 1963 (17) 858.	
3108-001			3108-006		

OCTAN-2-OL	$C_8H_{18}O$	130 Oil	OCTA-2,3-DIENE-5,7-DIYN-1-OL	C_8H_6O	118 -380°
CH$_3$.(CH$_2$)$_5$.CHOH.CH$_3$	Ber., 1942 (75) 502.		H.(C⋮C)$_2$.CH:C:CH.CH$_2$OH	JCS, 1966, 129.	
3108-002			3108-007		

OCTAN-3-OL	$C_8H_{18}O$	130 +10° Oil	OCTANAL	$C_8H_{16}O$	128 OI Oil
CH$_3$.(CH$_2$)$_4$.CHOH.CH$_2$.CH$_3$	JACS, 1951, 1848.		Me.(CH$_2$)$_6$.CHO	Chem. Abstr., 1951 (45) 4889.	
3108-003			3108-008		

OCT-2-EN-1-OL	$C_8H_{16}O$	128 OI Oil	OCT-2-ENAL	$C_8H_{14}O$	126 OI Oil
Me.(CH$_2$)$_4$.CH:CH.CH$_2$OH	J. Agr. Food Chem., 1970 (18) 538.		CH$_3$(CH$_2$)$_4$.CH:CH.CHO	Aust. J. C., 1969 (22) 1793.	
3108-004			3108-009		

OCTAN-2-ONE	$C_8H_{16}O$	128
		OI
		Oil
$Me.CO.(CH_2)_5.CH_3$	Helv. , 1939 (22) 382.	
3108-010		

OCTA-2,6-DIEN-4-YN-1,8-DOIC ACID DIME ESTER	$C_{10}H_{10}O_4$	194
		OI
		118^0C
MeOOC.$\overset{t}{CH:CH}$.C$\vdots$C.$\overset{t}{CH:CH}$.COOMe	JCS. 1957. 1607.	
3108-015		

OCTAN-3-ONE	$C_8H_{16}O$	128
		OI
		Oil
$CH_3.CH_2.CO.(CH_2)_4.CH_3$	Chem. Abstr., 1946 (40) 4480.	
3108-011		

AGROCYBIN	$C_8H_5O_2N$	147
		OI
		135^0C
$HOCH_2(C\vdots C)_3CONH_2$	JCS, 1958, 950.	
3108-016		

OCTANOIC ACID	$C_8H_{16}O_2$	144
		OI
		16^0C
$CH_3(CH_2)_6.COOH$	Ann. , 1845 (53) 399.	
3108-012		

DIATRETYNE-1	$C_8H_5O_3N$	163
		OI
		198^0C
$HOOC.OH:CH(C\vdots C)_2R$		
$R = CONH_2$	Proc. C. S. , 1960, 199.	
3108-017		

deleted		
3108-013		

DIATRETYNE-2	$C_8H_3O_2N$	145
		OI
		179^0C
$R = CN$	JCS, 1958, 950.	
3108-018		

OCTA-2,4,6-TRIENOIC ACID ME ESTER, 5-SMe-	$C_{10}H_{14}O_2S$	198
		OI
$Me.\overset{c}{CH:CH}.C(SMe):\overset{t}{CH}.\overset{c}{CH:CH}.COOMe$	Ann., 1966(694)149	
3108-014		

NONANE	C_9H_{20}	128
		OI
		Oil
$CH_3(CH_2)_7CH_3$	JACS, 1948, 2014.	
3109-001		

NONAN-1-OL	$C_9H_{20}O$	144 OI Oil		NONA-2,6-DIEN-1-OL	$C_9H_{16}O$	140 OI
$CH_3(CH_2)_7CH_2OH$		Chem. Abstr., 1937 (31) 1956.		Et.CH:CH.CH_2CH_2CH:CH.CH_2OH		Helv., 1944, 1561.
3109-002				3109-007		
NONAN-2-OL	$C_9H_{20}O$	144 Oil		NON-1-EN-3-OL	$C_9H_{18}O$	142 Oil
$CH_3(CH_2)_6CHOH.CH_3$		JCS, 1902, 1585.		$Me(CH_2)_5CHOH.CH:CH_2$		Helv., 1943, 1996.
3109-003				3109-008		
MARASIN, (-)-	C_9H_8O	132 -325^0e Oil		NON-2-ENE-4,6-DIYNE-1,9-DIOL	$C_9H_{10}O_2$	150 OI 58^0C
$H(C\vdots C)_2CH:C:CH.CH_2CH_2OH$		Arkiv. Kemi, 1959 (14) 475.		$HOCH_2CH_2(C\vdots C)_2CH:CH.CH_2OH$		JCS, 1963, 4160.
3109-004				3109-009		
MARASIN, (+)-	C_9H_8O	132 $+360^0e$ Oil		NON-4-ENE-6,8-DIYNE-1,2,3-TRIOL	$C_9H_{10}O_3$	166 -4^0e Oil
		JCS, 1963, 4120.		$H(C\vdots C)_2CH:\overset{t}{CH}.CHOH.CHOH.CH_2OH$		JCS, 1963, 4120.
3109-005				3109-010		
NON-2-ENE-4,6,8-TRIYN-1-OL	C_9H_6O	130 OI		NONA-3,5-DIYN-1,2,7-TRIOL	$C_9H_{12}O_3$	168
$H(C\vdots C)_3CH:\overset{t}{CH}.CH_2OH$		JCS, 1959, 2197.		Et. CHOH(C$\vdots$C)_2CHOH.CH_2OH		JCS, 1964, 1476.
3109-006				3109-011		

NONA-4, 6, 8-TRIYNE-1, 2, 3-TRIOL, -(2D, 3D)-	$C_9H_8O_3$	164 $+6^0$e 40^0C

H(C:C)$_3$CHOH.CHOH.CH$_2$OH

JCS, 1963, 4120.

3109-012

NONA-2, 6-DIEN-1-AL	$C_9H_{14}O$	138 OI Oil

Et.CH:CH.CH$_2$CH$_2$CH:CH.CHO

JACS, 1952, 4040.

3109-017

NONA-4, 6, 8-TRIYNE-1, 2, 3-TRIOL, -(2D, 3L)-	$C_9H_8O_3$	164 -8^0e

H(C:C)$_3$(CHOH)$_3$H

JCS, 1963, 4120.

3109-013

NON-2-ENE-4, 6, 8-TRIYN-1-AL	C_9H_4O	128 OI

H(C:C)$_3$CH:$\overset{t}{\text{CH}}$.CHO

JCS, 1959, 2197.

3109-018

NONANAL	$C_9H_{18}O$	142 OI Oil

Me(CH$_2$)$_7$CHO

J. Am. Pharm. Ass., 1935 (24) 38.

3109-014

NONAN-2-ONE	$C_9H_{18}O$	142 OI Oil

MeCO(CH$_2$)$_6$Me

Chem. Abstr., 1954 (48) 11732.

3109-019

NON-2-EN-1-AL	$C_9H_{16}O$	140 OI Oil

Me(CH$_2$)$_5$CH:CH.CHO

J. Agr. Food Chem., 1970 (18) 538.

3109-015

NONA-3, 5-DIYN-7-ONE, 1, 2-di-OH-	$C_9H_{10}O_3$	166 -30^0e 35^0C

Et.CO(C:C)$_2$CHOH.CH$_2$OH

JCS, 1966, 144.

3109-020

---

deleted

3109-016

NONANOIC ACID	$C_9H_{18}O_2$	158 OI 13^0C

Me(CH$_2$)$_7$COOH

Chem. Abstr., 1956 (60) 17341.

3109-021

DROSOPHILIN-E	$C_9H_8O_2$	148

$$H(C\vdots C)_2CH\vdots CH.CH_2CH_2COOH$$

JCS, 1960, 2257.

3109-022

AZELAIC ACID	$C_9H_{16}O_4$	188 OI 106^0C

$$HOOC(CH_2)_7COOH$$

Compt. Rend., 1957 (244) 2429.

3109-027

NONA-3,4-DIENE-6,8-DIYNOIC ACID ME ESTER	$C_{10}H_8O_2$	160 $+285^0e$ Oil

$$HC \equiv C - C \equiv C - CH = C = CHCH_2COOH$$

JCS, 1963, 4120.

3109-023

BIFORMYNE-1	$C_9H_6O_2$	146 41^0C

$$H(C\vdots C)_3 \text{ (epoxide) } CH_2OH$$

JCS, 1963, 2048.

3109-028

NONA-3,4-DIENE-6,8-DIYNOIC ACID	$C_9H_6O_2$	146 (+)

$$H(C\vdots C)_2CH\vdots C\vdots CH.CH_2COOH$$

JCS, 1963, 4120.

3109-024

	$C_{14}H_{13}NO_3$	243

$$H.(C\vdots C)_3CH\vdots CH.CONH.CH(COOH)CHMe_2$$

JCS, 1963, 2056.

3109-029

NON-2-ENE-4,6,8-TRIYNOIC ACID	$C_9H_4O_2$	144

$$H(C\vdots C)_3\overset{t}{CH}\vdots CHCOOH$$

JCS, 1963, 2056.

3109-025

---

deleted

3110-001

NON-7-EN-3,5-DIYNOIC ACID, -9-OH-	$C_9H_8O_3$	164 OI Oil

$$HOCH_2\overset{t}{CH}\vdots CH(C\vdots C)_2CH_2COOH$$

JCS, 1966, 135.

3109-026

---

deleted

3110-002

DECAN-1-OL	$C_{10}H_{22}O$	158 OI Oil
$Me(CH_2)_8CH_2OH$		Bull. Soc. Chim. Fr., 1953, 306.
3110-003		

MATRICARIANOL, dehydro-t-	$C_{10}H_8O$	144 OI 128^{0}C
trans isomer		JCS, 1966, 1216.
3110-008		

LACHNOPHYLLOL	$C_{10}H_{12}O$	148 OI 40^{0}C
$Pr.(C:C)_2CH:CH.CH_2OR$ (t) R = H		Ber., 1969 (102) 1682.
3110-004		

DEC-3-EN-1-OL, cis-	$C_{10}H_{20}O$	156 OI Oil
$Me(CH_2)_5CH:CH.CH_2CH_2OH$ (c)		Aust. J. C., 1966, 1495.
3110-009		

LACHNOPHYLLOL, Ac-	$C_{12}H_{14}O_2$	190 OI Oil
R = Ac		Ber., 1969 (102) 1682.
3110-005		

DEC-3-EN-1-OL, t-	$C_{10}H_{20}O$	156 OI Oil
trans isomer		Aust. J. C., 1966, 1495.
3110-010		

LACHNOPHYLLOL, Angeloyl-	$C_{15}H_{18}O_2$	230 OI Oil
R = Angeloyl		Ber., 1969 (102) 1682.
3110-006		

DEC-8-EN-4,6-DIYN-1-OL	$C_{10}H_{12}O$	148 OI Oil
$Me.CH:CH(C:C)_2CH_2CH_2CH_2OR$ (c) R = H		Ber., 1970 (103) 2853.
3110-011		

MATRICARIANOL, dehydro-cis-	$C_{10}H_8O$	144 OI
$Me(C:C)_3CH:CH.CH_2OH$ (c)		JCS, 1966, 139.
3110-007		

DEC-8-EN-4,6-DIYN-1-OL, Ac-	$C_{12}H_{14}O_2$	190 OI Oil
R = Ac		Ber., 1970 (103) 2853.
3110-012		

DEC-8-EN-4,6-DIYN-1-OL, i-Val-	$C_{15}H_{20}O_2$	232 OI Oil	MATRICARIANOL, i-Valeryl-t, c-	$C_{15}H_{18}O_2$	230 OI Oil
R = i-Valeryl		Ber., 1970 (103) 2853.	R = i-Valeryl (trans, cis-isomer)		Ber., 1970 (103) 2853.
3110-013			3110-018		

DECA-2,6-DIEN-4-YN-1-OL, Ac-	$C_{12}H_{16}O_2$	192 OI Oil	MATRICARIANOL, t, t-	$C_{10}H_{10}O$	146 OI 106^0C
Pr. $\overset{t}{C}H{:}CH.C{:}C.CH{:}CH.CH_2OAc$		Ber., 1970 (103) 3419.	R = H (trans, trans-isomer)		JCS, 1957, 1607.
3110-014			3110-019		

MATRICARIANOL, c, c-	$C_{10}H_{10}O$	146 OI Oil		$C_{12}H_{12}O_2$	188 OI Oil
Me. $\overset{c}{C}H{:}CH(C{:}C)_2\overset{c}{C}H{:}CH.CH_2OR$ R = H		Ber., 1969 (102) 1679.	R = Ac (trans, trans-isomer)		Ber., 1969 (102) 1682.
3110-015			3110-020		

MATRICARIANOL, Ac-c, c-	$C_{12}H_{12}O_2$	188 OI Oil	DECA-3,4-DIENE-6,8-DIYN-1-OL	$C_{10}H_{10}O$	146 $+340^0$e Oil
R = Ac (cis, cis-isomer)		Ber., 1969 (102) 1679.	Me$(C{:}C)_2CH{:}C{:}CH.CH_2CH_2OH$		JCS, 1966, 135.
3110-016			3110-021		

MATRICARIANOL, cis-trans-	$C_{10}H_{10}O$	146 OI 20^0C	DECA-4,5-DIENE-7,9-DIYN-1-OL	$C_{10}H_{10}O$	146 -290^0e Oil
R = H (cis, trans-isomer)		JCS, 1960, 691.	H. $(C{:}C)_2CH{:}C{:}CH.CH_2CH_2CH_2OH$		JCS, 1966, 129.
3110-017			3110-022		

DECA-2,6,8-TRIEN-4-YN-1-OL	$C_{10}H_{12}O$	148 OI 56^{0}C

$$\text{Me(CH:CH)}_2\overset{t}{\text{C}}\text{:C.}\overset{t}{\text{CH:CH.CH}}_2\text{OR}$$
R = H

Ber., 1969 (102) 1682.

3110-023

DECA-4,5-DIENE-7,9-DIYNE-1,3-DIOL	$C_{10}H_{10}O_2$	162 -210^{0}e Oil

$$\text{H(C:C)}_2\text{CH:C:CH.CHOH.CH}_2\text{CH}_2\text{OH}$$

JCS, 1966, 129.

3110-028

	$C_{12}H_{14}O_2$	190 OI 62^{0}C

R = Ac

Ber., 1969 (102) 1682.

3110-024

DEC-2-ENE-4,6,8-TRIYN-1,10-DIOL	$C_{10}H_8O_2$	160 OI 138^{0}C

$$\text{HOCH}_2\text{(C:C)}_3\overset{t}{\text{CH:CH.CH}}_2\text{OH}$$

JCS, 1959, 2197.

3110-029

	$C_{15}H_{18}O_2$	230 OI Oil

R = Angeloyl

Ber., 1969 (102) 1682.

3110-025

MATRICARIANOL, 10-OH-	$C_{10}H_{10}O_2$	162 OI 64^{0}C

$$\text{HOCH}_2\text{CH:CH}\overset{c}{(}\text{C:C)}_2\overset{c}{\text{CH:CH.CH}}_2\text{OH}$$

Ber., 1969 (102) 1679.

3110-030

DECA-2,4,6,8-TETRAENOL	$C_{10}H_{14}O$	150 OI 110^{0}C

$$\text{Me(}\overset{t}{\text{CH:CH)}}_4\text{CH}_2\text{OH}$$

JCS, 1966, 135.

3110-026

DECA-4,6,8-TRIYN-1,3-DIOL	$C_{10}H_{10}O_2$	162 (-) 55^{0}C

$$\text{Me(C:C)}_3\text{CHOH.CH}_2\text{CH}_2\text{OH}$$

JCS, 1969, 1096.

3110-031

DECA-4,6-DIYN-1-OL	$C_{10}H_{14}O$	150 OI Oil

$$\text{Pr.(C:C)}_2\text{CH}_2\text{CH}_2\text{CH}_2\text{OH}$$

Ber., 1970 (103) 2853.

3110-027

DECANAL	$C_{10}H_{20}O$	156 OI Oil

$$\text{Me(CH}_2\text{)}_8\text{CHO}$$

J. Pr. Chem., 1903 (68) 235.

3110-032

DEC-2-ENAL	$C_{10}H_{18}O$	154 OI Oil
$Me(CH_2)_6CH:CH.CHO$		Chem. Abstr., 1948 (42) 1388.
3110-033		

DECA-2,8-DIEN-4,6-DIYN-1-AL, -10-OH-	$C_{10}H_8O_2$	160 OI Oil
$HOCH_2.CH\overset{c}{:}CH.(C\vdots C)_2CH\overset{c}{:}CH.CHO$		Ber., 1969 (102) 1679.
3110-038		

DECA-2,4-DIEN-AL	$C_{10}H_{16}O$	152 OI Oil
$Me(CH_2)_4(CH:CH)_2CHO$		Chem. Abstr., 1954 (48) 388.
3110-034		

DECANONE, 2-	$C_{10}H_{20}O$	156 OI Oil
$Me.CO.(CH_2)_7Me$		Helv., 1932 (15) 1267.
3110-039		

DECA-2,6-DIEN-4-YNAL	$C_{10}H_{12}O$	148 OI Oil
$Pr.\overset{t}{CH:CH}.C\vdots C.\overset{t}{CH:CH}.CHO$		Ber., 1970 (103) 3422.
3110-035		

DEC-1-EN-4,6,8-TRIYN-3-ONE	$C_{10}H_6O$	142 OI Oil
$Me(C\vdots C)_3CO.CH:CH_2$		JCS, 1969, 1096.
3110-040		

DECA-2,8-DIEN-4,6-DIYN-1-AL	$C_{10}H_8O$	144 OI Oil
$Me.CH:CH(C\vdots C)_2CH:CH.CHO$		Arch. Pharm., 1970 (303) 912.
3110-036		

DECANOIC ACID	$C_{10}H_{20}O_2$	172 OI 31^0C
$Me(CH_2)_8COOH$		Ann., 1861 (118) 307.
3110-041		

DEC-2-EN-4,6,8-TRIYN-1-AL	$C_{10}H_6O$	142 OI 108^0C
$Me(C\vdots C)_3\overset{t}{CH:CH}.CHO$		JCS, 1960, 691.
3110-037		

LACHNOPHYLLUM ESTER, cis-	$C_{11}H_{12}O_2$	176 OI 33^0C
$Pr.(C\vdots C)_2CH\overset{c}{:}CH.COOMe$		Science, 1964 (146) 1461.
3110-042		

LACHNOPHYLLUM ESTER, t-	$C_{11}H_{12}O_2$	176 OI 16^0C

trans-isomer

Acta Chem. Scand.,
1954, 280.

3110-043

	$C_{12}H_{12}O_2S$	220 OI 71^0C

(cis, trans-isomer)

Ber., 1968, 2506.

3110-048

OBTUSILIC ACID	$C_{10}H_{18}O_2$	170 OI Oil

$Me(CH_2)_4CH:CH.CH_2CH_2COOH$

Bull. Chem. Soc. Jap.,
1937 (12) 226.

3110-044

	$C_{12}H_{12}O_2S$	220 OI 112^0C

(trans, cis-isomer)

Ber., 1966, 2096.

3110-049

DECA-2, 4-DIENOIC ACID	$C_{10}H_{16}O_2$	168 OI

$Me(CH_2)_4(CH:CH)_2COOR$
R = H

JCS, 1949, 3353.

3110-045

DECA-2, 6-DIEN-4, 8-DIYNOIC ACID ME ESTER, 7-SMe-t, c-	$C_{12}H_{12}O_2S$	220 OI

$ME.C:C.C(SMe)\overset{c}{:}CH.C:C.\overset{t}{CH:CH.COOMe}$

Ber., 1966, 2096.

3110-050

PELLITORINE	$C_{14}H_{25}NO$	223 OI 72^0C

R = NH.CH_2CHMe_2

JCS, 1969, 2477.

3110-046

SPILANTHOL	$C_{14}H_{25}ON$	223 OI

$Pr.(CH:CH)_2CH_2CH_2CO.NHCH_2CHMe_2$

JACS, 1956, 5084.

3110-051

DECA-2, 4-DIEN-6, 8-DIYNOIC ACID ME ESTER, 4-SMe-c, c-	$C_{12}H_{12}O_2S$	220 OI Oil

$Me(C:C)_2CH:\overset{c}{C}(SMe).\overset{c}{CH:CH.COOMe}$

Ber., 1968, 2506.

3110-047

DECA-2, 6-DIEN-4-YNOIC ACID ME ESTER	$C_{11}H_{14}O_2$	178 OI Oil

$Pr.CH:CH.C:C.CH:CH.COOMe$

Ber., 1969 (102) 1037.

3110-052

MATRICARIA ESTER, c, c–	$C_{11}H_{10}O_2$	174 OI	DECA-2, 4, 6-TRIEN-8-YNOIC ACID, 5-SMe-c, c, c–	$C_{11}H_{12}O_2S$	208 OI
Me. CH:CH.(C:C)$_2$CH:CH. COOMe			Me. C:C. CH:CH. C(SMe):CH. CH:CH. COOH		
	Ber., 1969 (102) 1682.			Ber., 1966, 2096.	
3110-053			3110-058		
MATRICARIA ESTER, t, c–	$C_{11}H_{10}O_2$	174 OI 2^{0}C	DECA-2, 4, 8-TRIEN-6-YNOIC ACID ME ESTER, 5-SMe-	$C_{12}H_{14}O_2S$	222 OI
(trans, cis-isomer) (acid)			Me. CH:CH. C:C. C(SMe):CH. CH:CH. COOMe		
	Ber., 1969 (102) 1682.				
3110-054			3110-059		
MATRICARIA ACID, t, t–	$C_{10}H_8O_2$	160 OI 175^{0}C	AFFININ	$C_{14}H_{23}ON$	221 OI 23^{0}C
(trans, trans-isomer)			Me. (CH:CH)$_2$CH$_2$CH$_2$CH:CH. CONHCH$_2$CHMe$_2$		
	Ber., 1967, 611.			JCS, 1963, 4970.	
3110-055			3110-060		
MATRICARIA ESTER, cis-dehydro–	$C_{11}H_8O_2$	172 OI 112^{0}C	DECA-2, 6, 7, 8-TETRAEN-4-YNOIC ACID, 9-SMe-	$C_{11}H_{10}O_2S$	206
Me(C:C)$_3$CH:CH. COOMe			Me(SMe)C:C:C:CH. C:C. CH:CH. COOH		
	Ber., 1966, 1830.			Ber., 1971 (104) 1329.	
3110-056			3110-061		
MATRICARIA ESTER, trans-dehydro–	$C_{11}H_8O_2$	172 OI 106^{0}C	MYRMICACIN	$C_{10}H_{20}O_3$	188 -21^{0}c 48^{0}C
trans-isomer			Me(CH$_2$)$_6$CHOR.CH$_2$COOH R = H		
	Acta Chem. Scand., 1954, 26.			Angew. Chem., 1971 (10) 124.	
3110-057			3110-062		

PYOLIPIC ACID	$C_{16}H_{30}O_7$	334	DEC-2-EN-4,6-DIYNOIC ACID	$C_{10}H_{10}O_3$	178
		Oil			OI
					155^0C

$$\overset{t}{HOCH_2CH_2CH_2(C\vdots C)_2CH:CH.COOH}$$

R = Rhamnosyl

Arch. Biochem.,
1946 (10) 165.

JCS, 1957, 1607.

3110-063

3110-068

DEC-8-EN-4,6-DIYNOIC ACID ME ESTER, 2-OAng-	$C_{16}H_{18}O_4$	274	DIATRETYNE-3	$C_{10}H_6O_3$	174
		Oil			OI

$$\overset{c}{\underset{*}{Me.CH:CH.(C\vdots C)_2CH_2CH(OAng).COOH}}$$

$$HOCH_2(C\vdots C)_3\overset{t}{CH:CH.COOH}$$

Ber., 1969 (102) 1682.

JCS, 1960, 691.

3110-064

3110-069

DECA-4,6-DIYNOIC ACID ME ESTER, 3-OAng-	$C_{16}H_{20}O_4$	276	DEC-4-ENE-6,8-DIYNOIC ACID, 2,10-diOH-	$C_{10}H_{10}O_4$	194
		Oil			

$$HOCH_2(C\vdots C)_2\overset{t}{CH:CH.}CH_2CHOH.COOH$$

* dihydro

Ber., 1970 (103) 561.

JCS, 1963, 3466.

3110-065

3110-070

DEC-2-ENOIC ACID, 9-OH-	$C_{10}H_{18}O_3$	186	DECA-4,6,8-TRIYNOIC ACID, 2,10-diOH-	$C_{10}H_6O_4$	192
		Oil			m.e. $+44^0e$
					m.e. 102^0C

$$Me.CHOH.(CH_2)_5\overset{t}{CH:CH.COOH}$$

$$HOCH_2(C\vdots C)_3CH_2CHOH.COOH$$

Nature, 1964 (201) 733.

JCS, 1963, 3466.

3110-066

3110-071

DEC-2-ENOIC ACID, 10-OH-	$C_{10}H_{18}O_3$	186	SEBACIC ACID	$C_{10}H_{18}O_4$	202
		OI			OI
		64^0C			134^0C

$$HOCH_2(CH_2)_6\overset{t}{CH:CH.COOH}$$

$$HOOC(CH_2)_8COOH$$

Tet., 1962, 177.

Ber., 1941(74)1617

3110-067

3110-072

DEC-2-EN-1,10-DIOIC ACID	$C_{10}H_{16}O_4$	200 OI

$$HOOC.CH:CH(CH_2)_6COOH$$

JACS, 1948, 4238.

3110-073

DEC-2-ENOIC ACID, 9-oxo-t-	$C_{10}H_{16}O_3$	184 OI 52^0C

$$Me.CO(CH_2)_5CH:CH.COOH$$

JCS, 1961, 3813.

3110-078

DECA-2,8-DIENE-4,6-DIYNE- -1,10-DIOIC ACID	$C_{10}H_6O_4$	190 OI 200^0C

$$HOOC.\overset{t}{C}H:CH(C:C)_2\overset{t}{C}H:CH.COOH$$

JCS, 1957, 1607.

3110-074

DECA-4,6,8-TRIYN-3-ONE, -1,2-epoxy-	$C_{10}H_6O_2$	158

$$Me(C:C)_3CO-$$

JCS, 1969, 1096.

3110-079

DECA-2,4,6-TRIYNDIOIC ACID -DIME ESTER	$C_{12}H_{10}O_4$	218 OI 45^0C

$$MeOOC.(C:C)_3CH_2CH_2COOMe$$

JCS, 1960, 691.

3110-075

	$C_{20}H_{38}O_5$	358

$$Me(CH_2)_6CHOH.CH_2COO\\Me(CH_2)_6\overset{t}{C}HCH_2COOH$$

JACS, 1949, 4124.

3110-080

DEC-2-ENE-4,6-DIYNDIOIC ACID -DIME ESTER	$C_{12}H_{12}O_4$	220 OI 57^0C

$$MeOOC.\overset{t}{C}H:CH.(C:C)_2CH_2CH_2COOMe$$

JCS, 1957, 1607.

3110-076

SERRATAMIC ACID	$C_{13}H_{25}NO_5$	285 -10^0e 138^0C

$$Me(CH_2)_6CHOH.CH_2CO.NHCH(COOH)CH_2OH$$

Biochem. J.,
1957 (67) 663.

3110-081

DEC-2-EN-4,6,8-TRIYNEDIOIC ACID	$C_{10}H_4O_4$	188 OI 0^0C

$$HOOC(C:C)_3\overset{t}{C}H:CH.COOH$$

JCS, 1963, 2056.

3110-077

UNDECANE	$C_{11}H_{24}$	156 OI Oil

$$Me(CH_2)_9Me$$

JACS, 1947, 2014.

3111-001

GALBANOLENE	$C_{11}H_{18}$	139 OI Oil		UNDEC-9-EN-2-OL	$C_{11}H_{22}O$	170 -5^0 Oil

GALBANOLENE $C_{11}H_{18}$ 139 OI Oil

$Me(CH_2)_4(CH:CH)_3H$

Bull. Soc. Chim. Fr.,
1967, 98.

3111-002

UNDEC-9-EN-2-OL $C_{11}H_{22}O$ 170 -5^0 Oil

$Me.CHOH.(CH_2)_7CH:CH_2$

Chem. Zentr.,
1911 (II) 1863.

3111-007

UNDECA-1,3,5,8-TETRAENE $C_{11}H_{16}$ 148 OI Oil

$Et.CH:CH.CH_2(CH:CH)_3H$

Chem. Comm.,
1970, 1093.

3111-003

UNDECA-5,6-DIENE-8,10--DIYN-1-OL $C_{11}H_{12}O$ 160 -180^0

$H(C{:}C)_2CH:C:CH(CH_2)_3CH_2OH$

JCS, 1966, 129.

3111-008

UNDECA-1-EN-5,7,9-TRIYNE $C_{11}H_{10}$ 142 OI Oil

$Me(C{:}C)_3CH_2CH_2CH:CH_2$

Ber., 1971 (104) 954.

3111-004

UNDECA-5,6-DIENE-8,10-DIYNE--1,3-DIOL $C_{11}H_{12}O_2$ 176

$H(C{:}C)_2CH:C:CH.CH_2CHOH.CH_2CH_2OH$

JCS, 1966, 129.

3111-009

UNDECA-1,3-DIEN-5,7,9-TRIYNE $C_{11}H_8$ 140 OI Oil

$Me(C{:}C)_3\overset{t}{C}H:CH.CH:CH_2$

Ber., 1971 (104) 954.

3111-005

UNDECA-5,6-DIENE-8,10-DIYNE--1,4-DIOL $C_{11}H_{12}O_2$ 176

$H(C{:}C)_2CH:C:CH.CHOH(CH_2)_2CH_2OH$

JCS, 1966, 129.

3111-010

UNDECAN-2-OL $C_{11}H_{24}O$ 172 Oil

$Me(CH_2)_8CHOH.Me$

JCS, 1902, 1585.

3111-006

UNDECYNE, 9-OMe- $C_{12}H_{22}O$ 182 Oil

$Me.CH(OMe).(CH_2)_7C{:}CH$

Chem. Ind., 1963, 122.

3111-011

UNDECAN-2-ONE	$C_{11}H_{22}O$	170 OI 14^{0}C
Me.CO.(CH$_2$)$_8$Me		JACS, 1938, 920.
3111-012		

DROSOPHILIN-C	$C_{11}H_8O_2$	172 OI 98^{0}C
HC:C.CH$_2$(C:C)$_2$CH:CH.CH$_2$COOH		JCS. 1960. 2257.
3111-017		

UNDEC-1-EN-10-ONE	$C_{11}H_{20}O$	168 OI Oil
MeCO.(CH$_2$)$_7$CH:CH$_2$		Chem. Zentr., 1911 (II) 1863.
3111-013		

DROSOPHILIN-D	$C_{11}H_8O_2$	172 (+) 21^{0}C
CH$_2$:C:CH.(C:C)$_2$CH:CH.CH$_2$COOH		JCS. 1960. 2257.
3111-018		

UNDECANOIC ACID	$C_{11}H_{22}O_2$	186 OI 30^{0}C
Me(CH$_2$)$_9$COOH		Chem. Abstr., 1948. (42)6140.
3111-014		

UNDEC-3,5,6-TRIENE-8.10- -DIYNOIC ACID	$C_{11}H_8O_2$	172 OI
H(C:C)$_2$CH:C:C.CH:CH.CH$_2$COOH		Science. 1957 (126) 1229.
3111-019		

UNDEC-10-ENOIC ACID	$C_{11}H_{20}O_2$	184 OI 24^{0}C
CH$_2$:CH.(CH$_2$)$_8$COOH		Chem. Abstr., 1955 (49) 11075.
3111-015		

NEMOTINIC ACID	$C_{11}H_{10}O_3$	190 +320^{0}r
H(C:C)$_2$CH:C:CH.CHOH.CH$_2$CH$_2$COOH		JCS, 1955, 4270.
3111-020		

UNDEC-10-YNOIC ACID	$C_{11}H_{18}O_2$	182 OI 39^{0}C
HC:C(CH$_2$)$_8$COOH		Chem. Abstr., 1955 (49) 11075.
3111-016		

UNDECA-2,4-DIEN-8,10-DIYNOIC ACID ISOBUTYLAMIDE	$C_{15}H_{19}NO$	229 OI
H(C:C)$_2$CH$_2$CH$_2$(CH:CH)$_2$CONHCH$_2$CHMe$_2$		Chem. Abstr., 1967 (66) 65025.
3111-021		

LABILOMYCIN	$C_{23}H_{34}O_8$	438 -237^0m 112^0C

OCOEt
|
Me   O   OCOCH.(CH:CH)$_3$COCH$_2$CH$_3$

MeO   OMe
OH

J. Antibiotics,
1964 (17) 200.

3111-022

DODEC-8-EN-1-OL, Ac-	$C_{14}H_{26}O_2$	226 OI Oil

Pr. CH:CH(CH$_2$)$_6$CH$_2$OAc

Nature, 1969 (224) 723.

3112-005

DODECA-1,11-DIEN-3,5,7,9--TETRAYNE	$C_{12}H_6$	150 OI

CH$_2$:CH(C:C)$_4$CH:CH$_2$

Acta Chem. Scand.,
1954 (8) 1944.

3112-001

DODECA-3,5,7-TRIEN-9,11--DIYN-1-OL	$C_{12}H_{12}O$	172 OI Oil

H(C:C)$_2$(CH:CH)$_3$CH$_2$CH$_2$OH

Ber., 1970 (103) 1879.

3112-006

DODECANOL	$C_{12}H_{26}O$	186 OI 23^0C

Me(CH$_2$)$_{10}$CH$_2$OH

JACS, 1943, 959.

3112-002

DODECA-3,6,8-TRIEN-1-OL	$C_{12}H_{20}O$	180 OI

Pr. CH:CH.CH:CH.CH$_2$CH:CH.CH$_2$CH$_2$OH

Nature, 1968 (219) 963.

3112-007

DODEC-7-EN-1-OL. Ac-c-	$C_{14}H_{26}O_2$	226 OI Oil

Bu. CH:CH(CH$_2$)$_5$CH$_2$OAc

Chem. Comm.,
1968, 792.

3112-003

DODECAN-1,12-DIOL	$C_{12}H_{26}O_2$	202 OI 80^0C

HOCH$_2$(CH$_2$)$_{10}$CH$_2$OH

Chem. Abstr.,
1956 (50) 13748.

3112-008

DODEC-7-EN-1-OL, Ac-t-	$C_{14}H_{26}O_2$	226 OI Oil

(trans-isomer)

Chem. Comm.,
1968, 792.

3112-004

DODECA-2,4,6,8-TETRAENE--1,12-DIOL	$C_{12}H_{18}O_3$	210 OI 126^0C

HOCH$_2$(CH:CH)$_4$CH$_2$CH$_2$CH$_2$OH

JCS, 1966, 135.

3112-009

DODECANAL	$C_{12}H_{24}O$	184 OI 45^0C
$Me(CH_2)_{10}CHO$	JACS, 1947, 2014.	
3112-010		

deleted

3112-015

DODEC-2-ENAL	$C_{12}H_{22}O$	182 OI Oil
$Me(CH_2)_8\overset{t}{CH}:CH.CHO$	Science, 1964 (144) 540.	
3112-011		

LINDERIC ACID	$C_{12}H_{22}O_2$	198 OI Oil
$Me(CH_2)_6CH:CH(CH_2)_2COOH$	J. Soc. Chem. Ind. Jap., 1937 (40) 285.	
3112-016		

---

deleted

3112-012

DODEC-9-ENOIC ACID	$C_{12}H_{22}O_2$	198 OI
$Et.CH:CH(CH_2)_7COOH$	Chem. Abstr., 1939 (33) 6239.	
3112-017		

DODECANAL, 3-oxo-	$C_{12}H_{22}O_2$	198 OI 8^0C
$Me(CH_2)_8COCH_2CHO$	Chem. Abstr., 1953 (47) 5356.	
3112-013		

DODECA-3,5,7,9,11-PENTAENOIC ACID	$C_{12}H_{14}O_2$	190
$H.(CH:CH)_5CH_2COOH$	JOC, 1971, 2621.	
3112-018		

DODECANOIC ACID	$C_{12}H_{24}O_2$	200 OI 44^0C
$Me(CH_2)_{10}COOH$	Ann., 1842 (41) 329.	
3112-014		

ODYSSIC ACID	$C_{12}H_{12}O_3$	204 $+300^0e$
$Me(C:C)_2CH:C:CH.CHOH.CH_2CH_2COOH$	JCS, 1957, 1607.	
3112-019		

SABINIC ACID $C_{12}H_{24}O_3$ 216
OI
84^0C

$HOCH_2(CH_2)_{10}COOH$

JCS, 1928, 2679.

3112-020

HERCULIN, neo- $C_{16}H_{25}NO$ 247
OI
70^0C

ttc
$Me(CH:CH)_3CH_2CH_2CH:CHCONHCH_2CHMe_2$

JOC, 1967, 1646.

3112-025

TRAUMATIC ACID $C_{12}H_{20}O_4$ 228
OI
165^0C

$HOOC.CH:CH(CH_2)_8COOH$

Science, 1939, 329.

3112-021

DODECA-2, 4-DIEN-8, 10-DIYNOIC $C_{16}H_{21}NO$ 243
ACID ISOBUTYLAMIDE   OI

$Me(C\vdots C)_2CH_2CH_2(CH:CH)_2CONHCH_2CHMe_2$

Chem. Abstr.,
1967 (66) 65018.

3112-026

DODECANEDIOIC ACID, 4-oxo- $C_{12}H_{20}O_5$ 244
OI
112^0C

$HOOC(CH_2)_2CO(CH_2)_7COOH$

Tet., 1970, 5215.

3112-022

TRIDECANE $C_{13}H_{28}$ 184
OI
Oil

$Me(CH_2)_{11}Me$

JACS, 1933, 3889.

3113-001

CERULENIN $C_{12}H_{17}NO_3$ 223

$$MeCH:CHCH_2CH_2CH:CHCH_2CO-\overset{\overset{O}{\diagup\!\diagdown}}{CH}-CH-CONH_2$$

J. Antibiotics,
1967, 349.

3112-023

TRIDEC-1-EN-3, 5, 7, 9, 11- $C_{13}H_6$ 162
-PENTAYNE   OI
Oil

$Me(C\vdots C)_5CH:CH_2$

Chem. Abstr.,
1968 (68) 2529.

3113-002

HERCULIN $C_{16}H_{29}NO$ 251
OI
60^0C

$Pr.CH:CH(CH_2)_4CH:CH.CONHCH_2CHMe_2$

JCS, 1952, 4338.

3112-024

TRIDECA-1, 11-DIEN-3, 5, 7, 9- $C_{13}H_8$ 164
-TETRAYNE   OI
Oil

$CH_2:CH(C\vdots C)_4CH:CH.Me$

Phytochem., 1971. 2227.

3113-003

TRIDECA-1, 3, 5-TRIEN-7, 9, 11--TRIYNE	$C_{13}H_{10}$	166 OI Oil	TRIDEC-2-EN-4, 6, 8-TRIYN-1-OL	$C_{13}H_{14}O$	186 OI
Me(C:C)$_3$(CH:CH)$_3$H			Bu. (C:C)$_3$CH:CH.CH$_2$OH		
		Ber., 1966, 1830.			JCS, 1966, 139.
3113-004			3113-009		

TRIDECA-1, 3, 11-TRIEN-5, 7, 9--TRIYNE	$C_{13}H_{10}$	166	TRIDECA-2, 8-DIENE-4, 6--DIYN-10-OL	$C_{13}H_{16}O$	188
Me.CH:CH(C:C)$_3$(CH:CH)$_2$H			Pr.CHOH.CH:CH(C:C)$_2$CH:CH.Me		
		Phytochem., 1971, 2227.			Arch. Pharm., 1970 (303) 912.
3113-005			3113-010		

AETHUSIN	$C_{13}H_{14}$	170 OI	TRIDECA-2, 12-DIEN-4, 6, 8, 10--TRIYN-1-OL	$C_{13}H_8O$	180 OI
Et.(CH:CH)$_2$(C:C)$_2$CH:CH.Me			CH$_2$:CH.(C:C)$_4$CH:CH.CH$_2$OR R = H		
		Arch. Pharm., 1970 (303) 912.			Phytochem., 1971 (10) 2227.
3113-006			3113-011		

TRIDECA-1, 3, 5, 11-TETRAEN-7,-9-DIYNE	$C_{13}H_{12}$	168 OI 71°C	TRIDECA-2, 12-DIEN-4, 6, 8, 10--TETRAYN-1-OL, Ac-	$C_{15}H_{10}O_2$	222 OI
Me.CH:CH(C:C)$_2$(CH:CH)$_3$H			R = Ac		
		Phytochem., 1971, 2227.			Phytochem., 1971, 2227.
3113-007			3113-012		

TRIDECA-2, 4, 6, 8-TETRAYNE	$C_{13}H_{12}$	168 OI	TRIDECA-3, 11-DIEN-5, 7, 9--TRIYN-2-OL, 1-Cl-	$C_{13}H_{11}OCl$	218
Me(C:C)$_4$(CH$_2$)$_3$Me			Me. CH:CH(C:C)$_3$CH:CH. CHOH.CH$_2$Cl		
		JCS, 1966, 139.			Chem. Abstr., 1967 (66) 65020.
3113-008			3113-013		

TRIDECA-8, 11-DIEN-1-OL, Ac-	$C_{15}H_{26}O_2$	238 OI	TRIDECA-4, 6, 8, 10-TETRAYN- -EN-1, 12, 12-TRIOL, 1-Glu-	$C_{19}H_{20}O_8$	376 Oil
Me. $\overset{t}{C}H{:}CH.CH_2CH\overset{c}{:}CH(CH_2)_6CH_2OAc$		Agr. Biol. Chem., 1971, 447.	$HOCH_2CHOH(C{:}C)_4CH{:}CH.CH_2OGlu$		Phytochem., 1971, 2233.
3113-014			3113-019		
TRIDECA-2, 10, 12-TRIEN-4, 6, -8-TRIYN-1-OL	$C_{13}H_{10}O$	182 OI	TRIDECA-2, 4, 6, 8-TETRAYN- -10, 11, 12, 13-TETROL	$C_{13}H_{12}O_4$	232
$H(CH{:}CH)_2(C{:}C)_3CH{:}CH.CH_2OR$ R = H		Phytochem., 1971, 2227.	$CH_3.(C{:}C)_4.CHOH.CHOH.CHOH.CH_2OH.$		JCS, 1966, 139.
3113-015			3113-020		
TRIDECA-2, 10, 12-TRIEN-4, 6, 8- -TRIYN-1-OL, Ac-	$C_{15}H_{12}O_2$	224 OI -15^0C	TRIDEC-2-EN-1-AL	$C_{13}H_{24}O$	196 OI Oil
R = Ac		Phytochem., 1971, 2227.	$Me(CH_2)_9CH{:}CH.CHO$		Chem. Zentr., 1938 (I) 1892.
3113-016			3113-021		
TRIDECA-1, 5-DIEN-7, 9, 11- -TRIYN-3, 4-DIOL	$C_{13}H_{12}O_2$	200 70^0C	TRIDECA-2, 4-DIEN-1-AL	$C_{13}H_{22}O$	194 OI Oil
$Me(C{:}C)_3\overset{t}{C}H{:}CH.CHOH.CHOH.CH{:}CH_2$		Ber., 1970 (103) 1879.	$Me(CH_2)_7(CH{:}CH)_2CHO$		Chem. Zentr., 1938 (I) 1892.
3113-017			3113-022		
SAFYNOL	$C_{13}H_{12}O_2$	200 -17^0m 112^0C	TRIDECA-2, 12-DIEN-4, 6, 8, 10- -TETRAYNAL	$C_{13}H_6O$	178 OI
$Me.\overset{t}{C}H{:}CH(C{:}C)_3\overset{t}{C}H{:}CH.CHOH.CH_2OH$		Phytochem., 1971, 1579.	$CH_2{:}CH(C{:}C)_4CH{:}CH.CHO$		Phytochem., 1971, 2227.
3113-018			3113-023		

TRIDECA-2, 10, 12-TRIENE-4, 6, 8--TRIYNAL, c, c-	$C_{13}H_8O$	180 OI	TRIDECAN-2-ONE	$C_{13}H_{26}O$	198 OI 28^0C

TRIDECA-2, 10, 12-TRIENE-4, 6, 8--TRIYNAL, c, c- $C_{13}H_8O$ 180 OI

$$CH_2:CH.CH:CH(C\vdots C)_3\overset{c}{CH}:CH.CHO$$
(c above first CH:CH)

Phytochem., 1971, 2227.

3113-024

TRIDECAN-2-ONE $C_{13}H_{26}O$ 198 OI 28^0C

$$MeCO(CH_2)_{10}Me$$

J. Pharm. Soc. Jap., 1937 (57) 920.

3113-029

TRIDECA-2, 10, 12-TRIEN-4, 6, 8--TRIYNAL, c, t- $C_{13}H_8O$ 180 OI

cis, trans-isomer

Ber., 1967, 1504.

3113-025

TRIDEC-2-EN-4, 6-DIYN-8-OL--10-ONE $C_{13}H_{16}O_2$ 204

$$Pr.CO.CH_2CHOH(C\vdots C)_2CH:CH.Me$$

Arch. Pharm., 1970 (303) 912.

3113-030

TRIDECANOIC ACID $C_{13}H_{26}O_2$ 214 OI 42^0C

$$Me(CH_2)_{11}COOH$$

Chem. Abstr., 1948 (42) 6140.

3113-026

TRIDECA-1, 3, 5-TRIEN-7, 9-DIYNE--10, 11-EPOXIDE $C_{13}H_{12}O$ 184

$$H(CH:CH)_3(C\vdots C)_2\overset{O}{CH-CHMe}$$

Ber., 1966, 1830.

3113-031

---

deleted

3113-027

TRIDEC-3-EN-5, 7, 9, 11--TETRAYNE-1, 2-EPOXIDE $C_{13}H_8O$ 180

$$Me(C\vdots C)_4CH:CH.\overset{O}{CH-CH_2}$$

Chem. Abstr., 1967 (66) 65023.

3113-032

MYCOMYCIN $C_{13}H_{10}O_2$ 198 -130^0e 75^0C

$$H(C\vdots C)_2CH:C:CH.(CH:CH)_2CH_2COOH$$

JACS, 1953, 1372.

3113-028

TRIDECA-1, 5-DIENE-7, 9, 11--TRIYNE-3, 4-EPOXIDE $C_{13}H_{10}O$ 182

$$Me(C\vdots C)_3CH:CH.\overset{O}{CH-CHCH:CH_2}$$

Ber., 1966, 1648.

3113-033

$C_{14}H_{10}O_2S$ 242 Me(C:C)$_2$CH:C(SO$_2$Me).(C:C)$_2$CH:CH$_2$ (t) Ber., 1967, 537. 3113-034	TETRADECAN-1-OL $C_{14}H_{30}O$ 214 OI 39^{0}C Me(CH$_2$)$_{12}$CH$_2$OH J. Pharm. Soc. Jap., 1937 (57) 196. 3114-001
$C_{14}H_{10}O_3S$ 258 Me(C:C)$_2$C—CH(C:C)$_2$CH:CH$_2$ O SO$_2$Me Ber., 1967, 537. 3113-035	TETRADECA-5-EN-8,10,12- -TRIYN-1-OL, Ac- $C_{16}H_{18}O_2$ 242 OI Me(C:C)$_3$CH$_2$CH:CH(CH$_2$)$_3$CH$_2$OAc (c) Ber., 1966, 1830. 3114-002
$C_{14}H_{10}O_4S$ 274 Me(C:C)$_2$C—CH(C:C)$_2$CH—CH$_2$ O O SO$_2$Me Ber., 1967, 537. 3113-036	TETRADECA-8-EN-2,4,6- -TRIYN-12-OL $C_{14}H_{16}O$ 200 Me(C:C)$_3$CH:CH.CH$_2$CH$_2$CHOH.Et (t) Ber., 1966, 1642. 3114-003
$C_{14}H_{10}OS$ 226 OI Me(C:C)$_2$C:CH(C:C)$_2$CH:CH$_2$ SOMe Ber., 1967, 537. 3113-037	TETRADEC-11-EN-1-OL, Ac- $C_{16}H_{30}O_2$ 254 OI Oil Et.CH:CH(CH$_2$)$_9$CH$_2$OH (c) Nature, 1970 (226) 1172. 3113-004
TRIDECA-1,3,5,7-TETRAEN- -9,11-DIYNE, 7-SMe- $C_{14}H_{14}S$ 214 OI Oil Me(C:C)$_2$CH:C(SMe).(CH:CH)$_3$H Ber., 1971 (104) 958. 3113-038	TETRADECA-4,6-DIEN-8,10,12- -TRIYN-1-OL $C_{14}H_{14}O$ 198 OI Me(C:C)$_3$(CH:CH)$_2$(CH$_2$)$_2$CH$_2$OR (t) R = H Ber., 1966, 1830. 3114-005

TETRADECA-4, 6-DIENE-8, 10, 12-
-TRIYN-1-OL, Ac- $C_{16}H_{16}O_2$ 240
OI

R = Ac

Ber., 1966, 2096.

3114-006

TETRADECA-4, 6-DIEN-8, 10, 12-
-TRIYN-1, 3-DIOL, 3-O-Ac- $C_{16}H_{16}O_3$ 256

R = H; R' = Ac

Ber., 1967, 611.

3114-011

TETRADECA-6, 12-DIEN-8, 10-
-DIYN-3-OL $C_{14}H_{18}O$ 202

38^0C

MeCH:CH(C$\vdots$C)$_2$CH:CH.CH$_2$CH$_2$CHOH. Et

Ber., 1966, 1642.

3114-007

TETRADECA-4, 6-DIEN-8, 10, 12-
-TRIYN-1, 3-DIOL, di-Ac- $C_{18}H_{18}O_4$ 298

R = R' = Ac   Phytochem., 1968, 270.

3114-012

TETRADECA-9, 12-DIEN-1-OL, Ac- $C_{16}H_{28}O_2$ 252
OI
Oil

Me. CH:CH.CH$_2$CH:CH(CH$_2$)$_7$CH$_2$OAc

Science, 1971 (171) 802.

3114-008

TETRADECA-4, 6, 12-TRIEN-8, 10-
-DIYN-1, 3-DIOL $C_{14}H_{16}O_2$ 216

MeCH:CH(C$\vdots$C)$_3$(CH:CH)$_2$CH.CH$_2$CH$_2$OR
OR'
R = R' = H

Chem. Abstr.,
1967 (66) 65020.

3114-013

TETRADECA-4, 6, 12-TRIEN-8, 10-
-DIYN-1-OL, Ac- $C_{16}H_{18}O_2$ 242

32^0C

MeCH:CH(C$\vdots$C)$_2$(CH:CH)$_2$CH$_2$CH$_2$CH$_2$OAc

Acta Chem. Scand.,
1966, 992.

3114-009

TETRADECA-4, 6, 12-TRIEN-
-8, 10-DIYN-1, 3-DIOL, 1-O-Ac- $C_{16}H_{18}O_3$ 258

R = Ac; R' = H

Chem. Abstr.,
1967 (66) 65020.

3114-014

TETRADECA-4, 6-DIEN-8, 10, 12-
-TRIYN-1, 3-DIOL $C_{14}H_{14}O_2$ 214

Me(C$\vdots$C)$_3$(CH:CH)$_2$CHOH'CH$_2$CH$_2$OR
R = R' = H

Phytochem., 1968, 270.

3114-010

TETRADECA-4, 6, 12-TRIEN-8, 10-
-DIYN-1, 3-DIOL, 3-O-Ac- $C_{16}H_{18}O_3$ 258

R = H; R' = Ac

Chem. Abstr.,
1967 (66) 65020.

3114-015

TETRADECA-4, 6, 12-TRIEN-8, 10--DIYN-1, 3-DIOL, di-Ac-	$C_{18}H_{20}O_4$	300
R = R' = Ac	Chem. Abstr., 1967 (66) 65020.	
3114-016		

TETRADECA-6-EN-1, 3-DIYN--13-ONE	$C_{14}H_{18}O$	202 OI
$H(C \vdots C)_2 CH_2 CH:CH(CH_2)_5 COMe$	Chem. Abstr., 1967 (66) 65023.	
3114-021		

TETRADECA-2, 8, 10-TRIEN-4, 6--DIYN-1, 14-DIOL	$C_{14}H_{16}O_2$	216 OI 132°C
$HOCH_2CH_2CH_2(CH\overset{t}{:}CH)_2(C \vdots C)_2 CH\overset{t}{:}CH.CH_2OH$	Ber., 1970 (103) 2095.	
3114-017		

TETRADEC-8-EN-2, 4, 6-TRIYN--12-ONE	$C_{14}H_{14}O$	198 OI 58°C
$Me(C \vdots C)_3 CH:CH.CH_2CH_2COEt$	Acta Chem. Scand., 1950. 1567.	
3114-022		

TETRADECA-6, 12-DIENE-8, 10--DIYN-1-OL, Ac-4, 5-epoxy-	$C_{16}H_{18}O_3$	258 Oil
$MeCH:CH(C \vdots C)_2 CH:CH.CH \overset{O}{\overbrace{\quad}} CH(CH_2)_3OAc$	Phytochem., 1971, 1877.	
3114-018		

MYRISTIC ACID	$C_{14}H_{28}O_2$	228 OI 54°C
$Me(CH_2)_{12}COOH$	JCS, 1950, 174.	
3114-023		

TETRADEC-6-EN-8, 10, 12--TRIYN-1-OL, 4, 5-epoxy-	$C_{14}H_{14}O_2$	214
$Me(C \vdots C)_3 CH:CH.CH \overset{O}{\overbrace{\quad}} CH(CH_2)_2CH_2OH$	Ber., 1966 (99) 1830.	
3114-019		

MACILENIC ACID	$C_{14}H_{26}O_2$	226 OI 70°C
$Me(CH_2)_{10}CH:CH.COOH$	Arch. Pharm., 1915 (253) 102.	
3114-024		

TETRADECANAL	$C_{14}H_{28}O$	212 OI 24°C
$Me(CH_2)_{12}CHO$	J. Am. Pharm. Ass., 1935 (24) 380.	
3114-020		

TSUZUIC ACID	$C_{14}H_{26}O_2$	226 OI 18°C
$Me(CH_2)_8CH:CH(CH_2)_2COOH$	Bull. Chem. Soc. Jap., 1937 (12) 433.	
3114-025		

444

TETRADEC-5-ENOIC ACID	$C_{14}H_{26}O_2$	226 OI 20^0C
$Me(CH_2)_7CH:CH(CH_2)_3COOH$	Chem. Abstr., 1936 (30) 315.	
3114-026		

TETRADECA-2, 4, 5-TRIENOIC ACID ME ESTER	$C_{15}H_{26}O_2$	238 -128^0 Oil
$Me(CH_2)_7CH:C:CH.\overset{t}{CH}:CH.COOMe$	JCS, 1970, 859.	
3114-031		

MYRISTOLEIC ACID	$C_{14}H_{26}O_2$	226 OI
$Me(CH_2)_3CH:CH(CH_2)_7COOR$ R = H	J. Soc. Chem. Ind., 1925 (44) 180.	
3114-027		

TETRADECA-2, 4, 6, 12-TETRAEN- -8, 10-DIYNOIC ACID ME ESTER	$C_{15}H_{14}O_2$	226 OI Oil
$MeCH:CH(C\vdots C)_2(CH:CH)_3COOMe$	Ber., 1966 (99) 3194.	
3114-032		

TETRADEC-9-ENOIC ACID -ET ESTER	$C_{16}H_{30}O_2$	254 OI Oil
R = Et	Nature, 1969 (221) 856.	
3114-028		

TETRADECA-5, 7, 9, 11, 13- -PENTAENOIC ACID	$C_{14}H_{18}O_2$	218 OI
$H(CH:CH)_5(CH_2)_3COOH$	JOC, 1971, 2621.	
3114-033		

MEGATOMIC ACID	$C_{14}H_{24}O_2$	224 OI
$Me(CH_2)_7(CH:CH)_2CH_2COOH$	JOC, 1970, 3152.	
3114-029		

TETRADECA-2, 4-DIENOIC ACID -ISOBUTYLAMIDE	$C_{18}H_{33}NO$	279 OI
$Me(CH_2)_8(CH:CH)_2CONHCH_2CHMe_2$	Chem. Abstr., 1967 (66) 65025.	
3114-034		

GOSHUYUIC ACID	$C_{14}H_{24}O_2$	224
$Me(CH_2)_4CH:CH.CH_2CH:CH(CH_2)_3COOH$	Chem. Pharm. Bull., 1972, 559.	
3114-030		

ANACYCLIN	$C_{18}H_{25}NO$	271 OI 118^0C
$Pr(C\vdots C)_2CH_2CH_2(CH:CH)_2CONHCH_2CHMe_2$	Chem. Ind., 1956, 406.	
3114-035		

ANACYCLIN, dehydro- $C_{18}H_{23}NO$ 269
OI

$MeCH:CH(C\vdots C)_2CH_2CH_2(CH:CH)_2CONHCH_2CHMe_2$

JCS, 1957, 2767.

3114-036

PHARBITIC ACID D $C_{50}H_{88}O_{30}$ 1168
-64^0
140^0C

R =

$$\begin{array}{c} | \\ Glu \\ _2 \diagup \quad \diagdown ^6 \\ Glu \qquad Rha \\ |^2 \qquad |^4 \\ Quinovosyl \quad Rha \\ |^3 \\ Rha \end{array}$$

Chem. Pharm. Bull.,
1971, 2394.

3114-041

MYRISTIC ACID, 3-OH- $C_{14}H_{28}O_3$ 244
-16^0c
74^0C

$Me(CH_2)_{10}CHOH.CH_2COOH$

JACS, 1953, 1035.

3114-037

TETRADECANOIC ACID, 9, 10-diOH- $C_{14}H_{28}O_4$ 260
OI
80^0C

$Me(CH_2)_3CHOH.CHOH(CH_2)_7COOH$

JCS, 1964, 5833.

3114-042

BUTOLIC ACID $C_{14}H_{28}O_3$ 244
-1^0c
58^0C

$Me(CH_2)_7CHOH(CH_2)_4COOH$

Tet., 1969, 3841.

3114-038

TETRADEC-2, 4, 6-TRIENOIC
ACID, 5-CH_2Ac- $C_{17}H_{26}O_4$ 294
OI
88^0C

$Me(CH_2)_6CH:CH.C(CH_2OAc):CH.CH:CH.COOH$

Aust. J. C., 1961 (14) 628.

3114-043

IPUROLIC ACID $C_{14}H_{28}O_4$ 260

100^0C

$PrCHOR(CH_2)_7CHOH.CH_2COOH$
R = H

Chem. Pharm. Bull.,
1971, 2394.

3114-039

TETRADEC-9-ENE-2, 4, 6-
-TRIYNDIOIC ACID $C_{14}H_{12}O_4$ 244
OI
120^0C

$HOOC(CH_2)_3CH:CH.CH_2(C\vdots C)_3COOH$

JCS, 1963, 2056.

3114-044

PHARBITIC ACID C $C_{44}H_{78}O_{26}$ 1022
-51^0
120^0C

R = .Glu^6—Rha^4—Quinovosyl
$|^2$
Glu^2—Rha

Chem. Pharm. Bull.,
1971, 2394.

3114-040

CORTICROCIN $C_{14}H_{14}O_4$ 246
OI
317^0C

$HOOC(CH:CH)_6COOH$

Acta Chem. Scand.,
1948 (2) 209.

3114-045

PENTADECANE	$C_{15}H_{32}$	212
		OI

Me(CH$_2$)$_{13}$Me

Chem. Zentr.,
1925 (I) 974.

deleted

3115-001

3115-006

PENTADECA-2, 9-DIENE- $C_{15}H_{20}$ 200
    -4, 6-DIYNE   OI
                                              Oil

Me(CH$_2$)$_4$CH:CHCH$_2$(C:C)$_2$CH:CHMe (c ... t)

Ber., 1968, 1889.

deleted

3115-002

3115-007

PENTADECA-2, 8. 10-TRIENE- $C_{15}H_{18}$ 198
    -4, 6-DIYNE   OI
                                                Oil

Me(CH$_2$)$_3$(CH:CH)$_2$(C:C)$_2$CH:CHMe (t ... t)

Ber., 1968 1889.

deleted

3115-003

3115-008

PENTADECA-2, 4, 6, 8-TETRAENE $C_{15}H_{24}$ 204
                                                    OI

Me(CH$_2$)$_5$(CH:CH)$_4$Me

Arch. Pharm.,
1970 (303) 912.

3115-004

PENTADECA-1, 8-DIEN-4, 6- $C_{15}H_{20}O$ 216
    -DIYN-3-OL

Me(CH$_2$)$_5$CH:CH(C:C)$_2$CHOH.CH:CH$_2$ (t)

Ber., 1968, 525.

3115-009

PENTADECA-2, 8, 10, 14-TETRAENE- $C_{15}H_{16}$ 196
    -4, 6-DIYNE   OI

CH$_2$:CHCH$_2$CH$_2$(CH:CH)$_2$(C:C)$_2$CH:CHMe

Aust. J. C., 1968, 2037.

3115-005

PENTADECA-2, 8-DIEN-4, 6- $C_{15}H_{20}O$ 216
    -DIYN-10-OL   Oil

Me(CH$_2$)$_4$CHOH.CH:CH(C:C)$_2$CH:CHMe

Ber., 1968 (101) 1163.

3115-010

PENTADECA-2, 9-DIEN-4, 6--DIYN-1-OL	$C_{15}H_{20}O$	216 OI Oil

$Me(CH_2)_4\overset{c}{CH}:CH.CH_2(C\vdots C)_2\overset{c}{CH}:CH.CH_2OH$

Ber., 1971 (104) 2030.

3115-011

PENTADECA-2, 9-DIEN-4, 6-DIYN--1, 8-DIOL, 1-Ac-	$C_{17}H_{22}O_3$	274 OI

$Me(CH_2)_4\overset{c}{CH}:CH.CHOH(C\vdots C)_2\overset{c}{CH}:CHCH_2OH$

Ber., 1971 (104) 1322.

3115-016

PENTADECA-1, 8, 10-TRIEN-4, 6--DIYN-3-OL	$C_{15}H_{18}O$	214 Oil

$Me(CH_2)_3(CH:CH)_2(C\vdots C)_2CHOH.CH:CH_2$

Ber., 1968, 1889.

3115-012

PENTADECA-5, 10-DIENAL	$C_{15}H_{26}O$	222 OI Oil

$Me(CH_2)_3CH:CH(CH_2)_3CH:CH(CH_2)_3CHO$

Ber., 1942 (75) 1830.

3115-017

PENTADECA-2, 8, 10-TRIEN-4, 6--DIYN-1-OL	$C_{15}H_{18}O$	214 OI Oil

$Me(CH_2)_3(\overset{t}{CH}:CH)(C\vdots C)_2\overset{t}{CH}:CHCH_2OH$

Ber., 1971 (104) 2030.

3115-013

PENTADECA-2, 9-DIEN-4, 6-DIYNAL	$C_{15}H_{18}O$	214 OI Oil

$Me(CH_2)_4\overset{c}{CH}:CHCH_2(C\vdots C)_2\overset{c}{CH}:CHCHO$

Ber., 1971 (104) 2030.

3115-018

PENTADECA-2, 8, 10-TRIEN-4, 6--DIYN-12-OL	$C_{15}H_{18}O$	214 34^0C

$PrCHOH.(\overset{t}{CH}:CH)_2(C\vdots C)_2CH:CHMe$

Ber., 1971 (104) 1362.

3115-014

PENTADECA-5, 7-DIEN-9, 11, 13--TRIYNAL	$C_{15}H_{14}O$	210

$Me(C\vdots C)_3(\overset{t}{CH}:CH)_2(CH_2)_3CHO$

Chem. Abstr., 1967 (66) 65023.

3115-019

---

deleted

3115-015

PENTADECA-2, 8, 10-TRIEN-4, 6--DIYNAL	$C_{15}H_{16}O$	212 OI Oil

$Me(CH_2)_3(\overset{t}{CH}:CH)_2(C\vdots C)_2\overset{t}{CH}:CHCHO$

Ber., 1971 (104) 2030.

3115-020

PENTADECA-5,7,13-TRIEN-9,11--DIYNAL	$C_{15}H_{16}O$	212

$\overset{t}{MeCH}:CH(C\vdots C)_2(CH:CH)_2(CH_2)_3CHO$

Chem. Abstr., 1967 (66) 65023.

3115-021

| PENTADECA-6,8,10,12--TETRAENAL | $C_{15}H_{22}O$ | 218 |

$Et(CH:CH)_4(CH_2)_4CHO$

Chem. Abstr., 1967 (66) 65023.

3115-022

| PENTADECA-1,8,10-TRIEN-4,6--DIYN-3-ONE | $C_{15}H_{16}O$ | 212 OI Oil |

$Me(CH_2)_3\overset{t}{(CH:CH)}_2(C\vdots C)_2COCH:CH_2$

Ber., 1968, 1889.

3115-023

| PENTADECA-7,13-DIEN-9,11--DIYN-4-ONE | $C_{15}H_{18}O$ | 214 OI Oil |

$MeCH:CH(C\vdots C)_2\overset{t}{CH}:CH.CH_2CH_2COPr$

Ber., 1971 (104) 1322.

3115-024

| PENTADECANOIC ACID | $C_{15}H_{30}O_2$ | 242 OI 53°C |

$Me(CH_2)_{13}COOH$

Arch. Pharm., 1931 384.

3115-025

| PENTADECANOIC ACID, 2-OH- | $C_{15}H_{30}O_3$ | 258 -1°e 90°C |

$Me(CH_2)_{12}CHOH.COOH$

Can. J. C., 1953 (31) 396.

3115-026

| CONVOLVULINOLIC ACID | $C_{15}H_{30}O_3$ | 258 50°C |

$Me(CH_2)_3CHOH(CH_2)_9COOH$

Planta Med., 1961 (9) 141.

3115-027

| PENTADECANOIC ACID, 15-OH- | $C_{15}H_{30}O_3$ | 258 OI 85°C |

$HOCH_2(CH_2)_{13}COOH$

JCS, 1963, 3505.

3115-028

| PENTADEC-8-EN-2-ONE | $C_{15}H_{26}O_3$ | 254 OI |

$Me(CH_2)_5CH:CH(CH_2)_5COCOOH$

Chem. Abstr., 1968 (68) 2529.

3115-029

| PENTADECANOIC ACID, 3-keto- | $C_{15}H_{28}O_3$ | 256 OI 90°C |

$Me(CH_2)_{10}COCH_2CH_2COOH$

Trans. Roy. Soc., 1931 (B220) 301.

3115-030

HEXADECANE	$C_{16}H_{34}$	226 OI 18^{0}C
$Me(CH_2)_{14}Me$		Chem. Abstr., 1956 (50) 11688.
3116-001		

HEXADEC-7-EN-1-OL	$C_{16}H_{32}O$	240 OI Oil
$Me(CH_2)_7\overset{c}{CH:CH}(CH_2)_5CH_2OR$ R = H		Nature, 1969 (221) 856.
3116-006		

HEXADECA-1,6,8-TRIEN--10,12,14-TRIYNE	$C_{16}H_{16}$	208 OI
$Me(C\overset{.}{:}C)_3\overset{t}{(CH:CH)}_2(CH_2)_3CH:CH_2$		Ber., 1965 (98) 2596.
3116-002		

HEXADEC-7-EN-1-OL, Ac-	$C_{18}H_{34}O_2$	282 OI Oil
R = Ac		Experientia, 1969 (25) 682.
3116-007		

deleted	
3116-003	

HEXADEC-9-EN-1-OL	$C_{16}H_{30}O$	240 OI Oil
$Me(CH_2)_5\overset{c}{CH:CH}(CH_2)_7CH_2OH$		Nature, 1969 (221) 856.
3116-008		

CETYL ALCOHOL	$C_{16}H_{34}O$	242 OI 50^{0}C
$Me(CH_2)_{14}CH_2OR$ R = H		Ann., 1941 (548) 270.
3116-004		

BOMBYKOL	$C_{16}H_{30}O$	236 OI Oil
$Pr.\overset{c}{CH:CH}.\overset{t}{CH:CH}(CH_2)_8CH_2OH$		Ann., 1962 (658) 65.
3116-009		

HEXADECYL ACETATE	$C_{18}H_{36}O_2$	384 OI Oil
R = Ac		JACS, 1966, 1305.
3116-005		

HEXADECA-7,14-DIEN-10,12--DIYN-1-OL	$C_{16}H_{22}O$	230 OI Oil
$Me.\overset{t}{CH:CH}(C\overset{.}{:}C)_2CH_2\overset{c}{CH:CH}(CH_2)_5CH_2OH$		Ber., 1970 (103) 2100.
3116-010		

HEXADECAN-1,16-DIOL	$C_{16}H_{34}O_2$	258 OI 90^{0}C	PALMITAMINE	$C_{16}H_{33}NO$	255 OI 105^{0}C
$HOCH_2(CH_2)_{14}CH_2OH$		Chem. Abstr., 1956 (50) 13748.	R = NH$_2$		JCS, 1956, 4163.
3116-011			3116-016		

HEXADECA-6,8-DIEN-10,12,14-TRIYNAL $C_{16}H_{16}O$ 224 OI

$Me(C{:}C)_3(\overset{t}{CH{:}CH})_2(CH_2)_4CHO$

Chem. Abstr., 1967 (66) 65023.

3116-012

PALMITIC ACID HEXADECYL-ESTER $C_{32}H_{64}O_2$ 480 OI Oil

R = .(CH$_2$)$_{14}$Me

JCS, 1926, 1463.

3116-017

HEXADECA-6,8,14-TRIEN-10,12-DIYNAL $C_{16}H_{18}O$ 226 OI

$Me(\overset{t}{CH{:}CH}(C{:}C)_2(\overset{t}{CH{:}CH})_2(CH_2)_4CHO$

Chem. Abstr., 1967 (66) 65023.

3116-013

HEXADEC-3-ENOIC ACID $C_{16}H_{30}O_2$ 254 OI 54^{0}C

$Me(CH_2)_{11}\overset{t}{CH{:}CH}.CH_2COOH$

Experientia, 1964 (20) 511.

3116-018

HEXADECA-7,9,11,13-TETRAENAL $C_{16}H_{26}O$ 234 OI

$Et(\overset{t}{CH{:}CH})_4(CH_2)_5CHO$

Chem. Abstr., 1967 (66) 65023.

3116-014

PALMITOLEIC ACID $C_{16}H_{30}O_2$ 254 OI 32^{0}C

$Me(CH_2)_5CH{:}CH(CH_2)_7COOH$

Biochem. J., 1956 (62) 222.

3116-019

PALMITIC ACID $C_{16}H_{32}O_2$ 256 OI 62^{0}C

$Me(CH_2)_{14}CO.R$
R = OH

Arch. Pharm., 1966 (299) 468.

3116-015

HEXADEC-11-ENOIC ACID $C_{16}H_{30}O_2$ 254 OI

$Me(CH_2)_3CH{:}CH(CH_2)_9COOH$

J. Biol. Chem., 1955 (213) 415.

3116-020

HEXADEC-12-ENOIC ACID	$C_{16}H_{30}O_2$	254 OI Oil		JALAPINOLIC ACID	$C_{16}H_{32}O_3$	272 68^0C
Pr. CH:CH(CH$_2$)$_{10}$COOH		Arch. Pharm., 1909 (247) 418.		Me(CH$_2$)$_4$CHOH(CH$_2$)$_9$COOH		Arch. Pharm., 1934, 841.
3116-021				3116-026		
HEXADECA-7, 10, 13-TRIENOIC ACID	$C_{16}H_{26}O_2$	250 OI		JUNIPERIC ACID	$C_{16}H_{32}O_3$	272 OI 94^0C
Et(CH:CH.CH$_2$)$_3$(CH$_2$)$_4$COOH		Chem. Abstr., 1946 (40) 225.		HOCH$_2$(CH$_2$)$_{14}$COOH		Chem. Abstr., 1953 (47) 4634.
3116-022				3116-027		
HEXADECANOIC ACID, D-2-OH-	$C_{16}H_{32}O_3$	272 -6^0c 45^0C		AMBRETTOLIC ACID	$C_{16}H_{30}O_3$	270 (α) 54^0C
Me(CH$_2$)$_{13}$CH(OH)COOH		Can. J. C., 1970 (48) 1985.		HOCH$_2$(CH$_2$)$_7$CH:CH(CH$_2$)$_5$COOH		Compt. Rend., 1945 (221) 205.
3116-023				3116-028		
PALMITIC ACID, D-3-OH-	$C_{16}H_{32}O_3$	272 -14^0c 80^0C		HEXADEC-9-ENOIC ACID, 12-OH-	$C_{16}H_{30}O_3$	270 m.e. $+6^0m$
Me(CH$_2$)$_{12}$CH(OH)CH$_2$COOH		Can. J. C., 1968 (46) 2628.		Me(CH$_2$)$_3$CHOHCH$_2$CH:CH(CH$_2$)$_7$COOH		JOC, 1966, 1477.
3116-024				3116-029		
HEXADECANOIC ACID, 8-OH-	$C_{16}H_{32}O_3$	272 $+1^0c$ 79^0C		HEXADEC-9-ENOIC ACID, 16-OH-	$C_{16}H_{30}O_3$	270 OI 18^0C
Me(CH$_2$)$_7$CHOH.(CH$_2$)$_6$COOH		Can. J. C., 1965 (43) 415.		HOCH$_2$(CH$_2$)$_5$CH:CH(CH$_2$)$_7$COOH		JCS, 1964, 5833.
3116-025				3116-030		

OPERCULINOLIC ACID $\qquad$ $C_{16}H_{32}O_4$ $\quad$ 288 $\\$ $\qquad$ 84^{0}C $\\$ $\\$ Me(CH$_2$)$_3$CH(OR)(CH$_2$)$_8$CHOHCH$_2$COOH $\\$ $\qquad$ R = H $\qquad$ Planta Med., 1961 (9) 141. $\\$ $\\$ 3116-031	USTILIC ACID B $\qquad$ $C_{16}H_{32}O_5$ $\quad$ 304 $\\$ $\\$ HOCH$_2$CHOH(CH$_2$)$_{12}$CHOH.COOH $\\$ $\qquad$ Can. J. C., 1953 (31) 1054. $\\$ $\\$ 3116-036
OPERCULINIC ACID $\qquad$ $C_{52}H_{92}O_{32}$ $\quad$ 1229 $\\$ $\qquad$ -49^{0}p $\\$ $\qquad$ 183^{0}C $\\$ $\\$ R = .Glu^2—Rha^4—Glu^4—Glu $\\$ $\qquad$ |3 $\qquad$ |6 $\\$ $\qquad$ Glu $\qquad$ Rha $\qquad$ Tet. Lett., 1971, 3232. $\\$ $\\$ 3116-032	THAPSIC ACID $\qquad$ $C_{16}H_{30}O_4$ $\quad$ 284 $\\$ $\qquad$ OI $\\$ $\qquad$ 125^{0}C $\\$ $\\$ HOOC(CH$_2$)$_{14}$COOH $\\$ $\qquad$ Monatsh., 1935 (66) 3. $\\$ $\\$ 3116-037
HEXADECANOIC ACID, 9, 10-di-OH- $\quad$ $C_{16}H_{32}O_4$ $\quad$ 288 $\\$ $\qquad$ m. e. 64^{0}C $\\$ $\\$ Me(CH$_2$)$_5$CHOH.CHOH(CH$_2$)$_7$COOH $\\$ $\qquad$ JCS, 1964, 5833. $\\$ $\\$ 3116-033	PAHUTOXIN $\qquad$ $C_{23}H_{46}NO_4^+$ $\quad$ 400 $\\$ $\qquad$ Cl: +3^{0}m $\\$ $\qquad$ Cl: 75^{0}C $\\$ $\\$ Me(CH$_2$)$_{12}$CH(OAc)CH$_2$COOCH$_2$CH$_2$N$^+$Me$_3$ $\\$ $\qquad$ Science, 1967 (155) 52. $\\$ $\\$ 3116-038
ALEURITIC ACID $\qquad$ $C_{16}H_{32}O_5$ $\quad$ 304 $\\$ $\qquad$ 102^{0}C $\\$ $\\$ HOCH$_2$(CH$_2$)$_5$CHOH.CHOH(CH$_2$)$_7$COOH $\\$ $\qquad$ Ind. Eng. Chem., $\\$ $\qquad$ 1938 (30) 333. $\\$ $\\$ 3116-034	deleted $\\$ $\\$ 3116-039
USTILIC ACID A $\qquad$ $C_{16}H_{32}O_4$ $\quad$ 288 $\\$ $\\$ HOCH$_2$(CH$_2$)$_{13}$CHOH.COOH $\\$ $\qquad$ Can. J. C., 1953 (31) 1054. $\\$ $\\$ 3116-035	deleted $\\$ $\\$ 3116-040

453

HEXADECA-1, 14-DIEN-10, 12--DIYNE-8, 9-EPOXIDE	$C_{16}H_{20}O$	228

$$MeCH:CH(C\vdots C)_2CH \overset{O}{\diagup\!\!\diagdown} CH(CH_2)_5CH:CH_2$$

Ber., 1966, 2096.

3116-041

HEPTADECA-2, 9-DIEN-4, 6-DIYNE	$C_{17}H_{24}$	228 OI Oil

$$Me(CH_2)_6\overset{c}{CH}:CH.CH_2(C\vdots C)_2\overset{t}{CH}:CHMe$$

Ber., 1971 (104) 1362.

3117-005

HEPTADECANE	$C_{17}H_{36}$	240 OI 23^0C

$$Me(CH_2)_{15}Me$$

Chem. Abstr., 1955 (49) 15720.

3117-001

HEPTADECA-1, 7, 9-TRIEN--3, 5-DIYNE	$C_{17}H_{22}$	226 OI

$$Me(CH_2)_5(CH:CH)_2(C\vdots C)_2CH:CH_2$$

Ber., 1968 (101) 525.

3117-006

HEPTADEC-5-ENE	$C_{17}H_{34}$	238 OI

$$Me(CH_2)_{10}CH:CH(CH_2)_3Me$$

Mar. Biol., 1971 (8) 183.

3117-002

HEPTADECA-1, 7, 9-TRIEN--11, 13-DIYNE	$C_{17}H_{22}$	226 OI Oil

$$Pr(C\vdots C)_2(CH:CH)_2(CH_2)_4CH:CH_2$$

Ber., 1969 (102) 1691.

3117-007

HEPTADEC-7-ENE	$C_{17}H_{34}$	238 OI

$$Me(CH_2)_8CH:CH(CH_2)_5Me$$

Mar. Biol., 1971 (8) 183.

3117-003

HEPTADECA-1, 7, 9-TRIEN-11, 13--15-TRIYNE, c-t-	$C_{17}H_{18}$	222 OI Oil

$$Me(C\vdots C)_3(\overset{t,\,c}{CH:CH})_2(CH_2)_4CH:CH_2$$

JCS, 1969, 1096.

3117-008

HEPTADECA-1, 8-DIEN-11, 13--DIYNE	$C_{17}H_{24}$	228 OI

$$Pr(C\vdots C)_2CH_2\overset{c}{CH:CH}(CH_2)_5CH:CH_2$$

Ber., 1968 (101) 532.

3117-004

HEPTADECA-1, 7, 9-TRIEN-11, --13, 15-TRIYNE, t, t-	$C_{17}H_{18}$	222 OI 18^0C

trans, trans-isomer

JCS, 1969, 1096.

3117-009

HEPTADECA-1,8,15-TRIEN-11,- -13-DIYNE, c,c-	$C_{17}H_{22}$	226 OI Oil

$$MeCH\overset{c}{:}CH(C\overset{.}{:}C)_2CH_2CH\overset{c}{:}CH(CH_2)_5CH:CH_2$$
*

Ber., 1968 (101) 532.

3117-010

HEPTADECA-1,7,9,15-TETRAEN- -11,13-DIYNE, t,t,t-	$C_{17}H_{20}$	224 OI 28^0C

*trans isomer

Phytochem., 1971, 2227.

3117-015

HEPTADECA-1,8,15-TRIEN-11,- -13-DIYNE, c,t-	$C_{17}H_{22}$	226 OI Oil

*trans isomer

Ber., 1969 (102) 1691.

3117-011

HEPTADECA-1,3,7,9,13-PENTAEN- -11-YNE	$C_{17}H_{22}$	226 OI Oil

$$PrCH:CH.C\overset{.}{:}C(CH\overset{t}{:}CH)_2(CH_2)_2(CH\overset{t}{:}CH)_2H$$

Ber., 1969 (102) 1037.

3117-016

HEPTADECA-2,8,10-TRIEN- -4,6-DIYNE	$C_{17}H_{22}$	226 OI Oil

$$Me(CH_2)_5(CH\overset{t}{:}CH)_2(C\overset{.}{:}C)_2CH:CHMe$$

Ber., 1971 (104) 1362.

3117-012

deleted

3117-017

HEPTADECA-1,7,9,13- -TETRAEN-11-YNE	$C_{17}H_{24}$	228 OI Oil

$$PrCH\overset{t}{:}CH.C\overset{.}{:}C(CH\overset{t}{:}CH)_2(CH_2)_4CH:CH_2$$

Ber., 1969 (102) 1037.

3117-013

deleted

3117-018

HEPTADECA-1,7,9,15-TETRAEN- -11,13-DYNE, t,t,c-	$C_{17}H_{20}$	224 OI Oil

$$MeCH\overset{c}{:}CH(C\overset{.}{:}C)_2(CH\overset{t}{:}CH)_2(CH_2)_4CH:CH_2$$
*

Ber., 1969 (102) 1034.

3117-014

deleted

3117-019

deleted

3117-020

HEPTADECA-2,8-DIEN-4,6-DIEN-10-OL $C_{17}H_{24}O$ 244 OI Oil

$Me(CH_2)_6CHOH.CH\overset{t}{:}CH(C\vdots C)_2CH\overset{t}{:}CHMe$

Ber., 1971 (104) 1362.

3117-025

HEPTADEC-9-EN-4,6-DIYN-1-OL $C_{17}H_{26}O$ 246 OI Oil

$Me(CH_2)_6CH\overset{c}{:}CHCH_2(C\vdots C)_2CH_2CH_2CH_2OH$

Ber., 1968 (101) 1163.

3117-021

HEPTADECA-2,9-DIEN-4,6-DIYN-1-OL, t,c- $C_{17}H_{24}O$ 244 OI Oil

$Me(CH_2)_6CH\overset{c}{:}CHCH_2(C\vdots C)_2CH\overset{t}{:}CHCH_2OH$

Ber., 1971 (104) 1362.

3117-026

HEPTADEC-9-EN-4,6-DIYN-8-OL $C_{17}H_{26}O$ 246 OI Oil

$Me(CH_2)_6CH\overset{c}{:}CHCHOH(C\vdots C)_2Pr$

Ber., 1971 (104) 1322.

3117-022

HEPTADECA-2,9-DIEN-4,6-DIYN-1-OL, c,c- $C_{17}H_{24}O$ 244 OI Oil

cis, cis-isomer

Ber., 1971 (104) 2030.

3117-027

FALCARINOL $C_{17}H_{24}O$ 244 -22^0 Oil

$Me(CH_2)_6CH:CHCH_2(C\vdots C)_2CHOH.CH:CH_2$

Chem. Comm., 1967, 439.
Tet., 1967, 465.

3117-023

HEPTADECA-2,9-DIEN-4,6-DIYN-8-OL $C_{17}H_{24}O$ 244 Oil

$Me(CH_2)_6CH\overset{c}{:}CH.CHOH(C\vdots C)_2CH\overset{c}{:}CHMe$

Ber., 1971 (104) 1322.

3117-028

HEPTADECA-2,8-DIEN-4,6-DIYN-1-OL, Ac- $C_{19}H_{24}O_2$ 284 OI Oil

$Me(CH_2)_5(CH:CH)_2(C\vdots C)_2CH:CHCH_2OAc$

Ber., 1968 (101) 525.

3117-024

HEPTADECA-9,16-DIEN-4,6-DIYN-3-OL $C_{17}H_{24}O$ 244 Oil

$CH_2:CH(CH_2)_5CH\overset{c}{:}CHCH_2(C\vdots C)_2CHOH.Et$

Ber., 1969 (102) 1691.

3117-029

HEPTADECA-1,8,10-TRIEN-4,6--DIYN-3-OL	$C_{17}H_{22}O$	242 Oil

$$Me(CH_2)_5(CH:CH)_2^t(C:C)_2CHOH.CH:CH_2$$

Ber., 1971 (104) 1362.

3117-030

CICUTOL	$C_{17}H_{22}O$	242 OI 66°C

$$Bu(CH:CH)_3(C:C)_2CH_2CH_2CH_2OH$$

JCS, 1953, 309.

3117-035

HEPTADECA-1,8,15-TRIEN--11,13-DIYN-10-OL, c,c-	$C_{17}H_{22}O$	242 Oil

$$MeCH\overset{c}{:}CH(C:C)_2CHOH.CH\overset{c}{:}CH(CH_2)_5CH:CH_2$$

Chem. Abstr., 1967 (66) 65019.

3117-031

HEPTADECA-8,10,16-TRIEN-4,6--DIYN-3-OL	$C_{17}H_{22}O$	242 Oil

$$CH_2:CH(CH_2)_4(CH\overset{t}{:}CH)_2(C:C)_2CHOH.Et$$

Ber., 1969 (102) 1691.

3117-036

HEPTADECA-1,9,15-TRIEN-11,13--DIYN-8-OL, c,c-	$C_{17}H_{22}O$	242

$$MeCH\overset{c}{:}CH(C:C)_2CH\overset{c}{:}CHCHOH(CH_2)_5CH:CH_2$$

Ber., 1966 (99) 590.

3117-032

HEPTADECA-1,7,9,13-TETRAEN--11-YN-3-OL	$C_{17}H_{24}O$	244 Oil

$$PrCH\overset{t}{:}CH.C:C(CH\overset{t}{:}CH)_2(CH_2)_3CHOH.CH:CH_2$$

Ber., 1969 (102) 1037.

3117-037

OENANTHETOL, cis-	$C_{17}H_{22}O$	242 OI 71°C

$$Me(CH_2)_5(CH:CH)_2(C:C)_2CH:CHCH_2OH$$

Aust. J. C., 1968, 2037.

3117-033

HEPTADECA-1,7,9,15-TETRAEN--11,13-DIYN-6-OL	$C_{17}H_{20}O$	240 41°C

$$MeCH\overset{t}{:}CH(C:C)_2(CH\overset{t}{:}CH)_2CHOR(CH_2)_3CH:CH_2$$
$$R = H$$

JCS, 1970, 314.

3117-038

HEPTADECA-2,9,16-TRIEN-4,6--DIYN-1-OL, c,c-	$C_{17}H_{22}O$	242 OI

$$CH_2:CH(CH_2)_5CH\overset{c}{:}CHCH_2(C:C)_2CH\overset{c}{:}CHCH_2OH$$

Ber., 1966, 2096.

3117-034

HEPTADECA-1,7,9,15-TETRAEN--11,13-DIYN-6-OL, Ac-	$C_{19}H_{22}O_2$	282 Oil

$$R = Ac$$

JCS, 1970, 314.

3117-039

HEPTADECA-2,7,9,13-TETRAEN--11-YN-1-OL	$C_{17}H_{24}O$	244 OI Oil

$$PrCH\overset{t}{:}CH.C\overset{t}{:}C(CH\overset{t}{:}CH)_2(CH_2)_3CH\overset{c}{:}CHCH_2OH$$

Ber., 1969,(102) 1037.

3117-040

PARINARIC ACID	$C_{18}H_{28}O_2$	276 OI 84^0C

$$CH_3CH_2(CH:CH)_4(CH_2)_7COOH$$

JCS, 1951, 291.

3117-044

HEPTADECA-2,8,10,16-TETRAEN--4,6-DIYN-1-OL	$C_{17}H_{20}O$	240 OI 77^0C

$$CH_2:CH(CH_2)_4(CH\overset{t}{:}CH)_2(C:C)_2CH\overset{c}{:}CHCH_2OH$$

Acta Chem. Scand., 1966 (20) 992.

3117-041

HEPTADECA-8-EN-4,6-DIYN--1,10-DIOL	$C_{17}H_{26}O_2$	262 44^0C

$$Me(CH_2)_6CHOH.CH\overset{t}{:}CH(C:C)_2CH_2CH_2CH_2OH$$

Ber., 1968 (101) 1163.

3117-045

HEPTADECA-4,8,10,16-TETRAEN--6-YN-3-OL	$C_{17}H_{24}O$	244 Oil

$$CH_2:CH(CH_2)_4(CH\overset{t}{:}CH)_2C:C.CH\overset{t}{:}CHCHOH.Et$$

Ber., 1969 (102) 1037.

3117-042

FALCARINDIOL	$C_{17}H_{24}O_2$	260

$$Me(CH_2)_6CH:CH.CHOH(C\overset{c}{:}C)_2CHOR.CH:CH_2$$
R = H

Ber., 1971 (104) 1957.

3117-046

HEPTADECA-2,4,8,10,16--PENTAEN-6-YN-1-OL	$C_{17}H_{22}O$	242 OI Oil

$$CH_2\overset{t}{:}CH(CH_2)_4(CH\overset{t}{:}CH)_2C\overset{t}{:}C(CH:CH)_2CH_2OH$$

Ann., 1970 (739) 135.

3117-043

FALCARINDIOL, 3-O-Ac-	$C_{19}H_{26}O_3$	302 Oil

R = Ac

JCS, 1969, 685.

3117-047

HEPTADECA-4,8,10,14,16--PENTAEN-6-YN-3-OL	$C_{17}H_{22}O$	242 Oil

$$H(CH\overset{t}{:}CH)_2CH_2CH_2(CH\overset{t}{:}CH)_2C:C.CH\overset{t}{:}CHCHOH.Et$$

Ber., 1969 (102) 1037.

3117-044

HEPTADECA-2,8-DIEN-4,6--DIYN-1,10-DIOL	$C_{17}H_{24}O_2$	260

$$Me(CH_2)_6CHOH.CH\overset{t}{:}CH(C:C)_2CH\overset{t}{:}CHCH_2OH$$

Ber., 1971 (104) 1362.

3117-048

HEPTADECA-2,9-DIEN-4,6--DIYN-1,8-DIOL	$C_{17}H_{24}O_2$	260 Oil

$$Me(CH_2)_6CH:CH.CHOH(C \overset{c}{:} C)_2CH:CHCH_2OR$$

R = H   Ber., 1971 (104) 1322.

3117-049

AVOCADENE	$C_{17}H_{34}O_3$	286 -7°c 68°C

$$CH_2:CH(CH_2)_{11}CHOH.CH_2CHOH.CH_2OR$$

R = H   Tet., 1969, 4623.

3117-054

	$C_{19}H_{26}O_3$	302 Oil

R = Ac

Ber., 1971 (104) 1322.

3117-050

AVOCADENE, 1-O-Ac-	$C_{19}H_{36}O_4$	328 -5°c 59°C

R = Ac

Phytochem., 1971, 1417.
Tet., 1969, 4623.

3117-055

HEPTADECA-8,10-DIEN-4,6--DIYN-1,12-DIOL	$C_{17}H_{24}O_2$	260 Oil

$$Me(CH_2)_4CHOH(CH:CH)_2(C:C)_2CH_2CH_2CH_2OH$$

Ber., 1969 (102) 3293.

3117-051

AVOCADYNE	$C_{17}H_{32}O_3$	284 -5°c 76°C

$$HC:C(CH_2)_{11}CHOH.CH_2CHOH.CH_2OR$$

R = H   Tet., 1969, 4623.

3117-056

OENANTHOTOXIN	$C_{17}H_{22}O_2$	258 +30° 87°C

$$PrCHOH.CH_2CH_2(CH:CH)_2(C:C)_2CH:CHCH_2OH$$

JCS, 1953, 309.

3117-052

AVOCADYNE, 1-O-Ac-	$C_{19}H_{34}O_4$	326 -5°c 73°C

R = Ac

Tet., 1969, 4623.

3117-057

CICUTOXIN	$C_{17}H_{22}O_2$	258 -15°e 54°C

$$PrCHOH(CH:CH)_3(C:C)_2CH_2CH_2CH_2OH$$

JCS, 1955, 1770.

3117-053

FALCARINONE, 1,2-Dihydro-	$C_{17}H_{24}O$	244 OI Oil

$$Me(CH_2)_6CH \overset{c}{:} CH.CH_2(C:C)_2CO.Et$$

Ber., 1971 (104) 1322.

3117-058

FALCARINONE	$C_{17}H_{22}O$	242 OI Oil	HEPTADECA-8, 10, 16-TRIEN-6- -YN-3-ONE	$C_{17}H_{24}O$	244 OI Oil

$Me(CH_2)_6CH\overset{c}{:}CH.CH_2(C\vdots C)_2COCH:CH_2$

Ber., 1971 (104) 1957.

$CH_2:CH(CH_2)_4(CH\overset{t}{:}CH)_2C\vdots C.CH_2CH_2CO.Et$

Ber., 1969 (102) 1037.

3117-059

3117-064

HEPTADEC-9, 16-DIEN-4, 6- -DIYN-3-ONE	$C_{17}H_{22}O$	242 OI Oil	HEPTADECA-8, 10, 16-TRIEN-4, 6- -DIYN-3-ONE	$C_{17}H_{20}O$	240 OI Oil

$CH_2:CH(CH_2)_5CH\overset{c}{:}CH.CH_2(C\vdots C)_2CO.Et$

Ber., 1969 (102) 1691.

$CH_2:CH(CH_2)_4(CH\overset{t}{:}CH)_2(C\vdots C)_2CO.Et$

Ber., 1969 (102) 1691.

3117-060

3117-065

HEPTADECA-1, 8, 16-TRIEN-4, 6- -DIYN-3-ONE	$C_{17}H_{20}O$	240 OI Oil	HEPTADECA-5, 7, 9, 15-TETRAEN- -11, 13-DIYN-4-ONE	$C_{17}H_{18}O$	238 OI Oil

$CH_2:CH(CH_2)_6CH\overset{t}{:}CH(C\vdots C)_2CO.CH:CH_2$

Ann., 1970 (739) 135.

$MeCH\overset{t}{:}CH(C\vdots C)_2(CH\overset{t}{:}CH)_3CO.Pr$

Ber., 1969 (102) 3293.

3117-061

3117-066

HEPTADECA-3, 9, 11-TRIEN-5, 7- -DIYN-2-ONE	$C_{17}H_{20}O$	240 OI	HEPTADECA-8, 10, 14, 16-TETRAEN- -6-YN-3-ONE	$C_{17}H_{22}O$	242 OI Oil

$Me(CH_2)_5(CH:CH)_2(C\vdots C)_2CH:CHCO.Me$

Ber., 1968, (101) 525.

$H(CH:CH)_2CH_2CH_2(CH\overset{t}{:}CH)_2C\vdots CCH_2CH_2CO.Et$

Ber., 1969 (102) 1037.

3117-062

3117-067

OENANTHETONE	$C_{17}H_{20}O$	240 OI 46^0C	HEPTADECA-1, 4, 8, 10, 16- -PENTAEN-6-YN-3-ONE	$C_{17}H_{20}O$	240 OI Oil

$MeCH:CH(C\vdots C)_2(CH:CH)_2CH_2CH_2CO.Pr$

JCS, 1953, 309.

$CH_2:CH(CH_2)_4(CH\overset{t}{:}CH)_2C\vdots C.CH\overset{t}{:}CHCO.CH:CH_2$

Ann., 1970 (739) 135.

3117-063

3117-068

FALCARINOLONE $\qquad$ $C_{17}H_{22}O_2$ $\qquad$ 258

$$Me(CH_2)_6CH\overset{c}{:}CH.CHOH.(C\vdots C)_2COCH:CH_2$$

Ber., 1971 (104) 1957.

3117-069

HEPTADECA-7,9-DIEN-11,13-
-DIYN-6-ONE, 17-OH- $\qquad$ $C_{17}H_{22}O_2$ $\qquad$ 258
OI
Oil

$$HO(CH_2)_3(C\vdots C)_2(CH\overset{t}{:}CH)_2CO(CH_2)_4Me$$

Ber., 1969 (102) 3293.

3117-074

AVOCADENONE ACETATE $\qquad$ $C_{19}H_{34}O_4$ $\qquad$ 326
$+20^0$
47^0C

$$CH_2:CH(CH_2)_{11}CO.CH_2CHOH.CH_2OAc$$

Tet., 1969, 4623.

3117-070

HEPTADECA-7,9-DIEN-11,13-
-DIYN-4-ONE, 17-OH- $\qquad$ $C_{17}H_{22}O_2$ $\qquad$ 258
OI
36^0C

$$HO(CH_2)_3(C\vdots C)_2(CH:CH)_2CH_2CH_2CO.Pr$$

Ber., 1969 (102) 3293.

3117-075

AVOCADYNONE ACETATE $\qquad$ $C_{19}H_{32}O_4$ $\qquad$ 324
$+26^0$
47^0C

$$HC\vdots C(CH_2)_{11}CO.CH_2CHOH.CH_2OAc$$

Tet., 1969, 4623.

3117-071

HEPTADECA-8,10,12,14-
-TETRAENAL $\qquad$ $C_{17}H_{26}O$ $\qquad$ 246
OI

$$Et(CH\overset{t}{:}CH)_4(CH_2)_6CHO$$

Chem. Abstr.,
1967 (66) 65023.

3117-076

HEPTADEC-9-EN-11,13-DIYN-
-4-ONE, 17-OH- $\qquad$ $C_{17}H_{24}O_2$ $\qquad$ 260
OI
87^0C

$$HOCH_2(CH_2)_2(C\vdots C)_2CH:CH(CH_2)_4CO.Pr$$

Ber., 1969 (102) 3293.

3117-072

HEPTADECA-2,8,10,16-TETRAEN-
-4,6-DIYNAL $\qquad$ $C_{17}H_{18}O$ $\qquad$ 238
OI

$$CH_2:CH(CH_2)_4(CH:CH)_2(C\vdots C)_2CH:CHCHO$$

Chem. Abstr.,
1967 (66) 65023.

3117-077

HEPTADECA-5,7,9-TRIEN-11,13-
-DIYN-4-ONE, 17-OH- $\qquad$ $C_{17}H_{20}O_2$ $\qquad$ 256
OI
79^0C

$$HO(CH_2)_3(C\vdots C)_2(CH\overset{t}{:}CH)_3CO.Pr$$

Ber., 1969 (102) 3293.

3117-073

HEPTADECA-2,4,8,10,16-
-PENTAEN-6-YNAL $\qquad$ $C_{17}H_{20}O$ $\qquad$ 240
OI
Oil

$$CH_2:CH(CH_2)_4(CH\overset{t}{:}CH)_2C\vdots C(CH\overset{t}{:}CH)_2CHO$$

Ann., 1970 (739) 135.

3117-078

HEPTADECA-2,8-DIEN-4,6--DIYNAL, 10-OH-	$C_{17}H_{22}O_2$	258
		Oil

$Me(CH_2)_6CHOH.CH:CH(C:C)_2CH:CH.CHO$

Ber., 1968 (101) 1163.

3117-079

HEPTADECA-10,16-DIEN-8--YNOIC ACID	$C_{17}H_{26}O_2$	262
		OI

$CH_2:CH(CH_2)_4C\overset{t}{H}:CH.C:C(CH_2)_6COOH$

Biochemistry,
1966 (5) 625.

3117-084

MARGARINIC ACID	$C_{17}H_{34}O_2$	270
		OI
		60^0C

$Me(CH_2)_{15}COOH$

Helv., 1942 (25) 649.

3117-080

HEPTADECA-10,16-DIEN-8--YNOIC ACID, 7-OH-	$C_{17}H_{26}O_3$	278
		Oil

$CH_2:CH(CH_2)_4C\overset{t}{H}:CH.C:C.CHOH(CH_2)_5COOH$

JOC, 1966, 528.

3117-085

HEPTADEC-9-ENOIC ACID	$C_{17}H_{32}O_2$	268
		OI
		12^0C

$Me(CH_2)_6CH:CH(CH_2)_7COOH$

Biochem. J.,
1960 (77) 64.

3117-081

HEPTADECA-7,9,15-TRIEN--11,13-DIYNOIC ACID ME ESTER	$C_{18}H_{20}O_2$	270
		OI
		Oil

$MeC\overset{t}{H}:CH(C:C)_2(C\overset{t}{H}:CH)_2(CH_2)_5COOMe$

Ber., 1970 (103) 2327.

3117-086

HEPTADEC-10-EN-8-YNOIC ACID	$C_{17}H_{28}O_2$	264
		OI
		34^0C

$Me(CH_2)_5C\overset{t}{H}:CH.C:C(CH_2)_6COOH$

Chem. Ind., 1967, 998.

3117-082

HEPTADECA-7,9,15-TRIEN-11,13--DIYNOIC ACID ME ESTER, 3-OH-	$C_{18}H_{22}O_3$	286
		42^0C

$MeCH:CH(C:C)_2(CH:CH)_2(CH_2)_3CHOH.CH_2COOMe$

Ber., 1970 (103) 2327.

3117-087

HEPTADEC-10-EN-8-YNOIC ACID, 7-OH-	$C_{17}H_{28}O_3$	280
		Oil

$Me(CH_2)_5CH:CH.C:C.CHOH(CH_2)_5COOH$

JOC, 1966, 528.

3117-083

STEARYL ALCOHOL	$C_{18}H_{38}O$	270
		OI
		59^0C

$Me(CH_2)_{16}CH_2OH$

Ann. Chim., 1952 (45) 502.

3118-001

OLEIC ALCOHOL	$C_{18}H_{36}O$	268 OI Oil
$Me(CH_2)_7\overset{c}{C}H:CH(CH_2)_7CH_2OH$		
	JACS. 1947. 1196.	
3118-002		

OCTADECA-8. 10. 14, 16-TETRAEN--12-YN-1-OL ACETATE, 3-keto-	$C_{20}H_{26}O_3$	314 OI 73^0C
$Me(\overset{t}{CH:CH})_2C:C(CH:CH)_2(CH_2)_1COCH_2CH_2OAc$		
	Ber. , 1966 (99) 142.	
3118-007		

ELAIDIC ALCOHOL	$C_{18}H_{36}O$	268 OI 36^0C
trans isomer		
	Compt. Rend. . 1927 (185) 279.	
3118-003		

STEARIC ALDEHYDE	$C_{18}H_{36}O$	268 OI 38^0C
$Me(CH_2)_{16}CHO$		
	Chem. Abstr. . 1942 (36) 6754.	
3118-008		

OCTADECANOL. d-2-	$C_{18}H_{38}O$	270 $+6^0c$ 56^0C
$Me(CH_2)_{15}CHOH. Me$		
	JACS. 1937. 858.	
3118-004		

OCTADECA-9. 17- DIENE-12. 14--DIYN-1- AL, 16-oxo-	$C_{18}H_{22}O_2$	270 OI
$CH_2:CHCO(C:C)_2CH_2\overset{c}{C}H:CH(CH_2)_7CHO$		
	JCS. 1966. 1220.	
3118-009		

OCTADECANOL. d-3-	$C_{18}H_{38}O$	270 56^0C
$Me(CH_2)_{11}CHOH. Et$		
	Chem. Abstr. . 1955 (49) 1142.	
3118-005		

STEARIC ACID	$C_{18}H_{36}O_2$	284 OI 69^0C
$Me(CH_2)_{16}COOH$		
	Ber. . 1879 (12) 1635.	
3118-010		

OCTADECAN-1. 18-DIOL	$C_{18}H_{38}O_2$	286 OI 99^0C
$HOCH_2(CH_2)_{16}CH_2OH$		
	JCS. 1952. 4393.	
3118-006		

OCTADECANOIC ACID, 3-OH-	$C_{18}H_{36}O_3$	300 -12^0c 84^0C
$CH_3(CH_2)_{14}CHOH. CH_2COOH$		
	Can. J. Chem. . 1964 (42) 830.	
3118-011		

OCTADECANOIC ACID, 18-OH-	$C_{18}H_{36}O_3$	300
		OI
$HOCH_2(CH_2)_{16}COOH$		
	Chem. Abstr., 1955 (49) 16472.	
3118-012		

SATIVIC ACID	$C_{18}H_{36}O_6$	348
		(α) 164^{0}C
$Me(CH_2)_4(CHOH)_2CH_2(CHOH)_2(CH_2)_7COOH$		
	JACS, 1922, 150.	
3118-017		

STEARIC ACID, 9, 10-diOH-c-	$C_{18}H_{36}O_4$	316
		133^{0}C
$Me(CH_2)_7CHOH.CHOH(CH_2)_7COOH$		
	JCS, 1942, 387.	
3118-013		

OCTADEC-5-ENOIC ACID	$C_{18}H_{31}O_2$	286
		OI
		44^{0}C
$Me(CH_2)_{11}\overset{t}{C}H:CH(CH_2)_3COOH$		
	Chem. Ind., 1970, 831.	
3118-018		

STEARIC ACID, 9, 10-diOH-t-	$C_{18}H_{36}O_4$	316
		96^{0}C
trans isomer		
	JCS, 1942, 387.	
3118-014		

PETROSELIC ACID	$C_{18}H_{31}O_2$	282
		OI
		32^{0}C
$Me(CH_2)_{10}\overset{c}{C}H:CH(CH_2)_4COOH$		
	JCS, 1954, 1808.	
3118-019		

ARTEMISIC ACID	$C_{18}H_{36}O_5$	332
		108^{0}C
$BuCHOH(CH_2)_3CHOH.CHOH(CH_2)_3COOH$		
	Chem. Abstr., 1951 (45) 9479.	
3118-015		

PETROSELIDIC ACID	$C_{18}H_{34}O_2$	282
		OI
		53^{0}C
trans isomer		
	JCS, 1954, 1808.	
3118-020		

PHLOIONOLIC ACID	$C_{18}H_{36}O_5$	332
		104^{0}C
$HOCH_2(CH_2)_7CHOH.CHOH(CH_2)_7COOH$		
	Helv., 1931 (14) 849.	
3118-016		

OLEIC ACID	$C_{18}H_{34}O_2$	282
		OI
		7^{0}C
$Me(CH_2)_7\overset{c}{C}H:CH(CH_2)_7COOH$		
	JCS, 1925, 175.	
3118-021		

ELAIDIC ACID	$C_{18}H_{34}O_2$	282 OI 43^0C	OCTADEC-9-ENOIC ACID, 5, 8, 12- -tri-OH- $\quad C_{18}H_{34}O_5 \quad$ 330

ELAIDIC ACID $\qquad C_{18}H_{34}O_2 \qquad$ 282
OI
43^0C

trans isomer

Chem. Abstr.,
1955 (49) 7639.

3118-022

OCTADEC-9-ENOIC ACID, 5, 8, 12- $\quad C_{18}H_{34}O_5 \quad$ 330
-tri-OH-

$$Me.(CH_2)_5.\overset{}{C}H.CH_2.\overset{t}{C}H:CH.\overset{}{C}H.CH_2.CH_2.\overset{}{C}H.(CH_2)_3.COOH$$
$$\quad\quad\quad\; OH \quad\quad\quad\quad OH \quad\quad\; OH$$

Phytochem.,
1971 (10) 631.

3118-027

OLEIC ACID, iso- $\qquad C_{18}H_{34}O_2 \qquad$ 282
OI
44^0C

$Me(CH_2)_6CH:CH(CH_2)_8COOH$

Chem. Abstr.,
1955 (49) 7267.

3118-023

LINOLEIC ACID, c, c- $\qquad C_{18}H_{32}O_2 \qquad$ 280
OI
Oil

$Me(CH_2)_4\overset{c}{C}H:CHCH_2\overset{c}{C}H:CH(CH_2)_7COOH$

Can. J. Chem.,
1965 (43) 2566.

3118-028

VACCENIC ACID, cis- $\qquad C_{18}H_{34}O_2 \qquad$ 282
OI
43^0C

$Me(CH_2)_5\overset{c}{C}H:CH(CH_2)_9COOH$

J. Biol. Chem.,
1955 (213) 425.

3118-024

LINOLEIC ACID, c, t- $\qquad C_{18}H_{32}O_2 \qquad$ 280
OI
Oil

cis, trans isomer

Chem. Ind., 1966, 1493.

3118-029

VACCENIC ACID, trans- $\qquad C_{18}H_{34}O_2 \qquad$ 282
OI
44^0C

trans isomer

Arch. Biochem. Biophys.,
1958 (76) 15.

3118-025

LINELAIDIC ACID $\qquad C_{18}H_{32}O_2 \qquad$ 280
OI
28^0C

trans, trans isomer

Can. J. C., 1963, 1888.

3118-030

OCTADEC-12-ENOIC ACID $\qquad C_{18}H_{34}O_2 \qquad$ 282

27^0C

$Me(CH_2)_4\overset{c}{C}H:CH(CH_2)_{10}COOH$

JCS, 1963, 489

3118-026

OCTADECA-10, 12-DIENOIC ACID $\quad C_{18}H_{32}O_2 \quad$ 280
OI
57^0C

$Me(CH_2)_4\overset{t}{(}CH:CH)_2(CH_2)_8COOH$

Can. J. C., 1965, 2566.

3118-031

OCTADECA-11, 13-DIEN-9- -YNOIC ACID	$C_{18}H_{28}O_2$	276
		45^0C

$$Bu(C\overset{t}{H}:CH)_2C\overset{\cdot}{:}C(CH_2)_7COOH$$

Aust. J. C. , 1960 (13) 488.

3118-032

OCTADECA-8, 10, 12-TRIENOIC ACID	$C_{18}H_{30}O_2$	278
		OI

trans, trans, cis-isomer

Can. J. C. , 1965, 3160.

3118-037

RANUNCULEIC ACID	$C_{18}H_{30}O_2$	278

$$Me(CH_2)_4(C\overset{c}{H}:CH.CH_2)_2CH_2C\overset{t}{H}:CH(CH_2)_3COOH$$

Khim. Prir. Soedin. ,
1970 (6) 167.

3118-033

PUNICIC ACID	$C_{18}H_{30}O_2$	278
		OI
		44^0C

$$Bu(C\overset{c,t,c}{H:CH})_3(CH_2)_7COOH$$

Chem. Ind. , 1966, 1551.

3118-038

LINOLENIC ACID, γ-	$C_{18}H_{30}O_2$	278
		OI

$$Bu(CH_2C\overset{c}{H}:CH)_3(CH_2)_4COOH$$

Chem. Ind. , 1966, 1551.

3118-034

ELAOSTEARIC, α-	$C_{18}H_{30}O_2$	278
		OI
		48^0C

cis, trans, trans-isomer

Can. J. C. , 1964, 560.

3118-039

OCTADECATRI-6, 10, 14-ENOIC ACID	$C_{18}H_{30}O_2$	278
		OI

$$Me(CH_2CH_2CH:CH)_3(CH_2)_4COOH$$

Chem. Abstr. ,
1954 (48) 1710.

3118-035

CATALPIC ACID	$C_{18}H_{30}O_2$	278
		OI
		32^0C

trans, trans, cis-isomer

JCS, 1966, 573.

3118-040

CALENDIC ACID	$C_{18}H_{30}O_2$	278
		OI
		78^0C

$$Me(CH_2)_4(C\overset{c,t,c}{H:CH})_3(CH_2)_6COOH$$

Chem. Ind. , 1966, 1551.

3118-036

LINOLENIC ACID, α-	$C_{18}H_{30}O_2$	278
		OI
		180^0C

$$Et(C\overset{c}{H}:CH.CH_2)_3(CH_2)_6COOH$$

Chem. Ind. ,
1966 (37) 1551.

3118-041

OCTADECATETRA-6, 9, 12, 15- -ENOIC ACID	$C_{18}H_{28}O_2$	276 OI

$$Et(\overset{c}{CH:CH}.CH_2)_4(CH_2)_3COOH$$

Biochem. Z.,
1963 (339) 212.

3118-042

TARIRIC ACID	$C_{18}H_{32}O_2$	280 OI 50^0C

$$Me(CH_2)_{10}C\!:\!C(CH_2)_4COOH$$

JCS, 1952, 5032.

3118-047

PARINARIC ACID	$C_{18}H_{28}O_2$	276 OI Oil

$$Et\overset{cttc}{(CH:CH)_4}(CH_2)_7COOH$$

Chem. Ind.,
1966 (37) 1551.

3118-043

CREPENYNIC	$C_{18}H_{30}O_2$	278 OI

$$Me(CH_2)_4C\!:\!C.CH_2\overset{c}{CH:CH}(CH_2)_7COOH$$

Chem. Comm.,
1967, 1055.

3118-048

SANTALBIC ACID	$C_{18}H_{30}^{i}O_2$	278 OI 40^0C

$$Me(CH_2)_5CH:CH.C\!:\!C(CH_2)_7COOH$$

JCS, 1955, 3782.

3118-049

OCTADECA-3, 9, 12, 15-TETRA- -ENOIC ACID	$C_{18}H_{28}O_2$	276 OI

$$Et(\overset{c}{CH:CH}.CH_2)_3(CH_2)_3\overset{t}{CH:CH}CH_2COOH$$

JCS, 1965, 907.

3118-045

ISANIC ACID	$C_{18}H_{26}O_2$	274 OI 42^0C

$$CH_2:CH(CH_2)_4(C\!:\!C)_2(CH_2)_7COOH$$

JCS, 1953, 1785.

3118-050

LABALLENIC ACID	$C_{18}H_{32}O_2$	280 (-)

$$\begin{array}{c}Me(CH_2)_{10}\\ \qquad\qquad C\!:\!C\!:\!CH(CH_2)_3COOH\\ H\end{array}$$

Tet. Lett., 1966, 4905.

3118-046

EXOCARPIC ACID	$C_{18}H_{26}O_2$	274 OI 42^0C

$$Bu\overset{t}{CH:CH}(C\!:\!C)_2(CH_2)_7COOH$$

Chem. Ind., 1966, 1533.

3118-051

OCTADEC-11-EN-9-YNOIC ACID, $C_{18}H_{30}O_3$ 294
   -8-OH-

$Me(CH_2)_5CH:CH.C\overset{t}{:}C.CHOH.(CH_2)_6COOH$

Chem. Ind., 1954, 249.

3118-052

DIMORPHECOLIC ACID, α- $C_{18}H_{32}O_3$ 296

10-trans, 12-cis isomer

Chem. Ind., 1966 (37) 1551.

3118-057

OCTADECA-11, 17-DIEN-9-YNOIC $C_{18}H_{28}O_3$ 292
    ACID, 8-OH-

Oil

$CH_2:CH(CH_2)_4\overset{t}{CH}:CH.C:C.CHOH(CH_2)_6COOH$

JOC, 1966, 528.

3118-053

HELENYNOLIC ACID $C_{18}H_{30}O_3$ 294

$Me(CH_2)_4C:C.\overset{t}{CH}:CH.CHOH(CH_2)_7COOH$

JOC, 1965, 610.

3118-058

ISANOLIC ACID $C_{18}H_{26}O_3$ 290

$PrCH:CH(C:C)_2CH_2CHOH(CH_2)_6COOH$

Chem. Abstr., 1942, 6852.

3118-054

RICINOLEIC ACID $C_{18}H_{34}O_3$ 298
                +4^0
                 5^{0}c

$Me(CH_2)_5CHOH.CH_2CH\overset{c}{:}CH(CH_2)_7COOH$

JOC, 1966, 1477.

3118-059

OCTADEC-12-ENOIC ACID, 9-OH- $C_{18}H_{64}O_3$ 298

$Me(CH_2)_4CH:CH(CH_2)_2CHOH(CH_2)_7COOH$

JCS, 1952, 1274.

3118-055

DENSIPOLIC ACID $C_{18}H_{32}O_3$ 296

Oil

$EtCH\overset{c}{:}CH(CH_2)_2CHOH.CH_2CH\overset{c}{:}CH(CH_2)_7COOH$

JOC, 1962, 3112.

3118-060

DIMORPHECOLIC ACID $C_{18}H_{32}O_3$ 296
                    m. e. +5^{0}c

$Me(CH_2)_4(\overset{t}{CH:CH})_2CHOH(CH_2)_7COOH$

Chem. Comm., 1965, 255.

3118-056

---

deleted

3118-061

CORIOLIC ACID	$C_{18}H_{32}O_3$	296	OCTADECA-9, 16-DIENOIC ACID, -11, OMe-12, 13-diOH-	$C_{19}H_{34}O_5$	326

Me. $(CH_2)_4$.CHOH.$\overset{t}{CH:CH}$.$\overset{c}{CH:CH}$.$(CH_2)_7$.COOH

Tet. Lett., 1966, 4329.

MeCH:CH$(CH_2)_2$(CHOH)$_2$CHOMe.CH:CH$(CH_2)_7$COOH

Helv., 1947, 1187.

3118-062

3118-067

ARTEMESIC ACID, β-	$C_{18}H_{32}O_3$	296	OCTADECANOIC ACID, 4-Oxo-	$C_{18}H_{34}O_3$	298 OI 97°C

trans, trans isomer

Chem. Ind., 1966 (37) 1551.

Me$(CH_2)_{13}$COCH$_2$CH$_2$COOH

Chem. Abstr., 1947 (41) 1472.

3118-063

3118-068

OCTADEC-9-ENOIC ACID, 18-OH-	$C_{18}H_{34}O_3$	298 OI	LACTARINIC ACID	$C_{18}H_{34}O_3$	298 OI 87°C

HOCH$_2$$(CH_2)_7$$\overset{t}{CH:CH}$$(CH_2)_7$COOH

Chem. Abstr., 1967 (66) 2159.

Me$(CH_2)_{12}$CO$(CH_2)_3$COOR
R = H

J. Biol. Chem., 1942 (142) 345.

3118-064

3118-069

KAMLOLENIC ACID, α-	$C_{18}H_{30}O_3$	278 OI 78°C	ACTINOMYCIN-J2	$C_{30}H_{58}O_3$	466 OI 81°C

HO$(CH_2)_4$(CH:CH)$_3$$(CH_2)_7$COOH

JCS, 1954, 2816.

R = .$(CH_2)_{11}$Me

Bull. Chem. Soc. Jap., 1949 (22) 121.

3118-065

3118-070

OCTADECA-10, 12-DIENOIC ACID, -9, 14-di-OH-	$C_{18}H_{32}O_4$	296 104°C	OCTADEC-9-ENOIC ACID, 13-Oxo-	$C_{18}H_{32}O_3$	296 OI

BuCHOH. (CH:CH)$_2$CHOH$(CH_2)_7$COOH

JACS, 1950, 124.

Me$(CH_2)_4$CO$(CH_2)_2$$\overset{t}{CH:CH}$$(CH_2)_7$COOH

Biochem. Biophys. Acta, 1970 (210) 353.

3118-066

3118-071

OCTADECA-9, 11-DIENOIC ACID, -13-Oxo- $\quad$ $C_{18}H_{30}O_3$ $\quad$ 294 $\quad$ OI $Me(CH_2)_4CO(C\overset{t}{H:}CH)_2(CH_2)_7COOH$ Biochem. Biophys. Acta, 1970 (210) 353. 3118-072	VERNOLIC ACID $\quad$ $C_{18}H_{32}O_3$ $\quad$ 296 $\quad$ -8^0 $\quad$ 22^0C $Me(CH_2)_4CH\overset{O}{-}CHCH_2C\overset{c}{H:}CH(CH_2)_7COOH$ JCS, 1954, 1611. 3118-077
LICANOIC ACID, α- $\quad$ $C_{18}H_{28}O_3$ $\quad$ 292 $\quad$ OI $\quad$ 75^0C $Bu(C\overset{c}{H:}CH)_3(CH_2)_4CO(CH_2)_2COOH$ Chem. Abstr., 1944 (38) 502. 3118-073	OCTADECA-3, 12-DIENOIC ACID, -9, 10-epoxy- $\quad$ $C_{18}H_{30}O_3$ $\quad$ 294 $Me(CH_2)_4C\overset{c}{H:}CHCH_2C\overset{O}{H}-CH(CH_2)_4C\overset{t}{H:}CHCH_2COOH$ J. Amer. Oil Chem. Soc., 1970 (47) 510. 3118-078
LICANOIC ACID, iso- $\quad$ $C_{18}H_{28}O_3$ $\quad$ 292 $\quad$ OI $\quad$ 97^0C cis, trans, cis isomer Biochem. J., 1935, 631. 3118-074	LINOLEIC ACID, 15, 16-epoxy- $\quad$ $C_{18}H_{30}O_3$ $\quad$ 294 $EtC\overset{O}{H}-CHCH_2(C\overset{c}{H:}CHCH_2)_2(CH_2)_6COOH$ Chem. Ind., 1966 (37) 1551. 3118-079
STEARIC ACID, 9, 10-epoxy- $\quad$ $C_{18}H_{34}O_3$ $\quad$ 298 $\quad$ 59^0C $Me(CH_2)_7C\underset{cis}{H-}CH(CH_2)_7COOH$ Chem. Ind., 1966 (37) 1551. 3118-075	OCTADEC-12-YNOIC ACID, -9, 10-epoxy- $\quad$ $C_{18}H_{30}O_3$ $\quad$ 294 $Me(CH_2)_4C:CCH_2C\overset{O}{H}-CH(CH_2)_7COOH$ Chem. Abstr., 1971 (74) 84000. 3118-080
CORONARIC ACID $\quad$ $C_{18}H_{32}O_3$ $\quad$ 296 $Me(CH_2)_4C\overset{c}{H:}CHCH_2C\overset{O}{H}-CH(CH_2)_7COOH$ Chem. Ind., 1966 (37) 1551. 3118-076	OCTADEC-9-ENOIC ACID, -18-Fluoro- $\quad$ $C_{18}H_{33}O_2F$ $\quad$ 300 $\quad$ OI $\quad$ 13^0C $FCH_2(CH_2)_8CH:CH(CH_2)_7COOH$ JACS, 1963, 622. 3118-081

OCTADEC-9-EN-DIOIC ACID	$C_{18}H_{32}O_4$	312 OI	

$$HOOC(CH_2)_7CH:CH(CH_2)_7COOH$$

Chem. Abstr.,
1956 (50) 806.

3118-082

NONADECANDIOIC ACID	$C_{19}H_{36}O_4$	328 OI

$$HOOC(CH_2)COOH$$

Helv., 1928 (11) 670.

3119-002

PHLOIONIC ACID	$C_{18}H_{34}O_6$	346 122^0C

$$HOOC(CH_2)_7(CHOH)_2(CH_2)_7COOH$$

Compt. Rend.,
1955 (240) 875.

3118-083

EICOSANE	$C_{20}H_{42}$	282 OI 37^0C

$$Me(CH_2)_{18}Me$$

Helv., 1950 (33) 249.

3120-001

SCABRIN	$C_{22}H_{35}NO$	329 OI Oil

$$PrCH:CH(CH_2)_2(CH:CH)_2(CH_2)_2$$
$$Me_2CHCH_2NHCO(CH:CH)_2 \Big)$$

JACS. 1951, 100.

3118-084

ARACHIDYL ALCOHOL	$C_{20}H_{42}O$	298 OI 66^0C

$$Me(CH_2)_{18}CH_2OH$$

Chem. Abstr..
1954 (48) 3252.

3120-002

OCTADECANOYL SOPHOROSIDE LACTONE, 17-OH-	$C_{34}H_{58}O_{16}$	690 104^0C

$$MeCH(CH_2)_{15}CO$$

diAc-sophorosyl

Chem. Comm.,
1967. 584.

3118-085

EICOSAN-2-OL	$C_{20}H_{12}$	298 $+4^0c$ 63^0C

$$Me(CH_2)_{17}CHOH.Me$$

Chem. Abstr.,
1954 (48) 5071.

3120-003

NONADECANE	$C_{19}H_{40}$	268 OI 33^0C

$$Me(CH_2)_{17}Me$$

Phytochem.,
1966 (5) 1029.

3119-001

EICOS-11-EN-1-OL	$C_{20}H_{40}O$	296 OI

$$Me(CH_2)_7CH:CH(CH_2)_9CH_2OH$$

JCS. 1936, 1750.

3120-004

ARACHIC ACID	$C_{20}H_{40}O_2$	312 OI 75^{0}C	EICOS-11-ENOIC ACID	$C_{20}H_{38}O_2$	310 OI 22^{0}C
$CH_3(CH_2)_{18}COOH$		Ann., 1854 (89) 1.	$Me(CH_2)_7\overset{c}{C}H{:}CH(CH_2)_9COOH$		Chem. Abstr., 1955 (49) 9939.
3120-005			3120-010		
MACILOLIC ACID	$C_{20}H_{40}O_3$	328 68^{0}C	EICOS-11-ENOIC ACID	$C_{20}H_{38}O_2$	310 OI 50^{0}C
$Me(CH_2)_{17}CHOH.COOH$		Arch. Pharm., 1915 (253) 102.	trans isomer		JCS. 1936. 1750.
3120-006			3120-011		
EICOSANOIC ACID, 20-OH-	$C_{20}H_{40}O_3$	328 OI	EICOSA-13-ENOIC ACID	$C_{20}H_{38}O_2$	310 OI
$HO(CH_2)_{19}COOH$		Chem. Abstr., 1955 (49) 16472.	$Me(CH_2)_5\overset{c}{C}H{:}CH(CH_2)_{11}COOH$		Lipids. 1970 (5) 430.
3120-007			3120-012		
GADOLEIC ACID	$C_{20}H_{38}O_2$	310 OI 23^{0}C	EICOSA-11. 14-DIENOIC ACID	$C_{20}H_{36}O_2$	308 OI
$Me(CH_2)_9\overset{c}{C}H{:}CH(CH_2)_7COOH$		JCS. 1952. 671.	$Me(CH_2)_4(CH{:}CH.CH_2)_2(CH_2)_8COOH$		Lipids. 1970 (5) 430.
3120-008			3120-013		
GADELAIDIC ACID	$C_{20}H_{30}O_2$	310 OI 54^{0}C	EICOSA-12, 14-DIENOIC ACID, -11-OH-	$C_{20}H_{36}O_3$	324 -10^{0}m 48^{0}C
trans isomer		JCS, 1952. 671.	$Me(CH_2)_4(\overset{c,t}{CH{:}CH})_2CHOH(CH_2)_9COOH$		Nature, 1966 (212) 38.
3120-009			3120-014		

EICOSA-5, 11, 14-TRIENOIC ACID	$C_{20}H_{34}O_2$	306 OI	HENEICOSANE	$C_{21}H_{44}$	296 OI 40^0C

$$Me(CH_2)_4(\overset{c}{CH:CH.CH_2})(CH_2)_3\overset{c}{CH:CH}(CH_2)_3COOH$$

Chem. Abstr.,
1968 (68) 113998.

$$Me(CH_2)_{19}Me$$

Phytochem.,
1966 (5) 1029.

3120-015

3121-001

ARACHIDONIC ACID	$C_{20}H_{32}O_2$	304 OI	HENEICOSA-1, 6, 9. 12. 15- -PENTAENE	$C_{21}H_{34}$	286 OI Oil

$$Me(CH_2)_4(\overset{c}{CH:CH.CH_2})_4(CH_2)_4COOH$$

Biochem. J., 1943 (37) 1.

$$Me(CH_2)_3(CH_2CH:CH)_4(CH_2)_3CH:CH_2$$

Chem. Comm.,
1971. 448.

3120-016

3121-002

EICOSA-11, 14, 17-TRIENOIC ACID	$C_{20}H_{34}O_2$	306 OI	HENEICOSA-1, 6, 9. 12, 15, 18- - HEXAENE	$C_{21}H_{32}$	284 OI Oil

$$Pr(\overset{c}{CH:CH.CH_2})_3(CH_2)_8COOH$$

Lipids, 1970 (5) 430.

$$Et(CH:CH.CH_2)_5(CH_2)_2CH:CH_2$$

Chem. Comm.,
1971. 448.

3120-017

3121-003

EICOSA-5, 11, 14, 17-TETRAENOIC ACID	$C_{20}H_{32}O_2$	304 OI	HENEICOSA-3, 6. 9, 12, 15. 18- - HEXAENE	$C_{21}H_{32}$	284 OI Oil

$$Et(CH:CH.CH_2)_3(CH_2)_3CH:CH(CH_2)_3COOH$$

Chem. Ind., 1967, 1326.

$$Et(CH:CH.CH_2)_6Et$$

Marine Biol.,
1971 (8) 183.

3120-018

3121-004

EICOSANDIOIC ACID	$C_{20}H_{38}O_4$	342 OI 126^0C	HENEICOSANDIOIC ACID	$C_{21}H_{40}O_4$	356 OI 115^0C

$$HOOC(CH_2)_{18}COOH$$

Chem. Abstr.,
1948 (42) 2117.

$$HOOC(CH_2)_{19}COOH$$

Helv., 1928 (11) 670.

3120-019

3121-005

DOCOSANE	$C_{22}H_{46}$	310 OI 45^0C	DOCOSANOIC ACID, 2,4-di-Me-	$C_{24}H_{48}O_2$	368
$Me(CH_2)_{20}Me$			$Me(CH_2)_{17}CHMe.CH_2CHMe.COOH$		
	Ber., 1882 (15) 1718.			Tet., 1964, p. 1955.	
3122-001			3122-006		
DOCOSANOL	$C_{22}H_{46}O$	326 OI 71^0C	CETOLIC ACID	$C_{22}H_{42}O_2$	338 OI 33^0C
$Me(CH_2)_{20}CH_2OH$			$Me(CH_2)_9CH:CH(CH_2)_9COOH$		
	Helv., 1942, 649.			Chem. Abstr., 1928 (22) 575.	
3122-002			3122-007		
DOCOS-13-EN-1-OL	$C_{22}H_{44}O$	324 OI	ERUCAIC ACID	$C_{22}H_{42}O_2$	338 OI 34^0C
$Me(CH_2)_7CH:CH(CH_2)_{11}CH_2OH$			$Me(CH_2)_7CH:\overset{c}{CH}(CH_2)_{11}COOH$		
	JCS. 1936. 1750.			Lipids. 1970 (5) 430.	
3122-003			3122-008		
DOCOSAN-1,22-DIOL	$C_{22}H_{46}O_2$	342 OI 106^0C	DOCOSA-5,13-DIENOIC ACID	$C_{22}H_{40}O_2$	336 OI
$HOCH_2(CH_2)_{20}CH_2OH$			$Me(CH_2)_7CH:\overset{c}{CH}(CH_2)_6\overset{c}{CH}:CH(CH_2)_3COOH$		
	Chem. Abstr., 1955 (49) 16471.			Lipids. 1966 (1) 73.	
3122-004			3122-009		
BEHENIC ACID	$C_{22}H_{44}O_2$	340 OI 80^0C	DOCOSA-13,16-DIENOIC ACID	$C_{22}H_{40}O_2$	336 OI
$Me(CH_2)_{20}COOH$			$Me(CH_2)_4(\overset{c}{CH}:CH.CH_2)_2(CH_2)_{10}COOH$		
	Ann., 1847 (64) 347.			Lipids, 1970 (5) 430.	
3122-005			3122-010		

DOCOSA-13, 16, 17-TRIENOIC ACID	$C_{22}H_{38}O_2$	334 OI	CLUPANODONIC ACID	$C_{22}H_{34}O_2$	330 OI Oil	
$Et(\overset{c}{CH}:CH.CH_2)_3(CH_2)_{10}COOH$			$PrC\overset{.}{:}C(CH_2)_5CH:CH(CH_2)_2(CH:CH.CH_2)_2CH_2COOH$			
	Lipids, 1970 (5) 430.			Bull. Chem. Soc. Jap., 1928 (3) 299.		
3122-011			3122-016			
ADRENIC ACID	$C_{22}H_{36}O_2$	332 OI	PHELLONIC ACID	$C_{22}H_{44}O_3$	356 OI 94⁰C	
$Bu(CH_2CH:CH)_4(CH_2)_5COOH$			$HOCH_2(CH_2)_{20}COOH$			
	Biochemistry, 1963 (2) 592.			Aust. J. C., 1955, 437.		
3122-012			3122-017			
DOCASA-4, 7, 10, 13, 16, 19-HEXAENOIC ACID	$C_{22}H_{32}O_2$	328 OI	PHELLOGENIC ACID	$C_{22}H_{42}O_4$	370 OI 125⁰C	
$Et(\overset{c}{CH}:CH.CH_2)_6CH_2COOH$			$HOOC(CH_2)_{20}COOH$			
	Biochem. J., 1963 (87) 263.			JACS. 1941. 617.		
3122-013			3122-018			
DOCOSA-4, 8, 12, 15, 18, 21-HEXAENOIC ACID	$C_{22}H_{32}O_2$	328 OI	TRICOSANE	$C_{23}H_{48}$	324 OI 47⁰C	
$H(CH:CH.CH_2)_3(CH_2CH:CH.CH_2)_2CH_2COOH$			$Me(CH_2)_{21}Me$			
	JCS. 1938, 427.			Phytochem., 1966 (5) 1029.		
3122-014			3123-001			
BEHENOLIC ACID	$C_{22}H_{40}O_2$	336 OI 58⁰C	TRICOSAN-12-OL	$C_{23}H_{48}O$	340 OI 75⁰C	
$Me(CH_2)_7C\overset{.}{:}C(CH_2)_{11}COOH$			$Me(CH_2)_{10}CHOH(CH_2)_{10}Me$			
	JCS, 1928, 2678.			Chem. Abstr., 1954, 3252.		
3122-015			3123-002			

TRICOSANDIOIC ACID	$C_{23}H_{44}O_4$	384 OI 128^{0}C
$HOOC(CH_2)_{21}COOH$		
Chem. Zentr., 1932 (I) 1544.		
3123-003		

CEREBRONIC ACID	$C_{24}H_{48}O_3$	384 +3^{0}p 100^{0}C
$Me(CH_2)_{21}CHOH.COOH$		
JCS, 1936, 283.		
3124-005		

TETRACOSANE	$C_{24}H_{50}$	338 OI 51^{0}C
$Me(CH_2)_{22}Me$		
J. Biol. Chem., 1953 (200) 319.		
3124-001		

TETRACOSANOIC ACID, 2,3-diOH-	$C_{24}H_{48}O_4$	400
$Me(CH_2)_{20}(CHOH)_2COOH$		
Chem. Abstr., 1952 (46) 6192.		
3124-006		

TETRACOSANOL	$C_{24}H_{50}O$	354 OI 77^{0}C
$Me(CH_2)_{22}CH_2OH$		
Chem. Abstr., 1950, (44) 8681.		
3124-002		

NERVONIC ACID	$C_{24}H_{46}O_2$	366 OI 43^{0}C
$Me(CH_2)_7CH:CH(CH_2)_{13}COOH$		
JCS, 1954, 448.		
3124-007		

TETRACOSANDIOL, 1,24-	$C_{24}H_{50}O_2$	370 OI 108^{0}C
$HOCH_2(CH_2)_{22}CH_2OH$		
Chem. Abstr., 1955 (44) 16471.		
3124-003		

TETRACOS-17-ENOIC ACID	$C_{24}H_{46}O_2$	366 OI
$Me(CH_2)_5CH:CH(CH_2)_{15}COOH$		
Helv., 1948, 377.		
3124-008		

LIGNOCERIC ACID	$C_{24}H_{48}O_2$	368 OI 82^{0}C
$Me(CH_2)_{22}COOH$		
Helv., 1948 (31) 977.		
3124-004		

TETRACOSANOIC ACID, 24-OH-	$C_{24}H_{48}O_3$	384 OI
$HOCH_2(CH_2)_{22}COOH$		
Chem. Abstr., 1955 (49) 16372.		
3124-009		

PENTACOSANE	$C_{25}H_{52}$	352 OI 54^0C		HEXACOSANDIOL, 1-Caffeoyl-26- -feruloyl-	$C_{45}H_{68}O_8$	736 OI 86^0C

PENTACOSANE $C_{25}H_{52}$ 352 OI 54°C

$Me(CH_2)_{23}Me$

JCS, 1952, 4580.

3125-001

HEXACOSANDIOL, 1-Caffeoyl-26--feruloyl- $C_{45}H_{68}O_8$ 736 OI 86°C

R = Caffeoyl
R' = Feruloyl

Chem. Ind., 1965, 1763.

3126-004

PENTACOSANOIC ACID $C_{25}H_{50}O_2$ 382 OI 84°C

$Me(CH_2)_{23}COOH$

Bull. Soc. Chim. Biol.,
1953 (35) 661.

3125-002

HEXACOSANDIOL, 1,26-di-caffeoyl- $C_{44}H_{66}O_8$ 722 OI 140°C

R = R' = Caffeoyl

Chem. Ind., 1965, 1763.

3126-005

HEXACOSANE $C_{26}H_{54}$ 366 OI 56°C

$Me(CH_2)_{24}Me$

Chem. Abstr.,
1952 (46) 4506.

3126-001

HEXACOSANDIOL, 1,26-di-feruloyl- $C_{46}H_{70}O_8$ 750 OI 92°C

R = R' = Feruloyl

Chem. Ind., 1965, 1763.

3126-006

HEXACOSANOL $C_{26}H_{54}O$ 382 OI 80°C

$Me(CH_2)_{24}CH_2OH$

Chem. Abstr.,
1956 (50) 11228.

3126-002

HEXACOSANOIC ACID $C_{26}H_{52}O_2$ 396 OI 88°C

$Me(CH_2)_{24}COOH$

Compt. Rend.,
1953 (237) 1804.

3126-007

HEXACOSAN-1, 26-DIOL $C_{26}H_{54}O_2$ 398 OI 110°C

$ROCH_2(CH_2)_{24}CH_2OR'$

R = R' = H

Chem. Ind., 1965, 1763.

3126-003

XIMENIC ACID $C_{26}H_{50}O_2$ 394 OI

$Me(CH_2)_7CH:CH(CH_2)_{15}COOH$

Chem. Abstr.,
1940 (34) 3521.

3126-008

CEROTIC ACID, 2-OH-	$C_{26}H_{52}O_3$	412 +2°p 104°C
$Me(CH_2)_{23}CHOH.COOH$		
	Biochem. J., 1953, 711.	
3126-009		

MYRISTONE	$C_{27}H_{54}O$	394 OI 77°C
$Me(CH_2)_{12}CO(CH_2)_{12}Me$		
	Biochem. J., 1933 (27) 1885.	
3127-004		

HEXACOSANOIC ACID, 26-OH-	$C_{26}H_{52}O_3$	412 OI
$HOCH_2(CH_2)_{24}COOH$		
	Chem. Abstr., 1955 (49) 16472.	
3126-010		

HEPTACOSANOIC ACID	$C_{27}H_{54}O_2$	410 OI 82°C
$CH_3(CH_2)_{24}CH_2CO_2H$		
	Compt. Rend., 1920 (170) 1328.	
3127-005		

HEPTACOSANE	$C_{27}H_{56}$	380 OI 59°C
$Me(CH_2)_{25}Me$		
	Phytochem., 1966 (5) 1029.	
3127-001		

OCTACOSANE	$C_{28}H_{58}$	394 OI 61°C
$Me(CH_2)_{26}Me$		
	Ann., 1941 (548) 270.	
3128-001		

HEPTACOSAN-1-OL	$C_{27}H_{56}O$	396 OI 76°C
$Me(CH_2)_{25}CH_2OH$		
	JCS, 1965, 3488.	
3127-002		

OCTACOSAN-1-OL	$C_{28}H_{58}$	410 OI 83°C
$Me(CH_2)_{26}CH_2OH$		
	Biochem. J., 1937 (31) 1981.	
3128-002		

HEPTACOSAN-14-OL	$C_{27}H_{56}O$	396 OI 81°C
$Me(CH_2)_{12}CHOH(CH_2)_{12}Me$		
	J. Am. Pharm. Ass., 1935 (24) 113.	
3127-003		

OCTACOSAN-1, 28-DIOL	$C_{28}H_{58}O_2$	426 OI 112°C
$HOCH_2(CH_2)_{26}CH_2OH$		
	Aust. J. C., 1955 (8) 432.	
3128-003		

OCTACOSANOIC ACID	$C_{28}H_{56}O_2$	424 OI 90^{0}C
$Me(CH_2)_{26}COOH$		JCS, 1929, 2444.
3128-004		

NONACOSAN-15-OL	$C_{29}H_{60}O$	424 OI 83^{0}C
$Me(CH_2)_{13}CHOH(CH_2)_{13}Me$		Compt. Rend., 1950 (230) 995.
3129-004		

OCTACOSANOIC ACID, 28-OH-	$C_{28}H_{56}O_3$	440
$HOCH_2(CH_2)_{26}COOH$		Chem. Abstr., 1955 (49) 16472.
3128-005		

GINNONE	$C_{29}H_{58}O$	422 OI 75^{0}C
$Me(CH_2)_{18}CO(CH_2)_8Me$		Biochem. J., 1931 (25) 2095.
3129-005		

NONACOSANE	$C_{29}H_{60}$	408 OI 64^{0}C
$Me(CH_2)_{27}Me$		Phytochem., 1966 (5) 1029.
3129-001		

DIMYRISTYL KETONE	$C_{29}H_{58}O$	422 OI 81^{0}C
$Me(CH_2)_{13}CO(CH_2)_{13}Me$		Biochem. J., 1931 (25) 606.
3129-006		

MONTANYL ALCOHOL	$C_{29}H_{60}O$	424 OI 84^{0}C
$Me(CH_2)_{27}CH_2OH$		JACS, 1932, 2935.
3129-002		

NONACOSAN-14-ONE, 16-OH-	$C_{29}H_{58}O_2$	438
$Me(CH_2)_{12}CHOH.CH_2CO(CH_2)_{12}Me$		J. Lipid Res., 1971 (12) 198.
3129-007		

GINNOL	$C_{29}H_{60}O$	424 83^{0}C
$Me(CH_2)_8CHOH(CH_2)_{18}Me$		Can. J. C., 1970, 67.
3129-003		

NONACOSAN-15-ONE, 17-OH-	$C_{29}H_{58}O_2$	438
$Me(CH_2)_{11}CHOH.CH_2CO(CH_2)_{13}Me$		J. Lipid Res., 1971 (12) 198.
3129-008		

TRIACONTANE.	$C_{30}H_{62}$	422 OI 66^0C
$Me(CH_2)_{28}Me$		
	Helv., 1951 (34) 382.	
3130-001		

TRIACONTANOIC ACID, 30-OH-	$C_{30}H_{60}O_3$	468 OI
$HOCH_2(CH_2)_{28}COOH$		
	Chem. Abstr., 1955 (49) 16472.	
3130-006		

TRIACONTANOL	$C_{30}H_{62}O$	438 OI 86^0C
$Me(CH_2)_{28}CH_2OH$		
	Chem. Abstr., 1956 (50) 13748.	
3130-002		

HENTRIACONTANE	$C_{31}H_{64}$	436 OI 68^0C
$Me(CH_2)_{29}Me$		
	Curr. Sci., 1934 (3) 250.	
3131-001		

MELISSIC ACID	$C_{30}H_{60}O_2$	452 OI 94^0C
$Me(CH_2)_{28}COOR$ $R = H$		
	J. Biol. Chem., 1936 (113) 487.	
3130-003		

HENTRIACONTAN-1-OL	$C_{31}H_{64}O$	452 OI 89^0C
$Me(CH_2)_{29}CH_2OH$		
	Chem. Abstr., 1971 (74) 20391.	
3131-002		

TRIACONTANOIC ACID -TETRACOSANOL ESTER	$C_{54}H_{108}O_2$	788 OI
$R = .CH_2(CH_2)_{22}Me$		
	Chem. Abstr., 1970 (73) 32275.	
3130-004		

HENTRIACONTAN-16-OL	$C_{31}H_{64}O$	452 OI 85^0C
$Me(CH_2)_{14}CHOH.(CH_2)_{14}Me$		
	Lipids, 1970 (5) 398.	
3131-003		

LUMEQUEIC ACID	$C_{30}H_{58}O_2$	450 OI
$Me(CH_2)_7CH:CH(CH_2)_{19}COOH$		
3130-005		

PALMITONE	$C_{31}H_{62}O$	450 OI 82^0C
$Me(CH_2)_{14}CO(CH_2)_{14}Me$		
	Bull. Soc. Chim. Biol., 1954 (36) 759.	
3131-004		

PALMITENONE	$C_{31}H_{60}O$	448 OI 40^{0}C

Me(CH$_2$)$_{14}$CO(CH$_2$)$_7$CH:CH(CH$_2$)$_5$Me

Bull. Soc. Chim. Biol.,
1954 (36) 759.

3131-005

DOTRIACONTANE	$C_{32}H_{66}$	450 OI 70^{0}C

Me(CH$_2$)$_{30}$Me

Biochem. J.,
1934 (28) 2189.

3132-001

PALMITONE, 10-OH-	$C_{31}H_{62}O_2$	466 97^{0}C

Me(CH$_2$)$_{14}$CO(CH$_2$)$_5$CHOH(CH$_2$)$_8$Me

Biochem. J.,
1937 (31) 1981.

3131-006

DOTRIACONTANOL	$C_{32}H_{66}O$	466 OI 89^{0}C

Me(CH$_2$)$_{30}$CH$_2$OH

Compt. Rend.,
1950 (230) 995.

3132-002

HENTRIACONTANE-14, 16-DIONE	$C_{31}H_{60}O_2$	464 OI

Me(CH$_2$)$_{14}$COCH$_2$CO(CH$_2$)$_{12}$Me

Phytochem.,
1971 (10) 487.

3131-007

DOTRIACONTANOIC ACID	$C_{32}H_{64}O_2$	480 OI 96^{0}C

Me(CH$_2$)$_{30}$COOH

Biochem. J.,
1934 (28) 2189.

3132-003

HENTRIACONTANE-14, 16-DIONE, -9-OH-	$C_{31}H_{60}O_3$	480 70^{0}C

Me(CH$_2$)$_{14}$COCH$_2$CO(CH$_2$)$_4$CHOH(CH$_2$)$_7$Me

Chem. Comm.,
1966, 226.

3131-008

TRITRIACONTANE	$C_{33}H_{68}$	464 OI 72^{0}C

Me(CH$_2$)$_{31}$Me

JACS, 1945, 881.

3133-001

HENTRIACONTANE-14, 16-DIONE, -8-OH-	$C_{31}H_{60}O_3$	480 70^{0}C

Me(CH$_2$)$_{14}$COCH$_2$CO(CH$_2$)$_5$CHOH(CH$_2$)$_6$Me

Chem. Comm.,
1966, 226.

3131-009

TRITRIACONTAN-16, 18-DIONE	$C_{33}H_{64}O_2$	492 OI 68^{0}C

Me(CH$_2$)$_{14}$COCH$_2$CO(CH$_2$)$_{14}$Me

Chem. Ind., 1962, 2032.

3133-002

TETRATRIACONTANOIC ACID	$C_{34}H_{68}O_2$	508 OI 98^0C
$Me(CH_2)_{32}COOH$		JCS, 1937, 999.
3134-001		

ISOBUTYL ACETATE	$C_6H_{12}O_2$	116 OI Oil
R = Ac		Biochem. J., 1950 (47) 55.
3194-002		

TETRATRIACONTANOL	$C_{34}H_{70}O$	494 OI 92^0C
$Me(CH_2)_{32}CH_2OH$		Biochem. J., 1934 (28) 2189.
3134-002		

ISOBUTANAL	C_4H_8O	72 OI Oil
Me_2CHCHO		Chem. Abstr., 1956 (50)15028.
3194-003		

PENTATRIACONTANE	$C_{35}H_{72}$	492 OI 74^0C
$Me(CH_2)_{33}Me$		JACS, 1905, 1467.
3135-001		

METHACROLEIN	C_4H_6O	70 OI Oil
$MeC(:CH_2)CHO$		JOC, 1941, 612.
3194-004		

HEPTATRIACONTANE	$C_{37}H_{76}$	520 OI
$Me(CH_2)_{35}Me$		Biochem. J., 1934 (28) 2189.
3137-001		

BUTYRIC ACID, iso-	$C_4H_8O_2$	88 OI Oil
$Me_2CHCOOH$		Chem. Abstr., 1956 (50) 1684.
3194-005		

ISOBUTANOL	$C_4H_{10}O$	74 OI Oil
Me_2CHCH_2OR R = H		Biochem. J., 1950 (47) 55.
3194-001		

ACRYLIC ACID, α-Me-	$C_4H_6O_2$	86 OI 16^0C
$MeC(:CH_2)COOH$		Bull. Soc. Chim. Fr., 1903 (29) 327.
3194-006		

ISOBUTYRIC ACID, 2, 2'-dithio-	$C_4H_8O_2S_2$	152 OI 62^{0}C		BUTANAL, 2-Me-	$C_5H_{10}O$	86 Oil
(HSCH$_2$)$_2$CH.COOH				EtCH(Me)CHO		
	J. Biol. Chem., 1948 (176) 657.				JACS, 1954, 1377.	
3194-007				3195-005		

ISOPENT-2-ENE	C_5H_{10}	70 OI Oil		PIVALIC ALDEHYDE	$C_5H_{10}O$	86 OI 6^{0}C
Me$_2$C:CH.Me				Me$_3$C.CHO		
	Can. J. Bot., 1955 (33) 363.				JACS, 1934, 444.	
3195-001				3195-006		

ISOAMYL ALCOHOL	$C_5H_{12}O$	88 OI Oil		BUTYRIC ACID, α-Me-	$C_5H_{10}O_2$	102 +18^{0}c 176^{0}C
Me$_2$CHCH$_2$CH$_2$OH				EtCH(Me)COOH		
	Rec. Trav. Chim., 1939 (58) 675.				JACS, 1948, 4238.	
3195-002				3195-007		

ISOVALERALDEHYDE	$C_5H_{10}O$	86 OI Oil		ISOVALERIC ACID	$C_5H_{10}O_2$	102 OI Oil
Me$_2$CHCH$_2$CHO				Me$_2$CHCH$_2$COOH		
	Ber.. 1934 (67) 1710.					
3195-003				3195-008		

TIGLALDEHYDE	C_5H_8O	84 OI Oil		ISOVALERIC ACID, α-OH-	$C_5H_{10}O_3$	118 -19^{0}c
MeCH:C(Me)CHO				Me$_2$CH.CHOH.COOH		
	Ann., 1854 (89) 345.				Chem. Abstr., 1948 (42) 2640.	
3195-004				3195-009		

ISOVALERIC ACID, 2, 3-diOH- $C_5H_{10}O_4$ 134 -12^0

$Me_2C(OH)CHOH.COOH$

JACS, 1954, 1085.

3195-010

ISOVALERIC ACID, α-keto- $C_5H_8O_3$ 116 OI 31^0C

$Me_2CHCOCOOH$

JACS, 1955, 4686.

3195-015

SENECIOIC ACID $C_5H_8O_2$ 100 OI 70^0C

$Me_2C:CH.COOH$

JCS, 1937, 584.

3195-011

NILIC ACID $C_5H_{10}O_3$ 118 Syrup

$MeCHOH.CH(Me)COOH$

J. Pharm. Soc. Jap.,
1922 (479) 1.

3195-016

TIGLIC ACID $C_5H_8O_2$ 100 OI 64^0C

$\overset{t}{MeCH:C(Me)COOR}$

R = H

JACS, 1948, 4238.

3195-012

ACETOLACTIC ACID $C_5H_8O_4$ 132

$$CH_3CO-\overset{Me}{\underset{OH}{C}}-COOH$$

Arch. Biochem. Biophys.,
1948 (17) 82.

3195-017

ANGELIC ACID $C_5H_8O_2$ 100 OI 46^0C

R = H
cis isomer

JACS, 1929, 2532.

3195-013

MESACONIC ACID $C_5H_6O_4$ 130 OI 205^0C

$$\underset{HOOC}{\overset{Me}{}}C=C\overset{COOH}{}$$

Proc. Nat. Acad. Sci.,
1958 (44) 1093.

3195-018

ANGELIC ACID ISOBUTYL ESTER $C_{10}H_{16}O_2$ 168 OI

R = $.CH_2CHMe_2$
cis isomer

Chem. Abstr.,
1971 (74) 15680.

3195-014

ITACONIC ACID $C_5H_6O_4$ 130 OI 163^0C

$$CH_2=C\overset{CH_2COOH}{\underset{COOH}{}}$$

Can. J. Microbiol.,
1955 (1) 749.

3195-019

CITRAMALIC ACID, L-	$C_5H_8O_5$	148 -30^0 110^0C
COOH | HO–C–CH₃ | CH₂COOH	Chem. Abstr., 1954 (48) 8885.	
3195-020		

PENTAN-2-ONE, 4-OH-4-Me-	$C_6H_{12}O_2$	116 OI Oil
CH₃COCH₂CMe₂ | OH	Experientia, 1964 (20) 390.	
3196-003		

ITATARTARIC	$C_5H_8O_6$	168
COOH | HO–C–CH₂OH | CH₂COOH	J. Biol. Chem., 1945 (161) 739.	
3195-021		

SYOYU-ALDEHYDE	$C_6H_8O_2$	112 OI
CH₃CO C=CH.CHO CH₃	Chem. Abstr., 1951 (45) 7015.	
3196-004		

ISOBUTYRIC ACID 2-ME- -BUTANYL ESTER	$C_9H_{18}O_2$	158 $+4^0$ Oil
Me₂CHCOOCH₂CHCH₂CH₃ CH₃	Chem. Ind.. 1959 1518.	
3195-022		

VALERIC ACID, d-β-Me-	$C_6H_{12}O_2$	116 $+10^0$ Oil
CH₃CH₂CHCH₂COOH CH₃	Compt. Rend., 1947 (225) 887.	
3196-005		

PENTAN-1-OL. 3-Me-	$C_6H_{14}O$	102 $+9^0C$ Oil
CH₃CH₂CHCH₂OH CH₃	JOC, 1941. 123.	
3196-001		

VALERIC ACID, γ-Me-	$C_6H_{12}O_2$	116 OI Oil
Me₂CHCH₂CH₂COOH	JACS, 1945, 1646.	
3196-006		

HEXAN-2-ONE, 4-Me-	$C_7H_{14}O$	114 Oil
CH₃COCH₂CHCH₂CH₃ CH₃	Chem. Abstr., 1963 (58) 8258.	
3196-002		

PYROTEREBIC ACID	$C_6H_{10}O_2$	114 OI Oil
Me₂C:CH.CH₂COOH	Ann., 1941 (546) 233.	
3196-007		

CAPROIC ACID, α-keto-iso-	$C_6H_{10}O_3$	130 OI

$$Me_2CHCH_2COCOOH$$

Acta Chem. Scand., 1957 (11) 1431.

3196-008

DICROTALIC ACID	$C_6H_{10}O_5$	162 OI 109°C

$$CH_3-\underset{\underset{CH_2COOH}{|}}{\overset{\overset{CH_2COOH}{|}}{C}}-OH$$

JACS, 1954, 1229.

3196-013

PENTANOIC ACID, 2,3-di-OH-3-Me-	$C_6H_{12}O_4$	148 -17°

COOH
OH
OH
CH₂CH₃

Chem. Comm., 1972, 398.

3196-009

GLUTARIC ACID, γ-Me-γ-OH-α--keto-	$C_6H_8O_6$	176

$$HOOC.\underset{\underset{OH}{|}}{\overset{\overset{CH_3}{|}}{C}}.CH_2COCOOH$$

Nature, 1955 (175) 703.

3196-014

PENT-2-ENOIC ACID, 4-Me-	$C_6H_{10}O_2$	114 OI Oil

$$Me_2CHCH:CHCOOH$$

JCS, 1929, 2505.

3196-010

GLUTARIC ACID, γ-methylene-α--keto-	$C_6H_6O_5$	158

$$CH_2=C\overset{\overset{CH_2COCOOH}{\diagup}}{\underset{\underset{COOH}{\diagdown}}{}}$$

Biochem. J., 1955 (59) 228.

3196-015

GLUTACONIC ACID, trans-β-Me-	$C_6H_8O_1$	144 OI 132°C

$$HOOC\overset{\diagup}{\underset{CH=C}{}}\overset{CH_3}{\underset{CH_2COOH}{}}$$

Tet. Lett., 1967, 2277.

3196-011

CITRIC ACID	$C_6H_8O_7$	192 OI anh. 153°C

$$HO-\underset{\underset{CH_2COOH}{|}}{\overset{\overset{CH_2COOH}{|}}{C}}-COOH$$

Chem. Eng. News, 1945 (23) 1952.

3196-016

ACONITIC ACID	$C_6H_6O_6$	174 OI 190°C

$$HOOC\overset{\diagup}{\underset{C=C}{}}\overset{CH_2COOH}{\underset{COOH}{}}$$

JACS, 1931, 3046.

3196-012

ISOCITRIC ACID, threo-DS-	$C_6H_8O_7$	192 +31°w 105°C

COOH
OH
HOOC
CH₂COOH

JACS, 1962, 309.

3196-017

486

ISOCITRIC ACID, erythro- LS-		$C_6H_8O_7$	192 $+38^0$e 246^0C	OCTANOIC ACID, 6- Me-	$C_9H_{18}O_2$	158

ISOCITRIC ACID, erythro- LS-

$C_6H_8O_7$ 192
$+38^0$e
246^0C

COOH
HO—
HOOC—
CH$_2$COOH

J. Biol. Chem.,
1962 (237) 1739.

3196-018

OCTANOIC ACID, 6- Me-

$C_9H_{18}O_2$ 158

Et . CH.(CH$_2$)$_4$COOH
|
CH$_3$

Nature, 1949, 622.

3197-004

TRICARBALLYLIC ACID

$C_6H_8O_6$ 176
OI
164^0C

CH$_2$COOH
|
CH.COOH
|
CH$_2$COOH

JACS, 1931, 3046.

3196-019

CAPSAICIN, nordihydro-

$C_{17}H_{27}NO_3$ 293
OI

NHCO(CH$_2$)$_n$CHMe$_2$

HO

OMe   n = 5

Agr. Biol. Chem.,
1970 (34) 248.

3197-005

ECHIMIDINIC ACID

$C_7H_{14}O_5$ 178
$+18^0$e

OH CH$_3$
| |
Me$_2$C.CH.C.COOH
| |
OH   OH

JCS, 1966, p. 1968.

3197-001

CAPSAICIN, dihydro-

$C_{18}H_{29}NO_3$ 307
OI

n = 6

Agr. Biol. Chem.,
1970 (34) 248.

3197-006

ADIPIC ACID, 2- Phospho- 4- OH-
-4- COOH-

$C_7H_{11}O_{11}P$ 301

OH
|
HOOCCHCH$_2$CCH$_2$COOH
| |
O   COOH
||
HO-P-OH
|
O

J. Bact., 1953 (66) 74.

3197-002

CAPSAICIN, homo- dihydro-

$C_{19}H_{31}NO_3$ 321

n = 7

Agr. Biol. Chem.,
1970 (34) 248.

3197-007

PHORBIC ACID

$C_8H_{12}O_8$ 236

OH OH
| |
HOOCCH$_2$CH$_2$CCH$_2$CHCOOH
|
COOH

Acta Chem. Scand.,
1965, 1705.

3197-003

NON- 2- EN- 1- AL, 8- Me-

$C_{10}H_{18}O$ 154
OI
Oil

Me$_2$CH(CH$_2$)$_4$CH:CH.CHO

J. Pr. Chem.,
1936 (144) 225.

3197-008

DICTYOPTERENE-A $C_{11}H_{18}$ 150
+77^{0}e

$$CH_3(CH_2)_3CH:CH\overset{t}{-}CH-CH:CH_2$$
with CH_2 cyclopropane

Tet. Lett., 1969, 3461.

3197-009

DODEC-2-ENAL, 2-Me- $C_{13}H_{24}O$ 196
OI
Oil

$$Me(CH_2)_8CH:\underset{CH_3}{C}.CHO$$

Chem. Zentr.,
1938 (I) 1892.

3197-014

DICTYOPTERENE-B $C_{11}H_{16}$ 148
-43^0
Oil

$$Et.(CH:CH)_2\overset{c}{C}H\overset{}{-}\underset{c}{C}H\,CH:CH_2$$
with CH_2 cyclopropane

Chem. Comm.,
1970, 1093.

3197-010

DODECANOIC ACID, (+)-10-Me- $C_{13}H_{26}O_2$ 214
+6^{0}c

$$Et.\underset{CH_3}{CH}(CH_2)_8COOH$$

Aust. J. C., 1971, 153.

3197-015

HYGROPHYLLINECIC ACID $C_{10}H_{16}O_6$ 232

$$CH_3CH:CH\underset{COOH}{-}CH\overset{OH}{-}CH\overset{Me}{-}\underset{COOH}{C}\overset{OH}{.}CH_3$$

JCS, 1965, 5707.

3197-011

TRIDECANOIC ACID, 12-Me- $C_{14}H_{28}O_2$ 228
OI

$$Me_2CH(CH_2)_{10}COOH$$

Aust. J. C., 1971, 153.

3197-016

TRICHODESMIC ACID $C_{10}H_{18}O_6$ 234

$$HOOCCH\underset{i-Pr}{-}\overset{Me}{\underset{OH}{C}}\overset{OH}{-}\underset{Me}{C}.COOH$$

JCS, 1969, 1386.

3197-012

TETRADECA-5,12-DIENE, 2-Me- $C_{15}H_{28}$ 208
OI
impure

$$MeCH:CH(CH_2)_5CH:CHCH_2CH_2CHMe_2$$

J. Am. Pharm. Ass.,
1924 (13) 898.

3197-017

DECA-2,4-DIENEDIOIC ACID, $C_{12}H_{18}O_5$ 242
-8-OH-2,7-diMe-
126^{0}C

Tet., 1967, 2417.

3197-013

TETRADECA-6,12-DIENE, 2-Me- $C_{15}H_{28}$ 208
OI
impure

$$MeCH:CH(CH_2)_4CH:CH(CH_2)_3CHMe_2$$

J. Am. Pharm. Ass.,
1924 (13) 898.

3197-018

deleted		
3197-019		

HEPTADECANE, 7-Me- $C_{18}H_{38}$ 254

$Me(CH_2)_5\overset{|}{C}H(CH_2)_9Me$
Me

Chem. Comm.,
1970, 1490.

3197-024

HEXADEC-1-ENE. 15-Me- $C_{17}H_{34}$ 238
OI
Oil

$Me_2CH(CH_2)_{12}CH:CH_2$

Chem. Abstr.,
1970 (73) 43892.

3197-020

HEPTADECANE, 8-Me- $C_{18}H_{38}$ 254

$Me(CH_2)_6\overset{|}{C}H(CH_2)_8Me$
Me

Chem. Comm.,
1970, 1490.

3197-025

HEXADEC-8-EN-1-OL. 14-Me- $C_{17}H_{34}O$ 254

R = CH_2OH

Et. $\overset{|}{C}H(CH_2)_4CH:CH(CH_2)_6R$
Me

JOC, 1971, 2902.

3197-021

HEPTADECANE, 2, 16-di-Me- $C_{19}H_{40}$ 268
OI
Oil

$Me_2CH(CH_2)_{13}CHMe_2$

Chem. Abstr.,
1970 (73) 43892.

3197-026

HEXADEC-8-ENOIC ACID
-ME ESTER, 14-Me- $C_{18}H_{34}O_2$ 282

R = COOMe

Science, 1969 (165) 905.

3197-022

MALVALIC ACID $C_{18}H_{32}O_2$ 280
OI
10^0C

$Me(CH_2)_7-\overset{\overset{CH_2}{\diagup\diagdown}}{C}=C-(CH_2)_6COOH$

Chem. Ind., 1961, 256.

3197-027

HEPTADECANE, 6-Me- $C_{18}H_{38}$ 254

$Me(CH_2)_4\overset{|}{C}H(CH_2)_{10}Me$
Me

Chem. Comm.,
1970, 1490.

3197-023

PROPYLURE $C_{18}H_{32}O_2$ 280
OI
110^0C

$Pr_2C:CHCH_2CH_2CH:CH(CH_2)_4OAc$

JCS, 1968, 2387.

3197-028

489

ROCCELLIC ACID	$C_{17}H_{32}O_4$	300 $+17^0$ 131^0C

$$Me(CH_2)_{11}CH.COOH$$
$$|$$
$$Me.CH.COOH$$

Z. Nat., 1969 (24B) 750.

3197-029

FOMENTARIC ACID	$C_{41}H_{80}O_4$	636 79^0C

$$(C_{18}H_{37})_2C.COOH$$
$$|$$
$$CH_3CHCOOH$$

Tet. Lett., 1967, 149.

3197-034

PEDICELLIC ACID	$C_{18}H_{34}O_4$	314 92^0C

$$Me(CH_2)_{12}CH.COOH$$
$$|$$
$$Me.CH.COOH$$

Tet., 1966, 1495.

3197-030

	$C_{44}H_{90}O$	634 OI 65^0C

$$(C_{21}H_{43})_2C.OH$$
$$|$$
$$CH_3$$

Chem. Abstr.,
1952 (46) 9262.

3197-035

RANGIFORMIC ACID	$C_{21}H_{38}O_6$	386 $+16^0$ 106^0C

$$Me(CH_2)_{13}CH.COOH$$
$$|$$
$$CH.COOH$$
$$|$$
$$CH_2COOH$$

Mono-Me ester (?)

J. Pharm. Soc. Jap.,
1946 (66A) 52.

3197-031

PHTHIOTRIOL-A	$C_{15}H_{32}O_3$	260

$$PrCHOH.CH_2CHOH.(CH_2)_4CH.CH_3$$
$$|$$
$$CHOH$$
$$|$$
$$CH_2CH_3$$

Chem. Abstr.,
1968 (69) 58781.

3197-036

CAPERATIC ACID	$C_{21}H_{38}O_7$	402 -4^0 132^0C
Mono-Me ester (?)		

$$Me(CH_2)_nCH.COOH$$
$$|$$
$$HO-C-COOH$$
$$|$$
$$CH_2COOH$$
$$n = 13$$

J. Pharm. Soc. Jap.,
1944 (64) 203.

3197-032

MYCOLIPENIC ACID	$C_{27}H_{52}O_2$	408 $+8^0r$

$$Me(CH_2)_{17}CH.CH_2CH.CH:C.COOH$$
$$|\quad\quad|\quad\quad|$$
$$Me\quad Me\quad Me$$

JCS, 1954, 1003.

3197-037

AGARICIC ACID	$C_{22}H_{40}O_7$	416 -9^0 142^0C

$$n = 15$$

Ann., 1907 (357) 145.

3197-033

PHTHIENOIC ACID	$C_{27}H_{52}O_2$	408 $+18^0$ 26^0C

$$Me(CH_2)_6CH.(CH_2)_8CH.\overset{t}{C}H:C.COOH$$
$$|\quad\quad\quad\quad|\quad\quad|$$
$$Me(CH_2)_4\quad Me\quad Me$$

JOC, 1957, 1284.

3197-038

MYCOCERANIC ACID $\quad C_{31}H_{62}O_2 \quad$ 466
$-7^0 r$
$30^0 C$

$$Me(CH_2)_{21}CH.CH_2CH.CH_2CH.COOH$$
$$\underset{Me}{|} \quad \underset{Me}{|} \quad \underset{Me}{|}$$

JCS, 1954, 1003.

3197-039

$C_{63}H_{122}O_5 \quad$ 958

$$HOOC(CH_2)_{16}CH:CH(CH_2)_{19}CHOH.\underset{\underset{C_{22}H_{45}}{|}}{CH}.COOH$$

Chem. Abstr.,
1967 (66) 28318.

3197-044

CORYNOMYCOLIC ACID $\quad C_{32}H_{64}O_3 \quad$ 436
$+7.5^0$
$70^0 C$

$$Me(CH_2)_{14}CHOH.\underset{\underset{COOH}{|}}{CH}.(CH_2)_{13}Me$$

Bull. Soc. Chim. Biol.,
1959 (41) 481.

3197-040

$C_{58}H_{112}O_5 \quad$ 888

$R = C_{20}H_{41}$

$$HOOC(CH_2)_{14}CH.CH:CH(CH_2)_{16}CHOH.\underset{\underset{R}{|}}{CH}.COOH$$

Chem. Abstr.,
1967 (66) 28318.

3197-045

CORYNOMYCOLENIC ACID $\quad C_{32}H_{62}O_3 \quad$ 434
$+9^0$
Oil

$$Me(CH_2)_5CH:CH(CH_2)_7CHOH.\underset{\underset{(CH_2)_{13}Me}{|}}{CH}.COOH$$

Biochim. Biophys. Acta.
1953 (11) 163.

3197-041

$C_{60}H_{116}O_5 \quad$ 916

$R = C_{22}H_{45}$

Chem. Abstr.,
1967 (66) 28318.

3197-046

CORYNINE $\quad C_{52}H_{104}O_4 \quad$ 792

$70^0 C$

$$MeCHOH(CH_2)_7\underset{\underset{Me(CH_2)_{14}}{|}}{CH}.CHOH\underset{\underset{Me}{|}}{(CH)}\underset{\underset{(CH_2)_{17}}{|}}{CH}.COOH$$

J. Pharm. Soc. Jap.,
1948 (68) 292.

3197-042

$C_{62}H_{122}O_3 \quad$ 914

$$Me(CH_2)_{17}CH:CH(CH_2)_{17}CHOH.\underset{\underset{C_{22}H_{45}}{|}}{CH}.COOH$$

Tet., 1966, 1113.

3197-047

$C_{57}H_{110}O_5 \quad$ 874

$$HOOC(CH_2)_{14}CH:CH(CH_2)_{17}CHOH.\underset{\underset{C_{20}H_{41}}{|}}{CH}.COOH$$

Chem. Abstr.,
1967 (66) 28318.

3197-043

HENTRIACONTANE, 3, 7, 11-tri-Me- $\quad C_{34}H_{70} \quad$ 478

$n = 17$

$$H(CH_2CH_2CH.CH_2)_3CH_2(CH_2)_nMe$$
$$\underset{CH_3}{|}$$

Tet., 1970, 307.

3197-048

TRITRIACONTANE, 3, 7. 11-tri-Me- $C_{36}H_{74}$ 506 n = 19 Tet., 1970, 307. 3197-049	TUBERCULOSTEARIC ACID $C_{19}H_{38}O_2$ 298 13^0C $Me(CH_2)_7CH(CH_2)_8COOH$ $\quad\quad\quad\;\; CH_3$ Chem. Comm., 1966, 855. 3197-054
PENTATRIACONTANE, 3. 7. 11- -tri-Me- n = 21 Tet., 1970. 307. 3197-050	LACTOBACILLIC ACID $C_{19}H_{36}O_2$ 296 34^0C $\quad\quad\quad\quad\; CH_2$ $Me(CH_2)_5CH\!-\!\!-\!CH(CH_2)_9COOH$ JACS. 1958, 5717. 3197-055
DOTRIACONTANE, 4. 8, 12-tri-Me- $C_{35}H_{72}$ 492 n = 17 $Me(CH_2CH_2CH.CH_2)_3CH_2(CH_2)_n Me$ $\quad\quad\quad\quad\;\; CH_3$ Tet., 1970, 307. 3197-051	STERCULIC ACID $C_{19}H_{34}O_2$ 294 OI 18^0C $\quad\quad\quad\;\; CH_2$ $Me(CH_2)_7C\!\!=\!\!C(CH_2)_7COOH$ Chem. Ind., 1958. 103. 3197-056
TETRATRIACONTANE, 4. 8. 12- -tri-Me- n = 19 Tet., 1970, 307. 3197-052	STERCULIC ACID, 2-OH- $C_{19}H_{34}O_3$ 310 $\quad\quad\quad\;\; CH_2$ $Me(CH_2)_7C\!\!=\!\!C(CH_2)_6CHOH.COOH$ Chem. Ind., 1967, 32. 3197-057
HEXATRIACONTANE, 4, 8, 12- -tri-Me- $C_{39}H_{80}$ 548 n = 21 Tet., 1970, 307. 3197-053	STERCULYNIC ACID $C_{19}H_{30}O_2$ 290 $\quad\quad\quad\;\; CH_2$ $HC\vdots C(CH_2)_7C\!\!=\!\!C(CH_2)_6COOH$ Tet. Lett., 1968, 2167. 3197-058

SELACHYL ALCOHOL $\quad C_{21}H_{42}O_3 \quad$ 342

6^0C

CH_2OR_1
|
$CHOR_2$
|
CH_2OR_3

$R_2 = R_3 = H$; R_1 = Oleoyl $\quad$ J. Biol. Chem. , 1944 (155) 447.

3198-001

OLEOPALMITOSTEARIN $\quad C_{55}H_{104}O_6 \quad$ 860

31^0C

R_1 = Oleoyl
R_2 = Stearoyl
R_3 = Palmitoyl

J. Soc. Chem. Ind., 1940 (59) 67

3198-006

TRILAURIN $\quad C_{39}H_{74}O_6 \quad$ 638 OI 49^0C

$R_1 = R_2 = R_3$ = Dodecanoyl

JACS, 1929, 870.

3198-002

OLEODISTEARIN $\quad C_{57}H_{108}O_6 \quad$ 888

44^0C

R_1 = Oleoyl
$R_2 = R_3$ = Stearoyl

JACS, 1943, 2143.

3198-007

TRIMYRISTIN $\quad C_{45}H_{86}O_6 \quad$ 722 OI 56^0C

$R_1 = R_2 = R_3$ = Tetradecanoyl

Chem. Abstr. , 1929 (23) 4676.

3198-003

TRISTEARIN $\quad C_{57}H_{110}O_6 \quad$ 890 OI 55^0C

$R_1 = R_2 = R_3$ = Stearoyl

JACS, 1955, 2646.

3198-008

OLEODIPALMITIN $\quad C_{53}H_{100}O_6 \quad$ 832

28^0C

$R_1 = R_2$ = Palmitoyl
R_3 = Oleoyl

JACS, 1944, 690.

3198-004

TRIOLEIN $\quad C_{57}H_{104}O_6 \quad$ 884 OI

$R_1 = R_2 = R_3$ = Oleoyl

J. Biol. Chem. , 1940 (132) 687.

3198-009

PALMITODISTEARIN $\quad C_{55}H_{106}O_6 \quad$ 862 OI 63^0C

$R_1 = R_3$ = Stearoyl
R_2 = Palmitoyl

JACS, 1932, 708.

3198-005

TRILINOLEIN $\quad C_{57}H_{98}O_6 \quad$ 878 OI

$R_1 = R_2 = R_3$ = Linoleoyl

JACS, 1944, 998.

3198-010

TRILINOLENIN $C_{57}H_{92}O_6$ 872 OI		**SORBO-1, 3-DIMYRISTIN** $C_{37}H_{66}O_6$ 606 OI 54^0C
$R_1 = R_2 = R_3 =$ Linolenoyl JACS, 1944, 998. 3198-011		$R_1 = R_3 =$ Myristoyl $R_2 =$ Sorboyl JCS, 1965, 5651. 3198-013
TRIVERNOLIN $C_{57}H_{98}O_9$ 932 $+2^0$ 25^0C $R_1 = R_2 = R_3 =$ Vernoloyl J. Am. Oil Chem. Soc., 1962 (39) 334. 3198-012		

Care must be exercised in using this class—reference to Class 30 may well be considered. The phenylpolyynes are sufficiently distinctive as a biogenetic group as to justify removing them from the relative anonymity of the C_6C_n class (04).

3201	Terminal Furans / γ-Lactones
3202	Nonterminal Furans
3203	Pyrans / δ-Lactones
3204	Dioxyspirocyclics
3205	Phenylpolyynes

FURAN, 2-pentyl- $C_9H_{14}O$ 138
OI
Oil

$(CH_2)_4Me$

Chem. Ind., 1966, 1926.

3201-001

ATRACYLODIN $C_{13}H_{10}O$ 182
OI
52^0C

$CH:CH.(C:C)_2CH:CHMe$

Chem. Pharm. Bull.,
1960 (8) 957.

3201-006

AVOCADENOFURAN $C_{17}H_{28}O$ 248
OI
Oil

$(CH_2)_{11}CH:CH_2$

Tet., 1969, 4623.

3201-002

MATRICARIALACTONE, c-dihydro- $C_{10}H_{10}O_2$ 162
OI

$CH.(CH:CH)_2Me$

Ber., 1967, 611.

3201-007

AVOCADYNOFURAN $C_{17}H_{26}O$ 246
OI
Oil

$(CH_2)_{11}C:CH$

Tet., 1969, 4623.

3201-003

MATRICARIALACTONE, c, c- $C_{10}H_8O_2$ 160
OI
98^0C

$CH.C:C.CH:CHMe$

Ber., 1966, 2096.

3201-008

AVOCADIENOFURAN $C_{17}H_{26}O$ 246
OI
Oil

$CH:CH.(CH_2)_9CH:CH_2$

Chem. Abstr., 1971,
(74) 115910.

3201-004

$C_{11}H_{10}O_4S$ 238
OI

$CH.C:C.CH:C(Me)SO_2Me$

Chem. Abstr., 1967 (66)
65021.

3201-009

AVOCADIENOFURAN, Iso- $C_{17}H_{26}O$ 246
OI
Oil

$CH_2CH:CH(CH_2)_8CH:CH_2$

Chem. Abstr., 1971
(74) 115910.

3201-005

$C_{10}H_{10}O_2$ 162

Oil

$C:C:CH.Pr$

Tet. Lett., 1970, 2465.

3201-010

FURAN, 2-(HEPTA-4, 6-DIEN-2-
-YNYLIDENE)-2, 5-dihydro- $C_{11}H_{10}O$ 158

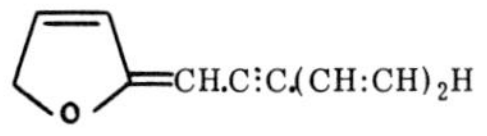

:CH.C:C:C(CH:CH)₂H

Chem. Abstr., 1967(66)
65023.

3201-011

$C_{13}H_{14}O_2$ 202

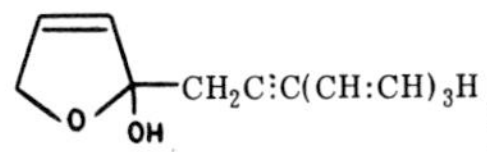

CH₂C:C(CH:CH)₃H
OH

Chem. Abstr., 1967(66)
65023.

3201-016

NEMOTIN $C_{11}H_8O_2$ 172
+380⁰e

-CH:C:CH(C:C)₂H

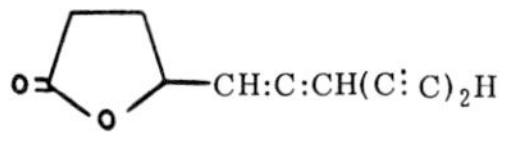

JCS, 1955, 4270.

3201-012

DODEC-6-EN-4-OLIDE $C_{12}H_{20}O_2$ 196

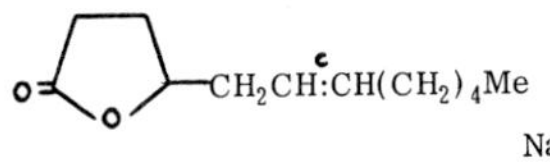

CH₂CH:CH(CH₂)₄Me

Nature, 1969 (221) 284.

3201-017

ODYSSIN $C_{12}H_{10}O_2$ 186
+360⁰e

-CH:C:CH(C:C)₂Me

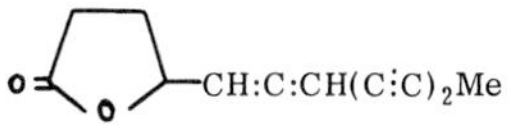

JCS, 1957, 1607.

3201-013

$C_{10}H_{10}O_2$ 162

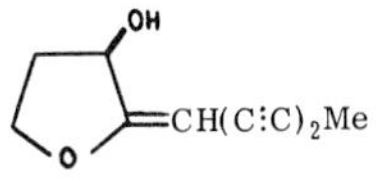

OH
:CH(C:C)₂Me

JCS, 1969, 1096.

3201-018

TRIDECA-2, 10, 12-TRIEN-6, 8-
-DIYN-4-OLIDE $C_{13}H_{10}O_2$ 198

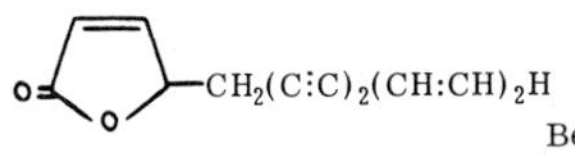

CH₂(C:C)₂(CH:CH)₂H

Ber., 1967, 1504.

3201-014

RHIZOSOLANIOL $C_{12}H_{10}O_6$ 250

139⁰C

OH OH

O   O

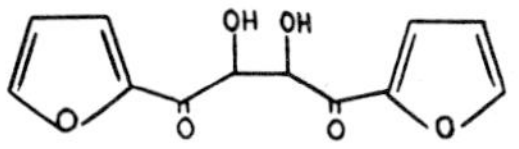

Experientia,
1971 (27) 1143.

3201-019

$C_{13}H_{10}O_2$ 198

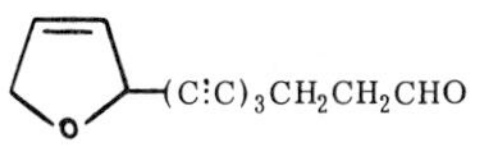

(C:C)₃CH₂CH₂CHO

Ber., 1967, 1504.

3201-015

DODECALACTONE, γ- $C_{12}H_{22}O_2$ 198

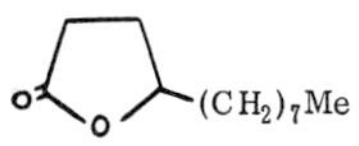

(CH₂)₇Me

Tet. Lett., 1972, 4635.

3201-020

FURAN, 2-(7-COOH-heptyl)- -5-hexyl- $C_{18}H_{30}O_3$ 294

Me$(CH_2)_5$ [furan ring] $(CH_2)_7$COOH

Tet. Lett. , 1966, 4249.
Chem. Comm. , 1966, 925.

3202-001

$C_{13}H_{12}O_2S$ 232
OI
42^0C

MeSCH:CHC:C. CH_2 [furan ring] CH:CHCHO

Ber. , 1969 (102) 4209.

3202-006

HEXADECANOIC ACID, -10-OH-6,9-oxido- $C_{16}H_{30}O_4$ 286
$+11^0$
43^0C

H$(CH_2)_6$CH(OH) [ring] $(CH_2)_4$COOH

Tet. Lett. , 1971, 4011.

3202-002

$C_{13}H_9OCl$ 216
88^0C

Me(C:C)$_3$·CH= [ring] CH_2Cl

Ber. , 1971 (104) 1962.

3202-007

WYEROL $C_{15}H_{16}O_4$ 260

EtCH:CHC:CCH(OH) [furan ring] CH:CHCOOMe

JCS. 1968, 2455.

3202-003

$C_{13}H_9OCl$ 216
70^0C

trans isomer

Ber. . 1971 (104) 1962.

3202-008

WYERONE $C_{15}H_{14}O_4$ 258
OI
64^0C

EtCH:CHC:C. CO. [furan ring] CH:CHCOOMe

JCS. 1968. 2455.

3202-004

PENTAN-5-OLIDE, 5-(pent-2-enyl)- $C_{10}H_{16}O_2$ 168
-30^0
Oil

[ring] CH_2CH:CHCH$_2$Me

Helv. , 1962 (45) 1250.

3203-001

WYERONE. dihydro- $C_{15}H_{16}O_4$ 260
OI

* dihydro

JCS. 1968, 2455.

3202-005

PYRONE, 5-(4-SMe-PENTA-1, 2, 3- -TRIENYL)-α- $C_{11}H_{10}O_2S$ 206
109^0C

[ring] CH:C:C:C(SMe)Me

Ann. , 1966, 149.

3203-002

ICHTHYOTHEREOL	$C_{14}H_{14}O_2$	214
		-40^0c
R = H		89^0C

CH:CH(C:C)$_3$Me

JCS, 1970, 314.

3203-003

	$C_{14}H_{16}O_2$	216
R = H		64^0C

CH:CH(C:C)$_2$CH:CHMe

Phytochem., 1971 (10) 1877.

3203-008

ICHTHYOTHEREOL, Ac-	$C_{16}H_{16}O_3$	256
		$+26^0$c
		64^0C
R = Ac		

JCS, 1970, 314.

3203-004

	$C_{16}H_{18}O_3$	258
		$+17^0$t
		76^0C
R = Ac		

JCS, 1970, 314.

3203-009

TRIDECA-3, 5-DIEN-7, 9, 11- -TRIYN-1, 5-OXIDE, 2-Cl-c-	$C_{13}H_9OCl$	216

CH(C:C)$_3$Me

Ber., 1966, 1648.

3203-005

HYPTOLIDE	$C_{18}H_{24}O_8$	368
		$+7^0$e
		88^0C

CH:CH.CH(OAc)CH$_2$CH(OAc)Me

JCS, 1964, 4167.

3203-010

TRIDECA-3, 5-DIEN-7, 9, 11- -TRIYNE-1, 5-OXIDE, 2-Cl-t-	$C_{13}H_9OCl$	216
		80^0C
trans isomer		

Ber., 1966, 1648.

3203-006

BORONOLIDE	$C_{18}H_{26}O_8$	370
		$+25^0$e
		90^0C

CH(OAc)CH(OAc)CH(OAc)Bu

Compt. Rend., 1971 (C273) 439.

3203-011

TRIDEC-5-EN-7, 9, 11-TRIYN- -1, 5-3, 4-DI-OXIDE, 2-Cl-	$C_{13}H_9O_2Cl$	232
		87^0C

CH(C:C)$_3$Me

Ber., 1966, 1648.

3203-007

LL-P880α	$C_{11}H_{18}O_4$	214
		-86^0m
		84^0C

CHOH(CH$_2$)$_3$Me

JOC, 1972, 2045.

3203-012

MASSOILACTONE $C_{10}H_{16}O_2$ 168
-93^0
Oil

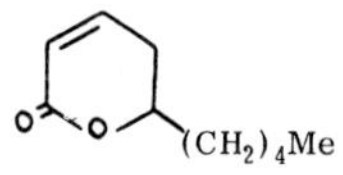

JCS, 1955, 2535.

3203-013

$C_{14}H_{20}O_2$ 220

Oil

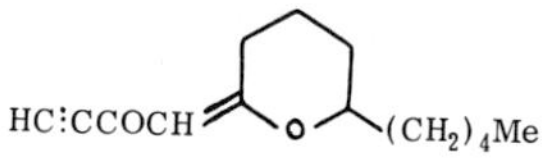

HC:CCOCH $(CH_2)_4$Me

Ber., 1971 (104) 2354.

3203-018

DODEC-2-EN-5-OLIDE $C_{12}H_{20}O_2$ 196

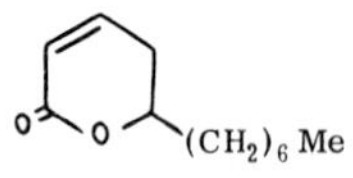

$(CH_2)_6$ Me

Aust. J. C., 1968 (21) 2819.

3203-014

$C_{12}H_{10}O_2$ 186

Oil

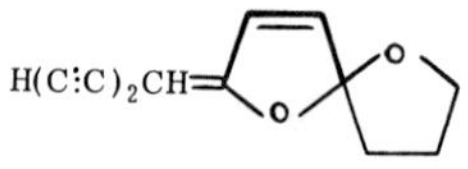

H(C:C)$_2$CH

Ber., 1970(103)1879

3204-001

$C_{18}H_{20}O_3$ 284
-46^{0}r
48^{0}C

AcO $(CH:CH)_2(C:C)_2CH:CHMe$

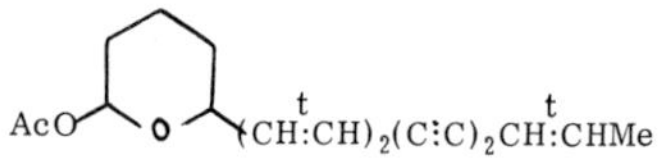

Ber., 1970 (103) 2100.

3203-015

$C_{12}H_8O_2$ 184

Oil

R = H

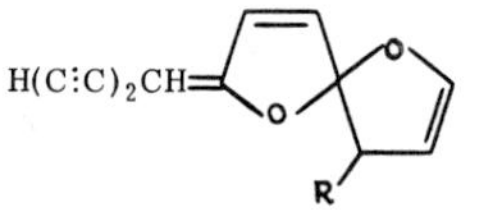

H(C:C)$_2$CH

R

Ber., 1970(103)1879

3204-002

$C_{14}H_{16}O_2$ 216

Oil

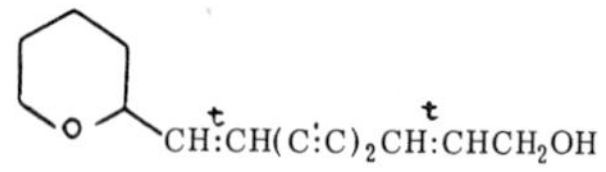

CH:CH(C:C)$_2$CH:CHCH$_2$OH

JCS, 1969, 830.

3203-016

$C_{14}H_{10}O_4$ 242

Oil

R = OAc

Ber., 1970(103)1879

3204-003

$C_{14}H_{18}O_2$ 218

Oil

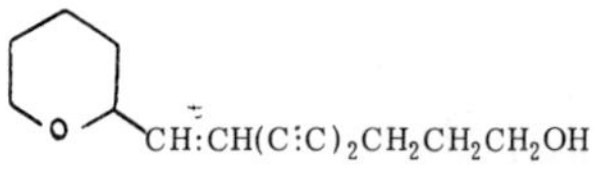

CH:CH(C:C)$_2$CH$_2$CH$_2$CH$_2$OH

JCS, 1969, 830.

3203-017

$C_{13}H_{12}O_2$ 200
-45^{0}r
49^{0}C

Me(C:C)$_2$CH

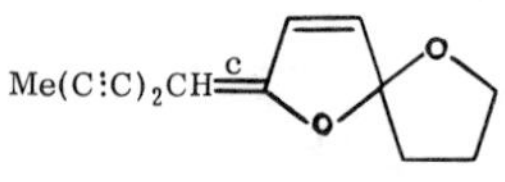

Ber., 1961(94)3193

3204-004

trans isomer	$C_{13}H_{12}O_2$	200 -104^0r amorph.	$R = H$ cis isomer	$C_{13}H_{10}O_2$ 198 -119^0r 76^0C
	Ber., 1961(94)3193			Ber., 1970(103)1881
3204-005			3204-010	

$C_{15}H_{14}O_4$ 258 amorph.

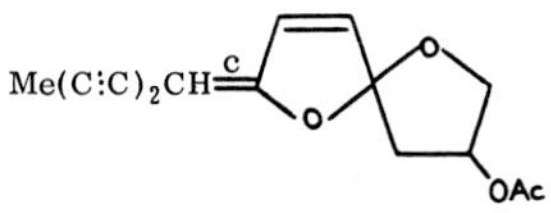

Me(C:C)$_2$CH$\overset{c}{=}$ (structure) OAc

Ber., 1965(98)2596

3204-006

$R = OAc$
cis isomer

$C_{15}H_{12}O_4$ 256 $+ 59^0r$ 91^0C

Ber., 1961(94)3193

3204-011

deleted

3204-007

$R = OAc$
trans isomer

$C_{15}H_{12}O_4$ 256 -868^0r 123^0C

Ber., 1961(94)3193

3204-012

Me(C:C)$_2$CH= (structure)

$C_{13}H_{12}O_3$ 216 126^0C

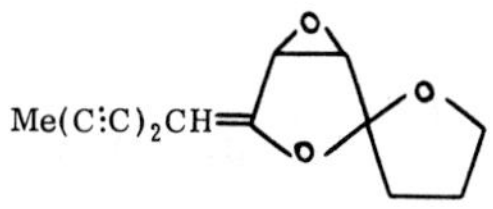

Ber., 1965(98)2596

3204-008

$R = O.COCH_2CHMe_2$

$C_{18}H_{18}O_4$ 298 66^0C

Ber., 1970(103)1879

3204-013

$R = H$
Me(C:C)$_2$CH$\overset{t}{=}$ (structure) R

$C_{13}H_{10}O_2$ 198 $- 444^0r$ 111^0C

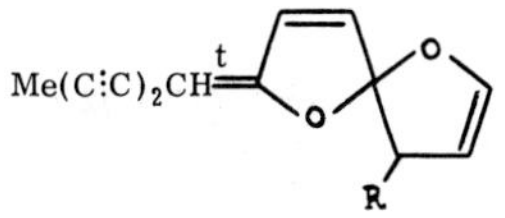

Ber., 1970(103)1881

3204-009

$R = H$
Me(C:C)$_2$CH= (structure) RO

$C_{13}H_{10}O_4$ 230 $- 123^0$ 139^0C

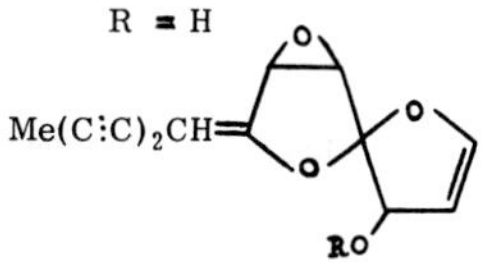

Ber., 1965(98)2596

3204-014

$C_{15}H_{12}O_5$ 272 154°C R = Ac Ann., 1963(668)51 3204-015	$C_{13}H_{14}O_3S$ 250 X = S→O cis, cis-isomer Ber., 1964(97)1179 3204-020
$C_{18}H_{18}O_5$ 314 impure R = isovaleryl Ber., 1970(103)1879 3204-016	$C_{13}H_{14}O_3S$ 250 X = S→O cis, trans-isomer Ber., 1964(97)1179 3204-021
$C_{18}H_{16}O_5$ 312 132°C R = Senecioyl Ber., 1970(103)1879 3204-017	$C_{13}H_{12}O_2S$ 232 MeS.CH:CH.C:C.CH= (c ... t) Fortschr. Chem. Org. Nat., 1967(25)38 3204-022
$C_{13}H_{14}O_2S$ 234 X = S MeX.CH:CH.C:C.CH= (c ... c) Ber., 1964(97)1179 3204-018	$C_{14}H_{14}O_2$ 214 0° 78°C Me(C:C)₂CH= (c) Ber., 1963(96)226 3204-023
$C_{13}H_{14}O_2S$ 234 +1°r 71°C X = S cis, trans-isomer Ber., 1964(97)1179 3204-019	$C_{14}H_{14}O_2$ 214 83°C trans isomer Ber., 1963(96)226 3204-024

deleted 3204-025	$C_{14}H_{16}O_2S$ 248 MeS.CH:CH.C:C.CH= Fortschr.Chem.Org. Nat., 1967(25) 36 3204-030
$C_{16}H_{16}O_4$ 272 127^0C R = Ac Me(C:C)$_2$CH= —OR Ber., 1966(99)2416 3204-026	FRUTESCIN $C_{15}H_{14}O_3$ 242 OI 67^0C OMe COOMe (C≡C)$_2$R R = Me Ber., 1962(95) 602. 3205-001
$C_{16}H_{16}O_4$ 272 R = Ac trans isomer Ber., 1963(96)226 3204-027	FRUTESCIN, desMe- $C_{14}H_{12}O_3$ 228 OI 68^0C R = H Ber.. 1962(95) 602. 3205-002
$C_{19}H_{22}O_4$ 314 88^0C R = isovaleryl cis isomer Ber., 1966(99)2416 3204-028	FRUTESCINONE $C_{15}H_{12}O_4$ 256 OI 121^0C OMe COOMe (C≡C)$_2$-Me O Ber., 1962(95) 602. 3205-003
$C_{19}H_{22}O_4$ 314 94^0C R = isovaleryl trans isomer Ber., 1966(99)2416 3204-029	FRUTESCINOL-LACTONE $C_{14}H_{10}O_3$ 226 188^0C OMe O O C≡C-C≡C-Me Ber., 1964(97) 1176. 3205-004

ARTEMIDIN	$C_{13}H_{12}O_2$	200 OI 48°C

Ber., 1970 (103) 2856.
Khim. Prir. Soedin.,
1970 (6) 467.

3205-005

HEXA-2, 4-DIYN-1-OL, -senecioyl-6-Ph-	$C_{17}H_{16}O_2$	252 OI 37°C

R = Senecioyl

Ber., 1966 (99) 2413.

3205-010

CAPILLARIN	$C_{13}H_{10}O_2$	198 OI 124°C

Ber., 1962 (95) 602.

3205-006

HEXA-2, 4-DIYNE, 1-(2-OH-Ph)-	$C_{12}H_{10}O$	170 OI Oil

R = H

Ber., 1971 (104) 1375.

3205-011

CAPILLIN	$C_{12}H_8O$	168 OI 81°C

J. Pharm. Soc. Jap.,
1956, 405.

3205-007

HEXA-2, 4-DIYN-1-OL, 1-(2-OH-Ph)-	$C_{12}H_{10}O_2$	186 Oil

R = OH

Ber., 1971 (104) 1375.

3205-012

HEXA-2, 4-DIYNE, 1-Ph-	$C_{12}H_{10}$	154 OI

Ber., 1966 (99) 2413.

3205-008

AGROPYRENE	$C_{12}H_{12}$	156 OI Oil

JCS, 1954, 1874.

3205-013

HEXA-2, 4-DIYN-1-OL, 6-Ph-	$C_{12}H_{10}O$	170 OI 93°C

R = H

Ber., 1966, 2413.

3205-009

CARLINA OXIDE	$C_{13}H_{10}O$	182 OI Oil

JACS, 1933, 3461.

3205-014

504

HEPT-2-EN-4,6-DIYN-1-AL, -7-Ph-

$C_{13}H_8O$

180
OI
64^0C

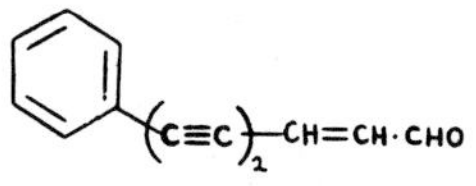

Chem. Abstr.,
1967 (66) 65019.

3205-015

HEPT-5-EN-1,3-DIYNE, -1(3-OH-Ph)-7-OAc-

$C_{15}H_{12}O_3$

240
OI

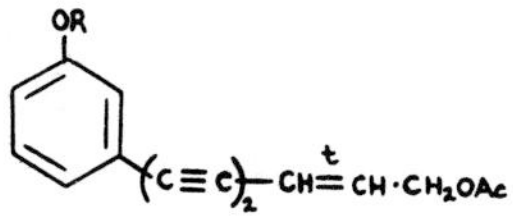

R = H

Ber., 1966 (99) 2413.

3205-018

HEPT-5-EN-1,3-DIYNE, -1-Ph-7-OH-

$C_{13}H_{10}O$

182
OI

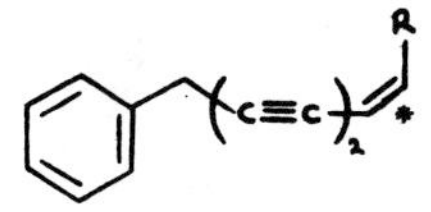

R = CH$_2$OH

Acta Chem. Scand.,
1954, 1757.

3205-016

HEPT-5-EN-1,3-DIYNE, -1-(3-OAc-Ph)-7-OAc-

$C_{17}H_{14}O_4$

282
OI

R = Ac

Ber., 1966 (99) 2413.

3205-019

HEPT-5-EN-1,3-DIYNE, -1-Ph-7-OAc-

$C_{15}H_{12}O_2$

224
OI

R = CH$_2$OAc
* trans

Acta Chem. Scand.,
1954, 1757.

3205-017

Included in this class are the acyclic sulphides, thienyls, and cyclic poly-sulphides. The thienyls are biogenetically related to the linear polyketides, as indicated by their cooccurrence with polyacetylenes.

3301	Acyclic sulphides
3302	Thiophenes
3303	Bithienyls
3304	Terthienyls
3305	Cyclic polysulphides

BISMETHYLTHIOMETHANE	$C_3H_8S_2$	108 OI Oil
MeS.CH$_2$SMe		
	Tet. Lett., 1967, 1681.	
3301-001		

BUTYL-PENT-1-ENYLDISULPHIDE	$C_9H_{18}S_2$	190 OI Oil
Me(CH$_2$)$_3$.S(CH$_2$)$_3$CH:CH$_2$		
	Chem. Abstr., 1946 (40) 6752.	
3301-006		

DIMETHYLSULPHIDE	C_2H_6S	62 OI Oil
MeSMe		
	Chem. Zentr.,1909 (I) 1564	
3301-002		

PROPENYL-SEC-BUTYL- DISULPHIDE	$C_7H_{14}S_2$	162 -18^0 Oil
Et.CH(Me)S.S CH:CHMe		
	Arch. Pharm., 1936, 461.	
3301-007		

DIMETHYLSULPHONE	$C_2H_6O_2S$	94 OI 168^{0}C
Me SO$_2$ Me		
	Acta Chem. Scand.. 1954 (8) 703.	
3301-003		

DIALLYLSULPHIDE	$C_6H_{10}S$	114 OI Oil
CH$_2$:CHCH$_2$(S)$_n$CH$_2$CH:CH$_2$ n = 1		
	Chem. Abstr., 1952, 3217.	
3301-008		

DIVINYLSULPHIDE	C_4H_6S	86 OI Oil
CH$_2$:CH.S. CH:CH$_2$		
	Chem. Abstr., 1953 (47) 3224.	
3301-004		

DIALLYLDISULPHIDE	$C_6H_{10}S_2$	146 OI Oil
n = 2		
	Arch. Pharm., 1892 (231) 434.	
3301-009		

PROPYL-ALLYLDISULPHIDE	$C_6H_{12}S_2$	148 OI Oil
MeCH$_2$CH$_2$S.S CH$_2$CH:CH$_2$		
	Arch. Pharm., 1892 (230) 434.	
3301-005		

DIALLYLTRISULPHIDE	$C_6H_{10}S_3$	178 OI Oil
n = 3		
	Arch. Pharm., 1892 (231) 434.	
3301-010		

DIALLYLTETRASULPHIDE	$C_6H_{10}S_4$	210 OI Oil		S-(3-OXO-UNDEC-4-ENYL)- THIOACETATE	$C_{13}H_{22}O_2S$	242 OI Oil

n = 4

$Me(CH_2)_5CH:CH.COCH_2CH_2S.COMe$

Chem. Abstr., 1948 (42) 8516.

Chem. Comm., 1971, 503.

3301-011

3301-016

BIS-(3-OXO-UNDECYL)-DISULPHIDE	$C_{22}H_{42}O_2S_2$	402 OI 67^0C	S-(3-OAc-UNDEC-5-ENYL)- THIOACETATE	$C_{15}H_{26}O_3S$ 286 -25^0t Oil

$(Me(CH_2)_7COCH_2CH_2S)_2$

$Me(CH_2)_4CH:CH.CH_2CH(OAc)$
$MeCOS.CH_2CH_2$

Chem. Comm., 1971, 503.

Chem. Comm., 1972, 326.

3301-012

3301-017

BIS-(3-OXO-UNDECYL)- TRISULPHIDE	$C_{22}H_{42}O_2S_3$	434 OI 61^0C	BIS-(3-OAc-UNDEC-5-ENYL)- DISULPHIDE	$C_{26}H_{50}O_4S_2$ 490 (-)

$(Me(CH_2)_7COCH_2CH_2S)_2S$

$(Me(CH_2)_4CH:CHCH_2CH(OAc)CH_2CH_2S)_2$

Chem. Comm., 1971, 1168.

Chem. Comm., 1972, 326.

3301-013

3301-018

BIS-(3-OXO-UNDECYL)- TETRASULPHIDE	$C_{22}H_{42}O_2S_4$	466 OI 33^0C	ALLICIN	$C_6H_{10}OS_2$ 162 Oil

$(Me(CH_2)_7COCH_2CH_2S.S)_2$

$CH_2:CHCH_2S(O)S.CH_2CH:CH_2$

Chem. Comm., 1971, 1168.

Angew., 1962 (1) 299.

3301-014

3301-019

S-(3-OXO-UNDECYL)- THIOACETATE	$C_{13}H_{24}O_2S$	244 OI Oil	JUNIPAL	C_8H_6OS 150 OI 80^0C

$Me(CH_2)_7COCH_2CH_2S.COMe$

OHC-S-C:C.Me

Chem. Comm., 1971, 503.

Biochem. J., 1955 (60) 255.

3301-015

3302-001

THIOPHENE, 2-Ac-3-OH-5-
-(prop-1-ynyl)-

$C_9H_8O_2S$ 180
OI
100^0C

RO   R = H

Ac   C:C.Me

Phytochem., 1971, 454.

3302-002

THIOPHENE, 2-(2-COOMe-ethenyl)- $C_{11}H_{12}O_2S$ 208
-5-(propenyl)-   OI

MeCH:CH   CH:CHCOOMe

Ann., 1966, 149.

3302-007

THIOPHENE, 2-Ac-3-OMe-5-
-(prop-1-ynyl)-

$C_{10}H_{10}O_2S$ 194
OI
90^0C

R = Me

Ber., 1962 (95) 2934.

3302-003

$C_{10}H_8O_2S$ 192
OI

Me   O   O

Ber., 1966, 1226.

3302-008

THIOPHENE, 2-(4-CHO-BUT-3-
EN-1-YNYL)-

C_9H_6OS 162
OI

C:C.CH:CH.CHO

Ber., 1966, 1642.

3302-004

$C_{14}H_{19}NOS$ 249
OI
105^0C

CH_2(CH:CH)_2CONH
Me_2CHCH_2

Lloydia, 1970 (33) 393.

3302-009

THIOPHENE, 2-(prop-1-ynyl)-5-
-pyruvoyl-

$C_{10}H_8O_2S$ 192
OI
Oil

Me.C:C   CO.CO.Me

JCS, 1969, 1813.

3302-005

THIOPHENE, 2-(1,3-butadiynyl)- $C_{12}H_6S$ 182
-5-(but-3-en-1-ynyl)-   OI
                                               Oil

H(C:C)_2   C:C.CH:CH_2

Ber., 1970 (103) 834.

3302-010

THIOPHENE, 2-Lactoyl-5-
-(Prop-1-ynyl)-

$C_{10}H_{10}O_2S$ 194

Oil

Me.C:C   CO.CHOH.Me

JCS, 1969, 1813.

3302-006

$C_{12}H_7OSCl$ 234
OI
Oil

H(C:C)_2   C:C.CO.CH_2Cl

Ber., 1970 (103) 834.

3302-011

$C_{12}H_9OSCl$ 236

$HC(C:C)_2$ [thiophene ring] $C:C.CHOH.CH_2Cl$

Ber., 1970 (103) 834.

3302-012

$C_{18}H_{22}O_3S$ 318
    $+1^0$

R = isovaleryl

Ber., 1966, 1642.

3302-017

$C_{12}H_8OS$ 200
    OI
    Oil

[thiophene ring] $C:C.CH\overset{c}{:}CH$ [furan ring]

Chem. Ind., 1964, 2101.

3302-013

THIOPHENE, 2-(6-OH-non-3-en-1- $C_{13}H_{14}O_2S$ 234
    -yn-7-onyl)- -35^0

[thiophene ring] $C:C.CH:CHCH_2CHOH.CO.Et$

Ber., 1966, 1642.

3302-018

THIOPHENE, 2-(7-OAc-Non-3-en-1- $C_{15}H_{18}O_2S$ 262
    -ynyl)-

[thiophene ring] $C:C.CH:CHCH_2CH_2CH(OAc)Et$

Ber., 1967, 611.

3302-014

THIOPHENE, 2-(nona-3,5-dien-1- $C_{13}H_{12}OS$ 216
    -yn-7-onyl)-   OI

[thiophene ring] $C:C.\overset{t,\,c}{(CH:CH)_2}CO.Et$

Ber., 1966, 1642.

3302-019

THIOPHENE, 2-(7-oxo-Non-3-en- $C_{12}H_{11}OS$ 218
    -1-ynyl)-   OI
    41^0C

[thiophene ring] $C:C.CH\overset{t}{:}CHCH_2CH_2CO.Et$

Ber., 1966, 1642.

3302-015

$C_{13}H_{12}OS$ 216
    OI

trans, trans isomer

Ber., 1966, 1642.

3302-020

THIOPHENE, 2-(5-OH-non-3-en- $C_{13}H_{14}O_2S$ 234
    -1-yn-7-onyl)-
    R = H

[thiophene ring] $C:C.CH:CH.CH(OR)CH_2CO.Et$

Ber., 1966, 1642.

3302-016

THIOPHENE, 2-(but-3-en-1-ynyl)- $C_{13}H_{10}S$ 198
    -5-(pent-3-en-1-ynyl)-   OI
    Oil

$MeCH:CHC:C$ [thiophene ring] $C:C.CH:CH_2$

Tet. Lett., 1965, 297.

3302-021

THIOPHENE, 2-(but-3-en-1-ynyl)-5-
-penta-1, 3-diynyl)-   C$_{13}$H$_8$S   196
OI
Oil

Me(C:C)$_2$ ⟨S⟩ C:C.CH:CH$_2$

Tet. Lett., 1969, 5111.

3302-022

THIOPHENE, 2-(4-Cl-3-OH-
-butynyl)-5-(penta-1, 3-diynyl)-   C$_{13}$H$_9$OSCl   248
R = H   Oil

Me(C:C)$_2$ ⟨S⟩ C:C.CHOR.CH$_2$Cl

Ber., 1970 (103) 834.

3302-027

THIOPHENE, 2-(4-OAc-3-Cl-
-butynyl)-5-pent-3-en-1-ynyl)-   C$_{15}$H$_{13}$O$_2$SCl   292

MeCH:CHC:C ⟨S⟩ C:C.CHCl.CH$_2$OAc

Chem. Abstr., 1967,
65023.

3302-023

THIOPHENE, 2-(3-OAc-4-chloro-
-but-1-ynyl)-5-(1, 3-pentadiynyl)-   C$_{15}$H$_{11}$O$_2$SCl   290
Oil

R = Ac

Ber., 1970 (103) 834.

3302-028

THIOPHENE, 2-(4-OAc-3-Cl-
-butynyl)-5-(penta-1, 3-diynyl)-   C$_{15}$H$_{11}$O$_2$SCl   290
R = Cl

Me(C:C)$_2$ ⟨S⟩ C:C.CHR.CH$_2$OAc

Chem. Abstr., 1967,
65023.

3302-024

THIOPHENE, 2-(hexa-3, 5-dien-1-
-ynyl)-5-(prop-1-ynyl)-   C$_{13}$H$_{10}$S   198
OI
Oil

MeC:C ⟨S⟩ C:C(CH:CH)$_2$H

Tet. Lett., 1965, 297.

3302-029

THIOPHENE, 2-(4-OAc-3-OH-
-butynyl)-5-(penta-1, 3-diynyl)-   C$_{15}$H$_{12}$O$_3$S   272
OI
Oil

R = OH

Ber., 1970, 834.

3302-025

THIOPHENE, 2-(hept-5-en-1, 3-
-diynyl)-5-(1-OH-2-OAc-ethyl)-   C$_{15}$H$_{14}$O$_3$S   274

MeCH:CH(C:C)$_2$ ⟨S⟩ CHOH.CH$_2$OAc

Ber., 1967, 1206.

3302-030

THIOPHENE, 2-(4-Cl-3-oxo-
-butynyl)-5-(penta-1, 3-diynyl)-   C$_{13}$H$_7$OSCl   246
OI

Me(C:C)$_2$ ⟨S⟩ C:C.CO.CH$_2$Cl

Ber., 1970 (103) 834.

3302-026

THIOPHENE, 2-CH$_2$OAc-5-
-(4-furyl-but-3-en-1-ynyl)-   C$_{15}$H$_{12}$O$_3$S   272
OI

AcOCH$_2$ ⟨S⟩ C:C.CH:CH ⟨O⟩

Ber., 1966, 135.

3302-031

512

THIOPHENE, 2-Ph-5-propynyl-	$C_{13}H_{10}S$	198 OI 42^0C	BITHIENYL, 5-(3-Cl-4-OAc-but-en-1-ynyl)-2, 2'-	$C_{14}H_9O_2S_2Cl$	308 OI

Ph—(thiophene)—C:C.Me

Acta Chem. Scand., 1958, 765.

3302-032

(bithienyl)—C:C.C(Cl):CHOAc

Ber., 1967, 1193.

3303-005

ARCTIC ACID $C_{12}H_8O_2S_2$ 248 OI

HOOC—(bithienyl)—C:C.Me

Chem. Abstr., 1971 (74) 108136.

3303-001

BITHIENYL, 5-(4-Cl-3-OH-butynyl)-2, 2'- $C_{12}H_9OS_2Cl$ 284 55^0C

(bithienyl)—C:C.CHOH.CH$_2$Cl

JCS, 1966, 1101.

3303-006

BITHIENYL, 5-(but-3-en-1-ynyl)-2, 2'- $C_{12}H_8S_2$ 216 OI Oil

(bithienyl)—C:C.CH:CH$_2$

JCS, 1966, 89.

3303-002

$C_{18}H_{16}O_2S_2$ 328 OI

ROCH$_2$—(bithienyl)—C:C.CH:CH$_2$

R = Angeloyl

Ber., 1970 (103) 834.

3303-007

BITHIENYL, 5-(4-OH-but-1-ynyl)-2, 2'- $C_{12}H_{10}OS_2$ 234 OI 67^0C

R = H

(bithienyl)—C:C.CH$_2$CH$_2$OR

JCS, 1965, 7109.

3303-003

$C_{18}H_{16}O_2S_2$ 328 OI

R = Tigloyl

Ber., 1970 (103) 834.

3303-008

BITHIENYL, 5-(4-OAc-but-1-ynyl)-2, 2'- $C_{14}H_{12}O_2S_2$ 276 OI 68^0C

R = Ac

JCS, 1965, 7109.

3303-004

$C_{17}H_{16}O_2S_2$ 316

R = i-Butyryl

Ber., 1970 (103) 834.

3303-009

$C_{18}H_{16}O_2S_2$ 328 R = Senecioyl Ber., 1970 (103) 834. 3303-010	$C_{18}H_{16}O_2S_3$ 360 OI 83^0C R = Angeloyl Ber., 1970 (103) 834. 3304-005
TERTHIENYL $C_{12}H_8S_3$ 248 OI 93^0C JCS, 1965, 7109. 3304-001	$C_{18}H_{16}O_2S_3$ 360 OI 61^0C R = Tigloyl Ber., 1970 (103) 834. 3304-006
TERTHIENYL, 2-CH$_2$OH-α- $C_{13}H_{10}OS_3$ 278 OI 150^0C R = H Tet. Lett., 1966, 4227. 3304-002	$C_{18}H_{16}O_2S_3$ 360 OI 71^0C R = Senecioyl Ber., 1970 (103) 834. 3304-007
$C_{15}H_{12}O_2S_3$ 320 OI 114^0C R = Ac Ber., 1970 (103) 834. 3304-003	TERTHIENYL, 2-CHO-α- $C_{13}H_8OS_3$ 276 OI Ind. J. C., 1970 (8) 761. 3304-008
$C_{17}H_{16}O_2S_3$ 348 OI 61^0C R = i-Butyryl Ber., 1970 (103) 834. 3304-004	BRUGIEROL $C_3H_6O_2S_2$ 138 OI 84^0C Tet. Lett., 1972, 203. 3305-001

BRUGIEROL, iso-	$C_3H_6O_2S_2$	138 OI Oil
* epimer		
	Tet. Lett., 1972, 203.	
3305-002		

DITHIACYCLOHEPTANONE, -3-hexyl-4, 5-	$C_{11}H_{20}OS_2$	232 -65^0t Oil
	Chem. Comm., 1971, 503.	
3305-004		

LIPOIC ACID, α-	$C_8H_{14}O_2S_2$	206 $+104^0$b 48^0C
	JACS, 1956, 5920.	
3305-003		

LENTHIONINE	$C_2H_4S_5$	188 OI
	Tet. Lett., 1966, 573.	
3305-005		

34 MACROLIDES

The macrolides are macrocyclic lactones constructed from a linear precursor of acetate or propionate origin. One structurally distinct subset of the macrolides is the polyene group which have been brought together here (3402). The miscellaneous macrolides (3499) include a variety of lactones that are most suitably placed here, although not all of them would earn the description of "macrolide" as originally conceived. A notable exclusion from this class is the group of macrolidic compounds recently discovered and designated generically as the "rifamycins." These compounds are constructed from an aromatic amino precursor, and as such will be included in Volume III.

3401	Non-Polyene Macrolides
3402	Polyene Macrolides
3499	Misc. Macrolide-like Compounds

METHYMYCIN

$C_{25}H_{43}NO_7$ 469
+61^{0}m
198^{0}C

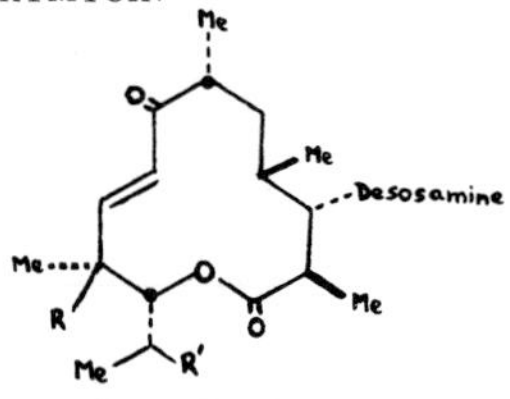

Tet. Lett., 1970, 1029.

R = OH: R' = H

3401-001

OLEADOMYCIN

$C_{35}H_{61}NO_{12}$ 687
-65^{0}m
110^{0}C

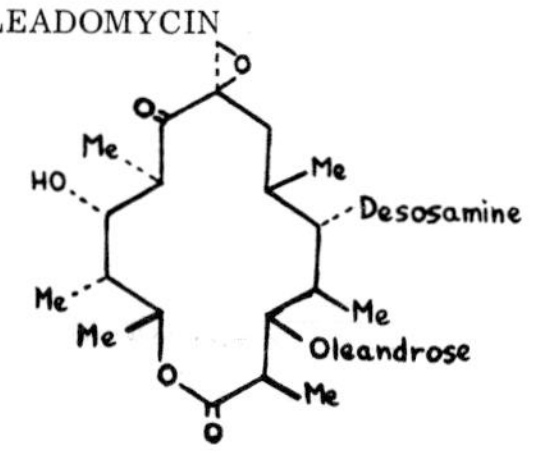

JACS, 1965, 1801.

3401-006

METHYMYCIN, neo-

$C_{25}H_{13}NO_7$ 469
+93^{0}c
157^{0}C

R = H: R' = OH

Tet., 1959 (4) 369.

3401-002

NARBONOLIDE

$C_{20}H_{32}O_5$ 352
+89^{0}m
126^{0}C

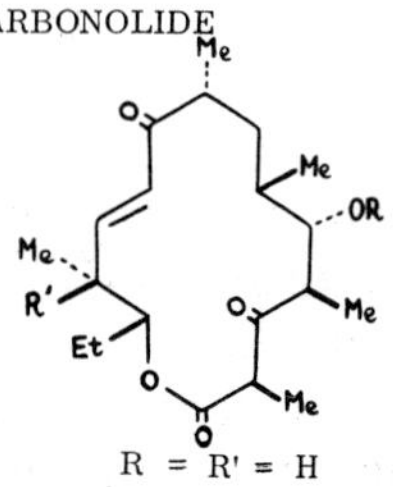

Chem. Comm.,
1971, 304.

R = R' = H

3401-007

NARBOMYCIN

$C_{28}H_{47}NO_7$ 509
+69^{0}c
114^{0}C

R = Desosaminyl
R' = H

Helv., 1962 (45) 4.

3401-008

CINEROMYCIN-B

$C_{17}H_{26}O_4$ 294

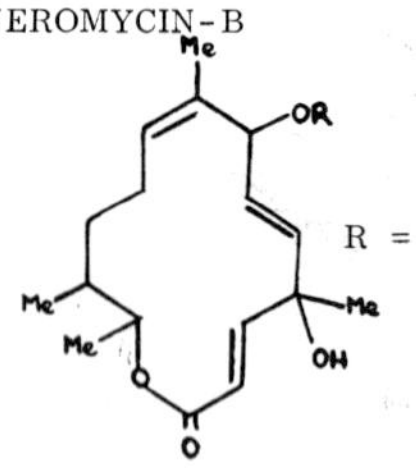

R = H

Chem. Pharm. Bull.,
1971. 655.
J. Antibiotics,
1966 (19) 56.

3401-004

PICROMYCIN

$C_{28}H_{47}NO_8$ 525
+6^{0}e
166^{0}C

R = Desosaminyl
R' = OH

JACS, 1968, 4748.

3401-009

ALBOCYCLINE

$C_{18}H_{28}O_4$ 308

ac: 84^{0}C

R = Me

Chem. Pharm. Bull.,
1971, 655.

3401-005

ERYTHROMYCIN-B

$C_{37}H_{67}NO_{12}$ 717
-78^{0}e
201^{0}C

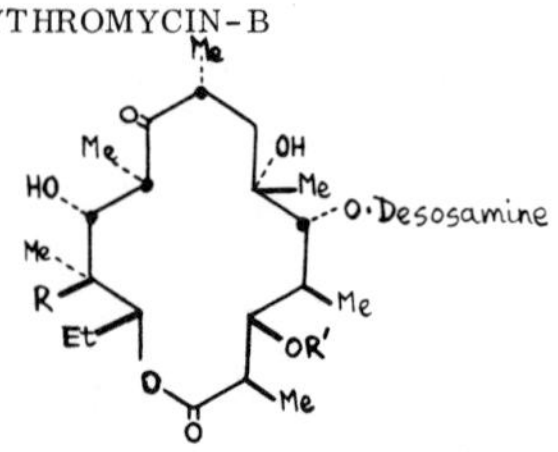

JACS, 1957, 6070.

R = H; R' = Cladinosyl

3401-010

ERYTHROMYCIN-C	$C_{36}H_{65}NO_{13}$ 719	
	121°C	
R = OH		
R' = Mycarosyl	Z. Nat., 1962 (17B) 852.	
3401-011		

MEGALOMICIN-C2	$C_{49}H_{86}N_2O_{17}$ 974	
R = Ac		
R' = CO.Et	U. S. Pat. 3632750 (1972).	
3401-016		

ERYTHROMYCIN-A	$C_{37}H_{67}NO_{13}$ 733	
	-78°e	
	191°C	
R' = Cladinosyl		
R = OH		
	Tet. Lett., 1965, 679.	
	Ber., 1963, 2867.	
3401-012		

KUJIMYCIN-A $C_{40}H_{70}O_{15}$ 790

114°C

R = Arcanosyl

J. Antibiotics,
1970 (23) 448.

3401-017

MEGALOMICIN-A $C_{44}H_{80}N_2O_{15}$ 876

-90°

255°C

JACS, 1969, 7506.

R = R' = H

3401-013

LANKAMYCIN	$C_{42}H_{72}O_{16}$ 832	
	-94°m	
	148°C	
R = Ac-arcanosyl		
	JACS, 1970 (92) 4129.	
3401-018		

MEGALOMICIN-B	$C_{46}H_{82}N_2O_{16}$ 918	
R = H		
R' = Ac		
	U. S. Pat. 3632750 (1972).	
3401-014		

NIDDAMYCIN $C_{40}H_{65}NO_{14}$ 783

-43°m

132°C

R = H

Arzneimittel Forsch,
1962 (12) 1191.

3401-019

MEGALOMICIN-C1	$C_{48}H_{84}N_2O_{17}$ 960	
R = R' = Ac	U. S. Pat. 3632750 (1972).	
3401-015		

MAGNAMYCIN-B	$C_{42}H_{67}NO_{15}$ 825	
	-35°c	
	142°C	
R = Ac		
	Antibiot. Chemoth.,	
	1957 (7) 349.	
3401-020		

MAGNAMYCIN-A	$C_{42}H_{67}NO_{16}$	841 -56^0c 208^0C

R = Ac

*epoxy (α)   JACS, 1965, 4662.

3401-021

LEUCOMYCIN-A5	$C_{39}H_{65}NO_{14}$	771 -52^0c 121^0C

R = H

R' = n-Butyryl   J. Antibiotics, 1968 (21A) 272.

3401-026

CHALCOMYCIN	$C_{35}H_{56}O_{14}$	700 -44^0e 122^0C

JACS, 1965 (87) 1801.

R = Me

3401-022

LEUCOMYCIN-A1	$C_{40}H_{67}NO_{14}$	785 -66^0c 132^0C

R = H

R' = isovaleryl   J. Antibiotics, 1968 (21A) 199.

3401-027

NEUTRAMYCIN	$C_{34}H_{54}O_{14}$	686 -35^0e 222^0C

R = H

Experientia, 1969, 12.

3401-023

LEUCOMYCIN-A8	$C_{39}H_{63}NO_{15}$	785 -58^0c 148^0C

R = R' = Ac

J. Antibiotics, 1968 (21A) 272.

3401-028

LEUCOMYCIN-A9	$C_{37}H_{61}NO_{14}$	743 -65^0c

R = H
R' = Ac

J. Antibiotics, 1968 (21A) 272.

3401-024

LEUCOMYCIN-A6	$C_{40}H_{65}NO_{15}$	799 -56^0c 136^0C

R = Ac

R' = Propionyl   J. Antibiotics, 1968 (21A) 272.

3401-029

LEUCOMYCIN-A7	$C_{38}H_{63}NO_{14}$	757 -65^0c

R = H

R' = Propionyl   J. Antibiotics, 1968 (21A) 272.

3401-025

LEUCOMYCIN-A4	$C_{41}H_{67}NO_{15}$	813 -50^0c 126^0C

R = Ac

R' = n-butyryl   J. Antibiotics, 1968 (21A) 272.

3401-030

LEUCOMYCIN-A3	$C_{42}H_{69}NO_{15}$	827 -58^0m 125^0C

R = Ac

R' = isovaleryl

Tet. Lett., 1967 (7) 609.

3401-031

ANTIBIOTIC YL-704C1 $C_{41}H_{67}NO_{16}$ 829 -69^0c 126^0C

J. Antibiotics, 1971 (24) 904.

3401-036

ANTIBIOTIC YL-704C2 $C_{40}H_{65}NO_{15}$ 799 -42^0c 116^0C

R = Propionyl

J. Antibiotics, 1971(24) 904.

3401-032

ANTIBIOTIC SF-837A3 $C_{41}H_{65}NO_{15}$ 811 -44^0e 122^0C

R = Propionyl

J. Antibiotics, 1971 (24) 526.

3401-037

ANTIBIOTIC YL-704B1 $C_{41}H_{67}NO_{5}$ 813 -42^0c 132^0C

R = R' = Propionyl

Tet. Lett., 1971, 435.

3401-033

ANTIBIOTIC SF-837A4 $C_{42}H_{67}NO_{15}$ 825 -40^0e 120^0C

R = Butyryl

J. Antibiotics. 1971 (24) 526.

3401-038

ANTIBIOTIC SF-837A2 $C_{42}H_{69}NO_{15}$ 827 -68^0e 125^0C

R = Propionyl

R' = n-Butyryl

J. Antibiotics, 1971 (24) 526.

3401-034

ANTIBIOTIC YL-704W1 $C_{43}H_{69}NO_{15}$ 839 -32^0c 160^0C

R = Isovaleryl

J. Antibiotics, 1971 (24) 904.

3401-039

ANTIBIOTIC YL-704A1 $C_{43}H_{71}NO_{15}$ 841 -50^0c 122^0C

R = Propionyl

R' = Isovaleryl

Tet. Lett., 1971, 435.

3401-035

FOROMACIDIN-A $C_{43}H_{74}N_2O_{14}$ 842 -91^0e 134^0C

R = H

JACS, 1965, 4660.

3401-040

FOROMACIDIN-B $C_{45}H_{76}N_2O_{15}$ 884
-80^0e
130^0C

R = Ac

JACS, 1965, 4660.

3401-041

PRIMYCIN $C_{55}H_{103}N_3O_{16}$ 1061

JACS, 1970, 5816.

3401-046

FOROMACIDIN-C $C_{46}H_{78}N_2O_{15}$ 898
-79^0e
128^0C

R = Propionyl

JACS, 1965, 4660.

3401-042

VENTURICIDIN-A $C_{41}H_{67}NO_{11}$ 749
$+119^0$c
141^0C

R = 3-O-carbamoyl-2-deoxy-
-rhamnosyl

Helv., 1968 (15) 1293.
Experientia,
1971 (27) 604.

3401-047

TYLOSIN $C_{45}H_{79}NO_{17}$ 905
-46^0m
130^0C

Tet. Lett., 1970, 4737.

3401-043

VENTURICIDIN-B $C_{40}H_{66}O_{10}$ 706
$+100^0$c
149^0C

R = 2-deoxy-rhamnosyl

Helv., 1968 (51) 1293.
Experientia, 1971 (27) 604.

3401-048

CIRRAMYCIN-A1 $C_{31}H_{51}NO_{10}$ 597

amorph.

J. Antibiotics,
1969 (22) 89.

R = H

3401-044

PIMARICIN $C_{33}H_{47}NO_{13}$ 681
$+250^0$m
200^0C

R = Mycosamine
R' = Me

Tet. Lett., 1966, 3551.

3402-001

ANTIBIOTIC B-58941 $C_{37}H_{59}O_{12}N$ 709
-88^0c
229^0C

R = cinerulosyl

Bull. Chem. Soc. Jap..
1970, 292.

3401-045

LUCENSOMYCIN $C_{36}H_{53}NO_{13}$ 723
$+50^0$
150^0C

R = Mycosamine
R' = C_4H_9

Tet. Lett., 1966, 3567.

3402-002

TETRIN-A $C_{34}H_{51}NO_{13}$ 681
$+8^0$p
$>350^0$C

OR
COOH
OH
OH
Me
OH
OH
Me
O
O
X

JACS, 1971, 3738.

X = H
R = Mycosamine

3402-003

MYCOTICIN-A $C_{36}H_{58}O_{10}$ 650

OH OH OH OH OH OH
Me OH Me
R O O OH

R = Me

JACS, 1967, 1535.

Me

3402-008

TETRIN-B $C_{34}H_{51}NO_{14}$ 697
$+43^0$m
amorph.

X = OH

JACS, 1971, 3747.

3402-004

MYCOTICIN-B $C_{37}H_{60}O_{10}$ 664

R = Et

JACS, 1967. 1535.

3402-009

RIMOCIDIN $C_{38}H_{63}NO_{13}$ 741

OH
OH
OH OH OH OH OH
Pr O O
Me OH
O

JACS, 1966, 4221.

(+ Mycosamine ?)

3402-005

NYSTATINOLIDE $C_{41}H_{64}O_{14}$ 880

OR
Me
Me
OH
OH OH OH OH OH
COOH
Me O O OH OH
OH

R = H

Tet. Lett., 1969. 5319.

3402-010

FILIPIN $C_{35}H_{58}O_{11}$ 654
-148^0m
195^0C

OH
X
OH
Me
OH
HO
OH OH OH OH
Me O O OH
HO Pentyl

X = H

Tet. Lett., 1964, 2615.

3402-006

NYSTATIN-A1 $C_{47}H_{75}NO_{17}$ 925

R = Mycosamine

Tet. Lett., 1971, 685.

3402-011

LAGOSIN $C_{35}H_{58}O_{12}$ 670
-160^0m
235^0C

X = OH

JCS, 1964, 862.

3402-007

CANDIDIN $C_{47}H_{71}NO_{17}$ 921

OMycosamine
Me
OH
Me OH OH
OH
COOH
Me O O OH O O OH
OH

Tet. Lett., 1971, 1987.

3402-012

AMPHOTERICIN-B $C_{47}H_{73}NO_{17}$ 923
-34^0m
170^0C

OMycosamine

Me
HO
Me
COOH
Me
OH OH OH OH OH
O
OH

Tet. Lett., 1970, 3909.

3402-013

ZEARALENONE $C_{18}H_{22}O_5$ 318
-170^0m
164^0C

OH
Me
O
HO
O

Chem. Comm., 1967, 761.

3499-005

BUNDLIN-A $C_{25}H_{33}NO_7$ 459
-221^0m
206^0C

Me
O
Me
NHCO.R'
Me
O
O
RO
OH
Me

R = H
R' = Ac

Agr. Biol. Chem.,
1971 (35) 27.

3499-001

CURVULARIN $C_{16}H_{20}O_5$ 292

OH
O
*
HO
O
Me

JCS, 1959, 3146.

3499-006

BUNDLIN-B $C_{27}H_{35}NO_8$ 501
-212^0m
214^0C

R = R' = Ac

Agr. Biol. Chem.,
1971 (35) 27.

3499-002

CURVULARIN, dehydro- $C_{16}H_{18}O_5$ 290
-80^0e
224^0C

*dehydro

JCS, 1967, 947.

3499-007

T-2636F $C_{25}H_{35}NO_7$ 461
-210^0f
178^0C

R = H; R' = .CHOH.Me

J. Antibiotics,
1971 (24) 23.

3499-003

COLLETODIOL $C_{14}H_{20}O_6$ 284

OH
OH
O
Me
O
O
Me
O

JCS, 1969, 911.

3499-008

T-2636D $C_{27}H_{37}NO_8$ 503
-226^0m
190^0C

R = Ac
R' = .CHOH.Me

J. Antibiotics,
1971 (24) 13.

3499-004

PYRENOPHORIN $C_{16}H_{20}O_6$ 308
-50^0c
175^0C

Me
O
O
O
O
O
Me

Tet. Lett., 1965, 4675.

3499-009

NONACTIN	$C_{40}H_{64}O_{12}$ 736 0^0 148^0C	

NONACTIN $C_{40}H_{64}O_{12}$ 736 0^0 148^0C

J. Antibiotics, 1965 (18A) 128.

$R_1 = R_2 = R_3 = R_4 = Me$

3499-010

BORRELIDIN $C_{28}H_{43}NO_6$ 489 -28^0e 145^0C

Experientia, 1966, 355.
Helv., 1967, 731.

3499-015

MONACTIN $C_{41}H_{66}O_{12}$ 750 $+2^0c$ 63^0C

$R_1 = Et; R_2 = R_3 = R_4 = Me$

J. Antibiotics. 1965 (18A) 128.

3499-011

DODEC-8-ENOIC ACID LACTONE, -11-OH-t- $C_{12}H_{20}O_2$ 196 $+73^0c$ Oil

Can. J. C., 1971 (49) 2029.

3499-016

DINACTIN $C_{42}H_{68}O_{12}$ 766 $+3^0c$ 67^0C

$R_1 = R_3 = Me$
$R_2 = R_4 = Et$

J. Antibiotics. 1965 (18A) 128.

3499-012

EXALTOLIDE $C_{15}H_{28}O_2$ 240 OI 31^0C

Ber. . 1927 (60) 902.

3499-017

TRINACTIN $C_{43}H_{70}O_{12}$ 780 $+2^0c$ 68^0C

$R_1 = Me;$
$R_2 = R_3 = R_4 = Et$

Helv., 1962, 129.

3499-013

AMBRETTOLIDE $C_{16}H_{28}O_2$ 252 OI Oil

Ber., 1927 (60) 902.

3499-018

TETRANACTIN $C_{44}H_{72}O_{12}$ 792 0^0 105^0C

$R_1 = R_2 = R_3 = R_4 = Et$

J. Antibiotics, 1971 (24) 418.

3499-014

CORIOLIDE $C_{18}H_{30}O_2$ 278 $+27^0m$ 40^0C

JOC, 1970, 1916.

3499-019

LASIODIPLODIN, des-O-Me- $C_{16}H_{22}O_4$ 278

128°C

R = H

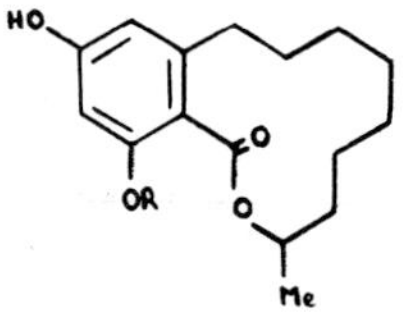

JCS, 1971, 1623.

3499-020

LASIODIPLODIN $C_{17}H_{24}O_4$ 292

184°C

R = Me

JCS, 1971, 1624.

3499-021

35   CARBOHYDRATES

The sugars are probably most significant in the field of natural products in their widespread occurrence as glycosides. A large number of the compounds isolated as "naturally occurring" are probably artifacts derived by hydrolysis during extraction of a parent glycoside. Many of the plant steroids and triterpenes occur as glycosides and are isolated as the aglycones (sapogenins); where the glycosides are themselves isolated, there can be some debate as to whether the saponins are "genuine" or not. Modern techniques of analysis, including the use of different hydrolysis enzymes, have been brought to bear on this problem. It has been the general policy in this handbook to regard aglycones as significant for classification purposes, and similarly the sugar moieties are included in this carbohydrate class even though they may have only been isolated as glycosides.

Included in this class are the cyclitols (3501) and aminocyclitols (3502). The latter are particularly significant because of their history as chemo- therapeutic agents (generally known as the "aminoglycoside antibiotics")

While some of the lower molecular weight oligosaccharides have been included here, no real attempt to cover the polymeric saccharides has been made.

3501 Cyclitols	3502 Amino-Cyclitols
3503 Pentoses	3504 Hexoses
3505 Heptoses	3506 Disaccharides
3507 Trisaccharides	3508 Oligosaccharides
3509 Amino-sugars	5310 Branched sugars

CYCLOHEXANE- TETROL,
-D-(+)-1, 4/2, 5-

$C_6H_{12}O_4$ 148
+22^{0}w
206^{0}C

Tet. Lett., 1966, 1527.

3501-001

QUERCITOL, 1-O-glucosyl-proto- $C_{12}H_{22}O_{10}$ 326

amorph.

R = Glucosyl

Chem. Pharm. Bull.,
1971, 1113.

3501-006

CONDURITOL- A

$C_6H_{10}O_4$ 146
OI
142^{0}C

Arch. Pharm., 1940
(278) 145.

3501-002

VIBURNITOL

$C_6H_{12}O_5$ 164
-50^{0}w
180^{0}C

JCS, 1952, 686.

3501-007

LEUCANTHEMITOL

$C_6H_{10}O_4$ 146
+101^{0}w
132^{0}C

Phytochem., 1966, 1091.

3501-003

INOSITOL, (D)-(+)-

$C_6H_{12}O_6$ 180
+68^{0}w
248^{0}C

Helv., 1936 (19) 1007.

$R_1 = R_3 = R_4 = H$

3501-008

QUERCITOL, (-)- Proto-

$C_6H_{12}O_5$ 164
-23^{0}w
233^{0}C

Aust. J. C., 1971, 431.

3501-004

INOSITOL 1-ME ETHER,
-(D)-(+)-

$C_7H_{14}O_6$ 194
+61^{0}w
207^{0}C

R_1 = Me
R_3 = H
R_4 = H

JCS, 1963, 5573.

3501-009

QUERCITOL, proto-

$C_6H_{12}O_5$ 164
+25^{0}w
235^{0}C

Helv., 1932, 948.

R = H

3501-005

PINITOL, D-

$C_7H_{14}O_6$ 194
+66^0
187^{0}C

R_1 = H
R_3 = Me
R_4 = H

Acta Chem. Scand.,
1956, 413.

3501-010

KASUGAMYCIN

R₁ = R₃ = H

$C_{14}H_{25}N_3O_9$ 379
+125⁰w
208⁰C

JACS, 1968, 6559.

3501-011

GALACTINOL

$C_{12}H_{22}O_{11}$ 342
+136⁰w
114⁰C

1-α-D-Galactosyl

JACS, 1953, 4507.

3501-016

INOSITOL, L-(-)-

$C_6H_{12}O_6$ 180
-65⁰w
238⁰C

Helv., 1936 (19) 1007.

3501-012

INOSITOL, 1-O-mannosyl-myo-

$C_{12}H_{22}O_{11}$ 342
+44⁰w
233⁰C

1-α-D-Mannosyl

Proc. C. S., 1963, 57.

3501-017

QUEBRACHITOL

$C_7H_{14}O_6$ 194
-80⁰w
191⁰C

Cf 012
2-O-Me ether

JCS, 1952, 686.

3501-013

MANNIMOSITOSE

$C_{18}H_{32}O_{14}$ 504
+77⁰
250⁰C

1-α-D-(6-O-α-D-mannosyl-
-mannosyl)-

Proc. C. S., 1963, 57.

3501-018

PINITOL, L-(-)-

$C_7H_{14}O_6$ 194
-65⁰w
186⁰C

Cf 012
3-O-Me ether

Compt. Rend., 1956
(243) 1913.

3501-014

BORNESITOL, (+)-D-

$C_7H_{14}O_6$ 194
+32⁰w
202⁰C

1-Me ether

JACS, 1954, 6130.

3501-019

INOSITOL, myo-

$C_6H_{12}O_6$ 180
0⁰
218⁰C

JACS, 1948, 4050.

3501-015

ONONITOL, D-

$C_7H_{14}O_6$ 194
+7⁰w
172⁰C

4-Me ether

Compt. Rend., 1962
(255) 1770.

3501-020

SEQUOYITOL	$C_7H_{14}O_6$	194
		235^0C
5-Me ether		
	Acta Chem. Scand., 1956, 413.	
3501-021		

SCYLLITOL	$C_6H_{12}O_6$	180
		OI
		350^0C

R = H

Acta Chem. Scand., 1955, 1097.

3501-026

DAMBONITOL	$C_8H_{16}O_6$	208
		0^0
		210^0C
1,3-di-Me ether	JCS, 1956, 480.	
3501-022		

SCYLLITOL, 1-O-Glu-	$C_{12}H_{22}O_{11}$	342
		amorph.

R = Glucosyl

Chem. Pharm. Bull., 1971, 1113.

3501-027

LIRIODENDRITOL	$C_8H_{16}O_6$	208
		-25^0w
		224^0C
1,4-di-Me ether	JCS, 1961, 4718.	
3501-023		

MYTILITOL	$C_7H_{14}O_6$	194
		266^0C

Acta Chem. Scand., 1957 (11) 506.

3501-028

PHYTIC ACID	$C_6H_{18}O_{24}P_6$	660

Hexa-phosphonyl
Myo-inositol

Biochem. J., 1936 (30) 252.

3501-024

LAMINITOL	$C_7H_{14}O_6$	194
		-3^0w
		266^0C

* epimer

Acta Chem. Scand., 1955 (10) 1323.

3501-029

BORNESITOL, (-)-L-	$C_7H_{14}O_6$	194
		-32^0w
		204^0C

JCS, 1953, 1192.

3501-025

STREPTAMINE	$C_6H_{14}N_2O_4$	178
		$>290^0C$

R = H

Ber., 1956 (89) 1152.

3502-001

STREPTIDINE $C_8H_{18}N_6O_4$ 262 R = .C(:NH)NH$_2$ JACS, 1963, 3896. 3502-002	**STREPTAMINE, 2-deoxy-** $C_6H_{14}N_2O_3$ 162 256°C R = H Tet. Lett., 1967, 2671. 3502-007
STREPTOMYCIN-A $C_{21}H_{39}N_7O_{12}$ 581 -95°w R = CHO R' = H Tet. Lett., 1968, 4725. 3502-003	**PAROMAMINE** $C_{12}H_{25}N_3O_7$ 323 +114°w J. Antibiotics, 1966 (19) 88. 3502-008
STREPTOMYCIN-B $C_{27}H_{49}N_7O_{17}$ 743 3HCl:-47°w 3HCl:190°C R = CHO R' = Mannosyl JACS, 1952, 5461. 3502-004	**NEOMYCIN-A** $C_{12}H_{26}N_4O_6$ 322 +123°w 250°C JACS, 1961, 3723. 3502-009
STREPTOMYCIN, dihydro- $C_{21}H_{41}N_7O_{12}$ 583 -88°w 260°C R = CH$_2$OH R' = H Chem. Pharm. Bull., 1954, 343. 3502-005	**RIBOSTAMYCIN** $C_{17}H_{34}N_4O_{10}$ 454 +42°w 193°C R=H Agr. Biol. Chem., 1970 (34) 980. 3502-010
STREPTOMYCIN, OH- $C_{21}H_{39}N_7O_{13}$ 597 -91°w JACS, 1951, 2290. 3502-006	**NEOMYCIN-B** $C_{23}H_{46}N_6O_{13}$ 614 +83° JACS, 1963, 1547. 3502-011

NEOMYCIN-C *epimer 3502-012	$C_{23}H_{46}N_6O_{13}$ 614 JACS, 1963, 1547.	KANAMYCIN-A R = OH R' = NH₂ 3502-017	$C_{18}H_{36}N_4O_{11}$ 484 +146° 263°C Tet. Lett., 1968, 1875.

KANAMYCIN-A
$C_{18}H_{36}N_4O_{11}$ 484
+146°
263°C

R = OH
R' = NH₂

Tet. Lett., 1968, 1875.

3502-017

PAROMOMYCIN-I
$C_{23}H_{45}N_5O_{14}$ 615
+64°w

R = H

JACS, 1967, 3364.

3502-013

KANAMYCIN-B
$C_{18}H_{37}N_5O_{10}$ 483
+135°w
170°C

R = R' = NH₂

J. Antibiotics,
1968 (21) 424.

3502-018

PAROMOMYCIN-II
$C_{23}H_{45}N_5O_{14}$ 615
+96°w

R = H

*epimer

JACS, 1967, 3364.

3502-014

KANAMYCIN-C
$C_{18}H_{36}N_4O_{11}$ 484
+126°w
270°C

R = NH₂
R' = OH

J. Antibiotics,
1968 (21) 162.

3502-019

PAROMOMYCIN, mannosyl-
$C_{29}H_{55}N_5O_{19}$ 777
+74°w
200°C

R = Mannosyl

J. Antibiotics,
1972 (25) 317.

3502-015

TOBRAMYCIN
$C_{18}H_{37}N_5O_9$ 467

Ant. Agents Chemother.,
1972 (1) 41.

3502-020

SISOMICIN
$C_{19}H_{37}N_5O_7$ 447
+189°w
199°C

Chem. Comm., 1971, 285.

3502-016

GENTAMICIN-A
$C_{18}H_{36}N_4O_{10}$ 468
+146°w
108°C

JACS, 1970, 1697.

3502-021

GENTAMICIN-D	$C_{19}H_{39}N_5O_7$	449

R = R' = H

JCS, 1971, 3126.

3502-022

GENTAMICIN-X	$C_{19}H_{38}N_4O_9$	466
		$+154^0w$

R = NH_2; R' = H;
R'' = OH

Belg. Pat. 768796 (1971).

3502-027

GENTAMICIN-C2	$C_{20}H_{41}N_5O_7$	463
		$+160^0w$
		107^0C

R = Me; R' = H

JCS, 1971. 3126.

3502-023

LIVIDOMYCIN-A	$C_{29}H_{55}N_5O_{18}$	761
		$+72^0w$
		197^0C

R = Mannosyl

J. Antibiotics. 1972 (25) 128.

3502-028

GENTAMICIN-C1	$C_{21}H_{43}N_5O_7$	477
		$+158^0w$
		94^0C

R = R' = Me

JCS. 1971, 3126.

3502-024

BUTIROSIN-A	$C_{21}H_{41}N_5O_{12}$	555
		$+26^0w$
		149^0C

Tet. Lett. , 1971. 2625.

3502-029

GENTAMICIN-B1	$C_{20}H_{40}N_4O_9$	480
		$+157^0w$

R = OH; R' = Me
R'' = NH_2

Belg. Pat. 768796 (1971).

3502-025

BUTIROSIN-B	$C_{21}H_{41}N_5O_{12}$	555
		$+33^0w$
		146^0C

*epimer

Tet. Lett. , 1971, 2625.

3502-030

GENTAMICIN-B	$C_{19}H_{38}N_4O_9$	466
		$+132^0w$

R = OH; R' = H;
R'' = NH_2

Belg. Pat. 768796 (1971).

3502-026

BLUENSIDINE	$C_8H_{16}N_4O_6$	264
		HCl: -1^0w
		HCl: 192^0C

R = H

J. Antibiotics. 1972 (25) 281.

3502-031

BLUENSOMYCIN	$C_{21}H_{39}N_5O_{14}$ 585	
R = [structure]	J. Antibiotics, 1972 (25) 281.	
3502-032		

ANTIBIOTIC A-396-I	$C_{19}H_{35}N_3O_3$ 513 +12^{0}w 190^{0}C	
R = R' = H	J. Antibiotics, 1970 (23) 569.	
3502-037		

ACTINOSPECTACIN	$C_{14}H_{24}N_2O_7$ 332 + 8^{0}w 66^{0}C	
[structure]	Chem. Comm., 1971 1541.	
3502-033		

VALIDAMINE	$C_7H_{15}NO_4$ 177 HCl: +57^0 HCl: 230^{0}C	
[structure]	J. Antibiotics, 1971 (24) 59.	
3502-038		

HYOSAMINE	$C_7H_{16}N_2O_3$ 176	
[structure]	Tet. Lett., 1967, 2671.	
3502-034		

ARABINOSE, D-	$C_5H_{10}O_5$ 150 -105^{0}w 160^{0}C	
[structure]	JCS, 1930, 553.	
3503-001		

DESTOMYCIN-A	$C_{20}H_{37}N_3O_{13}$ 527 +19^{0}w 190^{0}C	
[structure] R = H; R' = Me	J. Antibiotics, 1966 (19) 139.	
3502-035		

ARABINOSE, L-	$C_5H_{10}O_5$ 150 +105^{0}w 160^{0}C	
Enantiomer	JACS, 1937, 1124.	
3503-002		

HYGROMYCIN-B	$C_{20}H_{37}N_3O_{13}$ 527 +20^{0}w	
R = Me; R' = H	J. Antibiotics, 1970 (23)569.	
3502-036		

RIBOSE, D-	$C_5H_{10}O_5$ 150 -24^{0}w 87^{0}C	
[structure]	JACS, 1934, 748.	
3503-003		

XYLOSE, D-

$C_5H_{10}O_5$　　150
　　　　　　　+19°w
　　　　　　　144°C

CHO
—OH
HO—
—OH
CH₂OH

JCS, 1923, 620.

3503-004

ARABITOL, D-

$C_5H_{12}O_5$　　152
　　　　　　　+8°
　　　　　　　103°C

CH₂OH
HO—
—OH
—OH
CH₂OH

Acta Chem. Scand.,
1955 (9) 1234.

3503-009

RIBULOSE, D-

$C_5H_{10}O_5$　　150
　　　　　　　-16°w
　　　　　　　Syrup

CH₂OH
=O
—OH
—OH
CH₂OH

Helv., 1935 (18) 80.

3503-005

RIBITOL

$C_5H_{12}O_5$　　152
　　　　　　　102°C

CH₂OH
—OH
—OH
—OH
CH₂OH

Monatsh., 1938 (72) 168.

3503-010

XYLULOSE, L-

$C_5H_{10}O_5$　　150
　　　　　　　+35°w
　　　　　　　Syrup

CH₂OH
=O
—OH
HO—
CH₂OH

Ber., 1935 (68) 24.

3503-006

LYXURONIC ACID, D-

$C_5H_8O_6$　　164

COOH
HO—
—OH
—OH
CHO

Bull. Agr. Chem. Soc.
Jap., 1958 (22) 271.

3503-011

RIBOSE, 2-deoxy-D-

$C_5H_{10}O_4$　　134
　　　　　　　-55°w
　　　　　　　88°C

CHO
—OH
—OH
CH₂OH

JCS. 1949, 1363.

3503-007

GALACTOSE, D-

$C_6H_{12}O_6$　　180
　　　　　　　+80°w
　　　　　　　167°C

CHO
—OH
HO—
HO—
—OH　　R = H
CH₂OR

J. Biol. Chem.,
1953 (204) 169.

3504-001

CORDYCEPOSE

$C_5H_{10}O_4$　　134
　　　　　　　-26°e
　　　　　　　Syrup

CHO
—OH
—OH
CH₂OH

JACS, 1964. 948.

3503-008

GALACTOSE 6-ME ETHER

$C_7H_{14}O_6$　　194
　　　　　　　+77°w
　　　　　　　125°C

R = Me

Bull. Chem. Soc. Jap.,
1961 (34) 1048.

3504-002

GALACTOSE 6- PHOSPHATE $\quad$ $C_6H_{13}O_9P$ $\quad$ 260
$+36^0$w

R = PO(OH)$_2$

JCS, 1951, 980.

3504-003

GLUCOSE, D- $\quad$ $C_6H_{12}O_6$ $\quad$ 180
$+53^0$w
146^0C

CHO
—OH
HO—
—OH $\quad$ R = H
—OH
CH$_2$OR

Ann., 1909 (366) 277.

3504-008

GALACTOSE 1- PHOSPHATE $\quad$ $C_6H_{13}O_9P$ $\quad$ 260
$+143^0$

R = PO$_3$H$_2$

HOCH$_2$ O OR
HO OH
OH

JACS, 1950, 4824.

3504-004

GLUCOSE, 6- Ac - $\quad$ $C_8H_{14}O_7$ $\quad$ 222
$+48^0$w
133^0C

R = Ac

Nature, 1957 (179) 103.

3504-009

FLUORIDOSIDE $\quad$ $C_9H_{18}O_8$ $\quad$ 254
$+151^0$
86^0C

R = . CH(CH$_2$OH)$_2$

JCS, 1967, 1346.

3504-005

GLUCOSE, 1- O- Bu- $\quad$ $C_{10}H_{20}O_6$ $\quad$ 236
-38^0m
87^0C

R = n- Butyl

HOCH$_2$ O OR
HO OH
OH

Helv., 1969, 716.

3504-010

GLYCEROL, O-β-D-Galacto-
-furanosyl- $\quad$ $C_9H_{18}O_8$ $\quad$ 254
-78^0w

CH$_2$OH
—OH
H— —H
OH
HOCH$_2$
HO OH

Acta Chem, Scand.,
1967 (21) 2083.

3504-006

TULIPOSIDE- A $\quad$ $C_{11}H_{18}O_8$ $\quad$ 278
$+64^0$w
amorph.

R = . CO. C(:CH$_2$) CH$_2$CH$_2$OH

Tet. Lett., 1968, 701.

3504-011

GALACTOSE, L- $\quad$ $C_6H_{12}O_6$ $\quad$ 180
-18^0w
167^0C

CHO
HO —
—OH
—OH
HO —
CH$_2$OH

Nature, 1940 (146) 559.

3504-007

TULIPOSIDE- B $\quad$ $C_{11}H_{18}O_9$ $\quad$ 294
$+56^0$m
amorph.

R = . CO. C(:CH$_2$) CHOH. CH$_2$OH

Tet. Lett., 1968, 701.

3504-012

KARAKIN	$C_{15}H_{21}N_3O_{15}$	483
		$+5^0$
$R = .COCH_2CH_2NO_2$		122^0C

J. Sci. Food Agr.,
1951 (2) 54.

3504-013

ACOFRIOSE, L- $C_7H_{14}O_5$ 178 $+38^0w$ 114^0C

R = Me

Helv., 1955 (38) 499.

3504-018

MANNOSE, D- $C_6H_{12}O_6$ 180 $+15^0$ 133^0C

CHO
HO
RO
—OH
—OH   R = H
CH$_2$OH

JCS, 1950, 3292.

3504-014

TALOSE, 6-desoxy-L- $C_6H_{12}O_5$ 164 -20^0w 120^0C

CHO
—OH
—OH
—OH
HO
Me

Chem. Ind., 1963, 375.

3504-019

MANNOSE-3-ME ETHER $C_7H_{14}O_6$ 194

Syrup

R = Me

Biochem. J.,
1966 (98) 15.

3504-015

VALLAROSE, L- $C_7H_{14}O_5$ 178

CHO
—OH
MeO
HO
HO
Me

Helv., 1965 (48) 83.

3504-020

QUINOVOSE, D- $C_6H_{12}O_5$ 164 $+30^0w$ 140^0C

CHO
—OH
HO
—OH
—OH
Me

Ber., 1929 (62) 373.

3504-016

SORDAROSE $C_7H_{14}O_5$ 178 $+29^0w$

CHO
HO
—OH
—OMe
—OH
Me

Helv., 1971 (54) 1191.

3504-021

RHAMNOSE, L- $C_6H_{12}O_5$ 164 $+9^0w$ 93^0C

CHO
—OH
—OR
HO
HO   R = H
Me

Biochem. J.,
1948 (42) 251.

3504-017

FUCOSE, D- $C_6H_{12}O_5$ 164 $+89^0w$ 140^0C

CHO
—OH
RO
HO   R = H
—OH
Me

Ber., 1932 (65) 307.

3504-022

DIGITALOSE $C_7H_{14}O_5$ 178 +106°w 106°C R = Me Helv., 1949 (32) 163. 3504-023	**JAVOSE** $C_7H_{14}O_5$ 178 -40°w 113°C CHO, OMe, OH, OH, Me JCS, 1967, 1503. 3504-028
CURACOSE $C_7H_{14}O_5$ 178 +98°w 126°C CHO, OH, HO, MeO, OH, Me Carbohydrate Res., 1966 (2) 56. 3504-024	**ACOVENOSE** $C_7H_{14}O_5$ 178 Gum CHO, OH, OMe, OH, OH, Me Helv., 1950 (33) 485. 3504-029
FUCOSE, L- $C_6H_{12}O_5$ 164 -76°w 145°C CHO, HO, OH, OH, HO, Me Biochem. J., 1946 (40) 124. 3504-025	**DIGITOXOSE** $C_6H_{12}O_4$ 148 +46°w 111°C CHO, OR, OH, OH, Me R = H Helv., 1952 (35) 93. 3504-030
THEVETOSE, D- $C_7H_{14}O_5$ 178 +84°w 116°C CHO, OH, MeO, OH, OH, CH₂OH Helv., 1952 (35) 195. 3504-026	**CYMAROSE** $C_7H_{14}O_4$ 162 +55°w 93°C R = Me Helv., 1952 (35) 93. 3504-031
THEVETOSE, L- $C_7H_{14}O_5$ 178 -166°w 129°C enantiomer Helv., 1952 (35) 195. 3504-027	**OLIVOSE** $C_6H_{12}O_4$ 148 +45°w Oil CHO, HO, OH, OH, Me Tet., 1966, 2785. 3504-032

BOIVINOSE	$C_6H_{12}O_4$	148 -15^0a 101^{0}C

CHO
—
—OR   R = H
HO—
—OH
Me

Helv., 1953, 302.

3504-033

OLIVOMOSE	$C_7H_{14}O_4$	162 $+80^0$w 152^{0}C

CHO
—
HO—
MeO—
—OH
Me

Tet., 1966, 2785.

3504-038

SARMENTOSE	$C_7H_{14}O_4$	162 $+16^0$w 78^{0}C

R = Me

Helv., 2950 (33) 446.

3504-034

ABEQUOSE	$C_6H_{12}O_4$	148 -3^0w Gum

CHO
—OH
—
HO—
—OH
Me

Compt. Rend., 1958 (246) 2417.

3504-039

OLIOSE	$C_6H_{12}O_4$	148 $+46^0$w

CHO
—
RO—   R = H
HO—
—OH
Me

Tet. Lett., 1966, 1431.

3504-035

COLITOSE	$C_6H_{12}O_4$	148 $+4^0$w

CHO
HO—
—
—OH
HO—
Me

Compt. Rend., 1958 (246) 2417.

3504-040

DIGINOSE	$C_7H_{14}O_4$	162 $+56^0$w

R = Me

Helv., 1948 (31) 1630.

3504-036

ASCARYLOSE	$C_6H_{12}O_4$	148 -26^0w Oil

CHO
—OH
—
HO—
HO—
Me

Compt. Rend., 1958 (246) 2417.

3504-041

CHROMOSE-D	$C_8H_{14}O_5$	190 $+87^0$w 118^{0}C

R = Ac

Tet., 1966, 2785.

3504-037

TYVELOSE	$C_6H_{12}O_4$	148 $+25^0$w Oil

CHO
HO—
—
—OH
—OH
Me

Ann., 1959 (620) 8.

3504-042

CHALCOSE

CHO
—OH
MeO—
—OH
Me

$C_7H_{14}O_4$ 162
+76^{0}w
97^{0}C

Proc. C. S. , 1968, 279.

3504-043

SORBOSE, L-

CH_2OH
=O
HO—
—OH
HO—
CH_2OH

$C_6H_{12}O_6$ 180
-44^{0}w
159^{0}C

Adv. Carb. Chem. ,
1952 (7) 99.

3504-048

AMICETOSE

CHO

—OH
—OH
Me

$C_6H_{12}O_3$ 132
+44^{0}a
Oil

JACS, 1964, 3592.

3504-044

TAGATOSE, D-

CH_2OH
=O
HO—
HO—
—OH
CH_2OH

$C_6H_{12}O_6$ 180
-5^{0}w
134^{0}C

Adv. Carb. Chem. ,
1952 (7) 99.

3504-049

FRUCTOSE, D-

CH_2OH
=O
HO—
—OH
—OH
CH_2OH

$C_6H_{12}O_6$ 180
-93^{0}w
103^{0}C

Adv. Carb. Chem. ,
1952 (7) 53.

3504-045

GLUCOSONE

CHO
=O
HO—
—OH
—OH
CH_2OH

$C_6H_{10}O_6$ 178
Syrup

Science, 1956 (124) 171.

3504-050

FRUCTOSE 6-PHOSPHATE

R = PO_3H_2 ; R' = H

$C_6H_{13}O_9P$ 259
Ba^+ +4^{0}w

Arch. Biochem. Biophys.,
1944 (3) 33.

3504-046

HEXOSE, 2, 5-dioxo-

CHO
=O
HO—
—OH
=O
CH_2OH

$C_6H_{10}O_6$ 178
-89^{0}w
176^{0}C

Chem. Ind. , 1963, 1925.

3504-051

FRUCTOSE 1, 6-DIPHOSPHATE

R = R' = PO_3H_2

$C_6H_{14}O_{12}P_2$ 340
+4^{0}w

Arch. Biochem. Biophys.,
1944 (3) 33.

3504-047

ARABOHEXOSE, 6-deoxy-5-oxo-D-

$C_6H_{10}O_5$ 162
-4^{0}m
Oil

Tet. Lett. , 1967, 2285.

3504-052

DULCITOL	$C_6H_{14}O_6$	182 OI 188^0C

CH_2OH
—OH
HO—
HO—
—OH
CH_2OH

JACS. 1956, 2844.

3504-053

POLYGALITOL	$C_6H_{12}O_5$	164 $+43^0$w 142^0C

∗ epimer

JACS. 1947, 706.

3504-058

IDITOL, L- $C_6H_{14}O_6$ 182
                                  -3^0w
                                  74^0C

CH_2OH
—OH
HO—
—OH
HO—
CH_2OH

Bull. Soc. Chim. Fr.,
1905 (33) 264.

3504-054

GLUCONIC ACID $C_6H_{12}O_7$ 196
                                     $+6^0$w
                                     153^0C

COOH
—OH
HO—
—OH
—OH
CH_2OH

Ind. Eng. Chem.,
1940 (32) 1379.

3504-059

MANNITOL, D- $C_6H_{14}O_6$ 182
                                    -2^0w
                                    166^0C

CH_2OH
—OH
—OH
—OH
—OH
CH_2OH

J. Soc. Chem. Ind.,
1948 (67) 165.

3504-055

GLUCONIC ACID, 2-keto- $C_6H_{10}O_7$ 194
                                              152^0C

COOH
=O
HO—
—OH
—OH
CH_2OH

J. Biol. Chem.,
1953 (204) 43.

3504-060

SORBITOL, D- $C_6H_{14}O_6$ 182
                                    -2^0w
                                    97^0C

CH_2OH
—OH
HO—
—OH
—OH
CH_2OH

JACS, 1937, 2264.

3504-056

GLUCONIC ACID, 5-keto- $C_6H_{10}O_7$ 194

COOH
—OH
HO—
—OH
=O
CH_2OH

J. Ferm. Tech.,
1946 (24) 22.

3504-061

STYRACITOL $C_6H_{12}O_5$ 164
                                  -72^0w
                                  157^0C

HOCH_2
HO—
HO—   —OH
       —OH

JACS, 1943, 1477.

3504-057

GLUCONIC ACID, 2,5-diketo- $C_6H_8O_7$ 192

COOH
=O
—OH
HO—
=O
CH_2OH

J. Biol. Chem.,
1953 (204) 43.

3504-062

GLUCURONIC ACID $C_6H_{10}O_7$ 194
+36°w
165°C

CHO
—OH
HO—
—OH
—OH
COOH

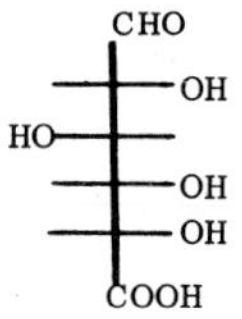

Natwiss., 1957 (179) 44.

3504-063

CINERULOSE-A $C_6H_{10}O_3$ 130

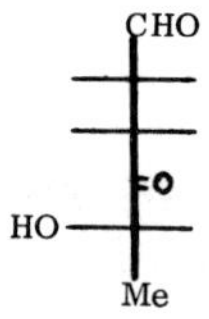

CHO
—
HO— =O
—
Me

Ant. Ag. Chemother.,
1970, 68.

3504-068

GULURONIC ACID $C_6H_{10}O_7$ 194

Syrup

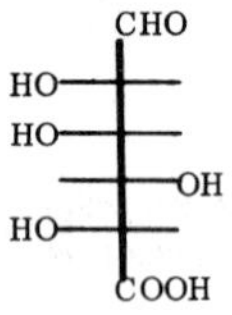

CHO
HO—
HO—
—OH
HO—
COOH

Z. Physiol. Ch.,
1955 (302) 186.

3504-064

MANNOHEPTULOSE $C_7H_{14}O_7$ 210
+30°w
152°C

CH₂OH
=O
HO—
HO—
—OH
—OH
CH₂OH

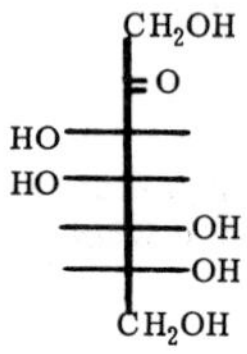

JACS, 1951, 730.

3505-001

GLUCARIC ACID $C_6H_{10}O_8$ 210
+21°w
125°C

COOH
—OH
HO—
—OH
—OH
COOH

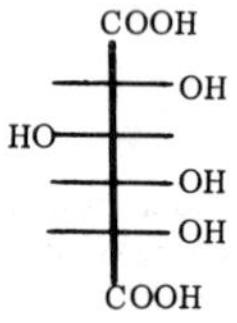

JCS, 1927, 3044.

3504-065

SEDOHEPTULOSE $C_7H_{14}O_7$ 210
+2°w
Syrup

CH₂OH
=O
HO—
—OH
—OH
—OH
CH₂OH

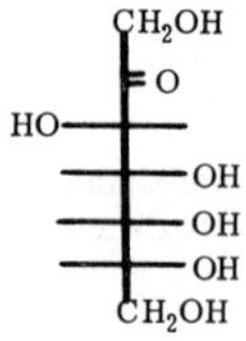

JACS, 1954, 2929.

3505-002

GALACTARIC ACID $C_6H_{10}O_8$ 210

214°C

COOH
—OH
HO—
HO—
—OH
COOH

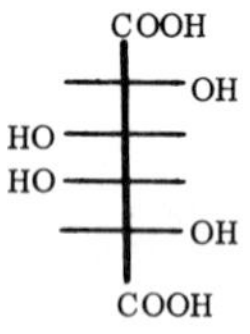

Nature, 1955 (49) 4201.

3504-066

CORIOSE $C_7H_{14}O_7$ 210
+22°w
170°C

CH₂OH
HO—
=O
—OH
—OH
—OH
CH₂OH

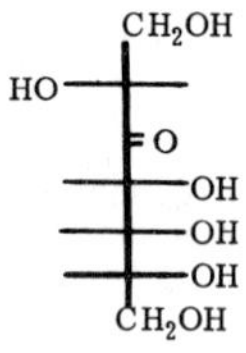

Tet., 1968, 6907.

3505-003

RHODINOSE $C_6H_{12}O_3$ 132
-12°

CHO
—
—
—OH
HO—
Me

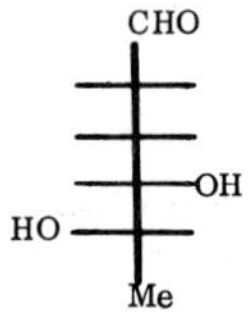

JACS, 1964, 3592.

3504-067

ANTIBIOTIC SF-666 $C_7H_{14}O_6$ 194
+38°
160°C

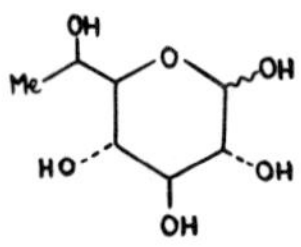

Chem. Abstr.,
1970 (73) 108247.

3505-004

KAMUSOL | $C_7H_{14}O_6$ | 194

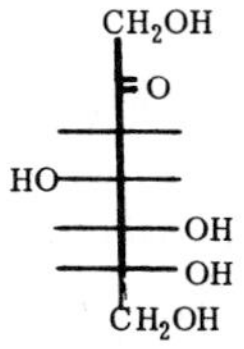

CH2OH
O
HO
OH
OH
CH2OH

Pak. J. Sci. Ind. Res., 1971 (14) 68.

3505-005

VOLEMITOL | $C_7H_{16}O_7$ | 212, $+2^0$w, 154^0C

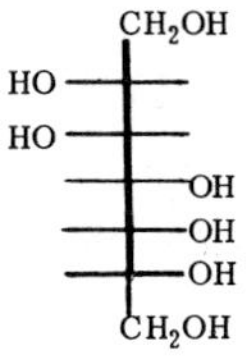

CH2OH
HO
HO
OH
OH
OH
CH2OH

Acta Chem. Scand., 1953 (7) 591.

3505-006

PERSEITOL | $C_7H_{16}O_7$ | 212, -1^0w, 188^0C

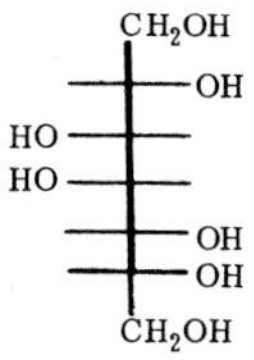

CH2OH
OH
HO
HO
OH
OH
CH2OH

JACS. 1948, 765.

3505-007

SIPHULITOL | $C_7H_{16}O_6$ | 196, -8^0w, 122^0C

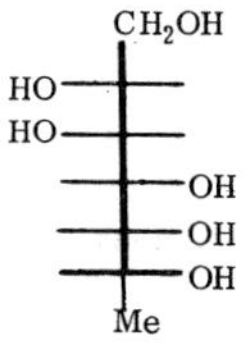

CH2OH
HO
HO
OH
OH
OH
Me

Acta Chem. Scand., 1962 (16) 543.

3505-008

ARABINOSE, 3-O-β-L--arabopyranosyl- | $C_{10}H_{18}O_9$ | 182, $+220^0$w

3β
Ara —— Ara

JCS, 1953, 4090.

3506-001

XYLOBIOSE | $C_{10}H_{18}O_9$ | 182, $+25^0$w, 185^0C

4β
Xyl —— Xyl

JACS. 1952, 3059.

3506-002

---

deleted

3506-003

---

deleted

3506-004

---

deleted

3506-005

UMBILICIN | $C_{11}H_{22}O_{10}$ | 314, -81^0w, 138^0C

HOCH2
HO
O
O
HO
OH
CH2OH
OH
OH
CH2OH

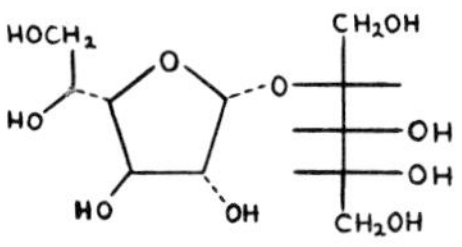

Acta Chem. Scand., 1962 (16) 2240.

3506-006

ARABINOPYRANOSE, 2-O-β-D--glucosyl- $C_{11}H_{20}O_{10}$ 312 $+23^0$w 2β Ara——Glu Acta Chem. Scand., 1969, 2914. 3506-007	MELIBIOSE, epi- $C_{12}H_{22}O_{11}$ 342 $+125^0$w 201^0C 6α Mann——Gal JCS, 1954, 295. 3506-012

VICIANOSE $C_{11}H_{20}O_{10}$ 312 $+40^0$w 210^0C 6α Glu——Ara Compt. Rend., 1910 (150) 180. 3506-008	PLANTEOBIOSE $C_{12}H_{22}O_{11}$ 342 $+125^0$w amorph. 6α Fruct——Gal Science, 1953 (117) 100. 3506-013

SAMBUBIOSE $C_{11}H_{20}O_{10}$ 312 202^0C 2α Glu——Xyl Acta Chem. Scand., 1969, 2213. 3506-009	SUCROSE $C_{12}H_{22}O_{11}$ 342 $+67^0$w 185^0C $\beta\ \alpha$ Fruct——Glu JACS, 1944, 1416. 3506-014

PRIMEVEROSE $C_{11}H_{20}O_{10}$ 312 -3^0w 210^0C 6β Glu——Xyl Ber., 1939 (72) 47. 3506-010	TURANOSE $C_{12}H_{22}O_{11}$ 342 $+76^0$w 157^0C 3α Fruct——Glu JOC, 1944, 470. 3506-015

ARABINOSE, 3-O-α-D--Galactopyranosyl- $C_{11}H_{20}O_{10}$ 312 $+152^0$w amorph. 3α Ara——Gal JCS, 1955, 269. 3506-011	TREHALOSE, α, α - $C_{12}H_{22}O_{11}$ 342 $+178^0$w 97^0C $\alpha\ \ \alpha$ Glu——Glu Adv. Carbohyd. Chem., 1963 (18) 201. 3506-016

KOJIBIOSE	$C_{12}H_{22}O_{11}$	342 +91^0

$$2\alpha$$
$$\text{Glu}\text{——}\text{Glu}$$

J. Agr. Chem. Soc. Jap. ,
1954 (28) 529.

3506-017

CELLOBIOSE	$C_{12}H_{22}O_{11}$	342 +35^{0}w 225^{0}C

$$4\beta$$
$$\text{Glu}\text{——}\text{Glu}$$

JACS 1953 (75) 1916.

3506-022

SOPHOROSE	$C_{12}H_{22}O_{11}$	342 +20^{0}w 196^{0}C

$$2\beta$$
$$\text{Glu}\text{——}\text{Glu}$$

JCS. 1960. 4213.

3506-018

LEUCROSE	$C_{12}H_{22}O_{11}$	342 +8^{0}w 162^{0}C

$$5\alpha$$
$$\text{Glu}\text{——}\text{Glu}$$

Natwiss. , 1964 (51) 163.

3506-023

SAKEBIOSE	$C_{12}H_{22}O_{11}$	342

$$3\alpha$$
$$\text{Glu}\text{——}\text{Glu}$$

J. Agr. Chem. Soc. Jap. ,
1955 (29) 861.

3506-019

MALTOSE, iso-	$C_{12}H_{22}O_{11}$	342 +120^{0}w

$$6\alpha$$
$$\text{Glu}\text{——}\text{Glu}$$

JACS. 1953, 5911.

3506-024

LAMINARIBIOSE	$C_{12}H_{22}O_{11}$	342 +20^{0}w 204^{0}C

$$3\beta$$
$$\text{Glu}\text{——}\text{Glu}$$

JCS. 1952, 1243.
Khim. Prir. Soedin. ,
1969. 595.

3506-020

GENTIOBIOSE	$C_{12}H_{22}O_{11}$	342 -11^{0}w 190^{0}C

$$6\beta$$
$$\text{Glu}\text{——}\text{Glu}$$

Ber. , 1961 (94) 2359.

3506-025

MALTOSE	$C_{12}H_{22}O_{11}$	342 +130^{0}w 102^{0}C

$$4\alpha$$
$$\text{Glu}\text{——}\text{Glu}$$

JACS. 1940, 1153.

3506-021

GALACTOSE, 3- β-Galactopyranosyl-	$C_{12}H_{22}O_{11}$	342 +62^{0}w 160^{0}C

$$3\beta$$
$$\text{Gal}\text{——}\text{Gal}$$

JCS. 1954. 2622.

3506-026

GALACTOSE, 4-α-Galactopyranosyl- $C_{12}H_{22}O_{11}$ 342

$+177^0$w

210^0C

4α
Gal——Gal

JACS, 1954, 1673.

3506-027

LACTOSE $C_{12}H_{22}O_{11}$ 342

(α) $+53^0$w

(α) 202^0C

4β
Glu——Gal

Chem. Revs. , 1926 (2) 85.

3506-032

GALACTOSE, 4-β-Galactopyranosyl- $C_{12}H_{22}O_{11}$ 342

$+68^0$w

204^0C

4β
Gal——Gal

Can. J. Chem. ,
1965 (43) 2508.

3506-028

MELIBIOSE $C_{12}H_{22}O_{11}$ 342

$+130^0$w

83^0C

6α
Glu——Gal

JACS, 1952, 5774.

3506-033

GALACTOSE, 6-β-Galactofuranosyl- $C_{12}H_{22}O_{11}$ 342

-28^0w

172^0C

Biochem. J. ,
1964 (90) 201.

3506-029

SCILLABIOSE $C_{12}H_{22}O_{10}$ 326

4β
Rha——Glu

Chem. Abstr. ,
1939 (33) 4202.

3506-034

AGAROBIOSE $C_{12}H_{22}O_9$ 310

-6^0w

J. Chem. Soc. Jap. ,
1944, 627.

3506-030

NEOHESPERIDOSE $C_{12}H_{22}O_{10}$ 326

-4^0w

191^0C

2α
Glu——Rha

Tet. , 1968, 4963.

3506-035

SOLABIOSE $C_{12}H_{22}O_{11}$ 342

$+41^0$w

200^0C

3β
Gal——Glu

JCS, 1963, 2848.

3506-031

RUTINOSE $C_{12}H_{22}O_{10}$ 326

-10^0e

190^0C

6β
Glu——Rha

Can. J. C. , 1959 (37) 1930.

3506-036

MANNOSE, 4-O-β-D-Mannosyl- $C_{12}H_{22}O_{11}$ 342
                                                      $+2^0$w
                                                      194^0C

4β
Mann———Mann

JACS, 1951, 4187.

3506-037

EVERNINOSE $C_{14}H_{26}O_{10}$ 354
                                      -74^0w
                                      200^0C

Chem. Comm. ,
1969. 1488.

3506-042

MANNITOL, 1-O-β-D-Glucosyl- $C_{12}H_{24}O_{11}$ 344
                                                       -18^0w
                                                       140^0C

CH$_2$O·Glu
HO
HO
        OH
        OH
CH$_2$OH

Acta Chem. Scand. ,
1955 (9) 168.

3506-038

ERYSCENOBIOSE $C_{12}H_{22}O_{10}$ 326
                                         $+22^0$w
                                         212^0C

HOCH$_2$
Glu·O
         OH

Tet. Lett. . 1966. 1703.

3506-043

MANNITOL, 3-O-β-D-Gluco- $C_{12}H_{24}O_{11}$ 344
    -pyranosyl-

CH$_2$OH
HO
Glu·O
        OH
        OH
CH$_2$OH

Acta Chem. Scand. .
1963 (17) 1348.

3506-039

CYMAROSE. 4-O-(2-O-Ac- $C_{16}H_{28}O_9$ 364
    -digitalosyl)- $+56^0$p
                                               177^0C

HO   Me
        OMe
            ·OAc
Me
        OMe
    OH

Chem. Pharm. Bull. ,
1972. 93.

3506-044

MANNITOL, 3-O-Galactofuranosyl- $C_{12}H_{24}O_{11}$ 344
                                                          -61^0w
                                                          162^0C

CH$_2$OH
HO
Gal·O
        OH
        OH
CH$_2$OH

Acta Chem. Scand. ,
1964 (18) 213.

3506-040

ASCLEPOBIOSE $C_{14}H_{26}O_8$ 322
                                     $+28^0$w
                                     101^0C

CHO
    OMe
    O
    OH   Me
CH$_3$ HO   OH
            OMe

Helv. , 1968, 311.

3506-045

ACACIABIURONIC ACID $C_{12}H_{20}O_{12}$ 356
                                              -9^0w
                                              118^0C

6β
Glu———Glucuronosyl

J. Biol. Chem. .
1938 (124) 207.

3506-041

MARSECTOBIOSE $C_{14}H_{26}O_8$ 322
                                      -10^0w
                                      136^0C

CHO
MeO
    O
    OH   Me
CH$_3$ HO   OH
            OMe

Helv. , 1970, 221.

3506-046

PACHYBIOSE $C_{14}H_{26}O_8$ 322
-9°w
160°C

Helv., 1968, 311.

3506-047

DEXTRANTRIOSE $C_{18}H_{32}O_{16}$ 504
+134°w

$\overset{6\alpha}{}\overset{6\alpha}{}$
Glu—Glu—Glu

J. Biol. Chem.,
1952 (196) 265.

3507-002

LILACINOBIOSE $C_{14}H_{26}O_8$ 322
+28°w
184°C

Helv., 1968, 738.

3506-048

MANNINOTRIOSE $C_{18}H_{32}O_{16}$ 504
+167°w
amorph.

$\overset{6\alpha}{}\overset{6\alpha}{}$
Glu—Gal—Gal

JCS. 1953. 567.

3507-003

VIMINOSE $C_{13}H_{24}O_8$ 308
+21°w
Syrup

Helv., 1968, 738.

3506-049

FRUCTOSE, α– Maltosyl- $C_{18}H_{32}O_{16}$ 504
+122°w
120°C

$\overset{2\alpha}{}\overset{4\alpha}{}$
Fruct—Glu—Glu
(β-fur)

JACS 1953, 1259.

3507-004

DREBYSSOBIOSE $C_{13}H_{24}O_8$ 308
+26°w
109°C

Helv., 1968, 668.

3506-050

GENTIANOSE $C_{18}H_{32}O_6$ 344
+31°w
210°C

$\overset{2\alpha}{}\overset{6\beta}{}$
Fruct—Glu—Glu
(β-fur)

J. Pharm. Chim.
1930 (10) 62.

3507-005

MALTOTRIOSE $C_{18}H_{30}O_{16}$ 504
+160°w
amorph.

$\overset{4\alpha}{}\overset{4\alpha}{}$
Glu—Glu—Glu

JACS 1954. 1671.

3507-001

RAFFINOSE $C_{18}H_{32}O_{16}$ 504
+105°w
80°C

$\overset{2\alpha}{}\overset{6\alpha}{}$
Fruct—Glu—Gal
(β-fur)

JCS, 1927, 1527.

3507-006

MELEZITOSE $C_{18}H_{32}O_{16}$ 504
$+88^0$w
154^0C

$$\overset{2\alpha}{Glu}\!-\!\overset{3\alpha}{Fruct}\!-\!Glu$$
(ß-fur)

Adv. Carbohyd. Chem.,
1953 (8) 277.

3507-007

STACHYOSE $C_{24}H_{42}O_{21}$ 366
$+148^0$w
168^0C

$$\overset{2\alpha}{Fruct}\!-\!\overset{6\alpha}{Glu}\!-\!\overset{6\alpha}{Gal}\!-\!Gal$$
(ß-fur)

JACS 1953, 3664.

3508-002

PLANTEOSE $C_{18}H_{32}O_{16}$ 504
$+107^0$w
124^0C

$$\overset{2\alpha}{Glu}\!-\!\overset{6\alpha}{Fruct}\!-\!Gal$$
(ß-fur)

Chem. Abstr.,
1971 (75) 16039.

3507-008

RAFFINOSE, fructosyl- $C_{24}H_{42}O_{21}$ 666
$+56^0$w
amorph.

$$\overset{2}{Fruct}\!-\!\overset{2\alpha}{Fruct}\!-\!\overset{6\alpha}{Glu}\!-\!Gal$$

Phytochem.,
1971 (10) 491.

3508-003

EVERTRIOSE $C_{21}H_{38}O_{14}$ 514
-42^0w
amorph.

Chem. Comm.,
1970. 911.

3507-009

LYCOTETRAOSE $C_{23}H_{10}O_{20}$ 636
$+2^0$w
188^0C

$$\overset{4\alpha\ \ 1}{Gal}\!-\!\overset{2\beta}{Glu}\!-\!Glu$$
$$|\ ^{3\beta}$$
$$Xyl$$

Ber., 1957 (90) 203.

3508-004

MANNITOL, 1,6-diglucosyl- $C_{18}H_{34}O_{16}$ 506
-14^0w
amorph.

Acta Chem. Scand.,
1954 (8) 1547.

3507-010

EVERTETROSE $C_{28}H_{50}O_{17}$ 658
-37^0w
amorph.

Chem. Comm.,
1971. 746.

3508-005

MALTOTETRAOSE $C_{24}H_{42}O_{21}$ 666
$+166^0$w

$$\overset{4\alpha}{Glu}\!-\!\overset{4\alpha}{Glu}\!-\!\overset{4\alpha}{Glu}\!-\!Glu$$

JACS. 1954, 1671.

3508-001

MALTOPENTAOSE $C_{30}H_{52}O_{26}$ 828
$+179^0$w

$$\overset{4\alpha}{Glu}\!-\!\overset{4\alpha}{Glu}\!-\!\overset{4\alpha}{Glu}\!-\!\overset{4\alpha}{Glu}\!-\!Glu$$

JACS. 1955, 1017.

3508-006

VERBASCOSE $C_{30}H_{52}O_{26}$ 828
+170^{0}w
253^{0}C

$\overset{2\alpha}{\text{Fruct}}\!-\!\overset{6\alpha}{\text{Glu}}\!-\!\overset{6\alpha}{\text{Gal}}\!-\!\overset{6\alpha}{\text{Gal}}\!-\!\text{Gal}$

Compt. Rend.,
1954 (239) 824.

3508-007

FUCOSAMINE, D- $C_6H_{13}NO_4$ 163
HCl: +91^{0}w
(dec.) HCl: 172^{0}C

CHO
NH$_2$
HO
HO
OH
CH$_3$

Biochem. J.,
1958 (70) 729.

3509-003

MALTOHEXOSE $C_{36}H_{62}O_{31}$ 990

$\overset{4\alpha}{\text{Glu}}\!-\!\overset{4\alpha}{(\text{Glu})_4}\!-\!\text{Glu}$

JACS, 1955. 5761.

3508-008

FUCOSAMINE, L- $C_6H_{13}NO_4$ 163
HCl: -92^{0}w
HCl: 192^{0}C

enantiomer

Nature. 1964 (202) 493.

3509-004

LYCOPOSE $C_{36}H_{62}O_{31}$ 990
+190^{0}w
270^{0}C

$\overset{2\alpha}{\text{Fruct}}\!-\!\overset{4\alpha}{(\text{Gal})_4}\!-\!\text{Gal}$

Acta Phytochim.,
1942 (13) 37.

3508-009

PNEUMOSAMINE $C_6H_{13}NO_4$ 163
HCl: -10^{0}w
HCl: 162^{0}C

CHO
H$_2$N
HO
HO
OH
CH$_3$

JCS, 1962. 5037.

3509-005

GLUCOSAMINE $C_6H_{13}NO_5$ 179
+48^{0}w
110^{0}C

CHO
NHR   R = H
HO
OH
OH
CH$_2$OH

Biochim. Biophys. Acta,
1958 (29) 522.

3509-001

TALOSAMINE $C_6H_{13}NO_5$ 179

HCl: 152^{0}C

CHO
H$_2$N
HO
HO
OH
CH$_2$OH

Tet. Lett., 1965, 2145.

3509-006

STREPTOZOTOCIN $C_8H_{15}N_3O_7$ 265

R = CO.NMe
NO

JACS, 1967, 4808.

3509-002

MANNOSAMINE $C_6H_{13}NO_5$ 179
HCl: -19^{0}w
HCl: 179^{0}C

CHO
H$_2$N
HO
OH
OH
CH$_{20}$H

Tet. Lett., 1965, 2143.
Nature, 1965 (206) 400.

3509-007

GULOSAMINE $C_6H_{13}NO_5$ 179

187^0C

CHO
—NHR
—OH
HO—
—OH   R = H
CH_2OH

JOC, 1961, 2153.

3509-008

GLUCOSE, 3-NH_2-3, 6-dideoxy- $C_6H_{13}NO_4$ 163

CHO
—OH
H_2N—
—OH
—OH
CH_3

JCS, 1962, 2758.

3509-013

GULOSAMINE. N-Me- $C_7H_{15}NO_5$ 193

HCl: 155^0C

R = Me

Tet. Lett. . 1966. 4187.

3509-009

RHODOSAMINE $C_8H_{17}NO_3$ 135
-48^0w
152^0C

CHO
—
—NMe_2
—OH
—OH
CH_3

Ber. , 1963 (96) 2925.

3509-014

CHONDROSINE $C_{12}H_{21}NO_{11}$ 355
$+39^0$w

CHO
—NH_2
Gal.O—
HO—
—OH
CH_2OH

JACS. 1962, 3029.

3509-010

DAUNOSAMINE $C_6H_{13}NO_3$ 147
HCl: -55^0w
HCl: 168^0C

CHO
—
—NH_2
—OH
HO—
CH_3

Chem. Comm. ,
1967. 973.

3509-015

KANOSAMINE $C_6H_{13}NO_5$ 179
NAc: $+43^0$w
NAc: 205^0C

CHO
—OH
R_2N—
—OH   R = H
—OH
CH_2OH

JACS, 1958, 4741.

3509-011

PEROSAMINE $C_6H_{13}NO_4$ 163

CHO
HO—
HO—
—NH_2
—OH
CH_3

Tet. Lett. , 1966. 5837.

3509-016

MYCAMINOSE $C_8H_{17}NO_4$ 191
HCl: $+31^0$w
HCl: 114^0C

R = Me

JCS, 1962, 2758.

3509-012

VIOSAMINE $C_6H_{13}NO_4$ 163

HCl: 135^0C

CHO
—OH
HO—
—NR_2 R = H
—OH
CH_3

JACS, 1964, 2939.

3509-017

AMOSAMINE $C_8H_{17}NO_4$ 191
HCl: $+46^0$w
HCl: 192^0C

R = Me

JACS, 1963, 1552.

3509-018

OSSAMINE $C_8H_{17}NO_2$ 159
-31^0w
Oil

✱ epimer

Tet. Lett., 1969, 1181.

3509-023

THOMOSAMINE $C_6H_{13}NO_4$ 163

R = H
✱ epimer

JACS, 1964, 2937.

3509-019

GLUCOSE, 6-NH_2-6-desoxy- $C_6H_{13}NO_5$ 179
HCl: $+50^0$w
HCl: 162^0C

CHO
—OH
HO—
—OH
—OH
CH_2NH_2

JACS, 1958, 2342.

3509-024

HOLOSAMINE, D- $C_7H_{15}NO_3$ 161

CHO
—OMe
—NH_2
—OH
CH_3

Tet., 1970, 1695.

3509-020

VANCOSAMINE $C_7H_{15}NO_3$ 161

CHO
Me——NH_2
—OH
HO—
CH_3

Chem. Comm., 1972, 361.

3509-025

HOLANTOSAMINE $C_7H_{15}NO_3$ 161

CHO
—OMe
H_2N—
HO—
CH_3

Bull. Soc. Chim. Fr., 1971, 864.

3509-021

GAROSAMINE $C_7H_{15}NO_4$ 177

CHO
—OH
MeHN—
HO——Me
CH_2OH

Angew. Chem. Int. Ed., 1971 (10) 660.

3509-026

FOROSAMINE $C_8H_{17}NO_2$ 159
$+84^0$m
60^0C

CHO

——NMe_2
—OH
CH_3

Tet. Lett., 1966, 5717.

3509-022

NEURAMINIC ACID, N-Glycollyl-
-8-O-Me- $C_{12}H_{21}NO_{10}$ 339

OH
COOH
HO
OH
MeOCH₂   NHCOCH₂OH

Biochim. Biophys. Acta, 1964 (83) 129.

3509-027

DESOSAMINE, D- $C_8H_{17}NO_3$ 175 83°C

CHO
— OH
Me_2N —
— OH
Me

Proc. Chem. Soc., 1963, 131.

3509-028

EVERMICOSE $C_7H_{14}O_4$ 162 +21°w 110°C

CHO
HO — Me
— OH
— OH
CH_3

Chem. Comm., 1971, 746.

3510-005

APIOSE, D- $C_5H_{10}O_5$ 150 +6°w Syrup

CHO
— OH
$HOCH_2$ — OH
CH_2OH

JCS, 1962, 2515.

3510-001

OLIVOMYCOSE $C_7H_{14}O_4$ 162 -13°w 109°C

CHO
Me — OH
RO —
HO —
CH_3 R = H

Tet. Lett., 1966, 2785.

3510-006

HAMMAMELOSE $C_6H_{12}O_6$ 180 -7°w Syrup

CHO
$HOCH_2$ — OH
— OH
— OH
CH_2OH

JCS, 1962, 3544.

3510-002

CHROMOSE-B $C_9H_{16}O_5$ 204 -26°w

R = Ac

Tet., 1966, 2785.

3510-007

NOGALOSE $C_{10}H_{20}O_5$ 220 +16°w 118°C

CHO
— OMe
Me — OMe
MeO —
HO —
Me

JOC, 1971, 2670.

3510-003

PILLAROSE $C_8H_{12}O_5$ 188 -17°w Syrup

Chem. Pharm. Bull., 1970, 1715.

3510-008

CLADINOSE $C_8H_{16}O_4$ 176 -23°w

CHO
Me — OMe
— OH
— OH
CH_3

JCS, 1966, 398.

3510-004

EVERNITROSE $C_8H_{15}NO_5$ 205 -5°e 90°C

CHO
O_2N — Me
MeO —
HO —
CH_3

Chem. Comm., 1971, 746.

3510-009

STREPTOSE $C_6H_{10}O_5$ 162

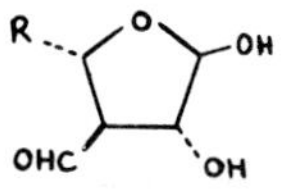

R = H

JACS, 1954, 3675.

3510-010

NOVIOSE $C_8H_{16}O_5$ 192
+20°e
129°C

CHO
——OH
——OH
MeO——
Me——OH
Me

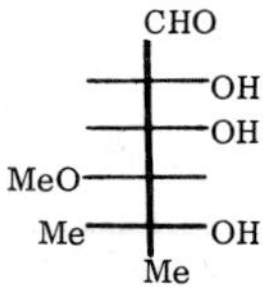

Biochim. Biophys. Acta,
1962 (57) 143.

3510-014

STREPTOSE, OH- $C_6H_{10}O_6$ 178

R = OH

Chem. Abstr.,
1953 (47) 4292e.

3510-011

ARCANOSE, L- $C_8H_{16}O_4$ 176
-21°e
97°C

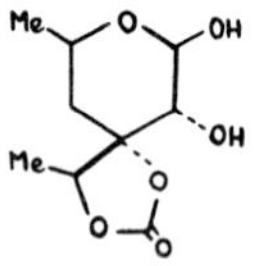

CHO
MeO——Me
——OH
HO——
Me

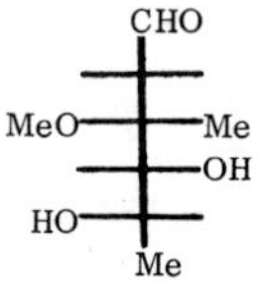

Helv., 1966 (49) 705.

3510-015

STREPTOSE, dihydro- $C_6H_{12}O_5$ 164
-70°w
135°C

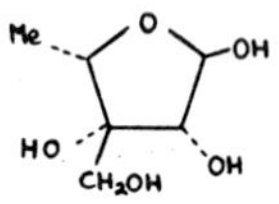

Chem. Ind.,
1964 (6) 238.

3510-012

MYCAROSE, L- $C_7H_{14}O_4$ 162
-31°w
128°C

CHO
HO——Me
HO——
HO——
Me

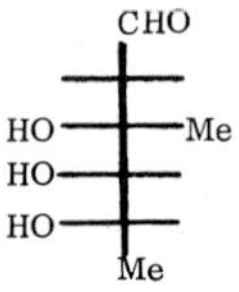

Tet., 1962, 1275.

3510-016

ALDGAROSE $C_9H_{14}O_6$ 218

Tet., 1967, 3893.

3510-013

39   MISCELLANEOUS

The structural/biogenetic classification system employed in this handbook has served to accommodate 97% of the compounds included in this volume. The remaining 3% (132 compounds) have been deposited by default into this miscellaneous class. The alternative of trying to force compounds into "near" classes was regarded as less satisfactory and likely to give rise to more problems in the future (thus, it seems unrealistic to regard a cyclohexanol as a cyclitol).

The miscellany is subclassified structurally according to the scheme below, those defying such simple classification being deposited in the 3999 subclass.

3901 Alicyclics	3902 Bezenoid Mono-aromatics
3903 Mono-heterocyclics	3904 Bicyclic fused heterocyclics
3905 Fused Benz-heterocyclics	3906 Polycyclics
3999 Truly Misc.	

CYCLOHEXANOL, 3-Me-	$C_7H_{14}O$	114

Oil

Helv., 1943, 1992.

3901-001

GRANDIDENTATIN	$C_{21}H_{28}O_9$	424

$-66^0 m$

$201^0 C$

O- p- Coumaroyl

O- Glu

JOC, 1962, 1806.

3901-006

CYCLOHEXANONE, 3-Me-	$C_7H_{12}O$	112

$+13^0$

Oil

JCS, 1907, 337.

3901-002

CYCLOHEXANE-1, 2-DIONE, 3-Me-	$C_7H_{10}O_2$	126

OI

$64^0 C$

Tet., 1963 (19) 2051.

3901-007

CYCLOHEXANDIOL, cis-1, 2-	$C_6H_{12}O_2$	116

OH

OR

R = H

Experientia, 1961, 31.

3901-003

UNDECANOIC ACID, 11-cyclohexyl-	$C_{17}H_{32}O_2$	268

$(CH_2)_{10}COOH$

Chem. Comm.,
1971, 1334.

3901-008

GRANDIDENTOSIDE	$C_{21}H_{28}O_{10}$	440

$196^0 C$

R = 2-O-Caffeoyl-Glucosyl-

Phytochem., 1970, 857.

3901-004

TRIDECANOIC ACID, 13-cyclohexyl-	$C_{19}H_{36}O_2$	296

$(CH_2)_{13}COOH$

Chem. Comm.,
1971, 1334.

3901-009

PURPUREIN	$C_{21}H_{28}O_9$	424

$-39^0 a$

$112^0 C$

R = 2-O-coumaroyl-glucosyl-

Phytochem., 1970, 853.

3901-005

CYCLOPENTADECANOL	$C_{15}H_{30}O$	226

$80^0 C$

$(CH_2)_5$ OH

$(CH_2)_6$

Helv., 1949 (32) 256.
JACS, 1942, 144.

3901-010

MUSCONE $C_{16}H_{30}O$ 238 -13^0 CO—$(CH_2)_{12}$ | | CH_2—CHMe JOC, 1971, 4124. 3901-011	CIVETONE $C_{17}H_{30}O$ 250 OI 32^{0}C $(CH_2)_7$ CH CO || CH CH_2 $(CH_2)_6$ JCS, 1963, 114. 3901-012

PHENYLACETIC ACID, α-Me-p-OH- $C_9H_{10}O_3$ 166
-36°c
142°C

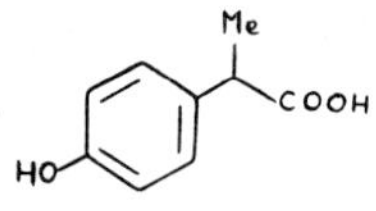

Aust. J. C. , 1964 (17) 379.

3902-001

ASPERRUBROL $C_{22}H_{24}O_3$ 336
OI

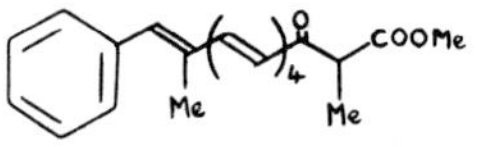

Compt. Rend. ,
1970 (271D) 795.

3902-005

TROPIC ACID, ℓ- $C_9H_{10}O_3$ 166
-79°w
125°C

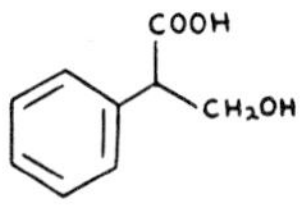

JACS. 1952. 253.

3902-002

PISCIDIC ACID $C_{11}H_{12}O_7$ 256
+41°w
186°C

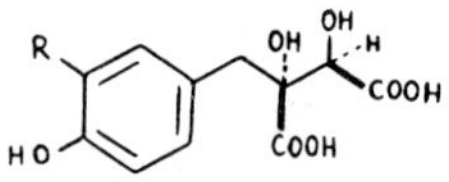

R = H

JACS, 1955, 5750.
Tet. Lett. , 1971, 3809.

3902-006

PENTANE-1, 3-DIONE, 4-Me-1-Ph- $C_{12}H_{14}O_2$ 190
OI
Oil

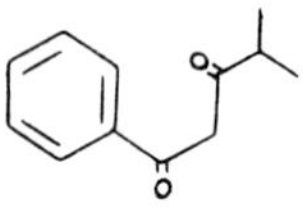

JCS, 1971, 3077.

3902-003

FUKIIK ACID $C_{11}H_{12}O_8$ 272
m. e. 189°C

R = OH

Tet. Lett. , 1971, 3809.

3902-007

CHAETOMIUM TRIACID $C_{14}H_{12}O_7$ 292
OI
170°C

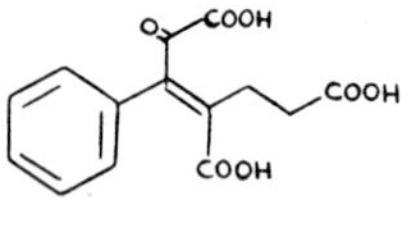

JCS, 1953, 2429.

3902-004

ANTIBIOTIC 1233A $\quad$ $C_{18}H_{28}O_5$ $\quad$ 324

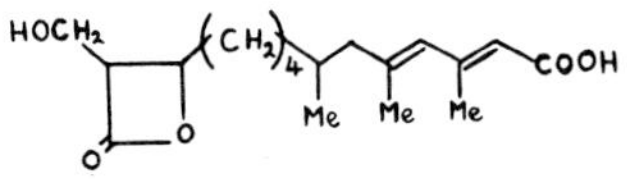

Chem. Comm.,
1970, 639.

3903-001

CITREOVIRIDIN $\quad$ $C_{23}H_{30}O_6$ $\quad$ 402

108^0C

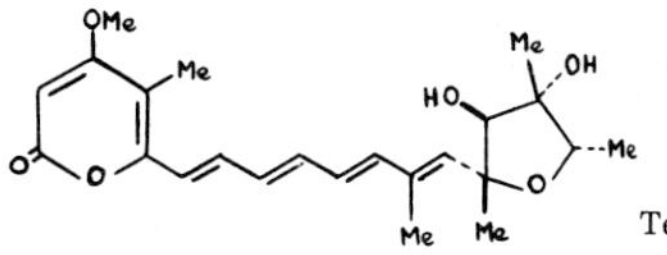

Tet. Lett., 1964, 1825.

3903-004

SARSAPIC ACID $\quad$ $C_6H_4O_6$ $\quad$ 172
OI
305^0C

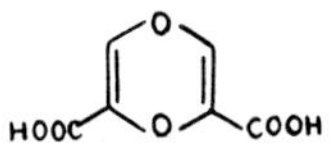

JCS, 1914, 205.

3903-002

ALTERNARIC ACID $\quad$ $C_{21}H_{30}O_8$ $\quad$ 410

138^0C

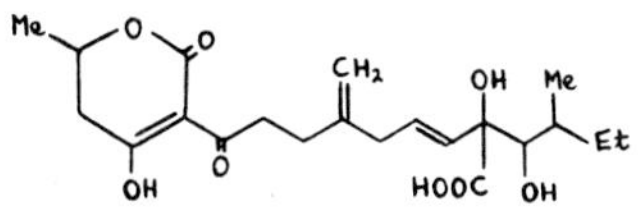

JCS, 1960, 1662.

3903-005

LAURENCIN $\quad$ $C_{17}H_{23}O_3Br$ $\quad$ 354
$+70^0$c
74^0C

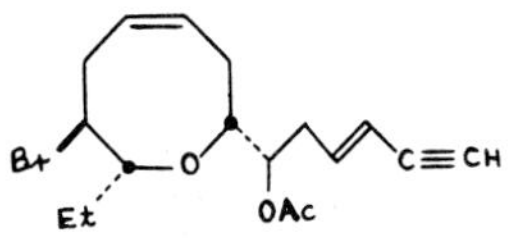

Tet., 1968, 4193.

3903-003

ALTENIN $\quad$ $C_9H_{14}O_6$ $\quad$ 218
(-)
Oil

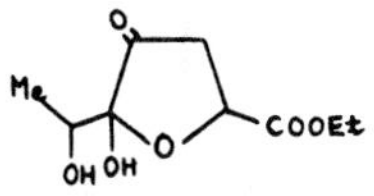

Bull. Chem. Soc. Jap.,
1967 (40) 345.

3903-006

LAUREFUCIN	$C_{15}H_{21}O_3Br$	328	HOP OXETONE I	$C_{11}H_{16}O_2$	180
		-80^0			
		108^0C			

R = H

Tet. Lett., 1972, 1.

Tet. Lett., 1967, 1715.

3904-001

3904-006

LAUREFUCIN, Ac-	$C_{17}H_{23}O_4Br$	370	HOP OXETONE II	$C_{11}H_{18}O_2$	182
		-126^0			
		Oil			

R = Ac

*dihydro

Tet. Lett., 1972, 1.

Tet. Lett., 1967, 1715.

3904-002

3904-007

LAUREATIN	$C_{15}H_{20}O_2Br_2$	(390)	SPHYDROFURAN	$C_8H_{14}O_5$	190
		$+96^0t$			
		82^0C			

Tet., 1970, 851.

J. Antibiotics, 1971 (24) 93.

3904-003

3904-008

LAUREATIN, iso-	$C_{15}H_{20}O_2Br_2$	(390)		$C_8H_{14}O_2$	142
		$+40^0t$			Oil
		83^0C			

Tet., 1970, 851.

Tet. Lett., 1967, 2459.

3904-004

3904-009

PATULIN	$C_7H_6O_4$	154	SCELERANECIC ACID DILACTONE	$C_{10}H_{14}O_5$	214
		OI			-9^0w
		111^0C			156^0C

Chem. Comm., 1967, 945.

JCS, 1963, 953.

3904-005

3904-010

BENZOFURAN	C_8H_6O	118 OI
R = H	Chem. Comm., 1967, 814.	
3905-001		

BENZOFURAN, 5-COOH-	$C_9H_6O_3$	162 OI 190^0C
R = COOH	Phytochem., 1971 (10) 1037.	
3905-006		

BENZOFURAN, 5-OH-	$C_8H_6O_2$	144 OI 60^0C
R = OH	Phytochem., 1971 (10) 1037.	
3905-002		

BENZOFURAN, 5-(2-OH-ethyl)-	$C_{10}H_{10}O_2$	162 OI
	Chem. Comm., 1967, 814.	
3905-007		

BENZOFURAN, 5-OMe-	$C_9H_8O_2$	148 OI 34^0C
R = OMe	Chem. Comm., 1967. 814.	
3905-003		

BENZOFURAN, 5-(1, 2-epoxy-propyl)-	$C_{11}H_{10}O_2$	174 Oil
	Chem. Comm., 1967, 814.	
3905-008		

BENZOFURAN, 5-CH$_2$OH-	$C_9H_8O_2$	148 OI 30^0C
R = CH$_2$OH	Chem. Comm., 1967, 814.	
3905-004		

BENZOFURAN, 5-(1, 2-diOH-propyl)-	$C_{11}H_{12}O_3$	192 -16^0 63^0C
	Phytochem., 1971 (10) 1037.	
3905-009		

BENZOFURAN, 5-CHO-	$C_9H_6O_2$	146 OI 36^0C
R = CHO	Chem. Comm., 1967, 814.	
3905-005		

BENZOFURAN, 5-(1-oxo-2-OH-propyl)-	$C_{11}H_{10}O_3$	190 OI Oil
	Chem. Comm., 1967, 814.	
3905-010		

BENZOFURAN, 2-isopropenyl-5,6--di-OMe-	$C_{13}H_{14}O_3$	218 OI 73^{0}C

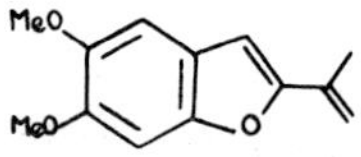

Tet., 1968, 2177.

3905-011

	$C_{12}H_{16}O_4$	224 0^{0}c 144^{0}C

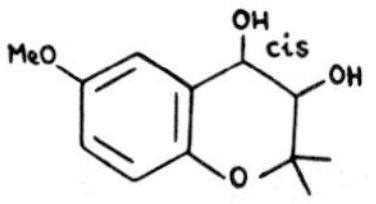

Helv., 1970, 1577.

3905-016

CHROMAN-4-ONE, 6-OH-2,2--di-Me- $C_{11}H_{12}O_3$ 192 OI 149^{0}C

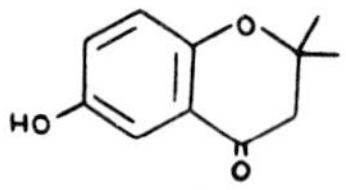

Helv., 1970, 1577.

3905-012

$C_{12}H_{16}O_4$ 224 -33^{0}c 77^{0}C

trans isomer

Helv., 1970, 1577.

3905-017

CHROMENE, 7-OMe-2,2-di-Me- $C_{12}H_{14}O_2$ 190 OI Oil

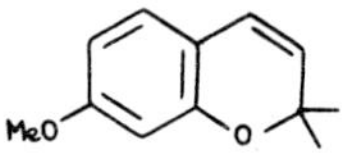

Tet. Lett., 1967, 2573.

3905-013

CHROMAN, 6-Ac-2,2-diMe-3-OH- $C_{13}H_{16}O_3$ 220

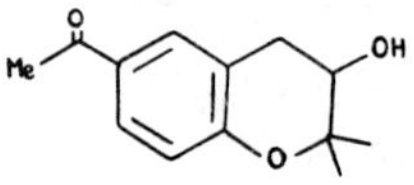

Chem. Abstr., 1971 (74) 126583.

3905-018

AGERATOCHROMENE $C_{13}H_{16}O_3$ 220 OI 47^{0}C

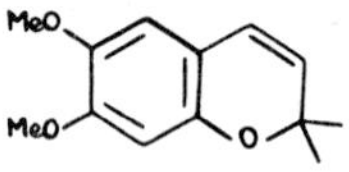

Acta Chem. Scand., 1955, 1725.

3905-014

deleted

3905-019

$C_{26}H_{32}O_6$ 440 -7^{0}c 154^{0}C

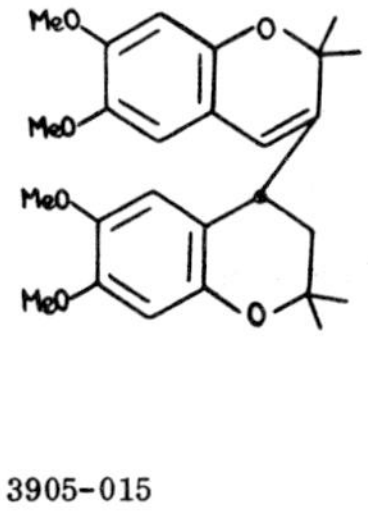

Tet. Lett., 1967, 2573.

3905-015

CHROMONE, 2-isopropyl- $C_{12}H_{12}O_2$ 188 OI Oil

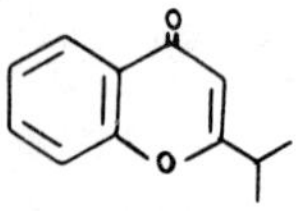

JCS, 1971, 3077.

3905-020

ROTTLERIN	$C_{30}H_{28}O_8$	516 OI 205^{0}C

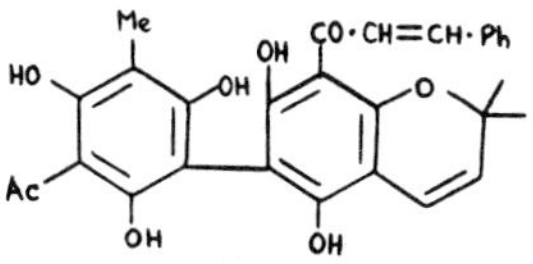

JCS, 1948, 113.

3905-021

CATALPALACTONE	$C_{15}H_{14}O_4$	258 OI 105^{0}C

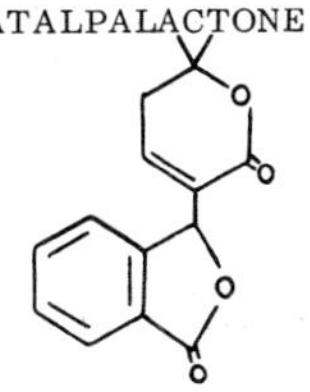

Chem. Pharm. Bull.,
1967, 786.

3905-026

ROSELLINIC ACID	$C_{11}H_{10}O_5$	222 207^{0}C

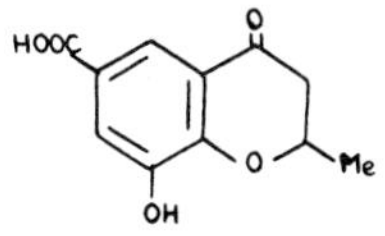

Agr. Biol. Chem.,
1964 (28) 431.

3905-022

GRAVOLENIC ACID	$C_{14}H_{16}O_6$	290 (-)

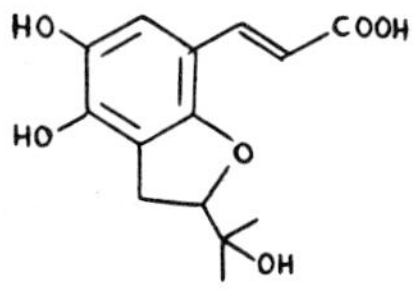

Chem. Abstr.,
1967 (66) 115519.

3905-027

SIPHULIN	$C_{24}H_{26}O_7$	426 OI 185^{0}C

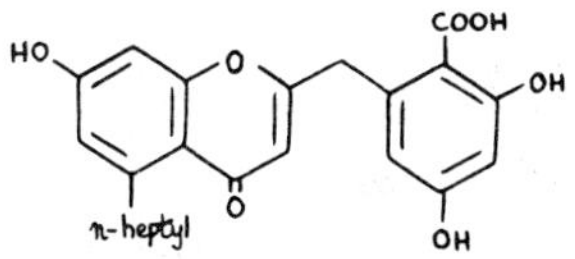

Acta Chem. Scand.,
1965, 1677.

3905-023

SCLERIN	$C_{13}H_{14}O_4$	234 +8^0 123^{0}C

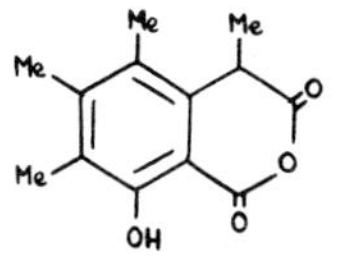

Tet. Lett., 1967, 745.

3905-028

deleted

3905-024

SCLEROLIDE	$C_{12}H_{14}O_4$	222 163^{0}C

Tet., 1970, 1085.

3905-029

MONOCHAETIN	$C_{18}H_{20}O_5$	316 +104^{0}C 203^{0}C

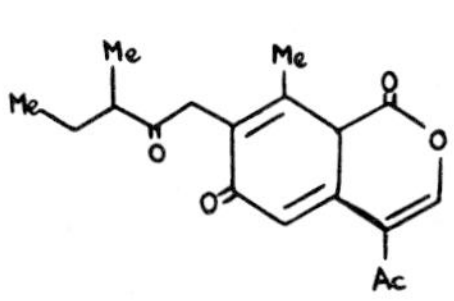

Can. J. C., 1970, 3654.

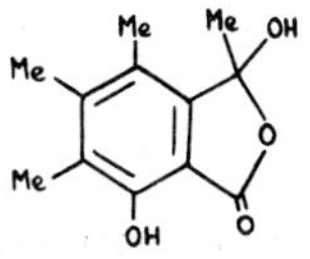

3905-025

	$C_{15}H_{20}O_2$	232 OI Oil

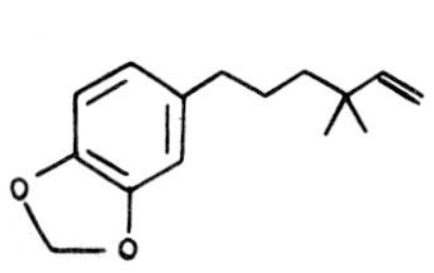

Tet. Lett., 1970, 4305.

3905-030

CYPERAQUINONE

$C_{14}H_{10}O_4$ 242
OI
182^0C

R = Me

Tet. Lett., 1969. 4669.

3906-001

CONICAQUINONE

$C_{13}H_8O_5$ 244
OI
190^0C

Tet. Lett., 1972, 241.

3906-006

CYPERAQUINONE, OH-

$C_{14}H_{10}O_5$ 258
OI
167^0C

R = CH_2OH

Tet. Lett., 1969, 4669.

3906-002

NAPHTHALIC ANHYDRIDE,
-4-OH-3-OMe-5-Ph-

$C_{19}H_{12}O_5$ 320
OI
261^0C

Phytochem., 1970. 1104.

3906-007

CYPERAQUINONE, dihydro-

$C_{14}H_{12}O_4$ 244
-35^0c
114^0C

Tet. Lett., 1969, 4669.

3906-003

RUBROMYCIN, γ-iso-

$C_{26}H_{18}O_{12}$ 522

295^0C

R = H

Ber., 1970 (103) 1709.

3906-008

CYPERAQUINONE, tetrahydro-

$C_{14}H_{14}O_4$ 246
$+210^0c$
139^0C

Tet. Lett., 1969, 4669.

3906-004

RUBROMYCIN, α-

$C_{27}H_{20}O_{12}$ 536

279^0C

R = Me

Ber., 1970 (103) 1709.

3906-009

CYPERAQUINONE, de-Me-

$C_{13}H_8O_4$ 228
OI
137^0C

Tet. Lett., 1969, 4669.

3906-005

RUBROMYCIN, γ -

$C_{26}H_{18}O_{12}$ 522

235^0C

Ber., 1970 (103) 1709.

3906-010

564

RUBROMYCIN, β-

$C_{27}H_{20}O_{12}$ 536

226^0C

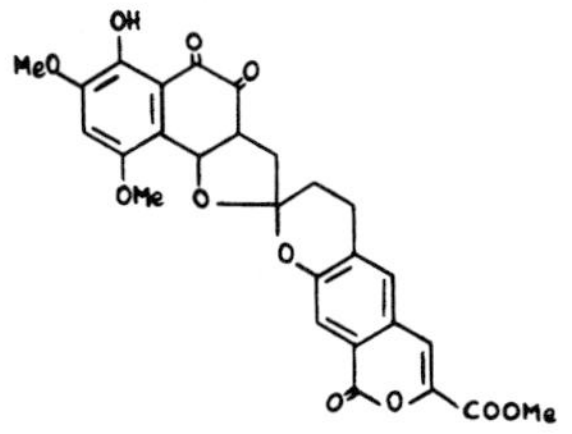

Ber. . 1970 (103) 1709.

3906-011

BYSSOCHLAMIC ACID

$C_{18}H_{20}O_6$ 332

186^0C

Et

Pr

JCS, 1965, 1787.

3906-016

THERMORUBIN-A

$C_{32}H_{24}O_{10}$ 568
OI
dec. $>200^0C$

MeOOC

HO

CH₂

Me

OH

OMe

OH

OMe

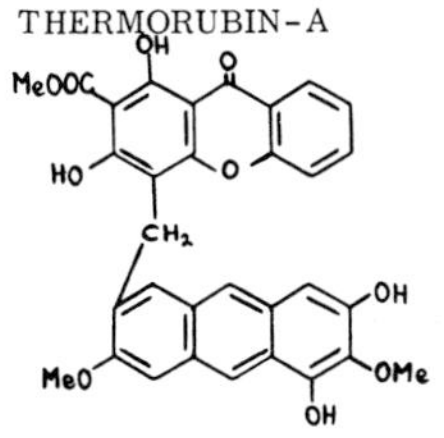

JACS, 1972. 3270.

3906-012

GLAUCANIC ACID

$C_{18}H_{20}O_6$ 332

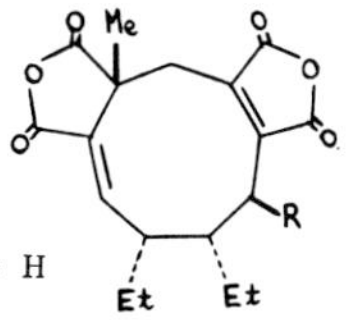

Me

R = H

Et

Et

R

Proc. C. S. . 1962. 385.

3906-017

BARAKOL

$C_{13}H_{12}O_4$ 232

165^0C

Me

OH

Me

O

OH

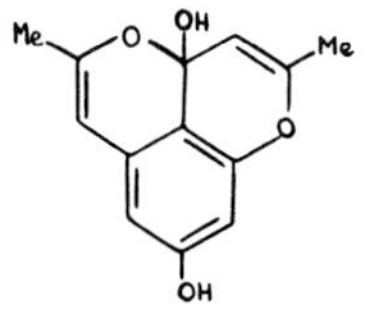

Chem. Comm. .
1969. 678.

3906-013

GLAUCONIC ACID

$C_{18}H_{20}O_7$ 348

202^0C

R = OH

Chem. Comm. .
1966. 772.

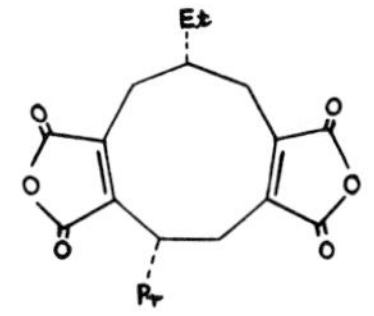

3906-018

BERGENIN

$C_{14}H_{16}O_9$ 328
-47^0w
232^0C

HO

Me

OH

O

OH

OH

CH_2OH

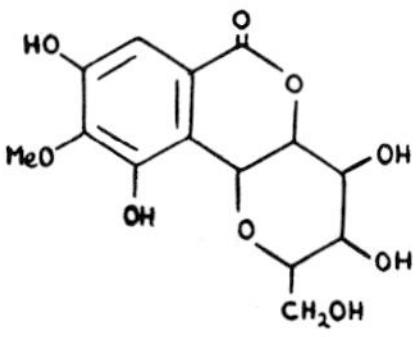

Chem. Rev. , 1964 (64) 247.

3906-014

RUBRATOXIN-A

$C_{26}H_{32}O_{11}$ 520

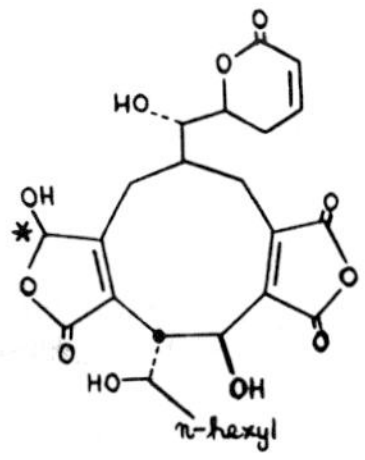

HO

OH

*

HO

OH

n-hexyl

Tet. Lett. . 1969. 367.

3906-019

BREVIFOLIC ACID

$C_{13}H_8O_8$ 292

250^0C

HO

HO

OH

O

O

COOH

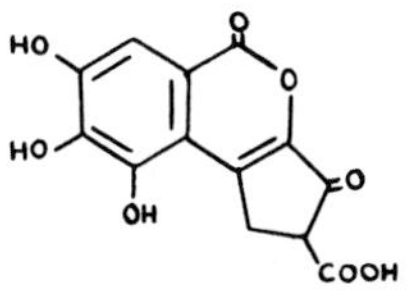

Ann. , 1954 (588) 211.

3906-015

RUBRATOXIN-B

$C_{26}H_{30}O_{11}$ 518
$+6^0a$
170^0C

*oxo

JACS, 1970, 6638.

3906-020

BRONIANONE	$C_{43}H_{58}O_6$	670

Chem. Comm.,
1969, 879.

3906-021

HAEMATOXYLIN	$C_{16}H_{14}O_6$	302

100^0C

R = OH

JCS, 1908, 496.

3906-026

PORTENTOL	$C_{16}H_{26}O_5$	310

$+21^0C$
260^0C

R = H

3906-022

BIKAVERIN	$C_{20}H_{14}O_8$	382

OI
322^0C

R = Me

JCS, 1971, 2788.

3906-027

PORTENTOL ACETATE	$C_{19}H_{28}O_6$	352

$+35^0c$
233^0C

R = Ac

Chem. Comm.,
1969. 162.

3906-023

BIKAVERIN, nor-	$C_{19}H_{12}O_8$	368

OI
$>350^0C$

R = H

JCS, 1971, 2788.

3906-028

CROMBENIN	$C_{16}H_{12}O_8$	332

Chem. Comm.,
1972, 392.

3906-024

AFLATOXIN-B1	$C_{17}H_{12}O_6$	312

-565^0c
265^0C

JACS. 1968, 5017.

3906-029

BRAZILIN	$C_{16}H_{14}O_5$	286

Ann., 1963 (667) 116.

R = H

3906-025

AFLATOXIN-B2	$C_{17}H_{14}O_6$	314

*dihydro

Nature, 1964, 1185.

3906-030

AFLATOXIN–B2A $C_{17}H_{14}O_8$ 346

240^{0}C

Chem. Ind. , 1968, 418.

3906-031

AFLATOXIN–GM1 $C_{17}H_{12}O_8$ 344

276^{0}C

Tet. , 1969, 1499.

3906-036

AFLATOXIN–M1 $C_{17}H_{12}O_7$ 328

-280^{0}f

299^{0}C

Tet. Lett. , 1966, 2799.

3906-032

AFLATOXIN G2A $C_{17}H_{14}O_8$ 346

190^{0}C

Chem. Ind. , 1968, 418.

3906-037

AFLATOXIN–M2 $C_{17}H_{14}O_7$ 330

293^{0}C

*dihydro

Tet. Lett. , 1966, 2799.

3906-033

AFLATOXIN–B3 $C_{16}H_{14}O_6$ 302

217^{0}C

Tet. , 1969, 1497.

3906-038

AFLATOXIN–G1 $C_{17}H_{12}O_7$ 328

-582^{0}c

258^{0}C

Nature, 1964, 1185.

3906-034

MIROESTROL $C_{20}H_{22}O_6$ 358

+301^{0}e

269^{0}C

Nature, 1964 (201) 1210.

3906-039

AFLATOXIN–G2 $C_{17}H_{14}O_7$ 330

*dihydro

JACS, 1963, 1706.

3906-035

deleted

3906-040

CONOCARPIN $C_{15}H_{16}O_8$ 324 +77[0]e 185[0]C JCS. 1970, 2127. R = H 3906-041	**RUBROPUNCTATIN** $C_{21}H_{22}O_5$ 354 156[0]C JCS. 1967. 751. R = $(CH_2)_4Me$ 3906-046
LEUCODRIN $C_{15}H_{16}O_8$ 324 -21[0]w 216[0]C R = H stereoisomer ? Tet., 1967. 1929. 3906-042	**MONASCORUBRIN** $C_{23}H_{26}O_5$ 382 135[0]C R = $(CH_2)_6Me$ Tet. Lett., 1960 (5) 24. 3906-047
LEUCOGLYCODRIN $C_{21}H_{26}O_{13}$ 486 221[0]C R = Glucosyl stereo ? Tet. Lett., 1966, 3773. 3906-043	**ROTIORIN** $C_{23}H_{24}O_5$ 380 +5080[0]c 246[0]C JCS. 1963. 3641. 3906-048
PIPTOSIDE $C_{17}H_{24}O_{12}$ 420 -28[0]w 230[0]C Tet. Lett., 1963, 1615. 3906-044	**ROTIORIN, (-)-7-epi-5-Cl-iso-** $C_{23}H_{23}ClO_5$ 414 Chem. Comm., 1970, 762. 3906-049
MONASCOFLAVIN $C_{21}H_{26}O_5$ 358 144[0]C JCS. 1961, 4579. 3906-045	**FLOCCOSIC ACID** $C_{14}H_{12}O_7$ 292 ~156[0]C Can. J. C., 1969, 1560. 3906-050

TABEBUIN $C_{31}H_{26}O_5$ 478
+94^{0}c
211^{0}C

JCS, 1967, 2100.

3906-051

MONENSIC ACID $C_{36}H_{62}O_{11}$ 670

104^0C

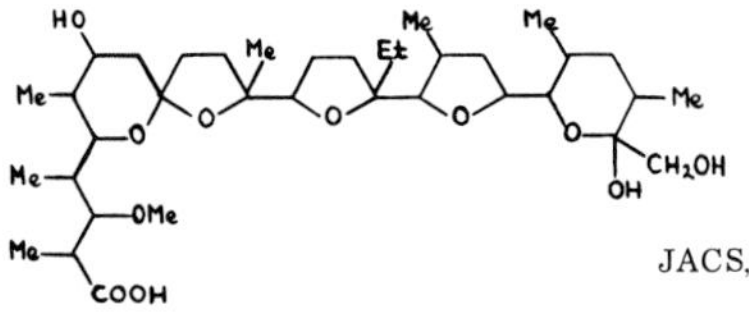

JACS, 1967, 5738.

3999-001

deleted

3999-006

GRISORIXIN $C_{40}H_{68}O_{10}$ 708
$+16^0a$
80^0C

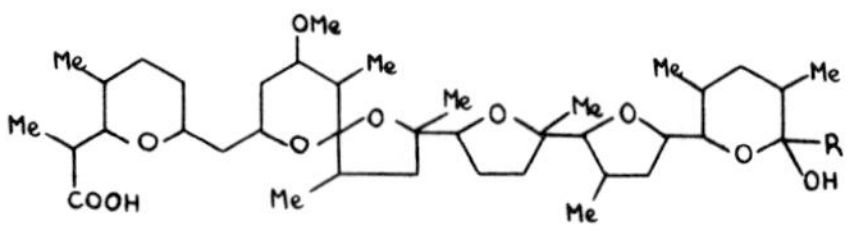

$R = CH_3$

Chem. Comm.,
1970, 1421.

3999-002

PEDERIN, pseudo- $C_{24}H_{43}NO_9$ 489

142^0C

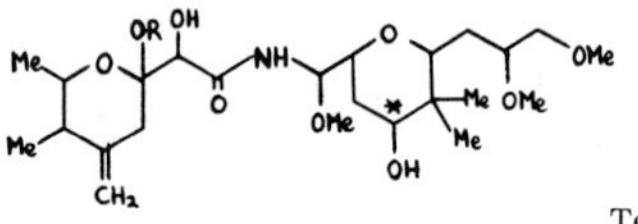

$R = H$

Tet. Lett., 1968, 6297.

3999-007

POLYETHERIN A $C_{40}H_{68}O_{11}$ 724
$+36^0c$
184^0C

$R = CH_2OH$

Chem. Comm.,
1968, 1541.

3999-003

PEDERIN $C_{25}H_{45}NO_9$ 503

113^0C

$R = Me$

Tet. Lett., 1968, 6297.

3999-008

ANTIBIOTIC X-206 $C_{45}H_{78}O_{13}$ 826

133^0C

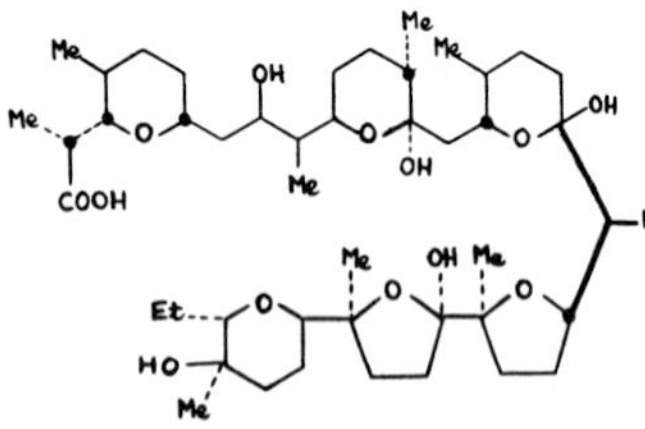

Chem. Comm.,
1971, 927.

3999-004

PEDERONE $C_{25}H_{43}NO_9$ 501
$+108^0e$

$R = Me$
*oxo

Tet. Lett., 1967, 4023.

3999-009

X-537A $C_{34}H_{54}O_8$ 590
-8^0m
112^0C

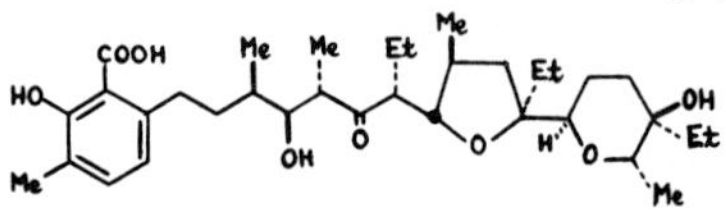

Chem. Comm.,
1971, 181.

3999-005

BULLATENONE $C_{12}H_{12}O_2$ 188
OI
69^0C

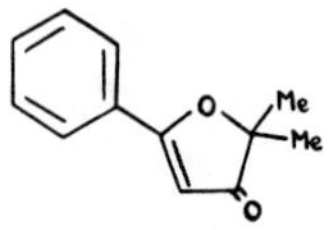

JCS, 1958, 3871.

3999-010

FUNICONE	$C_{19}H_{18}O_8$	374
		OI
		177^0C

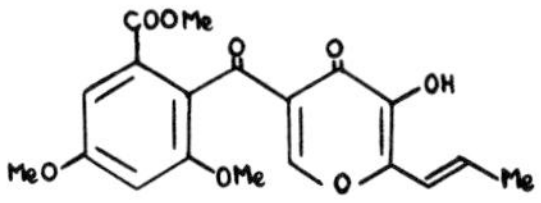

Tet., 1970, 2739.

3999-011

LUTEORETICULIN	$C_{19}H_{19}NO_5$	341
		185^0C

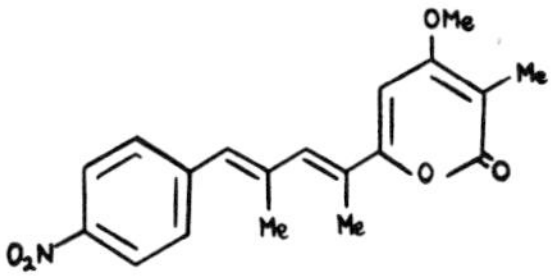

Tet. Lett., 1969, 355.

3999-014

ARENOL	$C_{21}H_{24}O_7$	388
		OI

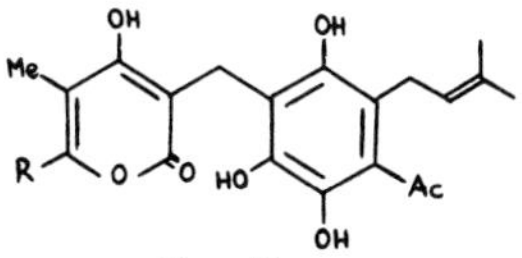

R = Me

Tet. Lett., 1971, 247.

3999-012

AUREOTHIN	$C_{22}H_{23}NO_6$	397
		$+51^0$
		158^0C

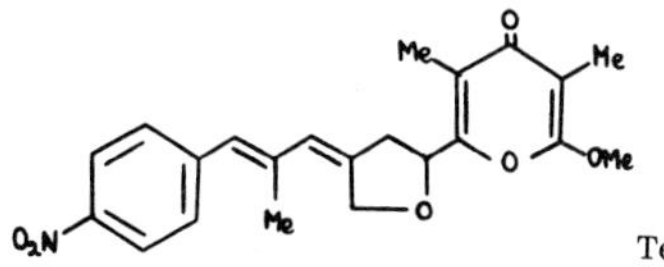

Tet. Lett., 1959 (16) 9.

3999-015

ARENOL, homo-	$C_{22}H_{26}O_7$	402
		OI

R = Et

Tet. Lett., 1971, 247.

3999-013

AUREOTHIN, neo-	$C_{28}H_{31}NO_6$	477

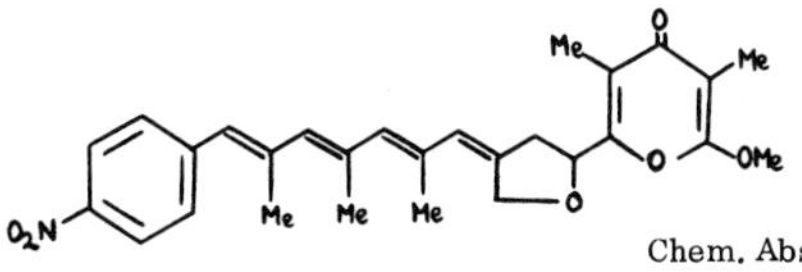

Chem. Abstr.,
1970 (73) 109622.

3999-016

INDICES GUIDE

In an attempt to cater for maximum access to this handbook three indices have been provided in addition to the structural guides.

The Molecular Weight Index provides a listing of the classification numbers in ascending molecular weight sequence.

The Molecular Formula Index alternatively provides a listing of accession numbers in ascending order of the number of C atoms in the molecular formula; within each carbon number, hydrogen, nitrogen, oxygen, etc., are similarly sequenced.

The Alphabetical Index provides access to the handbook by compound name and also serves to indicate alternative names for compounds. No attempt has been made to apply IUPAC nomenclature for compound names − those used in the index are in general the trivial names found commonly in the literature. In view of the structural organisation of this handbook there should be no special need to provide IUPAC nomenclature. It should be noted that in the alphabetical listing the common prefixes (iso, neo, dihydro, etc.) are placed *after* the parent name of the compound unless the prefix has some more meaningful connotation (e. g. cyclo- in the Buxus alkaloids). Thus:

4204-024	ABIETIC ACID
4204-025	ABIETIC ACID, DEHYDRO-
4204-041	ABIETIC ACID, 4-EPI-
4204-027	ABIETIC ACID, NEO-
4204-039	ABIETIC ACID, 15-OH-

and

4101-003	FARNESOL
4191-047	FARNESOL, CYCLO-

but

4417-001	CYCLOARTANOL (No such compound as ARTANOL)
4419-042	CYCLOBUXINE-D (Refers to the Cyclobuxane skeleton)

2111-014	ANACARDOL IV
3114-035	ANACYCLIN
3114-036	ANACYCLIN,DEHYDRO-
3005-042	ANCEPSENOLIDE
3005-043	ANCEPSENOLIDE,OH-
	ANCHUSIN
0503-006	ANDELIN
0203-006	ANDROSIN
3005-044	ANEMONIN
0301-003	ANETHOL
0399-001	ANETHOL GLYCOL
3195-013	ANGELIC ACID
3195-014	ANGELIC ACID,I-BUTYL ESTER
0501-036	ANGELICAL
0504-016	ANGELICIN
0508-005	ANGELICONE
0501-027	ANGELOL
0999-017	ANGOLENSIN
1203-030	ANGOPHOROL
2203-010	ANGUSTIFOLIONOL
2201-016	ANGUSTIONE
2201-017	ANGUSTIONE,DEHYDRO-
0103-002	ANISALDEHYDE
0103-014	ANISALDEHYDE,ORTHO-
0103-005	ANISALDEHYDE,PARA-
0104-007	ANISIC ACID
0104-008	ANISIC ACID ME ESTER
0102-002	ANISYL ALCOHOL
0399-003	ANISYL KETONE
1108-009	ANNULATIN
0301-005	ANOL ISOVALERYL ESTER
0399-017	ANOLOXIDE
0505-027	ANOMALIN,(+)-
0505-017	ANOMALIN,(-)-
2604-001	ANTHRAGALLOL
2604-004	ANTHRAGALLOL-1,2-DIME ETHER
2604-005	ANTHRAGALLOL-1,3-DIME ETHER
2604-006	ANTHRAGALLOL,2,3-DI-ME ETHER
2604-007	ANTHRAGALLOL TRI-ME ETHER
2604-002	ANTHRAGALLOL-2-MF-ETHER
2604-003	ANTHRAGALLOL 3-ME ETHER
2605-028	ANTHQUINONE,1,4,5,8-TETRA-OH-2,6-DI-ME-
2601-001	ANTHRAQUINONE
2601-003	ANTHRAQUINONE,2-CH2OH-
2601-004	ANTHRAQUINONE,2-CH2OAC-
2601-005	ANTHRAQUINONE,2-CH3-
2601-006	ANTHRAQUINONE,2-COOH-
2602-001	ANTHRAQUINONE,1-OH-
2602-003	ANTHRAQUINONE,2-OH-
2602-005	ANTHRAQUINONE,1-OH-2-CH3-
2602-008	ANTHRAQUINONE,3-OH-2-MF-
2602-002	ANTHRAQUINONE,1-OME-
2602-006	ANTHRAQUINONE,1-OME-2-ME-
2602-004	ANTHRAQUINONE,2-OME-
2602-009	ANTHRAQUINONE,3-OME-2-ME-
2605-006	ANTHRAQUINONE,1,7-DI-OH-6,8-DI-OME-3-ME-
2605-024	ANTHRAQUINONE,1,3,5-TRI-OH-8-OME-2-ME-
2606-002	ANTHRAQUINONE,1,2,6,7-TETRA-OH-8-OME-3-M
2605-019	ANTHRAQUINONE,1,3,5,8-TETRA-OH-6-ME-2-CL
2605-050	ANTHRAQUINONE,1,4,5,8-TETRA-OH-2,6-DI-ME
0605-002	ANTHRICIN = PODOPHYLLOTOXIN,DEOXY-
2699-012	ANTHRONE,2-CL-1,3,8-TRICH-6-ME-
0804-001	ANTIAROL
3903-001	ANTIBIOTIC 1233A
3502-037	ANTIBIOTIC A-396-I
3401-045	ANTIBIOTIC B-58941
3003-011	ANTIBIOTIC U-13,933
3505-004	ANTIBIOTIC SF-666
3401-034	ANTIBIOTIC SF-837A2
3401-037	ANTIBIOTIC SF-837A3
3401-038	ANTIBIOTIC SF837A4
3401-035	ANTIBIOTIV YL-704A1
3401-033	ANTIBIOTIC YL-704B1
3401-036	ANTIBIOTIC YL-704C1
3401-032	ANTIBIOTIC YL-704C2
3401-039	ANTIBIOTIC YL-704W1
3999-004	ANTIBIOTIC X-206
2301-026	ANZIAIC ACID
2204-026	APETALIC ACID
0902-011	APETALOLIDE
2509-004	APHININ-SM
2499-011	APHLIOL
2303-006	APHTHOSIN
1701-001	APIGENIDIN
1103-002	APIGENIN
1103-012	APIGENIN-5-GLUCOSIDE
1103-011	APIGENIN 5 ME ETHER
1103-006	APIGENIN-7-GLUCURONIDE
1103-004	APIGENIN,7-ME-GALACTURONOSYL-
1103-010	APIGENIN,7-O-GLUCOSIDO-XYLOSYL-
1103-007	APIGENIN,7-O-NEOHESPERIDOSYL-
1103-009	APIIN
0301-035	APIOL,DILL-
0301-019	APIOL,ISO-
3510-001	APIOSE-D
0203-006	APOCYNIN = ACETOVANILLONE
1199-031	APULEIDIN
1199-026	APULEIN
1199-025	APULEIN,5-O-DESME-
1108-027	APULEIRIN
1199-032	APULEISIN
1108-026	APULEITRIN
2703-048	AQUAYAMYCIN
3506-007	ARABINOPYRANOSE,2-GLUCOSYL-
3503-001	ARABINOSE, -
3503-002	ARABINOSE,L-
3506-011	ARABINOSE,3-ALPHA-GALACTOPYRANOSYL-
3506-009	ARABINOSE,3-BETA-ARABOPYRANOSYL-
3503-009	ARABITOL,D-
3504-052	ARABOHEXOSE,6-DEOXY-5-OXO-D-
3120-005	ARACHIC ACID
3120-016	ARACHIDONIC ACID
3120-002	ARACHIDYL ALCOHOL
2609-001	ARAROBINOL
0802-006	ARBUTIN
0802-008	ARBUTIN,2-O-CAFFEOYL-
0802-009	ARBUTIN,2'-O-GALLOYL-
0802-011	ARBUTIN,4-O-GALLOYL-
0802-010	ARBUTIN,6'-O-GALLOYL-
0802-015	ARBUTIN,METHYL-
0504-015	ARCHANGELICIN
0508-037	ARCHANGELIN
3303-001	ARCTIC ACID
0601-013	ARCTIGENIN
0601-014	ARCTIIN
2005-010	ARDISIAQUINONE-A
2005-011	ARDISIAQUINONE-B
2704-011	ARENICOCHROMINE
2605-001	ARECLATIN = COPAREOLATIN
2504-036	ARNEBIN-5
2504-037	ARNEBIN-6
2504-038	ARNEBIN-2
3999-012	ARENOL
3999-013	ARENOL,HOMO-
1204-001	AROMADENDRIN
1204-002	AROMADENDRIN,7-O-ME-
1204-004	AROMADENDRIN 7-RHAMNOSIDE
3205-005	ARTEMIDIN
3118-015	ARTEMISIC ACID
3118-063	ARTEMESIC ACID,BETA-
1106-087	ARTEMETIN
2301-033	ARTHONIOIC ACID
1199-009	ARTOCARPESIN
2403-016	ARTHOTHELIN
1105-039	ARTHRAXIN
1299-004	ARTOCARPANONE
1199-011	ARTOCARPETIN
1199-008	ARTOCARPETIN,NOR-
1199-012	ARTOCARPIN
1199-033	ARTOCARPIN,CYCLO-
1199-034	ARTOCARPIN DIME ETHER
1299-009	ARTOCARPIN,DIHYDRO-CYCLO-
1199-035	ARTOCARPIN,ISO-DIME-ETHER
2499-009	ARUGOSIN-A
2499-010	ARUGOSIN-B
2806-003	ASADANIN
2806-004	ASADANOL,EPI-
2806-005	ASADANOL,ISO-
0103-016	ASAFALDEHYDE
0601-030	ASARININ,(+)-
0301-031	ASARONE,ALPHA-
0601-031	ASARININ,(-)-
0301-032	ASARONE,BETA-
0301-033	ASARONE,GAMMA-
0301-034	ASARONE,3-OME-GAMMA-
0699-026	ASATONE
3005-013	ASCADIOL
3504-041	ASCARYLOSE
3506-045	ASCLEPOBIOSE
2106-018	ASCOCHITINE
3005-036	ASCORBIC ACID
2999-002	ASCOTOXIN = BREFELDIN A
1002-013	ASEBOGENIN
1002-014	ASEBOTIN
1003-011	ASPALATHIN
0499-008	ASPERENONE = ASPERYELLONE
2704-013	ASPEPGILLINE
3003-012	ASPERGILLUS LACTONE
3003-010	ASPERLINE,4-OH-6,7-DESOXY-
3902-005	ASPERRUBROL
3005-015	ASPERTETRONIN A
3005-016	ASPERTETRONIN B
2606-009	ASPERTHECIN
2403-042	ASPERTOXIN
2102-032	ASPERUGIN
2102-031	ASPERUGIN B
0499-008	ASPERYELLONE
2207-014	ASPIDIN
2299-006	ASPIDININ = PHLOROPYRONE
2203-003	ASPIDINOL
2399-003	ASTERRIC ACID
1206-004	ASTILBIN

0502-017 CHALEPENSIN	3401-044 CIRRAMYCIN-A1
0502-014 CHALEPIN	1103-041 CIRSIMARITIN
1599-007 CHAMAECYPARIN	1103-042 CIRSIMARITIN-4'-O-RUTINOSIDE
1401-007 CHANDALONE	3195-020 CITRAMALIC ACID,L-
2704-034 CHARTREUSIN	2604-045 CITREOROSEIN
2901-014 CHAULMOOGRIC ACID	2604-046 CITREOROSEIN,7-CL-
2901-017 CHAULMOOGRIC ACID,KETO-	2605-011 CITREOROSEIN,8-OH-
0301-009 CHAVIBETOL	2604-050 CITREOROSEIN-6,8-DI-ME ETHER
0402-002 CHAVICIC ACID	3903-004 CITREOVIRIDIN
0402-003 CHAVICIC ACID,ISO-	3196-016 CITRIC ACID
0301-001 CHAVICOL	3005-050 CITRIC ACID LACTONE,ALLO-ISO-
0199-011 CHEBULAGIC ACID	2104-024 CITRININ
0199-010 CHEBULINIC ACID	2104-004 CITRINONE,DE-COOH-DIHYDRO-
3002-014 CHELIDONIC ACID	2104-025 CITRINONE,DIHYDRO-
2702-014 CHELOCARDIN	1201-013 CITROMITIN
0339-012 CHICORIC ACID	1201-012 CITROMITIN,5-O-DES ME-
2502-005 CHIMAPHILIN	2199-008 CITROMYCETIN
1199-027 CHLORFLAVONIN	2703-004 CITROMYCINONE,ALPHA-
2502-018 CHLOROBIUMQUINONE	2703-003 CITROMYCINONE,GAMMA-
1902-006 CHLOROGENIC ACID	1299-001 CITRONETIN
2804-023 CHLOROPHORIN	1299-002 CITRONIN
2702-002 CHLOROTETRACYCLINE = AUREOMYCIN	3901-012 CIVETONE
3509-010 CHONDROSINE	3510-004 CLADINOSE
1902-018 CHORISMIC ACID	1402-021 CLADRASTIN
3905-018 CHROMAN,6-AC-2,2-DI-ME-3-OH-	1402-016 CLADRIN
2203-012 CHROMANONE,5-OH-2-ME-	0508-020 CLAUSENIDIN
3905-012 CHROMAN-4-ONE,6-OH-2,2-DI-ME-	0508-014 CLAUSENIN
2205-015 CHROMENE,6-AC-2,2-DI-ME-3-	2201-006 CLAVATOL
3905-013 CHROMENE,7-OME-2,2-DI-ME-	2605-007 CLAVORUBIN
2703-039 CHROMOCYCLINE = CHROCYCLOMYCINONE	0604-022 CLEISTANTHIN
2703-040 CHROMOCYCLOMYCIN	0604-021 CLEISTANTHIN-B
2703-039 CHROMOCYCLOMYCINONE	3003-007 CLIVONECIC ACID
2701-012 CHROMOMYCIN-A2	3122-016 CLUPANODONIC ACID
2701-011 CHROMOMYCIN-A3	2106-025 CNIDIUM LACTONE
2701-008 CHROMOMYCIN-A4	2106-019 CNILILIDE
2701-007 CHROMOMYCINONE	2106-020 CNILILIDE,NEO-
3905-020 CHROMONE,2-ISOPROPYL-	2301-007 COCCELLIC ACID = BARBATIC ACID
2203-013 CHROMONE,5-OH-2-ME-	2604-027 COELULATIN
3510-007 CHROMOSE B	2406-026 COELULATIN,DESOXY-
3504-037 CHROMOSE D	2503-002 COHULUPONE
1702-005 CHRYSANTHEMIN	2206-008 COHUMULONE
2699-001 CHRYSAROBIN	2206-009 COHUMULONE,4-DEOXY-
0507-005 CHRYSATROPIC ACID = SCOPOLETIN	2302-026 COLENSOIC ACID
1101-004 CHRYSIN	3504-040 COLITOSE
1101-008 CHRYSIN,GLUCURONIDE-	2302-029 COLLATIC ACID,ALPHA-
2704-028 CHRYSOAPHIN-SL-1	3499-008 COLLETODIOL
2704-029 CHRYSOAPHIN-SL2	0506-003 COLLININ
2704-030 CHRYSOAPHIN-SL-3	0604-017 COLLINUSIN
1105-021 CHRYSOERIOL	0504-005 COLUMBIANIDIN
1105-023 CHRYSOERIOL,7-O-GLUCOSYL-	0504-007 COLUMBIANIDIN OXIDE
1105-024 CHRYSOERIOL,7-O-GLUCURONOSYL-	0504-004 COLUMBIANIN
1105-025 CHRYSOERIOL,7-O-RUTINOSYL-	2206-010 COLUPULONE
2606-001 CHRYSO-OBTUSIN	2504-021 COMANTHERIN SULPHATE
2603-028 CHRYSOPHANEIN	2504-022 COMAPARVIN SULPHATE
2603-027 CHRYSOPHANIC ACID = CHRYSOPHANOL	2505-005 COMAPARVIN SULPHATE,6-OME-
2603-027 CHRYSOPHANOL	2505-006 COMAPARVIN-5-ME ETHER SULPHATE,6-OME-
1106-077 CHRYSOSPLENOL D	1108-016 COMERETOL
1199-023 CHRYSOSPLENOSIDE	3002-012 COMENIC ACID
1106-082 CHRYSOSPLENOSIDE B	3001-002 COMPOUND T
1199-017 CHRYSOSPLIN	3501-002 CONDURITOL
2607-001 CHRYSOTALUNIN	2301-031 CONFLUENTIN ACID
0507-003 CICHORIIN	2206-006 CONGLOMERONE
0605-004 CICUTIN = PICROPODOPHYLLIN,DEOXY-	3906-006 CONICAQUINONE
3117-035 CICUTOL	0604-012 CONIDENDRIN
3117-053 CICUTOXIN	0302-006 CONIFERIN
2901-020 CINEROLONE	0302-010 CONIFERIN,OME-
3401-004 CINEROMYCIN-B	0302-005 CONIFERYL ALCOHOL
2703-031 CINERUBINE-A	0699-019 CONIFERYL ALCOHOL,DEHYDRO-
0304-001 CINNAMIC ACID,CIS-	0399-008 CONIFERYL BENZOATE
0304-002 CINNAMIC ACID,TRANS-	3906-041 CONOCARPIN
0304-026 CINNAMIC ACID, 2-GLUCOSYLOXY-4-OME-	2302-020 CONSTICTIC ACID
0304-024 CINNAMIC ACID,2,3-DI-OH-	3115-027 CONVOLVULINOLIC ACID
0304-029 CINNAMIC ACID,2-OH-3-OME-	2605-001 COPAREOLATIN
0304-023 CINNAMIC ACID,3,4-DIOME-	2002-002 COPFININ
0304-027 CINNAMIC ACID, 2,4-DI-GLUCOSYLOXY-	2506-004 CORDEAUXIAQUINONE
0304-004 CINNAMIC ACID AMIDE,TRANS-	1902-016 CORDYCEPIC ACID
0239-013 CINNAMIC ACID BENZYL ESTER	3503-008 CORDYCEPOSE
0304-025 CINNAMIC ACID GLUCOSE ESTER,2,3-DI-OH-	1003-002 COREOPSIN
0304-003 CINNAMIC ACID ME ESTER	1205-016 COREOPSIN,ISO-
0304-014 CINNAMIC ACID ME ESTER,P-OME-	0199-009 CORILAGIN
0303-001 CINNAMIC ALDEHYDE	3118-062 CORIOLIC ACID
0303-002 CINNAMIC ALDEHYDE,DIHYDRO-	3499-019 CORIOLIDE
0303-003 CINNAMIC ALDEHYDE,2-OME-	3505-003 CORIOSE
0303-004 CINNAMIC ALDEHYDE,4-OME-	1106-095 CORNICULATUSIN
0303-007 CINNAMIC ALDEHYDE,3,4-METHLENEDIOXY-	1106-096 CORNICULATUSIN 3-GALACTOSIDE
0302-003 CINNAMYL ACETATE	3118-076 CORONARIC ACID
0302-002 CINNAMYL ALCOHOL	3114-045 CORTICROCIN
0302-001 CINNAMYL ALCOHOL,DIHYDRO-	0499-009 CORTISALIN
0399-007 CINNAMYL CINNAMATE	2405-007 CORYMBIFERIN
0401-004 CINNAMYLIDENE ACETONE	1107-006 CORYMBOSIN
0401-006 CINNAMYLIDENE ACETONE,3,4-METHYLENEDIOXY	3197-042 CORYNINE

3197-041	CORYNOMYCOLENIC ACID
3197-040	CORYNOMYCOLIC ACID
1103-005	COSMETIN = COSMOSIIN
1103-005	COSMOSIIN
2802-002	COTCIN
0402-006	COUMALIN,6-PH-
0304-006	COUMARIC ACID,ORTHO-
0304-008	COUMARIC ACID GLUCOSIDE,CIS-ORTHO-
0304-011	COUMARIC ACID,P-
0304-012	COUMARIC ACID ME ETHER,P-
0304-013	COUMARIC ACID ME ESTER,P-
0304-015	COUMARIC ACID ME ESTER,4-O-GERANYL-
0304-028	COUMARIC ACID, GLUCOSIDO-FURO-
0599-002	COUMARIN
0599-004	COUMARIN,4-HYDROXY-
0509-007	COUMARIN,6-OME-7,8-METHYLENEDIOXY-
0599-007	COUMARIN,8-OME-4-ME-
0508-012	COUMARIN,5-GERANYLOXY-7-OME-
0509-006	COUMARIN,6,7,8-TRIMETHOXY-
2805-009	COUMAROYL-FERULOYL-METHANE
1602-004	COUMESTAN,7-OH-11,12-DI-OME-
1602-001	COUMESTROL
1602-002	COUMESTROL DIME ETHER
0508-006	COUMURRAYIN
0101-006	CRESOL
3118-048	CREPENYNIC
2101-017	CRESOL,META-
0101-007	CRESOL,ORTHO-
0101-002	CRESOL METHYL ETHER,P-
0101-001	CRESOL,P-
0399-006	CROCATONE = LATIFOLONE
704-005	CROCEOMYCIN = RESISTOMYCIN
3906-024	CROMBENIN
1802-002	CROMPEONE
1901-017	CROTEPOXIDE
3104-011	CROTONIC ACID
3104-012	CROTONIC ACID,ISO-
0201-030	CROWEACIN
0499-011	CRYPTOCARYALACTONE
2201-035	CRYPTOCHLOROPHEIC ACID
2704-027	CRYPTOCLAUXIN
1599-001	CRYPTOMERIN A
1599-002	CRYPTOMERIN B
1599-006	CRYPTOMERIN,ISO-
1599-005	CRYPTOMERIN,NEO-
2903-008	CRYPTOSPORIOPSIN
2903-009	CRYPTOSPORIOPSIN,DES-CL-
2903-010	CRYPTOSPORIOPSIN,DES-CL-EPI-
2903-011	CRYPTOSPORIOPSIN,DIHYDRO-
1201-007	CRYPTOSTROBIN
0601-022	CUBEBIN
2001-007	CUMOQUINONE,PSEUDO-
1504-004	CUPRESSUFLAVONE
1504-005	CUPRESSUFLAVONE-7-ME ETHER
1504-002	CUPRESSUFLAVONE-7,7''-DIME ETHER
1504-006	CUPRESSUFLAVONE,4'',7,7''-TRIME ETHER
3504-024	CURACOSE
2805-010	CURCUMIN
2103-004	CURVIN
2108-006	CURVULARIN
2108-007	CURVULARIN,ALPHA BETA-DEHYDRO-
2103-003	CURVULIC ACID
2103-002	CURVULIN
2103-001	CURVULINIC ACID
2199-001	CURVULOL
1702-002	CYANIDIN
1702-003	CYANIDIN-3-ARABINOSIDE
1702-012	CYANIDIN-3-DIGLUCOSIDE
1702-010	CYANIDIN-3-RUTINOSIDE
1702-011	CYANIDIN-3-SOPHOROSIDE
1702-015	CYANIN
1399-006	CYANOMACLURIN
1199-037	CYCLOHETEROPHYLLIN
3901-003	CYCLOHEXANDIOL,CIS-1,2-
3901-007	CYCLOHEXANE-1,2-DIONE,3-ME-
3501-001	CYCLOHEXANE-TETROL,D-(+)-1,4/2,5-
3901-001	CYCLOHEXANOL,3-ME-
3901-002	CYCLOHEXANONE,3-ME-
1199-014	CYCLOMULBERRIN
1199-015	CYCLOMULBERROCHROMENE
2102-025	CYCLOPALDIC ACID
3901-010	CYCLOPENTADECANOL
2901-004	CYCLOPENTADECANONE,2,4,4-TRI-ME-
2901-001	CYCLOPENTANONE
2901-003	CYCLOPENT-2-ENONE,2-ME-
2102-023	CYCLOPOLIC ACID
3504-031	CYMAROSE
3506-044	CYMAROSE,4-O-(2-O-AC-DIGITALOSYL)-
1902-012	CYNARINE
2605-026	CYNODONTIN
3906-001	CYPERAQUINONE

3906-005	CYPERAQUINONE,DE-METHYL-
3906-003	CYPERAQUINONE,DIHYDRO-
3901-002	CYPERAQUINONE,HYDROXY-
3906-004	CYPERAQUINONE,TETRAHYDRO-
1205-014	CYRTOMINETIN
0302-007	DACRINOL
1902-002	DACTYLIFRIC ACID
1106-055	DACTYLIN
1401-023	DAIDZEIN
1401-024	DAIDZIN
0999-001	DALBERGICHROMENE
0902-017	DALBERGIN
0902-018	DALBERGIN,METHYL ETHER
0901-006	DALBERGIONE,(R)4-METHOXY-
0901-007	DALBERGIONE,(S)-4-OME-
0901-009	DALBERGIONE,(S)-4'-OH-4-OME-
0901-010	DALBERGIONE,S-DI-OME-
0901-008	DALBERGIONE,R-3,4-DIOME-
0901-004	DALBERGIONE QUINOL,R,-3,4-DIOME-
1801-016	DALPANOL
3501-022	DAMBONITOL
2603-019	DAMNACANTHAL
2603-018	DAMNACANTHAL,NOR-
2603-016	DAMNACANTHOL
0506-001	DAPHNETIN
0506-006	DAPHNETIN DI ME ETHER,3-(1,1-DMA)-
0506-002	DAPHNIN
0599-019	DAPHNORINE
0599-018	DAPHNORETIN
1199-001	DATISCETIN
3002-017	DAUCIC ACID
2703-024	DAUNOMYCIN
2703-026	DAUNOMYCIN,DIHYDRO-
2703-023	DAUNOMYCINONE
2703-025	DAUNOMYCINONE,DIHYDRO-
3509-015	DAUNOSAMINE
3110-032	DECANAL
3110-041	DECANOIC ACID
3110-003	DECAN-1-OL
3110-039	DECANONE,2-
3110-033	DEC-2-ENAL
3110-040	DEC-1-EN-4,6,8-TRIYN-3-ONE
3110-066	DEC-2-ENOIC ACID,9-OH-
3110-078	DEC-2-ENOIC ACID,9-OXO-
3110-067	DEC-2-ENOIC ACID,10-HYDROXY-T-
3110-073	DEC-2-ENE-1,10-DIOIC ACID
3110-076	DEC-2-ENE-4,6-DIYNE-DIOIC ACID DIMETHYL
3110-068	DEC-2-EN-4,6-DIYNOIC ACID
3110-038	DEC-2-ENE-4,6,8-TRIYN-1-AL,TRANS-
3110-029	DEC-2-ENE-4,6,8-TRIYN-1,10-DIOL,TRANS-
3110-077	DEC-2-EN-4,6,8-TRIYNEDIOIC ACID
3110-009	DEC-3-EN-1-OL,CIS-
3110-010	DEC-3-EN-1-OL,TRANS-
3110-070	DEC-4-ENE-6,8-DIYNOIC ACID,2,10-DIOH-
3110-011	DEC-8-EN-4,6-DIYN-1-OL
3110-013	DEC-8-EN-4,6-DIYN-1-OL,I-VAL-
3110-012	DEC-8-EN-4,6-DIYN-1-OL ACETATE
3110-064	DEC-8-EN-4,6-DIYNOIC ACID ME ESTER,2-OAN
3110-034	DECA-2,4-DIEN-AL
3197-013	DECA-2,4-DIENEDIOIC ACID,8-OH-2,7-DI-ME-
3110-045	DECA-2,4-DIENOIC ACID
3110-047	DECA-2,4-DIEN-6,8-DIYNOIC ACID ME ESTER,
3110-035	DECA-2,6-DIEN-4-YNAL
3110-014	DECA-2,6-DIEN-4-YN-1-OL,AC-
3110-052	DECA-2,6-DIEN-4-YNOIC ACID ME ESTER
3110-050	DECA-2,6-DIEN-4,8-DIYNOIC ACID ME ESTER,
3110-036	DECA-2,8-DIEN-4,6-DIYNAL
3110-038	DECA-2,8-DIEN-4,6-DIYNAL,10-OH-
3110-074	DECA-2,8-DIENE-4,6-DIYNE-1,10-DIOIC ACID
3110-021	DECA-3,4-DIENE-6,8-DIYN-1-OL
3110-022	DECA-4,5-DIENE-7,9-DIYN-1-OL
3110-028	DECA-4,5-DIENE-7,9-DIYNE-1,3-DIOL
3110-027	DECA-4,6-DIYN-1-OL
3110-065	DECA-4,6-DIYNOIC ACID ME ESTER,3-OANG-
3110-058	DECA-2,4,6-TRIEN-8-YNOIC ACID,5-SME-C,C,
3110-059	DECA-2,4,8-TRIEN-6-YNOIC ACID ME ESTER,5
3110-023	DECA-2,6,8-TRIEN-4-YN-1-OL
3110-026	DECA-2,4,6,8-TETRAEN-1-OL
3110-061	DECA-2,6,7,8-TETRAEN-4-YNOIC ACID,9-SME-
3110-075	DECA-2,4,6-TRIYNDIOIC ACID DIME ESTER
3110-031	DECA-4,6,8-TRIYN-1,3-DIOL
3110-071	DECA-4,6,8-TRIYNOIC ACID,2,10-DIOH-
3110-079	DECA-4,6,8-TRIYN-3-ONE,1,2-EPOXY-
2999-002	DECUMBIN = BREFELDIN A
0503-003	DECURSIN
0503-002	DECURSINOL
0503-004	DECURSINOL,3'-EPI-
0503-005	DECURSINOL,O-ANGELOYL-3'-EPI
2404-014	DECUSSATIN
1801-002	DEGUELIN
1801-003	DEGUELIN,DEHYDRO-

3120-017	EICOSA-11,14,17-TRIENOIC ACID
3120-018	EICOSA-5,11,14,17-TETRAENOIC ACID
3118-022	ELAIDIC ACID
3118-003	ELAIDIC ALCOHOL
3118-039	ELAOSTEARIC ACID,ALPHA-
0301-022	ELEMICIN
0301-023	ELEMICIN,ISO-
2503-049	ELEUTHERIN
2503-050	ELEUTHERIN,ISO-
2503-035	ELEUTHERINOL
2502-035	ELEUTHEROL
0105-001	ELLAGIC ACID
0105-003	ELLAGIC ACID 4,4'-DI-ME ETHER
2510-004	ELLIPTINONE
1801-018	ELLIPTONE
2704-017	ELSINOCHROME
2704-014	ELSINOCHROME A
2704-015	ELSINOCHROME B
2704-016	ELSINOCHROME C
2003-007	EMBELIN
2604-052	EMODIC ACID
2604-028	EMODIN
1703-003	EMPETRIN
2202-010	ENCECALIN
2202-009	ENCECALIN,DES-OME-
0203-014	ENCEOLIN,DES-OME-
2604-022	ENDOCROCIN
1204-005	ENGELITIN
1703-022	ENSATIN
0702-008	EPANORIN
1901-008	EPOXYDON
1104-020	EQUISETRIN
1405-001	EQUOL
2803-004	ERDIN,D,L-
2406-001	ERGOCHROME AA = SECALONIC ACID A
2406-004	ERGOCHROME AB = SECALONIC ACID C
2406-006	ERGOCHROME AC = ERGOCHRYSIN A
2406-005	ERGOCHROME AD
2406-003	ERGOCHROME BB = SECALONIC ACID B
2406-007	ERGOCHROME BC = ERGOCHRYSIN B
2406-006	ERGOCHROME BD
2406-005	ERGOCHROME CC = ERGOFLAVIN
2406-009	ERGOCHROME CD
2406-008	ERGOCHROME DD
2406-006	ERGOCHRYSIN A
2406-007	ERGOCHRYSIN B
2406-005	ERGOFLAVIN
2406-012	ERGOXANTHIN
0802-006	ERICOLIN = ARBUTIN
1205-004	ERIOCITRIN
1205-003	ERIODICTIN
1205-001	ERIODICTYOL
1205-002	ERIODICTYOL,7-O-GLUCOSYL
1205-005	ERIODICTYOL-5,3'-DIGLUCOSIDE
1205-007	ERIODICTYOL,HOMO-
1205-007	ERIODICTYONONE = ERIODICTYOL,HOMO-
2204-038	ERIOSTEMOIC ACID
2204-039	ERIOSTOIC ACID
1104-042	ERMANIN
1602-007	EROSNIN
3122-008	ERUCAIC ACID
3506-043	ERYSCENOBIOSE
2301-034	ERYTHRIN
3104-024	ERYTHRITOL,I-
3104-025	ERYTHRITOL,1-BETA-MANNOPYRANOSYL-MESO-
3104-024	ERYTHRITOL,MESO-
2605-020	ERYTHROGLAUCIN
2605-015	ERYTHROLACCIN
2604-020	ERYTHROLACCIN,DESOXY-
2605-002	ERYTHROLACCIN,ISO-
3401-012	ERYTHROMYCIN A
3401-010	ERYTHROMYCIN B
3401-011	ERYTHROMYCIN C
2506-013	ERYTHROSTOMINOL,DESOXY-
2506-015	ERYTHROSTOMINONE
2506-014	ERYTHROSTOMINONE,DESOXY-
0507-001	ESCULETIN
0507-006	ESCULETIN,6,7-DIMETHYL ETHER-
0507-002	ESCULIN
0301-002	ESDRAGOL
3102-002	ETHANOL
3005-045	ETHISOLIDE
3102-006	ETHYL ACETATE
3102-001	ETHYLENE
1103-031	EUCALYPTIN
0999-008	EUCOMIN
0999-007	EUCOMIN,4'-DES-ME-
0999-010	EUCOMIN,4'-DES-ME-5-O-ME-3,9-DIHYDRO-
0999-015	EUCOMNALIN
0999-009	EUCOMOL
0104-020	EUDESMIC ACID
0601-026	EUDESMIN
2203-006	EUGENIN
2203-007	EUGENITIN
2203-004	EUGENITOL
2203-005	EUGENITIN,ISO-
0301-010	EUGENOL
0399-019	EUGENOL-O-ANGELOYL ESTER,EPOXY-ISO-
0301-011	EUGENOL ME ETHER
0301-019	EUGENOL VICIANOSIDE
0399-018	EUGENOLOXIDE,ISO-
0301-013	EUGENOL,ISO-
0301-014	EUGENOL ME ETHER,ISO-
0301-021	EUGENOL,OME-
2203-001	EUGENONE
0301-019	EUGENYL ACETATE = ACETEUGENOL
1105-042	EUPAFOLIN
1104-050	EUPALIN
1104-049	EUPALITIN
2202-018	EUPARIN
2202-019	EUPARIN,DIHYDRO-
2202-023	EUPARIN,OME-
2202-024	EUPARIN,DIHYDRO-OME-
2202-027	EUPARONE ME ETHER
1105-050	EUPATILIN
1106-076	EUPATOLIN
1106-075	EUPATOLITIN
1105-049	EUPATORIN
0699-020	EUPOMATENE
1108-011	EUROPETIN
0199-010	EUTANNIN = CHEBULINIC ACID
2402-011	EUXANTHONE
3510-005	EVERMICOSE
2301-002	EVERNIC ACID
2101-006	EVERNINIC ACID
2101-032	EVERNINOCIN
2101-033	EVERNINONITROSE
3506-042	EVERNINOSE
3510-009	EVERNITROSE
3508-005	EVERTETROSE
3507-009	EVERTRIOSE
2202-008	EVODIONE
2202-003	EVODIONOL
2202-005	EVODIONOL,ALLO-
2202-007	EVODIONOL,ISO-
2202-004	EVODIONOL-7-ME-ETHER
2202-006	EVODIONOL-7-ME ETHER,ALLO-
3118-051	EXOCARPIC ACID
0902-020	EXOSTEMIN
0507-007	FABIATRIN
0601-034	FAGAROL
3117-046	FALCARINDIOL
3117-047	FALCARINDIOL,3-O-AC-
3117-023	FALCAPINOL
3117-069	FALCARINOLONE
3117-059	FALCARINONE
3117-058	FALCARINONE,1,2-DIHYDRO-
2604-051	FALLACINAL
0601-029	FARGESIN
0502-007	FELAMEDIN
1499-004	FERREIRIN
1499-005	FERREIRIN,HOMO-
1203-019	FARREROL
1299-009	FARREROL,PROTO-
2602-007	FERRUGINOL
1499-021	FERRUGONE
0903-023	FERRUOL A = MAMMEA B/BA
0599-005	FERULENOL
0304-019	FERULIC ACID
0304-021	FERULIC ACID,ISO-
0304-020	FERULIC ACID ME ESTER,4-O-GERANYL-
0304-022	FERULIC ACID CHOLINE ESTER,ISO-
0303-005	FERULYL ALDEHYDE
1601-023	FICIFOLINOL
1601-027	FICININ
1199-028	FICININE
1199-029	FICININE,ISO-
2299-002	FILICINIC ACID
3402-006	FILIPIN
2207-023	FILIXIC ACID BBB
2207-022	FILIXIC ACID PBB
2207-021	FILIXIC ACID PBP
1106-001	FISETIN
1106-002	FISETIN 3-ME-ETHER
1303-001	FISETINIDOL,(-)-
1303-002	FISETINIDOL,(+)-EPI-
1302-009	FISTUCACIDIN
1302-019	FLAVAN-3,4-DIOL,5,4'-DIOH-
1304-011	FLAVAN,(-)3,4,5,7,3',4',5'-HEPTA-OH-
1302-004	FLAVAN,DL-4'-5,7-TRI-OME-
1307-002	FLAVAN,(-)-4'-OH-7-OME-
1302-003	FLAVAN,(-)-4'-OH-7-OME-8-ME-

1102-001	GALANGIN
1102-003	GALANGIN-3-ME ETHER
0601-009	GALBACIN
3111-002	GALBANOLENE
0501-012	GALBANUM ACID
0601-006	GALBELGIN
0604-003	GALBULIN
0604-001	GALCATIN
0604-002	GALCATIN,ISO-
0601-007	GALGRAVIN
2604-010	GALIOSIN
2703-013	GALIRUBINONE = AKLAVINONE,7-DEOXY-
2703-001	GALIRUBINONE-B1
0104-016	GALLIC ACID
0199-014	GALLIC ACID,3-O-GALLOYL-
0104-017	GALLIC ACID,4-O-GLUCOSYL-
0104-018	GALLIC ACID GLU ESTER= GLUCOGALLIN
1304-004	GALLOCATECHIN,(+)-
1304-005	GALLOCATECHIN GALLATE,(-)-
1304-006	GALLOCATECHIN,EPI-(-)-
1304-008	GALLOCATECHIN,DL-
1304-007	GALLOCATECHIN GALLATE,EPI-(-)-
1105-014	GALUTEOLIN
1101-010	GAMATIN
2499-007	GAMBOGIC ACID
2204-021	GAMMANDOL
2302-009	GANGALEOIDIN
1601-026	GANGETIN
1199-036	GARCININ
1107-012	GARDENIN-A
1107-011	GARDENIN-C
1105-057	GARDENIN-D
1107-010	GARDENIN-E
3509-026	GAROSAMINE
2404-032	GARTANIN
2403-012	GARTANIN,8-DESOXY-
0104-025	GAULTHERIN
1501-006	GB-1
1501-001	GB-1A
1501-012	GB-2
0501-023	GEIJERIN
0501-024	GEIJERIN,DEHYDRO-
0301-020	GEIN
0501-008	GEIPARVARIN
1401-001	GENISTEIN
1401-002	GENISTIN
1103-020	GENKWANIN
1103-021	GENKWANIN,GLUCO-
1105-045	GENKWANIN,3'-OH-6-OME-
3502-021	GENTAMICIN A
3502-026	GENTAMICIN-B
3502-025	GENTAMICIN-B1
3502-024	GENTAMICIN-C1
3502-023	GENTAMICIN-C2
3502-022	GENTAMICIN-D
3502-027	GENTAMICIN-X
2404-010	GENTIACAULEIN
2404-009	GENTIAKOCHIANINE = SWERTIANIN
2403-019	GENTIANIN = GENTISIN
3507-005	GENTIANOSE
3506-025	GENTIOBIOSE
2403-021	GENTIOSIDE
2403-018	GENTISEIN
0104-030	GENTISIC ACID
0204-005	GENTISIC ACID,HOMO
0104-037	GENTISIC ACID ME ESTER,5-ME-3-NO2-
2403-019	GENTISIN
2403-020	GENTISIN,ISO-
0102-011	GENTISYL ALCOHOL,3-CL-
2001-003	GENTISYLQUINONE
2803-005	GEODIN
2802-017	GEODIN,DIHYDRO-
2803-001	GEODIN,(+)-BIS-DE-CL-
2803-002	GEODIN,(-)-BIS-DE-CL-
2399-010	GEODOXIN
0301-019	GEOSIDE
1701-003	GESNERIDINE = APIGENIDIN
1701-002	GESNERIN
0499-002	GINGEROL,6-
0499-004	GINGEROL,8-
0499-005	GINGEROL,10-
0499-003	GINGEROL,ME-6-
1505-007	GINKGETIN
1505-006	GINKGETIN,ISO-
2111-012	GINKGOL
2111-011	GINKGOL,HYDRO-
2111-002	GINKGOLIC ACID
2111-001	GINKGOLIC ACID,HYDRO-
3129-003	GINNOL
3129-005	GINNONE
0508-011	GLABRALACTONE
2102-024	GLADIOLIC ACID
2102-022	GLADIOLIC ACID,DIHYDRO-
3906-017	GLAUCANIC ACID
3906-018	GLAUCONIC ACID
0599-012	GLAUPALOL
1303-011	GLEDITSIN = MOLLISACACIDIN
1506-001	GLOBIFLORIN 3B1
1506-002	GLOBIFLORIN 3B2
2403-003	GLOBUXANTHONE
2301-029	GLOMELLIFERIC ACID
3504-065	GLUCARIC ACID,D-
1206-003	GLUCODISTYLIN
0104-032	GLUCOGALLIN
3504-059	GLUCONIC ACID,D-
3504-060	GLUCONIC ACID,2-KETO-
3504-061	GLUCONIC ACID,5-KETO-
3504-062	GLUCONIC ACID,2,5-DIKETO-
3509-001	GLUCOSAMINE,D-
3504-008	GLUCOSE,D-
3504-010	GLUCOSE,1-O-BUTYL-
3509-013	GLUCOSE,3-AMINO-3,6-DIDEOXY-
3504-009	GLUCOSE,6-O-ACETYL-
3509-024	GLUCOSE,6-AMINO-6-DESOXY-
3504-050	GLUCOSONE
0103-008	GLUCOVANILLIN = VANILLOSIDE
0102-008	GLUCOVANILLYL ALCOHOL = VANILLOLOSIDE
3504-063	GLUCURONIC ACID,D-
3105-007	GLUTACONIC ACID,TRANS-
3196-011	GLUTACONIC ACID,TRANS-BETA-ME-
2112-003	GLUTARENGOL
3105-004	GLUTARIC ACID
3105-006	GLUTARIC ACID,ALPHA-KETO-
3105-005	GLUTARIC ACID,ALPHA-OH-
3196-014	GLUTARIC ACID,GAMMA-ME-GAMMA-OH-ALPHA-KE
3196-015	GLUTARIC ACID,GAMMA-METHYLENE-ALPHA-KETO
3103-007	GLYCERALDEHYDE
3103-020	GLYCERIC ACID,L(-)-
3103-015	GLYCERIC ACID,2-PHOSPHO-
3103-003	GLYCEROL
3103-024	GLYCEROL,1,1-O-BETA-D-GALACTOFURANOSYL-
3504-006	GLYCEROL,O-BETA-D-GALACTOFURANOSYL-
0302-011	GLYCOL,C-SYRINGOYL-
3102-004	GLYCOLALDEHYDE
3102-008	GLYCOLIC ACID
0199-020	GLYCOSMIN
1002-010	GLYCYPHYLLIN
3102-009	GLYOXYLIC ACID
0601-035	GMELINOL
1102-008	GNAPHALIIN
1102-009	GNAPHALIIN 3-ME ETHER
1102-010	GNAPHALIIN,ISO-
2901-015	GORLIC ACID
3114-030	GOSHUYUIC ACID
1106-091	GOSSYPETIN
1106-115	GOSSYPETIN TETRA-ME ETHER
1106-105	GOSSYPETIN PENTA-ME ETHER
1702-006	GOSSYPICYANIN
1106-092	GOSSYPIN
1106-093	GOSSYPITRIN
2703-043	GRANATICIN
3901-006	GRANDIDENTATIN
3901-004	GRANDIDENTOSIDE
1099-015	GRANDIFLORONE
0104-021	GRANDIFOLINE
0501-038	GRAVELLIFERONE
0501-039	GRAVELLIFERONE,ME ETHER
0506-007	GRAVELLIFERONE,8-OME-
1105-006	GRAVEOBIOSIDE-A
1105-026	GRAVEOBIOSIDE B
3905-027	GRAVOLENIC ACID
2110-001	GREVILLOL
2101-025	GRIFOLIN
2803-007	GRISEOFULVIN
2803-008	GRISEOFULVIN,BROMO-
2803-006	GRISEOFULVIN,DECHLORO-
2803-009	GRISEOFULVIN,DEHYDRO-
2803-010	GRISEOFULVIN,DE-CL-6'-O-DES-ME-
2802-015	GRISEOPHENONE-A
2802-014	GRISEOPHENONE-B
2802-013	GRISEOPHENONE-C
2802-016	GRISEOPHENONE Y
2403-014	GRISEOXANTHONE C
3999-002	GRISORIXIN
0802-002	GUAIACOL
0201-004	GUAIACOL,ET-
0301-007	GUAIACOL,PROPYL-
0302-010	GUAIACYLGLYCEROL
0601-002	GUAIARETIC ACID
0601-001	GUAIARETIC ACID,NOR-DIHYDRO-
1106-011	GUAIJAVERIN
2402-005	GUANANDIN

3126-004	HEXACOSANDIOL,1-CAFFEOYL-26-FERULOYL-
3126-005	HEXACOSANDIOL,1,26-DI-CAFFEOYL-
3126-006	HEXACOSANDIOL,1,26-DI-FERULOYL-
3126-001	HEXACOSANE,N-
3126-007	HEXACOSANOIC ACID
3126-010	HEXACOSANOIC ACID,26-OH-
3126-002	HEXACOSAN-1-OL
3126-003	HEXACOSAN-1,26-DIOL
3116-001	HEXADECANE
3116-025	HEXADECANOIC ACID,8-OH-
3116-033	HEXADECANOIC ACID,9,10-DIOH-
3202-002	HEXADECANOIC ACID,10-3
3202-002	HEXADECANOIC ACID,10-OH-6,9-OXIDO-
3116-011	HEXADECAN-1,16-DIOL
3197-020	HEXADEC-1-ENE,15-ME-
3116-018	HEXADEC-3-ENOIC ACID
3116-006	HEXADEC-7-EN-1-OL
3116-007	HEXADEC-7-EN-1-OL,AC-
3197-021	HEXADEC-8-EN-1-OL,14-ME-
3197-022	HEXADEC-8-ENOIC ACID ME ESTER,14-ME-
3116-008	HEXADEC-9-EN-1-OL
3116-029	HEXADEC-9-ENOIC ACID,12-OH-
3116-030	HEXADEC-9-ENOIC ACID,16-OH-
3116-020	HEXADEC-11-ENOIC ACID
3116-021	HEXADEC-12-ENOIC ACID
3116-041	HEXADECA-1,14-DIEN-10,12-DIYNE-8,9-EPOXI
3116-012	HEXADECA-6,8-DIEN-10,12,14-TRIYNAL
3116-010	HEXADECA-7,14-DIEN-10,12-DIYN-1-OL
3116-002	HEXADECA-1,6,8-TRIEN-10,12,14-TRIYNE
3116-013	HEXADECA-6,8,14-TRIEN-10,12-DIYNAL
3116-022	HEXADECA-7,10,13-TRIENOIC ACID
3116-014	HEXADECA-7,9,11,13-TETRAENAL,ALL TRANS-
3116-005	HEXADECYL ACETATE
3106-004	HEXANOL
3106-005	HEXANAL
3196-002	HEXAN-2-ONE,4-ME-
3106-014	HEXANOIC ACID,3-OH-
3003-002	HEXANOIC ACID LACTONE,4-ME-5-OH-
3197-053	HEXATRIACONTANE,4,8,12-TRI-ME-
3504-051	HEXOSE,2,5-DIOXO-
2303-004	HIASIC ACID
2303-005	HIASCINNIC ACID = HIASIC ACID
1108-021	HIBISCETIN
1108-024	HIBISCETIN HEPTA-ME ETHER
1108-022	HIBISCITRIN
3005-009	HIBISCUSIC ACID
1599-003	HINOKIFLAVONE
1599-004	HINOKIFLAVONE,2,3-DIHYDRO-
0601-019	HINOKININ
0699-025	HINOKIOL
0699-012	HINOKIRESINOL
1703-024	HIRSUTIDIN
1703-025	HIRSUTIN
0404-017	HISPIDIN
1103-040	HISPIDULIN
3509-021	HOLANTOSAMINE
3509-022	HOLOSAMINE,D-
0101-003	HOMOCATECHOL
2699-013	HOMONATALOIN
2704-022	HOPEAPHENOL
3904-006	HOP OXETONE I
3904-007	HOP OXETONE II
0699-021	HORDATINE A
0699-022	HORDATINE B
1206-002	HULTENIN
2903-006	HULUPINIC ACID
2903-004	HULUPONE
2903-005	HUMULINIC ACID,PRE-
2903-001	HUMULINONE
2206-002	HUMULONE
2206-014	HUMULONE,4-DEOXY-
2206-021	HUMULONE,PRE-
1703-007	HYACIN
2901-013	HYDNOCARPIC ACID
2901-016	HYDNOCARPIC ACID,KETO-
2804-008	HYDRANGEIC ACID
2804-009	HYDRANGEIC ACID,DIHYDRO-
2804-010	HYDRANGENOL
2802-003	HYDROCOTOIN
2802-004	HYDROCOTOIN,METHYL-
0802-005	HYDROQUINONE
0802-012	HYDROQUINONE-ME ETHER-MONO
0802-014	HYDROQUINONE-DIME ETHER
0802-013	HYDROQUINONE-MONO ETHYL ETHER
0803-004	HYDROQUINONE,MEO-
3502-036	HYGROMYCIN B
3197-011	HYGROPHYLLINECIC ACID
1105-058	HYMENOXIN
3502-034	HYOSAMINE
2704-020	HYPERICIN
2704-021	HYPERICIN,PSEUDO-
1106-014	HYPERIN
1105-053	HYPOLAETIN-7-GLUCOSIDE
3203-010	HYPTOLIDE
1099-011	HYSSOPIN
2603-034	HYSTAZARIN-MONO-ME-ETHER
2603-017	IBERICIN
1104-039	ICARIIN
1104-033	ICARIIN,NOR-
1104-038	ICARITIN
1104-029	ICARITIN,ISO-ANHYDRO-
1104-030	ICARITIN,NOR-
1104-028	ICARITIN,NOR-ANHYDRO-
1499-018	ICHTHYNONE
3203-003	ICHTHYOTHEREOL
3203-004	ICHTHYOTHEREOL,AC-
1702-004	IDAEIN
3504-054	IDITOL,L-
2301-025	IMBRICARIC ACID
0506-013	IMPERATORIN
0508-024	IMPERATORIN,ISO-
0506-023	IMPERATORIN ME ETHER,EPOXY-ALLO-
3002-004	INNOVANOSIDE
2204-007	INOPHYLLOIDIC ACID
0902-013	INOPHYLLOLIDE,(+)-CIS-
0902-014	INOPHYLLOLIDE,(+)-TRANS-
0902-015	INOPHYLLOLIDE,(+)-DIHYDRO-
3501-008	INOSITOL,(D)-(+)-
3501-012	INOSITOL,L-
3501-012	INOSITOL,L-(-)-
3501-009	INOSITOL 1-ME-ETHER,(D)-(+)-
3501-015	INOSITOL,MYO-
3501-017	INOSITOL,1-O-MANNOSYL-MYO-
0799-001	INVOLUTIN
3114-039	IPUROLIC ACID
1403-002	IRIDIN
2608-002	IRIDOSKYRIN
1403-001	IRIGENIN
1401-030	IRISOLONE
1401-021	IRISOLIDONE
3118-050	ISANIC ACID
3118-054	ISANOLIC ACID
2604-054	ISLANDICIN
3195-002	ISOAMYL ALCOHOL
3194-003	ISOBUTANAL
3194-001	ISOBUTANOL
3194-002	ISOBUTYL ACETATE
3194-007	ISOBUTYRIC ACID,(2,2'-DITHIO)-
3196-017	ISOCITRIC ACID,THREO-DS-
3196-018	ISOCITRIC ACID,ERYTHRO-LS-
2104-015	ISOCOUMARIN,8-OH-3-ME-
2104-023	ISOCOUMARIN,8-OH-6,7-DI-OME-3-ME-
2104-016	ISOCOUMARIN,8-OH-6-OME-3-ME-
2104-021	ISOCOUMARIN,6,7,8-TRI-OH-3-ME-
2104-026	ISOCOUMARIN,7-COOH-3,4-DIHYDRO-8-OH-3-ME
2104-033	ISOCOUMARIN,7-AC-6,8-DI-OH-5-ME-
2104-012	ISOCOUMARIN,3-ME-4,8-DI-OH-3,4-DIHYDRO-
2104-011	ISOCOUMARIN,3-ME-3,8-DI-OH-3,4-DIHYDRO-
2507-003	ISOERHINOCHROME = SPINOCHROME-C
1405-009	ISOFLAVAN,2'-OH-4,7-DI-OME-
1499-014	ISOFLAVONE,2',4',5',6,7-PENTA-OME-
1402-010	ISOFLAVONE,3',4',7-TRIOH-
1403-003	ISOFLAVONE,3',6,7-TRI-OME-J',5'-METHYLEN
1402-022	ISOFLAVONE,3',4',6,7-TETRA-OME-
1401-028	ISOFLAVONE,4'-OH-7-OME-
1402-011	ISOFLAVONE,4',7-DIOH-3-OME-
1402-018	ISOFLAVANONE,6,7-DIOME-3',4'-METHYLENEDI
3195-001	ISOPENT-2-ENE
1106-043	ISORHAMNETIN
1106-053	ISORHAMNETIN-3-GALACTOSIDE-7-GLUCOSIDE
3195-003	ISOVALERALDEHYDE
3195-008	ISOVALERIC ACID
3195-009	ISOVALERIC ACID,ALPHA-OH-
3195-015	ISOVALERIC ACID,ALPHA-KETO-
3195-010	ISOVALERIC ACID,2,3-DIOH-
3195-019	ITACONIC ACID
3005-018	ITACONITIN
3195-021	ITATARTARIC
1102-002	IZALPININ
2404-027	JACAREUBIN
2403-011	JACAREUBIN,6-DES-OH-
1106-079	JACEIDIN
1106-080	JACEIN
1105-048	JACEOSIDE
3004-017	JACONECIC ACID
3116-026	JALAPINOLIC ACID
1499-017	JAMAICIN
2901-019	JASMONE
2999-004	JASMONIC ACID ISOLEUCINYL AMIDE
2999-005	JASMONIC ACID ISOLEUCINYL AMIDE,DIHYDRO-
2999-003	JASMONIC ACID LACTONE,5'-OH-
2505-014	JAVANICIN

2503-004	LAWSONE ME ETHER
1405-003	LAXIFLORAN
1303-018	LEBBECACIDIN
2301-001	LECANORIC ACID
1601-020	LEIOCARPIN
3305-005	LENTHIONINE
1104-023	LEPIDOSIDE
0702-010	LEPRAPIC ACID = LEPRAPINIC ACID
0702-010	LEPRAPINIC ACID
2203-009	LEPRARIC ACID
2111-010	LEPROSOL I,BETA-
2112-004	LEPROSOL-II,BETA-
2203-014	LEPTORUMOL
2203-015	LEPTORUMOLIN
1803-007	LEPTOSIDIN
1803-008	LEPTOSIN
2206-011	LEPTOSPERMONE
3501-003	LEUCANTHEMITOL
1303-019	LEUCOCYANIDIN,(+)-
1303-020	LEUCOCYANIDIN,(+)-
1303-021	LEUCOCYANIDIN GALLATE
1303-022	LEUCOCYANIDIN 4'-O-ME ETHER,3-O-GALACTOS
3906-042	LEUCODRIN
1303-017	LEUCOFISETINIDIN,(+)-
1502-004	LEUCOFISETINIDIN-(+)-CATECHIN
1303-013	LEUCOFISETINIDIN,(-)-
3906-043	LEUCOGLYCODRIN
0701-005	LEUCOMELONE
0701-006	LEUCOMELONE,PROTO-
3401-027	LEUCOMYCIN-A1
3401-031	LEUCOMYCIN A3
3401-030	LEUCOMYCIN-A4
3401-026	LEUCOMYCIN-A5
3401-029	LEUCOMYCIN-A6
3401-025	LEUCOMYCIN-A7
3401-028	LEUCOMYCIN-A8
3401-024	LEUCOMYCIN-A9
1302-016	LEUCOPELARGONIDIN
1599-012	LEUCOPELARGONIDIN DIMER
1599-017	LEUCOPELARGONIDIN DIOXAN DIMER
1599-027	LEUCOPELARGONIDIN TETRAMER
1304-010	LEUCOROBINETINIDIN
1506-007	LEUCOROBINETINIDIN-(+)-CATECHIN
1506-008	LEUCOROBINETINIDIN-(+)-GALLOCATECHIN
3506-023	LEUCROSE
0504-006	LIBANOTIN
3118-073	LICANIC ACID,ALPHA-
3118-074	LICANIC ACID,ISO-
3005-031	LICHESTERINIC ACID,(-)-
3005-027	LICHESTERINIC ACID,D-PROTO-
3005-028	LICHESTERINIC ACID,L-PROTO-
3005-029	LICHESTERINIC ACID,L-ALLO-PROTO-
2403-015	LICHEXANTHONE
2403-013	LICHEXANTHONE,NOR-
1405-008	LICORICIDIN
3124-004	LIGNOCERIC ACID
1103-054	LIGNOSIDE
2106-023	LIGUSTICUM LACTONE
2106-024	LIGUSTILIDE
3506-048	LILACINOBIOSE
0508-001	LIMETTIN
1106-100	LIMOCITRIN
1106-101	LIMOCITRIN,3-O-GLUCOSYL-
1106-107	LIMOCITROL
1106-108	LIMOCITROL,3-O-GLUCOSYL-
1106-109	LIMOCITROL,ISO-
1106-110	LIMOCITROL,3-O-GLUCOSYL-ISO-
0203-004	LINARHODIN = ACETOPHENONE,P-OME-
1103-025	LINARIN
3112-016	LINDERIC ACID
2903-015	LINDERONE
2903-016	LINDERONE,ME-
3118-030	LINELAIDIC ACID
3118-028	LINOLEIC ACID,C,C-
3118-029	LINOLEIC ACID,C,T-
3118-079	LINOLEIC ACID,15,16-EPOXY-
3118-041	LINOLENIC ACID,ALPHA-
3118-034	LINOLENIC ACID,GAMMA-
3305-003	LIPOIC ACID
1203-024	LIQUIRITIGENIN
1002-001	LIQUIRITIGENIN,ISO-
1002-003	LIQUIRITIGENIN,4'-O-ME-ISO-
1203-026	LIQUIRITIGENIN-7-GLUCOSIDE
1203-027	LIQUIRITIGENIN-7-DIGLUCOSIDE
1002-004	LIQUIRITIGENIN 4'-GLUCOSIDE,ISO-
1002-005	LIQUIRITIGENIN 4'-DIGLUCOSIDE,ISO-
1203-025	LIQUIRITIN
1002-002	LIQUIRITIN,ISO-
0603-004	LIRIODENDRIN
3501-023	LIRIODENDRITOL
0603-001	LIRIORESINOL-A
0603-002	LIRIORESINOL-B
0603-005	LIRIORESINOL C DI-ME ETHER
1602-010	LISETIN
3502-028	LIVIDOMYCIN-A
2204-041	LL-D253-ALPHA
2204-042	LL-D253-BETA
2204-040	LL-D253-GAMMA
3203-012	LL-P580-ALPHA
2101-027	LL-Z-1272 ALPHA
2101-026	LL-Z-1272 BETA
2101-029	LL-Z-1272 DELTA
2101-028	LL-Z-1272 EPSILON
2101-030	LL-Z-1272 GAMMA
2101-031	LL-Z-1272 ZETA
2302-023	LOBARIC ACID
2302-025	LOBARIDONE-NOR-
2506-016	LOMANDRONE
2507-006	LOMASTILONE
0505-002	LOMATIN
0505-005	LOMATIN,HEXANOYL-
0505-006	LOMATIN,OCTANOYL-
0505-007	LOMATIN,CIS-4-OCTENYL-
2503-008	LOMATIOL
2507-007	LOMAZARIN
1405-007	LONCHOCARPAN
1404-007	LONCHOCARPENIN
1404-006	LONCHOCARPIC ACID
1001-003	LONCHOCARPIN
2302-024	LOXODIN
1105-012	LUCENIN-1
3402-002	LUCENSOMYCIN
1602-003	LUCERNOL
2603-015	LUCIDIN
1105-060	LUCIDIN
1105-061	LUCIDIN DI ME ETHER
2903-013	LUCIDONE
2903-014	LUCIDONE,METHYL-
3130-005	LUMEQUEIC ACID
2206-015	LUPULONE
0105-008	LUTEOIC ACID
1105-001	LUTEOLIN
1105-003	LUTEOLIN,4'-O-GLUCOSYL-
1105-009	LUTEOLIN,7-O-GLUCO-GALACTOSYL-
1105-008	LUTEOLIN,7-O-GLUCO-RHAMNOSYL-
1105-004	LUTEOLIN,7-O-LAMINARIBIOSYL-
1105-007	LUTEOLIN,7-O-RUTINOSYL-
1105-002	LUTEOLIN 7-ME-ETHER
1105-015	LUTEOLIN 5,7-DIGLUCOSIDE
1105-062	LUTEOLIN-6-C-XYLOSIDE,7-O-RHAMNOSYL-
1702-001	LUTEOLINIDIN
3999-014	LUTEORETICULIN
1106-051	LUTEOSIDE
2608-015	LUTEOSKYRIN
0506-008	LUVANGETIN
3906-027	LYCOPERSIN = BIKAVERIN
3508-009	LYCOPOSE
1702-008	LYCORICYANIN
3508-004	LYCOTETRAOSE
0606-001	LYONIRESINOL,RACEMIC-
0602-002	LYONIRESINOL,(+)-2-ALPHA-O-RHAMNOSIDE
3503-011	LYXURONIC ACID,D-
1601-009	MAACKIAIN,D-
1601-011	MAACKIAIN,DL-
1601-007	MAACKIAIN,L-
0903-001	MAB-2 = MAMMEA B/AA
0902-007	MAB-3
0903-020	MAB-4
0902-010	MAB-5
2503-017	MACASSAR II
2503-018	MACASSAR III
3114-024	MACILENIC ACID
3120-006	MACILOLIC ACID
2404-028	MACLURAXANTHONE
2802-007	MACLURIN
2604-019	MACROSPORIN
2403-037	MACULATOXANTHONE
2607-005	MADAGASCARIN
2604-034	MADAGASCIN
2699-007	MADAGASCIN ANTHRONE
2003-013	MAESAQUINONE
1803-014	MAESOPSIN
3401-021	MAGNAMYCIN
3401-020	MAGNAMYCIN B
0602-004	MAGNOLIN
0699-003	MAGNOLOL
0504-001	MAJURIN
1801-021	MALACCOL
3104-023	MALEIC ACID,DIOH-
3104-018	MALIC ACID,D(+)-
3104-019	MALIC ACID,L-
1001-012	MALLOTUS CHALCONE

2602-010	PACHYBASIN
2602-011	PACHYBASIN ME ETHER
3506-047	PACHYBIOSE
1404-009	PACHYRRIZIN
1801-022	PACHYRRHIZONE
1801-023	PACHYRRIZONE,12A-OH-
1206-005	PADMATIN
1206-006	PADMATIN,GLUCOSYL-
2201-002	PAEONOL
3116-038	PAHUTOXIN
0304-033	PAJANEELIN
1803-012	PALASITRIN
2107-008	PALITANTIN
2609-006	PALMIDIN-A
2609-004	PALMIDIN-B
2609-005	PALMIDIN-C
2609-007	PALMIDIN-D
3116-016	PALMITAMINE
3131-005	PALMITENONE,CIS-
3116-015	PALMITIC ACID
3116-024	PALMITIC ACID,3-D-HYDROXY-
3116-017	PALMITIC ACID HEXADECYL ESTER
3198-005	PALMITODISTEARIN
3116-019	PALMITOLEIC ACID
3131-004	PALMITONE
3131-006	PALMITONE,10-OH-
2301-021	PALUDOSIC ACID
1104-013	PANASENOSIDE
1901-013	PANEPOXYDOL,NEO-
1901-014	PANEPOXYDOL,7-DEOXY-
1901-009	PANEPOXYDONE
1901-010	PANEPOXYDONE,ISO-
1901-011	PANEPOXYDONE,NEO-
1901-012	PANEPOXYDIONE
0508-030	PANGELINE
0508-031	PANGELINE,ANGELOYL-
2399-004	PANNARIC ACID
2302-021	PANNARIN
2199-010	PANOSIALIN-I
2199-011	PANOSIALIN-II
2199-012	PANOSIALIN-III
2204-005	PAPUANIC ACID
2204-006	PAPUANIC ACID,ISO-
2605-010	PAPULOSIN
2207-015	PARAASPIDIN
0601-020	PARABENZLACTONE
0402-010	PARACOTOIN
0402-011	PARACOTOIN,4-OME-
3003-001	PARASORBIC ACID
2604-053	PARIETINIC ACID
3118-043	PARINARIC ACID
2301-009	PARMELIN = ATRANORIN
3502-008	PAROMAMINE
3502-013	PAROMOMYCIN-I
3502-014	PAROMOMYCIN-II
3502-015	PAROMOMYCIN,MANNOSYL-
0303-006	PARVIFLORAL
1106-072	PATULETIN
3904-005	PATULIN
1106-073	PATULITRIN
0601-036	PAULOWNIN
0601-037	PAULOWNIN,ISO-
1103-043	PECTOLINARIGENIN
1103-044	PECTOLINARIN
1105-044	PEDALIIN
1105-043	PEDALITIN
3999-008	PEDERIN
3999-007	PEDERIN,PSEUDO-
3999-009	PEDERONE
3197-030	PEDICELLIC ACID
1001-014	PEDICELLIN
1001-013	PEDICIN
1201-010	PEDICIN,ISO-
1001-015	PEDICININ
1001-016	PEDICININ,METHYL-
2112-001	PELANDJUAIC ACID
1701-003	PELARGONIDIN
1701-005	PELARGONIDIN 3-RHAMNOGLUCOSIDE
1701-006	PELARGONIDIN 3-XYLOSYLGLUCOSIDE
1701-007	PELARGONIN
2804-011	PELLEPIPHYLLIN
3110-046	PELLITORINE
0698-003	PELTATIN,ALPHA-
0698-004	PELTATIN,BETA-
0698-005	PELTATIN A ME ETHER,BETA-
0698-006	PELTATIN A ME ETHER,BETA-5'-DES-OME-
1802-001	PELTOGYNAN-14-ONE,2,3,4-TRI-OH-8-OME-6,1
1802-009	PELTOGYNOL,(+)-
1802-002	PELTOGYNOL-B
1104-051	PENDULETIN
1104-052	PENDULIN

3005-012	PENICILLIC ACID
2608-013	PENICILLIOPSIN
0701-012	PENIOPHORIN
0701-012	PENIOPHORININ
3196-010	PENT-2-ENOIC ACID,4-ME-
3125-001	PENTACOSANE,N-
3125-002	PENTACOSANOIC ACID
3115-001	PENTADECANE,N-
3115-025	PENTADECANOIC ACID
3115-026	PENTADECANOIC ACID,2-OH-
3115-028	PENTADECANOIC ACID,15-OH-
3115-030	PENTADECANOIC ACID,3-KETO-
3115-029	PENTADEC-8-EN-2-ONE
3115-009	PENTADECA-1,8-DIEN-4,6-DIYN-3-OL,TRANS-
3115-010	PENTADECA-2,8-DIEN-4,6-DIYN-10-OL
3115-002	PENTADECA-2,9-DIEN-4,6-DIYNE
3115-011	PENTADECA-2,9-DIEN-4,6-DIYN-1-OL
3115-016	PENTADECA-2,9-DIEN-4,6-DIYN-1,8-DIOL,1-A
3115-018	PENTADECA-2,9-DIEN-4,4-DIYNAL
3115-019	PENTADECA-5,7-DIEN-9,11,13-TRIYNAL,T,T-
3115-017	PENTADECA-5,10-DIENAL
3115-024	PENTADECA-7,13-DIEN-9,11-DIYN-4-ONE
3115-012	PENTADECA-1,8,10-TRIEN-4,6-DIYN-3-ONE
3115-023	PENTADECA-1,8,10-TRIEN-4,6-DIYN-3-ONE
3115-003	PENTADECA-2,8,10-TRIEN-4,6-DIYNE
3115-013	PENTADECA-2,8,10-TRIEN-4,6-DIYN-1-OL
3115-014	PENTADECA-2,8,10-TRIEN-4,6-DIYN-12-OL
3115-020	PENTADECA-2,8,10-TRIEN-4,6-DIYNAL
3115-021	PENTADECA-5,7,13-TRIEN-9,11-DIYNAL,T,T-
3115-004	PENTADECA-2,4,6,8-TETRAENE
3115-005	PENTADECA-2,8,10,14-TETRA-ENE-4,6-DI-YNE
3115-022	PENTADECA-6,8,10,12-TETRAENAL
3105-001	PENTANAL
2805-004	PENTANE-1,3-DIOL,1,5-DI-PH-
3902-003	PENTANE-1,3-DIONE,4-ME-1-PH-
3196-009	PENTANOIC ACID,2,3-DI-OH-3-ME-
3196-001	PENTAN-1-OL,3-ME-
2805-003	PENTANOL,1,5-DI-PH-3-
3105-002	PENTAN-2-ONE
3196-003	PENTAN-2-ONE,4-OH-4-ME-
3135-001	PENTATRIACONTANE,N-
3197-050	PENTATRIACONTANE,3,7,11-TRI-ME-
1702-017	PEONIDIN
1702-018	PEONIDIN-3-ARABINOSIDE
1702-019	PEONIDIN-3-GALACTOSIDE
1702-021	PEONIDIN-3-XYLOSYLGLUCOSIDE
1702-022	PEONOL
2201-004	PEONOL,GLUCO-
2201-005	PEONOL PRIMEVEROSIDE
2301-027	PERLATOLIC ACID
3509-016	PEROSAMINE
3505-007	PERSEITOL
1106-044	PERSICARIN
1106-058	PERSICARIN-7-ME ETHER
2109-001	PERSOONOL
3003-006	PESTALOTIN
1106-031	PETIOLAROSIDE
0102-001	PERUVIN = BENZYL ALCOHOL
2704-012	PERYLENE-3,10-QUINONE,4,9-DIOH-
3118-019	PETROSELIC ACID
3118-020	PETROSELIDIC ACID
1099-005	PETROSTYRENE
1703-011	PETUNIDIN
1703-012	PETUNIDIN-3-ARABINOSIDE
1703-013	PETUNIDIN-3-GALACTOSIDE
1703-014	PETUNIDIN-3-MONOGLUCOSIDE
1703-016	PETUNIDIN-5-XYLOSIDE
1703-015	PETUNIDIN-3-XYLOSYLGLUCOSIDE
1703-017	PETUNIN
0502-010	PEUCEDANIN
0508-026	PEUCEDANIN,OXY-
0508-027	PEUCEDANIN,ISO-OXY-
0501-020	PEUCEDANOL
0504-013	PEUCENIDIN
2204-001	PEUCENIN
2204-004	PEUCENIN-DIME-ETHER,HETERO-
2204-002	PEUCENIN,HETERO-
2204-003	PEUCENIN-7-ME ETHER,HETERO-
0505-026	PEUFORMOSIN
1106-041	PEUMOSIDE
3114-040	PHARBITIC ACID C
3114-041	PHARBITIC ACID D
0304-017	PHASELIC ACID
1601-021	PHASEOLLIDIN
1601-022	PHASEOLLIN
0599-022	PHEBALIN
0501-034	PHEBALOSIN
1204-007	PHELLAMURETIN
1204-008	PHELLAMURIN
1104-034	PHELLATIN
1204-006	PHELLAVIN

3122-018	PHELLOGENIC ACID
3122-017	PHELLONIC ACID
0511-002	PHELLOPTERIN
1104-032	PHELLOZIDE
1204-009	PHELLOZIDE,DIHYDRO-
2701-022	PHENANTHRENE,1-OH-2,5,6,7-TETRA-OME-
2701-023	PHENANTHRENE,2-OH-1,7-DI-OME-5,6-METHYLE
2701-015	PHENANTHRENE,2-OH-3,5,7-TRI-OME-
2701-014	PHENANTHRENE,2,5-DI-OH-3,7-DI-OME-DIHYDO
2701-016	PHENANTHRENE,2,6-DI-OH-3,4,7-TRI-OME-
2701-017	PHENANTHRENE,2,6-DI-OH-3,4,7-TRI-OME-9,1
2701-020	PHENANTHRENE,4,7-DI-OH-2,3,6-TRI-OME-
2701-021	PHENANTHRENE,4,7-DI-OH-2,3,6-TRI-OME-DIH
2701-018	PHENANTHRENE,4,6,7-TRI-OH-2,3-DI-OME-
2701-019	PHENANTHRENE,4,6,7-TRI-OH-2,3-DI-OME-DIH
2701-024	PHENANTHRENE,1,2,7-TRI-OME-5,6-METHYLENE
2701-025	PHENANTHRENE QUINONE,1,3,4-TRI-OME-2,7-
2301-036	PHENARCTIN
0202-001	PHENETHYL ALCOHOL
0202-002	PHENETHYL ALCOHOL ME ETHER
0202-003	PHENETHYL ALCOHOL AMB-ESTER
0202-004	PHENETHYL ALCOHOL ISOVALERYL ESTER
0202-007	PHENETHYL ALCOHOL, 3,4-DI-OH-
0401-015	PHENETHYL KETONE, ME-
2704-007	PHENOCYCLINONE
0801-001	PHENOL
0802-003	PHENOL,M-OME-
0801-003	PHENOL,2,4-DI-CL-
0801-002	PHENOL,2,6-DI-BR-
0801-006	PHENOL,2-DECAPRENYL-
0802-003	PHENOL,2-DECAPRENYL-6-OME-
0801-005	PHENOL,2-NONAPRENYL-
0801-004	PHENOL,2-TETRAPRENYL-
0403-006	PHENOL,2(HEXA-2,4-DIYNYL)-
0403-007	PHENOL,2(1-OH-HEXA-2,4-DIYNUL)-
2105-006	PHENOL A
2105-007	PHENOL A ALDEHYDE
2303-020	PHENYLACETALDEHYDE
0204-001	PHENYLACETIC ACID
0204-005	PHENYLACETIC ACID, 3,4-DI-OH-
0204-003	PHENYLACETIC ACID,ORTHO-OH-
0204-004	PHENYLACETIC ACID,P-OH-
3901-001	PHENYLACETIC ACID,ALPHA-ME-P-OH-
0204-006	PHENYLGLYOXYLIC ACID,2,5-DIOH-
2201-015	PHENYLGLYOXYLIC ACID,2,4,5-TRIHYDROXY-
0302-001	PHENYLPROPYL ALCOHOL,GAMMA-
0304-010	PHENYLPYRUVIC ACID,P-OH-
1601-017	PHILENOPTERAN
1601-018	PHILENOPTERAN,9-O-ME-
0601-024	PHILLYGENOL
0601-025	PHILLYRIN
0701-007	PHLEBIARUBRONE
3118-083	PHLOIONIC ACID
3118-016	PHLOIONOLIC ACID
2201-013	PHLORACETOPHENONE-TRIME ETHER
2207-005	PHLORASPIN
0304-009	PHLORETIC ACID
1002-008	PHLORETIN
1002-009	PHLORIZIN
2299-001	PHLOROGLUCINOL
0201-008	PHLOROL,META-
0201-003	PHLOROL,PARA-
0201-010	PHLOROL ISOBUTYRYL ESTER,META-
0201-009	PHLOROL ME ETHER,META-
2299-004	PHLOROPYRONE
2005-001	PHOENICIN
2603-026	PHOMARIN
3197-003	PHORBIC ACID
2102-027	PHTHALIC ACID
2102-004	PHTHALIC ACID,3-OH-
2102-007	PHTHALIC ACID,3,5-DIOH-
2102-008	PHTHALIC ACID,4,5-DIOH-3-OME-
2102-009	PHTHALIC ACID,4,6-DIOH-3-OME-
2102-028	PHTHALIC ACID ME ESTER
2106-012	PHTHALIDE,N-BUTYL-
2102-026	PHTHALIDE,4-OH-
2102-011	PHTHALIDE,7-OH-5-OME-
2102-013	PHTHALIDE,5-OME-7-GLUCOSYLOXY-
2102-015	PHTHALIDE,7-OH-4,6-DIME-
2102-012	PHTHALIDE,5,7-DI-OME-
2102-016	PHTHALIDE,6-ME-4,5,7-TRIOH-
2199-013	PHTHALIDE,3-ISOBUTYLIDENYL-
2199-014	PHTHALIDE,3-ISOVALYLIDENYL-
2199-015	PHTHALIDE,3-ISOVALYLIDENYL-3,4-DIHYDRO-
3197-038	PHTHIENOIC ACID
2503-005	PHTHIOCOL
3197-036	PHTHIOTRIOL A
0601-003	PHYLLANTHIN
0698-011	PHYLLANTHIN,HYPO-
2804-016	PHYLLODULCINOL
2502-017	PHYLLOQUINONE

2604-036	PHYSCION
2604-038	PHYSCION,GLUCORHAMNOSYL-
2699-024	PHYSCION-9-ANTHRONE
2699-008	PHYSCION-10-ANTHRONE
2604-037	PHYSCIONIN
2302-027	PHYSODIC ACID
3501-024	PHYTIC ACID
2804-018	PICEATANNOL
2804-013	PICEID
2804-014	PICEID,4'-O-ME-
0203-003	PICEIN
0203-002	PICEOL = ACETOPHENONE, P-OH-
0203-003	PICEOSIDE = PICEIN
2399-018	PICROLICHENIC ACID
3401-009	PICROMYCIN
0605-005	PICROPODOPHYLLIN
0605-004	PICROPODOPHYLLIN,DEOXY-
0104-011	PICRORHIZIN
2703-042	PILLAROMYCIN A
2703-041	PILLAROMYCINONE
3510-004	PILLAROSE
1105-030	PILLOIN
2701-027	PILOQUINONE
2701-028	PILOQUINONE,4-OH-
0199-013	PILOPUBROSIN
3402-001	PIMARICIN
3107-010	PIMELIC ACID,ALPHA-KETO-
3107-011	PIMELIC ACID,ALPHA-KETO-GAMMA-OH-
0510-004	PIMPINELLIN
0511-001	PIMPINELLIN,ISO-
0702-009	PINASTRIC ACID
3501-010	PINITOL,D-
3501-011	PINITOL,L-(-)-
0508-007	PINNARIN
0508-035	PINNARIN,FURO-
0599-013	PINNATERIN
1105-041	PINNATIN
1202-001	PINOBANKSIN
1201-001	PINOCEMBRIN,(-)-
1201-003	PINOCEMBRIN,7-NEOHEPERIDOSYL-
1108-008	PINOMYRICETIN
1106-034	PINOQUERCETIN
0601-023	PINORESINOL
1201-005	PINOSTROBIN
2804-004	PINOSYLVIN
2804-005	PINOSYLVIN,MONO-METHYL ETHER
2804-007	PINOSYLVIN,DIHYDRO-MONO-ME-ETHER
2804-006	PINOSYLVIN-DIMETHYL ETHER
2402-014	PINSELIC ACID
2402-013	PINSELIN
0499-007	PIPATALINE
0402-004	PIPERIC ACID
0402-005	PIPERIC ACID,TETRAHYDRO-
0404-023	PIPEROLIDE
0103-007	PIPERONAL
0104-015	PIPERONYLIC ACID
3906-044	PIPTOSIDE
2199-004	PIROLAGENIN
1601-028	PISATIN
1499-020	PISCERYTHRONE
3902-006	PISCIDIC ACID
3195-006	PIVALIC ALDEHYDE
2301-028	PLANA ACID
3506-013	PLANTEOBIOSE
3507-008	PLANTEOSE
0899-009	PLASTOCHROMANOL-8
0802-018	PLASTOHYDROQUINONE ME ETHER,PHYTYL-
2001-015	PLASTOQUINONE,PHYTYL-
2001-011	PLASTOQUINONE-3
2001-012	PLASTOQUINONE-4
2001-013	PLASTOQUINONE-8
2001-014	PLASTOQUINONE-9
1205-023	PLATHYMENIN
0605-010	PLICATIC ACID
0605-011	PLICATINAPHTHOL
2503-046	PLUMBAGIN
2503-048	PLUMBAGIN,3-CHLORO-
2503-047	PLUMBAGIN ME ETHER
0601-028	PLUVIATILOL
0601-016	PLUVIATOLIDE
3509-005	PNEUMOSAMINE
1505-005	PODOCARPUSFLAVONE A
1505-010	PODOCARPUSFLAVONE B
0605-007	PODOPHYLLOTOXIN
0605-008	PODOPHYLLOTOXIN,DEHYDRO-
0605-006	PODOPHYLLOTOXIN,4'-DESMETHYL-
0605-002	PODOPHYLLOTOXIN,DEOXY-
0605-003	PODOPHYLLOTOXIN GLUCOSIDE,4'-DESME-DEOXY
0605-001	PODOPHYLLOTOXIN,7-DESOH-4-DESME-
0605-009	PODOPHYLLOTOXIN GLUCOSYL ESTER,DEOXY-
0602-003	PODORHIZOL

1499-002   PODOSPICATIN
1106-081   POLYCLADIN
1303-005   POLYDINE
3999-003   POLYETHERIN A
2405-005   POLYGALAXANTHONE A
2405-004   POLYGALAXANTHONE B
3504-058   POLYGALITOL
2003-014   POLYGONAQUINONE
0701-003   POLYPORIC ACID
1106-012   POLYSTACHOSIDE
2299-006   POLYSTICHININ = PHLOROPYRON
1402-003   POMIFERIN
1402-004   POMIFERIN,ISO-
1203-012   PONCIRIN
1106-005   PONGACHROMENE
1105-040   PONGAGLABRONE
1101-014   PONGAMIA GLABRA FURANO-FLAVONE
1001-018   PONGAMOL
1106-006   PONGAPIN
0199-018   POPULIN
0199-019   POPULIN,SALICOYL-
2201-003   POPULNEOL
1104-018   POPULNIN
0399-016   POPULOSIDE
1203-017   PORIOL
1203-018   PORIOLIN
2399-007   PORPHYRILIC ACID
3906-022   PORTENTOL
3906-023   PORTENTOL ACETATE
0501-040   PRANFERIN
0506-012   PRANGENIN
0508-025   PRANGOLARIN
0502-009   PRANGOSINE
0502-006   PRANTSCHIMGIN
1402-005   PRATENSEIN
1103-001   PRATOL
1104-002   PRATOLETIN
0507-004   PRENYLETIN
1902-019   PREPHENIC ACID
2702-015   PRETETRAMID,4-OH-6-ME-
1101-016   PRIMETIN
3506-010   PRIMEVEROSE
2002-003   PRIMIN
0104-032   PRIMULAVERIN = PRIMULAVEROSIDE
0104-035   PRIMVERIN = PRIMVEROSIDE
3401-046   PRIMYCIN
1599-009   PROANTHOCYANIDIN
1599-010   PROANTHOCYANIDIN
1506-003   PROANTHOCYANIDIN COOH
1502-005   PROCYANIDIN-B1
1502-006   PROCYANIDIN-B2
1502-007   PROCYANIDIN-B3
1502-008   PROCYANIDIN-B4
1599-025   PROCYANIDIN-(-)-EPICATECHIN
1599-011   PRODELPHINIDIN
3103-001   PROPANOL
1599-008   PROPELARGONIDIN
3301-007   PROPENYL-SEC-BUTYL DISULPHIDE
3103-005   PROPIONALDEHYDE
3103-010   PROPIONIC ACID
3103-026   PROPIONIC ACID,BETA-(DIME-SULPHONIUM HYD
3103-025   PROPIONIC ACID ME ESTER, BETA-SME-
0399-005   PROPIOSYRINGONE, ALPHA-OH-
0399-004   PROPIOVANILLONE, ALPHA-OH-
3301-005   PROPYL-ALLYL DISULPHIDE
3103-022   PROPYL MERCAPTAN
3103-023   PROPYLPHOSPHOSPHONIC ACID,(-)-C-1,2-EPOX
3197-028   PROPYLURE
2902-001   PROSTAGLANDIN-A1
2902-002   PROSTAGLANDIN-A1,19-OH-
2902-003   PROSTAGLANDIN-A2
2902-005   PROSTAGLANDIN A2,15-EPI-
2902-004   PROSTAGLANDIN A2,19-OH-
2902-006   PROSTAGLANDIN A2 METHYL ESTER,ACETYL-15-
2902-007   PROSTAGLANDIN-B1
2902-008   PROSTAGLANDIN-B1,19-OH-
2902-009   PROSTAGLANDIN-B2
2902-010   PROSTAGLANDIN-B2,19-OH-
2902-011   PROSTAGLANDIN E1
2902-012   PROSTAGLANDIN E2
2902-013   PROSTAGLANDIN E3
2902-014   PROSTAGLANDIN F1-ALPHA
2902-015   PROSTAGLANDIN F-2-ALPHA-
2902-016   PROSTAGLANDIN-F3ALPHA
2603-035   PROTETRONE
2699-025   PROTETRONE,ME-
2509-010   PROTOAPHIN-FB
0103-003   PROTOCATECHUALDEHYDE
0103-012   PROTOCATECHUALDEHYDE,5,6-DI-BR-
0104-009   PROTOCATECHUIC ACID
2302-004   PROTOCETRARIC ACID

2302-003   PROTOCETRARIC ACID,HYPTO-
2802-008   PROTOCOTOIN
2802-009   PROTOCOTOIN,METHYL-
2599-005   PROTODACTYNAPHIN-JC-1
2704-019   PROTOHYPERICIN
2207-006   PROTOKOSIN
1104-060   PRUDOMESTIN
1104-061   PRUDOMESTIN,3-O-ME-
1401-011   PRUNETIN
1401-012   PRUNETRIN
1203-003   PRUNIN
0402-009   PSILOTIN
0402-008   PSILOTININ
0502-013   PSORALENE
0506-021   PSORALENE,8-GERANYLOXY- = XANTHOTOXOL,O-
0511-006   PSORALENE,5-OME-8-GERANYLOXY-
0502-018   PSORALENE,4,5',8-TRI-ME-
0508-034   PSORALIDIN
2302-002   PSOROMIC ACID
2204-028   PTAEROCHROMENOL
2204-015   PTAEROCYCLIN
2204-014   PTAEROGLYCOL
2204-027   PTAEROXYLIN,ALLO-
2204-012   PTAEROXYLIN,DEHYDRO-
2204-013   PTAEROXYLINOL
1199-002   PTAEROXYLOL
2204-016   PTAEROXYLONE
1601-002   PTEROCARPAN,3-OH-9-OME-(+)-
1601-001   PTEROCARPAN,3-OH-9-OME-(-)-
1601-005   PTEROCARPAN,3-OH-4,9-DI-OME-
1601-015   PTEROCARPAN,3-OH,4-OME-8,9-METHYLENEDIOX
1601-016   PTEROCARPAN,3,4-DI-OME-8,9-METHYLENEDIOX
1601-003   PTEROCARPAN,3,9-DI-OME-
1601-019   PTEROCARPAN,(-)-3-OH-2-(3,3-DMA)-8,9-MET
           -DIOXY-
1601-006   PTEROCARPAN,3,4,9-TRI-OME-
1601-004   PTEROCARPEN,3,9-DI-OME-
1601-012   PTEROCARPIN
1601-014   PTEROCARPIN,(-)-2-OH-
2804-024   PTEROFURAN
2804-015   PTEROSTILBENE
0505-013   PTERYXIN
0505-014   PTERYXIN,EPOXY-
0505-009   PTERYXIN,ISO-
2604-060   PTILOMETRIC ACID
3001-004   PUBERULIC ACID
3001-005   PUBERULONIC ACID
1401-025   PUERARIN
0702-001   PULVIC ACID
0702-004   PULVIC ACID LACTONE
2107-009   PULVILLORIC ACID
0702-006   PULVINAMIDE
0999-011   PUNCTATIN
0999-012   PUNCTATIN,3,9-DIHYDRO-
0999-013   PUNCTATIN,4'-O-ME-
0999-014   PUNCTATIN,4'-O-ME-3,9-DIHYDRO-
3118-038   PUNICIC ACID
2608-009   PUNICOSKYRIN
3901-005   PURPUREIN
2604-008   PURPURIN
2604-009   PURPURIN,PSEUDO-
3001-008   PURPUROGALLIN
2704-008   PURPUROGENONE
2704-009   PURPUROGENONE,DEOXY-
3110-063   PYOLIPIC ACID
3499-009   PYRENOPHORIN
2901-021   PYRETHROLONE
2107-006   PYRICULOL
0803-001   PYROGALLOL
0803-002   PYROGALLOL-1-ME ETHER
0803-003   PYROGALLOL-1,3-DIME ETHER
0101-006   PYROGALLOL 1,3-DIME ETHER,5-ME-
0201-006   PYROGALLOL 1,3-DIME ETHER,5-ET-
0301-031   PYROGALLOL-1,3-DIME ETHER,5-PR-
0101-005   PYROGALLOL 1-ME ETHER,5-ME-
0301-027   PYROGALLOL-1-ME ETHER,5-PROPYL-
3002-011   PYRONE,3-CHO-2,3-DIHYDRO-GAMMA-
0402-007   PYRONE,2,3-DIHYDRO-3-ME-6-PH-GAMMA-
3203-002   PYRONE,5-(4-SME-PENTA-1,2,3-TRIENYL)-ALP
3003-005   PYRONE-6-COOH,3,4-DIME-ALPHA-
0802-007   PYROSIDE
3196-007   PYROTEREBIC ACID
2703-030   PYRROMYCIN
2703-029   PYRROMYCINONE,EPSILON-
2703-006   PYRROMYCINONE,ETA-
2703-005   PYRROMYCINONE,ETA-1-
2703-015   PYRROMYCINONE,ZETA-
3103-013   PYRUVIC ACID
3103-016   PYRUVIC ACID,HYDROXY-
3103-014   PYRUVIC ACID,PHOSPHO-ENOL-
2102-002   QUADRILINEATIN

2108-003	RUBRORDTIORIN
2608-016	RUBROSKYRIN
2608-014	RUGULOSIN
2608-001	RUGULOSIN,DIANHYDRO-
2203-011	RUPICOLON = SORDIDONE
1104-015	RUSTOSIDE
0507-021	RUTACULTIN
0502-016	RUTAMARIN
0506-009	RUTARETIN
1106-024	RUTIN
3506-036	RUTINOSE
0511-009	SABANDININ
0506-005	SABANDINONE
3112-020	SABINIC ACID
3113-018	SAFYNOL
0301-017	SAFROL
0301-018	SAFROL,ISO-
3506-019	SAKEBIOSE
1203-008	SAKURANETIN
1203-010	SAKURANETIN,L-ISO-
1203-009	SAKURANIN
1203-011	SAKURANIN,ISO-
1002-012	SAKURANIN,NEO-
2302-019	SALAZINIC ACID
1105-027	SALICAPRENE
0102-008	SALICIN
0199-017	SALICIN,6'-SALICOYL-
0199-002	SALICORTIN
0102-003	SALICOSIDE = SALIGENIN
0102-007	SALICYL ALCOHOL = SALIGENIN
0103-013	SALICYLALDEHYDE
0103-017	SALICYLALDEHYDE,4-OME-
0104-022	SALICYLIC ACID
0104-024	SALICYLIC ACID ME ESTER
0104-023	SALICYLIC ACID ME ETHER
0104-006	SALICYLIC ACID BETA-GLUCOSIDE
2102-003	SALICYLIC ACID,6-FORMYL-
2101-012	SALICYLIC ACID,6-ME-
0202-006	SALIDROSIDE
0102-007	SALIGENIN
0203-003	SALINIGRIN = PICEIN
1002-006	SALIPURPOL,ISO-
1203-007	SALIPURPOSIDE
1002-007	SALIPURPOSIDE,ISO-
0102-010	SALIREPIN
0102-002	SALIREPOL = GENTISYL ALCOHOL
0199-007	SALIREPOSIDE
0501-015	SAMARKANDINE
0501-006	SAMARKANDONE
3506-009	SAMBUBIOSE
0505-010	SAMIDIN
0505-011	SAMIDIN,DIHYDRO-
1402-002	SANTAL
3118-049	SANTALBIC ACID
1103-013	SAPONARETIN = VITEXIN,ISO-
1103-024	SAPONARIN
2801-003	SAPPANIN
2504-016	SAPTARANGI QUINONE A
2002-022	SARCODONTIC ACID
0201-011	SARISAN
2901-005	SARKOMYCIN
3504-034	SARMENTOSE
3903-002	SARSAPIC ACID
3118-017	SATIVIC ACID
0601-018	SAVININ
1402-012	SAYANEDINE
3118-084	SCABRIN
1404-008	SCANDENIN
1401-016	SCANDINONE
1107-013	SCAPOSIN
2701-033	SCHIZANDRIN
2399-005	SCHIZOPELTIC ACID
1505-012	SCIADOPITYSIN
1505-015	SCIADOPITYSIN,2,3-DIHYDRO-
3506-034	SCILLABIOSE
3904-010	SCLERANECIC ACID
3905-028	SCLERIN
2802-006	SCLERDIN
3905-029	SCLEROLIDE
2503-045	SCLERONE
2104-034	SCLEROTININ A
2104-035	SCLEROTININ B
2108-004	SCLEROTIORIN
2108-005	SCLEROTIORIN,7-EPI-
0507-005	SCOPOLETIN
0507-010	SCOPOLETIN,ISO-
0507-009	SCOPOLETIN,O-GERANYL-
0507-008	SCOPOLETIN,O-(3,3DMA)-
2402-010	SCRIBLITIFOLIC ACID
1103-032	SCUTELLAREIN
1103-039	SCUTELLAREIN,4'-O-ME-
1103-037	SCUTELLAREIN,7-O-GLUCOBIOSYL-
1103-034	SCUTELLAREIN, 7-O-GLUCOSYL-
1103-035	SCUTELLAREIN, 7-O-GLUCURONOSYL-
1103-038	SCUTELLARIN
3501-026	SCYLLITOL
3501-027	SCYLLITOL,1-O-GLUCOSYL-
3110-072	SEBACIC ACID
2406-001	SECALONIC ACID A
2406-003	SECALONIC ACID B
2406-004	SECALONIC ACID C
2406-002	SECALONIC ACID D
2106-026	SEDANONIC ACID
2106-021	SEDANONIC ACID ANHYDRIDE
3506-006	SEDOHEPTITOL = VOLEMITOL,D-
3505-002	SEDOHEPTULOSE
2301-019	SEKIKAIC ACID
2301-023	SEKIKAIC ACID,HOMO-
3198-001	SELACHYL ALCOHOL
0505-004	SELINIDIN
1203-007	SELINONE
3195-011	SENECIOIC ACID
1901-015	SENEOL
1901-016	SENEPOXIDE
2609-008	SENNIDINE A
2609-012	SENNOSIDE-A
2609-013	SENNOSIDE-B
2609-014	SENNOSIDE-C
3001-006	SEPEDONIN
3001-007	SEPEDONIN,ANHYDRO-
0699-016	SEQUIRIN B
0699-014	SEQUIRIN C
1505-003	SEQUOIAFLAVONE
3501-021	SEQUOYITOL
1102-004	SERICETIN
1199-021	SERPYLLIN
3110-081	SERRATAMIC ACID
0601-033	SESAMIN,(-)-
0601-032	SESAMIN,(+-)- = FAGAROL
0601-030	SESAMIN,EPI- = ASARININ
0803-005	SESAMOL
0699-024	SESAMOLIN
0698-010	SESANGOLIN
0502-008	SESELIFLORIN
0505-001	SESELIN
0505-021	SESELIN,3'-ISOVALERYLOXY-
0505-019	SESELIN,3'-SENECIOYLOXY-
2204-024	SESELIRIN
2002-020	SHANORELLIN
1902-001	SHIKIMIC ACID
1902-003	SHIKIMIC ACID,DIHYDRO-
1902-004	SHIKIMIC ACID,5-DEHYDRO-
2504-031	SHIKONIN
2504-032	SHIKONIN,O-AC-
2504-033	SHIKONIN,O-ISOBUTYRYL-
2504-034	SHIKONIN,O-SENECIOYL-
2302-012	SHIRIN
2503-043	SHINANOLONE
2503-044	SHINANOLONE,ISO-
1702-016	SHISONIN
0499-001	SHOGAOL
0508-008	SIBIRICIN
2101-024	SICCANIN
2101-023	SICCANOCHROMENE E
2101-022	SICCANOCHROMENE,PRE-
2101-021	SICCANOCHROMENIC ACID
2101-020	SICCANOCHROMENIC ACID,PRE-
1299-010	SILYBIN
1299-011	SILYCHRISTIN
1299-012	SILYDIANIN
0304-030	SINAPIC ACID
0304-031	SINAPIN
0303-008	SINAPYL ALDEHYDE
1204-003	SINENSIN
3905-023	SIPHULIN
3505-008	SIPHULITOL
3502-016	SISOMICIN
1401-014	SISSOTRIN
0501-003	SKIMMIN
2608-003	SKYRIN
2608-011	SKYRIN,OXY-
2608-010	SKYRINOL
0502-011	SMIRNIORIN
0502-012	SMIRNIORIDIN
3506-031	SOLABIOSE
0899-010	SOLANACHROMENE
2505-015	SOLANIOL,(+)-
2605-031	SOLORINIC ACID
2605-030	SOLORINIC ACID,NOR-
1601-010	SOPHOJAPONICIN
1002-019	SOPHORADIN
1002-020	SOPHORADOCHROMENE

1104-014	SOPHORAFLAVANOLOSIDE
1203-032	SOPHORANOCHROMENE
1203-031	SOPHORANONE
1401-005	SOPHORICOBIOSIDE
1401-004	SOPHORICOSIDE
1499-003	SOPHOROL
3506-018	SOPHOROSE
2603-025	SORANJIDIOL
1103-036	SORBARIN
3106-010	SORBIC ACID
2205-001	SORBICILLIN
1103-033	SORBIFOLIN
2204-029	SORBIFOLIN
3504-056	SORBITOL,D-
3198-013	SORBO-1,3-DIMYRISTIN
3504-048	SORBOSE,L-
3504-021	SORDAROSE
2203-011	SORDIDONE
2503-026	SORIGENIN,ALPHA-
2502-033	SORIGENIN,BETA-
2503-027	SORININ,ALPHA-
2502-034	SORININ,BETA-
1505-002	SOTETSUFLAVONE
3106-007	SOYANAL
2101-005	SPARASSOL
2204-018	SPATHELIACHROMENE
2204-017	SPATHELIACHROMENE,ME-ISO-
1401-003	SPHAEROBIOSIDE
2301-017	SPHAEROPHORIN
0507-020	SPHONDIN
3904-008	SPHYDROFURAN
1501-005	SPICATASIDE
3005-032	SPICULISPORIC ACID
3110-051	SPILANTHOL
1106-078	SPINACETIN
2506-010	SPINOCHROME A
2506-001	SPINOCHROME B
2506-001	SPINOCHROME-B1= SPINOCHROME-B
2507-003	SPINOCHROME C
2507-001	SPINOCHROME D
2508-001	SPINOCHROME E
2507-003	SPINOCHROME-F = SPINOCHROME-C
2506-010	SPINOCHROME-M = SPINOCHROME-A
2506-001	SPINOCHROME-M2= SPINOCHROME-B
2506-001	SPINOCHROME-N = SPINOCHROME-B
2506-001	SPINOCHROME-P1= SPINOCHROME-B
2506-006	SPINOCHROME-S
2507-003	SPINONE-A = SPINOCHROME-C
1105-019	SPINOSIDE
2004-001	SPINULOSIN
1901-006	SPINULOSIN-HYDRATE
1901-005	SPINULOSIN-QUINOL,DIHYDRO-
1901-007	SPINULOSIN-QUINOL-HYDRATE
1106-114	SPIRAEOSIDE
2301-012	SQUAMATIC ACID
3508-002	STACHYOSE
3118-010	STEARIC ACID
3118-013	STEARIC ACID,9,10-DI-OH-CIS-
3118-014	STEARIC ACID,9,10-DI-OH-TRANS-
3118-075	STEARIC ACID,9,10-EPOXY-
3118-008	STEARIC ALDEHYDE
3118-011	STEARYL ALCOHOL
1299-003	STEPPOGENIN
1299-005	STEPPOSIDE
3197-056	STERCULIC ACID
3197-057	STERCULIC ACID,2-OH-
3197-058	STERCULYNIC ACID
2403-039	STERIGMATOCYSTIN
2403-040	STERIGMATOCYSTIN,METHYL ETHER
2404-033	STERIGMATOCYSTIN,5-OME-
2403-038	STERIGMATOCYSTIN,DES-ME-
2403-041	STERIGMATOCYSTIN,DIHYDRO-O-ME-
2301-020	STERNOSPORIC ACID
2302-018	STICTIC ACID
2302-017	STICTIC ACID,NOR-
2804-001	STILBENE,TRANS-
2804-002	STILBENE,4-OXY-
2804-003	STILBENE,4-METHOXY-
2804-021	STILBENE,3,4,5,3',5'-PENTAOXY-
2804-017	STILBENE-5-GLUCOSIDE,3,5,4'-TRI-OH-3'-OM
1003-005	STILLOPSIDIN
3001-001	STIPITATIC ACID
3001-003	STIPITATONIC ACID
2399-008	STREPSILIN
3502-001	STREPTAMINE
3501-007	STREPTAMINE,2-DEOXY-
3502-002	STREPTIDINE
3502-003	STREPTOMYCIN-A
3502-004	STREPTOMYCIN-B
3502-005	STREPTOMYCIN,DIHYDRO-
3502-006	STREPTOMYCIN,OH-

3510-010	STREPTOSE
3510-012	STREPTOSE,DIHYDRO-
3510-011	STREPTOSE,OH-
3509-002	STREPTOZOTOCIN
2805-017	STRIATOL
1202-004	STROBOBANKSIN
1101-009	STROBOCHRYSIN
1201-006	STROBOPININ
2503-042	STYPANDRONE
0399-007	STYRACIN = CINNAMYL CINNAMATE
3504-057	STYRACITOL
0201-001	STYRENE
0201-002	STYRENE,PARA-OH-
0201-005	STYRENE,3-OME-4-HYDROXY-
0501-022	SUBEROSIN
0501-019	SUBEROSIN,DEMETHYL-
3104-015	SUCCINIC ACID
3506-014	SUCROSE
1105-055	SUDACHITIN
1103-048	SUDACHITIN,DE-OME-
0699-015	SUGIRESINOL
0505-016	SUKDORFIN
2802-011	SULOCHRIN
1803-010	SULPHUREIN
1803-009	SULPHURETIN
1803-011	SULPHURETIN 6-DIGLUCOSIDE
1801-017	SUMATROL
0903-008	SURANGIN A
0903-009	SURANGIN B
0601-021	SVENTENIN ACETATE
2404-031	SWERCHIRIN
1105-020	SWERTIAJAPONIN
2404-009	SWERTIANIN
2404-011	SWERTIANIN,ME-
2404-007	SWERTIANIN,NOR-
2404-012	SWERTININ
1103-022	SWERTISIN
2299-003	SYCARPIC ACID
3004-002	SYLVAN
2404-006	SYMPHOXANTHONE
3196-004	SYOYU-ALDEHYDE
0104-019	SYRINGAIC ACID
0103-010	SYRINGIC ALDEHYDE,GLUCO-
0302-009	SYRINGENIN
0603-003	SYRINGERESINOL,DL-
1108-013	SYRINGETIN
0104-033	SYRINGIC ACID
0302-009	SYRINGIN
0103-009	SYRINGYL ALDEHYDE
3499-002	T-2636A = BUNDLIN B
3499-001	T-2636C = BUNDLIN A
3499-004	T-2636D
3499-003	T-2636F
3906-051	TABEBUIN
1101-019	TACHROSIN
3504-049	TAGATOSE,D-
1106-070	TAGETIIN
0601-017	TAIWANIN A
0604-018	TAIWANIN C
0604-023	TAIWANIN E
0604-024	TAIWANIN-E ME ETHER
1501-002	TALBOTAFLAVONE
3509-006	TALOSAMINE
3504-019	TALOSE,6-DESOXY-L-
1106-060	TAMARIXETIN
1106-061	TAMARIXETIN 3-ME ETHER
1106-062	TAMARIXIN
1106-063	TAMARIXIN
1104-062	TAMBULIN
1103-052	TANGERETIN
1103-051	TANGERETIN,5-O-DESME-
3118-047	TARIRIC ACID
3104-021	TARTARIC ACID,D-
3104-022	TARTARIC ACID,L-
3103-017	TARTRONIC ACID
2206-003	TASMANONE
0803-006	TAXICATIGENIN
0803-007	TAXICATIN
1206-001	TAXIFOLIN
0604-008	TAXIRESINOL,ISO-
0604-009	TAXIRESINOL 6-ME ETHER,ISO-
0104-013	TECOMIN
1101-006	TECTOCHRYSIN
2510-015	TECTOL
2510-018	TECTOL,DEHYDRO-
2510-016	TECTOL,TETRAHYDRO-
2510-017	TECTOL,TETRAHYDRO-DIME-ETHER
2605-027	TECTOLEAF QUINONE
2601-002	TECTOQUINONE
1401-019	TECTORIDIN
1401-017	TECTORIGENIN

1401-018	TECTORIGENIN,7-O-ME-
1401-022	TECTORIGENIN,7,4'-DIME-
2604-049	TELOSCHISTIN
2303-003	TENUIORIN
1801-025	TEPHROSIN
1801-004	TEPHROSIN,ISO-
1302-006	TERACACIDIN,(-)-
1302-007	TERACACIDIN,ISO-
1106-104	TERNATIN
2702-006	TERRAMYCIN = TETRACYCLINE,OXY-
2702-013	TERRAMYCIN X
1901-003	TERREIC ACID
2999-006	TERREIN
1901-002	TERREMUTIN
3005-033	TERRESTRIC ACID
3304-001	TERTHIENYL
3304-002	TERTHIENYL,2-HYDROXYMETHYL-ALPHA-
3304-008	TERTHIENYL,2-CHO-ALPHA-
3003-009	TETRACETIC ACID LACTONE
3124-001	TETRACOSANE,N-
3124-002	TETRACOSANOL,1-
3124-003	TETRACOSANDIOL,1,24-
3124-009	TETRACOSANOIC ACID,24-OH-
3124-006	TETRACOSANOIC ACID,2,3-DIOH-
3124-008	TETRACOS-17-ENOIC ACID
2702-001	TETRACYCLINE
2702-012	TETRACYCLINE,2-AC-2-DECARBOXAMIDO-
2702-010	TETRACYCLINE,ANHYDRO-DESME-CHLOR-
2702-011	TETRACYCLINE,4-AMINO-DE-DIME-AMINO-ANHYD
2702-003	TETRACYCLINE,BROMO-
2702-008	TETRACYCLINE,7-CL-DEHYDRO-
2702-007	TETRACYCLINE,7-CL-5-OH-
2702-005	TETRACYCLINE,7-CHLORO-6-DEMETHYL-
2702-004	TETRACYCLINE,6-DEMETHYL
2702-006	TETRACYCLINE,OXY-
3114-042	TETRADECANOIC ACID,9,10-DI-OH-
3114-001	TETRADECAN-1-OL
3114-026	TETRADEC-5-ENOIC ACID
3114-002	TETRADEC-5-EN-8,10,12-TRIYN-1-OL ACETATE
3114-021	TETRADEC-6-EN-1,3-DIYN-13-ONE
3114-019	TETRADEC-6-EN-8,10,12-TRIYN-1-OL,4,5-EPO
3114-003	TETRADEC-8-EN-2,4,6-TRIYN-12-OL,TRANS-
3114-022	TETRADEC-8-EN-2,4,6-TRIYN-12-ONE
3114-028	TETRADEC-9-ENOIC ACID ET ESTER
3114-044	TETRADEC-9-EN-2,4,6-TRIYNDIOIC ACID
3114-004	TETRADEC-11-EN-1-OL,AC-
3114-034	TETRADECA-2,4-DIEN-1-OIC ACID ISOBUTYLAM
3114-005	TETRADECA-4,6-DIEN-8,10,12-TRIYN-1-OL
3114-006	TETRADECA-4,6-DIENE-8,10,12-TRIYNE,1-OAC
3114-010	TETRADECA-4,6-DIEN-8,10,12-TRIYN-1,3-DIO
3114-011	TETRADECA-4,6-DIEN-8,10,12-TRIYNE,1-OH-3
3114-012	TETRADECA-4,6-DIEN-8,10,12-TRIYN-1,3-DIO
3197-017	TETRADECA-5,12-DIEN,2-ME-
3197-018	TETRADECA-6,12-DIENE,2-ME-
3114-007	TETRADECA-6,12-DIEN-8,10-DIYNE,3-OH-
3114-018	TETRADECA-6,12-DIEN-8,10-DIYN-1-OL,AC--4,5-EPOXY-
3114-008	TETRADECA-9,12-DIEN-1-OL,AC-
3114-043	TETRADEC-2,4,6-TRIENOIC ACID,5-ACETOXYME
3114-031	TETRADECA-2,4,5-TRIENOIC ACID ME ESTER
3114-017	TETRADECA-2,8,10-TRIEN-4,6-DIYN-1,14-DIO
3114-009	TETRADECA-4,6,12-TRIEN-8,10-DIYN-OL ACET
3114-013	TETRADECA-4,6,12-TRIEN-8,10-DIYN-1,3-DIO
3114-014	TETRADECA-4,6,12-TRIEN-8,10-DIYNE,1-OAC-
3114-015	TETRADECA-4,6,12-TRIEN-8,10-DIYNE,1-OH-3
3114-016	TETRADECA-4,6,12-TRIEN-8,10-DIYNE,1,3-DI
3114-032	TETRADECA-2,4,6,12-TETRAEN-8,10-DIYNOIC
3114-033	TETRADECA-5,7,9,11,13-PENTAENOIC ACID
3499-014	TETRANACTIN
2703-045	TETRANGOMYCIN
2703-044	TETRANGULOL
2901-008	TETRAPHYLLIN-A
2901-009	TETRAPHYLLIN-B
3197-052	TETRATRIACONTANE,4,8,12-TRI-ME-
3134-001	TETRATRIACONTANOIC ACID
3134-002	TETRATRIACONTAN-1-OL
3005-039	TETRENOLIN
3402-003	TETRIN-A
3402-004	TETRIN-B
3005-005	TETRONIC ACID,GAMMA-METHYL-
1103-003	THALICTIIN
2301-014	THAMNOLIC ACID
2301-015	THAMNOLIC ACID,HAEMA-
2301-013	THAMNOLIC ACID,HYPO-
0599-023	THAMNOSIN
0501-025	THAMNOSMIN
3116-037	THAPSIC ACID
1599-018	THEAFLAVIN
1599-019	THEAFLAVIN-2A
1599-020	THEAFLAVIN-2B
1599-021	THEAFLAVIN-3

1599-022	THEAFLAVIN,ISO-
1599-023	THEAFLAVIC ACID
1599-024	THEAFLAVIC ACID,EPI-
0701-008	THELEPHORIC ACID
1902-020	THEOGALLIN
3906-012	THERMORUBIN-A
3504-026	THEVETOSE,D-
3504-027	THEVETOSE,L-
2403-017	THIOPHANIC ACID
3302-021	THIOPHENE,2-(BUT-3-EN-1-YNYL)-5-(PENT-3-
3302-022	THIOPHENE,2-(BUT-3-EN-1-YNYL)-5-(PENTA-1
3302-004	THIOPHENE,2-(4-CHO-BUT-3-EN-1-YNYL)-
3302-030	THIOPHENE,2-(HEPT-5-EN-1,3-DIYNYL)-5-(1-
3302-029	THIOPHENE,2-(HEXA-3,5-DIEN-1-YNYL)-5-(PR
3302-006	THIOPHENE,2-LACTOYL-5-(PROP-1-YNYL)-
3302-032	THIOPHENE,2-PH-5-PROPYNYL-
3302-005	THIOPHENE,2-(PROP-1-YNYL)-5-PYRUVOYL-
3302-019	THIOPHENE,2-T,C-(NONA-3,5-DIENE-1-YN-7-O
3302-020	THIOPHENE,2-T,T-(NONA-3,5-DIEN-1-YN-7-ON
3302-007	THIOPHENE,2-(2-CARBOMETHOXY-ETHENYL)-5-(
3302-023	THIOPHENE,2-(4-ACETOXY-3-CL-BUT-L-YNYL)-
3302-024	THIOPHENE,2-(4-ACETOXY-3-CL-BUT-L-YNYL)-
3302-016	THIOPHENE,2-(5-OH-NON-3-EN-1-YN-7-ONYL)-
3302-018	THIOPHENE,2-(6-OH-NON-3-EN-1-YN-7-ONYL)-
3302-014	THIOPHENE,2-(7-OAC-NON-3-EN-1-YNYL)-
3302-015	THIOPHENE,2-(NON-3-EN-1-YN-7-ONYL)-
3302-002	THIOPHENE,2-AC-3-OH-5-PROP-1-YNYL-
3302-003	THIOPHENE,2-AC-3-OME-5-PROP-1-YNYL-
0499-010	THITSIOL
0606-003	THOMASIC ACID
0606-004	THOMASIDIOIC ACID
3509-019	THOMOSAMINE
0602-001	THUJAPLICATENE,GAMMA-
0602-002	THUJAPLICATIN ME ETHER,DI-OH-
3503-007	THYMINOSE = RIBOSE,2-DEOXY-
3195-004	TIGLALDEHYDE
3195-012	TIGLIC ACID
1103-055	TILIANIN
1104-026	TILIROSIDE
1499-001	TLATLANCUAYIN
3502-020	TOBRAMYCIN
0899-001	TOCOPHEROL,ALPHA-
0899-002	TOCOPHEROL,BETA-
0899-004	TOCOPHEROL,DELTA-
0899-006	TOCOPHEROL,EPSILON- = TOCOTRIENOL,BETA-
0899-003	TOCOPHEROL,GAMMA-
0899-005	TOCOPHEROL,ZETA-1- = TOCOTRIENOL,ALPHA-
0899-011	TOCOPHEROL DIMER,ALPHA-
2001-016	TOCOPHEROLQUINONE,ALPHA-
0899-005	TOCOTRIENOL,ALPHA-
0899-006	TOCOTRIENOL,BETA-
0899-007	TOCOTRIENOL,GAMMA-
0899-008	TOCOTRIENOL,DELTA-
0508-002	TODDACULINE
0508-004	TODDALOLACTONE
2001-002	TOLUQUINONE,PARA-
2003-005	TOLUQUINONE,6-OH-4-OME-
0902-016	TOMENTOLIDE A
0903-022	TOMENTOLIDE B
2503-025	TORACHRYSONE
1101-005	TORINGIN
2206-012	TORQUATONE
2403-031	TOVOXANTHONE
1801-005	TOXICAROL
1801-006	TOXICAROL,DEHYDRO-
1499-013	TOXICAROL ISOFLAVONE
0203-017	TOXOL
2701-011	TOYOMYCIN = CHROMOMYCIN-A3
0699-017	TRACHELOSIDE
0699-018	TRACHELOSIDE,NOR-
0508-015	TRACHYPHYLLIN
1106-036	TRANSILIN
1106-112	TRANSILITIN
3112-021	TRAUMATIC ACID
3506-016	TREHALOSE,ALPHA,ALPHA-
0203-016	TREMETONE,(-)-
0203-015	TREMETONE,DEHYDRO-
2202-020	TREMETONE,OH-
0203-018	TREMETONE,OME-
0199-003	TREMULACIN
3003-008	TRIACETIC LACTONE,ME-
3130-001	TRIACONTANE,N-
3130-006	TRIACONTANOIC ACID,30-OH-
3130-004	TRIACONTANOIC ACID TETRACOSANOL ESTER
3130-002	TRIACONTAN-1-OL
2599-010	TRIANELLINONE
2207-026	TRIASPIDIN
3196-019	TRICARBALLYLIC ACID
0199-008	TRICHOCARPIN
3197-012	TRICHODESMIC ACID
1107-002	TRICIN

MOLECULAR WEIGHT INDEX

ID	MW	C	H	□	Other
3102-001	28	C 2	H 4		
3101-003	30	C 1	H 2	□ 1	
3101-001	32	C 1	H 4	□ 1	
3102-003	44	C 2	H 4	□ 1	
3101-004	46	C 1	H 2	□ 2	
3102-002	46	C 2	H 6	□ 1	
3101-002	48	C 1	H 4		S 1
3104-001	56	C 4	H 8		
3103-005	58	C 3	H 6	□ 1	
3103-008	58	C 3	H 6	□ 1	
3103-002	58	C 3	H 6	□ 1	
3103-001	60	C 3	H 8	□ 1	
3102-004	60	C 2	H 4	□ 2	
3102-005	60	C 2	H 4	□ 2	
3301-002	62	C 2	H 6		S 1
3004-001	68	C 4	H 4	□ 1	
3195-001	70	C 5	H 10		
3194-004	70	C 4	H 6	□ 1	
3194-003	72	C 4	H 8	□ 1	
3104-008	72	C 4	H 8	□ 1	
3103-021	72	C 3	H 4	□ 2	
3104-026	72	C 4	H 8	□ 1	
3103-006	72	C 3	H 4	□ 2	
3104-003	72	C 4	H 8	□ 1	
3104-004	72	C 4	H 8	□ 1	
3106-002	74	C 6	H 2		
3104-002	74	C 4	H 10	□ 1	
3102-009	74	C 2	H 2	□ 3	
3103-010	74	C 3	H 6	□ 2	
3194-001	74	C 4	H 10	□ 1	
3103-022	76	C 3	H 8		S 1
3102-008	76	C 2	H 4	□ 3	
3102-007	78	C 2	H 3	□ 2	F 1
3106-001	82	C 6	H 10		
3004-002	82	C 5	H 6	□ 1	
2901-001	84	C 5	H 8	□ 1	
3195-004	84	C 5	H 8	□ 1	
3195-003	86	C 5	H 10	□ 1	
3195-005	86	C 5	H 10	□ 1	
3195-006	86	C 5	H 10	□ 1	
3301-004	86	C 4	H 6		S 1
3194-006	86	C 4	H 6	□ 2	
3104-007	86	C 4	H 6	□ 2	
3105-001	86	C 5	H 10	□ 1	
3104-011	86	C 4	H 6	□ 2	
3104-012	86	C 4	H 6	□ 2	
3105-002	86	C 5	H 10	□ 1	
3104-010	88	C 4	H 8	□ 2	
3102-006	88	C 4	H 8	□ 2	
3104-009	88	C 4	H 8	□ 2	
3104-006	88	C 4	H 8	□ 2	
3103-013	88	C 3	H 4	□ 3	
3195-002	88	C 5	H 12	□ 1	
3194-005	88	C 4	H 8	□ 2	
3104-005	90	C 4	H 10	□ 2	
3103-012	90	C 3	H 6	□ 3	
3103-011	90	C 3	H 3	□ 3	
3102-010	90	C 2	H 2	□ 4	
3103-009	90	C 3	H 6	□ 3	
3103-007	90	C 3	H 6	□ 3	
3103-003	92	C 3	H 8	□ 3	
3301-003	94	C 2	H 6	□ 2	S 1
0801-001	94	C 6	H 6	□ 1	
3004-004	96	C 5	H 4	□ 2	
2901-003	96	C 6	H 8	□ 1	
3005-004	96	C 5	H 4	□ 2	
3005-003	98	C 5	H 6	□ 2	
3106-006	98	C 6	H 10	□ 1	
3004-003	98	C 5	H 6	□ 2	
3107-001	100	C 7	H 16		
3106-005	100	C 6	H 12	□ 1	
3106-004	100	C 6	H 12	□ 1	
3004-010	100	C 5	H 8	□ 2	
3195-011	100	C 5	H 8	□ 2	
3195-013	100	C 5	H 8	□ 2	
3195-012	100	C 5	H 8	□ 2	
3196-001	102	C 6	H 14	□ 1	
3195-008	102	C 5	H 10	□ 2	
3195-007	102	C 5	H 10	□ 2	
3106-003	102	C 6	H 14	□ 1	
3105-003	102	C 5	H 10	□ 2	
3103-016	104	C 3	H 4	□ 4	
3103-019	104	C 3	H 4	□ 4	
0201-001	104	C 8	H 8		
3103-020	106	C 3	H 6	□ 4	
0101-001	108	C 7	H 8	□ 1	
0102-001	108	C 7	H 8	□ 1	
0101-007	108	C 7	H 8	□ 1	
3301-001	108	C 3	H 8		S 2
2101-017	108	C 7	H 8	□ 1	
2001-001	108	C 6	H 4	□ 2	
3004-005	110	C 6	H 6	□ 2	
0802-005	110	C 6	H 6	□ 2	
0802-001	110	C 6	H 6	□ 2	
2901-007	112	C 6	H 8	□ 2	
3106-010	112	C 6	H 8	□ 2	
3004-007	112	C 5	H 4	□ 4	
3107-006	112	C 7	H 12	□ 1	
3107-005	112	C 7	H 12	□ 1	
3107-007	112	C 7	H 12	□ 1	
3003-001	112	C 6	H 8	□ 2	
3002-001	112	C 5	H 4	□ 3	
3196-004	112	C 6	H 8	□ 2	
3901-002	112	C 7	H 12	□ 1	
3901-001	114	C 7	H 14	□ 1	
3196-007	114	C 6	H 10	□ 2	
3301-008	114	C 6	H 10		S 1
3196-010	114	C 6	H 10	□ 2	
3196-002	114	C 7	H 14	□ 1	
3107-004	114	C 7	H 14	□ 1	
3104-014	114	C 4	H 6	□ 4	
3005-005	114	C 5	H 6	□ 3	
3106-009	114	C 6	H 10	□ 2	
3106-007	114	C 6	H 10	□ 2	
3104-013	114	C 4	H 6	□ 4	
3005-001	114	C 5	H 6	□ 3	
3107-008	114	C 7	H 14	□ 1	
3106-008	116	C 6	H 12	□ 2	
3107-002	116	C 7	H 16	□ 1	
3107-003	116	C 7	H 16	□ 1	
3104-016	116	C 4	H 4	□ 4	
3196-003	116	C 6	H 12	□ 2	
3195-015	116	C 5	H 8	□ 3	
3196-006	116	C 6	H 12	□ 2	
3196-005	116	C 6	H 12	□ 2	
3194-002	116	C 6	H 12	□ 2	
3901-003	116	C 6	H 12	□ 2	
3905-001	118	C 8	H 6	□ 1	
3905-020	118	C12	H 12	□ 2	
3195-016	118	C 5	H 10	□ 3	
3195-009	118	C 5	H 10	□ 3	
3108-007	118	C 8	H 6	□ 1	
3104-015	118	C 4	H 6	□ 4	
3103-018	118	C 3	H 2	□ 5	
3103-017	120	C 3	H 4	□ 5	
0203-020	120	C 8	H 8	□ 1	
0203-009	120	C 8	H 8	□ 1	
0201-002	120	C 8	H 8	□ 1	
0201-003	122	C 8	H 10	□ 3	
0201-008	122	C 8	H 10	□ 1	
0202-001	122	C 8	H 10	□ 1	
0103-001	122	C 7	H 6	□ 2	
0104-001	122	C 7	H 6	□ 2	
0103-013	122	C 7	H 6	□ 2	
0101-002	122	C 8	H 10	□ 1	
3104-024	122	C 4	H 10	□ 4	
3103-023	122	C 3	H 7	□ 3	P 1
0899-012	122	C 8	H 10	□ 1	
0899-014	122	C 8	H 10	□ 1	
0899-013	122	C 8	H 10	□ 1	
0899-015	122	C 8	H 10	□ 1	
0899-016	122	C 8	H 10	□ 1	
2001-002	122	C 7	H 6	□ 2	
2101-013	122	C 7	H 8	□ 2	
2101-018	124	C 7	H 8	□ 2	
0802-012	124	C 7	H 8	□ 2	
0802-004	124	C 7	H 8	□ 2	
0802-002	124	C 7	H 8	□ 2	
3106-011	124	C 7	H 8	□ 2	
3106-012	124	C 7	H 8	□ 2	
3004-006	124	C 7	H 8	□ 2	
0101-003	124	C 7	H 8	□ 2	
0102-007	124	C 7	H 8	□ 2	
3004-009	126	C 6	H 6	□ 3	

Code	MW	C	H				
3002-003	126	C 6	H 6	□ 3			
3005-006	126	C 6	H 6	□ 3			
3002-011	126	C 6	H 6	□ 3			
3108-009	126	C 8	H 14	□ 1			
2901-004	126	C 8	H 14	□ 1			
0803-001	126	C 6	H 6	□ 3			
3901-007	126	C 7	H 10	□ 2			
2299-001	126	C 6	H 6	□ 3			
2599-001	128	C10	H 8				
3108-008	128	C 8	H 16	□ 1			
3003-002	128	C 7	H 12	□ 2			
3109-018	128	C 9	H 4	□ 1			
3004-011	128	C 6	H 8	□ 3			
3108-011	128	C 8	H 16	□ 1			
3108-010	128	C 8	H 16	□ 1			
3108-004	128	C 8	H 16	□ 1			
3002-002	128	C 5	H 4	□ 4			
3109-001	128	C 9	H 20				
3004-014	128	C 8	H 16	□ 1			
3108-005	128	C 8	H 16	□ 1			
3108-003	130	C 8	H 18	□ 1			
3106-015	130	C 6	H 10	□ 3			
3105-007	130	C 5	H 6	□ 4			
3003-003	130	C 6	H 10	□ 3			
3108-001	130	C 8	H 18	□ 1			
3109-006	130	C 9	H 6	□ 1			
3107-009	130	C 7	H 14	□ 2			
3108-002	130	C 8	H 18	□ 1			
3195-018	130	C 5	H 6	□ 4			
3195-019	130	C 5	H 6	□ 4			
3196-008	130	C 6	H 10	□ 3			
3195-017	132	C 5	H 8	□ 4			
3109-004	132	C 9	H 8	□ 1			
3109-005	132	C 9	H 8	□ 1			
3104-020	132	C 4	H 4	□ 5			
3106-014	132	C 6	H 12	□ 3			
3104-017	132	C 4	H 4	□ 5			
2299-010	132	C13	H 12	□ 4			
3504-067	132	C 6	H 12	□ 3			
3504-044	132	C 6	H 12	□ 3			
0303-001	132	C 9	H 8	□ 1			
0303-002	134	C 9	H 10	□ 1			
0302-002	134	C 9	H 10	□ 1			
0301-001	134	C 9	H 12	□ 1			
3104-018	134	C 4	H 6	□ 5			
3104-019	134	C 4	H 6	□ 5			
3103-025	134	C 5	H 10	□ 2	S 1		
3195-010	134	C 5	H 10	□ 4			
3503-008	134	C 5	H 10	□ 4			
3503-007	134	C 5	H 10	□ 4			
3509-014	135	C 8	H 17	N 1	□ 3		
3105-004	136	C 5	H 8	□ 4			
0204-001	136	C 8	H 8	□ 2			
0202-002	136	C 9	H 12	□ 1			
0201-009	136	C 9	H 12	□ 1			
0203-001	136	C 8	H 8	□ 2			
0302-001	136	C 9	H 12	□ 1			
0103-002	136	C 8	H 8	□ 2			
0103-014	136	C 8	H 8	□ 2			
2201-001	136	C 8	H 8	□ 2			
2001-005	136	C 8	H 8	□ 2			
2001-006	136	C 8	H 8	□ 2			
2001-004	136	C 8	H 8	□ 2			
2001-003	138	C 7	H 6	□ 3			
2002-001	138	C 7	H 6	□ 3			
2101-015	138	C 8	H 10	□ 2			
2101-014	138	C 8	H 10	□ 2			
0103-003	138	C 7	H 6	□ 3			
0102-002	138	C 8	H 10	□ 2			
0101-004	138	C 8	H 10	□ 2			
0202-005	138	C 8	H 10	□ 2			
0104-022	138	C 7	H 6	□ 3			
0104-004	138	C 7	H 6	□ 3			
3109-017	138	C 9	H 14	□ 1			
3305-002	138	C 3	H 6	□ 2	S 2		
3201-001	138	C 9	H 14	□ 1			
3305-001	138	C 3	H 6	□ 2	S 2		
0802-013	138	C 8	H 10	□ 2			
0802-014	138	C 8	H 10	□ 2			
0803-005	139	C 7	H 6	□ 3			
0803-004	140	C 7	H 8	□ 3			
0803-002	140	C 7	H 8	□ 3			
3002-010	140	C 7	H 8	□ 3			
3111-005	140	C11	H 8				
3109-007	140	C 9	H 16	□ 1			
3003-008	140	C 7	H 8	□ 3			
3109-015	140	C 9	H 16	□ 1			
2901-005	140	C 7	H 8	□ 3			
0102-003	140	C 7	H 8	□ 3			
0102-009	140	C 7	H 8	□ 3			
3005-007	141	C 6	H 7	N 1	□ 3		
3002-007	142	C 6	H 6	□ 4			
3002-008	142	C 6	H 6	□ 4			
3002-005	142	C 6	H 6	□ 4			
3110-040	142	C10	H 6	□ 1			
3004-008	142	C 6	H 6	□ 4			
3109-008	142	C 9	H 18	□ 1			
3109-019	142	C 9	H 18	□ 1			
3109-014	142	C 9	H 18	□ 1			
3111-004	142	C11	H 10				
3110-037	142	C10	H 6	□ 1			
3002-016	142	C 6	H 6	□ 4			
3904-009	142	C 8	H 14	□ 2			
3905-002	144	C 8	H 6	□ 2			
3004-015	144	C 7	H 12	□ 3			
3110-008	144	C10	H 8	□ 1			
3109-025	144	C 9	H 4	□ 2			
3110-036	144	C10	H 8	□ 1			
3109-003	144	C 9	H 20	□ 1			
3109-002	144	C 9	H 20	□ 1			
3110-007	144	C10	H 8	□ 1			
3108-012	144	C 8	H 16	□ 2			
3002-006	144	C 6	H 8	□ 4			
3196-011	144	C 6	H 8	□ 4			
3108-018	145	C 8	H 3	N 1	□ 2	□ 2	
3110-017	146	C10	H 10	□ 1			
3109-024	146	C 9	H 9	□ 2			
3110-021	146	C10	H 10	□ 1			
3106-016	146	C 6	H 10	□ 4			
3105-006	146	C 5	H 6	□ 5			
3109-028	146	C 9	H 6	□ 2			
3110-015	146	C10	H 10	□ 1			
3110-022	146	C10	H 10	□ 1			
3110-019	146	C10	H 10	□ 1			
3301-009	146	C 6	H 10	S 2			
3501-002	146	C 6	H 10	□ 4			
3501-003	146	C 6	H 10	□ 4			
3905-005	146	C 9	H 6	□ 2			
0401-004	146	C10	H 10	□ 1			
0599-002	146	C 9	H 6	□ 2			
0304-004	147	C 9	H 9	N 1	□ 1		
3509-015	147	C 6	H 13	N 1	□ 3		
3108-016	147	C 8	H 5	N 1	□ 2		
3110-004	148	C10	H 12	□ 1			
3110-035	148	C10	H 12	□ 1			
3111-002	148	C11	H 18				
3104-023	148	C 4	H 4	□ 6			
3110-011	148	C10	H 12	□ 1			
3105-005	148	C 5	H 8	□ 5			
3110-023	148	C10	H 12	□ 1			
3109-022	148	C 9	H 8	□ 2			
3111-003	148	C11	H 16				
3004-013	148	C 9	H 8	□ 2			
3001-010	148	C11	H 16				
3905-004	148	C 9	H 8	□ 2			
3905-003	148	C 9	H 8	□ 2			
0304-002	148	C 9	H 8	□ 2			
0304-001	148	C 9	H 8	□ 2			
0401-003	148	C10	H 12	□ 1			
0599-001	148	C 9	H 8	□ 2			
0301-003	148	C10	H 12	□ 1			
0301-002	148	C10	H 12	□ 1			
3504-035	148	C 6	H 12	□ 4			
3504-030	148	C 6	H 12	□ 4			
3195-020	148	C 5	H 8	□ 5			
3197-010	148	C11	H 16				
3504-039	148	C 6	H 12	□ 4			
3504-040	148	C 6	H 12	□ 4			
3504-041	148	C 6	H 12	□ 4			
3504-042	148	C 6	H 12	□ 4			
3504-033	148	C 6	H 12	□ 4			
3504-032	148	C 6	H 12	□ 4			
3196-009	148	C 6	H 12	□ 4			
3501-001	148	C 6	H 12	□ 4			
3301-005	148	C 6	H 12	S 2			
3503-002	150	C 5	H 10	□ 5			
3302-001	150	C 8	H 6	□ 1	S 1		
3503-006	150	C 5	H 10	□ 5			
3503-005	150	C 5	H 10	□ 5			
3503-003	150	C 5	H 10	□ 5			
3503-004	150	C 5	H 10	□ 5			
3197-009	150	C11	H 18				
3503-001	150	C 5	H 10	□ 5			
0301-008	150	C 9	H 10	□ 2			
0201-005	150	C 9	H 10	□ 2			
0203-003	150	C 9	H 10	□ 2			
0103-007	150	C 8	H 6	□ 3			
3510-001	150	C 5	H 10	□ 5			
3001-009	150	C11	H 18				
3104-022	150	C 4	H 6	□ 6			
3110-027	150	C10	H 14	□ 1			
3005-013	150	C 7	H 8	□ 4			
3112-001	150	C12	H 6				
3104-021	150	C 4	H 6	□ 6			
3109-009	150	C 9	H 10	□ 2			
2102-026	150	C 8	H 6	□ 3			
2199-002	150	C 9	H 10	□ 2			
2002-002	152	C 8	H 8	□ 3			
2002-019	152	C 8	H 8	□ 3			
2101-012	152	C 8	H 8	□ 3			

Code	MW	Formula
2201-007	152	C 8 H 8 O 3
2001-007	152	C 9 H 10 O 2
3110-034	152	C10 H 16 O 1
3103-026	152	C 5 H 12 O 3 S 1
0103-017	152	C 8 H 8 O 3
0103-004	152	C 8 H 8 O 3
0203-004	152	C 8 H 8 O 3
0201-004	152	C 9 H 12 O 2
0301-007	152	C 9 H 12 O 2
0204-004	152	C 8 H 8 O 3
0204-003	152	C 8 H 8 O 3
0204-002	152	C 8 H 8 O 3
0104-005	152	C 8 H 8 O 3
0104-007	152	C 8 H 8 O 3
0104-023	152	C 8 H 8 O 3
0104-024	152	C 8 H 8 O 3
3503-010	152	C 5 H 12 O 5
3194-007	152	C 4 H 8 O 2 S 2
3503-009	152	C 5 H 12 O 5
3197-008	154	C10 H 18 O 1
3205-008	154	C12 H 10
3501-004	154	C 6 H 12 O 5
0104-027	154	C 7 H 6 O 4
0104-033	154	C 7 H 6 O 4
0104-030	154	C 7 H 6 O 4
0104-028	154	C 7 H 6 O 4
0104-009	154	C 7 H 6 O 4
0202-007	154	C 8 H 10 O 3
0102-004	154	C 8 H 10 O 3
0101-005	154	C 8 H 10 O 3
3003-010	154	C 8 H 10 O 3
3109-016	154	C10 H 18 O 1
3110-033	154	C10 H 18 O 1
2101-016	154	C 8 H 10 O 3
1901-003	154	C 7 H 6 O 4
1901-001	154	C 7 H 6 O 4
1901-002	154	C 7 H 8 O 4
3904-005	154	C 7 H 6 O 4
0803-003	154	C 8 H 10 O 3
0803-006	154	C 8 H 10 O 3
2299-002	154	C 8 H 10 O 3
1901-008	156	C 7 H 8 O 4
3111-001	156	C11 H 24
3110-039	156	C10 H 20 O 1
3110-032	156	C10 H 20 O 1
3110-010	156	C10 H 20 O 1
3110-009	156	C10 H 20 O 1
3003-004	156	C 7 H 8 O 4
3005-014	156	C 8 H 12 O 3
3110-026	156	C10 H 14 O 1
3205-013	156	C12 H 12
3201-011	158	C11 H 10 O 1
3196-015	158	C 6 H 6 O 5
3197-004	158	C 9 H 18 O 2
3109-021	158	C 9 H 18 O 2
3110-003	158	C10 H 22 O 1
3110-079	158	C10 H 6 O 2
3108-013	158	C 9 H 18 O 2
3002-009	158	C 6 H 6 O 5
2501-001	158	C11 H 10 O 1
3509-022	159	C 8 H 17 N 1 O 2
3509-023	159	C 8 H 17 N 1 O 2
3110-029	160	C10 H 8 O 2
3111-008	160	C11 H 12 O 1
3109-023	160	C10 H 8 O 2
3106-017	160	C 6 H 8 O 5
3110-055	160	C10 H 8 O 2
3110-038	160	C10 H 8 O 2
3201-008	160	C10 H 8 O 2
3509-021	161	C 7 H 15 N 1 O 3
3509-025	161	C 7 H 15 N 1 O 3
3509-020	161	C 7 H 15 N 1 O 3
3510-016	162	C 7 H 14 O 4
3510-010	162	C 6 H 10 O 5
3905-006	162	C 9 H 6 O 3
3905-007	162	C10 H 10 O 2
3510-006	162	C 7 H 14 O 4
3510-005	162	C 7 H 14 O 4
3201-010	162	C10 H 10 O 2
3201-018	162	C10 H 10 O 2
3301-019	162	C 6 H 10 O 1 S 2
3504-052	162	C 6 H 10 O 5
3201-007	162	C10 H 10 O 2
3196-013	162	C 6 H 10 O 5
3502-007	162	C 6 H 14 N 2 O 3
3301-007	162	C 7 H 14 S 2
3504-036	162	C 7 H 14 O 4
3504-034	162	C 7 H 14 O 4
3504-031	162	C 7 H 14 O 4
3504-038	162	C 7 H 14 O 4
3302-004	162	C 9 H 6 O 1 S 1
3504-043	162	C 7 H 14 O 4
3110-031	162	C10 H 10 O 2
3110-030	162	C10 H 10 O 2
3113-002	162	C13 H 6
3110-028	162	C10 H 10 O 2
0801-003	162	C 6 H 4 O 1 CL2
0501-001	162	C 9 H 6 O 3
0599-004	162	C 9 H 6 O 3
0301-018	162	C10 H 10 O 2
0301-017	162	C10 H 10 O 2
0304-003	162	C10 H 10 O 2
0303-004	162	C10 H 10 O 2
0303-003	162	C10 H 10 O 2
3108-017	163	C 8 H 5 N 1 O 3
3509-005	163	C 6 H 13 N 1 O 4
3509-004	163	C 6 H 13 N 1 O 4
3509-013	163	C 6 H 13 N 1 O 4
3509-003	163	C 6 H 13 N 1 O 4
3509-016	163	C 6 H 13 N 1 O 4
3509-017	163	C 6 H 13 N 1 O 4
3509-019	163	C 6 H 13 N 1 O 4
3510-012	164	C 6 H 12 O 5
2901-019	164	C11 H 16 O 1
3109-012	164	C 9 H 8 O 3
3109-013	164	C 9 H 8 O 3
3109-026	164	C 9 H 8 O 3
3113-003	164	C13 H 8
0304-006	164	C 9 H 8 O 3
0304-011	164	C 9 H 8 O 3
0399-003	164	C10 H 12 O 2
0301-013	164	C10 H 12 O 2
0301-010	164	C10 H 12 O 2
0301-011	164	C10 H 12 O 2
3504-057	164	C 6 H 12 O 5
3504-058	164	C 6 H 12 O 5
3501-007	164	C 6 H 12 O 5
3504-016	164	C 6 H 12 O 5
3504-025	164	C 6 H 12 O 5
3504-022	164	C 6 H 12 O 5
3503-011	164	C 5 H 8 O 6
3501-005	164	C 6 H 12 O 5
3504-017	164	C 6 H 12 O 5
3504-019	164	C 6 H 12 O 5
0203-005	166	C 9 H 10 O 3
0104-008	166	C 9 H 10 O 3
0104-015	166	C 8 H 6 O 4
0401-002	166	C10 H 14 O 2
0304-009	166	C 9 H 10 O 3
0304-005	166	C 9 H 10 O 3
0103-006	166	C 9 H 10 O 3
3113-004	166	C13 H 10
3113-005	166	C13 H 10
2901-020	166	C10 H 14 O 2
3109-010	166	C 9 H 10 O 3
3109-009	166	C 9 H 10 O 3
3902-001	166	C 9 H 10 O 3
3902-002	166	C 9 H 10 O 3
2201-002	166	C 9 H 10 O 3
2102-027	166	C 8 H 6 O 4
2102-003	166	C 8 H 6 O 4
2201-008	166	C 9 H 10 O 3
2003-005	168	C 8 H 8 O 4
2003-004	168	C 8 H 8 O 4
2003-003	168	C 8 H 8 O 4
2003-002	168	C 8 H 8 O 4
2003-001	168	C 8 H 8 O 4
2002-023	168	C 8 H 8 O 4
2101-001	168	C 8 H 8 O 4
3111-013	168	C11 H 20 O 1
3103-014	168	C 3 H 5 O 6 P 1
3003-009	168	C 8 H 8 O 4
3004-016	168	C 9 H 12 O 4
3109-011	168	C 9 H 12 O 3
3110-045	168	C10 H 16 O 2
3113-008	168	C13 H 12
3003-005	168	C 8 H 8 O 4
3113-007	168	C13 H 12
2901-010	168	C10 H 16 O 2
0101-006	168	C 9 H 12 O 3
0104-010	168	C 8 H 8 O 4
0104-029	168	C 8 H 8 O 4
0204-005	168	C 8 H 8 O 4
0204-006	168	C 8 H 8 O 4
3203-001	168	C10 H 16 O 2
3203-013	168	C10 H 16 O 2
3205-007	168	C12 H 8 O 1
3195-021	168	C 5 H 8 O 6
3195-014	168	C10 H 16 O 2
3205-009	170	C12 H 10 O 1
3205-012	170	C12 H 10 O 1
0104-016	170	C 7 H 6 O 5
0104-036	170	C 7 H 6 O 5
0103-008	170	C 7 H 6 O 5
3110-044	170	C10 H 18 O 2
3111-012	170	C11 H 22 O 1
3005-012	170	C 8 H 10 O 4
3108-006	170	C10 H 18 O 2
3113-006	170	C13 H 14
3111-007	170	C11 H 22 O 1

2101-019	170	C 8	H 10	☐ 4						
2502-026	170	C11	H 10	☐ 2						
2901-006	171	C 5	H 8	☐ 2	CL2					
3111-006	172	C11	H 24	☐ 1						
3111-017	172	C11	H 8	☐ 2						
3110-056	172	C11	H 8	☐ 2						
3110-057	172	C11	H 8	☐ 2						
3106-013	172	C 8	H 12	☐ 2	S 1					
3110-041	172	C10	H 20	☐ 2						
3112-006	172	C12	H 12	☐ 1						
3111-019	172	C11	H 8	☐ 2						
3111-018	172	C11	H 8	☐ 2						
2502-003	172	C11	H 8	☐ 2						
2502-004	172	C11	H 8	☐ 2						
2502-039	172	C11	H 8	☐ 2						
1902-004	172	C 7	H 8	☐ 5						
0402-006	172	C10	H 8	☐ 2						
3201-012	172	C11	H 8	☐ 2						
3903-002	172	C 6	H 4	☐ 6						
3905-008	174	C11	H 10	☐ 2						
3196-012	174	C 6	H 6	☐ 6						
0102-011	174	C 7	H 7	☐ 3	CL1					
1902-001	174	C 7	H 10	☐ 5						
2503-003	174	C10	H 6	☐ 3						
2502-024	174	C11	H 10	☐ 2						
2502-001	174	C11	H 10	☐ 2						
2503-039	174	C10	H 6	☐ 3						
2503-021	174	C10	H 6	☐ 3						
3005-050	174	C 6	H 6	☐ 6						
3110-069	174	C10	H 6	☐ 3						
3110-054	174	C11	H 10	☐ 2						
3107-010	174	C 7	H 10	☐ 5						
3110-053	174	C11	H 10	☐ 2						
3509-028	175	C 8	H 17	N 1	☐ 3					
3510-004	176	C 8	H 16	☐ 4						
3510-015	176	C 8	H 16	☐ 4						
3111-010	176	C11	H 12	☐ 2						
3111-009	176	C11	H 12	☐ 2						
3005-036	176	C 6	H 8	☐ 6						
3110-042	176	C11	H 12	☐ 2						
3110-043	176	C11	H 12	☐ 2						
2503-037	176	C10	H 8	☐ 3						
2001-008	176	C11	H 12	☐ 2						
2104-015	176	C10	H 8	☐ 3						
1902-003	176	C 7	H 12	☐ 5						
2203-013	176	C10	H 8	☐ 3						
0501-002	176	C10	H 8	☐ 3						
0303-007	176	C10	H 8	☐ 3						
0302-003	176	C11	H 12	☐ 2						
3196-014	176	C 6	H 8	☐ 6						
3196-019	176	C 6	H 8	☐ 6						
3502-034	176	C 7	H 16	N 2	☐ 3					
3502-038	177	C 7	H 15	N 1	☐ 4					
3509-026	177	C 7	H 15	N 1	☐ 4					
3510-011	178	C 6	H 10	☐ 6						
3301-010	178	C 6	H 10	S 3						
3504-020	178	C 7	H 14	☐ 5						
3197-001	178	C 7	H 14	☐ 5						
3504-023	178	C 7	H 14	☐ 5						
3504-026	178	C 7	H 14	☐ 5						
3504-029	178	C 7	H 14	☐ 5						
3504-021	178	C 7	H 14	☐ 5						
3504-027	178	C 7	H 14	☐ 5						
3504-024	178	C 7	H 14	☐ 5						
3504-050	178	C 6	H 10	☐ 6						
3504-028	178	C 7	H 14	☐ 5						
3504-051	178	C 6	H 10	☐ 6						
3504-018	178	C 7	H 14	☐ 5						
3502-001	178	C 6	H 14	N 2	☐ 4					
0303-005	178	C10	H 10	☐ 3						
0301-028	178	C11	H 14	☐ 2						
0304-012	178	C10	H 10	☐ 3						
0304-013	178	C10	H 10	☐ 3						
0401-001	178	C11	H 14	☐ 2						
0301-011	178	C11	H 14	☐ 2						
0301-014	178	C11	H 14	☐ 2						
0201-011	178	C10	H 10	☐ 3						
0506-001	178	C 9	H 6	☐ 4						
0507-001	178	C 9	H 6	☐ 4						
2104-014	178	C10	H 10	☐ 3						
2102-015	178	C10	H 10	☐ 3						
2203-012	178	C10	H 10	☐ 3						
2503-045	178	C10	H 10	☐ 3						
3110-052	178	C11	H 14	☐ 2						
3110-068	178	C10	H 10	☐ 3						
3113-023	178	C13	H 6	☐ 1						
2901-021	178	C11	H 14	☐ 2						
3509-011	179	C 6	H 13	N 1	☐ 5					
3509-008	179	C 6	H 13	N 1	☐ 5					
3509-007	179	C 6	H 13	N 1	☐ 5					
3509-001	179	C 6	H 13	N 1	☐ 5					
3509-006	179	C 6	H 13	N 1	☐ 5					
3509-024	179	C 6	H 13	N 1	☐ 5	
3904-006	180	C11	H 16	☐ 2	
3510-002	180	C 6	H 12	☐ 6	
3113-032	180	C13	H 8	☐ 1	
3113-025	180	C13	H 8	☐ 1	
3113-011	180	C13	H 8	☐ 1	
2804-001	180	C14	H 12		
3005-017	180	C11	H 16	☐ 2	
3113-024	180	C13	H 8	☐ 1	
3002-012	180	C 6	H 4	☐ 5	
3112-007	180	C12	H 20	☐ 1	
2203-002	180	C10	H 12	☐ 3	
2102-028	180	C 9	H 18	☐ 4	
2102-011	180	C 9	H 8	☐ 4	
2201-006	180	C10	H 12	☐ 3	
0203-008	180	C10	H 12	☐ 3	
0401-007	180	C10	H 12	☐ 3	
0304-010	180	C 9	H 8	☐ 4	
0304-024	180	C 9	H 8	☐ 4	
0304-016	180	C 9	H 8	☐ 4	
0302-005	180	C10	H 12	☐ 3	
3501-012	180	C 6	H 12	☐ 6	
3504-048	180	C 6	H 12	☐ 6	
3504-007	180	C 6	H 12	☐ 6	
3501-008	180	C 6	H 12	☐ 6	
3504-045	180	C 6	H 12	☐ 6	
3504-049	180	C 6	H 12	☐ 6	
3504-001	180	C 6	H 12	☐ 6	
3504-008	180	C 6	H 12	☐ 6	
3205-015	180	C13	H 8	☐ 1	
3504-014	180	C 6	H 12	☐ 6	
3302-002	180	C 9	H 8	☐ 2	S 1
3501-015	180	C 6	H 12	☐ 6	
3501-026	180	C 6	H 12	☐ 6	
3506-002	182	C10	H 18	☐ 9	
3504-055	182	C 6	H 14	☐ 6	
3504-054	182	C 6	H 14	☐ 6	
3205-016	182	C13	H 10	☐ 1	
3504-056	182	C 6	H 14	☐ 6	
3506-001	182	C10	H 18	☐ 9	
3504-053	182	C 6	H 14	☐ 6	
3201-006	182	C13	H 10	☐ 1	
3302-010	182	C12	H 6	S 1	
3205-014	182	C13	H 10	☐ 1	
0301-027	182	C10	H 14	☐ 3	
0304-018	182	C 9	H 10	☐ 4	
0399-001	182	C10	H 14	☐ 3	
0201-007	182	C10	H 14	☐ 3	
0201-006	182	C10	H 14	☐ 3	
0204-007	182	C 8	H 6	☐ 5	
0104-012	182	C 9	H 10	☐ 4	
0104-031	182	C 9	H 10	☐ 4	
0104-034	182	C 9	H 10	☐ 4	
0103-009	182	C 9	H 10	☐ 4	
2101-008	182	C 9	H 10	☐ 4	
2101-007	182	C 9	H 10	☐ 4	
2102-017	182	C 9	H 10	☐ 4	
2102-030	182	C 8	H 6	☐ 5	
2101-006	182	C 9	H 10	☐ 4	
2104-001	182	C10	H 14	☐ 3	
2102-004	182	C 8	H 6	☐ 5	
2101-004	182	C 9	H 10	☐ 4	
2002-024	182	C 9	H 10	☐ 4	
2002-020	182	C 9	H 10	☐ 4	
2999-006	182	C10	H 14	☐ 3	
3111-011	182	C12	H 22	☐ 1	
3113-015	182	C13	H 10	☐ 1	
3001-001	182	C 8	H 6	☐ 5	
3112-011	182	C12	H 22	☐ 1	
3113-033	182	C13	H 10	☐ 1	
3005-045	182	C 9	H 10	☐ 4	
3111-016	182	C11	H 18	☐ 2	
3904-007	182	C11	H 18	☐ 2	
2299-003	182	C10	H 14	☐ 3	
3111-015	184	C11	H 20	☐ 2	
3110-078	184	C10	H 16	☐ 3	
3003-012	184	C 9	H 12	☐ 4	
3002-014	184	C 7	H 4	☐ 6	
3113-016	184	C13	H 28		
3113-031	184	C13	H 12	☐ 1	
3112-010	184	C12	H 24	☐ 1	
2004-001	184	C 8	H 8	☐ 5	
2102-029	184	C 8	H 8	☐ 5	
1901-004	184	C 8	H 8	☐ 5	
3204-002	184	C12	H 8	☐ 2	
0804-001	184	C 9	H 12	☐ 4	
3204-001	186	C12	H 10	☐ 2	
3201-013	186	C12	H 10	☐ 2	
3205-012	186	C12	H 10	☐ 2	
3004-012	186	C 8	H 10	☐ 5	
3103-015	186	C 3	H 7	☐ 7	P 1
3113-009	186	C13	H 14	☐ 1	

ID	MW	C	H				
3110-066	186	C10	H 18	O 3			
3110-067	186	C10	H 18	O 3			
3111-014	186	C11	H 22	O 2			
3112-002	186	C12	H 26	O 1			
2502-005	186	C12	H 10	O 2			
2501-006	186	C12	H 10	O 2			
0504-016	186	C11	H 6	O 3			
0502-013	186	C11	H 6	O 3			
0402-007	188	C12	H 12	O 2			
2503-046	188	C11	H 8	O 3			
2503-040	188	C11	H 8	O 3			
2503-005	188	C11	H 8	O 3			
2502-025	188	C12	H 12	O 2			
2503-004	188	C11	H 8	O 3			
3110-016	188	C12	H 12	O 2			
3109-027	188	C 9	H 16	O 4			
3110-020	188	C12	H 12	O 2			
3110-062	188	C10	H 20	O 3			
3110-077	188	C10	H 4	O 4			
3113-010	188	C13	H 16	O 1			
3005-051	188	C 8	H 12	O 5			
3305-005	188	C 2	H 4	S 5			
2106-023	188	C12	H 12	O 2			
1901-005	188	C 8	H 12	O 5			
2199-013	188	C12	H 12	O 2			
3510-008	188	C 8	H 12	O 5			
3909-010	188	C12	H 12	O 2			
3905-010	190	C11	H 10	O 3			
3905-013	190	C12	H 14	O 2			
3904-008	190	C 8	H 14	O 5			
3902-003	190	C12	H 14	O 2			
2106-022	190	C12	H 14	O 2			
2106-024	190	C12	H 14	O 2			
2001-009	190	C12	H 14	O 2			
2104-018	190	C11	H 10	O 3			
1902-017	190	C 7	H 10	O 6			
3301-006	190	C 9	H 18	S 2			
3504-037	190	C 8	H 14	O 5			
3110-024	190	C12	H 14	O 2			
3110-074	190	C10	H 6	O 4			
3110-005	190	C12	H 14	O 2			
3111-020	190	C11	H 10	O 3			
3107-011	190	C 7	H 10	O 6			
3005-009	190	C 6	H 6	O 7			
3110-012	190	C12	H 14	O 2			
3112-018	190	C12	H 14	O 2			
2503-041	190	C11	H 10	O 3			
0402-008	190	C11	H 10	O 3			
0401-006	190	C11	H 10	O 3			
0599-007	190	C11	H 10	O 3			
0507-011	190	C10	H 6	O 4			
3509-018	191	C 8	H 17	N 1	O 4		
3509-012	191	C 8	H 17	N 1	O 4		
3510-014	192	C 8	H 16	O 5			
3905-009	192	C11	H 12	O 3			
3905-012	192	C11	H 12	O 5			
0507-010	192	C10	H 8	O 4			
0507-005	192	C10	H 8	O 4			
0201-010	192	C12	H 16	O 2			
0304-014	192	C11	H 12	O 3			
0301-025	192	C11	H 12	O 3			
0301-024	192	C11	H 12	O 3			
0301-029	192	C11	H 12	O 3			
0301-030	192	C11	H 12	O 3			
2503-043	192	C11	H 12	O 3			
2503-044	192	C11	H 12	O 3			
3005-044	192	C10	H 8	O 4			
3110-071	192	C10	H 8	O 4			
3110-014	192	C12	H 16	O 2			
3196-018	192	C 6	H 8	O 7			
3302-005	192	C10	H 8	O 2	S 1		
3196-017	192	C 6	H 8	O 7			
3504-062	192	C 6	H 8	O 7			
3302-008	192	C10	H 8	O 2	S 1		
3196-016	192	C 6	H 8	O 7			
1902-016	192	C 7	H 12	O 6			
2104-017	192	C11	H 12	O 3			
2106-021	192	C12	H 16	O 2			
1902-005	192	C 7	H 12	O 6			
3509-009	193	C 7	H 15	N 1	O 5		
3302-006	194	C10	H 10	O 2	S 1		
3501-019	194	C 7	H 14	O 6			
3501-025	194	C 7	H 14	O 6			
3501-020	194	C 7	H 14	O 6			
3302-003	194	C10	H 10	O 2	S 1		
3504-015	194	C 7	H 14	O 6			
3504-063	194	C 6	H 10	O 7			
3504-002	194	C 7	H 14	O 6			
3501-010	194	C 7	H 14	O 6			
3504-064	194	C 6	H 10	O 7			
3504-060	194	C 6	H 10	O 7			
3504-061	194	C 6	H 10	O 7			
3501-029	194	C 7	H 14	O 6			
3501-021	194	C 7	H 14	O 6			
3505-005	194	C 7	H 14	O 6			
3501-013	194	C 7	H 14	O 6			
3501-028	194	C 7	H 14	O 6			
3501-014	194	C 7	H 14	O 6			
3505-004	194	C 7	H 14	O 6			
3501-009	194	C 7	H 14	O 6			
2102-014	194	C10	H 10	O 4			
2106-019	194	C12	H 18	O 2			
2106-020	194	C12	H 18	O 2			
2201-017	194	C11	H 14	O 3			
2102-002	194	C10	H 10	O 4			
2102-012	194	C10	H 10	O 4			
2102-010	194	C10	H 10	O 4			
2106-025	194	C12	H 18	O 2			
2104-012	194	C10	H 10	O 4			
2104-011	194	C10	H 10	O 4			
2104-013	194	C10	H 10	O 4			
3110-070	194	C10	H 10	O 4			
3113-022	194	C13	H 22	O 1			
3108-015	194	C10	H 10	O 4			
2504-002	194	C10	H 10	O 4			
0304-021	194	C10	H 10	O 4			
0304-019	194	C10	H 10	O 4			
0304-029	194	C10	H 10	O 4			
0401-005	194	C11	H 14	O 3			
0304-032	194	C10	H 10	O 4			
0301-021	194	C11	H 14	O 3			
0399-004	196	C10	H 12	O 4			
0301-026	196	C11	H 16	O 3			
0103-011	196	C10	H 12	O 4			
0103-016	196	C10	H 12	O 4			
2804-002	196	C14	H 12	O 1			
3113-021	196	C13	H 24	O 1			
2901-011	196	C17	H 20	O 2			
3002-013	196	C 6	H 4	O 6			
3115-005	196	C15	H 16				
2101-002	196	C10	H 12	O 4			
2101-010	196	C 9	H 8	O 5			
1901-014	196	C11	H 16	O 3			
2101-011	196	C10	H 12	O 4			
2107-005	196	C19	H 28	O 3			
2201-011	196	C10	H 12	O 4			
2101-009	196	C10	H 12	O 4			
2003-006	196	C10	H 12	O 4			
2201-016	196	C11	H 16	O 3			
2105-006	196	C11	H 16	O 3			
2102-001	196	C 9	H 8	O 5			
2102-006	196	C 9	H 8	O 5			
2101-005	196	C10	H 12	O 4			
2199-001	196	C10	H 12	O 4			
3505-008	196	C 7	H 16	O 6			
3504-059	196	C 6	H 12	O 7			
3203-014	196	C12	H 20	O 2			
3302-022	196	C13	H 8	S 1			
3201-017	196	C12	H 20	O 2			
3197-014	196	C13	H 24	O 1			
3302-029	198	C13	H 10	S 1			
3201-015	198	C13	H 10	O 2			
3201-014	198	C13	H 10	O 2			
3205-006	198	C13	H 10	O 2			
3302-032	198	C13	H 10	S 1			
3204-010	198	C13	H 10	O 2			
3204-009	198	C13	H 10	O 2			
3201-020	198	C12	H 22	O 2			
3302-021	198	C13	H 10	S 1			
2104-002	198	C10	H 14	O 4			
2102-007	198	C 8	H 6	O 6			
2201-015	198	C 8	H 6	O 6			
3112-017	198	C12	H 22	O 2			
3112-016	198	C12	H 22	O 2			
3114-022	198	C14	H 14	O 1			
3112-013	198	C12	H 22	O 2			
3001-004	198	C 8	H 6	O 6			
3113-029	198	C13	H 26	O 1			
3108-014	198	C10	H 14	O 2	S 1		
3114-005	198	C14	H 14	O 1			
2802-001	198	C13	H 10	O 2			
3113-028	198	C13	H 10	O 2			
3115-003	198	C15	H 18				
0404-008	198	C13	H 10	O 2			
0104-019	198	C 9	H 10	O 5			
0203-015	200	C13	H 12	O 2			
0404-007	200	C13	H 12	O 2			
3113-018	200	C13	H 12	O 2			
3113-017	200	C13	H 12	O 2			
3115-002	200	C15	H 20				
3112-014	200	C12	H 24	O 2			
3114-003	200	C14	H 16	O 1			

ID	MW	C	H	□	Extra
3002-015	200	C 7	H 4	□ 7	
3110-073	200	C10	H 16	□ 4	
3205-005	200	C13	H 12	□ 2	
3302-013	200	C12	H 8	□ 1	S 1
3204-005	200	C13	H 12	□ 2	
3204-004	200	C13	H 12	□ 2	
3201-016	202	C13	H 14	□ 2	
3114-021	202	C14	H 18	□ 1	
3114-007	202	C14	H 18	□ 1	
3110-072	202	C10	H 18	□ 4	
3112-008	202	C12	H 26	□ 2	
0203-014	202	C13	H 14	□ 2	
0203-012	202	C13	H 14	□ 2	
0301-004	202	C14	H 18	□ 1	
0203-016	202	C13	H 14	□ 2	
0508-022	202	C11	H 6	□ 4	
2202-027	202	C13	H 14	□ 2	
2202-015	202	C13	H 14	□ 2	
1901-006	202	C 8	H 10	□ 6	
2199-014	202	C13	H 14	□ 2	
2502-027	202	C12	H 10	□ 3	
2503-047	202	C12	H 10	□ 3	
2503-022	202	C12	H 10	□ 3	
2502-028	202	C12	H 10	□ 3	
2504-009	204	C11	H 8	□ 4	
2504-023	204	C11	H 8	□ 4	
1901-007	204	C 8	H 12	□ 6	
2199-015	204	C13	H 16	□ 2	
0203-010	204	C13	H 16	□ 2	
0501-036	204	C11	H 8	□ 4	
3002-017	204	C 7	H 8	□ 7	
3115-004	204	C15	H 24		
3112-019	204	C12	H 12	□ 3	
3113-030	204	C13	H 16	□ 2	
3510-007	204	C 9	H 16	□ 5	
3510-009	205	C 8	H 15	N 1 □ 5	
3305-003	206	C 8	H 14	□ 2	S 2
3203-002	206	C11	H 10	□ 2	S 1
3001-007	206	C11	H 10	□ 4	
3110-061	206	C11	H 10	□ 2	S 1
0507-006	206	C11	H 10	□ 4	
0508-001	206	C11	H 10	□ 4	
0202-004	206	C13	H 18	□ 2	
0202-003	206	C13	H 18	□ 2	
0301-019	206	C12	H 14	□ 3	
2104-010	206	C11	H 10	□ 4	
2203-006	206	C11	H 10	□ 4	
2203-004	206	C11	H 10	□ 4	
2203-005	206	C11	H 10	□ 4	
2505-001	206	C10	H 6	□ 5	
2505-007	206	C10	H 6	□ 5	
2601-001	208	C14	H 8	□ 2	
2104-021	208	C10	H 8	□ 5	
2002-003	208	C12	H 16	□ 3	
2206-007	208	C12	H 16	□ 3	
2104-006	208	C11	H 12	□ 4	
1901-012	208	C11	H 12	□ 4	
0301-023	208	C12	H 16	□ 3	
0301-022	208	C12	H 16	□ 3	
0399-006	208	C11	H 12	□ 4	
0301-031	208	C12	H 16	□ 3	
0509-001	208	C10	H 8	□ 5	
0301-033	208	C12	H 16	□ 3	
0301-032	208	C12	H 16	□ 3	
0304-023	208	C11	H 12	□ 4	
0303-008	208	C11	H 12	□ 4	
2903-007	208	C12	H 16	□ 3	
3001-003	208	C 9	H 4	□ 6	
3110-058	208	C11	H 12	□ 2	S 1
3005-039	208	C11	H 12	□ 4	
2901-018	208	C12	H 16	□ 3	
3116-002	208	C16	H 16		
2999-003	208	C12	H 16	□ 3	
3115-007	208	C15	H 28		
3115-006	208	C15	H 28		
3501-023	208	C 8	H 16	□ 6	
3501-022	208	C 8	H 16	□ 6	
3197-018	208	C15	H 28		
3197-017	208	C15	H 28		
3302-007	208	C11	H 12	□ 2	S 1
3301-011	210	C 6	H 10		S 4
3504-066	210	C 6	H 10	□ 8	
3505-001	210	C 7	H 14	□ 7	
3505-002	210	C 7	H 14	□ 7	
3505-003	210	C 7	H 14	□ 7	
3504-065	210	C 6	H 10	□ 8	
3115-019	210	C15	H 14	□ 1	
3005-048	210	C11	H 14	□ 4	
3112-009	210	C12	H 18	□ 3	
3001-002	210	C10	H 10	□ 5	
3005-033	210	C11	H 14	□ 4	
2804-003	210	C15	H 14	□ 1	
0302-008	210	C11	H 14	□ 4	
2201-012	210	C11	H 14	□ 9	
2106-026	210	C12	H 18	□ 3	
2106-027	210	C12	H 18	□ 3	
2201-013	210	C11	H 14	□ 4	
1901-009	210	C11	H 14	□ 4	
2103-001	210	C10	H 10	□ 5	
1901-010	210	C11	H 14	□ 4	
2105-001	210	C10	H 10	□ 5	
1901-011	210	C11	H 14	□ 4	
2501-005	210	C15	H 14	□ 1	
1001-001	210	C15	H 14	□ 1	
2501-004	212	C15	H 16	□ 1	
2501-002	212	C15	H 16	□ 1	
2401-002	212	C13	H 8	□ 3	
2401-001	212	C13	H 8	□ 3	
1901-013	212	C11	H 16	□ 4	
0199-012	212	C14	H 12	□ 2	
0104-020	212	C10	H 12	□ 5	
0506-010	212	C11	H 16	□ 4	
2804-004	212	C14	H 12	□ 2	
2801-004	212	C13	H 8	□ 3	
3003-011	212	C10	H 12	□ 5	
3115-023	212	C15	H 16	□ 1	
3115-001	212	C15	H 32		
3115-021	212	C15	H 16	□ 1	
3005-049	212	C11	H 16	□ 4	
3115-020	212	C15	H 16	□ 1	
3114-020	212	C14	H 28	□ 1	
3505-006	212	C 7	H 16	□ 7	
3505-007	212	C 7	H 16	□ 7	
3203-003	214	C14	H 14	□ 2	
3203-012	214	C11	H 18	□ 4	
3204-023	214	C14	H 14	□ 2	
3204-024	214	C14	H 14	□ 2	
3197-015	214	C13	H 26	□ 2	
3115-014	214	C15	H 18	□ 1	
3115-013	214	C15	H 18	□ 1	
3005-008	214	C10	H 14	□ 5	
3115-024	214	C15	H 18	□ 1	
3115-012	214	C15	H 18	□ 1	
3114-019	214	C14	H 14	□ 2	
3114-001	214	C14	H 30	□ 1	
3003-006	214	C11	H 18	□ 4	
3005-038	214	C 9	H 10	□ 6	
3005-011	214	C 9	H 10	□ 6	
3114-010	214	C14	H 14	□ 2	
3113-038	214	C14	H 14		S 1
3113-026	214	C13	H 26	□ 2	
3115-018	214	C15	H 18	□ 1	
0302-010	214	C10	H 14	□ 5	
2501-003	214	C15	H 18	□ 1	
2502-036	214	C14	H 14	□ 2	
2502-029	216	C13	H 12	□ 3	
2502-030	216	C13	H 12	□ 3	
2502-033	216	C12	H 8	□ 4	
0402-010	216	C12	H 8	□ 4	
0507-020	216	C12	H 8	□ 4	
0508-036	216	C12	H 8	□ 4	
0508-023	216	C21	H 8	□ 4	
3114-017	216	C14	H 16	□ 2	
3115-011	216	C15	H 20	□ 1	
3115-010	216	C15	H 20	□ 1	
3115-009	216	C15	H 20	□ 1	
3112-020	216	C12	H 24	□ 3	
3114-013	216	C14	H 16	□ 2	
3204-008	216	C13	H 12	□ 3	
3203-006	216	C13	H 9	□ 1	CL1
3302-020	216	C13	H 12	□ 1	S 1
3203-008	216	C14	H 16	□ 2	
3203-005	216	C13	H 9	□ 1	CL1
3302-019	216	C13	H 12	□ 1	S 1
3303-002	216	C12	H 8		S 2
3202-007	216	C13	H 9	□ 1	CL1
3202-008	216	C13	H 9	□ 1	CL1
3203-016	216	C14	H 16	□ 2	
2202-018	216	C13	H 12	□ 3	
2202-019	218	C13	H 14	□ 3	
2202-020	218	C13	H 14	□ 3	
2202-020	218	C13	H 14	□ 3	
3203-017	218	C14	H 18	□ 2	
3510-013	218	C 9	H 14	□ 6	
3302-015	218	C13	H 14	□ 1	S 1
3905-011	218	C13	H 14	□ 3	
3903-006	218	C 9	H 14	□ 6	
3114-033	218	C14	H 18	□ 2	
3110-075	218	C12	H 10	□ 4	
3115-022	218	C15	H 22	□ 1	
3113-013	218	C13	H 11	□ 1	CL1
2801-003	218	C12	H 10	□ 4	

ID	MW	C	H	N	O	Other
0402-004	218	C12	H 10		□ 4	
0402-003	218	C12	H 10		□ 4	
0402-002	218	C12	H 10		□ 4	
0301-005	218	C14	H 18		□ 2	
0203-017	218	C13	H 14		□ 3	
0203-013	218	C14	H 18		□ 2	
2504-001	218	C12	H 10		□ 4	
2503-017	218	C13	H 14		□ 3	
2504-008	218	C12	H 10		□ 4	
2504-010	218	C12	H 10		□ 4	
2504-024	218	C12	H 10		□ 4	
2504-012	218	C12	H 10		□ 4	
2504-003	218	C12	H 10		□ 4	
2505-008	220	C11	H 8		□ 5	
0302-012	220	C14	H 20		□ 2	
3110-049	220	C12	H 12		□ 2	S 1
3110-048	220	C12	H 12		□ 2	S 1
3110-076	220	C12	H 12		□ 4	
3110-047	220	C12	H 12		□ 2	S 1
3110-050	220	C12	H 12		□ 2	S 1
3001-008	220	C11	H 8		□ 5	
3510-003	220	C10	H 20		□ 5	
3905-014	220	C13	H 16		□ 3	
3203-018	220	C14	H 20		□ 2	
2104-032	220	C11	H 8		□ 5	
2203-007	220	C12	H 12		□ 4	
3905-018	220	C13	H 16		□ 3	
3110-060	221	C14	H 23	N 1	□ 1	
3113-012	222	C15	H 10		□ 2	
3115-017	222	C15	H 26		□ 1	
3110-059	222	C12	H 14		□ 2	S 1
3117-009	222	C17	H 18			
3117-008	222	C17	H 18			
3905-022	222	C11	H 10		□ 5	
3905-029	222	C12	H 14		□ 4	
2203-014	222	C12	H 14		□ 4	
2104-004	222	C12	H 14		□ 4	
2104-031	222	C11	H 10		□ 5	
2102-024	222	C11	H 10		□ 5	
2104-026	222	C11	H 10		□ 5	
2104-022	222	C11	H 10		□ 5	
3504-009	222	C 8	H 14		□ 7	
0301-035	222	C12	H 14		□ 4	
0509-007	222	C11	H 8		□ 5	
0301-037	222	C12	H 14		□ 4	
0301-036	222	C12	H 14		□ 4	
0402-005	222	C12	H 14		□ 4	
0509-005	222	C11	H 10		□ 5	
0509-003	222	C11	H 10		□ 5	
0510-001	222	C11	H 10		□ 5	
2506-001	222	C10	H 6		□ 6	
2506-007	222	C10	H 6		□ 6	
2601-002	222	C15	H 10		□ 2	
2503-048	222	C11	H 7		□ 3	CL1
1101-001	222	C15	H 10		□ 2	
3110-051	223	C14	H 25	N 1	□ 1	
3110-046	223	C14	H 25	N 1	□ 1	
3112-023	223	C12	H 17	N 1	□ 3	
3001-005	224	C 9	H 4		□ 7	
3113-016	224	C15	H 12		□ 2	
3001-006	224	C11	H 12		□ 5	
3114-030	224	C14	H 24		□ 2	
3114-029	224	C14	H 24		□ 2	
3116-012	224	C16	H 16		□ 1	
3117-015	224	C17	H 20			
2901-012	224	C14	H 24		□ 2	
3117-014	224	C17	H 20			
2602-001	224	C14	H 8		□ 3	
2602-003	224	C14	H 8		□ 3	
0304-030	224	C11	H 12		□ 5	
3905-017	224	C12	H 16		□ 4	
3905-016	224	C12	H 16		□ 4	
3205-017	224	C15	H 12		□ 2	
1902-018	224	C10	H 8		□ 6	
2106-001	224	C12	H 16		□ 4	
2105-007	224	C12	H 16		□ 4	
2102-022	224	C11	H 12		□ 5	
2105-003	224	C10	H 8		□ 4	
2104-019	224	C11	H 12		□ 5	
2203-003	224	C12	H 16		□ 4	
2201-014	226	C11	H 14		□ 5	
1902-019	226	C10	H 10		□ 6	
2105-002	226	C10	H 10		□ 6	
2102-006	226	C10	H 10		□ 6	
2102-005	226	C10	H 10		□ 6	
3901-010	226	C15	H 30		□ 1	
3205-004	226	C14	H 10		□ 3	
0399-005	226	C11	H 14		□ 5	
0506-011	226	C12	H 8		□ 4	
2503-013	226	C14	H 10		□ 3	
2503-014	226	C14	H 10		□ 3	
2502-006	226	C15	H 14		□ 2	
3112-005	226	C14	H 26		□ 2.	
3117-012	226	C17	H 22			
3116-013	226	C16	H 18		□ 1	
3112-003	226	C14	H 26		□ 2	
3113-037	226	C14	H 10		□ 1	S 1
3117-007	226	C17	H 22			
3117-011	226	C17	H 22			
3116-001	226	C16	H 34			
3114-027	226	C14	H 26		□ 2	
3114-026	226	C14	H 26		□ 2	
3112-004	226	C14	H 26		□ 2	
3117-016	226	C17	H 22			
2804-005	226	C15	H 14		□ 2	
3114-032	226	C15	H 14		□ 2	
3114-024	226	C14	H 26		□ 2	
3114-025	226	C14	H 26		□ 2	
3117-006	226	C17	H 22			
3117-010	226	C17	H 22			
0104-037	227	C 9	H 9	N 1	□ 6	
0404-010	228	C14	H 12		□ 3	
0505-001	228	C14	H 12		□ 3	
0504-001	228	C14	H 12		□ 3	
0503-001	228	C14	H 12		□ 3	
0502-018	228	C14	H 12		□ 3	
3117-004	228	C17	H 24			
3116-041	228	C16	H 20		□ 1	
3117-005	228	C17	H 24			
3112-021	228	C12	H 20		□ 4	
3005-019	228	C10	H 12		□ 6	
3005-020	228	C10	H 12		□ 6	
2804-007	228	C15	H 16		□ 2	
3113-027	228	C14	H 28		□ 2	
3114-023	228	C14	H 28		□ 2	
2804-012	228	C14	H 12		□ 3	
3117-013	228	C17	H 24			
2402-003	228	C13	H 8		□ 4	
2502-008	228	C15	H 16		□ 2	
2402-011	228	C13	H 8		□ 4	
3197-016	228	C14	H 28		□ 2	
3205-002	228	C14	H 12		⊐ 3	
2102-008	228	C 9	H 8		□ 7	
2102-009	228	C 9	H 8		□ 7	
3906-005	228	C13	H 8		□ 4	
3111-021	229	C15	H 19	N 1	□ 1	
3110-025	230	C15	H 18		□ 2	
2801-001	230	C14	H 14		□ 3	
3116-010	230	C16	H 22		□ 1	
2903-010	230	C10	H 11		□ 4	CL1
3110-006	230	C15	H 18		□ 2	
2903-009	230	C10	H 11		□ 4	CL1
3110-018	230	C15	H 18		□ 2	
2204-032	230	C13	H 10		□ 4	
3204-014	230	C13	H 10		□ 4	
2503-042	230	C13	H 10		□ 4	
0501-028	230	C14	H 14		□ 3	
0501-004	230	C14	H 14		□ 3	
0501-019	230	C14	H 14		□ 3	
0404-009	230	C14	H 14		□ 3	
0404-011	230	C14	H 14		□ 3	
0404-005	230	C14	H 14		□ 3	
0404-016	230	C13	H 10		□ 4	
0404-006	232	C14	H 16		□ 3	
0404-012	232	C14	H 16		□ 3	
0203-019	232	C14	H 16		□ 3	
0203-018	232	C14	H 16		□ 3	
2504-003	232	C13	H 12		□ 4	
2503-018	232	C14	H 16		□ 3	
2503-023	232	C13	H 12		□ 4	
2503-019	232	C13	H 12		□ 4	
3305-004	232	C11	H 20		□ 1	S 2
3197-011	232	C10	H 16		□ 6	
3204-022	232	C13	H 12		□ 2	S 1
3203-002	232	C13	H 9		□ 2	CL1
3905-030	232	C15	H 20		□ 2	
3202-006	232	C13	H 12		□ 2	S 1
2205-001	232	C14	H 16		□ 3	
2202-026	232	C14	H 16		□ 3	
2202-010	232	C14	H 16		□ 3	
3110-013	232	C15	H 20		□ 2	
3113-020	232	C13	H 12		□ 4	
3005-021	232	C 9	H 12		□ 5	
3004-017	232	C10	H 16		□ 6	
3906-013	232	C13	H 12		□ 4	
3116-014	234	C16	H 26		□ 1	
2203-003	234	C13	H 14		□ 4	
2204-040	234	C13	H 14		□ 4	
2104-033	234	C12	H 10		□ 5	
2104-005	234	C13	H 14		□ 4	
3302-016	234	C13	H 14		□ 2	S 1
3204-018	234	C13	H 14		□ 2	S 1

Reg. No.	MW	Formula		Reg. No.	MW	Formula
3302-018	234	C13 H 14 O 2 S 1		3116-004	242	C15 H 34 O 1
3204-019	234	C13 H 14 O 2 S 1		2104-007	242	C11 H 11 O 4 CL1
3905-028	234	C13 H 14 O 4		1405-001	242	C15 H 14 O 3
3302-011	234	C12 H 7 O 1 S 1 CL1		2201-003	242	C15 H 14 O 3
3303-003	234	C12 H 10 O 1 S 2		2104-008	242	C11 H 11 O 4 CL1
3197-012	234	C10 H 18 O 6		3204-003	242	C14 H 10 O 4
2505-009	234	C12 H 10 O 5		3301-016	242	C13 H 22 O 2 S 1
2505-002	234	C12 H 10 O 5		3906-001	242	C14 H 10 O 4
0203-011	234	C14 H 18 O 3		3197-013	242	C12 H 18 O 5
0399-017	234	C14 H 18 O 3		3205-001	242	C15 H 14 O 3
0509-006	236	C21 H 12 O 5		2503-011	242	C15 H 14 O 3
2601-005	236	C15 H 8 O 3		2503-009	242	C15 H 14 O 3
2604-001	236	C14 H 8 O 5		2402-001	242	C14 H 10 O 4
3504-010	236	C10 H 20 O 6		2503-016	242	C15 H 14 O 3
3302-012	236	C12 H 9 O 1 S 1 CL1		2503-015	242	C14 H 10 O 4
3197-003	236	C 8 H 12 O 8		2402-015	242	C14 H 10 O 4
2104-023	236	C12 H 12 O 5		2402-004	242	C14 H 10 O 4
3116-009	236	C16 H 30 O 1		2503-006	242	C15 H 14 O 3
3117-002	238	C17 H 34		2502-009	242	C16 H 18 O 2
3117-066	238	C17 H 18 O 1		2502-037	242	C15 H 14 O 3
3117-003	238	C17 H 34		0404-022	242	C14 H 10 O 4
3114-031	238	C15 H 26 O 2		0502-009	243	C14 H 13 N 1 O 3
3116-003	238	C17 H 34		3109-029	243	C14 H 13 N 1 O 3
3117-077	238	C17 H 18 O 1		3112-026	243	C16 H 21 N 1 O 1
3113-014	238	C15 H 26 O 2		2804-018	244	C14 H 12 O 4
2206-002	238	C13 H 18 O 4		3117-027	244	C17 H 24 O 1
2206-006	238	C13 H 18 O 4		3117-028	244	C17 H 24 O 1
2206-005	238	C13 H 18 O 4		3117-037	244	C17 H 24 O 1
2104-020	238	C12 H 14 O 5		3117-025	244	C17 H 24 O 1
2102-025	238	C11 H 10 O 6		3117-064	244	C17 H 24 O 1
2103-002	238	C12 H 14 O 5		3117-023	244	C17 H 24 O 1
3901-011	238	C16 H 30 O 1		3117-026	244	C17 H 24 O 1
3197-020	238	C17 H 34		3112-024	244	C12 H 20 O 5
3201-009	238	C11 H 10 O 4 S 1		3114-037	244	C14 H 28 O 3
2602-008	238	C15 H 10 O 3		3117-058	244	C17 H 24 O 1
2602-002	238	C15 H 10 O 4		2802-002	244	C14 H 12 O 4
2602-010	238	C 5 H 10 O 3		2804-022	244	C14 H 12 O 4
2602-005	238	C15 H 10 O 3		3117-040	244	C17 H 24 O 1
2603-023	238	C15 H 10 O 3		3117-042	244	C17 H 24 O 1
2507-001	238	C10 H 6 O 7		3114-044	244	C14 H 12 O 4
2602-004	238	C15 H 10 O 3		3117-029	244	C17 H 24 O 1
2601-003	238	C15 H 10 O 3		3114-038	244	C14 H 28 O 3
0399-013	238	C16 H 14 O 2		0501-030	244	C15 H 16 O 3
0301-034	238	C13 H 18 O 4		0501-018	244	C15 H 16 O 3
1101-002	238	C15 H 10 O 3		0501-022	244	C15 H 16 O 3
1001-002	240	C16 H 16 O 2		0503-008	244	C14 H 12 O 4
2502-010	240	C16 H 16 O 2		0504-003	244	C14 H 12 O 4
2502-007	240	C16 H 16 O 2		0504-002	244	C14 H 12 O 4
2699-001	240	C15 H 12 O 3		0507-017	244	C14 H 12 O 4
2503-012	240	C15 H 12 O 3		0404-021	244	C14 H 12 O 4
2603-007	240	C14 H 8 O 4		2403-028	244	C13 H 8 O 5
2603-001	240	C14 H 8 O 4		2403-018	244	C13 H 8 O 5
2402-002	240	C14 H 8 O 4		2502-035	244	C14 H 12 O 4
2503-010	240	C15 H 12 O 3		3906-003	244	C14 H 12 O 4
3205-018	240	C15 H 12 O 3		3906-004	244	C13 H 8 O 5
2103-003	240	C11 H 12 O 6		3301-015	244	C13 H 24 O 2 S 1
1302-001	240	C16 H 16 O 2		3906-004	246	C14 H 14 O 4
2203-011	240	C11 H 9 O 4 CL1		3201-005	246	C17 H 26 O 1
2102-023	240	C11 H 12 O 6		3302-026	246	C13 H 7 O 1 S 1 CL1
3117-062	240	C17 H 20 O 1		3201-003	246	C17 H 26 O 1
3117-061	240	C17 H 20 O 1		3201-004	246	C17 H 26 O 1
3117-078	240	C17 H 20 O 1		2503-010	246	C14 H 14 O 4
3117-038	240	C17 H 20 O 1		2503-025	246	C14 H 14 O 4
2805-003	240	C17 H 20 O 1		2503-036	246	C14 H 14 O 4
3117-063	240	C17 H 20 O 1		2503-026	246	C13 H 10 O 5
3114-006	240	C16 H 16 O 2		0404-017	246	C13 H 10 O 5
3116-006	240	C16 H 32 O 1		0402-011	246	C13 H 10 O 5
3117-041	240	C17 H 20 O 1		0303-006	246	C15 H 18 O 3
3116-008	240	C16 H 30 O 1		0511-001	246	C13 H 10 O 5
3117-065	240	C17 H 20 O 1		0507-016	246	C14 H 14 O 4
3117-001	240	C17 H 36		0507-004	246	C14 H 14 O 4
3117-068	240	C17 H 20 O 1		0510-004	246	C13 H 10 O 5
2804-006	240	C16 H 16 O 2		0599-008	246	C13 H 10 O 5
3117-044	242	C17 H 22 O 1		0503-004	246	C14 H 14 O 4
3117-031	242	C17 H 22 O 1		0503-002	246	C14 H 14 O 4
3117-030	242	C17 H 22 O 1		0505-002	246	C14 H 14 O 4
3115-008	242	C17 H 34 O 1		2802-005	246	C13 H 10 O 5
2805-001	242	C10 H 18 O 2		3114-045	246	C14 H 14 O 4
3117-043	242	C17 H 22 O 1		3117-022	246	C17 H 26 O 1
3117-033	242	C17 H 22 O 1		3117-021	246	C17 H 26 O 1
3117-032	242	C17 H 22 O 1		3117-076	246	C17 H 26 O 1
3117-067	242	C17 H 22 O 1		2202-023	246	C14 H 14 O 4
3117-035	242	C17 H 22 O 1		2204-033	246	C13 H 10 O 5
3117-036	242	C17 H 22 O 1		2204-035	246	C13 H 10 O 5
3117-034	242	C17 H 22 O 1		3112-027	247	C16 H 25 N 1 O 1
3114-002	242	C16 H 18 O 2		2202-017	248	C14 H 16 O 4
3113-034	242	C14 H 10 O 2 S 1		2202-005	248	C14 H 16 O 4
3115-025	242	C15 H 30 O 2		2202-011	248	C14 H 16 O 4
3114-009	242	C16 H 18 O 2		2202-003	248	C14 H 16 O 4
3117-060	242	C17 H 22 O 1		2202-024	248	C14 H 16 O 4
3117-059	242	C17 H 22 O 1		2202-007	248	C14 H 16 O 4

ID	MW	Formula
2107-006	248	C14 H 16 O14
0404-013	248	C14 H 16 O 4
2503-001	248	C13 H 12 O 5
2505-010	248	C12 H 8 O 6
2505-012	248	C12 H 8 O 6
2505-004	248	C12 H 8 O 6
3303-001	248	C12 H 8 O 2 S 2
3201-002	248	C17 H 28 O 1
3302-027	248	C13 H 9 O 1 S 1 CL1
3304-001	248	C12 H 8 S 3
3204-030	248	C14 H 16 O 2 S 1
3302-009	249	C14 H 19 N 1 O 1 S 1
3112-025	249	C16 H 27 N 1 O 1
3116-022	250	C16 H 26 O 2
3901-012	250	C17 H 30 O 1
3201-019	250	C12 H 10 O 6
3204-020	250	C13 H 14 O 3 S 1
3204-021	250	C13 H 14 O 3 S 1
2506-009	250	C12 H 10 O 6
2506-008	250	C12 H 10 O 6
2506-005	250	C12 H 10 O 6
2506-002	250	C12 H 10 O 6
0801-002	250	C 6 H 4 O 1 BR2
0302-007	250	C15 H 22 O 3
2104-024	250	C13 H 14 O 5
2003-017	250	C14 H 18 O 4
2106-017	250	C13 H 14 O 5
2202-002	250	C14 H 18 O 4
2202-001	250	C14 H 18 O 4
3112-024	251	C16 H 29 N 1 O 1
2901-013	252	C16 H 28 O 2
3114-008	252	C16 H 28 O 2
2204-041	252	C13 H 16 O 5
2101-025	252	C22 H 32 O 2
2203-001	252	C13 H 16 O 5
1399-002	252	C16 H 12 O 3
2206-004	252	C14 H 20 O 4
2206-003	252	C14 H 20 O 4
2107-007	252	C14 H 20 O 4
0699-012	252	C17 H 16 O 2
2602-006	252	C16 H 12 O 3
2602-011	252	C16 H 12 O 3
2601-006	252	C15 H 8 O 4
2602-009	252	C16 H 12 O 3
3205-010	252	C17 H 16 O 2
3905-024	254	C15 H 10 O 4
3504-005	254	C 9 H 18 O 8
3197-021	254	C17 H 34 O 1
3504-006	254	C 9 H 18 O 8
3197-025	254	C18 H 48
3197-024	254	C18 H 38
3197-023	254	C18 H 38
2603-005	254	C15 H 10 O 4
2603-002	254	C15 H 10 O 4
2603-003	254	C15 H 10 O 4
2603-027	254	C15 H 10 O 4
2603-026	254	C15 H 10 O 4
2602-007	254	C15 H 10 O 4
2603-011	254	C15 H 10 O 4
2701-026	254	C15 H 10 O 4
2508-001	254	C10 H 6 O 8
2603-021	254	C15 H 10 O 4
2603-008	254	C15 H 10 O 4
2603-009	254	C15 H 10 O 4
2603-034	254	C15 H 10 O 4
2603-024	254	C15 H 10 O 4
2603-025	254	C15 H 10 O 4
0502-017	254	C16 H 14 O 3
2107-008	254	C14 H 22 O 4
1401-023	254	C15 H 10 O 4
3116-020	254	C16 H 30 O 2
3116-021	254	C16 H 30 O 2
3116-019	254	C16 H 30 O 2
3116-018	254	C16 H 30 O 2
3114-028	254	C16 H 30 O 2
3117-019	254	C18 H 38
3103-004	254	C 9 H 18 O 8
3117-018	254	C18 H 38
3114-004	254	C16 H 30 O 2
3116-039	254	C17 H 34 O 1
3117-017	254	C18 H 38
3115-029	254	C15 H 26 O 3
3103-024	254	C 9 H 18 O 8
1099-013	254	C16 H 14 O 3
0999-001	254	C16 H 14 O 3
1101-004	254	C15 H 10 O 4
0999-016	254	C15 H 10 O 4
0901-007	254	C16 H 14 O 3
0901-006	254	C16 H 14 O 3
1101-016	254	C15 H 10 O 4
3116-016	255	C16 H 33 N 1 O 1
1701-001	255	C15 H 11 O 4 +
1301-002	256	C16 H 16 O 3
2105-004	256	C11 H 12 O 7
1302-002	256	C16 H 16 O 3
2204-012	256	C15 H 12 O 4
2903-013	256	C15 H 12 O 4
3114-011	256	C16 H 16 O 3
3005-034	256	C12 H 16 O 6
2804-008	256	C15 H 12 O 4
3116-015	256	C16 H 32 O 2
2804-015	256	C16 H 16 O 3
2804-010	256	C15 H 12 O 4
3115-030	256	C15 H 28 O 3
3117-073	256	C17 H 20 O 2
2805-004	256	C17 H 20 O 2
1002-001	256	C15 H 12 O 4
1201-001	256	C15 H 12 O 4
1203-024	256	C15 H 12 O 4
0999-003	256	C16 H 16 O 3
0901-001	256	C16 H 16 O 3
0502-003	256	C14 H 14 O 4
0502-001	256	C14 H 14 O 4
0302-011	256	C12 H 16 O 3
2504-014	256	C15 H 12 O 4
2699-006	256	C15 H 12 O 4
2503-035	256	C15 H 12 O 4
2502-038	256	C16 H 16 O 3
2604-008	256	C14 H 8 O 5
2504-015	256	C14 H 8 O 5
2503-007	256	C16 H 16 O 3
2403-035	256	C14 H 8 O 5
2701-013	256	C16 H 16 O 3
3204-012	256	C15 H 12 O 4
3205-003	256	C15 H 12 O 4
3204-011	256	C15 H 12 O 4
3902-006	256	C11 H 12 O 7
3203-004	256	C16 H 16 O 3
3203-009	258	C16 H 18 O 3
3905-026	258	C15 H 14 O 4
3906-002	258	C14 H 10 O 5
3204-006	258	C15 H 14 O 4
3204-007	258	C15 H 14 O 4
3202-004	258	C15 H 14 O 4
2403-004	258	C14 H 10 O 5
2403-032	258	C14 H 10 O 5
2503-008	258	C15 H 14 O 4
2504-017	258	C14 H 10 O 5
2403-013	258	C14 H 10 O 5
2403-006	258	C14 H 10 O 5
2403-027	258	C14 H 10 O 5
2403-020	258	C14 H 10 O 5
2403-001	258	C14 H 10 O 5
2403-029	258	C14 H 10 O 5
2403-019	258	C14 H 10 O 5
0404-015	258	C15 H 14 O 4
0404-023	258	C15 H 14 O 4
0501-034	258	C15 H 14 O 4
0501-032	258	C15 H 14 O 4
0502-010	258	C15 H 14 O 4
0501-025	258	C15 H 14 O 4
0501-024	258	C15 H 14 O 4
0506-012	258	C15 H 14 O 4
0506-008	258	C15 H 14 O 4
0507-019	258	C15 H 14 O 4
0508-021	258	C15 H 14 O 4
0508-017	258	C15 H 14 O 4
2802-003	258	C15 H 14 O 4
3117-075	258	C17 H 22 O 2
3114-018	258	C16 H 18 O 3
2804-019	258	C15 H 14 O 4
2804-011	258	C16 H 18 O 3
3117-074	258	C17 H 22 O 2
3117-053	258	C17 H 22 O 2
2804-026	258	C15 H 14 O 4
3113-035	258	C14 H 10 O 3 S 1
3114-018	258	C16 H 18 O 3
3117-069	258	C17 H 22 O 2
3114-014	258	C16 H 18 O 3
3115-027	258	C15 H 30 O 3
2801-005	258	C14 H 10 O 5
3115-026	258	C15 H 30 O 3
3117-052	258	C17 H 22 O 2
3115-028	258	C15 H 30 O 3
2804-009	258	C15 H 14 O 4
3116-011	258	C16 H 34 O 2
3117-079	258	C17 H 22 O 2
2204-010	258	C15 H 14 O 4
2204-018	258	C15 H 14 O 4
2204-027	258	C15 H 14 O 4
2204-016	258	C14 H 10 O 5
2001-010	258	C17 H 22 O 2

ID	Mass	C	H	O	Extra
1302-018	258	C20	H 22	O 6	
3504-046	259	C 6	H 13	O 9	P 1
3197-036	260	C15	H 32	O 3	
3504-003	260	C 6	H 13	O 9	P 1
3202-003	260	C15	H 16	O 4	
3504-004	260	C 6	H 13	O 9	P 1
3202-005	260	C15	H 16	O 4	
2204-036	260	C14	H 12	O 5	
2204-002	260	C15	H 16	O 4	
2199-004	260	C17	H 24	O 2	
2204-001	260	C15	H 16	O 4	
3117-072	260	C17	H 24	O 2	
3114-039	260	C14	H 28	O 4	
3117-051	260	C17	H 24	O 2	
3117-046	260	C17	H 24	O 2	
3114-042	260	C14	H 28	O 4	
3117-049	260	C17	H 24	O 2	
3117-048	260	C17	H 24	O 2	
2804-021	260	C14	H 12	O 5	
0508-014	260	C14	H 12	O 5	
0802-016	260	C 7	H 4	O 2	CL4
0507-018	260	C14	H 12	O 5	
0507-012	260	C15	H 16	O 4	
0507-014	260	C15	H 16	O 4	
0507-013	260	C15	H 16	O 4	
0506-004	260	C15	H 16	O 4	
0501-023	260	C15	H 16	O 4	
0501-033	260	C15	H 16	O 4	
0501-035	260	C15	H 16	O 4	
0501-031	260	C15	H 16	O 4	
0599-012	260	C15	H 16	O 4	
2404-034	260	C13	H 8	O 6	
2801-002	260	C15	H 16	O 4	
2504-005	260	C14	H 12	O 5	
2404-022	260	C13	H 8	O 6	
2503-002	260	C14	H 12	O 5	
2404-007	260	C13	H 8	O 6	
2404-029	260	C13	H 8	O 6	
1106-077	260	C18	H 16	O 8	
1101-003	262	C17	H 10	O 3	
2504-006	262	C14	H 14	O 5	
2505-003	262	C14	H 14	O 5	
0404-024	262	C15	H 18	O 4	
0399-019	262	C15	H 18	O 4	
0506-009	262	C14	H 14	O 5	
0505-008	262	C14	H 14	O 5	
0504-008	262	C14	H 14	O 5	
3117-045	262	C17	H 26	O 2	
3117-084	262	C17	H 26	O 2	
2802-007	262	C13	H 10	O 6	
2202-004	262	C15	H 18	O 4	
2109-001	262	C17	H 26	O 2	
2202-012	262	C15	H 18	O 4	
2202-006	262	C15	H 18	O 4	
3502-002	262	C 8	H 18	N 6 O 4	
3302-014	262	C15	H 18	O 2	S 1
3502-031	264	C 8	H 16	N 4 O 6	
2203-008	264	C14	H 16	O 5	
2202-025	264	C14	H 16	O 5	
2903-008	264	C10	H 10	O 4	CL2
2903-006	264	C15	H 20	O 4	
3117-082	264	C17	H 28	O 2	
3005-018	264	C14	H 14	O 5	
0501-020	264	C14	H 16	O 5	
0399-018	264	C15	H 20	O 4	
0399-007	264	C18	H 16	O 2	
2506-006	264	C12	H 8	O 7	
2506-003	264	C12	H 8	O 7	
2506-010	264	C12	H 8	O 7	
3509-002	265	C 8	H 15	N 3 O 7	
2507-002	266	C12	H 10	O 7	
2505-011	266	C12	H 10	O 5	
2699-020	266	C17	H 14	O 3	
0699-003	266	C18	H 18	O 2	
0699-025	266	C18	H 18	O 2	
2901-016	266	C16	H 26	O 2	
3005-047	266	C15	H 22	O 4	
3005-046	266	C15	H 22	O 4	
2903-011	266	C10	H 12	O 4	CL2
1399-003	266	C17	H 14	O 3	
2104-034	266	C15	H 22	O 4	
2206-011	266	C15	H 22	O 4	
2104-025	266	C13	H 14	O 6	
1602-001	268	C15	H 8	O 5	
2206-019	268	C14	H 20	O 5	
1401-026	268	C16	H 12	O 4	
2103-004	268	C13	H 16	O 6	
1401-028	268	C16	H 12	O 4	
3119-001	268	C19	H 40		
3117-081	268	C17	H 32	O 2	
3118-003	268	C18	H 36	O 1	
3118-002	268	C18	H 36	O 1	
3117-020	268	C19	H 40		
3118-008	268	C18	H 36	O 1	
0104-014	268	C14	H 22	N 1 O 4	+
2603-010	268	C16	H 12	O 4	
2603-022	268	C16	H 12	O 4	
2603-012	268	C16	H 12	O 4	
2603-006	268	C16	H 12	O 4	
2508-002	268	C11	H 8	O 8	
3901-008	268	C17	H 32	O 2	
3197-026	268	C19	H 40		
1101-006	268	C16	H 12	O 4	
0902-017	268	C16	H 12	O 4	
1103-001	268	C16	H 12	O 4	
1101-009	268	C16	H 12	O 4	
3114-036	269	C18	H 23	N 1 O 1	
3117-086	270	C18	H 20	O 2	
3116-028	270	C16	H 30	O 3	
3116-029	270	C16	H 30	O 3	
3116-030	270	C16	H 30	O 3	
3118-005	270	C18	H 38	O 1	
3118-009	270	C18	H 22	O 2	
3118-004	270	C18	H 38	O 1	
3118-001	270	C18	H 38	O 1	
2903-014	270	C16	H 14	O 4	
2903-017	270	C17	H 18	O 3	
3117-080	270	C17	H 34	O 2	
1199-030	270	C15	H 10	O 5	
1101-011	270	C15	H 10	O 5	
1102-001	270	C15	H 10	O 5	
1001-022	270	C16	H 14	O 4	
1103-002	270	C15	H 10	O 5	
1201-007	270	C16	H 14	O 4	
1201-006	270	C16	H 14	O 4	
1201-005	270	C16	H 14	O 4	
1201-004	270	C16	H 14	O 4	
1002-003	270	C16	H 14	O 4	
1099-004	270	C16	H 14	O 4	
1001-009	270	C16	H 14	O 4	
1001-007	270	C16	H 14	O 4	
1001-011	270	C16	H 14	O 4	
1001-006	270	C17	H 18	O 3	
1104-001	270	C15	H 10	O 5	
1001-004	270	C17	H 18	O 3	
1001-005	270	C17	H 18	O 3	
2603-018	270	C15	H 8	O 5	
2604-002	270	C15	H 10	O 5	
2604-011	270	C15	H 10	O 5	
2604-003	270	C15	H 10	O 5	
2604-026	270	C15	H 10	O 5	
2604-020	270	C15	H 10	O 5	
2603-015	270	C15	H 10	O 5	
2699-024	270	C16	H 14	O 4	
2604-054	270	C15	H 10	O 5	
2604-055	270	C15	H 10	O 5	
2403-036	270	C15	H 10	O 5	
2503-034	270	C16	H 14	O 4	
2603-029	270	C15	H 10	O 5	
2604-028	270	C15	H 10	O 5	
2699-008	270	C16	H 14	O 4	
0105-002	270	C14	H 16	O 6	
0901-009	270	C14	H 16	O 4	
0508-034	270	C16	H 14	O 4	
0508-024	270	C16	H 14	O 4	
0506-013	270	C16	H 14	O 4	
1301-003	270	C17	H 18	O 3	
1601-001	270	C16	H 14	O 4	
1803-009	270	C15	H 10	O 5	
1601-002	270	C16	H 14	O 4	
1302-003	270	C17	H 18	O 3	
2399-008	270	C15	H 10	O 5	
1401-001	270	C15	H 10	O 5	
1402-010	270	C15	H 16	O 5	
1702-001	271	C15	H 11	O 5	+
1701-003	271	C15	H 11	O 5	+
3114-035	271	C18	H 25	N 1 O 1	
2901-008	271	C12	H 17	N 1 O 6	
3116-027	272	C16	H 32	O 3	
3116-026	272	C16	H 32	O 3	
2806-027	272	C15	H 12	O 5	
3116-025	272	C16	H 32	O 3	
3116-024	272	C16	H 32	O 3	
3116-023	272	C16	H 32	O 3	
2204-017	272	C16	H 16	O 4	
2101-003	272	C12	H 16	O 7	
2403-002	272	C15	H 12	O 5	
2204-031	272	C15	H 12	O 5	
1205-015	272	C15	H 12	O 5	
2403-005	272	C15	H 12	O 5	
1405-002	272	C16	H 16	O 4	
2403-007	272	C15	H 12	O 5	

Code	MW	Formula			
2403-008	272	C15	H 12	O 5	
0802-006	272	C12	H 16	O 7	
2802-005	272	C16	H 16	O 4	
2403-014	272	C15	H 12	O 5	
2699-014	272	C15	H 12	O 5	
2403-022	272	C15	H 12	O 5	
2701-014	272	C16	H 16	O 4	
2503-050	272	C16	H 6	O 4	
2802-004	272	C16	H 16	O 4	
2403-034	272	C15	H 12	O 5	
2504-019	272	C15	H 12	O 5	
2504-018	272	C15	H 12	O 5	
2403-033	272	C15	H 12	O 5	
2503-049	272	C16	H 16	O 4	
1202-001	272	C15	H 12	O 5	
1002-006	272	C15	H 12	O 5	
0999-017	272	C16	H 16	O 4	
1003-001	272	C15	H 12	O 5	
1203-001	272	C15	H 12	O 5	
1203-033	272	C15	H 12	O 5	
3902-007	272	C11	H 12	O 8	
3302-031	272	C15	H 12	O 3	S 1
3204-027	272	C16	H 16	O 4	
3302-025	272	C15	H 12	O 3	S 1
3204-015	272	C15	H 12	O 5	
3204-026	272	C16	H 16	O 4	
3302-030	274	C15	H 14	O 3	S 1
1002-008	274	C15	H 14	O 5	
2505-017	274	C14	H 10	O 6	
2404-023	274	C14	H 10	O 6	
2506-012	274	C14	H 10	O 6	
2404-009	274	C14	H 10	O 6	
2404-042	274	C14	H 10	O 6	
2504-025	274	C16	H 18	O 4	
2505-021	274	C14	H 10	O 6	
2404-030	274	C14	H 10	O 6	
2802-006	274	C15	H 14	O 5	
2404-018	274	C14	H 10	O 6	
2404-001	274	C14	H 10	O 6	
2404-008	274	C14	H 10	O 6	
2804-028	274	C16	H 18	O 4	
0499-006	274	C18	H 26	O 2	
0404-018	274	C15	H 14	O 5	
0802-017	274	C 8	H 6	O 2	CL4
0404-019	274	C15	H 14	O 5	
0599-013	274	C16	H 18	O 4	
0506-006	274	C16	H 18	O 4	
0507-021	274	C16	H 18	O 4	
0508-007	274	C16	H 18	O 4	
0508-006	274	C16	H 18	O 4	
0508-002	274	C16	H 18	O 4	
2005-001	274	C14	H 10	O 6	
1399-004	274	C16	H 12	O 5	
1302-011	274	C15	H 14	O 5	
2202-021	274	C15	H 14	O 4	
1302-019	274	C15	H 14	O 5	
1303-001	274	C15	H 14	O 5	
1303-002	274	C15	H 14	O 5	
2204-028	274	C15	H 14	O 5	
2204-003	274	C16	H 18	O 4	
1302-005	274	C15	H 14	O 5	
1302-012	274	C15	H 14	O 5	
2204-013	274	C15	H 14	O 5	
1302-013	274	C15	H 14	O 5	
1302-014	274	C15	H 14	O 5	
1302-015	274	C15	H 14	O 5	
2204-011	274	C15	H 14	O 5	
3118-051	274	C18	H 26	O 2	
3118-050	274	C18	H 26	O 2	
3113-036	274	C14	H 10	O 4	S 1
3110-064	274	C16	H 18	O 4	
3115-016	274	C17	H 27	O 3	
3118-045	276	C18	H 28	O 2	
3118-043	276	C18	H 28	O 2	
3110-065	276	C16	H 20	O 4	
3005-035	276	C11	H 16	O 8	
3118-042	276	C18	H 28	O 2	
3005-015	276	C16	H 20	O 4	
3118-032	276	C18	H 28	O 2	
2204-043	276	C15	H 16	O 5	
2106-018	276	C15	H 16	O 5	
2204-030	276	C15	H 16	O 5	
2202-013	276	C15	H 16	O 5	
2204-037	276	C14	H 12	O 6	
2104-009	276	C11	H 10	O 4	CL2
2204-021	276	C15	H 16	O 5	
2204-022	276	C15	H 16	O 5	
2202-022	276	C15	H 16	O 5	
0599-010	276	C14	H 12	O 6	
0510-003	276	C15	H 16	O 5	
0599-009	276	C14	H 12	O 6	
0404-020	276	C15	H 16	O 5	
0499-001	276	C17	H 24	O 3	
2505-020	276	C14	H 12	O 6	
2505-013	276	C14	H 12	O 6	
2504-007	276	C15	H 16	O 5	
2405-009	276	C13	H 8	O 7	
3304-008	276	C13	H 8	O 1	S 3
3303-004	276	C14	H 12	O 2	S 2
3504-011	278	C11	H 18	O 8	
3304-002	278	C13	H 10	O 1	S 3
2506-011	278	C13	H 10	O 7	
0499-008	278	C20	H 22	O 1	
0501-037	278	C15	H 18	O 5	
2108-001	278	C16	H 22	O 4	
2103-005	278	C15	H 18	O 5	
2107-009	278	C15	H 18	O 5	
3118-037	278	C18	H 30	O 2	
3118-036	278	C18	H 30	O 2	
3118-049	278	C18	H 30	O 2	
3118-038	278	C18	H 30	O 2	
3118-034	278	C18	H 30	O 2	
3118-035	278	C18	H 30	O 2	
3118-065	278	C18	H 30	O 3	
3118-040	278	C18	H 30	O 2	
3118-048	278	C18	H 30	O 2	
3118-033	278	C18	H 30	O 2	
3118-039	278	C18	H 30	O 2	
3118-041	278	C18	H 30	O 2	
3117-085	278	C17	H 26	O 3	
2901-015	278	C18	H 30	O 2	
3114-034	279	C18	H 33	N 1	O 1
3118-046	280	C18	H 32	O 2	
3118-047	280	C18	H 32	O 2	
3118-029	280	C18	H 32	O 2	
3118-028	280	C18	H 32	O 2	
2901-014	280	C18	H 32	O 2	
3118-031	280	C18	H 32	O 2	
2999-002	280	C16	H 24	O 4	
3118-030	280	C18	H 32	O 2	
3117-088	280	C18	H 32	O 2	
3117-083	280	C17	H 28	O 3	
2903-005	280	C16	H 24	O 4	
2206-012	280	C16	H 24	O 4	
2103-006	280	C15	H 16	O 5	
2003-012	280	C16	H 24	O 4	
2199-006	280	C13	H 12	O 7	
0304-022	280	C15	H 22	N 1	O 4
2507-005	280	C13	H 12	O 7	
2805-007	280	C19	H 20	O 2	
2507-003	280	C12	H 8	O 8	
2601-004	280	C17	H 12	O 4	
2507-004	280	C13	H 12	O 7	
3197-028	280	C18	H 32	O 2	
3197-027	280	C18	H 32	O 2	
3205-019	282	C17	H 14	O 4	
3197-022	282	C18	H 34	O 2	
2508-003	282	C12	H 10	O 8	
2508-004	282	C12	H 10	O 8	
2805-006	282	C19	H 22	O 2	
0902-018	282	C17	H 14	O 4	
1402-013	282	C16	H 10	O 5	
2206-017	282	C15	H 22	O 5	
2206-018	282	C15	H 22	O 5	
1601-004	282	C17	H 14	O 4	
3118-024	282	C18	H 34	O 2	
3118-023	282	C18	H 34	O 2	
3118-021	282	C18	H 34	O 2	
3118-020	282	C18	H 34	O 2	
3118-019	282	C18	H 34	O 2	
3116-007	282	C18	H 34	O 2	
3118-022	282	C18	H 34	O 2	
3116-040	282	C18	H 34	O 2	
3120-001	282	C20	H 42		
3118-025	282	C18	H 34	O 2	
3117-039	282	C19	H 22	O 2	
3118-026	282	C18	H 34	O 2	
1101-015	282	C17	H 14	O 4	
1103-011	284	C16	H 12	O 5	
1206-017	284	C16	H 14	O 5	
1103-023	284	C16	H 12	O 5	
1101-013	284	C16	H 12	O 5	
1103-020	284	C16	H 12	O 5	
1102-003	284	C16	H 12	O 5	
1199-003	284	C16	H 12	O 5	
0999-007	284	C16	H 12	O 5	
1001-024	284	C17	H 16	O 4	
1201-008	284	C17	H 16	O 4	
1102-002	284	C16	H 12	O 5	
1001-023	284	C17	H 16	O 4	
1101-017	284	C16	H 12	O 5	
3117-024	284	C19	H 24	O 2	

Code	MW	C	H	O	Other
3104-025	284	C10	H 20	O 9	
3117-056	284	C17	H 32	O 3	
3116-037	284	C16	H 30	O 4	
3121-004	284	C21	H 32		
3121-003	284	C21	H 32		
3118-010	284	C18	H 36	O 2	
1601-007	284	C16	H 12	O 5	
1601-003	284	C17	H 16	O 4	
1602-003	284	C15	H 8	O 6	
1401-013	284	C16	H 12	O 5	
1601-009	284	C16	H 12	O 5	
1601-011	284	C16	H 12	O 5	
1402-011	284	C16	H 12	O 5	
1602-008	284	C15	H 8	O 6	
0901-010	284	C17	H 16	O 4	
0901-008	284	C17	H 16	O 4	
0506-022	284	C17	H 16	O 4	
0508-035	284	C17	H 16	O 4	
0104-003	284	C13	H 16	O 7	
0104-002	284	C13	H 16	O 7	
0999-002	284	C17	H 16	O 4	
0103-015	284	C13	H 16	O 7	
0399-008	284	C17	H 16	O 4	
2805-005	284	C19	H 24	O 2	
2604-005	284	C16	H 12	O 5	
2604-004	284	C16	H 12	O 5	
2604-036	284	C16	H 12	O 5	
2604-019	284	C16	H 12	O 5	
2603-019	284	C16	H 10	O 5	
2804-025	284	C16	H 12	O 5	
2604-035	284	C16	H 12	O 5	
2603-031	284	C15	H 8	O 6	
2603-020	284	C15	H 8	O 6	
2604-016	284	C16	H 12	O 5	
2604-015	284	C16	H 12	O 5	
2604-006	284	C16	H 12	O 4	
3203-015	284	C14	H 16	O 2	
3303-006	284	C12	H 9	O 1	S 2 CL1
3499-008	284	C14	H 20	O 6	
3110-081	285	C13	H 25	N 1 O 5	
3118-018	286	C18	H 34	O 2	
3118-006	286	C18	H 38	O 2	
3121-002	286	C21	H 34		
2903-015	286	C16	H 14	O 5	
3117-087	286	C18	H 22	O 3	
3117-054	286	C17	H 34	O 3	
3906-025	286	C16	H 14	O 5	
3301-017	286	C15	H 26	O 3	S 1
3202-002	286	C16	H 30	O 4	
2605-009	286	C15	H 10	O 6	
2804-024	286	C16	H 14	O 5	
2605-004	286	C15	H 10	O 6	
2605-026	286	C15	H 10	O 6	
2604-027	286	C15	H 10	O 6	
2605-002	286	C15	H 10	O 6	
2701-015	286	C17	H 18	O 4	
2804-016	286	C16	H 14	O 5	
2704-002	286	C19	H 10	O 3	
2604-045	286	C15	H 10	O 6	
2605-001	286	C15	H 10	O 6	
2605-015	286	C15	H 10	O 6	
2603-016	286	C16	H 12	O 5	
2605-016	286	C15	H 10	O 6	
2605-018	286	C15	H 10	O 6	
2701-018	286	C16	H 14	O 5	
0699-015	286	C17	H 18	O 4	
0699-013	286	C17	H 18	O 4	
0508-026	286	C16	H 14	O 5	
0508-029	286	C16	H 14	O 5	
0508-030	286	C16	H 14	O 5	
0508-027	286	C16	H 14	O 5	
0102-008	286	C13	H 18	O 7	
0508-025	286	C16	H 14	O 5	
0901-004	286	C17	H 18	O 4	
0506-020	286	C16	H 14	O 4	
0506-016	286	C16	H 14	O 5	
0506-015	286	C16	H 14	O 5	
0499-011	286	C17	H 18	O 4	
0506-014	286	C16	H 14	O 5	
0802-015	286	C13	H 18	O 7	
2402-014	286	C15	H 10	O 6	
2403-009	286	C16	H 14	O 5	
1405-009	286	C17	H 18	O 4	
1803-001	286	C15	H 10	O 6	
2105-005	286	C19	H 26	O 2	
2403-030	286	C16	H 14	O 5	
2403-015	286	C16	H 14	O 5	
1803-005	286	C15	H 10	O 6	
2403-023	286	C16	H 14	O 5	
1402-001	286	C15	H 10	O 6	
2204-004	286	C17	H 20	O 4	
1199-001	286	C15	H 10	O 6	
1199-008	286	C15	H 10	O 6	
1203-008	286	C16	H 14	O 5	
1201-009	286	C16	H 14	O 5	
1202-004	286	C16	H 14	O 5	
1203-010	286	C16	H 14	O 5	
1104-003	286	C15	H 10	O 6	
1104-002	286	C15	H 10	O 6	
1299-001	286	C16	H 14	O 5	
1103-032	286	C15	H 10	O 6	
1105-001	286	C15	H 10	O 6	
1106-001	286	C15	H 10	O 6	
1202-002	286	C16	H 14	O 5	
1099-002	286	C17	H 18	O 4	
1701-008	287	C15	H 11	O 6	+
1702-002	287	C15	H 11	O 6	+
2901-009	287	C12	H 17	N 1 O 7	
2504-031	288	C16	H 16	O 5	
2701-019	288	C16	H 16	O 5	
2505-018	288	C15	H 12	O 6	
2699-018	288	C15	H 12	O 6	
2801-007	288	C15	H 12	O 6	
2504-004	288	C16	H 16	O 5	
2704-001	288	C19	H 12	O 3	
2504-026	288	C16	H 16	O 5	
2404-011	288	C16	H 12	O 6	
2404-012	288	C15	H 12	O 6	
1399-006	288	C15	H 12	O 6	
1803-014	288	C15	H 12	O 6	
2404-010	288	C15	H 12	O 6	
2404-005	288	C15	H 12	O 6	
1401-011	288	C16	H 12	O 5	
2404-031	288	C15	H 12	O 6	
2404-019	288	C15	H 12	O 6	
2404-015	288	C15	H 12	O 6	
1003-006	288	C15	H 12	O 6	
1299-003	288	C15	H 12	O 6	
1205-019	288	C15	H 12	O 6	
1205-023	288	C15	H 12	O 6	
1004-001	288	C15	H 12	O 6	
1205-001	288	C15	H 12	O 6	
1002-013	288	C16	H 16	O 5	
1002-011	288	C16	H 16	O 5	
1203-013	288	C15	H 12	O 6	
1204-001	288	C15	H 12	O 6	
1002-021	288	C15	H 12	O 6	
1003-007	288	C15	H 12	O 6	
1207-001	288	C15	H 12	O 6	
1206-009	288	C15	H 12	O 6	
1003-005	288	C15	H 12	O 6	
0499-007	288	C19	H 28	O 2	
0501-026	288	C15	H 12	O 6	
0508-011	288	C16	H 16	O 5	
0508-005	288	C16	H 16	O 5	
3116-031	288	C16	H 32	O 4	
3116-035	288	C16	H 32	O 4	
3116-033	288	C16	H 32	O 4	
3118-089	290	C19	H 30	O 2	
3118-054	290	C18	H 26	O 3	
0508-003	290	C16	H 18	O 5	
0508-008	290	C16	H 18	O 5	
0508-010	290	C16	H 18	O 5	
0507-008	290	C15	H 16	O 4	
0702-004	290	C18	H 10	O 4	
0506-005	290	C10	H 6	O 4	
1399-001	290	C15	H 14	O 6	
2108-007	290	C16	H 18	O 5	
2204-014	290	C15	H 14	O 6	
1302-016	290	C15	H 14	O 6	
1303-003	290	C15	H 14	O 6	
1302-007	290	C15	H 14	O 6	
1302-006	290	C15	H 14	O 6	
1303-015	290	C15	H 14	O 6	
1303-007	290	C15	H 14	O 6	
1303-008	290	C15	H 14	O 6	
1303-014	290	C15	H 14	O 6	
1302-008	290	C15	H 14	O 6	
1303-006	290	C15	H 14	O 6	
1303-011	290	C15	H 14	O 6	
1303-012	290	C15	H 14	O 6	
1303-013	290	C15	H 14	O 6	
1303-018	290	C15	H 14	O 6	
1302-009	290	C15	H 14	O 6	
1303-016	290	C15	H 14	O 6	
1304-002	290	C15	H 14	O 6	
1304-001	290	C15	H 14	O 6	
1303-017	290	C15	H 14	O 6	
2699-019	290	C15	H 14	O 6	
2699-012	290	C15	H 11	O 4	CL1
2801-006	290	C15	H 14	O 6	
2504-036	290	C16	H 18	O 5	

2505-014	290	C15	H 14	□ 6						1401-031	298	C17	H 14	□ 5		
3302-024	290	C15	H 11	□ 2	S 1	CL1				2111-014	298	C21	H 30	□ 1		
3302-028	290	C15	H 11	□ 2	S 1	CL1				1802-003	298	C16	H 10	□ 6		
3905-027	290	C14	H 16	□ 6						1402-012	298	C17	H 14	□ 5		
3906-050	292	C14	H 12	□ 7						2107-004	298	C19	H 22	□ 3		
3902-004	292	C14	H 12	□ 7						1601-012	298	C17	H 14	□ 5		
3906-015	292	C13	H 8	□ 8						1399-005	298	C17	H 14	□ 5		
3302-023	292	C15	H 13	□ 2	S 1	CL1				2604-044	298	C17	H 14	□ 5		
2505-015	292	C15	H 16	□ 6						2604-051	298	C16	H 10	□ 6		
2506-016	292	C15	H 16	□ 6						2701-025	298	C17	H 14	□ 5		
2506-004	292	C14	H 12	□ 7						2805-014	298	C19	H 22	□ 3		
2110-001	292	C19	H 32	□ 2						2701-023	298	C17	H 14	□ 5		
2202-008	292	C16	H 20	□ 5						2604-018	298	C16	H 10	□ 6		
2108-002	292	C17	H 24	□ 4						2604-007	298	C17	H 14	□ 5		
2108-006	292	C16	H 20	□ 5						2805-013	298	C19	H 22	□ 3		
0701-002	292	C18	H 12	□ 4						1106-007	298	C15	H 10	□ 7		
0701-001	292	C18	H 12	□ 4						1103-024	298	C17	H 14	□ 5		
3005-002	292	C11	H 16	□ 9						1106-008	298	C15	H 10	□ 7		
3118-073	292	C18	H 28	□ 3						1103-048	298	C17	H 14	□ 7		
3118-053	292	C18	H 28	□ 3						1199-002	300	C16	H 12	□ 6		
3118-074	292	C18	H 28	□ 3						1103-039	300	C16	H 12	□ 6		
1101-010	292	C18	H 12	□ 4						1105-021	300	C16	H 12	□ 6		
1102-005	292	C18	H 12	□ 4						1104-040	300	C16	H 12	□ 6		
1101-014	292	C18	H 12	□ 4						1199-011	300	C16	H 12	□ 6		
3197-005	293	C17	H 27	N 1	□ 3					1103-040	300	C16	H 12	□ 6		
3504-012	294	C11	H 18	□ 9						1302-004	300	C18	H 20	□ 4		
3401-004	294	C17	H 26	□ 4						1104-004	300	C16	H 12	□ 6		
3202-001	294	C18	H 30	□ 3						1099-011	300	C16	H 12	□ 6		
3118-079	294	C18	H 30	□ 3						1103-033	300	C16	H 12	□ 6		
3118-078	294	C18	H 30	□ 3						1104-036	300	C16	H 12	□ 6		
3114-043	294	C17	H 26	□ 4						1203-019	300	C17	H 16	□ 5		
3118-080	294	C18	H 30	□ 3						1099-014	300	C18	H 20	□ 4		
3118-058	294	C18	H 30	□ 3						1099-003	300	C18	H 20	□ 4		
3118-052	294	C18	H 30	□ 3						1105-002	300	C16	H 12	□ 6		
3118-072	294	C18	H 30	□ 3						1105-029	300	C16	H 12	□ 6		
3118-087	294	C19	H 34	□ 2						1106-003	300	C16	H 12	□ 6		
3005-016	294	C16	H 22	□ 5						1106-002	300	C16	H 12	□ 6		
0499-002	294	C17	H 26	□ 4						2805-012	300	C19	H 24	□ 3		
0103-012	294	C 7	H 4	□ 3	BR2					2701-016	300	C17	H 16	□ 5		
1001-018	294	C18	H 14	□ 4						2605-020	300	C16	H 12	□ 6		
2204-042	294	C15	H 18	□ 6						2604-047	300	C16	H 12	□ 6		
2402-008	294	C18	H 14	□ 2						2604-009	300	C15	H 8	□ 7		
2003-007	294	C17	H 26	□ 4						2604-052	300	C15	H 8	□ 7		
2901-017	294	C18	H 30	□ 3						2604-048	300	C16	H 12	□ 6		
1602-002	296	C17	H 12	□ 5						2805-011	300	C19	H 24	□ 3		
2402-009	296	C18	H 16	□ 4						2701-020	300	C17	H 16	□ 5		
1602-005	296	C16	H 8	□ 6						2605-005	300	C16	H 12	□ 6		
2402-005	296	C18	H 16	□ 4						2604-049	300	C16	H 12	□ 6		
1601-013	296	C17	H 12	□ 5						2604-017	300	C16	H 12	□ 6		
0404-014	296	C19	H 20	□ 3						2903-016	300	C17	H 16	□ 5		
0699-009	296	C18	H 16	□ 4						2605-017	300	C16	H 12	□ 6		
0699-010	296	C18	H 16	□ 4						2603-017	300	C17	H 14	□ 5		
0304-017	296	C13	H 12	□ 8						2605-012	300	C16	H 12	□ 6		
0699-008	296	C18	H 16	□ 4						2503-033	300	C17	H 16	□ 5		
0699-007	296	C18	H 16	□ 4						2401-003	300	C15	H 8	□ 7		
0699-006	296	C18	H 16	□ 4						2402-013	300	C16	H 12	□ 6		
3118-066	296	C18	H 32	□ 4						1401-017	300	C16	H 12	□ 6		
3118-057	296	C18	H 32	□ 3						1402-002	300	C16	H 12	□ 6		
3118-085	296	C19	H 36	□ 2						1803-004	300	C16	H 12	□ 6		
3118-071	296	C18	H 32	□ 3						2111-013	300	C21	H 32	□ 1		
3118-056	296	C18	H 32	□ 3						1402-005	300	C16	H 12	□ 6		
3003-007	296	C10	H 12	□ 4						1803-007	300	C16	H 12	□ 6		
3118-063	296	C18	H 32	□ 3						1001-019	300	C18	H 20	□ 4		
3121-001	296	C21	H 44							0506-023	300	C17	H 16	□ 5		
3005-026	296	C17	H 28	□ 4						0901-003	300	C18	H 20	□ 4		
3005-037	296	C17	H 28	□ 4						0510-005	300	C17	H 16	□ 5		
3118-076	296	C18	H 32	□ 3						0999-010	300	C17	H 16	□ 5		
3118-077	296	C18	H 32	□ 3						0202-006	300	C14	H 20	□ 7		
3118-060	296	C18	H 32	□ 3						0104-006	300	C13	H 16	□ 8		
3120-004	296	C20	H 40	□ 1						0511-002	300	C17	H 16	□ 5		
3118-062	296	C18	H 32	□ 3						0511-007	300	C17	H 16	□ 5		
3901-009	296	C19	H 36	□ 2						1001-015	300	C16	H 12	□ 6		
3204-013	298	C18	H 18	□ 4						3118-012	300	C18	H 36	□ 3		
3118-059	298	C18	H 34	□ 3						3114-016	300	C18	H 20	□ 4		
3120-003	298	C20	H 42	□ 1						3118-081	300	C18	H 33	□ 2	F 1	
3120-002	298	C20	H 42	□ 1						3118-011	300	C18	H 36	□ 3		
3118-075	298	C18	H 34	□ 3						3197-029	300	C17	H 32	□ 4		
3118-064	298	C18	H 34	□ 3						3197-002	301	C 7	H 11	□11	P 1	+
3005-024	298	C17	H 30	□ 4						1702-017	301	C16	H 13	□ 6		
3118-055	298	C18	H 34	□ 3						2404-014	302	C16	H 14	□ 6		
3118-084	298	C19	H 38	□ 2						1802-009	302	C16	H 14	□ 6		
3118-069	298	C18	H 34	□ 3						1802-010	302	C16	H 14	□ 6		
3118-068	298	C18	H 34	□ 3						2404-016	302	C16	H 14	□ 6		
3114-012	298	C18	H 18	□ 4						2111-012	302	C21	H 34	□ 1		
0203-002	298	C14	H 18	□ 7						2404-020	302	C16	H 14	□ 6		
0902-021	298	C17	H 14	□ 5						1802-008	302	C16	H 14	□ 6		
0501-021	298	C19	H 22	□ 3						1802-007	302	C16	H 14	□ 6		
0501-005	298	C19	H 22	□ 3						2404-003	302	C16	H 14	□ 6		
0501-038	298	C19	H 22	□ 3						2404-002	302	C16	H 14	□ 6		
1499-003	298	C16	H 10	□ 6						2404-004	302	C16	H 14	□ 6		
1402-016	298	C17	H 14	□ 5						1499-004	302	C16	H 14	□ 6		

ID	Mass	Formula
1405-003	302	C17 H 18 O 5
1405-004	302	C17 H 18 O 5
2404-013	302	C16 H 14 O 6
3906-026	302	C16 H 14 O 6
3906-038	302	C16 H 14 O 6
3117-050	302	C19 H 26 O 3
3117-047	302	C19 H 26 O 3
0699-016	302	C17 H 18 O 5
0699-014	302	C17 H 18 O 5
0601-001	302	C18 H 22 O 4
0105-001	302	C14 H 6 O 8
0999-004	302	C17 H 18 O 5
0102-010	302	C13 H 18 O 8
2802-008	302	C16 H 14 O 6
2606-006	302	C15 H 10 O 7
2605-011	302	C15 H 10 O 7
2605-021	302	C15 H 10 O 7
2701-017	302	C17 H 18 O 5
2701-021	302	C17 H 18 O 5
1104-057	302	C15 H 10 O 7
1299-004	302	C16 H 14 O 6
1205-007	302	C16 H 14 O 6
1205-008	302	C16 H 14 O 6
1205-021	302	C16 H 14 O 6
1003-009	302	C16 H 14 O 6
1106-009	302	C15 H 10 O 7
1202-005	302	C16 H 14 O 6
1199-016	302	C15 H 10 O 7
1204-002	302	C16 H 14 O 6
1108-017	302	C15 H 10 O 7
1703-001	303	C15 H 11 O 7 +
2405-007	304	C15 H 12 O 7
2199-005	304	C18 H 24 O 4
2405-010	304	C15 H 12 O 7
2405-008	304	C15 H 12 O 7
2202-014	304	C17 H 20 O 5
2111-011	304	C21 H 36 O 1
1208-001	304	C15 H 12 O 7
1206-001	304	C15 H 12 O 7
1003-012	304	C15 H 12 O 7
1302-010	304	C16 H 16 O 6
2803-010	304	C16 H 16 O 6
2802-013	304	C16 H 16 O 6
2703-044	304	C19 H 12 O 4
2699-016	304	C16 H 16 O 6
2604-039	304	C15 H 9 O 5 CL1
0506-017	304	C16 H 16 O 6
0701-007	304	C19 H 12 O 4
3116-036	304	C16 H 32 O 5
3116-034	304	C16 H 32 O 5
3120-018	304	C20 H 32 O 2
3120-016	304	C20 H 32 O 2
3120-017	306	C20 H 34 O 2
3120-015	306	C20 H 34 O 2
0702-005	306	C18 H 10 O 5
1001-003	306	C20 H 18 O 3
2505-016	306	C15 H 14 O 7
2703-046	306	C19 H 14 O 4
1105-040	306	C18 H 10 O 5
1303-024	306	C15 H 14 O 7
1303-023	306	C15 H 14 O 7
1304-004	306	C15 H 14 O 7
1303-020	306	C15 H 14 O 7
1304-006	306	C15 H 14 O 7
1303-019	306	C15 H 14 O 7
1304-008	306	C15 H 14 O 7
2104-003	306	C15 H 14 O 7
1304-010	306	C15 H 14 O 7
2204-015	306	C15 H 14 O 7
2204-044	306	C16 H 18 O 6
2005-002	306	C14 H 10 O 8
0702-006	307	C18 H 13 N 1 O 4
3197-006	307	C18 H 29 N 1 O 3
3499-009	308	C16 H 20 O 6
3303-005	308	C14 H 9 O 2 S 2 CL1
3506-049	308	C13 H 24 O 8
3506-050	308	C13 H 24 O 8
3401-005	308	C18 H 28 O 4
0499-003	308	C18 H 28 O 4
0508-009	308	C16 H 20 O 6
0508-004	308	C16 H 20 O 6
0702-001	308	C18 H 12 O 5
2107-002	308	C16 H 20 O 6
2199-009	308	C14 H 12 O 8
2805-008	308	C19 H 16 O 4
3120-013	308	C20 H 36 O 2
3120-010	310	C20 H 38 O 2
3120-012	310	C20 H 38 O 2
3120-011	310	C20 H 38 O 2
3120-008	310	C20 H 38 O 2
3118-088	310	C19 H 34 O 3
3122-001	310	C22 H 46
3120-009	310	C20 H 30 O 2
2403-031	310	C18 H 14 O 5
2403-011	310	C18 H 14 O 5
2106-014	310	C21 H 26 O 2
2499-011	310	C14 H 14 O 8
1602-006	310	C17 H 10 O 6
2403-026	310	C18 H 14 O 5
2403-038	310	C17 H 10 O 6
1402-020	310	C17 H 10 O 6
0102-006	310	C 8 H 8 O 3 BR2
3906-022	310	C17 H 26 O 5
3506-030	310	C12 H 22 O 9
3506-010	312	C11 H 20 O10
3506-007	312	C11 H 20 O10
3506-008	312	C11 H 20 O10
3506-009	312	C11 H 20 O10
3506-011	312	C11 H 20 O10
3906-029	312	C17 H 12 O 6
3204-017	312	C18 H 16 O 5
0508-015	312	C19 H 20 O 4
0508-018	312	C19 H 20 O 4
0501-039	312	C20 H 24 O 3
0501-006	312	C19 H 20 O 4
0508-039	312	C19 H 20 O 4
1602-004	312	C17 H 12 O 6
2403-025	312	C18 H 16 O 5
2402-006	312	C18 H 16 O 5
1402-017	312	C18 H 16 O 5
2403-024	312	C18 H 16 O 5
2403-003	312	C18 H 16 O 5
3120-005	312	C20 H 40 O 2
3118-082	312	C18 H 32 O 4
2604-025	312	C17 H 12 O 6
2805-016	312	C20 H 24 O 3
2701-024	312	C18 H 16 O 5
2805-015	312	C20 H 24 O 3
1199-006	312	C18 H 16 O 5
1199-005	312	C18 H 16 O 5
1103-030	312	C18 H 16 O 5
1103-050	312	C18 H 16 O 7
1103-049	312	C18 H 16 O 7
2503-051	313	C14 H 10 O 4 CL2
2604-050	314	C17 H 14 O 6
2604-022	314	C16 H 10 O 7
2604-011	314	C16 H 10 O 7
2704-012	314	C20 H 10 O 4
2701-022	314	C18 H 18 O 5
2605-006	314	C17 H 14 O 6
2604-053	314	C16 H 10 O 7
2604-057	314	C17 H 14 O 6
2505-019	314	C15 H 22 O 7
2805-002	314	C18 H 18 O 5
2604-056	314	C17 H 14 O 6
1203-021	314	C18 H 18 O 5
1203-020	314	C18 H 18 O 5
1104-035	314	C17 H 14 O 6
1106-069	314	C15 H 10 O 8
1105-030	314	C17 H 14 O 6
1103-041	314	C17 H 14 O 6
1103-043	314	C17 H 14 O 6
1104-043	314	C17 H 14 O 6
1104-042	314	C17 H 14 O 6
1102-008	314	C17 H 14 O 6
1105-022	314	C17 H 14 O 6
1102-010	314	C17 H 14 O 6
1103-045	314	C17 H 14 O 6
3118-007	314	C20 H 26 O 3
1601-028	314	C17 H 14 O 6
2202-016	314	C19 H 22 O 4
2106-014	314	C21 H 30 O 2
2111-020	314	C21 H 30 O 2
1601-015	314	C17 H 14 O 6
2106-012	314	C21 H 30 O 2
2106-011	314	C21 H 30 O 2
1802-002	314	C16 H 10 O 7
1401-021	314	C17 H 14 O 6
2302-022	314	C16 H 10 O 7
1601-014	314	C17 H 14 O 6
2106-015	314	C21 H 30 O 2
2111-019	314	C21 H 30 O 2
2405-006	314	C16 H 10 O 7
2403-010	314	C12 H 16 O 5
1602-011	314	C16 H 10 O 7
1401-020	314	C17 H 14 O 6
2111-009	314	C21 H 30 O 2
2402-007	314	C18 H 18 O 5
2106-013	314	C21 H 30 O 2
2399-007	314	C16 H 10 O 7
1601-006	314	C18 H 18 O 5

ID	MW	C	H	O	Other
1401-018	314	C17	H 14	O 6	
0802-007	314	C14	H 18	O 8	
0304-015	314	C20	H 26	O 3	
1001-016	314	C17	H 14	O 6	
0999-005	314	C17	H 14	O 6	
0104-025	314	C14	H 18	O 8	
1002-015	314	C18	H 18	O 5	
0999-011	314	C17	H 14	O 6	
0799-001	314	C17	H 14	O 6	
0502-015	314	C19	H 22	O 4	
0502-014	314	C19	H 22	O 4	
0999-015	314	C17	H 14	O 6	
1099-015	314	C19	H 22	O 4	
0902-019	314	C17	H 14	O 6	
0103-005	314	C14	H 18	O 8	
3204-016	314	C18	H 18	O 5	
3906-030	314	C17	H 14	O 6	
3204-029	314	C19	H 22	O 4	
3204-028	314	C19	H 22	O 4	
3197-030	314	C18	H 34	O 4	
3506-006	314	C11	H 22	O10	
1702-023	315	C17	H 15	O 6	+
2404-024	316	C17	H 16	O 6	
2111-018	316	C21	H 32	O 2	
1499-005	316	C17	H 16	O 6	
2404-021	316	C17	H 16	O 6	
2399-004	316	C16	H 12	O 7	
1601-017	316	C17	H 16	O 6	
1405-006	316	C17	H 16	O 6	
2404-017	316	C17	H 16	O 6	
2111-008	316	C21	H 32	O 2	
3905-025	316	C18	H 20	O 5	
3303-009	316	C17	H 16	O 2	S 2
0803-007	316	C14	H 20	O 8	
0102-005	316	C14	H 20	O 8	
0999-012	316	C17	H 16	O 6	
0901-005	316	C18	H 20	O 5	
0511-003	316	C17	H 16	O 6	
0999-006	316	C17	H 16	O 6	
1099-005	316	C18	H 20	O 5	
1099-004	316	C18	H 20	O 5	
3118-013	316	C18	H 36	O 4	
3118-014	316	C18	H 36	O 4	
1104-045	316	C16	H 12	O 7	
1104-046	316	C16	H 12	O 7	
1205-014	316	C17	H 16	O 6	
1105-042	316	C16	H 12	O 7	
1106-039	316	C16	H 12	O 7	
1106-112	316	C16	H 12	O 7	
1106-060	316	C16	H 12	O 7	
1106-037	316	C16	H 12	O 7	
1105-043	316	C16	H 12	O 7	
1104-054	316	C16	H 12	O 7	
1106-043	316	C16	H 12	O 7	
1106-034	316	C16	H 12	O 7	
1205-009	316	C17	H 16	O 6	
1106-035	316	C16	H 12	O 7	
1105-046	316	C16	H 12	O 7	
2606-002	316	C16	H 12	O 7	
2802-009	316	C17	H 16	O 6	
2704-013	316	C20	H 12	O 4	
2606-001	316	C16	H 12	O 7	
1703-011	317	C16	H 13	O 7	+
2102-019	318	C17	H 18	O 6	
2111-006	318	C21	H 34	O 2	
2111-007	318	C21	H 34	O 2	
2003-018	318	C19	H 26	O 4	
2301-001	318	C16	H 14	O 7	
2111-016	318	C21	H 34	O 2	
2204-023	318	C17	H 18	O 6	
2405-011	318	C16	H 14	O 7	
2111-017	318	C21	H 34	O 2	
2803-006	318	C17	H 18	O 6	
2704-003	318	C20	H 14	O 4	
2604-042	318	C16	H 11	O 5	CL1
2903-002	318	C19	H 26	O 4	
2801-008	318	C16	H 14	O 7	
2606-010	318	C15	H 10	O 8	
2509-001	318	C20	H 14	O 4	
2604-040	318	C16	H 11	O 5	CL1
2606-009	318	C15	H 10	O 8	
1299-009	318	C17	H 18	O 6	
1206-008	318	C16	H 14	O 7	
1108-001	318	C15	H 10	O 8	
1106-091	318	C15	H 10	O 8	
1206-007	318	C16	H 14	O 7	
1206-005	318	C16	H 14	O 7	
1206-002	318	C16	H 14	O 7	
0506-018	318	C17	H 18	O 6	
0506-024	318	C17	H 18	O 6	
0501-040	318	C18	H 22	O 5	
3302-017	318	C18	H 22	O 3	S 1
3499-005	318	C18	H 22	O 5	
3906-007	320	C19	H 12	O 5	
3304-003	320	C15	H 12	O 2	S 3
0499-009	320	C21	H 20	O 3	
0511-009	320	C11	H 8	O 5	
0105-008	320	C14	H 8	O 9	
1208-002	320	C15	H 12	O 8	
1304-009	320	C16	H 16	O 7	
1303-025	320	C16	H 16	O 7	
2604-046	320	C15	H 9	O 6	CL1
3003-013	320	C17	H 20	O 6	
2605-010	320	C15	H 9	O 6	CL1
2605-005	320	C15	H 9	O 6	CL1
1602-007	320	C18	H 8	O 6	
2102-020	320	C17	H 20	O 6	
2111-005	320	C21	H 36	O 2	
2111-015	320	C21	H 36	O 2	
3197-007	321	C19	H 31	N 1 O 3	
3506-045	322	C14	H 26	O 8	
3502-009	322	C12	H 26	N 4 O 6	
3506-048	322	C14	H 26	O 8	
3506-047	322	C14	H 26	O 8	
3506-046	322	C14	H 26	O 8	
1601-022	322	C20	H 18	O 4	
2199-008	322	C14	H 10	O 7	
2107-001	322	C16	H 18	O 7	
2003-008	322	C19	H 30	O 4	
2703-045	322	C19	H 14	O 5	
2604-061	322	C18	H 10	O 6	
2507-006	322	C15	H 14	O 8	
1304-011	322	C15	H 14	O 8	
0199-014	322	C14	H 10	O 9	
0599-014	322	C18	H 10	O 6	
0604-005	322	C20	H 18	O 4	
0499-004	322	C19	H 30	O 4	
1001-008	322	C20	H 18	O 4	
2999-004	323	C18	H 29	N 1 O 4	
3502-008	323	C12	H 25	N 3 O 7	
3906-042	324	C15	H 16	O 8	
3906-041	324	C15	H 16	O 8	
3903-001	324	C18	H 28	O 5	
2699-007	324	C20	H 20	O 4	
2507-007	324	C15	H 16	O 8	
0508-016	324	C20	H 20	O 4	
0604-006	324	C20	H 30	O 4	
0701-003	324	C18	H 12	O 6	
1002-017	324	C20	H 20	O 4	
1203-028	324	C20	H 20	O 4	
1203-030	324	C20	H 20	O 4	
2107-003	324	C16	H 20	O 7	
2204-020	324	C20	H 20	O 4	
2403-039	324	C18	H 12	O 6	
1601-021	324	C20	H 20	O 4	
3005-028	324	C19	H 32	O 4	
3122-003	324	C22	H 44	O 1	
3005-027	324	C19	H 32	O 4	
3005-029	324	C19	H 32	O 4	
3117-071	324	C19	H 32	O 4	
3120-014	324	C20	H 36	O 3	
3123-001	324	C23	H 48		
3005-031	324	C19	H 32	O 4	
2999-005	325	C18	H 31	N 1 O 4	
3005-030	326	C19	H 34	O 4	
3005-025	326	C19	H 34	O 4	
3117-070	326	C19	H 34	O 4	
3118-067	326	C19	H 34	O 5	
3117-057	326	C19	H 34	O 4	
3122-002	326	C22	H 46	O 1	
1499-001	326	C18	H 14	O 6	
2302-013	326	C19	H 18	O 5	
2204-019	326	C20	H 22	O 4	
2404-027	326	C18	H 14	O 6	
1103-051	326	C19	H 18	O 7	
1106-004	326	C18	H 14	O 6	
1103-031	326	C19	H 18	O 5	
0699-004	326	C19	H 18	O 5	
0302-004	326	C16	H 22	O 7	
0304-008	326	C15	H 18	O 8	
0304-007	326	C15	H 18	O 8	
0604-004	326	C20	H 22	O 4	
0304-033	326	C15	H 18	O 8	
0508-019	326	C20	H 22	O 4	
0501-009	326	C19	H 18	O 5	
0501-008	326	C19	H 18	O 5	
3501-006	326	C12	H 22	O10	
3506-036	326	C12	H 22	O10	
3506-035	326	C12	H 22	O10	
3506-034	326	C12	H 22	O10	
3506-043	326	C12	H 22	O10	
0304-031	327	C16	H 25	N 1 O 6	

ID	Mass	C	H	O	Notes
0508-038	328	C19	H 20	O 5	
0601-002	328	C20	H 24	O 4	
0203-006	328	C15	H 20	O 8	
0504-005	328	C19	H 20	O 5	
0504-006	328	C19	H 20	O 5	
0503-003	328	C19	H 20	O 5	
0503-005	328	C19	H 20	O 5	
0505-003	328	C19	H 20	O 5	
0505-004	328	C19	H 20	O 5	
0506-007	328	C20	H 24	O 4	
0506-003	328	C20	H 24	O 4	
0508-020	328	C19	H 20	O 5	
0508-012	328	C20	H 24	O 4	
0507-009	328	C20	H 24	O 4	
0507-015	328	C20	H 24	O 4	
0999-013	328	C18	H 16	O 6	
0902-020	328	C18	H 16	O 6	
3303-008	328	C18	H 16	O 2	S 2
3906-014	328	C14	H 16	O 9	
3303-007	328	C18	H 16	O 2	S 2
3303-010	328	C18	H 16	O 2	S 2
3906-034	328	C17	H 12	O 7	
3904-001	328	C15	H 21	O 3	BR1
3906-032	328	C17	H 12	O 7	
1202-003	328	C18	H 16	O 6	
1105-038	328	C18	H 16	O 6	
1401-022	328	C18	H 16	O 6	
1199-018	328	C18	H 16	O 6	
1101-018	328	C18	H 16	O 6	
1102-007	328	C18	H 16	O 6	
1102-009	328	C18	H 16	O 6	
1103-047	328	C18	H 16	O 6	
1402-018	328	C18	H 16	O 6	
1104-044	328	C18	H 16	O 6	
2404-026	328	C18	H 16	O 6	
2404-025	328	C18	H 16	O 6	
1601-016	328	C18	H 16	O 6	
2201-004	328	C15	H 20	O 8	
1802-001	328	C17	H 12	O 7	
2404-038	328	C18	H 16	O 6	
2101-022	328	C22	H 32	O 2	
2201-009	328	C15	H 20	O 8	
2404-006	328	C18	H 16	O 6	
1402-021	328	C18	H 16	O 6	
3117-055	328	C19	H 36	O 4	
3120-006	328	C20	H 40	O 3	
3120-007	328	C20	H 40	O 3	
3119-002	328	C19	H 36	O 4	
3122-014	328	C22	H 32	O 2	
3122-013	328	C22	H 32	O 2	
3005-032	328	C17	H 28	O 6	
2605-008	328	C17	H 12	O 7	
2604-023	328	C17	H 12	O 7	
2604-058	328	C18	H 16	O 6	
3118-086	329	C22	H 35	N 1 O 1	
3118-027	330	C18	H 34	O 5	
3005-023	330	C16	H 26	O 7	
3122-016	330	C22	H 34	O 2	
2504-027	330	C18	H 18	O 6	
2605-003	330	C16	H 10	O 8	
2803-002	330	C17	H 14	O 7	
2504-032	330	C18	H 18	O 6	
2803-001	330	C17	H 14	O 7	
2605-013	330	C16	H 10	O 8	
2605-007	330	C16	H 10	O 8	
2504-029	330	C21	H 22	O 6	
1499-002	330	C17	H 14	O 7	
1601-018	330	C18	H 18	O 6	
1801-004	330	C17	H 14	O 7	
2405-005	330	C17	H 14	O 7	
2106-009	330	C22	H 34	O 2	
1802-005	330	C17	H 14	O 7	
1104-060	330	C17	H 14	O 7	
1104-047	330	C17	H 14	O 7	
1102-011	330	C17	H 14	O 7	
1104-049	330	C17	H 14	O 7	
1106-064	330	C17	H 14	O 7	
1105-047	330	C17	H 14	O 7	
1201-010	330	C18	H 18	O 6	
1106-057	330	C17	H 14	O 7	
1106-061	330	C17	H 14	O 7	
1105-045	330	C17	H 14	O 7	
1104-059	330	C17	H 14	O 7	
1106-056	330	C17	H 14	O 7	
1106-042	330	C17	H 14	O 7	
1107-002	330	C17	H 14	O 7	
3906-033	330	C17	H 14	O 7	
3906-035	330	C17	H 14	O 7	
0999-014	330	C18	H 18	O 6	
1001-013	330	C18	H 18	O 6	
0502-006	330	C19	H 22	O 5	
0502-005	330	C19	H 22	O 5	
0599-011	330	C18	H 18	O 6	
1703-018	331	C17	H 15	O 7	+
2301-002	332	C17	H 16	O 7	
2405-012	332	C17	H 16	O 7	
2206-009	332	C20	H 28	O 4	
2405-003	332	C17	H 16	O 7	
2405-002	332	C17	H 16	O 7	
2405-001	332	C17	H 16	O 7	
0104-018	332	C13	H 16	O10	
0104-017	332	C13	H 16	O10	
0501-007	332	C19	H 24	O 5	
3906-016	332	C18	H 20	O 6	
3906-024	332	C16	H 12	O 8	
3502-033	332	C14	H 24	N 2 O 7	
3906-017	332	C18	H 20	O 6	
1108-011	332	C16	H 12	O 8	
1108-012	332	C16	H 12	O 8	
1106-095	332	C16	H 12	O 8	
1106-094	332	C16	H 12	O 8	
1106-072	332	C16	H 12	O 8	
1108-009	332	C16	H 12	O 8	
1405-005	332	C18	H 20	O 6	
1107-008	332	C16	H 12	O 8	
1405-007	332	C18	H 20	O 6	
1108-008	332	C16	H 12	O 8	
2506-014	332	C17	H 16	O 7	
2604-043	332	C17	H 13	O 5	CL1
2903-003	332	C20	H 28	O 4	
2802-011	332	C17	H 16	O 7	
2903-004	332	C20	H 28	O 4	
3122-012	332	C22	H 36	O 2	
3118-016	332	C18	H 36	O 5	
3118-015	332	C18	H 36	O 5	
3122-011	334	C22	H 38	O 2	
3110-063	334	C16	H 30	O 7	
2902-003	334	C20	H 30	O 4	
2902-005	334	C20	H 30	O 4	
2506-013	334	C17	H 18	O 7	
2902-009	334	C20	H 30	O 4	
1108-021	334	C15	H 10	O 9	
0511-004	334	C17	H 18	O 7	
0999-008	334	C17	H 14	O 5	
2003-010	334	C20	H 30	O 4	
2301-003	334	C16	H 14	O 8	
1902-002	336	C16	H 16	O 8	
2003-009	336	C20	H 32	O 4	
1801-020	336	C19	H 12	O 6	
2102-018	336	C17	H 20	O 7	
0599-015	336	C19	H 12	O 6	
0702-003	336	C20	H 16	O 5	
1001-012	336	C21	H 20	O 4	
1401-006	336	C20	H 16	O 5	
1401-029	336	C21	H 20	O 4	
1105-041	336	C19	H 12	O 6	
1201-011	336	C21	H 20	O 4	
1106-006	336	C19	H 12	O 6	
1404-009	336	C19	H 12	O 6	
2902-007	336	C20	H 32	O 4	
2802-016	336	C17	H 17	O 5	CL1
2902-001	336	C20	H 32	O 4	
2699-015	336	C16	H 16	O 8	
2699-017	336	C16	H 16	O 8	
3122-009	336	C22	H 40	O 2	
3122-015	336	C22	H 40	O 2	
3122-010	336	C22	H 40	O 2	
3902-005	336	C22	H 24	O 3	
1199-028	337	C20	H 19	N 1 O 4	
1199-029	337	C20	H 19	N 1 O 4	
1199-009	338	C20	H 18	O 5	
1203-029	338	C21	H 22	O 4	
1499-011	338	C20	H 18	O 6	
3122-007	338	C22	H 42	O 2	
3122-008	338	C22	H 44	O 2	
3124-001	338	C24	H 50		
2703-047	338	C19	H 14	O 6	
2604-041	338	C15	H 8	O 5	CL2
2805-009	338	C20	H 18	O 5	
2604-034	338	C20	H 18	O 5	
2605-040	338	C18	H 10	O 7	
2605-027	338	C19	H 14	O 6	
2802-014	338	C16	H 15	O 6	CL1
0508-037	338	C21	H 22	O 4	
0508-032	338	C21	H 22	O 4	
1002-018	338	C21	H 22	O 4	
0199-015	338	C14	H 10	O10	
1601-027	338	C19	H 14	O 6	
2403-040	338	C19	H 14	O 6	
1499-019	338	C19	H 14	O 6	
2503-038	338	C16	H 18	O 8	
1902-007	338	C16	H 18	O 6	

Code	MW	C	H			Extra
3509-027	339	C12	H 21	N 1	O10	
3504-047	340	C 6	H 14	O12	P 2	
2001-011	340	C23	H 32	O 2		
2403-041	340	C19	H 16	O 6		
0105-003	340	C16	H 10	O 8		
0105-004	340	C16	H 10	O 8		
0601-008	340	C20	H 20	O 5		
0701-005	340	C18	H 12	O 7		
0601-009	340	C20	H 20	O 5		
0604-002	340	C21	H 24	O 4		
0604-007	340	C20	H 20	O 5		
0604-001	340	C21	H 24	O 4		
0507-002	340	C15	H 16	O 9		
0507-003	340	C15	H 16	O 9		
0699-002	340	C20	H 20	O 5		
0506-021	340	C21	H 24	O 4		
0506-002	340	C15	H 16	O 9		
2605-042	340	C18	H 12	O 7		
2605-043	340	C18	H 12	O 7		
3122-005	340	C22	H 44	O 2		
3123-002	340	C23	H 48	O 1		
1203-007	340	C20	H 20	O 5		
3999-014	341	C19	H 19	N 1	O 5	
3506-015	342	C12	H 22	O11		
3506-032	342	C12	H 22	O11		
3506-014	342	C12	H 22	O11		
3506-033	342	C12	H 22	O11		
3506-013	342	C12	H 22	O11		
3506-012	342	C12	H 22	O11		
3506-024	342	C12	H 22	O11		
3501-016	342	C12	H 22	O11		
3506-016	342	C12	H 22	O11		
3501-017	342	C12	H 22	O11		
3506-023	342	C12	H 22	O11		
3198-001	342	C21	H 42	O 3		
3506-022	342	C12	H 22	O11		
3506-037	342	C12	H 22	O11		
3506-017	342	C12	H 22	O11		
3506-031	342	C12	H 22	O11		
3506-026	342	C12	H 22	O11		
3506-025	342	C12	H 22	O11		
3506-027	342	C12	H 22	O11		
3506-029	342	C12	H 22	O11		
3506-028	342	C12	H 22	O11		
3506-020	342	C12	H 22	O11		
3506-019	342	C12	H 12	O11		
3501-027	342	C12	H 22	O11		
3506-021	342	C12	H 22	O11		
3506-018	342	C12	H 22	O11		
1199-019	342	C19	H 18	O 6		
1402-022	342	C19	H 18	O 6		
1199-020	342	C19	H 18	O 6		
1303-010	342	C22	H 18	O10		
3122-004	342	C22	H 46	O 2		
3120-019	342	C20	H 38	O 4		
2806-004	342	C19	H 18	O 6		
2806-005	342	C19	H 18	O 6		
2701-029°	342	C19	H 18	O 6		
2604-060	342	C18	H 14	O 7		
0899-018	342	C12	H 8	O 2		BR2
0501-010	342	C19	H 18	O 6		
0501-003	342	C15	H 16	O 8		
0304-025	342	C15	H 18	O 9		
0302-006	342	C16	H 22	O 8		
0699-005	342	C20	H 22	O 5		
2404-043	342	C19	H 16	O 6		
1801-001	342	C19	H 18	O 6		
2503-032	342	C19	H 18	O 6		
2402-010	342	C19	H 18	O 6		
2111-004	342	C22	H 30	O 3		
2101-024	342	C22	H 30	O 3		
2101-023	342	C22	H 30	O 3		
1601-005	342	C19	H 18	O 6		
2204-029	342	C20	H 22	O 5		
2102-013	342	C15	H 18	O 9		
2111-003	344	C22	H 32	O 3		
2207-002	344	C18	H 16	O 7		
2204-039	344	C20	H 24	O 5		
2204-038	344	C20	H 24	O 5		
2207-003	344	C18	H 16	O 7		
2112-002	344	C23	H 36	O 2		
2207-001	344	C18	H 16	O 7		
2003-015	344	C21	H 28	O 4		
0199-004	344	C14	H 16	O10		
0104-013	344	C15	H 20	O 9		
0304-020	344	C21	H 28	O 4		
0499-010	344	C23	H 36	O 2		
0505-005	344	C20	H 24	O 5		
0504-007	344	C19	H 20	O 6		
0508-013	344	C20	H 24	O 5		
0103-010	344	C15	H 20	O 9		
2803-003	344	C18	H 16	O 7		
2605-022	344	C17	H 12	O 8		
2806-003	344	C19	H 20	O 6		
1106-099	344	C17	H 12	O 8		
1104-063	344	C18	H 16	O 7		
1205-022	344	C19	H 20	O 6		
1106-066	344	C18	H 16	O 7		
1106-067	344	C18	H 16	O 7		
1199-007	344	C18	H 16	O 7		
1104-053	344	C18	H 16	O 7		
1105-050	344	C18	H 16	O 7		
1104-062	344	C18	H 16	O 7		
1104-061	344	C18	H 16	O 7		
1104-051	344	C18	H 16	O 7		
1105-049	344	C18	H 16	O 7		
3506-040	344	C12	H 24	O11		
3906-036	344	C17	H 12	O 8		
3507-005	344	C18	H 32	O 6		
3506-038	344	C12	H 24	O11		
3506-039	344	C12	H 24	O11		
1703-024	345	C18	H 17	O 7		+
2405-004	346	C18	H 18	O 7		
1802-006	346	C17	H 10	O 9		
1901-016	346	C18	H 18	O 7		
2301-006	346	C18	H 18	O 7		
2301-004	346	C18	H 18	O 7		
2111-002	346	C22	H 34	O 3		
2112-003	346	C23	H 38	O 2		
2206-014	346	C21	H 30	O 4		
3906-037	346	C17	H 14	O 8		
3906-031	346	C17	H 14	O 8		
1106-078	346	C17	H 14	O 8		
1106-074	346	C17	H 14	O 8		
1106-075	346	C17	H 14	O 8		
1106-100	346	C17	H 14	O 8		
1106-103	346	C17	H 14	O 8		
1108-013	346	C17	H 14	O 8		
1106-097	346	C17	H 14	O 8		
2504-039	346	C18	H 18	O 7		
2510-002	346	C22	H 18	O 4		
0504-009	346	C19	H 22	O 6		
0504-011	346	C18	H 18	O 7		
0502-011	346	C18	H 18	O 7		
0604-008	346	C19	H 22	O 6		
3118-083	346	C18	H 34	O 6		
3118-017	348	C18	H 36	O 6		
0604-015	348	C20	H 12	O 6		
0502-008	348	C18	H 20	O 5		S 1
0604-018	348	C20	H 12	O 6		
2510-001	348	C22	H 18	O 4		
2504-037	348	C18	H 20	O 7		
2506-015	348	C17	H 16	O 8		
3304-004	348	C17	H 16	O 2		S 3
3906-018	348	C18	H 20	O 7		
2111-010	348	C23	H 40	O 2		
2206-008	348	C20	H 28	O 5		
2111-001	348	C22	H 36	O 3		
2399-003	348	C17	H 16	O 8		
2102-021	348	C19	H 24	O 6		
2299-004	348	C19	H 26	O 6		
1601-020	350	C21	H 18	O 5		
2902-010	350	C20	H 30	O 5		
2803-009	350	C17	H 15	O 6		CL1
2902-004	350	C20	H 30	O 5		
2902-013	350	C20	H 30	O 5		
0502-007	350	C21	H 18	O 5		
0599-020	350	C20	H 14	O 6		
0601-017	350	C20	H 14	O 6		
0499-005	350	C21	H 34	O 4		
1402-006	350	C20	H 14	O 6		
1402-015	350	C21	H 18	O 5		
0402-009	352	C17	H 20	O 8		
0999-009	352	C17	H 16	O 6		
0601-018	352	C20	H 16	O 6		
0599-018	352	C19	H 12	O 7		
0502-016	352	C21	H 24	O 5		
0702-010	352	C20	H 16	O 6		
0702-009	352	C20	H 16	O 6		
2902-012	352	C20	H 32	O 5		
2701-027	352	C24	H 20	O 5		
2902-008	352	C20	H 32	O 5		
2802-015	352	C17	H 17	O 6		CL1
2605-039	352	C18	H 16	O 8		
2902-016	352	C20	H 32	O 5		
2803-007	352	C17	H 17	O 6		CL1
2902-002	352	C20	H 32	O 5		
2502-002	352	C18	H 24	O 7		
1601-019	352	C21	H 20	O 5		
1801-020	352	C20	H 16	O 6		
1801-021	352	C19	H 12	O 7		
3401-007	352	C20	H 32	O 5		

3906-023	352	C19	H 28	O 6		
3125-001	352	C25	H 52			
3903-003	354	C17	H 23	O 3	BR1	
3506-042	354	C12	H 26	O10		
3906-046	354	C21	H 22	O 5		
2403-042	354	C19	H 14	O 7		
2106-007	354	C22	H 26	O 4		
2404-033	354	C19	H 14	O 7		
1902-006	354	C16	H 18	O 9		
1902-009	354	C16	H 18	O 9		
1902-010	354	C16	H 18	O 9		
2902-011	354	C20	H 34	O 5		
2605-033	354	C20	H 18	O 6		
2703-002	354	C20	H 18	O 6		
2703-003	354	C20	H 18	O 6		
2902-015	354	C20	H 34	O 5		
0699-001	354	C21	H 22	O 6		
0601-032	354	C20	H 18	O 6		
0698-001	354	C21	H 22	O 5		
0508-033	354	C21	H 22	O 5		
0601-034	354	C20	H 18	O 6		
0507-007	354	C16	H 18	O 9		
1002-016	354	C21	H 22	O 5		
0601-019	354	C20	H 18	O 6		
0601-030	354	C20	H 18	O 6		
0105-006	354	C17	H 22	O 8		
0601-031	354	C20	H 18	O 6		
0698-007	354	C20	H 18	O 6		
0698-008	354	C20	H 18	O 6		
0601-033	354	C20	H 18	O 6		
0511-008	354	C21	H 22	O 5		
1203-023	354	C21	H 22	O 5		
3124-002	354	C24	H 50	O 1		
3509-010	355	C12	H 21	N 1 O11		
3506-041	356	C12	H 20	O12		
3122-017	356	C22	H 44	O 3		
3121-005	356	C21	H 40	O 4		
1499-015	356	C19	H 16	O 7		
1204-007	356	C20	H 20	O 6		
1102-006	356	C16	H 12	O 6		
1404-001	356	C19	H 16	O 7		
1403-003	356	C19	H 16	O 7		
1108-018	356	C19	H 16	O 7		
0601-022	356	C20	H 20	O 6		
0304-026	356	C16	H 20	O 9		
0604-003	356	C22	H 28	O 4		
0601-038	356	C12	H 24	O 5		
0601-016	356	C20	H 20	O 6		
0698-009	356	C20	H 20	O 6		
0601-028	356	C20	H 20	O 6		
0604-012	356	C20	H 20	O 6		
2902-014	356	C20	H 36	O 5		
2703-007	356	C19	H 16	O 7		
2605-028	356	C19	H 16	O 7		
2806-002	356	C21	H 24	O 5		
2701-033	356	C20	H 20	O 6		
2106-002	356	C22	H 28	O 4		
2101-026	356	C23	H 32	O 3		
2106-008	358	C22	H 30	O 4		
2106-003	358	C22	H 30	O 4		
2106-004	358	C22	H 30	O 4		
2302-002	358	C18	H 14	O 8		
2302-001	358	C18	H 14	O 8		
2106-006	358	C22	H 30	O 4		
2399-005	358	C19	H 18	O 7		
2701-030	358	C15	H 18	O 7		
2504-033	358	C20	H 22	O 6		
2806-001	358	C21	H 26	O 5		
0699-019	358	C20	H 22	O 6		
0601-011	358	C20	H 22	O 6		
0601-023	358	C20	H 22	O 6		
1001-014	358	C20	H 22	O 6		
0903-007	358	C21	H 26	O 5		
1107-006	358	C19	H 18	O 7		
1304-007	358	C22	H 18	O11		
1105-054	358	C19	H 18	O 7		
1104-056	358	C19	H 18	O 7		
1304-005	358	C22	H 18	O11		
1105-060	358	C18	H 14	O 8		
1107-009	358	C19	H 18	O 8		
1104-064	358	C19	H 18	O 7		
3110-080	358	C20	H 38	O 5		
3906-039	358	C20	H 22	O 6		
3906-045	358	C21	H 26	O 5		
3304-005	360	C18	H 16	O 2	S 3	
3304-006	360	C18	H 16	O 2	S 3	
3304-007	360	C18	H 16	O 2	S 3	
1105-056	360	C18	H 16	O 8		
1106-083	360	C18	H 16	O 8		
1106-098	360	C18	H 16	O 8		
1199-031	360	C18	H 16	O 8		
1106-085	360	C18	H 16	O 8		
1403-001	360	C18	H 16	O 8		
1199-022	360	C18	H 16	O 8		
1108-014	360	C18	H 16	O 8		
1106-079	360	C18	H 16	O 8		
0601-005	360	C20	H 24	O 6		
0399-014	360	C18	H 16	O 8		
0604-010	360	C20	H 24	O 6		
0604-009	360	C20	H 24	O 6		
2704-011	360	C21	H 12	O 6		
2301-007	360	C19	H 20	O 7		
2403-016	361	C14	H 7	O 5	CL3	
2206-013	362	C21	H 30	O 5		
2302-021	362	C18	H 15	O 6	CL1	
1901-017	362	C18	H 18	O 8		
2003-011	362	C22	H 34	O 4		
2206-020	362	C21	H 30	O 5		
2604-024	362	C17	H 11	O 7	CL1	
3005-042	362	C22	H 34	O 4		
2604-059	362	C18	H 15	O 6	CL1	
0699-020	362	C20	H 18	O 4		
0601-004	362	C20	H 26	O 6		
0604-023	364	C20	H 12	O 7		
0604-016	364	C21	H 16	O 6		
0503-007	364	C24	H 28	O 3		
2509-003	364	C20	H 12	O 7		
2299-005	364	C20	H 28	O 6		
2109-002	364	C18	H 17	O 6	CL1	
1402-008	364	C21	H 16	O 6		
3506-044	364	C16	H 28	O 9		
3126-001	366	C26	H 54			
3124-002	366	C24	H 46	O 2		
1402-007	366	C21	H 18	O 6		
1404-010	366	C20	H 14	O 7		
1801-023	366	C20	H 14	O 7		
2605-041	366	C20	H 14	O 7		
0604-017	366	C21	H 18	O 6		
0501-011	366	C24	H 30	O 3		
0599-005	366	C24	H 30	O 3		
0801-004	366	C26	H 38	O 1		
0304-005	366	C17	H 18	O 9		
3124-007	366	C24	H 46	O 2		
3122-006	368	C24	H 48	O 2		
3124-004	368	C24	H 48	O 2		
0511-006	368	C22	H 24	O 5		
0508-031	368	C21	H 20	O 6		
0903-022	368	C22	H 24	O 5		
2605-044	368	C20	H 16	O 7		
2701-028	368	C21	H 20	O 6		
2805-010	368	C21	H 20	O 6		
2605-036	368	C20	H 16	O 7		
2605-034	368	C21	H 20	O 6		
1801-022	368	C20	H 16	O 7		
2302-006	368	C17	H 14	O 5	CL2	
1902-008	368	C17	H 20	O 7		
1103-052	368	C20	H 20	O 7		
1104-065	368	C20	H 20	O 7		
3906-028	368	C19	H 12	O 8		
3203-010	368	C18	H 24	O 8		
2104-027	369	C20	H 19	N 1	O 6	
2101-021	370	C23	H 30	O 4		
2399-006	370	C22	H 26	O 5		
3904-002	370	C17	H 23	O 4	BR1	
1104-066	370	C19	H 18	O 8		
1106-090	370	C19	H 14	O 8		
2605-030	370	C20	H 18	O 7		
2605-029	370	C20	H 18	O 7		
2703-008	370	C20	H 18	O 7		
2703-041	370	C20	H 18	O 7		
2605-035	370	C20	H 18	O 7		
2703-004	370	C20	H 18	O 7		
2504-028	370	C21	H 22	O 6		
2504-034	370	C21	H 20	O 6		
0903-021	370	C22	H 26	O 5		
0699-024	370	C20	H 18	O 7		
0702-012	370	C18	H 10	O 9		
0505-007	370	C22	H 26	O 5		
0601-029	370	C21	H 22	O 6		
0509-002	370	C16	H 18	O10		
0601-020	370	C20	H 18	O 7		
0105-007	370	C17	H 22	O 9		
0601-036	370	C20	H 18	O 7		
0601-037	370	C20	H 18	O 7		
3124-003	370	C24	H 50	O 2		
3122-018	370	C22	H 42	O 4		
0302-009	372	C17	H 24	O 9		
0602-001	372	C20	H 20	O 7		
0601-013	372	C21	H 24	O 6		
0702-002	372	C19	H 14	O 5		
0903-001	372	C22	H 28	O 5		
0903-003	372	C22	H 28	O 5		

ID	MW	C	H	N	O	Other
0903-002	372	C22	H 28		O 5	
0601-006	372	C22	H 28		O 5	
0601-007	372	C22	H 28		O 5	
0505-006	372	C22	H 28		O 5	
0702-011	372	C18	H 12		O 9	
0601-024	372	C21	H 24		O 6	
2701-031	372	C20	H 20		O 7	
2605-032	372	C20	H 20		O 7	
2605-045	372	C18	H 12		O 9	
2504-038	372	C21	H 24		O 6	
2701-032	372	C20	H 20		O 7	
1499-014	372	C20	H 20		O 7	
1105-051	372	C20	H 20		O 7	
1106-088	372	C19	H 16		O 7	
1104-030	372	C20	H 20		O 7	
2112-001	372	C24	H 36		O 8	
2101-028	372	C23	H 32		O 4	
2101-020	372	C23	H 32		O 4	
2301-016	372	C20	H 20		O 7	
2302-017	372	C18	H 12		O 9	
2301-009	374	C19	H 18		O 8	
2301-008	374	C19	H 18		O 8	
2106-005	374	C22	H 30		O 5	
2301-011	374	C20	H 22		O 7	
2302-003	374	C18	H 14		O 9	
1105-058	374	C19	H 18		O 8	
1107-013	374	C19	H 18		O 9	
1499-012	374	C19	H 18		O 8	
1105-057	374	C19	H 18		O 8	
1108-019	374	C19	H 18		O 8	
1106-086	374	C19	H 18		O 8	
1106-104	374	C19	H 18		O 8	
1106-115	374	C19	H 18		O 8	
1106-081	374	C19	H 18		O 8	
1108-015	374	C19	H 18		O 8	
1199-024	374	C19	H 18		O 8	
1206-010	374	C20	H 22		O 7	
1199-017	374	C19	H 18		O 8	
2511-001	374	C22	H 14		O 6	
2509-002	374	C22	H 14		O 6	
2599-011	374	C22	H 14		O 6	
2510-004	374	C22	H 14		O 6	
2510-003	374	C22	H 14		O 6	
2510-006	374	C22	H 14		O 6	
2510-005	374	C22	H 14		O 6	
0903-013	374	C21	H 26		O 6	
0903-010	374	C21	H 26		O 6	
0601-039	374	C20	H 22		O 7	
3999-011	374	C19	H 18		O 8	
0601-010	376	C20	H 24		O 7	
0903-004	376	C22	H 28		O 5	
0604-011	376	C20	H 24		O 7	
0501-027	376	C20	H 24		O 7	
2704-005	376	C22	H 16		O 6	
2703-001	376	C22	H 16		O 6	
1106-107	376	C18	H 16		O 9	
1199-032	376	C18	H 16		O 9	
1106-109	376	C18	H 16		O 9	
2399-001	376	C18	H 16		O 9	
2204-024	376	C19	H 20		O 6	S 1
1801-009	376	C22	H 16		O 6	
2206-021	376	C22	H 32		O 5	
2002-022	376	C22	H 32		O 5	
3113-019	376	C19	H 20		O 8	
2502-031	378	C19	H 22		O 8	
1901-015	378	C19	H 22		O 8	
1801-010	378	C22	H 18		O 6	
1801-008	378	C22	H 18		O 6	
1106-005	378	C22	H 18		O 6	
1499-017	378	C22	H 18		O 6	
1402-009	378	C22	H 18		O 6	
1199-027	378	C18	H 15		O 7	CL1
1402-019	378	C22	H 18		O 6	
2903-001	378	C21	H 30		O 6	
2703-005	378	C21	H 14		O 7	
2605-023	378	C27	H 11		O 8	CL1
0604-024	378	C21	H 14		O 7	
0604-025	378	C21	H 14		O 7	
0698-002	378	C21	H 14		O 7	
3501-011	379	C14	H 25	N 3	O 9	
3127-001	380	C27	H 56			
3906-048	380	C23	H 24		O 5	
0604-019	380	C21	H 16		O 7	
2804-023	380	C24	H 28		O 4	
2504-022	380	C17	H 16		O 8	S 1
3005-043	380	C22	H 36		O 5	
1499-016	380	C22	H 20		O 6	
1404-002	380	C22	H 20		O 6	
2101-032	380	C15	H 18		O 7	CL2
2403-012	380	C23	H 24		O 5	
2702-015	381	C20	H 15	N 1	O 7	
2605-037	382	C21	H 18		O 7	
2106-028	382	C21	H 18		O 7	
1602-010	382	C21	H 18		O 7	
1801-024	382	C20	H 14		O 8	
2203-009	382	C18	H 18		O 8	
0599-006	382	C24	H 30		O 4	
0501-014	382	C24	H 30		O 4	
0501-013	382	C24	H 30		O 4	
3906-047	382	C23	H 26		O 5	
3126-002	382	C26	H 54		O 1	
3125-002	382	C25	H 50		O 2	
3906-027	382	C20	H 14		O 8	
2603-035	383	C19	H 13	N 1	O 8	
2605-031	384	C21	H 20		O 7	
2803-004	384	C16	H 10		O 7	CL2
2504-030	384	C22	H 24		O 6	
3124-009	384	C24	H 48		O 3	
0510-002	384	C17	H 20		O10	
0509-004	384	C17	H 20		O10	
0605-010	384	C21	H 20		O 7	
0599-003	384	C24	H 32		O 4	
0698-010	384	C21	H 20		O 7	
0605-011	384	C20	H 16		O 8	
1104-067	384	C20	H 20		O 8	
1499-020	384	C21	H 20		O 7	
1205-022	384	C22	H 24		O 6	
3124-005	384	C24	H 48		O 3	
3116-005	384	C18	H 36		O 2	
3123-003	384	C23	H 44		O 4	
1105-061	386	C20	H 18		O 8	
1104-038	386	C21	H 22		O 7	
1106-089	386	C20	H 18		O 8	
1106-106	386	C20	H 18		O 8	
0601-026	386	C22	H 26		O 6	
0601-027	386	C22	H 26		O 6	
0508-022	386	C21	H 22		O 7	
0903-023	386	C23	H 30		O 5	
0901-002	386	C17	H 18		O 4	
0502-012	386	C21	H 22		O 7	
0505-009	386	C21	H 22		O 7	
0504-010	386	C21	H 22		O 7	
0504-013	386	C21	H 22		O 7	
0505-015	386	C21	H 22		O 7	
0505-013	386	C21	H 22		O 7	
0505-016	386	C21	H 22		O 7	
3197-031	386	C21	H 38		O 6	
2703-019	386	C20	H 18		O 8	
2703-018	386	C20	H 18		O 8	
2703-049	396	C20	H 18		O8	
2703-017	386	C20	H 18		O 8	
2003-019	386	C24	H 34		O 4	
2302-018	386	C19	H 14		O 9	
2302-019	388	C18	H 12		O10	
2301-018	388	C21	H 24		O 7	
1602-009	388	C22	H 12		O 7	
2203-015	388	C18	H 23		O 9	
2302-010	388	C19	H 19		O 6	CL1
2204-026	388	C22	H 28		O 6	
2504-021	388	C16	H 13		O 8	S 1 NA1
2504-035	388	C21	H 24		O 7	
3999-012	388	C21	H 24		O 7	
0505-011	388	C21	H 24		O 7	
0505-012	388	C21	H 24		O 7	
0505-016	388	C21	H 24		O 7	
0504-012	388	C21	H 24		O 7	
0903-020	388	C22	H 28		O 6	
0903-011	388	C22	H 28		O 6	
0904-014	388	C22	H 28		O 6	
0903-012	388	C22	H 28		O 6	
0903-015	388	C22	H 28		O 6	
0903-019	388	C22	H 28		O 6	
1106-105	388	C20	H 20		O 8	
1106-087	388	C20	H 20		O 8	
1199-021	388	C20	H 20		O 8	
1108-016	388	C20	H 20		O 8	
1108-023	390	C19	H 18		O 9	
1199-025	390	C19	H 18		O 9	
1108-026	390	C19	H 18		O 9	
1104-028	390	C20	H 22		O 8	
1201-012	390	C20	H 22		O 8	
1107-010	390	C19	H 18		O 9	
0399-002	390	C21	H 26		O 7	
0199-039	390	C20	H 22		O 8	
0902-008	390	C24	H 22		O 5	
3904-003	390	C15	H 20		O 2	BR2
3904-004	390	C15	H 20		O 2	BR1
2511-002	390	C22	H 14		O 7	
2804-013	390	C20	H 22		O 8	
2902-006	390	C23	H 34		O 5	
2108-005	390	C21	H 23		O 5	CL1
2108-004	390	C21	H 23		O 5	CL1

ID	MW	C	H	N	O	Misc
2299-006	390	C21	H 26		□ 7	
1601-025	390	C25	H 26		□ 4	
2399-002	390	C18	H 14		□10	
3201-012	390	C19	H 18		□ 9	
2101-027	390	C23	H 31		□ 3	CL1
1601-024	390	C25	H 26		□ 4	
2112-004	390	C26	H 46		□ 2	
1801-003	392	C23	H 20		□ 6	
1801-013	392	C23	H 20		□ 6	
1601-023	392	C25	H 28		□ 4	
2701-001	392	C19	H 20		□ 9	
2703-006	392	C22	H 16		□ 7	
2704-006	392	C22	H 16		□ 7	
0501-029	392	C20	H 24		□ 8	
0902-001	392	C24	H 24		□ 5	
1101-019	392	C23	H 20		□ 6	
1404-003	394	C23	H 22		□ 6	
1404-004	394	C22	H 18		□ 7	
0604-014	394	C22	H 18		□ 7	
0604-020	394	C22	H 18		□ 7	
2199-007	394	C19	H 22		□ 9	
1801-011	394	C22	H 18		□ 7	
2302-014	394	C19	H 14		□ 5	CL2
2503-024	394	C19	H 22		□ 9	
2005-004	394	C20	H 26		□ 8	
1801-002	394	C23	H 22		□ 6	
1801-012	394	C23	H 22		□ 6	
2404-028	394	C23	H 22		□ 6	
3128-001	394	C28	H 58			
3127-004	394	C27	H 54		□ 1	
3126-008	394	C26	H 50		□ 2	
2403-017	395	C14	H 6		□ 5	CL4
2404-039	396	C23	H 24		□ 6	
2404-032	396	C23	H 24		□ 6	
3126-007	396	C26	H 52		□ 2	
3127-003	396	C27	H 56		□ 1	
3127-002	396	C27	H 56		□ 1	
0701-008	396	C20	H 12		□ 9	
0501-017	396	C24	H 28		□ 5	
0899-008	396	C27	H 40		□ 2	
2605-038	396	C22	H 20		□ 7	
2703-013	396	C22	H 20		□ 7	
2803-008	397	C17	H 17		□ 6	BR
3999-015	397	C22	H 23	N 1	□ 6	
3126-003	398	C26	H 54		□ 2	
2703-023	398	C21	H 18		□ 8	
2803-005	398	C17	H 12		□ 7	CL2
0501-016	398	C24	H 30		□ 5	
0605-004	398	C22	H 22		□ 7	
0605-002	398	C22	H 22		□ 7	
0698-006	398	C22	H 22		□ 7	
2106-029	398	C21	H 18		□ 8	
2699-025	399	C20	H 17	N 1	□ 8	
2703-025	400	C21	H 20		□ 8	
2802-017	400	C17	H 14		□ 7	CL2
2206-010	400	C25	H 36		□ 4	
2301-005	400	C17	H 14		□ 7	CL2
0698-003	400	C21	H 20		□ 8	
0605-006	400	C21	H 20		□ 8	
0501-015	400	C24	H 32		□ 5	
0903-024	400	C24	H 32		□ 5	
3124-006	400	C24	H 48		□ 4	
3116-038	400	C23	H 46	N 1	□ 4	+
0902-014	402	C25	H 22		□ 5	
0601-035	402	C22	H 26		□ 7	
0903-017	402	C22	H 26		□ 7	
0903-018	402	C22	H 26		□ 7	
0505-014	402	C21	H 22		□ 8	
0899-004	402	C27	H 46		□ 2	
0902-013	402	C25	H 22		□ 5	
0902-016	402	C25	H 22		□ 5	
2302-011	402	C20	H 21		□ 6	CL1
2302-020	402	C19	H 14		□10	
2302-007	402	C17	H 13		□ 5	CL3
2703-033	402	C20	H 18		□ 9	
3999-013	402	C22	H 26		□ 7	
3301-012	402	C22	H 42		□ 2	S 2
3197-032	402	C21	H 38		□ 7	
3903-004	402	C23	H 30		□ 6	
1105-059	402	C21	H 22		□ 8	
1107-007	402	C21	H 22		□ 3	
2104-028	403	C20	H 18	N 1	□ 6	CL1
2207-016	404	C21	H 24		□ 8	
2114-001	404	C27	H 48		□ 2	
2101-030	404	C23	H 29		□ 4	CL1
2302-004	404	C18	H 12		□11	
2301-015	404	C19	H 16		□10	
1199-026	404	C20	H 20		□ 9	
1401-007	404	C25	H 24		□ 5	
1108-027	404	C20	H 20		□ 9	
1401-010	404	C25	H 24		□ 5	
1401-008	404	C25	H 24		□ 5	
1401-009	404	C25	H 24		□ 5	
1107-011	404	C20	H 20		□ 9	
1104-029	404	C21	H 24		□ 8	
1201-013	404	C21	H 24		□ 8	
1301-001	404	C21	H 24		□ 8	
1102-004	404	C25	H 24		□ 5	
2510-012	404	C24	H 20		□ 6	
2804-014	404	C21	H 24		□ 8	
0903-016	404	C22	H 28		□ 7	
0902-009	404	C25	H 24		□ 5	
0902-010	404	C25	H 24		□ 5	
0902-015	404	C25	H 24		□ 5	
0902-004	406	C25	H 26		□ 5	
0902-003	406	C25	H 26		□ 5	
0902-002	406	C25	H 26		□ 5	
1099-007	406	C25	H 26		□ 5	
0199-017	406	C20	H 22		□ 9	
0199-016	406	C20	H 22		□ 9	
0199-013	406	C20	H 22		□ 9	
0199-008	406	C20	H 22		□ 9	
0199-007	406	C20	H 22		□ 9	
0902-005	406	C25	H 26		□ 5	
1099-010	406	C25	H 26		□ 5	
2701-002	406	C20	H 22		□ 9	
2101-029	406	C23	H 31		□ 4	CL1
2301-013	406	C19	H 18		□10	
2204-034	408	C19	H 20		□10	
2001-012	408	C28	H 40		□ 2	
1801-007	408	C23	H 20		□ 7	
2301-010	408	C19	H 17		□ 8	CL1
0701-010	408	C26	H 16		□ 5	
1099-006	408	C25	H 20		□ 5	
0504-004	408	C20	H 24		□ 9	
1404-005	408	C23	H 20		□ 7	
1499-018	408	C23	H 20		□ 7	
1199-036	408	C22	H 16		□ 8	
1499-021	408	C23	H 20		□ 7	
3197-038	408	C27	H 52		□ 2	
3197-037	408	C27	H 52		□ 2	
3129-001	408	C29	H 60			
3128-002	410	C28	H 58		□ 1	
3903-005	410	C21	H 30		□ 8	
3127-005	410	C27	H 54		□ 2	
1499-013	410	C23	H 22		□ 7	
0899-007	410	C28	H 42		□ 2	
0605-008	410	C22	H 18		□ 8	
0899-005	410	C28	H 42		□ 2	
1801-018	410	C23	H 22		□ 7	
1801-017	410	C23	H 24		□ 7	
2404-040	410	C24	H 26		□ 6	
1801-015	410	C23	H 22		□ 7	
1801-014	410	C23	H 22		□ 7	
1801-006	410	C23	H 22		□ 7	
1801-025	410	C23	H 22		□ 7	
1801-005	410	C23	H 22		□ 7	
1801-004	410	C23	H 22		□ 7	
2505-005	410	C18	H 18		□ 9	S 1
2702-014	411	C22	H 21	N 1	□ 7	
2703-016	412	C22	H 20		□ 8	
2805-018	412	C26	H 36		□ 4	
2703-015	412	C22	H 20		□ 8	
2903-012	412	C21	H 32		□ 8	
2703-014	412	C22	H 20		□ 8	
2302-009	412	C18	H 14		□ 7	CL2
2003-016	412	C23	H 36		□ 4	
2106-030	412	C21	H 16		□ 9	
0601-021	412	C22	H 20		□ 8	
0501-012	412	C25	H 32		□ 5	
1405-C08	412	C25	H 32		□ 5	
1105-039	412	C21	H 16		□ 9	
3126-010	412	C26	H 52		□ 3	
3126-009	412	C26	H 52		□ 3	
3906-049	414	C23	H 23		□ 5	CL1
0605-005	414	C22	H 22		□ 8	
0605-007	414	C22	H 22		□ 8	
0698-004	414	C22	H 22		□ 8	
2206-015	414	C26	H 30		□ 4	
2206-016	414	C26	H 38		□ 4	
2108-003	414	C23	H 23		□ 5	CL1
2001-015	414	C28	H 46		□ 2	
2703-036	414	C21	H 18		□ 9	
2399-010	415	C17	H 12		□ 8	CL2
2301-020	416	C23	H 28		□ 7	
2301-036	416	C21	H 20		□ 9	
2204-025	416	C24	H 32		□ 6	
2301-025	416	C23	H 28		□ 7	
2301-017	416	C23	H 28		□ 7	
2603-013	416	C21	H 20		□ 9	
2603-028	416	C21	H 20		□ 9	
2604-031	416	C21	H 20		□ 9	

ID	MW	C	H	N	□	Other
2604-030	416	C21	H 20		□ 9	
0602-004	416	C23	H 28		□ 7	
0602-003	416	C22	H 24		□ 8	
0902-012	416	C26	H 24		□ 5	
0899-003	416	C28	H 48		□ 2	
0902-011	416	C26	H 24		□ 5	
0899-002	416	C28	H 48		□ 2	
3197-033	416	C22	H 40		□ 7	
1106-111	416	C21	H 20		□ 9	
1401-025	416	C21	H 20		□ 9	
1101-005	416	C21	H 20		□ 9	
1431-024	416	C21	H 20		□ 9	
1103-053	416	C21	H 20		□ 9	
1499-009	416	C26	H 24		□ 5	
1701-002	417	C21	H 21		□ 9	+
2104-029	417	C21	H 20	N 1	□ 6	CL1
2301-019	418	C22	H 26		□ 8	
2003-013	418	C26	H 42		□ 4	
2207-009	418	C22	H 26		□ 8	
2207-011	418	C22	H 26		□ 8	
1104-005	418	C20	H 18		□10	
1201-002	418	C21	H 22		□ 9	
1401-016	418	C26	H 26		□ 5	
1203-025	418	C21	H 22		□ 9	
1107-012	418	C21	H 22		□ 9	
1203-026	418	C21	H 22		□ 9	
1199-015	418	C25	H 22		□ 6	
0603-003	418	C22	H 26		□ 8	
0502-004	418	C20	H 24		□ 9	
0502-002	418	C20	H 24		□ 9	
0511-005	418	C22	·H 26		□ 8	
0601-003	418	C24	H 34		□ 6	
1002-004	418	C21	H 22		□ 9	
0603-001	418	C22	H 26		□ 8	
1002-002	418	C21	H 22		□ 9	
0606-001	418	C22	H 26		□ 8	
0603-002	418	C22	H 26		□ 8	
2699-003	418	C21	H 22		□ 9	
2699-002	418	C21	H 22		□ 9	
2702-011	419	C19	H 16	N 2	□ 7	CL1
2104-030	419	C20	H 18	N 1	□ 7	CL1
1702-003	419	C20	H 19		□10	+
2301-014	420	C19	H 16		□11	
2701-007	420	C21	H 24		□ 9	
2804-017	420	C21	H 24		□ 9	
2804-020	420	C21	H 24		□ 9	
0602-002	420	C21	H 24		□ 9	
0901-011	420	C25	H 24		□ 6	
1402-004	420	C25	H 24		□ 6	
1499-006	420	C25	H 24		□ 6	
1601-026	420	C26	H 28		□ 5	
1199-014	420	C25	H 24		□ 6	
1499-008	420	C25	·H 24		□ 6	
1402-003	420	C25	H 24		□ 6	
1199-013	420	C25	H 24		□ 6	
3906-044	420	C17	H 24		□12	
3129-006	422	C29	H 58		□ 1	
3130-001	422	C30	H 62			
3129-005	422	C29	H 58		□ 1	
1199-010	422	C25	H 26		□ 6	
1299-014	422	C24	H 22		□ 7	
1303-005	422	C20	H 22		□10	
1002-010	422	C21	H 26		□ 9	
0902-006	422	C25	H 26		□ 6	
0902-007	422	C25	H 26		□ 6	
0701-011	422	C27	H 18		□ 5	
0605-010	422	C20	H 22		□10	
1099-008	422	C25	H 26		□ 6	
2302-008	422	C16	H 10		□ 5	CL4
2404-037	422	C19	H 18		□11	
2404-035	422	C19	H 18		□11	
2002-005	424	C28	H 40		□ 3	
2404-041	424	C25	H 28		□ 6	
2499-010	424	C25	H 28		□ 6	
2499-009	424	C25	H 28		□ 6	
0899-005	424	C29	H 44		□ 2	
0701-012	424	C26	H 16		□ 6	
0599-017	424	C23	H 20		□ 8	
0802-010	424	C19	H 20		□11	
0802-009	424	C19	H 20		□11	
0802-011	424	C19	H 20		□11	
1299-006	424	C25	H 28		□ 6	
3129-004	424	C29	H 60		□ 1	
3128-004	424	C28	H 56		□ 2	
3129-003	424	C29	H 60		□ 1	
3129-002	424	C29	H 60		□ 1	
3901-005	424	C21	H 28		□ 9	
3901-006	424	C21	H 28		□ 9	
2505-006	424	C19	H 20		□ 9	S 1
2806-006	426	C27	H 38		□ 4	
3905-023	426	C24	H 26		□ 7	
0505-027	426	C24	H 26		□ 7	
0505-026	426	C24	H 26		□ 7	
0505-017	426	C24	H 26		□ 7	
0505-020	426	C24	H 26		□ 7	
0505-019	426	C24	H 26		□ 7	
0505-018	426	C24	H 26		□ 7	
0504-015	426	C24	H 26		□ 7	
0503-006	426	C24	H 26		□ 7	
0505-021	428	C24	H 28		□ 7	
0698-005	428	C23	H 24		□ 8	
3128-003	428	C28	H 58		□ 2	
2703-032	428	C22	H 20		□ 9	
2703-029	428	C22	H 20		□ 9	
2703-035	428	C22	H 20		□ 9	
2703-028	428	C22	H 20		□ 9	
2703-027	428	C22	H 20		□ 9	
1599-023	428	C21	H 16		□10	
1599-024	428	C21	H 16		□10	
2302-015	428	C19	H 15		□ 5	CL3
2207-004	428	C24	H 28		□ 7	
2301-026	430	C24	H 30		□ 7	
1101-007	430	C21	H 20		□ 9	
1401-027	430	C22	H 22		□ 9	
2606-007	430	C17	H 14		□ 7	
2702-004	430	C21	H 22	N 2	□ 8	
2606-003	430	C17	H 14		□ 7	
0903-005	430	C24	H 30		□ 7	
0802-018	430	C29	H 50		□ 2	
0698-011	430	C24	H 30		□ 7	
0903-006	430	C24	H 30		□ 7	
0899-001	430	C29	H 50		□ 2	
0504-014	430	C24	H 30		□ 7	
0699-023	432	C24	H 32		□ 7	
0399-015	432	C22	H 24		□ 9	
0606-003	432	C22	H 24		□ 9	
2699-005	432	C21	H 20		□10	
2604-029	432	C21	H 20		□10	
2603-030	432	C21	H 20		□10	
1108-024	432	C22	H 24		□ 9	
1401-004	432	C21	H 20		□10	
1103-003	432	C21	H 20		□10	
1108-020	432	C22	H 24		□ 9	
1103-036	432	C21	H 20		□10	
1103-012	432	C21	H 20		□10	
1103-013	432	C21	H 20		□10	
1104-010	432	C21	H 20		□10	
1103-015	432	C21	H 20		□10	
1103-005	432	C21	H 20		□10	
1401-002	432	C21	H 20		□10	
1104-019	432	C21	H 20		□10	
1803-010	432	C21	H 20		□10	
2204-005	432	C25	H 36		□ 6	
2204-006	432	C25	H 36		□ 6	
2207-010	432	C23	H 28		□ 8	
2207-017	432	C23	H 28		□ 8	
2301-021	432	C23	H 28		□ 8	
2207-012	432	C23	H 28		□ 8	
2207-005	432	C23	H 28		□ 8	
1702-004	433	C21	H 21		□10	+
1701-004	433	C21	H 21		□10	+
1108-025	434	C21	H 22		□10	
1199-033	434	C26	H 26		□ 6	
1106-062	434	C16	H 11		□10	S1 K1
1499-010	434	C26	H 26		□ 6	
1106-012	434	C20	H 18		□11	
1106-013	434	C20	H 18		□11	
1204-005	434	C21	H 22		□10	
1204-004	434	C21	H 22		□10	
1106-019	434	C20	H 18		□11	
1106-011	434	C20	H 18		□11	
1203-016	434	C21	H 22		□10	
1106-044	434	C16	H 11		□10	S1 K1
1404-006	434	C26	H 26		□ 6	
1205-016	434	C21	H 22		□10	
1205-017	434	C21	H 22		□10	
1203-003	434	C21	H 22		□10	
1203-002	434	C21	H 22		□10	
1203-015	434	C21	H 22		□10	
1404-008	434	C26	H 26		□ 6	
1205-003	434	C21	H 22		□10	
1106-034	434	C20	H 18		□11	
1499-007	434	C26	H 26		□ 6	
2699-011	434	C21	H 22		□10	
3002-004	434	C21	H 22		□10	
0802-008	434	C21	H 22		□10	
1001-017	434	C21	H 22		□10	
1003-003	434	C21	H 22		□12	
1003-002	434	C21	H 22		□10	
1002-007	434	C21	H 22		□10	
3197-041	434	C32	H 62		□ 3	
3301-013	434	C22	H 42		□ 2	S 3

621

0702-008	435	C25	H 25	N 1	O 6					1002-023	448	C21	H 20	O11			
1703-002	435	C20	H 19	O11	+					0399-016	448	C22	H 24	O10			
1299-008	436	C26	H 28	O 6						0699-026	448	C24	H 32	O 8			
1199-012	436	C26	H 28	O 6						1703-012	449	C21	H 21	O11	+		
1002-009	436	C21	H 24	O10						1703-016	449	C21	H 21	O11	+		
0701-013	436	C27	H 16	O 6						1702-004	449	C21	H 21	O11	+		
3131-001	436	C31	H 64							1702-005	449	C21	H 21	O11	+		
3197-040	436	C32	H 64	O 3						1703-005	449	C21	H 21	O11	+		
2806-010	436	C27	H 36	O 4						3502-022	449	C19	H 39	N 5	O 7		
2806-008	436	C27	H 36	O 4						3132-001	450	C32	H 66				
2404-036	436	C20	H 20	O11						3130-005	450	C30	H 58	O 2			
2503-030	436	C21	H 24	O10						3131-004	450	C31	H 62	O 1			
2503-029	436	C21	H 24	O10						1299-005	450	C21	H 22	O11			
2503-028	438	C21	H 26	O10						1205-002	450	C21	H 22	O11			
2301-034	438	C20	H 22	O11						1205-020	450	C21	H 32	O11			
2806-009	438	C28	H 38	O 4						1203-014	450	C21	H 22	O11			
2806-011	438	C28	H 38	O 4						1206-004	450	C21	H 22	O11			
3129-007	438	C29	H 58	O 2						1401-012	450	C22	H 22	O10			
3129-008	438	C29	H 56	O 2						1204-003	450	C21	H 22	O11			
3130-002	438	C30	H 62	O 1						1204-010	450	C21	H 22	O11			
0599-016	438	C24	H 22	O 8						1003-008	450	C21	H 22	O11			
1099-009	438	C26	H 30	O 6						0199-020	450	C22	H 26	O10			
1099-012	438	C26	H 30	O 6						1002-022	450	C21	H 22	O11			
1299-007	438	C26	H 30	O 6						1002-014	450	C22	H 26	O10			
3111-022	438	C23	H 34	O 8						1002-012	450	C22	H 26	O10			
1205-013	440	C26	H 32	O 6						2502-017	450	C31	H 46	O 2			
0903-008	440	C27	H 36	O 5						2510-015	450	C30	H 26	O 4			
3128-005	440	C28	H 56	O 3						1003-011	452	C21	H 24	O11			
3901-004	440	C21	H 28	O10						1303-009	452	C21	H 24	O11			
3905-015	440	C26	H 32	O 6						3131-003	452	C31	H 64	O 1			
2806-007	440	C28	H 40	O 4						3131-002	452	C31	H 64	O 1			
2805-017	442	C28	H 42	O 4						3130-003	452	C30	H 60	O 2			
0505-023	442	C24	H 26	O 8						3502-010	454	C17	H 34	N 4	O10		
1304-003	442	C22	H 18	O10						2510-016	454	C30	H 30	O 4			
1303-004	442	C22	H 18	O10						2003-020	454	C29	H 42	O 4			
2302-016	442	C20	H 17	O 5	CL3					2302-023	456	C25	H 28	O 8			
2399-009	442	C25	H 30	O 9						2302-024	456	C25	H 28	O 8			
2302-026	442	C25	H 30	O 7						2302-012	456	C19	H 17	O 6	CL3		
2702-012	443	C23	H 25	N 1	O 8					2703-039	456	C24	H 24	O 9			
2703-038	444	C22	H 20	O10						2510-007	458	C26	H 18	O 7			
2606-004	444	C18	H 16	O 7						2606-005	458	C19	H 18	O 7			
2702-001	444	C22	H 24	N 2	O 8					2301-029	458	C25	H 30	O 8			
2703-043	444	C22	H 20	O10						1303-021	458	C22	H 18	O11			
2301-027	444	C25	H 32	O 7						1203-032	458	C30	H 34	O 4			
1101-008	444	C21	H 18	O10						0505-025	458	C24	H 26	O 9			
0505-022	444	C23	H 24	O 7						0505-024	458	C24	H 26	O 9			
0606-004	446	C22	H 22	O10						0301-020	458	C21	H 30	O11			
0104-026	446	C19	H 26	O12						1002-020	458	C30	H 34	O 4			
0603-005	446	C24	H 30	O 8						2702-013	459	C23	H 25	N 1	O 9		
1103-029	446	C22	H 22	O10						3499-001	459	C25	H 33	N 1	O 7		
1199-004	446	C22	H 22	O10						2699-009	460	C30	H 36	O 1			
1601-008	446	C22	H 22	O10						2699-010	460	C30	H 36	O 4			
1401-014	446	C22	H 22	O10						2702-006	460	C22	H 24	N 2	O 9		
1103-006	446	C21	H 18	O11						0701-009	460	C25	H 16	O 9			
1103-021	446	C22	H 22	O10						1002-019	460	C30	H 36	O 4			
1103-022	446	C22	H 22	O10						0203-007	460	C20	H 28	O12			
1203-018	446	C22	H 24	O10						1203-031	460	C30	H 36	O 4			
1601-010	446	C22	H 22	O10						1103-004	460	C22	H 20	O 4			
1103-055	446	C22	H 22	O10						1101-012	460	C21	H 18	O11			
2001-016	446	C29	H 50	O 3						2201-005	460	C20	H 28	O12			
2301-022	446	C24	H 30	O 8						2207-018	460	C25	H 32	O 8			
2301-023	446	C24	H 30	O 8						2207-007	460	C25	H 32	O 8			
2207-013	446	C24	H 30	O 8						2207-008	460	C25	H 32	O 8			
2604-037	446	C22	H 22	O10						2301-024	460	C25	H 32	O 8			
2603-032	446	C21	H 18	O11						2207-006	460	C25	H 32	O 8			
2702-010	447	C21	H 20	N 2	O 7	CL1				2301-035	460	C25	H 32	O 8			
3502-016	447	C19	H 37	N 5	O 7					2207-015	460	C25	H 32	O 8			
3131-005	448	C31	H 60	O 1						2207-014	460	C25	H 32	O 8			
2699-013	448	C22	H 24	O10						3499-003	461	C25	H 36	N 1	O 7		
2510-018	448	C30	H 24	O 4						2101-031	462	C25	H 31	O 6	CL1		
2003-014	448	C28	H 48	O 4						2403-037	462	C28	H 30	O 6			
1105-011	448	C21	H 20	O11						1199-035	462	C28	H 30	O 6			
1104-008	448	C21	H 20	O11						1105-033	462	C22	H 22	O11			
1104-018	448	C21	H 20	O11						1105-031	462	C22	H 22	O11			
1106-045	448	C21	H 20	O11						1103-038	462	C21	H 18	O12			
1404-007	448	C27	H 28	O 6						1104-009	462	C21	H 18	O12			
1203-011	448	C22	H 24	O10						1104-041	462	C22	H 22	O11			
1203-009	448	C22	H 24	O10						1104-037	462	C22	H 22	O11			
1105-010	448	C21	H 20	O11						1803-008	462	C22	H 22	O11			
1106-058	448	C17	H 13	O10	S1	K1				1401-019	462	C22	H 22	O11			
1104-007	448	C21	H 20	O11						1105-023	462	C22	H 22	O11			
1106-027	448	C21	H 20	O11						1105-018	462	C22	H 22	O11			
1104-006	448	C21	H 20	O11						1105-020	462	C22	H 22	O11			
1105-016	448	C21	H 20	O11						1105-034	462	C22	H 22	O11			
1106-018	448	C21	H 20	O11						1103-035	462	C21	H 18	O12			
1803-002	448	C21	H 20	O11						2605-024	462	C22	H 22	O11			
1803-003	448	C21	H 20	O11						1702-019	463	C22	H 23	O11	+		
1105-003	448	C21	H 20	O11						1703-019	463	C22	H 23	O11	+		
1103-034	448	C21	H 20	O11						1702-020	463	C22	H 23	O11	+		
1803-006	448	C21	H 20	O11						3502-023	463	C20	H 41	N 5	O 7		
1105-014	448	C21	H 20	O11						3131-007	464	C31	H 60	O 2			

622

3133-001	464	C33	H 68					
1107-001	464	C21	H 20	O12				
1803-013	464	C21	H 20	O12				
1205-006	464	C22	H 24	O11				
1106-015	464	C21	H 20	O12				
1104-058	464	C21	H 20	O12				
1106-026	464	C21	H 20	O16				
1105-053	464	C21	H 20	O12				
1106-038	464	C22	H 22	O11				
1108-003	464	C21	H 20	O12				
1106-014	464	C21	H 20	O12				
1106-114	464	C21	H 20	O12				
1199-034	464	C28	H 32	O 6				
2605-025	464	C20	H 20	N 2	O11			
1003-010	464	C22	H 24	O11				
2702-005	465	C21	H 22	N 2	O 8	CL1		
1703-003	465	C21	H 21	O12	+			
1703-004	465	C21	H 21	O12	+			
1206-003	466	C21	H 22	O12				
1302-017	466	C22	H 26	O11				
0506-019	466	C22	H 26	O11				
3301-014	466	C22	H 42	O 2	S 4			
3131-006	466	C31	H 62	O 2				
3502-026	466	C19	H 38	N 4	O 9			
3502-027	466	C19	H 38	N 4	O 9			
3197-039	466	C31	H 62	O 2				
3132-002	466	C32	H 66	O 1				
3118-070	466	C30	H 58	O 3				
3502-020	467	C18	H 37	N 5	O 9			
3130-006	468	C30	H 60	O 3				
3502-021	468	C18	H 36	N 4	O10			
0601-015	468	C26	H 28	O 8				
2303-001	468	C24	H 20	O10				
2499-008	468	C29	H 40	O 5				
0702-007	469	C28	H 23	N 1	O 6			
3401-001	469	C25	H 43	N 1	O 7			
3401-002	469	C25	H 43	N 1	O 7			
2302-027	470	C26	H 30	O 8				
1206-006	470	C22	H 24	O12				
2301-028	472	C27	H 36	O 7				
2302-005	472	C22	H 16	O12				
2301-030	472	C26	H 32	O 8				
2606-008	472	C19	H 16	O 8				
2704-023	472	C26	H 14	O 9				
1599-015	474	C31	H 22	O 5				
0399-012	474	C22	H 18	O12				
0104-035	476	C20	H 28	O13				
0104-032	476	C20	H 28	O13				
1104-050	476	C23	H 24	O11				
1105-036	476	C23	H 24	O11				
1105-024	476	C22	H 20	O12				
1103-046	476	C23	H 24	O11				
2702-008	476	C22	H 21	N 2	O 8	CL1		
3502-024	477	C21	H 43	N 5	O 7			
3999-016	477	C28	H 31	N 1	O 6			
3197-048	478	C34	H 70					
2599-002	478	C31	H 26	O 5				
2609-001	478	C30	H 22	O 6				
1106-063	478	C22	H 22	O12				
1104-055	478	C22	H 22	O12				
1106-046	478	C22	H 22	O11				
1106-036	478	C22	H 22	O12				
1106-017	478	C21	H 18	O13				
1105-044	478	C22	H 22	O11				
2503-031	478	C23	H 26	O11				
1703-014	479	C22	H 23	O12	+			
1703-013	479	C22	H 23	O12	+			
2702-002	479	C22	H 23	N 2	O 8	CL1		
1108-004	480	C21	H 20	O13				
1108-006	480	C21	H 20	O13				
1108-002	480	C21	H 20	O13				
1108-028	480	C21	H 20	O13				
1106-092	480	C21	H 20	O13				
1106-093	480	C21	H 20	O12				
1106-070	480	C21	H 20	O13				
1106-071	480	C21	H 20	O13				
2199-010	480	C21	H 36	O 8	S 2			
2199-011	480	C21	H 36	O 8	S 2			
3502-025	480	C20	H 40	N 4	O 9			
3132-003	480	C32	H 64	O 2				
3131-009	480	C31	H 60	O 3				
3131-008	480	C31	H 60	O 3				
3116-017	480	C32	H 64	O 2				
2303-002	482	C25	H 22	O10				
2303-004	482	C25	H 22	O10				
1299-011	482	C25	H 22	O10				
1299-012	482	C25	H 22	O10				
1299-010	482	C25	H 22	O10				
1303-022	482	C22	H 26	O12				
2510-017	482	C32	H 34	O 4				
3005-041	482	C26	H 42	O 8				
3504-013	483	C15	H 21	N 3	O15			
3502-018	483	C18	H 37	N 5	O10			
3502-017	484	C18	H 36	N 4	O11			
3502-019	484	C18	H 36	N 4	O11			
2303-005	484	C24	H 20	O11				
0599-022	484	C30	H 28	O 6				
0599-023	484	C30	H 28	O 6				
3906-043	486	C21	H 26	O13				
2999-001	486	C28	H 38	O 7				
2703-048	486	C25	H 26	O10				
1599-016	488	C32	H 24	O 5				
3499-015	489	C28	H 43	N 1	O 6			
3301-018	490	C26	H 50	O 4	S 2			
1106-068	490	C24	H 26	O11				
1104-044	492	C23	H 24	O12				
1106-059	492	C23	H 24	O12				
1105-048	492	C23	H 24	O12				
1106-076	492	C23	H 24	O12				
1107-004	492	C23	H 24	O12				
2002-006	492	C33	H 48	O 3				
1107-003	492	C23	H 24	O12				
3133-002	492	C33	H 64	O 2				
3197-051	492	C35	H 72					
3135-001	492	C35	H 72					
2605-014	492	C22	H 20	O13				
0104-011	492	C20	H 28	O14				
0105-005	492	C22	H 20	O13				
0399-010	492	C24	H 28	O11				
1703-021	493	C23	H 25	O12	+			
1703-020	493	C23	H 25	O12	+			
1106-073	494	C22	H 22	O13				
1106-096	494	C22	H 22	O13				
1108-010	494	C22	H 22	O13				
2702-007	494	C22	H 23	N 2	O 9	CL1		
2609-004	494	C30	H 22	O 7				
2609-005	494	C30	H 22	O 7				
3134-002	494	C34	H 70	O 1				
2199-012	494	C22	H 38	O 8	S 2			
2605-047	495	C24	H 17	N 1	O11			
2605-049	496	C24	H 16	O12				
2303-003	496	C26	H 24	O10				
1108-022	496	C21	H 30	O14				
2207-020	498	C28	H 34	O 8				
2502-019	498	C35	H 46	O 2				
0903-009	498	C29	H 38	O 7				
2301-031	500	C28	H 36	O 8				
3999-009	501	C25	H 43	N 1	O 9			
3499-002	501	C27	H 35	N 1	O 8			
1199-037	502	C30	H 30	O 7				
3499-004	503	C27	H 37	N 1	O 8			
3999-008	503	C25	H 45	N 1	O 9			
3501-018	504	C18	H 32	O14				
3507-007	504	C18	H 32	O16				
3507-008	504	C18	H 32	O16				
3507-006	504	C18	H 32	O16				
3507-001	504	C18	H 30	O16				
3507-002	504	C18	H 32	O16				
3507-003	504	C18	H 32	O16				
3507-004	504	C18	H 32	O16				
0304-027	504	C21	H 28	O14				
2704-024	504	C27	H 20	O10				
2704-020	504	C30	H 16	O 8				
2704-019	506	C30	H 18	O 8				
2607-001	506	C30	H 18	O 8				
2607-002	506	C30	H 18	O 8				
2608-001	506	C30	H 18	O 8				
0199-001	506	C21	H 14	O15				
3197-049	506	C36	H 74					
3507-010	506	C18	H 34	O16				
1107-005	506	C23	H 22	O13				
1106-016	506	C23	H 22	O13				
1104-052	506	C24	H 26	O12				
1106-102	508	C23	H 24	O13				
3134-001	508	C34	H 68	O 2				
2609-008	508	C30	H 20	O 8				
3401-008	509	C28	H 47	N 1	O 7			
2608-013	510	C30	H 22	O 8				
2609-006	510	C30	H 22	O 8				
2609-003	510	C30	H 22	O 8				
2609-002	510	C30	H 22	O 8				
1205-010	510	C28	H 34	O15				
1205-011	510	C28	H 34	O15				
0199-019	510	C27	H 26	O10				
0399-011	510	C25	H 34	O11				
2502-034	510	C23	H 26	O13				
2207-019	512	C28	H 36	O 8				
2302-028	512	C28	H 32	O 9				
3502-037	513	C19	H 35	N 3	O 3			
3507-009	514	C21	H 38	O14				
2302-025	514	C23	H 26	O 6				
0599-019	514	C25	H 22	O12				

No.	MW	C	H	N	O	Other
2509-006	514	C28	H 18		O10	
3905-021	516	C30	H 28		O 8	
1902-014	516	C25	H 24		O12	
1902-013	516	C25	H 24		O12	
1299-013	516	C30	H 28		O 8	
1902-015	516	C25	H 24		O12	
1902-012	516	C25	H 24		O12	
1902-011	516	C25	H 24		O12	
3906-020	518	C26	H 30		O11	
2509-005	518	C28	H 22		O10	
2604-032	518	C27	H 30		O14	
3906-019	520	C26	H 32		O11	
3137-001	520	C37	H 76			
3197-052	520	C37	H 76			
0601-012	520	C26	H 32		O11	
0104-038	522	C28	H 44	N 1	O 8	+
3906-008	522	C26	H 18		O12	
3906-010	522	C26	H 18		O12	
2607-003	522	C30	H 18		O 9	
2608-008	522	C30	H 18		O 9	
2608-005	522	C30	H 18		O 9	
2608-004	522	C30	H 18		O 9	
1106-080	522	C24	H 26		O13	
1199-023	522	C24	H 26		O13	
1403-002	522	C24	H 26		O13	
1106-084	522	C24	H 36		O13	
2003-021	522	C34	H 50		O 4	
2702-003	523	C22	H 23	N 2	O 8	BR1
2609-007	524	C31	H 24		O 8	
2609-010	524	C30	H 20		O 9	
1502-001	524	C32	H 28		O 7	
1106-101	524	C23	H 24		O14	
0199-002	524	C20	H 24		O10	
2502-022	524	C25	H 32		O12	
3401-009	525	C28	H 47	N 1	O 8	
2204-007	526	C32	H 46		O 6	
2204-008	526	C32	H 46		O 6	
2302-029	526	C29	H 34		O 9	
2204-009	526	C32	H 46		O 6	
3502-036	527	C20	H 37	N 3	O13	
3502-035	527	C20	H 37	N 3	O13	
2703-009	527	C28	H 33	N 1	O 9	
2703-024	527	C27	H 29	N 1	O10	
2704-028	528	C30	H 24		O 9	
2704-009	528	C29	H 20		O10	
2704-029	528	C30	H 24		O 9	
2704-030	528	C30	H 24		O 9	
2504-016	528	C33	H 52		O 5	
2005-010	528	C30	H 40		O 8	
2301-032	528	C29	H 36		O 9	
2005-011	528	C30	H 40		O 8	
2301-033	528	C29	H 36		O 9	
0199-003	528	C27	H 28		O11	
1105-017	528	C21	H 20		O14	S 1
2499-001	530	C33	H 38		O 6	
2704-026	530	C28	H 18		O11	
2509-007	530	C28	H 18		O11	
0701-004	532	C32	H 20		O 8	
0601-014	534	C27	H 34		O11	
0601-025	534	C27	H 34		O11	
2704-018	534	C29	H 26		O10	
2603-004	534	C25	H 26		O13	
1104-034	534	C26	H 30		O12	
3197-050	534	C38	H 78			
3906-009	536	C27	H 20		O12	
3906-011	536	C27	H 20		O12	
1106-082	536	C25	H 28		O13	
1204-006	536	C26	H 32		O12	
1204-008	536	C26	H 32		O12	
0699-018	536	C26	H 32		O12	
2605-048	537	C26	H 19	N 1	O12	
2609-011	538	C31	H 22		O 9	
2608-003	538	C30	H 18		O10	
2607-004	538	C30	H 18		O10	
2608-006	538	C30	H 18		O10	
2609-008	538	C30	H 18		O10	
2607-005	538	C30	H 18		O10	
2608-007	538	C30	H 18		O10	
2608-002	538	C30	H 18		O10	
0199-006	538	C27	H 22		O18	
1106-108	538	C24	H 26		O14	
1504-001	538	C30	H 18		O10	
1106-110	538	C24	H 26		O14	
1505-001	538	C30	H 18		O10	
1599-003	538	C30	H 18		O10	
2605-046	539	C25	H 17	N 1	O13	
2703-042	540	C28	H 28		O11	
1599-004	540	C30	H 20		O10	
1502-002	540	C32	H 28		O 8	
1501-004	540	C30	H 20		O10	
1501-003	540	C30	H 20		O10	
1501-002	540	C30	H 20		O10	
0604-022	540	C28	H 28		O11	
2503-027	540	C24	H 28		O14	
0604-021	542	C27	H 26		O12	
1501-001	542	C30	H 22		O10	
2699-021	542	C30	H 22		O10	
2608-014	542	C30	H 22		O10	
2704-033	542	C30	H 22		O10	
2703-037	543	C27	H 29	N 1	O11	
2703-020	543	C28	H 33	N 1	O10	
2704-008	544	C29	H 20		O11	
2704-014	544	C30	H 24		O10	
1599-017	544	C30	H 20		O10	
2499-002	544	C33	H 36		O 7	
2499-003	544	C33	H 36		O 7	
3906-040	544	C29	H 20		O11	
2499-004	546	C33	H 38		O 7	
1506-001	546	C30	H 26		O10	
1506-002	546	C30	H 26		O10	
2509-008	546	C28	H 18		O12	
2510-008	546	C30	H 26		O10	
2704-017	546	C30	H 26		O10	
2704-015	546	C30	H 26		O10	
2704-025	546	C29	H 22		O11	
2607-025	546	C30	H 26		O10	
2704-031	546	C30	H 26		O10	
2704-032	546	C30	H 26		O10	
0605-003	546	C27	H 30		O12	
2603-014	548	C26	H 28		O13	
2704-016	548	C30	H 28		O10	
3197-053	548	C39	H 80			
3999-007	549	C29	H 43	N 1	O 9	
2510-011	550	C30	H 20		O11	
0699-017	550	C27	H 34		O12	
0699-021	550	C28	H 38	N 8	O 4	
2802-012	552	C33	H 38		O 8	
1599-001	552	C31	H 20		O10	
1505-004	552	C31	H 20		O10	
1505-003	552	C31	H 20		O10	
1505-002	552	C31	H 20		O10	
1505-005	552	C31	H 20		O10	
1103-017	552	C28	H 24		O12	
1599-005	552	C31	H 20		O10	
1599-006	552	C31	H 20		O10	
1504-005	552	C31	H 20		O10	
1503-001	552	C31	H 20		O10	
2403-021	552	C25	H 28		O14	
2207-029	554	C32	H 42		O 8	
2608-009	554	C30	H 18		O11	
2608-011	554	C30	H 18		O11	
3502-029	555	C21	H 41	N 5	O12	
3502-030	555	C21	H 41	N 5	O12	
2599-009	558	C33	H 18		O 9	
1501-006	558	C30	H 22		O11	
1501-007	558	C30	H 20		O11	
1599-014	558	C30	H 22		O11	
1501-009	558	C30	H 20		O11	
2002-007	560	C38	H 56		O 3	
2499-006	560	C33	H 36		O 8	
2499-005	560	C33	H 36		O 8	
1502-003	562	C30	H 26		O11	
1502-004	562	C30	H 26		O11	
1599-012	562	C30	H 26		O11	
1506-004	562	C30	H 26		O11	
1506-006	562	C30	H 26		O11	
1506-005	562	C30	H 26		O11	
2704-027	562	C29	H 22		O12	
2604-012	564	C26	H 28		O14	
2699-004	564	C27	H 32		O13	
2599-003	564	C30	H 28		O11	
2599-006	564	C30	H 28		O11	
2704-021	564	C32	H 20		O10	
1104-023	564	C26	H 28		O14	
1201-003	564	C27	H 32		O13	
1104-027	564	C26	H 28		O14	
1599-022	564	C29	H 24		O12	
1103-009	564	C26	H 28		O14	
1599-004	564	C29	H 24		O12	
1103-016	564	C26	H 28		O14	
1103-010	564	C26	H 28		O14	
1701-006	565	C26	H 29		O14	+
1702-007	565	C26	H 29		O14	+
1505-009	566	C32	H 22		O10	
1505-007	566	C32	H 22		O10	
1505-006	566	C32	H 22		O10	
1599-007	566	C32	H 22		O10	
1503-002	566	C32	H 22		O10	
1505-010	566	C32	H 22		O10	
1504-003	566	C32	H 22		O10	
1504-002	566	C32	H 22		O10	
1599-002	566	C32	H 22		O10	

Code	MW	C	H	N	O	Extra
2101-033	566	C23	H30	N1	O11	CL2
2502-020	566	C40	H54		O2	
1505-008	568	C32	H24		O10	
2509-009	568	C32	H24		O10	
3906-012	568	C32	H24		O10	
3203-011	570	C18	H26		O8	
2511-003	570	C32	H26		O10	
2608-010	570	C30	H18		O12	
2510-013	570	C30	H18		O12	
2299-007	570	C37	H62		O4	
0599-021	570	C32	H26		O10	
1501-010	572	C31	H22		O11	
1501-012	574	C30	H22		O12	
1401-005	574	C27	H30		O14	
2608-015	574	C30	H22		O12	
2608-016	574	C30	H22		O12	
2510-010	574	C30	H22		O12	
2699-023	574	C30	H22		O12	
2699-022	574	C30	H22		O12	
1599-025	576	C30	H24		O12	
1506-007	578	C30	H26		O12	
1104-022	578	C27	H30		O14	
1103-008	578	C27	H30		O14	
1599-013	578	C30	H26		O12	
1502-008	578	C30	H26		O12	
1502-005	578	C30	H26		O12	
1401-003	578	C27	H30		O14	
1599-008	578	C30	H26		O12	
1502-007	578	C30	H26		O12	
1502-006	578	C30	H26		O12	
1103-007	578	C27	H30		O14	
1401-015	578	C27	H30		O14	
2604-033	578	C27	H30		O14	
2604-013	578	C27	H30		O14	
0605-009	578	C28	H34		O13	
1701-005	579	C27	H31		O14	+
1505-011	580	C33	H24		O10	
1105-006	580	C26	H28		O15	
1105-005	580	C26	H28		O15	
1106-025	580	C26	H28		O15	
1104-015	580	C26	H28		O15	
1105-062	580	C26	H28		O15	
1203-027	580	C27	H32		O14	
1504-006	580	C33	H24		O10	
1104-012	580	C26	H28		O15	
1505-014	580	C33	H24		O10	
1505-012	580	C33	H24		O10	
1203-004	580	C27	H32		O14	
1203-005	580	C27	H32		O14	
0699-022	580	C29	H40	N8	O5	
1002-005	580	C27	H32		O14	
0899-017	580	C12	H5		O2	BR5
2599-004	580	C30	H28		O12	
2599-005	580	C30	H28		O12	
2502-011	580	C41	H56		O2	
1702-006	581	C26	H29		O15	+
1702-008	581	C26	H29		O15	+
3502-003	581	C21	H39	N7	O12	
1505-013	582	C33	H26		O10	
3502-005	583	C21	H41	N7	O12	
3502-032	585	C21	H39	N5	O14	
2703-030	585	C30	H35	N1	O11	
2607-009	590	C34	H22		O10	
3999-005	590	C34	H54		O8	
2003-022	590	C39	H58		O4	
1402-014	590	C28	H30		O14	
1506-003	590	C31	H26		O12	
1502-009	592	C31	H28		O12	
1103-025	592	C28	H32		O14	
1103-027	592	C28	H32		O14	
1103-026	592	C28	H32		O14	
1506-008	594	C30	H26		O13	
1803-012	594	C27	H30		O15	
1203-012	594	C28	H34		O14	
1803-011	594	C27	H30		O15	
1106-041	594	C27	H30		O15	
1505-015	594	C34	H26		O10	
1105-008	594	C27	H30		O15	
1103-018	594	C27	H30		O15	
1104-011	594	C27	H30		O15	
1103-019	594	C27	H30		O15	
1105-027	594	C27	H30		O15	
1502-010	594	C30	H26		O13	
1105-007	594	C27	H30		O15	
1104-026	594	C30	H26		O13	
1106-047	594	C27	H30		O15	
1104-016	594	C27	H30		O15	
1105-026	594	C27	H30		O15	
1599-009	594	C30	H26		O13	
1504-004	594	C34	H26		O10	
1599-010	594	C30	H26		O13	
1299-002	594	C28	H34		O14	
2604-014	594	C27	H30		O15	
2604-038	594	C28	H32		O14	
2604-010	594	C26	H26		O16	
1701-007	595	C27	H31		O15	+
1702-010	595	C27	H31		O15	+
1702-009	595	C27	H31		O15	+
1702-021	595	C27	H31		O15	+
1205-004	596	C27	H32		O15	
1205-018	596	C27	H32		O15	
1106-033	596	C26	H28		O16	
1106-029	596	C26	H28		O16	
1203-006	596	C27	H32		O15	
2504-020	596	C27	H32		O15	
1003-004	596	C27	H32		O15	
3401-044	597	C31	H51	N1	O10	
3502-006	597	C21	H39	N7	O13	
2608-012	598	C32	H22		O12	
1104-031	598	C26	H30		O12	
2299-008	598	C39	H66		O4	
2005-005	600	C35	H52		O8	
1103-014	604	C27	H30		O15	
1105-032	608	C28	H32		O15	
1105-019	608	C28	H32		O15	
1105-025	608	C28	H32		O15	
1105-037	608	C28	H32		O15	
2603-033	608	C27	H28		O16	
1105-015	610	C27	H30		O16	
1599-011	610	C30	H26		O14	
1104-021	610	C27	H30		O16	
1105-004	610	C27	H30		O16	
1106-024	610	C27	H30		O16	
1103-037	610	C26	H30		O16	
1106-030	610	C27	H30		O16	
1105-009	610	C27	H30		O16	
1106-031	610	C27	H30		O16	
1105-012	610	C27	H30		O16	
1104-014	610	C27	H30		O16	
1104-013	610	C27	H30		O16	
1104-020	610	C27	H30		O16	
1106-023	610	C27	H30		O16	
2406-005	610	C30	H26		O14	
1702-013	611	C27	H31		O16	+
1703-006	611	C27	H31		O16	+
1702-012	611	C27	H31		O16	+
1702-011	611	C27	H31		O16	+
1702-015	611	C27	H31		O16	+
1703-015	611	C27	H31		O16	+
1205-005	612	C27	H32		O16	
2609-016	612	C30	H19		O8	CL3
0599-024	612	C31	H36	N2	O11	
3502-011	614	C23	H46	N6	O13	
3502-012	614	C23	H46	N6	O13	
3502-013	615	C23	H45	N5	O14	
3502-014	615	C23	H45	N5	O14	
2704-010	616	C33	H32	N2	O10	
1103-042	622	C29	H34		O1	K1
1103-044	622	C29	H34		O15	
1106-052	624	C28	H32		O16	
1105-028	624	C28	H32		O16	
1106-049	624	C28	H32		O16	
1106-051	624	C28	H32		O16	
1105-035	624	C28	H32		O16	
2406-006	624	C31	H28		O14	
2406-007	624	C31	H28		O14	
2406-012	624	C31	H28		O14	
1702-022	625	C28	H33		O16	+
1106-022	626	C27	H30		O17	
1106-021	626	C27	H30		O17	
1106-026	626	C27	H30		O17	
1106-020	626	C27	H30		O17	
2299-009	626	C41	H70		O4	
1703-028	627	C27	H31		O17	+
1703-007	627	C27	H31		O17	+
2002-008	628	C43	H64		O3	
2499-007	628	C38	H44		O8	
1105-055	630	C18	H16		O8	
2502-021	634	C45	H62		O2	
2510-014	634	C32	H26		O14	
2607-019	634	C36	H26		O11	
3197-035	634	C44	H90		O1	
3197-034	636	C41	H80		O4	
3508-004	636	C23	H40		O20	
1103-028	636	C28	H28		O17	
0701-006	636	C32	H28		O14	
1106-065	638	C29	H34		O16	
1203-022	638	C30	H38		O15	
3198-002	638	C39	H74		O6	
2607-024	638	C36	H30		O11	
2406-001	638	C32	H30		O14	
2406-002	638	C32	H30		O14	

Code	Mass	C	H	N	O	Notes
2406-003	638	C32	H 30		O14	
2406-004	638	C32	H 30		O14	
2207-021	640	C34	H 40		O12	
2704-034	640	C32	H 32		O14	
1106-055	640	C28	H 32		O17	
1106-053	640	C28	H 32		O17	
1106-048	640	C28	H 32		O17	
1703-017	641	C28	H 33		O17	+
1108-005	642	C27	H 30		O18	
1108-007	642	C27	H 30		O18	
2704-004	642	C32	H 34		O14	
2406-011	642	C31	H 30		O15	
2303-006	646	C34	H 30		O13	
2609-018	646	C30	H 18		O 8	CL4
2609-017	646	C30	H 18		O 8	CL4
2502-012	648	C46	H 64		O 2	
3402-008	650	C36	H 58		O10	
2607-015	652	C36	H 28		O12	
3402-006	654	C35	H 58		O11	
2207-022	654	C35	H 42		O12	
2207-025	654	C35	H 42		O12	
2207-024	654	C35	H 42		O12	
1703-023	655	C29	H 35		O17	+
2406-009	656	C32	H 32		O15	
2003-023	658	C44	H 36		O 4	
3508-005	658	C28	H 50		O17	
3501-024	660	C 6	H 18		O24	P 6
2502-018	662	C46	H 62		O 3	
2510-009	662	C34	H 30		O14	
3402-009	664	C37	H 60		O10	
3508-003	666	C24	H 42		O21	
3508-002	666	C24	H 42		O21	
3508-001	666	C24	H 42		O21	
0606-002	666	C28	H 38		O12	
2607-023	668	C36	H 28		O13	
2704-007	668	C35	H 24		O14	
2207-023	668	C36	H 44		O12	
2207-027	668	C36	H 44		O12	
2207-026	668	C36	H 44		O12	
1703-025	669	C30	H 37		O17	+
2607-021	670	C36	H 30		O13	
3402-007	670	C35	H 58		O12	
3999-001	670	C36	H 62		O11	
3906-021	670	C43	H 58		O 6	
2406-010	674	C32	H 34		O16	
2406-008	674	C32	H 34		O16	
2607-007	678	C38	H 30		O12	
1104-033	680	C32	H 40		O16	
2001-013	680	C48	H 22		O 2	
3402-003	681	C34	H 51	N 1	O13	
3402-001	681	C33	H 47	N 1	O13	
2703-010	684	C36	H 48	N 2	O11	
2599-010	684	C39	H 24		O12	
2106-016	684	C43	H 56		O 7	
2607-010	686	C38	H 38		O12	
3401-023	686	C34	H 54		O14	
3401-006	687	C35	H 61	N 1	O12	
0204-008	688	C29	H 36		O19	
0104-021	691	C36	H 53	N 1	O12	
1104-039	694	C33	H 42		O16	
2002-009	696	C48	H 72		O 3	
1104-032	696	C32	H 40		O17	
2607-012	696	C38	H 32		O13	
1702-016	697	C36	H 37		O18	+
3402-004	697	C34	H 51	N 1	O14	
1204-009	698	C32	H 42		O17	
3401-022	700	C35	H 56		O14	
2607-016	700	C38	H 36		O13	
2703-021	700	C36	H 48	N 2	O12	
2005-006	700	C43	H 56		O 4	
2502-022	702	C50	H 70		O 2	
1501-005	702	C36	H 30		O15	
1801-016	704	C34	H 40		O16	
0801-005	706	C51	H 78		O 1	
3999-002	708	C40	H 68		O10	
3401-048	708	C40	H 66		O10	
3401-045	709	C37	H 59	N 1	O12	
1103-054	714	C34	H 34		O17	
2607-018	714	C38	H 34		O14	
2607-006	714	C38	H 34		O14	
2703-034	716	C36	H 28	N 2	O13	
1599-019	716	C36	H 28		O16	
1599-020	716	C36	H 28		O16	
2502-013	716	C51	H 72		O 2	
3401-010	717	C37	H 67	N 1	O12	
2502-015	718	C51	H 74		O 2	
2607-011	718	C38	H 38		O14	
3401-011	719	C36	H 65	N 1	O13	
1501-008	720	C36	H 30		O16	
1501-011	722	C36	H 32		O16	
3126-005	722	C44	H 66		O 8	
3198-003	722	C45	H 86		O 6	
3402-002	723	C36	H 53	N 1	O13	
3999-003	724	C40	H 68		O11	
2003-024	726	C49	H 74		O 4	
2509-004	726	C36	H 38		O16	
2607-014	730	C38	H 38		O15	
2607-017	732	C38	H 36		O15	
2607-013	732	C38	H 36		O15	
3401-012	733	C37	H 67	N 1	O13	
2607-022	734	C38	H 38		O15	
3126-004	736	C45	H 68		O 8	
3499-010	736	C40	H 64		O12	
1104-025	740	C33	H 40		O19	
3402-005	741	C38	H 63	N 1	O13	
2599-005	742	C36	H 38		O17	
0603-004	742	C34	H 46		O18	
3401-024	743	C37	H 61	N 1	O14	
3502-004	743	C27	H 49	N 7	O17	
2607-026	746	C38	H 34		O16	
2599-008	746	C44	H 26		O12	
2607-020	748	C38	H 36		O16	
0899-010	748	C53	H 80		O 2	
2001-014	748	C53	H 80		O 2	
3401-047	749	C41	H 67	N 1	O11	
3126-006	750	C46	H 70		O 8	
3499-011	750	C41	H 66		O12	
2002-013	750	C52	H 78		O 3	
0899-009	750	C53	H 82		O 2	
2607-008	750	C38	H 38		O16	
2005-007	754	C47	H 62		O 8	
1104-017	756	C33	H 40		O20	
1702-014	757	C33	H 41		O20	+
3401-025	757	C38	H 63	N 1	O14	
3502-028	761	C29	H 55	N 5	O18	
2002-010	764	C53	H 80		O 3	
3499-012	766	C42	H 68		O12	
1703-022	769	C38	H 41		O17	+
1106-040	770	C34	H 42		O20	
2502-023	770	C55	H 78		O 2	
3401-026	771	C39	H 65	N 1	O14	
1106-032	772	C33	H 40		O21	
1104-024	772	C33	H 40		O21	
1106-113	772	C33	H 40		O21	
0801-006	774	C56	H 86		O 1	
3502-015	777	C29	H 55	N 5	O19	
2002-016	779	C53	H 81	N 1	O 3	
3499-013	780	C43	H 70		O12	
2509-010	780	C36	H 44		O19	
3401-019	783	C40	H 65	N 1	O14	
2502-014	784	C56	H 80		O 2	
3401-027	785	C40	H 67	N 1	O14	
3401-028	785	C39	H 63	N 1	O15	
2502-016	786	C56	H 82		O 2	
1106-050	786	C34	H 42		O21	
3130-004	788	C54	H108		O 2	
3401-017	790	C40	H 70		O15	
3197-042	792	C52	H104		O 4	
3499-014	792	C44	H 72		O12	
2003-025	794	C54	H 82		O 4	
3401-029	799	C40	H 65	N 1	O15	
3401-032	799	C40	H 65	N 1	O15	
1106-054	802	C34	H 42		O22	
0802-003	804	C57	H 88		O 2	
2703-011	810	C42	H 58	N 2	O14	
3401-037	811	C41	H 65	N 1	O15	
3401-033	813	C41	H 67	N 1	O15	
3401-030	813	C41	H 67	N 1	O15	
2703-031	815	C42	H 41	N 1	O16	
2002-014	818	C57	H 86		O 3	
2002-015	820	C27	H 88		O 3	
2005-008	822	C52	H 70		O 8	
3401-038	825	C42	H 67	N 1	O15	
3401-034	825	C42	H 67	N 1	O15	
3999-004	826	C45	H 78		O13	
2003-026	826	C59	H 90		O 4	
3401-034	827	C42	H 69	N 1	O15	
3401-031	827	C42	H 69	N 1	O15	
3508-007	828	C30	H 52		O26	
3508-006	828	C30	H 52		O26	
3401-036	829	C41	H 67	N 1	O16	
3401-018	832	C42	H 72		O16	
3198-004	832	C53	H100		O 6	
2002-011	832	C58	H 88		O 3	
2002-012	834	C58	H 90		O 3	
2005-009	836	C53	H 72		O 8	
3401-039	839	C43	H 69	N 1	O15	
3401-021	841	C42	H 67	N 1	O16	
3401-035	841	C43	H 71	N 1	O15	
3401-040	842	C43	H 74	N 2	O14	
2002-017	847	C58	H 89	N 1	O 3	
2003-031	848	C58	H 88		O 4	

ID	C	H	O	Other
3101-003	C 1	H 2	O 1	
3101-004	C 1	H 2	O 2	
3101-001	C 1	H 4	O 1	
3101-002	C 1	H 4	S 1	
3102-009	C 2	H 2	O 2	
3102-010	C 2	H 2	O 4	
3102-007	C 2	H 3	O 2	F 1
3102-001	C 2	H 4		
3102-003	C 2	H 4	O 1	
3102-004	C 2	H 4	O 2	
3102-005	C 2	H 4	O 2	
3102-008	C 2	H 4	O 3	
3305-005	C 2	H 4	S 5	
3102-002	C 2	H 6	O 1	
3301-003	C 2	H 6	O 2	S 1
3301-002	C 2	H 6	S 1	
3103-018	C 3	H 2	O 5	
3103-011	C 3	H 3	O 3	
3103-021	C 3	H 4	O 2	
3103-006	C 3	H 4	O 2	
3103-013	C 3	H 4	O 3	
3103-019	C 3	H 4	O 4	
3103-016	C 3	H 4	O 4	
3103-017	C 3	H 4	O 5	
3103-014	C 3	H 5	O 6	P 1
3103-005	C 3	H 6	O 1	
3103-008	C 3	H 6	O 1	
3103-002	C 3	H 6	O 1	
3103-010	C 3	H 6	O 2	
3305-002	C 3	H 6	O 2	S 2
3305-001	C 3	H 6	O 2	S 2
3103-012	C 3	H 6	O 3	
3103-007	C 3	H 6	O 3	
3103-009	C 3	H 6	O 3	
3103-020	C 3	H 6	O 4	
3103-023	C 3	H 7	O 3	P 1
3103-015	C 3	H 7	O 7	P 1
3103-001	C 3	H 8	O 1	
3103-003	C 3	H 8	O 3	
3103-022	C 3	H 8	S 1	
3301-001	C 3	H 8	S 2	
3004-001	C 4	H 4	O 1	
3104-016	C 4	H 4	O 4	
3104-020	C 4	H 4	O 5	
3104-017	C 4	H 4	O 5	
3104-023	C 4	H 4	O 6	
3194-004	C 4	H 6	O 1	
3194-006	C 4	H 6	O 2	
3104-012	C 4	H 6	O 2	
3104-011	C 4	H 6	O 2	
3104-007	C 4	H 6	O 2	
3104-015	C 4	H 6	O 4	
3104-014	C 4	H 6	O 4	
3104-013	C 4	H 6	O 4	
3104-018	C 4	H 6	O 5	
3104-019	C 4	H 6	O 5	
3104-022	C 4	H 6	O 6	
3104-021	C 4	H 6	O 6	
3301-004	C 4	H 6	S 1	
3104-001	C 4	H 8		
3104-026	C 4	H 8	O 1	
3104-008	C 4	H 8	O 1	
3104-003	C 4	H 8	O 1	
3104-004	C 4	H 8	O 1	
3194-003	C 4	H 8	O 1	
3194-005	C 4	H 8	O 2	
3104-006	C 4	H 8	O 2	
3104-009	C 4	H 8	O 2	
3102-006	C 4	H 8	O 2	
3104-010	C 4	H 8	O 2	
3194-007	C 4	H 8	O 2	S 2
3194-001	C 4	H 10	O 1	
3104-002	C 4	H 10	O 1	
3104-005	C 4	H 10	O 2	
3104-024	C 4	H 10	O 4	
3004-004	C 5	H 4	O 2	
3005-004	C 5	H 4	O 2	
3002-001	C 5	H 4	O 3	
3002-002	C 5	H 4	O 4	
3004-007	C 5	H 4	O 4	
3004-002	C 5	H 6	O 1	
3005-003	C 5	H 6	O 2	
3004-003	C 5	H 6	O 2	
3005-005	C 5	H 6	O 3	
3005-001	C 5	H 6	O 3	
3105-007	C 5	H 6	O 4	
3195-018	C 5	H 6	O 4	
3195-019	C 5	H 6	O 4	
3105-006	C 5	H 6	O 5	
2901-001	C 5	H 8	O 1	
3195-004	C 5	H 8	O 1	
3195-011	C 5	H 8	O 2	
3195-013	C 5	H 8	O 2	
3195-012	C 5	H 8	O 2	
3004-010	C 5	H 8	O 2	
2901-006	C 5	H 8	O 2	CL2
3195-015	C 5	H 8	O 3	
3195-017	C 5	H 8	O 4	
3105-004	C 5	H 8	O 4	
3105-005	C 5	H 8	O 5	
3195-020	C 5	H 8	O 5	
3503-011	C 5	H 8	O 6	
3195-001	C 5	H 8	O 6	
3195-001	C 5	H 10		
3195-005	C 5	H 10	O 1	
3195-006	C 5	H 10	O 1	
3195-003	C 5	H 10	O 1	
3105-002	C 5	H 10	O 1	
3105-001	C 5	H 10	O 1	
3105-003	C 5	H 10	O 2	
3195-008	C 5	H 10	O 2	
3195-007	C 5	H 10	O 2	
3103-025	C 5	H 10	O 2	S 1
3195-016	C 5	H 10	O 3	
3195-009	C 5	H 10	O 3	
2602-010	C 5	H 10	O 3	
3195-010	C 5	H 10	O 4	
3503-007	C 5	H 10	O 4	
3503-008	C 5	H 10	O 4	
3503-001	C 5	H 10	O 5	
3503-003	C 5	H 10	O 5	
3503-006	C 5	H 10	O 5	
3503-002	C 5	H 10	O 5	
3503-005	C 5	H 10	O 5	
3503-004	C 5	H 10	O 5	
3510-001	C 5	H 10	O 5	
3195-002	C 5	H 12	O 1	
3103-026	C 5	H 12	O 3	S 1
3503-010	C 5	H 12	O 5	
3503-009	C 5	H 12	O 5	
3106-002	C 6	H 2		
0801-002	C 6	H 4	O 1	BR2
0801-003	C 6	H 4	O 1	CL2
2001-001	C 6	H 4	O 2	
3002-012	C 6	H 4	O 5	
3002-013	C 6	H 4	O 6	
3903-002	C 6	H 4	O 6	
0801-001	C 6	H 6	O 1	
0802-005	C 6	H 6	O 2	
0802-001	C 6	H 6	O 2	
3004-005	C 6	H 6	O 2	
3002-003	C 6	H 6	O 3	
3005-006	C 6	H 6	O 3	
3004-009	C 6	H 6	O 3	
3002-011	C 6	H 6	O 3	
0803-001	C 6	H 6	O 3	
2299-001	C 6	H 6	O 3	
3004-008	C 6	H 6	O 4	
3002-005	C 6	H 6	O 4	
3002-016	C 6	H 6	O 4	
3002-008	C 6	H 6	O 4	
3002-007	C 6	H 6	O 4	
3002-009	C 6	H 6	O 5	
3196-015	C 6	H 6	O 5	
3196-012	C 6	H 6	O 6	
3005-050	C 6	H 6	O 6	
3005-009	C 6	H 6	O 7	
3005-007	C 6	H 7	N 1 O 3	
2901-003	C 6	H 8	O 1	
3003-001	C 6	H 8	O 2	
3106-010	C 6	H 8	O 2	
2901-007	C 6	H 8	O 2	
3196-004	C 6	H 8	O 2	
3004-011	C 6	H 8	O 3	
3002-006	C 6	H 8	O 4	
3196-011	C 6	H 8	O 4	
3106-017	C 6	H 8	O 5	
3005-036	C 6	H 8	O 6	
3196-014	C 6	H 8	O 6	
3196-019	C 6	H 8	O 6	
3504-062	C 6	H 8	O 7	
3196-016	C 6	H 8	O 7	
3196-018	C 6	H 8	O 7	
3196-017	C 6	H 8	O 7	
3106-001	C 6	H 10		
3106-006	C 6	H 10	O 1	
3301-019	C 6	H 10	O 1	S 2
3196-007	C 6	H 10	O 2	
3196-010	C 6	H 10	O 2	
3106-009	C 6	H 10	O 2	
3106-007	C 6	H 10	O 2	
3003-003	C 6	H 10	O 2	
3106-015	C 6	H 10	O 3	
3196-008	C 6	H 10	O 3	
3501-003	C 6	H 10	O 4	
3501-002	C 6	H 10	O 4	
3106-016	C 6	H 10	O 4	
3504-052	C 6	H 10	O 5	
3196-013	C 6	H 10	O 5	
3510-010	C 6	H 10	O 5	
3510-011	C 6	H 10	O 6	
3504-051	C 6	H 10	O 6	
3504-050	C 6	H 10	O 6	
3504-064	C 6	H 10	O 7	
3504-060	C 6	H 10	O 7	
3504-061	C 6	H 10	O 7	
3504-063	C 6	H 10	O 7	
3504-065	C 6	H 10	O 8	
3504-066	C 6	H 10	O 8	
3301-008	C 6	H 10	S 1	
3301-009	C 6	H 10	S 2	
3301-010	C 6	H 10	S 3	
3301-011	C 6	H 10	S 4	
3106-005	C 6	H 12	O 1	
3106-004	C 6	H 12	O 1	
3106-008	C 6	H 12	O 2	
3194-002	C 6	H 12	O 2	
3196-005	C 6	H 12	O 2	
3196-006	C 6	H 12	O 2	
3196-003	C 6	H 12	O 2	
3901-003	C 6	H 12	O 2	
3504-067	C 6	H 12	O 3	
3504-044	C 6	H 12	O 3	
3106-014	C 6	H 12	O 3	
3504-030	C 6	H 12	O 4	
3504-040	C 6	H 12	O 4	
3504-039	C 6	H 12	O 4	
3196-009	C 6	H 12	O 4	
3504-041	C 6	H 12	O 4	
3504-032	C 6	H 12	O 4	
3501-001	C 6	H 12	O 4	
3504-033	C 6	H 12	O 4	
3504-035	C 6	H 12	O 4	
3504-042	C 6	H 12	O 4	
3501-004	C 6	H 12	O 5	
3504-017	C 6	H 12	O 5	
3501-005	C 6	H 12	O 5	
3501-007	C 6	H 12	O 5	
3504-019	C 6	H 12	O 5	
3504-022	C 6	H 12	O 5	
3504-058	C 6	H 12	O 5	
3504-016	C 6	H 12	O 5	
3504-057	C 6	H 12	O 5	
3504-025	C 6	H 12	O 5	
3510-012	C 6	H 12	O 5	
3510-002	C 6	H 12	O 6	
3501-008	C 6	H 12	O 6	
3501-015	C 6	H 12	O 6	
3504-007	C 6	H 12	O 6	
3504-048	C 6	H 12	O 6	
3504-008	C 6	H 12	O 6	
3504-001	C 6	H 12	O 6	
3501-026	C 6	H 12	O 6	
3504-045	C 6	H 12	O 6	
3501-012	C 6	H 12	O 6	
3504-049	C 6	H 12	O 6	

3504-014	C 6 H 12 O 6
3504-059	C 6 H 12 O 7
3301-005	C 6 H 12 S 2
3509-015	C 6 H 13 N 1 O 3
3509-003	C 6 H 13 N 1 O 4
3509-016	C 6 H 13 N 1 O 4
3509-017	C 6 H 13 N 1 O 4
3509-019	C 6 H 13 N 1 O 4
3509-005	C 6 H 13 N 1 O 4
3509-013	C 6 H 13 N 1 O 4
3509-004	C 6 H 13 N 1 O 4
3509-007	C 6 H 13 N 1 O 5
3509-006	C 6 H 13 N 1 O 5
3509-011	C 6 H 13 N 1 O 5
3509-008	C 6 H 13 N 1 O 5
3509-001	C 6 H 13 N 1 O 5
3509-024	C 6 H 13 N 1 O 5
3504-046	C 6 H 13 O 9 P 1
3504-003	C 6 H 13 O 9 P 1
3504-004	C 6 H 13 O 9 P 1
3502-007	C 6 H 14 N 2 O 3
3502-001	C 6 H 14 N 2 O 4
3196-001	C 6 H 14 O 1
3106-003	C 6 H 14 O 1
3504-055	C 6 H 14 O 6
3504-054	C 6 H 14 O 6
3504-056	C 6 H 14 O 6
3504-053	C 6 H 14 O 6
3504-047	C 6 H 14 O 12 P 2
3501-024	C 6 H 18 O 24 P 6
0802-016	C 7 H 4 CL 2
0103-012	C 7 H 4 O 3 BR 2
3002-014	C 7 H 4 O 6
3002-015	C 7 H 4 O 7
0104-001	C 7 H 6 O 2
0103-013	C 7 H 6 O 2
0103-001	C 7 H 6 O 2
2001-002	C 7 H 6 O 2
2002-001	C 7 H 6 O 3
2001-003	C 7 H 6 O 3
0103-003	C 7 H 6 O 3
0104-022	C 7 H 6 O 3
0104-004	C 7 H 6 O 3
0803-005	C 7 H 6 O 3
0104-009	C 7 H 6 O 4
0104-027	C 7 H 6 O 4
0104-033	C 7 H 6 O 4
0104-030	C 7 H 6 O 4
0104-028	C 7 H 6 O 4
1901-003	C 7 H 6 O 4
1901-001	C 7 H 6 O 4
3904-005	C 7 H 6 O 4
0104-016	C 7 H 6 O 5
0104-036	C 7 H 6 O 5
0103-008	C 7 H 6 O 5
0102-011	C 7 H 7 O 2 CL1
0102-001	C 7 H 8 O 1
0101-007	C 7 H 8 O 1
0101-001	C 7 H 8 O 1
2101-017	C 7 H 8 O 1
2101-013	C 7 H 8 O 2
2101-018	C 7 H 8 O 2
0101-003	C 7 H 8 O 2
0102-007	C 7 H 8 O 2
0802-004	C 7 H 8 O 2
0802-012	C 7 H 8 O 2
0802-002	C 7 H 8 O 2
3106-012	C 7 H 8 O 2
3106-011	C 7 H 8 O 2
3004-006	C 7 H 8 O 2
3003-008	C 7 H 8 O 3
3002-010	C 7 H 8 O 3
2901-005	C 7 H 8 O 3
0803-002	C 7 H 8 O 3
0803-004	C 7 H 8 O 3
0102-003	C 7 H 8 O 3
0102-009	C 7 H 8 O 3
3003-004	C 7 H 8 O 4
3005-013	C 7 H 8 O 4
1901-002	C 7 H 8 O 4
1901-008	C 7 H 8 O 4
1902-004	C 7 H 8 O 5
3002-017	C 7 H 8 O 7
3901-007	C 7 H 10 O 2
3107-010	C 7 H 10 O 5
1902-001	C 7 H 10 O 5
1902-002	C 7 H 10 O 6
3107-011	C 7 H 10 O 6
3197-002	C 7 H 11 O 11 P 1
3107-005	C 7 H 12 O 1
3107-007	C 7 H 12 O 1
3107-006	C 7 H 12 O 1
3901-002	C 7 H 12 O 1
3003-003	C 7 H 12 O 2
3004-015	C 7 H 12 O 3
1902-003	C 7 H 12 O 5
1902-005	C 7 H 12 O 6
1902-016	C 7 H 12 O 6
3107-004	C 7 H 14 O 1
3107-008	C 7 H 14 O 1
3901-001	C 7 H 14 O 1
3196-002	C 7 H 14 O 1
3107-009	C 7 H 14 O 2
3504-034	C 7 H 14 O 4
3504-031	C 7 H 14 O 4
3504-043	C 7 H 14 O 4
3504-038	C 7 H 14 O 4
3504-036	C 7 H 14 O 4
3510-005	C 7 H 14 O 4
3510-016	C 7 H 14 O 4
3510-006	C 7 H 14 O 4
3504-018	C 7 H 14 O 5
3504-024	C 7 H 14 O 5
3504-021	C 7 H 14 O 5
3504-028	C 7 H 14 O 5
3504-027	C 7 H 14 O 5
3197-001	C 7 H 14 O 5
3504-026	C 7 H 14 O 5
3504-020	C 7 H 14 O 5
3504-023	C 7 H 14 O 5
3504-029	C 7 H 14 O 5
3501-025	C 7 H 14 O 6
3501-009	C 7 H 14 O 6
3501-010	C 7 H 14 O 6
3501-019	C 7 H 14 O 6
3501-021	C 7 H 14 O 6
3505-004	C 7 H 14 O 6
3504-002	C 7 H 14 O 6
3501-020	C 7 H 14 O 6
3504-015	C 7 H 14 O 6
3505-005	C 7 H 14 O 6
3501-013	C 7 H 14 O 6
3501-028	C 7 H 14 O 6
3501-029	C 7 H 14 O 6
3501-014	C 7 H 14 O 6
3505-001	C 7 H 14 O 7
3505-002	C 7 H 14 O 7
3301-007	C 7 H 14 S 2
3509-025	C 7 H 15 N 1 O 3
3509-021	C 7 H 15 N 1 O 3
3509-020	C 7 H 15 N 1 O 3
3502-038	C 7 H 15 N 1 O 4
3509-026	C 7 H 15 N 1 O 4
3509-009	C 7 H 15 N 1 O 5
3107-001	C 7 H 16
3502-034	C 7 H 16 N 2 O 3
3107-002	C 7 H 16 O 1
3107-003	C 7 H 16 O 1
3505-008	C 7 H 16 O 6
3505-006	C 7 H 16 O 7
3505-007	C 7 H 16 O 7
3108-018	C 8 H 3 N 1 O 2
3108-016	C 8 H 5 N 1 O 2
3108-017	C 8 H 5 N 1 O 3
3108-007	C 8 H 6 O 1
3905-001	C 8 H 6 O 1
3302-001	C 8 H 6 O 1 S 1
3905-002	C 8 H 6 O 2
0802-017	C 8 H 6 O 2 CL4
2102-026	C 8 H 6 O 3
0103-007	C 8 H 6 O 3
0104-015	C 8 H 6 O 4
2102-027	C 8 H 6 O 4
2102-003	C 8 H 6 O 4
2102-004	C 8 H 6 O 5
2102-030	C 8 H 6 O 5
0204-007	C 8 H 6 O 5
3001-001	C 8 H 6 O 5
3001-004	C 8 H 6 O 6
2201-015	C 8 H 6 O 6
2102-007	C 8 H 6 O 6
0201-001	C 8 H 8
0203-020	C 8 H 8 O 1
0203-009	C 8 H 8 O 1
0201-002	C 8 H 8 O 1
0203-001	C 8 H 8 O 2
0204-001	C 8 H 8 O 2
0103-014	C 8 H 8 O 2
0103-002	C 8 H 8 O 2
2001-006	C 8 H 8 O 2
2001-005	C 8 H 8 O 2
2201-001	C 8 H 8 O 2
2001-004	C 8 H 8 O 2
2002-002	C 8 H 8 O 3
2201-007	C 8 H 8 O 3
2002-019	C 8 H 8 O 3
2101-012	C 8 H 8 O 3
0103-004	C 8 H 8 O 3
0103-017	C 8 H 8 O 3
0204-003	C 8 H 8 O 3
0204-002	C 8 H 8 O 3
0204-004	C 8 H 8 O 3
0203-004	C 8 H 8 O 3
0104-005	C 8 H 8 O 3
0104-007	C 8 H 8 O 3
0104-024	C 8 H 8 O 3
0104-023	C 8 H 8 O 3
0102-006	C 8 H 8 O 3 BR2
0104-029	C 8 H 8 O 4
0104-010	C 8 H 8 O 4
0204-005	C 8 H 8 O 4
0204-006	C 8 H 8 O 4
2003-001	C 8 H 8 O 4
2003-005	C 8 H 8 O 4
2002-023	C 8 H 8 O 4
2003-002	C 8 H 8 O 4
2003-003	C 8 H 8 O 4
2003-004	C 8 H 8 O 4
2101-001	C 8 H 8 O 4
3003-005	C 8 H 8 O 4
3003-009	C 8 H 8 O 4
1901-004	C 8 H 8 O 5
2102-029	C 8 H 8 O 5
2004-001	C 8 H 8 O 5
0201-008	C 8 H 10 O 1
0202-001	C 8 H 10 O 1
0101-002	C 8 H 10 O 1
0899-013	C 8 H 10 O 1
0899-012	C 8 H 10 O 1
0899-014	C 8 H 10 O 1
0899-015	C 8 H 10 O 1
0899-016	C 8 H 10 O 1
0802-014	C 8 H 10 O 2
0802-013	C 8 H 10 O 2
0101-004	C 8 H 10 O 2
0102-002	C 8 H 10 O 2
0202-005	C 8 H 10 O 2
2101-015	C 8 H 10 O 2
2101-014	C 8 H 10 O 2
2101-016	C 8 H 10 O 3
0201-003	C 8 H 10 O 3
0202-007	C 8 H 10 O 3
0102-004	C 8 H 10 O 3
0101-005	C 8 H 10 O 3
0803-006	C 8 H 10 O 3
0803-003	C 8 H 10 O 3
3003-010	C 8 H 10 O 3
2299-002	C 8 H 10 O 3
3005-012	C 8 H 10 O 4
2101-019	C 8 H 10 O 4
3004-012	C 8 H 10 O 5
1901-006	C 8 H 10 O 5
3106-013	C 8 H 12 O 2 S 1
3005-014	C 8 H 12 O 3
3005-051	C 8 H 12 O 5
1901-005	C 8 H 12 O 5
3510-008	C 8 H 12 O 5
1901-007	C 8 H 12 O 6
3197-003	C 8 H 12 O 8
2901-004	C 8 H 14 O 1
3108-009	C 8 H 14 O 1
3904-009	C 8 H 14 O 2
3305-003	C 8 H 14 O 2 S 2
3504-037	C 8 H 14 O 5
3904-008	C 8 H 14 O 5
3504-009	C 8 H 14 O 7
3510-009	C 8 H 15 N 1 O 5
3509-027	C 8 H 15 N 3 O 7
3502-031	C 8 H 16 N 4 O 6
3108-010	C 8 H 16 O 1
3108-005	C 8 H 16 O 1
3108-008	C 8 H 16 O 1
3108-011	C 8 H 16 O 1
3108-004	C 8 H 16 O 1
3004-014	C 8 H 16 O 1
3108-012	C 8 H 16 O 2
3510-015	C 8 H 16 O 4
3510-004	C 8 H 16 O 4
3510-014	C 8 H 16 O 5
3501-022	C 8 H 16 O 6
3501-023	C 8 H 16 O 6
3509-022	C 8 H 17 N 1 O 2
3509-023	C 8 H 17 N 1 O 2
3509-028	C 8 H 17 N 1 O 3
3509-014	C 8 H 17 N 1 O 3
3509-012	C 8 H 17 N 1 O 4
3509-018	C 8 H 17 N 1 O 4
3502-002	C 8 H 18 N 6 O 4
3108-003	C 8 H 18 O 1
3108-002	C 8 H 18 O 1
3108-001	C 8 H 18 O 1
3109-018	C 9 H 4 O 1
3109-025	C 9 H 4 O 2
3001-003	C 9 H 4 O 6
3001-005	C 9 H 4 O 7
3109-006	C 9 H 6 O 1
3302-004	C 9 H 6 O 1 S 1

628

3111-018	C11	H	8	□	2	
3110-056	C11	H	8	□	2	
3111-019	C11	H	8	□	2	
3111-017	C11	H	8	□	2	
3110-057	C11	H	8	□	2	
2502-004	C11	H	8	□	2	
2502-039	C11	H	8	□	2	
2502-003	C11	H	8	□	2	
3201-012	C11	H	8	□	2	
2503-046	C11	H	8	□	3	
2503-005	C11	H	8	□	3	
2503-004	C11	H	8	□	3	
2503-040	C11	H	8	□	3	
2504-009	C11	H	8	□	4	
2504-023	C11	H	8	□	4	
0501-036	C11	H	8	□	4	
0511-009	C11	H	8	□	5	
0509-007	C11	H	8	□	5	
2505-008	C11	H	8	□	5	
3001-008	C11	H	8	□	5	
2104-032	C11	H	8	□	5	
2508-002	C11	H	8	□	8	
2203-011	C11	H	9	□	4	CL1
3111-004	C11	H	10			
2501-001	C11	H	10	□	1	
3201-011	C11	H	10	□	1	
3905-008	C11	H	10	□	2	
2502-026	C11	H	10	□	2	
2502-024	C11	H	10	□	2	
2502-001	C11	H	10	□	2	
3110-053	C11	H	10	□	2	
3110-054	C11	H	10	□	2	
3110-061	C11	H	10	□	2	S 1
3203-002	C11	H	10	□	2	S 1
3905-010	C11	H	10	□	3	
3111-020	C11	H	10	□	3	
2503-041	C11	H	10	□	3	
2104-018	C11	H	10	□	3	
0599-007	C11	H	10	□	3	
0401-006	C11	H	10	□	3	
0402-008	C11	H	10	□	3	
0507-006	C11	H	10	□	4	
0508-001	C11	H	10	□	4	
2203-006	C11	H	10	□	4	
2203-005	C11	H	10	□	4	
2203-004	C11	H	10	□	4	
2104-010	C11	H	10	□	4	
3001-007	C11	H	10	□	4	
2104-009	C11	H	10	□	4	CL2
3201-009	C11	H	10	□	4	S 1
3905-022	C11	H	10	□	5	
2102-024	C11	H	10	□	5	
2104-026	C11	H	10	□	5	
2104-031	C11	H	10	□	5	
2104-022	C11	H	10	□	5	
0509-003	C11	H	10	□	5	
0509-005	C11	H	10	□	5	
0510-001	C11	H	10	□	5	
2102-025	C11	H	10	□	6	
2104-008	C11	H	11	□	4	CL1
2104-007	C11	H	11	□	4	CL1
3111-008	C11	H	12	□	1	
3111-009	C11	H	12	□	2	
3111-010	C11	H	12	□	2	
3110-043	C11	H	12	□	2	
3110-042	C11	H	12	□	2	
2001-008	C11	H	12	□	2	
0302-003	C11	H	12	□	2	
3110-058	C11	H	12	□	2	S 1
3302-007	C11	H	12	□	2	S 1
3905-009	C11	H	12	□	3	
0301-025	C11	H	12	□	3	
0301-024	C11	H	12	□	3	
0301-030	C11	H	12	□	3	
0301-029	C11	H	12	□	3	
0304-014	C11	H	12	□	3	
2104-017	C11	H	12	□	3	
2503-044	C11	H	12	□	3	
2503-043	C11	H	12	□	3	
1901-012	C11	H	12	□	4	
2104-006	C11	H	12	□	4	
0303-008	C11	H	12	□	4	
0304-023	C11	H	12	□	4	
0399-006	C11	H	12	□	4	
3005-039	C11	H	12	□	4	
3001-006	C11	H	12	□	5	
0304-030	C11	H	12	□	5	
2102-022	C11	H	12	□	5	
2104-019	C11	H	12	□	5	
3905-012	C11	H	12	□	5	
2103-003	C11	H	12	□	6	
2102-023	C11	H	12	□	6	
2105-004	C11	H	12	□	7	
3902-006	C11	H	12	□	7	
3902-007	C11	H	12	□	8	
0301-028	C11	H	14	□	2	
0401-001	C11	H	14	□	2	
0301-011	C11	H	14	□	2	
0301-014	C11	H	14	□	2	
2901-021	C11	H	14	□	2	
3110-052	C11	H	14	□	2	
0301-021	C11	H	14	□	3	
0401-005	C11	H	14	□	3	
2201-017	C11	H	14	□	3	
1901-009	C11	H	14	□	4	
1901-010	C11	H	14	□	4	
1901-011	C11	H	14	□	4	
2201-013	C11	H	14	□	4	
0302-008	C11	H	14	□	4	
3005-048	C11	H	14	□	4	
3005-033	C11	H	14	□	4	
0399-005	C11	H	14	□	5	
2201-014	C11	H	14	□	5	
2201-012	C11	H	14	□	9	
3111-003	C11	H	16			
3001-010	C11	H	16			
3197-010	C11	H	16			
2901-019	C11	H	16	□	1	
3005-017	C11	H	16	□	2	
3904-006	C11	H	16	□	2	
2105-006	C11	H	16	□	3	
2201-016	C11	H	16	□	3	
1901-014	C11	H	16	□	3	
0301-026	C11	H	16	□	3	
0506-010	C11	H	16	□	4	
1901-013	C11	H	16	□	4	
3005-049	C11	H	16	□	4	
3005-035	C11	H	16	□	8	
3005-002	C11	H	16	□	9	
3001-009	C11	H	18			
3111-002	C11	H	18			
3197-009	C11	H	18			
3904-007	C11	H	18	□	2	
3111-016	C11	H	18	□	2	
3003-006	C11	H	18	□	4	
3203-012	C11	H	18	□	4	
3504-011	C11	H	18	□	8	
3504-012	C11	H	18	□	9	
3111-013	C11	H	20	□	1	
3305-004	C11	H	20	□	1	S 2
3111-015	C11	H	20	□	2	
3506-008	C11	H	20	□	10	
3506-007	C11	H	20	□	10	
3506-010	C11	H	20	□	10	
3506-011	C11	H	20	□	10	
3506-009	C11	H	20	□	10	
3111-012	C11	H	22	□	1	
3111-007	C11	H	22	□	1	
3111-014	C11	H	22	□	2	
3506-006	C11	H	22	□	10	
3111-001	C11	H	24			
3111-006	C11	H	24	□	1	
0899-017	C12	H	5	□	2	BR5
3112-001	C12	H	6			
3302-010	C12	H	6	S 1		
3302-011	C12	H	7	□	1	S 1 CL1
3205-007	C12	H	8	□	1	
3302-013	C12	H	8	□	1	S 1
3204-002	C12	H	8	□	2	
0899-018	C12	H	8	□	2	BR2
3303-001	C12	H	8	□	2	S 2
0506-011	C12	H	8	□	4	
0508-036	C12	H	8	□	4	
0507-020	C12	H	8	□	4	
0402-010	C12	H	8	□	4	
2502-033	C12	H	8	□	4	
2505-010	C12	H	8	□	6	
2505-004	C12	H	8	□	6	
2505-012	C12	H	8	□	6	
2506-006	C12	H	8	□	7	
2506-010	C12	H	8	□	7	
2506-003	C12	H	8	□	7	
2507-003	C12	H	8	□	8	
3303-002	C12	H	8	S 2		
3304-001	C12	H	8	S 3		
3302-012	C12	H	9	□	1	S 1 CL1
3303-006	C12	H	9	□	1	S 2 CL1
3205-008	C12	H	10			
3205-009	C12	H	10	□	1	
3205-011	C12	H	10	□	1	
3303-003	C12	H	10	□	1	S 2
3201-013	C12	H	10	□	2	
3205-012	C12	H	10	□	2	
3204-001	C12	H	10	□	2	
2502-005	C12	H	10	□	2	
2501-006	C12	H	10	□	2	
2503-022	C12	H	10	□	3	
2502-028	C12	H	10	□	3	
2503-047	C12	H	10	□	3	
2502-027	C12	H	10	□	3	
2504-008	C12	H	10	□	4	
2504-001	C12	H	10	□	4	
2504-024	C12	H	10	□	4	
2504-010	C12	H	10	□	4	
2504-012	C12	H	10	□	4	
2504-003	C12	H	10	□	4	
0402-004	C12	H	10	□	4	
0402-003	C12	H	10	□	4	
0402-002	C12	H	10	□	4	
3110-075	C12	H	10	□	4	
2801-003	C12	H	10	□	4	
2505-011	C12	H	10	□	5	
2505-002	C12	H	10	□	5	
2505-009	C12	H	10	□	5	
2104-033	C12	H	10	□	5	
2506-002	C12	H	10	□	6	
2506-005	C12	H	10	□	6	
2506-008	C12	H	10	□	6	
2506-009	C12	H	10	□	6	
3201-019	C12	H	10	□	6	
2507-002	C12	H	10	□	7	
2508-004	C12	H	10	□	8	
2508-003	C12	H	10	□	8	
3205-013	C12	H	12			
3112-006	C12	H	12	□	1	
3110-016	C12	H	12	□	2	
3110-020	C12	H	12	□	2	
3905-020	C12	H	12	□	2	
2502-025	C12	H	12	□	2	
2106-023	C12	H	12	□	2	
2199-013	C12	H	12	□	2	
0402-007	C12	H	12	□	2	
3999-010	C12	H	12	□	2	
3110-049	C12	H	12	□	2	S 1
3110-047	C12	H	12	□	2	S 1
3110-050	C12	H	12	□	2	S 1
3110-048	C12	H	12	□	2	S 1
3112-019	C12	H	12	□	3	
3110-076	C12	H	12	□	4	
2203-007	C12	H	12	□	4	
2104-023	C12	H	12	□	5	
3506-019	C12	H	12	□	11	
3902-003	C12	H	14	□	2	
3905-013	C12	H	14	□	2	
2106-024	C12	H	14	□	2	
2001-009	C12	H	14	□	2	
2106-022	C12	H	14	□	2	
3110-012	C12	H	14	□	2	
3110-024	C12	H	14	□	2	
3112-018	C12	H	14	□	2	
3110-005	C12	H	14	□	2	
3110-059	C12	H	14	□	2	S 1
0301-019	C12	H	14	□	3	
0402-005	C12	H	14	□	4	
0301-037	C12	H	14	□	4	
0301-036	C12	H	14	□	4	
0301-035	C12	H	14	□	4	
2203-014	C12	H	14	□	4	
2104-004	C12	H	14	□	4	
3905-029	C12	H	14	□	4	
2104-020	C12	H	14	□	5	
2103-002	C12	H	14	□	5	
2106-021	C12	H	16	□	2	
0201-010	C12	H	16	□	2	
3110-014	C12	H	16	□	2	
2901-018	C12	H	16	□	3	
2999-003	C12	H	16	□	3	
2903-007	C12	H	16	□	3	
0301-023	C12	H	16	□	3	
0301-022	C12	H	16	□	3	
0301-033	C12	H	16	□	3	
0301-032	C12	H	16	□	3	
0301-031	C12	H	16	□	3	
0302-011	C12	H	16	□	3	
2002-003	C12	H	16	□	3	
2206-007	C12	H	16	□	3	
2106-001	C12	H	16	□	4	
2105-007	C12	H	16	□	4	
2203-003	C12	H	16	□	4	
3905-016	C12	H	16	□	4	
3905-017	C12	H	16	□	4	
2403-010	C12	H	16	□	5	
3005-034	C12	H	16	□	6	

2101-003	C12 H 16 O 7
0802-006	C12 H 16 O 7
3112-023	C12 H 17 N 1 O 3
2901-008	C12 H 17 N 1 O 6
2901-009	C12 H 17 N 1 O 7
2106-025	C12 H 18 O 2
2106-019	C12 H 18 O 2
2106-020	C12 H 18 O 2
2106-026	C12 H 18 O 3
2106-027	C12 H 18 O 3
3112-009	C12 H 18 O 3
3197-013	C12 H 18 O 5
3112-007	C12 H 20 O 1
3203-014	C12 H 20 O 2
3201-017	C12 H 20 O 2
3112-021	C12 H 20 O 4
3112-022	C12 H 20 O 5
3506-041	C12 H 20 O12
3509-027	C12 H 21 N 1 O10
3509-010	C12 H 21 N 1 O11
3112-011	C12 H 22 O 1
3111-011	C12 H 22 O 1
3112-013	C12 H 22 O 2
3112-017	C12 H 22 O 2
3112-016	C12 H 22 O 2
3201-020	C12 H 22 O 2
3506-030	C12 H 22 O 9
3506-036	C12 H 22 O10
3506-035	C12 H 22 O10
3506-034	C12 H 22 O10
3506-043	C12 H 22 O10
3501-006	C12 H 22 O10
3506-028	C12 H 22 O11
3506-016	C12 H 22 O11
3501-017	C12 H 22 O11
3506-029	C12 H 22 O11
3506-015	C12 H 22 O11
3506-020	C12 H 22 O11
3506-014	C12 H 22 O11
3506-031	C12 H 22 O11
3506-032	C12 H 22 O11
3506-013	C12 H 22 O11
3501-016	C12 H 22 O11
3506-033	C12 H 22 O11
3506-025	C12 H 22 O11
3506-024	C12 H 22 O11
3506-012	C12 H 22 O11
3506-023	C12 H 22 O11
3506-017	C12 H 22 O11
3506-022	C12 H 22 O11
3506-026	C12 H 22 O11
3506-021	C12 H 22 O11
3501-027	C12 H 22 O11
3506-018	C12 H 22 O11
3506-037	C12 H 22 O11
3506-027	C12 H 22 O11
3112-010	C12 H 24 O 1
3112-014	C12 H 24 O 2
3112-020	C12 H 24 O 3
0601-038	C12 H 24 O 5
3506-040	C12 H 24 O11
3506-038	C12 H 24 O11
3506-039	C12 H 24 O11
3502-008	C12 H 25 N 3 O 7
3502-009	C12 H 26 N 4 O 6
3112-002	C12 H 26 O 1
3112-008	C12 H 26 O 2
3506-042	C12 H 26 O10
3113-002	C13 H 6
3113-023	C13 H 6 O 1
3302-026	C13 H 7 O 1 S 1 CL1
3113-003	C13 H 8
3113-011	C13 H 8 O 1
3113-024	C13 H 8 O 1
3113-025	C13 H 8 O 1
3113-032	C13 H 8 O 1
3205-015	C13 H 8 O 1
3304-008	C13 H 8 O 1 S 3
2801-004	C13 H 8 O 3
2401-002	C13 H 8 O 3
2401-001	C13 H 8 O 3
2402-003	C13 H 8 O 4
2402-011	C13 H 8 O 4
3906-005	C13 H 8 O 4
3906-006	C13 H 8 O 5
2403-018	C13 H 8 O 5
2403-028	C13 H 8 O 5
2404-034	C13 H 8 O 6
2404-022	C13 H 8 O 6
2404-029	C13 H 8 O 6
2404-007	C13 H 8 O 6
2405-009	C13 H 8 O 7
3906-015	C13 H 8 O 8
3302-022	C13 H 8 S 1
3202-008	C13 H 9 O 1 CL1
3203-006	C13 H 9 O 1 CL1
3202-007	C13 H 9 O 1 CL1
3203-005	C13 H 9 O 1 CL1
3302-027	C13 H 9 O 1 S 1 CL1
3203-007	C13 H 9 O 2 CL1
3113-004	C13 H 10
3113-005	C13 H 10
3113-015	C13 H 10 O 1
3113-033	C13 H 10 O 1
3201-006	C13 H 10 O 1
3205-016	C13 H 10 O 1
3205-014	C13 H 10 O 1
3304-002	C13 H 10 O 1 S 3
3201-015	C13 H 10 O 2
3204-010	C13 H 10 O 2
3205-006	C13 H 10 O 2
3201-014	C13 H 10 O 2
3204-009	C13 H 10 O 2
2802-001	C13 H 10 O 2
3113-028	C13 H 10 O 2
0404-008	C13 H 10 O 2
0404-016	C13 H 10 O 4
3204-014	C13 H 10 O 4
2503-042	C13 H 10 O 4
2204-032	C13 H 10 O 4
2204-033	C13 H 10 O 5
2204-035	C13 H 10 O 5
2503-026	C13 H 10 O 5
0404-017	C13 H 10 O 5
0402-011	C13 H 10 O 5
0510-004	C13 H 10 O 5
0511-001	C13 H 10 O 5
0599-008	C13 H 10 O 5
2802-010	C13 H 10 O 5
2802-007	C13 H 10 O 6
2506-011	C13 H 10 O 7
3302-029	C13 H 10 S 1
3302-032	C13 H 10 S 1
3302-021	C13 H 10 S 1
3113-013	C13 H 11 O 1 CL1
3113-008	C13 H 12
3113-007	C13 H 12
3113-031	C13 H 12 O 1
3302-019	C13 H 12 O 1 S 1
3302-020	C13 H 12 O 1 S 1
3204-004	C13 H 12 O 2
3204-005	C13 H 12 O 2
3205-005	C13 H 12 O 2
3113-018	C13 H 12 O 2
3113-017	C13 H 12 O 2
0203-015	C13 H 12 O 2
0404-007	C13 H 12 O 2
3202-006	C13 H 12 O 2 S 1
3204-022	C13 H 12 O 2 S 1
3204-008	C13 H 12 O 3
2502-030	C13 H 12 O 3
2502-029	C13 H 12 O 3
2202-018	C13 H 12 O 3
2299-010	C13 H 12 O 4
2503-019	C13 H 12 O 4
2504-013	C13 H 12 O 4
2503-023	C13 H 12 O 4
3906-013	C13 H 12 O 4
3113-020	C13 H 12 O 4
2503-001	C13 H 12 O 5
2507-004	C13 H 12 O 7
2507-005	C13 H 12 O 7
2199-006	C13 H 12 O 7
0304-017	C13 H 12 O 8
3113-006	C13 H 14
3113-009	C13 H 14 O 1
3302-015	C13 H 14 O 1 S 1
3201-016	C13 H 14 O 2
0203-014	C13 H 14 O 2
0203-012	C13 H 14 O 2
0203-016	C13 H 14 O 2
2202-027	C13 H 14 O 2
2199-014	C13 H 14 O 2
2202-015	C13 H 14 O 2
3302-016	C13 H 14 O 2 S 1
3204-018	C13 H 14 O 2 S 1
3204-019	C13 H 14 O 2 S 1
3302-018	C13 H 14 O 2 S 1
3905-011	C13 H 14 O 3
2202-019	C13 H 14 O 3
2202-009	C13 H 14 O 3
2202-020	C13 H 14 O 3
0203-017	C13 H 14 O 3
2503-017	C13 H 14 O 3
3204-021	C13 H 14 O 3 S 1
3204-020	C13 H 14 O 3 S 1
3905-028	C13 H 14 O 4
2203-010	C13 H 14 O 4
2104-005	C13 H 14 O 4
2204-040	C13 H 14 O 4
2104-024	C13 H 14 O 5
2106-017	C13 H 14 O 5
2104-025	C13 H 14 O 6
3113-010	C13 H 16 O 1
3113-030	C13 H 16 O 2
2199-015	C13 H 16 O 2
0203-010	C13 H 16 O 2
3905-014	C13 H 16 O 3
3905-018	C13 H 16 O 3
2204-041	C13 H 16 O 5
2203-001	C13 H 16 O 5
2103-004	C13 H 16 O 6
0103-015	C13 H 16 O 7
0104-003	C13 H 16 O 7
0104-002	C13 H 16 O 7
0104-006	C13 H 16 O 8
0104-018	C13 H 16 O10
0104-017	C13 H 16 O10
0202-004	C13 H 18 O 2
0202-003	C13 H 18 O 2
0301-034	C13 H 18 O 4
2206-005	C13 H 18 O 4
2206-002	C13 H 18 O 4
2206-006	C13 H 18 O 4
0102-008	C13 H 18 O 7
0802-015	C13 H 18 O 7
0102-010	C13 H 18 O 8
3113-022	C13 H 22 O 1
3301-016	C13 H 22 O 2 S 1
3197-014	C13 H 24 O 1
3113-021	C13 H 24 O 1
3301-015	C13 H 24 O 2 S 1
3506-049	C13 H 24 O 8
3506-050	C13 H 24 O 8
3110-081	C13 H 25 N 1 O 5
3113-029	C13 H 26 O 1
3113-026	C13 H 26 O 2
3197-015	C13 H 26 O 2
3113-001	C13 H 28
2403-017	C14 H 6 O 5 CL4
0105-001	C14 H 6 O 8
2403-016	C14 H 7 O 5 CL3
2601-001	C14 H 8 O 2
2602-001	C14 H 8 O 3
2602-003	C14 H 8 O 3
2603-001	C14 H 8 O 4
2603-007	C14 H 8 O 4
2402-002	C14 H 8 O 4
2604-008	C14 H 8 O 5
2403-035	C14 H 8 O 5
2604-001	C14 H 8 O 5
2504-015	C14 H 8 O 5
0105-008	C14 H 8 O 9
3303-005	C14 H 9 O 2 S 2 CL1
3113-037	C14 H 10 O 1 S 1
3113-034	C14 H 10 O 2 S 1
3205-004	C14 H 10 O 3
2503-013	C14 H 10 O 3
2503-014	C14 H 10 O 3
3113-035	C14 H 10 O 3 S 1
2503-015	C14 H 10 O 4
2402-004	C14 H 10 O 4
2402-001	C14 H 10 O 4
2402-015	C14 H 10 O 4
3204-003	C14 H 10 O 4
3906-001	C14 H 10 O 4
0404-022	C14 H 10 O 4
2503-051	C14 H 10 O 4 CL2
3113-036	C14 H 10 O 4 S 1
2801-005	C14 H 10 O 5
2403-001	C14 H 10 O 5
2403-029	C14 H 10 O 5
2403-006	C14 H 10 O 5
2403-032	C14 H 10 O 5
2403-004	C14 H 10 O 5
2403-019	C14 H 10 O 5
2504-017	C14 H 10 O 5
2403-020	C14 H 10 O 5
2403-027	C14 H 10 O 5
2403-013	C14 H 10 O 5
3906-002	C14 H 10 O 5
2204-016	C14 H 10 O 5
2005-001	C14 H 10 O 6

Registry	C	H	nH	O	nO	Other
2506-012	C14	H	10	□	6	
2404-001	C14	H	10	□	6	
2404-009	C14	H	10	□	6	
2404-008	C14	H	10	□	6	
2505-021	C14	H	10	□	6	
2404-042	C14	H	10	□	6	
2404-023	C14	H	10	□	6	
2404-030	C14	H	10	□	6	
2404-018	C14	H	10	□	6	
2505-017	C14	H	10	□	6	
2199-008	C14	H	10	□	7	
2005-002	C14	H	10	□	8	
0199-014	C14	H	10	□	9	
0199-015	C14	H	10	□	10	
2804-001	C14	H	12			
2804-002	C14	H	12	□	1	
2804-004	C14	H	12	□	2	
0199-012	C14	H	12	□	2	
3303-004	C14	H	12	□	2	S 2
3205-002	C14	H	12	□	3	
0404-010	C14	H	12	□	3	
0505-001	C14	H	12	□	3	
0504-001	C14	H	12	□	3	
0503-001	C14	H	12	□	3	
0502-018	C14	H	12	□	3	
2804-012	C14	H	12	□	3	
2804-022	C14	H	12	□	4	
2802-002	C14	H	12	□	4	
3114-044	C14	H	12	□	4	
2804-018	C14	H	12	□	4	
0503-008	C14	H	12	□	4	
0504-003	C14	H	12	□	4	
0504-002	C14	H	12	□	4	
0507-017	C14	H	12	□	4	
0404-021	C14	H	12	□	4	
3906-003	C14	H	12	□	4	
2502-035	C14	H	12	□	4	
2503-002	C14	H	12	□	5	
2504-005	C14	H	12	□	5	
0507-018	C14	H	12	□	5	
0508-014	C14	H	12	□	5	
2804-021	C14	H	12	□	5	
2204-036	C14	H	12	□	5	
2204-037	C14	H	12	□	6	
0599-010	C14	H	12	□	6	
0599-009	C14	H	12	□	6	
2505-013	C14	H	12	□	6	
2505-020	C14	H	12	□	6	
2506-004	C14	H	12	□	7	
3906-050	C14	H	12	□	7	
3902-004	C14	H	12	□	7	
2199-009	C14	H	12	□	8	
0502-009	C14	H	13	N 1 □	3	
3109-029	C14	H	13	N 1 □	3	
3114-005	C14	H	14	□	1	
3114-022	C14	H	14	□	1	
3114-010	C14	H	14	□	2	
3114-019	C14	H	14	□	2	
3203-003	C14	H	14	□	2	
3204-023	C14	H	14	□	2	
3204-024	C14	H	14	□	2	
2502-036	C14	H	14	□	2	
2801-001	C14	H	14	□	3	
0501-028	C14	H	14	□	3	
0501-004	C14	H	14	□	3	
0501-019	C14	H	14	□	3	
0404-009	C14	H	14	□	3	
0404-011	C14	H	14	□	3	
0404-005	C14	H	14	□	3	
0502-003	C14	H	14	□	4	
0502-001	C14	H	14	□	4	
0503-004	C14	H	14	□	4	
0503-002	C14	H	14	□	4	
0505-002	C14	H	14	□	4	
0507-016	C14	H	14	□	4	
0507-004	C14	H	14	□	4	
3114-045	C14	H	14	□	4	
2503-025	C14	H	14	□	4	
2503-010	C14	H	14	□	4	
2503-036	C14	H	14	□	4	
3906-004	C14	H	14	□	4	
2202-023	C14	H	14	□	4	
2505-003	C14	H	14	□	5	
2504-006	C14	H	14	□	5	
3005-018	C14	H	14	□	5	
0505-008	C14	H	14	□	5	
0504-008	C14	H	14	□	5	
0506-009	C14	H	14	□	5	
2499-011	C14	H	14	□	8	
3113-038	C14	H	14	S 1		
3114-003	C14	H	16	□	1	
3114-017	C14	H	16	□	2	
3114-013	C14	H	16	□	2	
3203-016	C14	H	16	□	2	
3203-015	C14	H	16	□	2	
3203-008	C14	H	16	□	2	
3204-030	C14	H	16	□	2	S 1
2503-018	C14	H	16	□	3	
0404-006	C14	H	16	□	3	
0404-012	C14	H	16	□	3	
0203-019	C14	H	16	□	3	
0203-018	C14	H	16	□	3	
2205-001	C14	H	16	□	3	
2202-010	C14	H	16	□	3	
2202-026	C14	H	16	□	3	
2202-007	C14	H	16	□	4	
2202-005	C14	H	16	□	4	
2202-024	C14	H	16	□	4	
2202-011	C14	H	16	□	4	
2202-017	C14	H	16	□	4	
2202-003	C14	H	16	□	4	
0404-013	C14	H	16	□	4	
0901-009	C14	H	16	□	4	
0501-020	C14	H	16	□	5	
2203-008	C14	H	16	□	5	
2202-025	C14	H	16	□	5	
0105-002	C14	H	16	□	6	
3905-027	C14	H	16	□	6	
3906-014	C14	H	16	□	9	
0199-004	C14	H	16	□	10	
2107-006	C14	H	16	□	14	
0301-004	C14	H	18	□	1	
3114-007	C14	H	18	□	1	
3114-021	C14	H	18	□	1	
3114-033	C14	H	18	□	2	
0301-005	C14	H	18	□	2	
0203-013	C14	H	18	□	2	
3203-017	C14	H	18	□	2	
0203-011	C14	H	18	□	3	
0399-017	C14	H	18	□	3	
2003-017	C14	H	18	□	4	
2202-001	C14	H	18	□	4	
2202-002	C14	H	18	□	4	
0203-002	C14	H	18	□	7	
0802-007	C14	H	18	□	8	
0104-025	C14	H	18	□	8	
0103-005	C14	H	18	□	8	
3302-009	C14	H	19	N 1 □	1	S 1
3203-018	C14	H	20	□	2	
0302-012	C14	H	20	□	2	
2206-004	C14	H	20	□	4	
2206-003	C14	H	20	□	4	
2107-007	C14	H	20	□	4	
2206-019	C14	H	20	□	5	
3499-008	C14	H	20	□	6	
0202-006	C14	H	20	□	7	
0102-005	C14	H	20	□	8	
0803-007	C14	H	20	□	8	
0104-014	C14	H	22	N 1 □	4	+
2107-008	C14	H	22	□	4	
3110-060	C14	H	23	N 1 □	1	
3502-033	C14	H	24	N 2 □	7	
2901-012	C14	H	24	□	2	
3114-030	C14	H	24	□	2	
3114-029	C14	H	24	□	2	
3110-051	C14	H	25	N 1 □	1	
3110-046	C14	H	25	N 1 □	1	
3501-011	C14	H	25	N 3 □	9	
3114-027	C14	H	26	□	2	
3114-024	C14	H	26	□	2	
3114-025	C14	H	26	□	2	
3114-026	C14	H	26	□	2	
3112-003	C14	H	26	□	2	
3112-004	C14	H	26	□	2	
3112-005	C14	H	26	□	2	
3506-048	C14	H	26	□	8	
3506-047	C14	H	26	□	8	
3506-046	C14	H	26	□	8	
3506-045	C14	H	26	□	8	
3114-020	C14	H	28	□	1	
3113-027	C14	H	28	□	2	
3114-023	C14	H	28	□	2	
3197-016	C14	H	28	□	2	
3114-037	C14	H	28	□	3	
3114-038	C14	H	28	□	3	
3114-039	C14	H	28	□	4	
3114-042	C14	H	28	□	4	
3114-001	C14	H	30	□	1	
2601-005	C15	H	8	□	3	
2601-006	C15	H	8	□	4	
2603-018	C15	H	8	□	4	
1602-001	C15	H	8	□	5	
2604-041	C15	H	8	□	5	CL2
2603-020	C15	H	8	□	6	
2603-031	C15	H	8	□	6	
1602-003	C15	H	8	□	6	
1602-008	C15	H	8	□	6	
2604-052	C15	H	8	□	7	
2604-009	C15	H	8	□	7	
2401-003	C15	H	8	□	7	
2604-039	C15	H	9	□	5	CL1
2604-046	C15	H	9	□	6	CL1
2605-010	C15	H	9	□	6	CL1
2605-019	C15	H	9	□	6	CL1
2601-002	C15	H	10	□	2	
3113-012	C15	H	10	□	2	
1101-001	C15	H	10	□	2	
1101-002	C15	H	10	□	3	
2601-003	C15	H	10	□	3	
2602-008	C15	H	10	□	3	
2603-023	C15	H	10	□	3	
2602-004	C15	H	10	□	3	
2602-005	C15	H	10	□	3	
2603-011	C15	H	10	□	4	
2603-024	C15	H	10	□	4	
2603-008	C15	H	10	□	4	
2603-025	C15	H	10	□	4	
2603-002	C15	H	10	□	4	
2602-002	C15	H	10	□	4	
2603-003	C15	H	10	□	4	
2603-021	C15	H	10	□	4	
2603-026	C15	H	10	□	4	
2603-034	C15	H	10	□	4	
2603-009	C15	H	10	□	4	
2602-007	C15	H	10	□	4	
2603-027	C15	H	10	□	4	
2701-026	C15	H	10	□	4	
2603-005	C15	H	10	□	4	
1101-004	C15	H	10	□	4	
0999-016	C15	H	10	□	4	
1101-016	C15	H	10	□	4	
1401-023	C15	H	10	□	4	
3905-024	C15	H	10	□	4	
1803-009	C15	H	10	□	5	
1401-001	C15	H	10	□	5	
1103-002	C15	H	10	□	5	
1199-030	C15	H	10	□	5	
1101-011	C15	H	10	□	5	
1102-001	C15	H	10	□	5	
1104-001	C15	H	10	□	5	
2603-015	C15	H	10	□	5	
2403-036	C15	H	10	□	5	
2604-028	C15	H	10	□	5	
2604-020	C15	H	10	□	5	
2603-029	C15	H	10	□	5	
2604-002	C15	H	10	□	5	
2604-003	C15	H	10	□	5	
2604-026	C15	H	10	□	5	
2604-011	C15	H	10	□	5	
2399-008	C15	H	10	□	5	
2604-055	C15	H	10	□	5	
2604-054	C15	H	10	□	5	
2605-009	C15	H	10	□	6	
2605-016	C15	H	10	□	6	
2605-004	C15	H	10	□	6	
2402-014	C15	H	10	□	6	
2605-018	C15	H	10	□	6	
2604-045	C15	H	10	□	6	
2605-001	C15	H	10	□	6	
2604-027	C15	H	10	□	6	
2605-026	C15	H	10	□	6	
2605-002	C15	H	10	□	6	
2605-015	C15	H	10	□	6	
1104-002	C15	H	10	□	6	
1104-003	C15	H	10	□	6	
1103-032	C15	H	10	□	6	
1199-001	C15	H	10	□	6	
1106-001	C15	H	10	□	6	
1105-001	C15	H	10	□	6	
1199-008	C15	H	10	□	6	
1803-001	C15	H	10	□	6	
1803-005	C15	H	10	□	6	
1402-001	C15	H	10	□	6	
1106-007	C15	H	10	□	7	
1104-057	C15	H	10	□	7	
1199-016	C15	H	10	□	7	
1106-008	C15	H	10	□	7	
1106-009	C15	H	10	□	7	
1108-017	C15	H	10	□	7	
2605-011	C15	H	10	□	7	
2606-006	C15	H	10	□	7	
2605-021	C15	H	10	□	7	
2606-010	C15	H	10	□	8	

Ref	C	H	O	Other
2606-009	C15	H 10	O 8	
1108-001	C15	H 10	O 8	
1106-091	C15	H 10	O 8	
1106-069	C15	H 10	O 8	
1108-021	C15	H 10	O 9	
3302-024	C15	H 11	O 2	S 1 CL1
3302-028	C15	H 11	O 2	S 1 CL1
1701-001	C15	H 11	O 4	+
2699-012	C15	H 11	O 4	CL1
1701-003	C15	H 11	O 5	+
1702-001	C15	H 11	O 5	+
1701-008	C15	H 11	O 6	+
1702-002	C15	H 11	O 6	+
1703-001	C15	H 11	O 7	+
3205-017	C15	H 12	O 2	
3113-016	C15	H 12	O 2	
3304-003	C15	H 12	O 2	S 3
3205-018	C15	H 12	O 3	
2503-012	C15	H 12	O 3	
2503-010	C15	H 12	O 3	
2699-001	C15	H 12	O 3	
3302-031	C15	H 12	O 3	S 1
3302-025	C15	H 12	O 3	S 1
3204-012	C15	H 12	O 4	
3204-011	C15	H 12	O 4	
3205-003	C15	H 12	O 4	
2699-006	C15	H 12	O 4	
2503-035	C15	H 12	O 4	
2504-014	C15	H 12	O 4	
2804-008	C15	H 12	O 4	
2804-010	C15	H 12	O 4	
2903-013	C15	H 12	O 4	
2204-012	C15	H 12	O 4	
1201-001	C15	H 12	O 4	
1203-024	C15	H 12	O 4	
1002-001	C15	H 12	O 4	
1203-003	C15	H 12	O 4	
1203-033	C15	H 12	O 5	
1002-006	C15	H 12	O 5	
1202-001	C15	H 12	O 5	
1003-001	C15	H 12	O 5	
2204-031	C15	H 12	O 5	
1205-015	C15	H 12	O 5	
2403-022	C15	H 12	O 5	
2699-014	C15	H 12	O 5	
2504-018	C15	H 12	O 5	
2806-027	C15	H 12	O 5	
2504-019	C15	H 12	O 5	
2403-014	C15	H 12	O 5	
2403-008	C15	H 12	O 5	
2403-034	C15	H 12	O 5	
2403-005	C15	H 12	O 5	
2403-033	C15	H 12	O 5	
2403-002	C15	H 12	O 5	
2403-007	C15	H 12	O 5	
3204-015	C15	H 12	O 5	
2404-019	C15	H 12	O 6	
2801-007	C15	H 12	O 6	
2505-018	C15	H 12	O 6	
2404-015	C15	H 12	O 6	
2699-018	C15	H 12	O 6	
2404-005	C15	H 12	O 6	
2404-010	C15	H 12	O 6	
2404-031	C15	H 12	O 6	
2404-012	C15	H 12	O 6	
1399-006	C15	H 12	O 6	
1299-003	C15	H 12	O 6	
1205-023	C15	H 12	O 6	
1206-009	C15	H 12	O 6	
1207-001	C15	H 12	O 6	
1205-019	C15	H 12	O 6	
1803-014	C15	H 12	O 6	
1003-007	C15	H 12	O 6	
1205-001	C15	H 12	O 6	
1003-006	C15	H 12	O 6	
1003-005	C15	H 12	O 6	
1002-021	C15	H 12	O 6	
1203-013	C15	H 12	O 6	
1204-001	C15	H 12	O 6	
1004-001	C15	H 12	O 6	
0501-026	C15	H 12	O 6	
1003-012	C15	H 12	O 7	
1208-001	C15	H 12	O 7	
1206-001	C15	H 12	O 7	
2405-010	C15	H 12	O 7	
2405-007	C15	H 12	O 7	
2405-008	C15	H 12	O 7	
1208-002	C15	H 12	O 8	
3302-023	C15	H 13	O 2	S 1 CL1
2804-003	C15	H 14	O 1	
2501-005	C15	H 14	O 1	
1001-001	C15	H 14	O 1	
3115-019	C15	H 14	O 1	
3114-032	C15	H 14	O 2	
2804-005	C15	H 14	O 2	
2502-006	C15	H 14	O 2	
2503-011	C15	H 14	O 3	
2502-037	C15	H 14	O 3	
2503-009	C15	H 14	O 3	
2503-006	C15	H 14	O 3	
2503-016	C15	H 14	O 3	
3205-001	C15	H 14	O 3	
1405-001	C15	H 14	O 3	
2201-003	C15	H 14	O 3	
3302-030	C15	H 14	O 3	S 1
3204-007	C15	H 14	O 4	
3905-026	C15	H 14	O 4	
3204-006	C15	H 14	O 4	
3202-004	C15	H 14	O 4	
2204-010	C15	H 14	O 4	
2204-018	C15	H 14	O 4	
2202-021	C15	H 14	O 4	
2204-027	C15	H 14	O 4	
2802-003	C15	H 14	O 4	
2804-009	C15	H 14	O 4	
2904-026	C15	H 14	O 4	
2503-008	C15	H 14	O 4	
2804-019	C15	H 14	O 4	
0501-025	C15	H 14	O 4	
0501-024	C15	H 14	O 4	
0501-034	C15	H 14	O 4	
0501-032	C15	H 14	O 4	
0502-010	C15	H 14	O 4	
0506-008	C15	H 14	O 4	
0506-012	C15	H 14	O 4	
0507-019	C15	H 14	O 4	
0508-017	C15	H 14	O 4	
0508-021	C15	H 14	O 4	
0404-015	C15	H 14	O 4	
0404-023	C15	H 14	O 4	
0404-018	C15	H 14	O 5	
0404-019	C15	H 14	O 5	
2802-006	C15	H 14	O 5	
2204-028	C15	H 14	O 5	
2204-013	C15	H 14	O 5	
1302-015	C15	H 14	O 5	
1302-014	C15	H 14	O 5	
1302-013	C15	H 14	O 5	
1302-011	C15	H 14	O 5	
1302-019	C15	H 14	O 5	
1303-002	C15	H 14	O 5	
1303-001	C15	H 14	O 5	
1302-005	C15	H 14	O 5	
1302-012	C15	H 14	O 5	
2204-011	C15	H 14	O 5	
1002-008	C15	H 14	O 5	
1304-002	C15	H 14	O 5	
1302-006	C15	H 14	O 6	
1304-001	C15	H 14	O 6	
1303-017	C15	H 14	O 6	
1303-018	C15	H 14	O 6	
1302-007	C15	H 14	O 6	
1303-015	C15	H 14	O 6	
1303-016	C15	H 14	O 6	
1302-008	C15	H 14	O 6	
1303-015	C15	H 14	O 6	
1302-009	C15	H 14	O 6	
1303-007	C15	H 14	O 6	
1303-006	C15	H 14	O 6	
1303-008	C15	H 14	O 6	
1302-016	C15	H 14	O 6	
1303-013	C15	H 14	O 6	
1399-001	C15	H 14	O 6	
1303-012	C15	H 14	O 6	
1303-011	C15	H 14	O 6	
2204-014	C15	H 14	O 6	
2505-014	C15	H 14	O 6	
2699-019	C15	H 14	O 6	
2801-004	C15	H 14	O 6	
2505-016	C15	H 14	O 6	
2505-016	C15	H 14	O 7	
1304-006	C15	H 14	O 7	
1304-008	C15	H 14	O 7	
1303-023	C15	H 14	O 7	
1304-004	C15	H 14	O 7	
1303-024	C15	H 14	O 7	
1304-010	C15	H 14	O 7	
1303-020	C15	H 14	O 7	
1303-019	C15	H 14	O 7	
2204-015	C15	H 14	O 7	
1304-011	C15	H 14	O 8	
2507-006	C15	H 14	O 8	
3115-005	C15	H 16		
3115-023	C15	H 16	O 1	
3115-020	C15	H 16	O 1	
3115-021	C15	H 16	O 1	
2501-002	C15	H 16	O 1	
2501-004	C15	H 16	O 1	
2804-007	C15	H 16	O 2	
2502-008	C15	H 16	O 2	
0501-030	C15	H 16	O 3	
0501-022	C15	H 16	O 3	
0501-018	C15	H 16	O 3	
0501-023	C15	H 16	O 4	
0501-031	C15	H 16	O 4	
0501-033	C15	H 16	O 4	
0501-035	C15	H 16	O 4	
0506-004	C15	H 16	O 4	
0507-008	C15	H 16	O 4	
0507-012	C15	H 16	O 4	
0507-014	C15	H 16	O 4	
0507-013	C15	H 16	O 4	
0599-012	C15	H 16	O 4	
2801-002	C15	H 16	O 4	
2204-002	C15	H 16	O 4	
2204-001	C15	H 16	O 4	
3202-005	C15	H 16	O 4	
3202-003	C15	H 16	O 4	
2202-013	C15	H 16	O 5	
1402-010	C15	H 16	O 5	
2204-030	C15	H 16	O 5	
2204-022	C15	H 16	O 5	
2204-021	C15	H 16	O 5	
2106-018	C15	H 16	O 5	
2204-043	C15	H 16	O 5	
2202-022	C15	H 16	O 5	
2103-006	C15	H 16	O 5	
2504-007	C15	H 16	O 5	
0510-003	C15	H 16	O 5	
0404-020	C15	H 16	O 5	
2505-015	C15	H 16	O 6	
2506-016	C15	H 16	O 6	
2507-007	C15	H 16	O 8	
0501-003	C15	H 16	O 8	
3906-041	C15	H 16	O 8	
3906-042	C15	H 16	O 8	
0506-002	C15	H 16	O 9	
0507-003	C15	H 16	O 9	
0507-002	C15	H 16	O 9	
3115-003	C15	H 18		
3115-013	C15	H 18	O 1	
3115-014	C15	H 18	O 1	
3115-024	C15	H 18	O 1	
3115-012	C15	H 18	O 1	
3115-018	C15	H 18	O 1	
2501-003	C15	H 18	O 1	
3110-025	C15	H 18	O 2	
3110-018	C15	H 18	O 2	
3110-006	C15	H 18	O 2	
3302-014	C15	H 18	O 2	S 1
0303-006	C15	H 18	O 3	
0404-024	C15	H 18	O 4	
0399-019	C15	H 18	O 4	
2202-006	C15	H 18	O 4	
2202-004	C15	H 18	O 4	
2202-012	C15	H 18	O 4	
2103-005	C15	H 18	O 5	
2107-009	C15	H 18	O 5	
0501-037	C15	H 18	O 5	
2204-042	C15	H 18	O 6	
2701-030	C15	H 18	O 7	
2101-032	C15	H 18	O 7	CL2
0304-033	C15	H 18	O 8	
0304-008	C15	H 18	O 8	
0304-007	C15	H 18	O 9	
0304-025	C15	H 18	O 9	
2102-013	C15	H 18	O 9	
3111-021	C15	H 19	N 1 O 1	
3115-002	C15	H 20		
3115-009	C15	H 20	O 1	
3115-011	C15	H 20	O 1	
3115-010	C15	H 20	O 1	
3110-013	C15	H 20	O 2	
3905-030	C15	H 20	O 2	
3904-004	C15	H 20	O 2	BR1
3904-003	C15	H 20	O 2	BR2
2903-006	C15	H 20	O 2	
0399-018	C15	H 20	O 4	
0203-006	C15	H 20	O 8	
2201-009	C15	H 20	O 8	
2201-004	C15	H 20	O 8	
0104-013	C15	H 20	O 9	

ID	C	H	N	O	Other
0103-010	C15	H 20		O 9	
3504-013	C15	H 21	N 3	O15	
3904-001	C15	H 21		O 3	BR1
0304-022	C15	H 22	N 1	O 4	
3115-022	C15	H 22		O 1	
0302-007	C15	H 22		O 3	
3005-047	C15	H 22		O 4	
3005-046	C15	H 22		O 4	
2206-011	C15	H 22		O 4	
2104-034	C15	H 22		O 4	
2206-018	C15	H 22		O 5	
2206-017	C15	H 22		O 5	
2505-019	C15	H 22		O 7	
3115-004	C15	H 24			
3115-017	C15	H 26		O 1	
3114-031	C15	H 26		O 2	
3113-014	C15	H 26		O 2	
3115-029	C15	H 26		O 3	
3301-017	C15	H 26		O 3	S 1
3197-017	C15	H 28			
3197-018	C15	H 28			
3115-007	C15	H 28			
3115-006	C15	H 28			
3115-030	C15	H 28		O 3	
3901-010	C15	H 30		O 1	
3115-025	C15	H 30		O 2	
3115-026	C15	H 30		O 3	
3115-027	C15	H 30		O 3	
3115-028	C15	H 30		O 3	
3115-001	C15	H 32			
3197-036	C15	H 32		O 3	
3116-004	C15	H 34		O 1	
2503-050	C16	H 6		O 4	
1602-005	C16	H 8		O 6	
1402-013	C16	H 10		O 5	
2603-019	C16	H 10		O 5	
2302-008	C16	H 10		O 5	CL4
1802-003	C16	H 10		O 6	
1499-003	C16	H 10		O 6	
2604-051	C16	H 10		O 6	
2604-018	C16	H 10		O 6	
2604-053	C16	H 10		O 7	
2604-021	C16	H 10		O 7	
2604-022	C16	H 10		O 7	
2405-006	C16	H 10		O 7	
2302-022	C16	H 10		O 7	
1802-002	C16	H 10		O 7	
1602-011	C16	H 10		O 7	
2399-007	C16	H 10		O 7	
2803-004	C16	H 10		O 7	CL2
2605-003	C16	H 10		O 8	
2605-013	C16	H 10		O 8	
2605-007	C16	H 10		O 8	
0105-003	C16	H 10		O 8	
0105-004	C16	H 10		O 8	
2604-040	C16	H 11		O 5	CL1
2604-042	C16	H 11		O 5	CL1
1106-044	C16	H 11		O10	S1 K1
1106-062	C16	H 11		O10	S1 K1
2602-011	C16	H 12		O 3	
2602-009	C16	H 12		O 3	
2602-006	C16	H 12		O 3	
1399-002	C16	H 12		O 3	
1401-028	C16	H 12		O 4	
1401-026	C16	H 12		O 4	
2603-022	C16	H 12		O 4	
2604-006	C16	H 12		O 4	
2603-010	C16	H 12		O 4	
2603-012	C16	H 12		O 4	
2603-006	C16	H 12		O 4	
1101-009	C16	H 12		O 4	
1103-001	C16	H 12		O 4	
1101-006	C16	H 12		O 4	
0902-017	C16	H 12		O 4	
0999-007	C16	H 12		O 5	
1103-023	C16	H 12		O 5	
1102-002	C16	H 12		O 5	
1101-017	C16	H 12		O 5	
1103-011	C16	H 12		O 5	
1102-003	C16	H 12		O 5	
1199-003	C16	H 12		O 5	
1103-020	C16	H 12		O 5	
1101-013	C16	H 12		O 5	
2604-015	C16	H 12		O 5	
2604-019	C16	H 12		O 5	
2604-005	C16	H 12		O 5	
2604-004	C16	H 12		O 5	
2804-025	C16	H 12		O 5	
2604-016	C16	H 12		O 5	
2604-036	C16	H 12		O 5	
2603-016	C16	H 12		O 5	
2604-035	C16	H 12		O 5	
1601-009	C16	H 12		O 5	
1399-004	C16	H 12		O 5	
1402-011	C16	H 12		O 5	
1601-011	C16	H 12		O 5	
1401-013	C16	H 12		O 5	
1401-011	C16	H 12		O 5	
1601-007	C16	H 12		O 5	
1402-005	C16	H 12		O 6	
1401-017	C16	H 12		O 6	
2404-011	C16	H 12		O 6	
2402-013	C16	H 12		O 6	
1402-002	C16	H 12		O 6	
1803-004	C16	H 12		O 6	
1803-007	C16	H 12		O 6	
2604-049	C16	H 12		O 6	
2604-047	C16	H 12		O 6	
2604-048	C16	H 12		O 6	
2605-005	C16	H 12		O 6	
2605-020	C16	H 12		O 6	
2604-047	C16	H 12		O 6	
2605-017	C16	H 12		O 6	
2605-012	C16	H 12		O 6	
1099-011	C16	H 12		O 6	
1102-006	C16	H 12		O 6	
1105-021	C16	H 12		O 6	
1104-040	C16	H 12		O 6	
1105-029	C16	H 12		O 6	
1199-002	C16	H 12		O 6	
1104-036	C16	H 12		O 6	
1105-002	C16	H 12		O 6	
1106-003	C16	H 12		O 6	
1106-002	C16	H 12		O 6	
1103-040	C16	H 12		O 6	
1103-039	C16	H 12		O 6	
1103-033	C16	H 12		O 6	
1199-011	C16	H 12		O 6	
1104-004	C16	H 12		O 6	
1001-015	C16	H 12		O 6	
1106-043	C16	H 12		O 7	
1105-042	C16	H 12		O 7	
1106-112	C16	H 12		O 7	
1105-046	C16	H 12		O 7	
1106-034	C16	H 12		O 7	
1106-035	C16	H 12		O 7	
1105-043	C16	H 12		O 7	
1104-054	C16	H 12		O 7	
1106-039	C16	H 12		O 7	
1104-045	C16	H 12		O 7	
1104-046	C16	H 12		O 7	
1106-060	C16	H 12		O 7	
1106-037	C16	H 12		O 7	
2606-002	C16	H 12		O 7	
2606-001	C16	H 12		O 7	
2399-004	C16	H 12		O 7	
1108-011	C16	H 12		O 8	
1107-008	C16	H 12		O 8	
1108-009	C16	H 12		O 8	
1108-008	C16	H 12		O 8	
1108-012	C16	H 12		O 8	
1106-072	C16	H 12		O 8	
1106-094	C16	H 12		O 8	
1106-095	C16	H 12		O 8	
3906-024	C16	H 12		O 8	
1702-017	C16	H 13		O 6	+
1703-011	C16	H 13		O 7	+
2504-021	C16	H 13		O 8	S 1 NA1
0399-013	C16	H 14		O 2	
0999-001	C16	H 14		O 3	
0901-007	C16	H 14		O 3	
0901-006	C16	H 14		O 3	
0502-017	C16	H 14		O 3	
1099-013	C16	H 14		O 3	
1099-001	C16	H 14		O 4	
1201-005	C16	H 14		O 4	
1001-007	C16	H 14		O 4	
1001-009	C16	H 14		O 4	
1201-004	C16	H 14		O 4	
1201-006	C16	H 14		O 4	
1001-011	C16	H 14		O 4	
1002-003	C16	H 14		O 4	
1001-022	C16	H 14		O 4	
1201-007	C16	H 14		O 4	
0506-013	C16	H 14		O 4	
0506-020	C16	H 14		O 4	
0508-024	C16	H 14		O 4	
0508-034	C16	H 14		O 4	
2699-024	C16	H 14		O 4	
2903-014	C16	H 14		O 4	
2699-008	C16	H 14		O 4	
2503-034	C16	H 14		O 4	
1601-002	C16	H 14		O 4	
1601-001	C16	H 14		O 4	
2403-015	C16	H 14		O 5	
2403-023	C16	H 14		O 5	
2403-009	C16	H 14		O 5	
2403-030	C16	H 14		O 5	
2804-016	C16	H 14		O 5	
2804-024	C16	H 14		O 5	
2903-015	C16	H 14		O 5	
2701-018	C16	H 14		O 5	
0508-029	C16	H 14		O 5	
0508-030	C16	H 14		O 5	
0508-027	C16	H 14		O 5	
0508-026	C16	H 14		O 5	
0508-025	C16	H 14		O 5	
0506-016	C16	H 14		O 5	
0506-015	C16	H 14		O 5	
0506-014	C16	H 14		O 5	
1206-017	C16	H 14		O 5	
1202-004	C16	H 14		O 5	
1203-010	C16	H 14		O 5	
1201-009	C16	H 14		O 5	
1299-001	C16	H 14		O 5	
1202-002	C16	H 14		O 5	
1203-008	C16	H 14		O 5	
3906-025	C16	H 14		O 5	
3906-026	C16	H 14		O 6	
3906-038	C16	H 14		O 6	
1204-002	C16	H 14		O 6	
1299-004	C16	H 14		O 6	
1003-009	C16	H 14		O 6	
1205-008	C16	H 14		O 6	
1205-021	C16	H 14		O 6	
1205-007	C16	H 14		O 6	
1202-005	C16	H 14		O 6	
2802-008	C16	H 14		O 6	
1499-004	C16	H 14		O 6	
2404-016	C16	H 14		O 6	
2404-002	C16	H 14		O 6	
2404-013	C16	H 14		O 6	
2404-020	C16	H 14		O 6	
1802-009	C16	H 14		O 6	
1802-010	C16	H 14		O 6	
1802-008	C16	H 14		O 6	
1802-007	C16	H 14		O 6	
2404-014	C16	H 14		O 6	
2404-003	C16	H 14		O 6	
2404-004	C16	H 14		O 6	
2301-001	C16	H 14		O 7	
2405-011	C16	H 14		O 7	
2801-008	C16	H 14		O 7	
1206-005	C16	H 14		O 7	
1206-002	C16	H 14		O 7	
1206-008	C16	H 14		O 7	
1206-007	C16	H 14		O 7	
2301-003	C16	H 14		O 7	
2802-014	C16	H 15		O 6	CL1
3116-002	C16	H 16			
3116-012	C16	H 16		O 1	
3114-006	C16	H 16		O 2	
2502-007	C16	H 16		O 2	
2804-006	C16	H 16		O 2	
2502-010	C16	H 16		O 2	
1302-001	C16	H 16		O 2	
1001-002	C16	H 16		O 2	
0901-001	C16	H 16		O 3	
0999-003	C16	H 16		O 3	
1302-002	C16	H 16		O 3	
1301-002	C16	H 16		O 3	
2701-013	C16	H 16		O 3	
2804-015	C16	H 16		O 3	
2502-038	C16	H 16		O 3	
2503-007	C16	H 16		O 3	
3114-011	C16	H 16		O 3	
3203-004	C16	H 16		O 3	
3204-027	C16	H 16		O 4	
3204-026	C16	H 16		O 4	
2503-049	C16	H 16		O 4	
2802-004	C16	H 16		O 4	
2802-005	C16	H 16		O 4	
2701-014	C16	H 16		O 4	
0999-017	C16	H 16		O 4	
2204-017	C16	H 16		O 4	
1405-002	C16	H 16		O 4	
0508-011	C16	H 16		O 4	
0508-005	C16	H 16		O 5	
2701-019	C16	H 16		O 5	
2504-004	C16	H 16		O 5	
2504-026	C16	H 16		O 5	
2504-031	C16	H 16		O 5	
1002-011	C16	H 16		O 5	

635

No.	C	H	O	n	Notes
2903-017	C17	H 18	O	3	
1302-003	C17	H 18	O	3	
1301-003	C17	H 18	O	3	
1001-005	C17	H 18	O	3	
1001-004	C17	H 18	O	3	
1001-006	C17	H 18	O	3	
0901-004	C17	H 18	O	4	
1099-002	C17	H 18	O	4	
0699-013	C17	H 18	O	4	
0499-011	C17	H 18	O	4	
0901-002	C17	H 18	O	4	
0699-015	C17	H 18	O	4	
2701-015	C17	H 18	O	4	
1405-009	C17	H 18	O	4	
1405-004	C17	H 18	O	5	
1405-003	C17	H 18	O	5	
2701-017	C17	H 18	O	5	
2701-021	C17	H 18	O	5	
0699-014	C17	H 18	O	5	
0699-016	C17	H 18	O	5	
0999-004	C17	H 18	O	5	
0506-018	C17	H 18	O	6	
0506-024	C17	H 18	O	6	
2803-006	C17	H 18	O	6	
2204-023	C17	H 18	O	6	
2102-019	C17	H 18	O	6	
1299-009	C17	H 18	O	6	
2506-013	C17	H 18	O	7	
0511-004	C17	H 18	O	7	
0304-028	C17	H 18	O	9	
3117-014	C17	H 20			
3117-015	C17	H 20			
3117-041	C17	H 20	O	1	
3117-061	C17	H 20	O	1	
3117-062	C17	H 20	O	1	
3117-063	C17	H 20	O	1	
3117-068	C17	H 20	O	1	
3117-038	C17	H 20	O	1	
3117-078	C17	H 20	O	1	
3117-065	C17	H 20	O	1	
2805-003	C17	H 20	O	1	
2805-004	C17	H 20	O	2	
2901-011	C17	H 20	O	2	
3117-073	C17	H 20	O	2	
2204-004	C17	H 20	O	4	
2202-014	C17	H 20	O	5	
1902-008	C17	H 20	O	6	
2102-020	C17	H 20	O	6	
3003-013	C17	H 20	O	6	
2102-018	C17	H 20	O	7	
0402-009	C17	H 20	O	8	
0510-002	C17	H 20	O	10	
0509-004	C17	H 20	O	10	
3117-011	C17	H 22			
3117-010	C17	H 22			
3117-012	C17	H 22			
3117-006	C17	H 22			
3117-016	C17	H 22			
3117-007	C17	H 22			
3117-036	C17	H 22	O	1	
3117-031	C17	H 22	O	1	
3117-035	C17	H 22	O	1	
3117-043	C17	H 22	O	1	
3117-032	C17	H 22	O	1	
3117-059	C17	H 22	O	1	
3117-067	C17	H 22	O	1	
3117-030	C17	H 22	O	1	
3117-033	C17	H 22	O	1	
3117-060	C17	H 22	O	1	
3117-034	C17	H 22	O	1	
3117-044	C17	H 22	O	1	
3117-075	C17	H 22	O	2	
3117-079	C17	H 22	O	2	
3117-069	C17	H 22	O	2	
3117-074	C17	H 22	O	2	
3117-052	C17	H 22	O	2	
3117-053	C17	H 22	O	2	
2001-010	C17	H 22	O	2	
0105-006	C17	H 22	O	8	
0105-007	C17	H 22	O	9	
3903-003	C17	H 23	O	3	BR1
3904-002	C17	H 23	O	4	BR1
3117-005	C17	H 24			
3117-013	C17	H 24			
3117-004	C17	H 24			
3117-025	C17	H 24	O	1	
3117-029	C17	H 24	O	1	
3117-037	C17	H 24	O	1	
3117-026	C17	H 24	O	1	
3117-028	C17	H 24	O	1	
3117-027	C17	H 24	O	1	
3117-040	C17	H 24	O	1	
3117-042	C17	H 24	O	1	
3117-058	C17	H 24	O	1	
3117-023	C17	H 24	O	1	
3117-064	C17	H 24	O	1	
3117-051	C17	H 24	O	2	
3117-048	C17	H 24	O	2	
3117-049	C17	H 24	O	2	
3117-046	C17	H 24	O	2	
3117-072	C17	H 24	O	2	
2199-004	C17	H 24	O	2	
0499-001	C17	H 24	O	3	
2108-002	C17	H 24	O	4	
0302-009	C17	H 24	O	9	
3906-044	C17	H 24	O	12	
3201-003	C17	H 26	O	1	
3201-005	C17	H 26	O	1	
3201-004	C17	H 26	O	1	
3117-076	C17	H 26	O	1	
3117-021	C17	H 26	O	1	
3117-022	C17	H 26	O	1	
3117-045	C17	H 26	O	2	
3117-084	C17	H 26	O	2	
2109-001	C17	H 26	O	2	
3117-085	C17	H 26	O	3	
3114-043	C17	H 26	O	4	
2003-007	C17	H 26	O	4	
3401-004	C17	H 26	O	4	
0499-002	C17	H 26	O	4	
3906-022	C17	H 26	O	5	
3197-005	C17	H 27	N 1 O	3	
3115-016	C17	H 27	O	3	
3201-002	C17	H 28	O	1	
3117-082	C17	H 28	O	2	
3117-083	C17	H 28	O	3	
3005-026	C17	H 28	O	4	
3005-037	C17	H 28	O	4	
3005-032	C17	H 28	O	6	
3901-012	C17	H 30	O	1	
3005-024	C17	H 30	O	4	
3901-008	C17	H 32	O	2	
3117-081	C17	H 32	O	2	
3117-056	C17	H 32	O	3	
3197-029	C17	H 32	O	4	
3197-020	C17	H 34			
3117-003	C17	H 34			
3116-003	C17	H 34			
3117-002	C17	H 34			
3502-010	C17	H 34	N 4 O	10	
3197-021	C17	H 34	O	1	
3116-039	C17	H 34	O	1	
3117-080	C17	H 34	O	2	
3117-054	C17	H 34	O	3	
3117-001	C17	H 36			
1602-007	C18	H 8	O	6	
0702-004	C18	H 10	O	4	
0702-005	C18	H 10	O	5	
1105-040	C18	H 10	O	5	
0599-014	C18	H 10	O	6	
2604-061	C18	H 10	O	6	
2605-040	C18	H 10	O	7	
0702-012	C18	H 10	O	9	
0701-001	C18	H 12	O	4	
0701-002	C18	H 12	O	4	
1101-010	C18	H 12	O	4	
1101-014	C18	H 12	O	4	
1102-005	C18	H 12	O	4	
0702-001	C18	H 12	O	5	
0701-003	C18	H 12	O	6	
2403-039	C18	H 12	O	6	
0701-005	C18	H 12	O	7	
2605-042	C18	H 12	O	7	
2605-043	C18	H 12	O	7	
2605-045	C18	H 12	O	9	
0702-011	C18	H 12	O	9	
2302-017	C18	H 12	O	9	
2302-019	C18	H 12	O	10	
2302-004	C18	H 12	O	11	
0702-006	C18	H 13	N 1 O	4	
2402-008	C18	H 14	O	2	
1001-018	C18	H 14	O	4	
2403-031	C18	H 14	O	5	
2403-011	C18	H 14	O	5	
2403-026	C18	H 14	O	5	
2404-027	C18	H 14	O	6	
1499-001	C18	H 14	O	6	
1106-004	C18	H 14	O	6	
2604-060	C18	H 14	O	7	
2302-009	C18	H 14	O	7	CL2
2302-001	C18	H 14	O	8	
2302-002	C18	H 14	O	8	
1105-060	C18	H 14	O	8	
2302-003	C18	H 14	O	9	
2399-002	C18	H 14	O	10	
2302-021	C18	H 15	O	6	CL1
2604-059	C18	H 15	O	6	CL1
1199-027	C18	H 15	O	7	CL1
0399-007	C18	H 16	O	2	
3303-010	C18	H 16	O	2	S 2
3303-008	C18	H 16	O	2	S 2
3303-007	C18	H 16	O	2	S 2
3304-005	C18	H 16	O	2	S 3
3304-006	C18	H 16	O	2	S 3
3304-007	C18	H 16	O	2	S 3
0699-009	C18	H 16	O	4	
0699-010	C18	H 16	O	4	
0699-008	C18	H 16	O	4	
0699-007	C18	H 16	O	4	
0699-006	C18	H 16	O	4	
2402-009	C18	H 16	O	4	
2402-005	C18	H 16	O	4	
2402-006	C18	H 16	O	5	
2403-025	C18	H 16	O	5	
2403-024	C18	H 16	O	5	
2403-003	C18	H 16	O	5	
3204-017	C18	H 16	O	5	
1103-030	C18	H 16	O	5	
1199-006	C18	H 16	O	5	
1402-017	C18	H 16	O	5	
1199-005	C18	H 16	O	5	
2701-024	C18	H 16	O	5	
2604-058	C18	H 16	O	6	
1402-018	C18	H 16	O	6	
1401-022	C18	H 16	O	6	
1402-021	C18	H 16	O	6	
1103-047	C18	H 16	O	6	
1101-018	C18	H 16	O	6	
1104-044	C18	H 16	O	6	
1202-003	C18	H 16	O	6	
1105-038	C18	H 16	O	6	
1199-018	C18	H 16	O	6	
1102-009	C18	H 16	O	6	
1102-007	C18	H 16	O	6	
1601-016	C18	H 16	O	6	
2404-006	C18	H 16	O	6	
2404-025	C18	H 16	O	6	
2404-026	C18	H 16	O	6	
2404-038	C18	H 16	O	6	
0902-020	C18	H 16	O	6	
0999-013	C18	H 16	O	6	
2207-001	C18	H 16	O	7	
2207-003	C18	H 16	O	7	
2207-002	C18	H 16	O	7	
1105-050	C18	H 16	O	7	
1104-051	C18	H 16	O	7	
1103-050	C18	H 16	O	7	
1104-053	C18	H 16	O	7	
1104-061	C18	H 16	O	7	
1104-062	C18	H 16	O	7	
1104-063	C18	H 16	O	7	
1105-049	C18	H 16	O	7	
1199-007	C18	H 16	O	7	
1103-049	C18	H 16	O	7	
1106-067	C18	H 16	O	7	
1106-066	C18	H 16	O	7	
2803-003	C18	H 16	O	7	
2606-004	C18	H 16	O	7	
2605-039	C18	H 16	O	8	
1106-077	C18	H 16	O	8	
1106-079	C18	H 16	O	8	
1106-098	C18	H 16	O	8	
1105-056	C18	H 16	O	8	
1403-001	C18	H 16	O	8	
1199-022	C18	H 16	O	8	
1105-055	C18	H 16	O	8	
1106-083	C18	H 16	O	8	
1199-031	C18	H 16	O	8	
1108-014	C18	H 16	O	8	
1106-085	C18	H 16	O	8	
0399-014	C18	H 16	O	8	
1199-032	C18	H 16	O	9	
1106-107	C18	H 16	O	9	
1106-109	C18	H 16	O	9	
2399-001	C18	H 16	O	9	
2109-002	C18	H 17	O	6	CL1
1703-024	C18	H 17	O	7	+
0699-025	C18	H 18	O	2	
0699-003	C18	H 18	O	2	
3204-013	C18	H 18	O	4	
3114-012	C18	H 18	O	4	
3204-016	C18	H 18	O	5	
1002-015	C18	H 18	O	5	

ID	C	H	O	Other
2402-007	C18	H 18	O 5	
1601-006	C18	H 18	O 5	
1203-020	C18	H 18	O 5	
1203-021	C18	H 18	O 5	
2805-002	C18	H 18	O 5	
2701-022	C18	H 18	O 5	
2504-032	C18	H 18	O 6	
2504-027	C18	H 18	O 6	
1201-010	C18	H 18	O 6	
1601-018	C18	H 18	O 6	
0599-011	C18	H 18	O 6	
0999-014	C18	H 18	O 6	
1001-013	C18	H 18	O 6	
0504-011	C18	H 18	O 7	
0502-011	C18	H 18	O 7	
1901-016	C18	H 18	O 7	
2405-004	C18	H 18	O 7	
2301-006	C18	H 18	O 7	
2301-004	C18	H 18	O 7	
2504-039	C18	H 18	O 7	
2203-009	C18	H 18	O 8	
1901-017	C18	H 18	O 8	
2505-005	C18	H 18	O 9	S 1
3117-086	C18	H 20	O 2	
3114-016	C18	H 20	O 4	
0901-003	C18	H 20	O 4	
1001-019	C18	H 20	O 4	
1099-003	C18	H 20	O 4	
1099-014	C18	H 20	O 4	
1302-004	C18	H 20	O 4	
1099-005	C18	H 20	O 5	
0901-005	C18	H 20	O 5	
1099-004	C18	H 20	O 5	
3905-025	C18	H 20	O 5	
0502-008	C18	H 20	O 5	S 1
3906-017	C18	H 20	O 6	
3906-016	C18	H 20	O 6	
1405-007	C18	H 20	O 6	
1405-005	C18	H 20	O 6	
3906-018	C18	H 20	O 7	
2504-037	C18	H 20	O 7	
3118-009	C18	H 22	O 2	•
3117-087	C18	H 22	O 3	•
3302-017	C18	H 22	O 3	S 1
0601-001	C18	H 22	O 4	•
0501-040	C18	H 22	O 5	
3499-005	C18	H 22	O 5	
3114-036	C18	H 23	N 1 O 1	
2203-015	C18	H 23	O 9	
2199-005	C18	H 24	O 4	
2502-002	C18	H 24	O 7	
3203-010	C18	H 24	O 8	
3114-035	C18	H 25	N 1 O 1	
3118-051	C18	H 26	O 2	
3118-050	C18	H 26	O 2	
0499-006	•C18	H 26	O 2	
3118-054	C18	H 26	O 3	
3203-011	C18	H 26	O 8	
3118-032	C18	H 28	O 2	
3118-043	C18	H 28	O 2	
3118-045	C18	H 28	O 2	
3118-042	C18	H 28	O 2	
3118-073	C18	H 28	O 3	
3118-074	C18	H 28	O 3	
3118-053	C18	H 28	O 3	
3401-005	C18	H 28	O 4	
0499-003	C18	H 28	O 4	
3903-001	C18	H 28	O 5	
3197-006	C18	H 29	N 1 O 3	
2999-004	C18	H 29	N 1 O 4	
2901-015	C18	H 30	O 2	
3118-039	C18	H 30	O 2	
3118-036	C18	H 30	O 2	
3118-041	C18	H 30	O 2	
3118-040	C18	H 30	O 2	
3118-037	C18	H 30	O 2	
3118-034	C18	H 30	O 2	
3118-049	C18	H 30	O 2	
3118-035	C18	H 30	O 2	
3118-033	C18	H 30	O 2	
3118-038	C18	H 30	O 2	
3118-048	C18	H 30	O 2	
3118-065	C18	H 30	O 3	
3118-078	C18	H 30	O 3	
3118-052	C18	H 30	O 3	
3118-080	C18	H 30	O 3	
3118-079	C18	H 30	O 3	
3118-072	C18	H 30	O 3	
3118-058	C18	H 30	O 3	
2901-017	C18	H 30	O 3	
3202-001	C18	H 30	O 3	
3507-001	C18	H 30	O16	
2999-005	C18	H 31	N 1 O 4	
2901-014	C18	H 32	O 2	
3197-028	C18	H 32	O 2	
3197-027	C18	H 32	O 2	
3118-029	C18	H 32	O 2	
3118-030	C18	H 32	O 2	
3118-046	C18	H 32	O 2	
3118-047	C18	H 32	O 2	
3117-088	C18	H 32	O 2	
3118-028	C18	H 32	O 2	
3118-031	C18	H 32	O 2	
3118-056	C18	H 32	O 3	
3118-077	C18	H 32	O 3	
3118-076	C18	H 32	O 3	
3118-057	C18	H 32	O 3	
3118-071	C18	H 32	O 3	
3118-060	C18	H 32	O 3	
3118-062	C18	H 32	O 3	
3118-066	C18	H 32	O 4	
3118-082	C18	H 32	O 4	
3507-005	C18	H 32	O 6	
3501-018	C18	H 32	O14	
3507-007	C18	H 32	O16	
3507-006	C18	H 32	O16	
3507-004	C18	H 32	O16	
3507-008	C18	H 32	O16	
3507-002	C18	H 32	O16	
3507-003	C18	H 32	O16	
3114-034	C18	H 33	N 1 O 1	
3118-081	C18	H 33	O 2	F 1
3116-007	C18	H 34	O 2	
3118-019	C18	H 34	O 2	
3118-023	C18	H 34	O 2	
3118-022	C18	H 34	O 2	
3118-025	C18	H 34	O 2	
3118-020	C18	H 34	O 2	
3118-024	C18	H 34	O 2	
3116-040	C18	H 34	O 2	
3118-018	C18	H 34	O 2	
3118-021	C18	H 34	O 2	
3197-022	C18	H 34	O 2	
3118-059	C18	H 34	O 3	
3118-068	C18	H 34	O 3	
3118-069	C18	H 34	O 3	
3118-055	C18	H 34	O 3	
3118-075	C18	H 34	O 3	
3118-064	C18	H 34	O 3	
3197-030	C18	H 34	O 4	
3118-027	C18	H 34	O 5	
3118-083	C18	H 34	O 6	
3507-010	C18	H 34	O16	
3502-021	C18	H 36	N 4 O10	
3502-017	C18	H 36	N 4 O11	
3502-019	C18	H 36	N 4 O11	
3118-002	C18	H 36	O 1	
3118-003	C18	H 36	O 1	
3118-008	C18	H 36	O 1	
3118-010	C18	H 36	O 2	
3116-005	C18	H 36	O 2	
3118-012	C18	H 36	O 3	
3118-011	C18	H 36	O 3	
3118-013	C18	H 36	O 4	
3118-014	C18	H 36	O 4	
3118-016	C18	H 36	O 5	
3118-015	C18	H 36	O 5	
3118-017	C18	H 36	O 6	
3502-020	C18	H 37	N 5 O 9	
3502-018	C18	H 37	N 5 O10	
3197-024	C18	H 38		
3197-023	C18	H 38		
3117-017	C18	H 38		
3117-018	C18	H 38		
3117-019	C18	H 38		
3118-004	C18	H 38	O 1	
3118-005	C18	H 38	O 1	
3118-001	C18	H 38	O 1	
3118-006	C18	H 38	O 2	
3197-025	C18	H 48		
2704-002	C19	H 10	O 3	
2704-001	C19	H 12	O 3	
2703-044	C19	H 12	O 4	
0701-007	C19	H 12	O 4	
3906-007	C19	H 12	O 5	
0599-015	C19	H 12	O 6	
1801-020	C19	H 12	O 6	
1105-041	C19	H 12	O 6	
1404-009	C19	H 12	O 6	
1106-006	C19	H 12	O 6	
1801-021	C19	H 12	O 7	
0599-018	C19	H 12	O 7	
3906-028	C19	H 12	O 7	
2603-035	C19	H 13	N 1 O 8	
2703-046	C19	H 14	O 4	
2703-045	C19	H 14	O 5	
0702-002	C19	H 14	O 5	
2302-014	C19	H 14	O 5	CL2
1601-027	C19	H 14	O 6	
2403-040	C19	H 14	O 6	
2605-027	C19	H 14	O 6	
2703-047	C19	H 14	O 6	
1499-019	C19	H 14	O 6	
2403-042	C19	H 14	O 7	
2404-033	C19	H 14	O 7	
1106-090	C19	H 14	O 8	
2302-018	C19	H 14	O 9	
2302-020	C19	H 14	O10	
2302-015	C19	H 15	O 5	CL3
2702-011	C19	H 16	N 2 O 7	CL1
2805-008	C19	H 16	O 4	
2403-041	C19	H 16	O 6	
2404-043	C19	H 16	O 6	
2605-028	C19	H 16	O 7	
2703-007	C19	H 16	O 7	
1108-018	C19	H 16	O 7	
1403-003	C19	H 16	O 7	
1404-001	C19	H 16	O 7	
1499-015	C19	H 16	O 7	
1106-088	C19	H 16	O 8	
2606-008	C19	H 16	O 8	
2301-015	C19	H 16	O10	
2301-014	C19	H 16	O11	
2302-012	C19	H 17	O 6	CL3
2301-010	C19	H 17	O 8	CL1
2302-013	C19	H 18	O 5	
1103-031	C19	H 18	O 5	
0699-004	C19	H 18	O 5	
0501-009	C19	H 18	O 5	
0501-008	C19	H 18	O 5	
0501-010	C19	H 18	O 6	
1199-019	C19	H 18	O 6	
1402-022	C19	H 18	O 6	
1199-020	C19	H 18	O 6	
2503-032	C19	H 18	O 6	
1601-005	C19	H 18	O 6	
2402-010	C19	H 18	O 6	
1801-001	C19	H 18	O 6	
2701-029	C19	H 18	O 6	
2806-005	C19	H 18	O 6	
2806-004	C19	H 18	O 6	
2606-005	C19	H 18	O 7	
2399-005	C19	H 18	O 7	
1104-056	C19	H 18	O 7	
1104-064	C19	H 18	O 7	
1107-006	C19	H 18	O 7	
1103-051	C19	H 18	O 7	
1105-054	C19	H 18	O 7	
1108-019	C19	H 18	O 8	
1106-081	C19	H 18	O 8	
1199-017	C19	H 18	O 8	
1107-009	C19	H 18	O 8	
1108-015	C19	H 18	O 8	
1106-086	C19	H 18	O 8	
1199-024	C19	H 18	O 8	
1104-066	C19	H 18	O 8	
1106-115	C19	H 18	O 8	
1105-058	C19	H 18	O 8	
1105-057	C19	H 18	O 8	
1499-012	C19	H 18	O 8	
1106-104	C19	H 18	O 8	
2301-009	C19	H 18	O 8	
2301-008	C19	H 18	O 8	
3999-011	C19	H 18	O 8	
3201-012	C19	H 18	O 9	
1199-025	C19	H 18	O 9	
1107-010	C19	H 18	O 9	
1107-013	C19	H 18	O 9	
1108-023	C19	H 18	O 9	
1108-026	C19	H 18	O 9	
2301-013	C19	H 18	O10	
2404-037	C19	H 18	O11	
2404-035	C19	H 18	O11	
3999-014	C19	H 19	N 1 O 5	
2302-010	C19	H 19	O 6	CL1
2805-007	C19	H 20	O 3	
0404-014	C19	H 20	O 3	
0501-006	C19	H 20	O 4	
0508-039	C19	H 20	O 4	
0508-015	C19	H 20	O 4	
0508-018	C19	H 20	O 4	

637

0508-020	C19	H 20	□ 5			0604-015	C20	H 12	□ 6				1204-007	C20	H 20	□ 6		
0508-038	C19	H 20	□ 5			0604-023	C20	H 12	□ 7				0698-009	C20	H 20	□ 6		
0505-003	C19	H 20	□ 5			2509-003	C20	H 12	□ 7				0604-012	C20	H 20	□ 6		
0505-004	C19	H 20	□ 5			0701-008	C20	H 12	□ 9				0601-016	C20	H 20	□ 6		
0504-006	C19	H 20	□ 5			2509-001	C20	H 14	□ 4				0601-028	C20	H 20	□ 6		
0504-005	C19	H 20	□ 5			2704-004	C20	H 14	□ 4				0601-022	C20	H 20	□ 6		
0503-003	C19	H 20	□ 5			0601-017	C20	H 14	□ 6				2701-033	C20	H 20	□ 6		
0503-005	C19	H 20	□ 5			0599-020	C20	H 14	□ 6				2701-032	C20	H 20	□ 7		
0504-007	C19	H 20	□ 6			1402-006	C20	H 14	□ 6				2701-031	C20	H 20	□ 7		
2806-003	C19	H 20	□ 6			1404-010	C20	H 14	□ 7				2605-032	C20	H 20	□ 7		
1205-022	C19	H 20	□ 6			2605-041	C20	H 14	□ 7				0602-001	C20	H 20	□ 7		
2204-024	C19	H 20	□ 6	S 1		1801-023	C20	H 14	□ 7				1104-065	C20	H 20	□ 7		
2301-007	C19	H 20	□ 7			1801-014	C20	H 14	□ 8				1103-052	C20	H 20	□ 7		
3113-019	C19	H 20	□ 8			3906-027	C20	H 14	□ 8				1499-014	C20	H 20	□ 7		
2701-001	C19	H 20	□ 9			2702-015	C20	H 15	N 1	□ 7			1105-051	C20	H 20	□ 7		
2505-006	C19	H 20	□ 9	S 1		1401-014	C20	H 16	□ 5				1104-030	C20	H 20	□ 7		
2204-034	C19	H 20	□10			0702-003	C20	H 16	□ 5				2301-016	C20	H 20	□ 7		
0802-011	C19	H 20	□11			0702-010	C20	H 16	□ 6				1199-021	C20	H 20	□ 8		
0802-010	C19	H 20	□11			0702-009	C20	H 16	□ 6				1106-087	C20	H 20	□ 8		
0802-009	C19	H 20	□11			0601-018	C20	H 16	□ 6				1108-016	C20	H 20	□ 8		
2805-006	C19	H 22	□ 2			1801-019	C20	H 16	□ 6				1104-067	C20	H 20	□ 8		
3117-039	C19	H 22	□ 2			1801-022	C20	H 16	□ 7				1106-105	C20	H 20	□ 8		
2805-013	C19	H 22	□ 3			2605-036	C20	H 16	□ 7				1107-011	C20	H 20	□ 9		
2805-014	C19	H 22	□ 3			2605-044	C20	H 16	□ 7				1199-026	C20	H 20	□ 9		
0501-005	C19	H 22	□ 3			0605-011	C20	H 16	□ 8				1108-027	C20	H 20	□ 9		
0501-021	C19	H 22	□ 3			2699-025	C20	H 17	N 1	□ 8			2404-036	C20	H 20	□11		
0501-038	C19	H 22	□ 3			2302-016	C20	H 17	□ 5	CL3			2302-011	C20	H 21	□ 6	CL1	
2107-004	C19	H 22	□ 3			2104-028	C20	H 18	N 1	□ 6	CL1		0499-008	C20	H 22	□ 1		
2202-016	C19	H 22	□ 4			2104-030	C20	H 18	N 1	□ 7	CL1		0508-019	C20	H 22	□ 4		
0502-015	C19	H 22	□ 4			1001-003	C20	H 18	□ 3				0604-004	C20	H 22	□ 4		
0502-014	C19	H 22	□ 4			1001-008	C20	H 18	□ 4				2204-019	C20	H 22	□ 4		
1099-015	C19	H 22	□ 4			0604-005	C20	H 18	□ 4				2204-029	C20	H 22	□ 5		
3204-028	C19	H 22	□ 4			0699-020	C20	H 18	□ 4				0699-005	C20	H 22	□ 5		
3204-029	C19	H 22	□ 4			1601-022	C20	H 18	□ 4				0699-019	C20	H 22	□ 6		
0502-006	C19	H 22	□ 5			2604-034	C20	H 18	□ 5				0601-011	C20	H 22	□ 6		
0502-005	C19	H 22	□ 5			2805-009	C20	H 18	□ 5				1001-014	C20	H 22	□ 6		
0504-009	C19	H 22	□ 6			1199-009	C20	H 18	□ 5				0601-023	C20	H 22	□ 6		
0604-008	C19	H 22	□ 6			1499-011	C20	H 18	□ 6				2504-033	C20	H 22	□ 6		
1901-015	C19	H 22	□ 8			2605-033	C20	H 18	□ 6				1302-018	C20	H 22	□ 6		
2502-031	C19	H 22	□ 8			2703-003	C20	H 18	□ 6				3906-039	C20	H 22	□ 6		
2199-007	C19	H 22	□ 9			2703-002	C20	H 18	□ 6				1206-010	C20	H 22	□ 7		
2503-024	C19	H 22	□ 9			0698-008	C20	H 18	□ 6				2301-011	C20	H 22	□ 7		
2805-005	C19	H 24	□ 2			0698-007	C20	H 18	□ 6				0601-039	C20	H 22	□ 7		
3117-024	C19	H 24	□ 2			0601-032	C20	H 18	□ 6				0199-018	C20	H 22	□ 8		
2805-012	C19	H 24	□ 3			0601-019	C20	H 18	□ 6				1104-028	C20	H 22	□ 8		
2805-011	C19	H 24	□ 3			0601-034	C20	H 18	□ 6				1201-012	C20	H 22	□ 8		
0501-007	C19	H 24	□ 5			0601-030	C20	H 18	□ 6				2804-013	C20	H 22	□ 8		
2102-021	C19	H 24	□ 6			0601-031	C20	H 18	□ 6				2701-002	C20	H 22	□ 9		
2105-005	C19	H 26	□ 2			0601-033	C20	H 18	□ 6				0199-017	C20	H 22	□ 9		
3117-050	C19	H 26	□ 3			0601-020	C20	H 18	□ 7				0199-016	C20	H 22	□ 9		
3117-047	C19	H 26	□ 3			0699-024	C20	H 18	□ 7				0199-013	C20	H 22	□ 9		
2003-018	C19	H 26	□ 4			0601-037	C20	H 18	□ 7				0199-008	C20	H 22	□ 9		
2903-002	C19	H 26	□ 4			0601-036	C20	H 18	□ 7				0199-007	C20	H 22	□ 9		
2299-004	C19	H 26	□ 6			2605-030	C20	H 18	□ 7				0605-010	C20	H 22	□10		
0104-026	C19	H 26	□12			2605-029	C20	H 18	□ 7				1303-005	C20	H 22	□10		
0499-007	C19	H 28	□ 2			2605-035	C20	H 18	□ 7				2301-034	C20	H 22	□11		
2107-005	C19	H 28	□ 3			2703-004	C20	H 18	□ 7				0501-039	C20	H 24	□ 3		
3906-023	C19	H 28	□ 6			2703-041	C20	H 18	□ 7				2805-016	C20	H 24	□ 3		
3118-089	C19	H 30	□ 2			2703-008	C20	H 18	□ 7				2805-015	C20	H 24	□ 3		
2003-008	C19	H 30	□ 4			2703-019	C20	H 18	□ 8				0506-003	C20	H 24	□ 4		
0499-004	C19	H 30	□ 4			2703-017	C20	H 18	□ 8				0506-007	C20	H 24	□ 4		
3197-007	C19	H 31	N 1	□ 3		2703-018	C20	H 18	□ 8				0507-009	C20	H 24	□ 4		
2110-001	C19	H 32	□ 2			1106-106	C20	H 18	□ 8				0507-015	C20	H 24	□ 4		
3117-071	C19	H 32	□ 4			1106-089	C20	H 18	□ 8				0508-012	C20	H 24	□ 4		
3005-031	C19	H 32	□ 4			1105-061	C20	H 18	□ 8				0601-002	C20	H 24	□ 4		
3005-028	C19	H 32	□ 4			2703-033	C20	H 18	□ 9				0508-013	C20	H 24	□ 5		
3005-029	C19	H 32	□ 4			1104-005	C20	H 18	□10				0505-005	C20	H 24	□ 5		
3005-027	C19	H 32	□ 4			1106-011	C20	H 18	□11				2204-039	C20	H 24	□ 5		
3118-087	C19	H 34	□ 2			1106-012	C20	H 18	□11				2204-038	C20	H 24	□ 5		
3118-088	C19	H 34	□ 3			1106-019	C20	H 18	□11				0604-010	C20	H 24	□ 6		
3117-070	C19	H 34	□ 4			1106-013	C20	H 18	□11				0604-009	C20	H 24	□ 6		
3117-057	C19	H 34	□ 4			1106-010	C20	H 18	□11				0601-005	C20	H 24	□ 6		
3005-025	C19	H 34	□ 4			2703-049	C20	H 18	□8				0604-011	C20	H 24	□ 7		
3005-030	C19	H 34	□ 4			1199-028	C20	H 19	N 1	□ 4			0601-010	C20	H 24	□ 7		
3118-067	C19	H 34	□ 5			1199-029	C20	H 19	N 1	□ 4			0501-027	C20	H 24	□ 7		
3502-037	C19	H 35	N 3	□ 3		2104-027	C20	H 19	N 1	□ 6			0501-029	C20	H 24	□ 8		
3901-009	C19	H 36	□ 2			1702-003	C20	H 19	□10	+			0502-002	C20	H 24	□ 9		
3118-085	C19	H 36	□ 2			1703-002	C20	H 19	□11	+			0502-004	C20	H 24	□ 9		
3119-002	C19	H 36	□ 4			2605-025	C20	H 20	N 2	□11			0504-004	C20	H 24	□ 9		
3117-055	C19	H 36	□ 4			2699-007	C20	H 20	□ 4				0199-002	C20	H 24	□10		
3502-016	C19	H 37	N 5	□ 7		1601-021	C20	H 20	□ 4				0304-015	C20	H 26	□ 3		
3502-026	C19	H 38	N 4	□ 9		2204-020	C20	H 20	□ 4				3118-007	C20	H 26	□ 3		
3502-027	C19	H 38	N 4	□ 9		1203-028	C20	H 20	□ 4				0601-004	C20	H 26	□ 6		
3118-084	C19	H 38	□ 2			1203-030	C20	H 20	□ 4				2005-004	C20	H 26	□ 8		
3502-022	C19	H 39	N 5	□ 7		0508-016	C20	H 20	□ 4				2206-009	C20	H 28	□ 4		
3197-026	C19	H 40				1002-017	C20	H 20	□ 4				2903-003	C20	H 28	□ 4		
3119-001	C19	H 40				0699-002	C20	H 20	□ 5				2903-004	C20	H 28	□ 5		
3117-020	C19	H 40				0601-008	C20	H 20	□ 5				2206-008	C20	H 28	□ 5		
2704-012	C20	H 12	□ 4			0601-009	C20	H 20	□ 5				2299-005	C20	H 28	□ 6		
2704-013	C20	H 12	□ 4			0604-007	C20	H 20	□ 5				2201-005	C20	H 28	□12		
0604-018	C20	H 12	□ 6			1203-007	C20	H 20	□ 5				0203-007	C20	H 28	□12		

Reg. No.	C	H	N	O	Other
2003-015	C21	H 28		O 4	
0304-020	C21	H 28		O 4	
3901-005	C21	H 28		O 9	
3901-006	C21	H 28		O 9	
3901-004	C21	H 28		O10	
0304-027	C21	H 28		O14	
2111-014	C21	H 30		O 1	
2106-010	C21	H 30		O 2	
2111-009	C21	H 30		O 2	
2106-011	C21	H 30		O 2	
2111-020	C21	H 30		O 2	
2106-012	C21	H 30		O 2	
2111-019	C21	H 30		O 2	
2106-015	C21	H 30		O 2	
2106-013	C21	H 30		O 2	
2206-014	C21	H 30		O 4	
2206-013	C21	H 30		O 5	
2206-020	C21	H 30		O 5	
2903-001	C21	H 30		O 6	
3903-005	C21	H 30		O 8	
0301-020	C21	H 30		O11	
1108-022	C21	H 30		O14	
3121-004	C21	H 32			
3121-003	C21	H 32			
2111-013	C21	H 32		O 1	
2111-018	C21	H 32		O 2	
2111-008	C21	H 32		O 2	
2903-012	C21	H 32		O 8	
1205-020	C21	H 32		O11	
3121-002	C21	H 34			
2111-012	C21	H 34		O 1	
2111-017	C21	H 34		O 2	
2111-016	C21	H 34		O 2	
2111-006	C21	H 34		O 2	
2111-007	C21	H 34		O 2	
0499-005	C21	H 34		O 4	
2111-011	C21	H 36		O 1	
2111-005	C21	H 36		O 2	
2111-015	C21	H 36		O 2	
2199-011	C21	H 36		O 8	S 2
2199-010	C21	H 36		O 9	S 2
3197-031	C21	H 38		O 6	
3197-032	C21	H 38		O 7	
3507-009	C21	H 38		O14	
3502-032	C21	H 39	N 5	O14	
3502-003	C21	H 39	N 7	O12	
3502-006	C21	H 39	N 7	O13	
3121-005	C21	H 40		O 4	
3502-029	C21	H 41	N 5	O12	
3502-030	C21	H 41	N 5	O12	
3502-005	C21	H 41	N 7	O12	
3198-001	C21	H 42		O 3	
3502-024	C21	H 43	N 5	O 7	
3121-001	C21	H 44			
1602-009	C22	H 12		O 7	
2510-003	C22	H 14		O 6	
2509-002	C22	H 14		O 6	
2510-005	C22	H 14		O 6	
2510-004	C22	H 14		O 6	
2510-006	C22	H 14		O 6	
2511-001	C22	H 14		O 6	
2599-011	C22	H 14		O 6	
2511-002	C22	H 14		O 7	
2704-005	C22	H 16		O 6	
2703-001	C22	H 16		O 6	
1801-009	C22	H 16		O 6	
2704-006	C22	H 16		O 7	
2703-006	C22	H 16		O 7	
1199-036	C22	H 16		O 8	
2302-005	C22	H 16		O12	
2510-002	C22	H 18		O 4	
2510-001	C22	H 18		O 4	
1801-008	C22	H 18		O 6	
1801-010	C22	H 18		O 6	
1106-005	C22	H 18		O 6	
1499-017	C22	H 18		O 6	
1402-009	C22	H 18		O 6	
1402-010	C22	H 18		O 6	
1404-004	C22	H 18		O 7	
1801-011	C22	H 18		O 7	
0604-014	C22	H 18		O 7	
0604-020	C22	H 18		O 7	
0605-008	C22	H 18		O 8	
1304-003	C22	H 18		O10	
1303-004	C22	H 18		O10	
1303-010	C22	H 18		O10	
1304-007	C22	H 18		O11	
1304-005	C22	H 18		O11	
1303-021	C22	H 18		O11	
0399-012	C22	H 18		O12	
1103-004	C22	H 20		O 4	
1404-002	C22	H 20		O 6	
1499-016	C22	H 20		O 6	
2605-038	C22	H 20		O 7	
2703-013	C22	H 20		O 7	
2703-016	C22	H 20		O 8	
2703-014	C22	H 20		O 8	
2703-015	C22	H 20		O 8	
0601-021	C22	H 20		O 8	
2703-032	C22	H 20		O 9	
2703-035	C22	H 20		O 9	
2703-028	C22	H 20		O 9	
2703-027	C22	H 20		O 9	
2703-029	C22	H 20		O 9	
2703-038	C22	H 20		O10	
2703-043	C22	H 20		O10	
1105-024	C22	H 20		O12	
2605-014	C22	H 20		O13	
0105-005	C22	H 20		O13	
2702-014	C22	H 21	N 1	O 7	
2702-008	C22	H 21	N 2	O 8	CL1
0605-004	C22	H 22		O 7	
0698-006	C22	H 22		O 7	
0605-002	C22	H 22		O 7	
0605-005	C22	H 22		O 8	
0698-004	C22	H 22		O 8	
0605-007	C22	H 22		O 8	
1401-027	C22	H 22		O 9	
1601-010	C22	H 22		O10	
1103-022	C22	H 22		O10	
1103-055	C22	H 22		O10	
1601-008	C22	H 22		O10	
1199-004	C22	H 22		O10	
1401-014	C22	H 22		O10	
1401-012	C22	H 22		O10	
1103-029	C22	H 22		O10	
1103-021	C22	H 22		O10	
0606-004	C22	H 22		O10	
2604-037	C22	H 22		O10	
2605-024	C22	H 22		O11	
1105-033	C22	H 22		O11	
1104-041	C22	H 22		O11	
1105-020	C22	H 22		O11	
1106-046	C22	H 22		O11	
1106-038	C22	H 22		O11	
1105-018	C22	H 22		O11	
1105-031	C22	H 22		O11	
1105-023	C22	H 22		O11	
1105-044	C22	H 22		O11	
1104-037	C22	H 22		O11	
1401-019	C22	H 22		O11	
1105-034	C22	H 22		O11	
1803-008	C22	H 22		O11	
1104-055	C22	H 22		O12	
1106-063	C22	H 22		O12	
1106-036	C22	H 22		O12	
1108-020	C22	H 22		O13	
1106-073	C22	H 22		O13	
1106-096	C22	H 22		O13	
3999-015	C22	H 23	N 1	O 6	
2702-003	C22	H 23	N 2	O 8	BR1
2702-002	C22	H 23	N 2	O 8	CL1
2702-007	C22	H 23	N 2	O 9	CL1
1702-020	C22	H 23		O11	+
1703-019	C22	H 23		O11	+
1702-019	C22	H 23		O11	+
1703-014	C22	H 23		O12	+
1703-013	C22	H 23		O12	+
2702-001	C22	H 24	N 2	O 8	
2702-006	C22	H 24	N 2	O 9	
3902-005	C22	H 24		O 3	
0511-006	C22	H 24		O 5	
0903-022	C22	H 24		O 5	
1205-012	C22	H 24		O 6	
2504-030	C22	H 24		O 6	
0602-003	C22	H 24		O 8	
0606-003	C22	H 24		O 9	
0399-015	C22	H 24		O 9	
1108-020	C22	H 24		O 9	
1108-024	C22	H 24		O 9	
1203-018	C22	H 24		O10	
1203-011	C22	H 24		O10	
1203-009	C22	H 24		O10	
0399-016	C22	H 24		O10	
2699-013	C22	H 24		O10	
1003-010	C22	H 24		O11	
1205-006	C22	H 24		O11	
1206-006	C22	H 24		O12	
2106-007	C22	H 26		O 4	
2399-006	C22	H 26		O 5	
0505-007	C22	H 26		O 5	
0903-021	C22	H 26		O 5	
0601-026	C22	H 26		O 6	
0601-027	C22	H 26		O 6	
0903-018	C22	H 26		O 7	
0903-017	C22	H 26		O 7	
0601-035	C22	H 26		O 7	
3999-013	C22	H 26		O 7	
0511-005	C22	H 26		O 8	
0606-001	C22	H 26		O 8	
0603-003	C22	H 26		O 8	
0603-002	C22	H 26		O 8	
0603-001	C22	H 26		O 8	
2207-011	C22	H 26		O 8	
2301-019	C22	H 26		O 8	
2207-009	C22	H 26		O 8	
1002-014	C22	H 26		O10	
0199-020	C22	H 26		O10	
1002-012	C22	H 26		O10	
0506-019	C22	H 26		O11	
1302-017	C22	H 26		O11	
1303-022	C22	H 26		O12	
0604-003	C22	H 28		O 4	
2106-002	C22	H 28		O 4	
0601-006	C22	H 28		O 5	
0601-007	C22	H 28		O 5	
0903-001	C22	H 28		O 5	
0903-003	C22	H 28		O 5	
0903-004	C22	H 28		O 5	
0903-002	C22	H 28		O 5	
0505-006	C22	H 28		O 5	
0903-019	C22	H 28		O 6	
0903-020	C22	H 28		O 6	
0903-012	C22	H 28		O 6	
0903-011	C22	H 28		O 6	
0903-015	C22	H 28		O 6	
0904-014	C22	H 28		O 6	
2204-026	C22	H 28		O 6	
0903-016	C22	H 28		O 7	
2111-004	C22	H 30		O 3	
2101-023	C22	H 30		O 3	
2101-024	C22	H 30		O 3	
2106-003	C22	H 30		O 4	
2106-008	C22	H 30		O 4	
2106-004	C22	H 30		O 4	
2106-006	C22	H 30		O 4	
2106-005	C22	H 30		O 5	
2101-025	C22	H 32		O 2	
2101-022	C22	H 32		O 2	
3122-014	C22	H 32		O 2	
3122-013	C22	H 32		O 2	
2111-003	C22	H 32		O 3	
2002-022	C22	H 32		O 5	
2206-021	C22	H 32		O 5	
2106-009	C22	H 34		O 2	
3122-016	C22	H 34		O 2	
2111-002	C22	H 34		O 3	
2003-011	C22	H 34		O 4	
3005-042	C22	H 34		O 4	
3118-086	C22	H 35	N 1	O 1	
3122-012	C22	H 36		O 2	
2111-001	C22	H 36		O 3	
3005-043	C22	H 36		O 5	
3122-011	C22	H 38		O 2	
2199-012	C22	H 38		O 8	S 2
3122-010	C22	H 40		O 2	
3122-015	C22	H 40		O 2	
3122-009	C22	H 40		O 2	
3197-033	C22	H 40		O 7	
3122-007	C22	H 42		O 2	
3301-012	C22	H 42		O 2	S 2
3301-013	C22	H 42		O 2	S 3
3301-014	C22	H 42		O 2	S 4
3122-018	C22	H 42		O 4	
3122-003	C22	H 44		O 1	
3122-005	C22	H 44		O 2	
3122-008	C22	H 44		O 2	
3122-017	C22	H 44		O 3	
3122-001	C22	H 46			
3122-002	C22	H 46		O 1	
3122-004	C22	H 46		O 2	
1101-019	C23	H 20		O 6	
1801-003	C23	H 20		O 6	
1801-013	C23	H 20		O 6	
1499-018	C23	H 20		O 7	
1801-007	C23	H 20		O 7	
1404-005	C23	H 20		O 7	
1499-021	C23	H 20		O 7	
0599-017	C23	H 20		O 8	
1801-012	C23	H 22		O 6	
1404-003	C23	H 22		O 6	
1801-002	C23	H 22		O 6	
2404-028	C23	H 22		O 6	

Index	C	H	N	O	Other
1801-014	C23	H 22		O 7	
1801-025	C23	H 22		O 7	
1801-015	C23	H 22		O 7	
1801-006	C23	H 22		O 7	
1801-004	C23	H 22		O 7	
1801-018	C23	H 22		O 7	
1499-013	C23	H 22		O 7	
1801-005	C23	H 22		O 7	
1106-016	C23	H 22		O13	
1107-005	C23	H 22		O13	
2108-003	C23	H 23		O 5	CL1
3906-049	C23	H 23		O 5	CL1
3906-048	C23	H 24		O 5	
2403-012	C23	H 24		O 5	
2404-039	C23	H 24		O 6	
2404-032	C23	H 24		O 6	
1801-017	C23	H 24		O 7	
0505-022	C23	H 24		O 7	
0698-005	C23	H 24		O 8	
1103-046	C23	H 24		O11	
1105-036	C23	H 24		O11	
1104-050	C23	H 24		O11	
1106-076	C23	H 24		O12	
1106-059	C23	H 24		O12	
1104-048	C23	H 24		O12	
1107-004	C23	H 24		O12	
1105-048	C23	H 24		O12	
1107-003	C23	H 24		O12	
1106-102	C23	H 24		O13	
1106-101	C23	H 24		O14	
2702-012	C23	H 25	N 1	O 8	
2702-013	C23	H 25	N 1	O 9	
1703-020	C23	H 25		O12	+
1703-021	C23	H 25		O12	+
3906-047	C23	H 26		O 5	
2302-025	C23	H 26		O 6	
2503-031	C23	H 26		O11	
2502-034	C23	H 26		O13	
2301-017	C23	H 28		O 7	
2301-025	C23	H 28		O 7	
2301-020	C23	H 28		O 7	
0602-004	C23	H 28		O 7	
2207-017	C23	H 28		O 8	
2207-012	C23	H 28		O 8	
2207-010	C23	H 28		O 8	
2207-005	C23	H 28		O 8	
2301-021	C23	H 28		O 8	
2101-030	C23	H 29		O 4	CL1
2101-033	C23	H 30	N 1	O11	CL2
2101-021	C23	H 30		O 4	
0903-023	C23	H 30		O 5	
3903-004	C23	H 30		O 6	
2101-027	C23	H 31		O 3	CL1
2101-029	C23	H 31		O 4	CL1
2001-011	C23	H 32		O 2	
2101-026	C23	H 32		O 3	
2101-028	C23	H 32		O 4	
2101-020	C23	H 32		O 4	
2902-006	C23	H 34		O 5	
3111-022	C23	H 34		O 8	
2112-002	C23	H 36		O 2	
0499-010	C23	H 36		O 2	
2003-016	C23	H 36		O 4	
2112-003	C23	H 38		O 2	
2111-010	C23	H 40		O 2	
3508-004	C23	H 40		O20	
3123-003	C23	H 44		O 4	
3502-014	C23	H 45	N 5	O14	
3502-013	C23	H 45	N 5	O14	
3116-038	C23	H 46	N 1	O 4	+
3502-011	C23	H 46	N 6	O13	
3502-012	C23	H 46	N 6	O13	
3123-001	C23	H 48			
3123-002	C23	H 48		O 1	
2605-049	C24	H 16		O12	
2605-047	C24	H 17	N 1	O11	
2701-027	C24	H 20		O 5	
2510-012	C24	H 20		O 6	
2303-001	C24	H 20		O10	
2303-005	C24	H 20		O11	
0902-008	C24	H 22		O 5	
1299-014	C24	H 22		O 7	
0599-016	C24	H 22		O 8	
0902-001	C24	H 24		O 5	
2703-039	C24	H 24		O 9	
2404-040	C24	H 26		O 6	
0505-020	C24	H 26		O 7	
0505-019	C24	H 26		O 7	
0505-018	C24	H 26		O 7	
0505-017	C24	H 26		O 7	
0505-027	C24	H 26		O 7	
0505-026	C24	H 26		O 7	
0504-015	C24	H 26		O 7	
0503-006	C24	H 26		O 7	
3905-023	C24	H 26		O 7	
0505-023	C24	H 26		O 8	
0505-025	C24	H 26		O 9	
0505-024	C24	H 26		O 9	
1106-068	C24	H 26		O11	
1104-052	C24	H 26		O12	
1403-002	C24	H 26		O13	
1106-080	C24	H 26		O13	
1199-023	C24	H 26		O13	
1106-110	C24	H 26		O14	
1106-108	C24	H 26		O14	
0503-007	C24	H 28		O 3	
2804-023	C24	H 28		O 4	
0501-017	C24	H 28		O 5	
0505-021	C24	H 28		O 7	
2207-004	C24	H 28		O 7	
0399-010	C24	H 28		O11	
2503-027	C24	H 28		O14	
0599-005	C24	H 30		O 3	
0501-011	C24	H 30		O 3	
0501-014	C24	H 30		O 4	
0501-013	C24	H 30		O 4	
0599-006	C24	H 30		O 4	
0501-016	C24	H 30		O 5	
0903-005	C24	H 30		O 7	
0903-006	C24	H 30		O 7	
0504-014	C24	H 30		O 7	
0698-011	C24	H 30		O 7	
2301-026	C24	H 30		O 7	
2207-013	C24	H 30		O 8	
2301-022	C24	H 30		O 8	
2301-023	C24	H 30		O 8	
0603-005	C24	H 30		O 8	
0599-003	C24	H 32		O 4	
0903-024	C24	H 32		O 5	
0501-015	C24	H 32		O 5	
2204-025	C24	H 32		O 6	
0699-023	C24	H 32		O 7	
0699-026	C24	H 32		O 8	
2003-019	C24	H 34		O 4	
0601-003	C24	H 34		O 6	
2112-001	C24	H 36		O 8	
1106-084	C24	H 36		O13	
3508-001	C24	H 42		O21	
3508-003	C24	H 42		O21	
3508-002	C24	H 42		O21	
3124-008	C24	H 46		O 2	
3124-007	C24	H 46		O 2	
3124-004	C24	H 48		O 2	
3122-006	C24	H 48		O 2	
3124-005	C24	H 48		O 3	
3124-009	C24	H 48		O 3	
3124-006	C24	H 48		O 4	
3124-001	C24	H 50			
3124-002	C24	H 50		O 1	
3124-003	C24	H 50		O 2	
0701-009	C25	H 16		O 9	
2605-046	C25	H 17	N 1	O13	
1099-006	C25	H 20		O 5	
0902-013	C25	H 22		O 5	
0902-014	C25	H 22		O 5	
0902-016	C25	H 22		O 5	
1199-015	C25	H 22		O 6	
1299-011	C25	H 22		O10	
1299-012	C25	H 22		O10	
1299-013	C25	H 22		O10	
2303-002	C25	H 22		O10	
2303-004	C25	H 22		O10	
0599-019	C25	H 22		O12	
0902-015	C25	H 24		O 5	
0902-010	C25	H 24		O 5	
0902-009	C25	H 24		O 5	
1102-004	C25	H 24		O 5	
1401-010	C25	H 24		O 5	
1401-009	C25	H 24		O 5	
1401-007	C25	H 24		O 5	
1401-008	C25	H 24		O 5	
1199-013	C25	H 24		O 6	
1499-008	C25	H 24		O 6	
1199-014	C25	H 24		O 6	
1402-003	C25	H 24		O 6	
1402-004	C25	H 24		O 6	
0901-011	C25	H 24		O 6	
1902-015	C25	H 24		O12	
1902-014	C25	H 24		O12	
1902-012	C25	H 24		O12	
1902-013	C25	H 24		O12	
1902-011	C25	H 24		O12	
0702-008	C25	H 25	N 1	O 6	
1601-025	C25	H 26		O 4	
1601-024	C25	H 26		O 4	
1099-007	C25	H 26		O 5	
0902-004	C25	H 26		O 5	
0902-003	C25	H 26		O 5	
0902-005	C25	H 26		O 5	
0902-002	C25	H 26		O 5	
0902-006	C25	H 26		O 6	
0902-007	C25	H 26		O 6	
1199-010	C25	H 26		O 6	
1099-008	C25	H 26		O 6	
2703-048	C25	H 26		O10	
2603-004	C25	H 26		O13	
1601-023	C25	H 28		O 4	
1299-006	C25	H 28		O 6	
2404-041	C25	H 28		O 6	
2499-009	C25	H 28		O 6	
2499-010	C25	H 28		O 6	
2302-024	C25	H 28		O 8	
2302-023	C25	H 28		O 8	
1106-082	C25	H 28		O13	
2403-021	C25	H 28		O14	
2302-026	C25	H 30		O 7	
2301-029	C25	H 30		O 8	
2399-009	C25	H 30		O 9	
2101-031	C25	H 31		O 6	CL1
1405-008	C25	H 32		O 5	
0501-012	C25	H 32		O 5	
2301-027	C25	H 32		O 7	
2301-035	C25	H 32		O 8	
2207-008	C25	H 32		O 8	
2207-006	C25	H 32		O 8	
2207-007	C25	H 32		O 8	
2301-024	C25	H 32		O 8	
2207-014	C25	H 32		O 8	
2207-015	C25	H 32		O 8	
2207-018	C25	H 32		O 8	
2502-032	C25	H 32		O12	
3499-001	C25	H 33	N 1	O 7	
0399-011	C25	H 34		O11	
3499-003	C25	H 36	N 1	O 7	
2206-010	C25	H 36		O 4	
2204-005	C25	H 36		O 6	
2204-006	C25	H 36		O 6	
3401-001	C25	H 43	N 1	O 7	
3401-002	C25	H 43	N 1	O 7	
3999-009	C25	H 43	N 1	O 9	
3999-008	C25	H 45	N 1	O 9	
3125-002	C25	H 50		O 2	
3125-001	C25	H 52			
2704-023	C26	H 14		O 9	
0701-010	C26	H 16		O 5	
0701-012	C26	H 16		O 6	
2510-007	C26	H 18		O 8	
3906-008	C26	H 18		O12	
3906-010	C26	H 18		O12	
2605-048	C26	H 19	N 1	O12	
0902-012	C26	H 24		O 5	
0902-011	C26	H 24		O 5	
1499-009	C26	H 24		O 5	
2303-003	C26	H 24		O10	
1401-016	C26	H 26		O 5	
1404-006	C26	H 26		O 6	
1199-033	C26	H 26		O 6	
1499-010	C26	H 26		O 6	
1404-008	C26	H 26		O 6	
1499-007	C26	H 26		O 6	
2604-010	C26	H 26		O16	
1601-026	C26	H 28		O 5	
1299-008	C26	H 28		O 6	
1199-012	C26	H 28		O 6	
0601-015	C26	H 28		O 8	
2603-014	C26	H 28		O13	
2604-012	C26	H 28		O14	
1103-009	C26	H 28		O14	
1104-027	C26	H 28		O14	
1103-010	C26	H 28		O14	
1104-023	C26	H 28		O14	
1103-016	C26	H 28		O14	
1105-006	C26	H 28		O15	
1105-062	C26	H 28		O15	
1105-005	C26	H 28		O15	
1104-015	C26	H 28		O15	
1106-025	C26	H 28		O15	
1104-012	C26	H 28		O15	
1106-029	C26	H 28		O16	
1106-033	C26	H 28		O16	
1702-007	C26	H 29		O14	+

ID	C	H	N	O	Flag
1701-006	C26	H 29		O14	+
1702-006	C26	H 29		O15	+
1702-008	C26	H 29		O15	+
2206-015	C26	H 30		O 4	
1099-012	C26	H 30		O 6	
1099-009	C26	H 30		O 6	
1299-007	C26	H 30		O 6	
2302-027	C26	H 30		O 8	
3906-020	C26	H 30		O11	
1104-031	C26	H 30		O12	
1104-034	C26	H 30		O12	
1103-037	C26	H 30		O16	
1205-013	C26	H 32		O 6	
3905-015	C26	H 32		O 6	
2301-030	C26	H 32		O 8	
3906-019	C26	H 32		O11	
0601-012	C26	H 32		O11	
0699-018	C26	H 32		O12	
1204-006	C26	H 32		O12	
1204-008	C26	H 32		O12	
2805-018	C26	H 36		O 4	
0801-004	C26	H 38		O 1	
2206-016	C26	H 38		O 4	
2003-013	C26	H 42		O 4	
3005-041	C26	H 42		O 8	
2112-004	C26	H 46		O 2	
3126-008	C26	H 50		O 2	
3301-018	C26	H 50		O 4	S 2
3126-007	C26	H 52		O 2	
3126-010	C26	H 52		O 3	
3126-009	C26	H 52		O 3	
3126-001	C26	H 54			
3126-002	C26	H 54		O 1	
3126-003	C26	H 54		O 2	
2605-023	C27	H 11		O 8	CL1
0701-013	C27	H 16		O 6	
0701-011	C27	H 18		O 5	
2704-024	C27	H 20		O10	
3906-009	C27	H 20		O12	
3906-011	C27	H 20		O12	
0199-006	C27	H 22		O18	
0199-019	C27	H 26		O10	
0604-021	C27	H 26		O12	
1404-007	C27	H 28		O 6	
0199-003	C27	H 28		O11	
2603-033	C27	H 28		O16	
2703-024	C27	H 29	N 1	O10	
2703-037	C27	H 29	N 1	O11	
0605-003	C27	H 30		O12	
2604-032	C27	H 30		O14	
2604-033	C27	H 30		O14	
2604-013	C27	H 30		O14	
1103-008	C27	H 30		O14	
1401-015	C27	H 30		O14	
1104-022	C27	H 30		O14	
1401-003	C27	H 30		O14	
1401-005	C27	H 30		O14	
1103-007	C27	H 30		O14	
1106-047	C27	H 30		O15	
1103-014	C27	H 30		O15	
1105-026	C27	H 30		O15	
1105-027	C27	H 30		O15	
1803-011	C27	H 30		O15	
1104-016	C27	H 30		O15	
1103-018	C27	H 30		O15	
1103-019	C27	H 30		O15	
1104-011	C27	H 30		O15	
1803-012	C27	H 30		O15	
1105-008	C27	H 30		O15	
1105-007	C27	H 30		O15	
1106-041	C27	H 30		O15	
2604-014	C27	H 30		O15	
1105-004	C27	H 30		O16	
1104-013	C27	H 30		O16	
1106-031	C27	H 30		O16	
1104-014	C27	H 30		O16	
1106-030	C27	H 30		O16	
1105-009	C27	H 30		O16	
1104-020	C27	H 30		O16	
1104-021	C27	H 30		O16	
1106-023	C27	H 30		O16	
1105-015	C27	H 30		O16	
1105-012	C27	H 30		O16	
1106-024	C27	H 30		O16	
1106-020	C27	H 30		O17	
1106-022	C27	H 30		O17	
1106-021	C27	H 30		O17	
1106-028	C27	H 30		O17	
1108-007	C27	H 30		O18	
1108-005	C27	H 30		O18	
1701-005	C27	H 31		O14	+
1702-009	C27	H 31		O15	+
1701-007	C27	H 31		O15	+
1702-021	C27	H 31		O15	+
1702-010	C27	H 31		O15	+
1702-015	C27	H 31		O16	+
1703-015	C27	H 31		O16	+
1703-006	C27	H 31		O16	+
1702-011	C27	H 31		O16	+
1702-012	C27	H 31		O16	+
1702-013	C27	H 31		O16	+
1703-008	C27	H 31		O17	+
1703-007	C27	H 31		O17	+
1201-003	C27	H 32		O13	
2699-004	C27	H 32		O13	
1203-027	C27	H 32		O14	
1203-004	C27	H 32		O14	
1203-005	C27	H 32		O14	
1002-005	C27	H 32		O14	
1003-004	C27	H 32		O15	
1203-006	C27	H 32		O15	
1205-018	C27	H 32		O15	
1205-004	C27	H 32		O15	
2504-020	C27	H 32		O15	
1205-005	C27	H 32		O16	
0601-014	C27	H 34		O11	
0601-025	C27	H 34		O11	
0699-017	C27	H 34		O12	
3499-002	C27	H 35	N 1	O 8	
2806-008	C27	H 36		O 4	
2806-010	C27	H 36		O 4	
0903-008	C27	H 36		O 5	
2301-028	C27	H 36		O 7	
3499-004	C27	H 37	N 1	O 8	
2806-006	C27	H 38		O 4	
0899-008	C27	H 40		O 2	
0899-004	C27	H 46		O 2	
2114-001	C27	H 48		O 2	
3502-004	C27	H 49	N 7	O17	
3197-037	C27	H 52		O 2	
3197-038	C27	H 52		O 2	
3127-004	C27	H 54		O 1	
3127-005	C27	H 54		O 2	
3127-001	C27	H 56			
3127-003	C27	H 56		O 1	
3127-002	C27	H 56		O 1	
2002-015	C27	H 88		O 3	
2509-006	C28	H 18		O10	
2509-007	C28	H 18		O11	
2704-026	C28	H 18		O11	
2509-008	C28	H 18		O12	
2509-005	C28	H 22		O10	
0702-007	C28	H 23	N 1	O 6	
1103-017	C28	H 24		O12	
0604-022	C28	H 28		O11	
2703-042	C28	H 28		O11	
1103-028	C28	H 28		O17	
1199-035	C28	H 30		O 6	
2403-037	C28	H 30		O 6	
1402-014	C28	H 30		O14	
3999-016	C28	H 31	N 1	O 6	
1199-034	C28	H 32		O 6	
2302-028	C28	H 32		O 9	
1103-027	C28	H 32		O14	
1103-025	C28	H 32		O14	
1103-026	C28	H 32		O14	
2604-038	C28	H 32		O14	
1105-019	C28	H 32		O15	
1105-025	C28	H 32		O15	
1105-032	C28	H 32		O15	
1105-037	C28	H 32		O15	
1105-028	C28	H 32		O16	
1106-051	C28	H 32		O16	
1106-052	C28	H 32		O16	
1105-035	C28	H 32		O16	
1106-049	C28	H 32		O16	
1106-048	C28	H 32		O17	
1106-053	C28	H 32		O17	
1106-055	C28	H 32		O17	
2703-009	C28	H 33	N 1	O 9	
2703-020	C28	H 33	N 1	O10	
1702-022	C28	H 33		O16	+
1703-017	C28	H 33		O17	+
2207-020	C28	H 34		O 8	
0605-009	C28	H 34		O13	
1299-002	C28	H 34		O14	
1203-012	C28	H 34		O14	
1205-010	C28	H 34		O15	
1205-011	C28	H 34		O15	
2207-019	C28	H 36		O 8	
2301-031	C28	H 36		O 8	
0699-021	C28	H 38	N 8	O 4	
2806-009	C28	H 38		O 4	
2806-011	C28	H 38		O 4	
2999-001	C28	H 38		O 7	
0606-002	C28	H 38		O12	
2001-012	C28	H 40		O 2	
2002-005	C28	H 40		O 3	
2806-007	C28	H 40		O 4	
0899-006	C28	H 42		O 2	
0899-007	C28	H 42		O 2	
2805-017	C28	H 42		O 4	
3499-015	C28	H 43	N 1	O 6	
0104-038	C28	H 44	N 1	O 8	+
2001-015	C28	H 46		O 2	
3401-008	C28	H 47	N 1	O 7	
3401-009	C28	H 47	N 1	O 8	
0899-002	C28	H 48		O 2	
0899-003	C28	H 48		O 2	
2003-014	C28	H 48		O 4	
3508-005	C28	H 50		O17	
3128-004	C28	H 56		O 2	
3128-005	C28	H 56		O 3	
3128-001	C28	H 58			
3128-002	C28	H 58		O 1	
3128-003	C28	H 58		O 2	
2704-009	C29	H 20		O10	
2704-008	C29	H 20		O11	
3906-040	C29	H 20		O11	
2704-025	C29	H 22		O11	
2704-027	C29	H 22		O12	
1599-018	C29	H 24		O12	
1599-022	C29	H 24		O12	
2704-018	C29	H 26		O10	
1103-042	C29	H 34		O 1	K1
2302-029	C29	H 34		O 9	
1103-044	C29	H 34		O15	
1106-065	C29	H 34		O16	
1703-023	C29	H 35		O17	+
2301-033	C29	H 36		O 9	
2301-032	C29	H 36		O 9	
0204-008	C29	H 36		O19	
0903-009	C29	H 38		O 7	
0699-022	C29	H 40	N 8	O 5	
2499-008	C29	H 40		O 5	
2003-020	C29	H 42		O 4	
3999-007	C29	H 43	N 1	O 9	
0899-005	C29	H 44		O 2	
0899-001	C29	H 50		O 2	
0802-018	C29	H 50		O 2	
2001-016	C29	H 50		O 3	
3502-028	C29	H 55	N 5	O18	
3502-015	C29	H 55	N 5	O19	
3129-008	C29	H 56		O 2	
3129-005	C29	H 58		O 1	
3129-006	C29	H 58		O 1	
3129-007	C29	H 58		O 2	
3129-001	C29	H 60			
3129-002	C29	H 60		O 1	
3129-003	C29	H 60		O 1	
3129-004	C29	H 60		O 1	
2704-020	C30	H 16		O 8	
2608-001	C30	H 18		O 8	
2607-001	C30	H 18		O 8	
2607-002	C30	H 18		O 8	
2704-019	C30	H 18		O 8	
2609-018	C30	H 18		O 8	CL4
2609-017	C30	H 18		O 8	CL4
2607-003	C30	H 18		O 9	
2608-004	C30	H 18		O 9	
2608-008	C30	H 18		O 9	
2608-005	C30	H 18		O 9	
2607-005	C30	H 18		O10	
2608-006	C30	H 18		O10	
2608-003	C30	H 18		O10	
2608-007	C30	H 18		O10	
2609-008	C30	H 18		O10	
2607-004	C30	H 18		O10	
2608-002	C30	H 18		O10	
1505-001	C30	H 18		O10	
1599-003	C30	H 18		O10	
1504-001	C30	H 18		O10	
2608-011	C30	H 18		O11	
2608-009	C30	H 18		O11	
2608-010	C30	H 18		O12	
2510-013	C30	H 18		O12	
2609-016	C30	H 19		O 8	CL3
2609-009	C30	H 20		O 8	
2609-010	C30	H 20		O 9	
1501-003	C30	H 20		O10	
1599-004	C30	H 20		O10	
1501-002	C30	H 20		O10	
1599-017	C30	H 20		O10	

ID	C	H	O	extra
1501-004	C30	H 20	O10	
1501-007	C30	H 20	O11	
1501-009	C30	H 20	O11	
2510-011	C30	H 20	O11	
2609-001	C30	H 22	O 6	
2609-004	C30	H 22	O 7	
2609-005	C30	H 22	O 7	
2609-003	C30	H 22	O 8	
2608-013	C30	H 22	O 8	
2609-006	C30	H 22	O 8	
2609-002	C30	H 22	O 8	
2699-021	C30	H 22	O10	
2608-014	C30	H 22	O10	
2704-033	C30	H 22	O10	
1501-001	C30	H 22	O10	
1501-006	C30	H 22	O11	
1599-014	C30	H 22	O11	
1501-012	C30	H 22	O12	
2510-010	C30	H 22	O12	
2699-023	C30	H 22	O12	
2608-015	C30	H 22	O12	
2608-016	C30	H 22	O12	
2699-022	C30	H 22	O12	
2510-018	C30	H 24	O 4	
2704-028	C30	H 24	O 9	
2704-029	C30	H 24	O 9	
2704-030	C30	H 24	O 9	
2704-014	C30	H 24	O10	
1599-025	C30	H 24	O12	
2510-015	C30	H 26	O 4	
2704-017	C30	H 26	O10	
2510-008	C30	H 26	O10	
2704-032	C30	H 26	O10	
2704-015	C30	H 26	O10	
2607-025	C30	H 26	O10	
2704-031	C30	H 26	O10	
1506-002	C30	H 26	O10	
1506-001	C30	H 26	O10	
1506-005	C30	H 26	O11	
1506-006	C30	H 26	O11	
1599-012	C30	H 26	O11	
1502-003	C30	H 26	O11	
1506-004	C30	H 26	O11	
1502-004	C30	H 26	O11	
1502-008	C30	H 26	O12	
1502-007	C30	H 26	O12	
1502-005	C30	H 26	O12	
1502-006	C30	H 26	O12	
1599-013	C30	H 26	O12	
1506-007	C30	H 26	O12	
1599-008	C30	H 26	O12	
1506-008	C30	H 26	O13	
1599-010	C30	H 26	O13	
1502-010	C30	H 26	O13	
1599-009	C30	H 26	O13	
1104-026	C30	H 26	O13	
1599-011	C30	H 26	O14	
2406-005	C30	H 26	O14	
0599-023	C30	H 28	O 6	
0599-022	C30	H 28	O 6	
1299-013	C30	H 28	O 8	
3905-021	C30	H 28	O 8	
2704-016	C30	H 28	O10	
2599-003	C30	H 28	O11	
2599-006	C30	H 28	O11	
2599-007	C30	H 28	O12	
2599-004	C30	H 28	O12	
2510-016	C30	H 30	O 4	
1199-037	C30	H 30	O 7	
1203-032	C30	H 34	O 4	
1002-020	C30	H 34	O 4	
2703-030	C30	H 35	N 1 O11	
2699-009	C30	H 36	O 1	
2699-010	C30	H 36	O 4	
1002-019	C30	H 36	O 4	
1203-031	C30	H 36	O 4	
1703-025	C30	H 37	O17	+
1203-022	C30	H 38	O15	
2005-010	C30	H 40	O 8	
2005-011	C30	H 40	O 8	
3508-006	C30	H 52	O26	
3508-007	C30	H 52	O26	
3130-005	C30	H 58	O 2	
3118-070	C30	H 58	O 3	
3130-003	C30	H 60	O 2	
3130-006	C30	H 60	O 3	
3130-001	C30	H 62		
3130-002	C30	H 62	O 1	
1503-001	C31	H 20	O10	
1505-005	C31	H 20	O10	
1505-004	C31	H 20	O10	
1505-002	C31	H 20	O10	
1599-006	C31	H 20	O10	
1599-005	C31	H 20	O10	
1505-003	C31	H 20	O10	
1504-005	C31	H 20	O10	
1599-001	C31	H 20	O10	
1599-015	C31	H 22	O 5	
2609-011	C31	H 22	O 9	
1501-010	C31	H 22	O11	
2609-007	C31	H 24	O 8	
2599-002	C31	H 26	O 5	
1506-003	C31	H 26	O12	
1502-009	C31	H 28	O12	
2406-007	C31	H 28	O14	
2406-006	C31	H 28	O14	
2406-012	C31	H 28	O14	
2406-011	C31	H 30	O15	
0599-024	C31	H 36	O11	N 2
2502-017	C31	H 46	O 2	
3401-044	C31	H 51	N 1 O10	
3131-005	C31	H 60	O 1	
3131-007	C31	H 60	O 2	
3131-008	C31	H 60	O 3	
3131-009	C31	H 60	O 3	
3131-004	C31	H 62	O 1	
3197-039	C31	H 62	O 2	
3131-006	C31	H 62	O 2	
3131-001	C31	H 64		
3131-002	C31	H 64	O 1	
3131-003	C31	H 64	O 1	
0701-004	C32	H 20	O 8	
2704-021	C32	H 20	O10	
1505-006	C32	H 22	O10	
1505-009	C32	H 22	O10	
1505-010	C32	H 22	O10	
1504-002	C32	H 22	O10	
1504-003	C32	H 22	O10	
1599-002	C32	H 22	O10	
1503-002	C32	H 22	O10	
1505-007	C32	H 22	O10	
2608-012	C32	H 22	O12	
1599-016	C32	H 24	O 5	
1505-008	C32	H 24	O10	
3906-012	C32	H 24	O10	
2509-009	C32	H 24	O10	
2511-003	C32	H 26	O10	
0599-021	C32	H 26	O10	
2510-014	C32	H 26	O14	
1502-001	C32	H 28	O 7	
1502-002	C32	H 28	O 8	
0701-006	C32	H 28	O14	
2406-004	C32	H 30	O14	
2406-002	C32	H 30	O14	
2406-003	C32	H 30	O14	
2406-001	C32	H 30	O14	
2704-034	C32	H 32	O14	
2406-009	C32	H 32	O15	
2510-017	C32	H 34	O 4	
2704-004	C32	H 34	O14	
2406-008	C32	H 34	O16	
2406-010	C32	H 34	O16	
1104-033	C32	H 40	O16	
1104-032	C32	H 40	O17	
2207-029	C32	H 42	O 8	
1204-009	C32	H 42	O17	
2204-007	C32	H 46	O 6	
2204-009	C32	H 46	O 6	
2204-008	C32	H 46	O 6	
3197-041	C32	H 62	O 3	
3132-003	C32	H 64	O 2	
3116-017	C32	H 64	O 2	
3197-040	C32	H 64	O 3	
3132-001	C32	H 66		
3132-002	C32	H 66	O 1	
2599-008	C33	H 18	O 9	
1504-006	C33	H 24	O10	
1505-012	C33	H 24	O10	
1505-011	C33	H 24	O10	
1505-014	C33	H 24	O10	
1505-013	C33	H 26	O10	
2704-010	C33	H 32	N 2 O10	
2499-002	C33	H 36	O 7	
2499-003	C33	H 36	O 7	
2499-005	C33	H 36	O 8	
2499-006	C33	H 36	O 8	
2499-001	C33	H 38	O 6	
2499-004	C33	H 38	O 7	
2802-012	C33	H 38	O 8	
1104-025	C33	H 40	O19	
1104-017	C33	H 40	O20	
1106-032	C33	H 40	O21	
1104-024	C33	H 40	O21	
1106-113	C33	H 40	O21	
1702-014	C33	H 41	O20	+
1104-039	C33	H 42	O22	
3402-001	C33	H 47	N 1 O13	
2002-006	C33	H 48	O 3	
2504-016	C33	H 52	O 5	
3133-002	C33	H 64	O 2	
3133-001	C33	H 68		
2607-009	C34	H 22	O10	
1505-015	C34	H 26	O10	
1504-004	C34	H 26	O10	
2303-006	C34	H 30	O13	
2510-009	C34	H 30	O13	
1103-054	C34	H 34	O17	
2207-021	C34	H 40	O12	
1801-016	C34	H 40	O16	
1106-040	C34	H 42	O20	
1106-050	C34	H 42	O21	
1106-054	C34	H 42	O22	
0603-004	C34	H 46	O18	
2003-021	C34	H 50	O 4	
3402-003	C34	H 51	N 1 O13	
3402-004	C34	H 51	N 1 O14	
3999-005	C34	H 54	O 8	
3401-023	C34	H 54	O14	
3134-001	C34	H 68	O 2	
3197-048	C34	H 70		
3134-002	C34	H 70	O 1	
2704-007	C35	H 24	O14	
2207-024	C35	H 42	O12	
2207-025	C35	H 42	O12	
2207-022	C35	H 42	O12	
2502-019	C35	H 46	O 2	
2005-005	C35	H 52	O 8	
3401-022	C35	H 56	O 7	
3402-006	C35	H 58	O11	
3402-007	C35	H 58	O12	
3401-021	C35	H 61	N 1 O12	
3135-001	C35	H 72		
3197-051	C35	H 72		
2607-019	C36	H 26	O11	
2703-034	C36	H 28	N 2 O13	
2607-015	C36	H 28	O12	
2607-023	C36	H 28	O13	
1599-020	C36	H 28	O16	
1599-019	C36	H 28	O16	
2607-024	C36	H 30	O11	
2607-021	C36	H 30	O13	
1501-005	C36	H 30	O15	
1501-008	C36	H 30	O16	
1501-011	C36	H 32	O16	
1702-016	C36	H 37	O18	+
2509-004	C36	H 38	O16	
2599-005	C36	H 38	O17	
2207-026	C36	H 44	O12	
2207-027	C36	H 44	O12	
2207-023	C36	H 44	O12	
2509-010	C36	H 44	O19	
2703-010	C36	H 48	N 2 O11	
2703-021	C36	H 48	N 2 O12	
0104-021	C36	H 53	N 1 O12	
3402-002	C36	H 53	N 1 O13	
3402-008	C36	H 58	O10	
3999-001	C36	H 62	O11	
3508-008	C36	H 62	O31	
3508-009	C36	H 62	O37	
3401-011	C36	H 65	N 1 O13	
3197-049	C36	H 74		
3401-045	C37	H 59	N 1 O12	
3402-009	C37	H 60	O10	
3401-024	C37	H 61	N 1 O14	
2299-007	C37	H 62	O 4	
3401-010	C37	H 67	N 1 O12	
3401-012	C37	H 67	N 1 O13	
3137-001	C37	H 76		
3197-052	C37	H 76		
2607-007	C38	H 30	O12	
2607-012	C38	H 32	O13	
2607-006	C38	H 34	O14	
2607-018	C38	H 34	O14	
2607-026	C38	H 34	O16	
2607-016	C38	H 36	O13	
2607-017	C38	H 36	O15	
2607-013	C38	H 36	O15	
2607-020	C38	H 36	O16	
2607-010	C38	H 38	O12	
2607-011	C38	H 38	O14	
2607-022	C38	H 38	O15	
2607-014	C38	H 38	O15	

ID	C	H	N	O
2607-008	C38	H 38		O16
1703-022	C38	H 41		O17 +
2499-007	C38	H 44		O 8
2002-007	C38	H 56		O 3
3402-005	C38	H 63	N 1	O13
3401-025	C38	H 63	N 1	O14
3197-050	C38	H 78		
2599-010	C39	H 24		O12
2003-022	C39	H 58		O 4
3401-028	C39	H 63	N 1	O15
3401-026	C39	H 65	N 1	O14
2299-008	C39	H 66		O 4
3198-002	C39	H 74		O 6
3197-053	C39	H 80		
2502-020	C40	H 54		O 2
3499-010	C40	H 64		O12
3401-019	C40	H 65	N 1	O14
3401-032	C40	H 65	N 1	O15
3401-029	C40	H 65	N 1	O15
3401-048	C40	H 66		O10
3401-027	C40	H 67	N 1	O14
3999-002	C40	H 68		O10
3999-003	C40	H 68		O11
3401-017	C40	H 70		O15
0199-009	C41	H 32		O28
0199-011	C41	H 32		O28
0199-010	C41	H 34		O28
0199-005	C41	H 34		O28
2502-011	C41	H 56		O 2
3402-010	C41	H 64		O14
3401-037	C41	H 65	N 1	O15
3499-011	C41	H 66		O12
3401-047	C41	H 67	N 1	O11
3401-030	C41	H 67	N 1	O15
3401-033	C41	H 67	N 1	O15
3401-036	C41	H 67	N 1	O16
2299-009	C41	H 70		O 4
3197-034	C41	H 80		O 4
2609-012	C42	H 38		O20
2609-013	C42	H 38		O20
2609-015	C42	H 40		O19
2609-014	C42	H 40		O19
2703-031	C42	H 41	N 1	O16
1703-009	C42	H 47		O23 +
1703-010	C42	H 47		O23 +
2703-011	C42	H 58	N 2	O14
3401-020	C42	H 67	N 1	O15
3401-038	C42	H 67	N 1	O15
3401-021	C42	H 67	N 1	O16
3499-012	C42	H 68		O12
3401-031	C42	H 69	N 1	O15
3401-034	C42	H 69	N 1	O15
3401-018	C42	H 72		O16
1599-021	C43	H 32		O20
2106-016	C43	H 56		O 7
2005-006	C43	H 56		O 8
3906-021	C43	H 58		O 6
2002-008	C43	H 64		O 3
3401-039	C43	H 69	N 1	O15
3499-013	C43	H 70		O12
3401-035	C43	H 71	N 1	O15
3401-040	C43	H 74	N 2	O14
2599-008	C44	H 26		O12
2003-023	C44	H 36		O 4
3126-005	C44	H 66		O 8
3499-014	C44	H 72		O12
3114-040	C44	H 78		O26
3401-013	C44	H 80	N 2	O15
3197-035	C44	H 90		O 1
2502-021	C45	H 62		O 2
3126-004	C45	H 68		O 8
3401-041	C45	H 76	N 2	O15
3999-004	C45	H 78		O13
3401-043	C45	H 79	N 1	O17
3198-003	C45	H 86		O 6
2502-018	C46	H 62		O 3
2502-012	C46	H 64		O 2
3126-006	C46	H 70		O 8
3401-042	C46	H 78	N 2	O15
3401-014	C46	H 82	N 2	O16
2207-028	C47	H 56		O16
2005-007	C47	H 62		O 2
2701-006	C47	H 66		O22
3402-012	C47	H 71	N 1	O17
3402-013	C47	H 73	N 1	O17
3402-011	C47	H 75	N 1	O17
2001-013	C48	H 22		O 2
2703-040	C48	H 64		O21
2703-012	C48	H 68	N 2	O16
2703-022	C48	H 68	N 2	O17
2701-008	C48	H 68		O22
2002-009	C48	H 72		O 3
3401-015	C48	H 84	N 2	O17
2003-024	C49	H 74		O 4
3401-016	C49	H 86	N 2	O17
2502-022	C50	H 70		O 2
3114-041	C50	H 88		O30
2502-013	C51	H 72		O 2
2502-015	C51	H 74		O 2
0801-005	C51	H 78		O 1
2005-008	C52	H 70		O 8
2701-009	C52	H 76		O24
2002-013	C52	H 78		O 3
3116-032	C52	H 92		O32
3197-042	C52	H104		O 4
2005-009	C53	H 72		O 8
2001-014	C53	H 80		O 2
0899-010	C53	H 80		O 2
2002-010	C53	H 80		O 3
2002-016	C53	H 81	N 1	O 3
0899-009	C53	H 82		O 2
3198-004	C53	H100		O 6
2003-025	C54	H 82		O 4
3130-004	C54	H108		O 2
2502-023	C55	H 78		O 2
3401-046	C55	H103	N 3	O16
3198-006	C55	H104		O 6
3198-005	C55	H106		O 6
2704-022	C56	H 42		O12
2502-014	C56	H 80		O 2
2701-004	C56	H 80		326
2502-016	C56	H 82		O 2
2701-005	C56	H 82		O25
0801-006	C56	H 86		O 1
2701-011	C57	H 82		O26
2701-010	C57	H 84		O25
2002-014	C57	H 86		O 3
0802-003	C57	H 88		O 2
3198-011	C57	H 92		O 6
3198-010	C57	H 98		O 6
3198-012	C57	H 98		O 9
3198-009	C57	H104		O 6
3198-007	C57	H108		O 6
3197-043	C57	H110		O 5
3198-008	C57	H110		O 6
2701-003	C58	H 84		O26
2002-011	C58	H 88		O 3
2003-031	C58	H 88		O 4
2002-017	C58	H 89	N 1	O 3
2002-012	C58	H 90		O 3
0899-011	C58	H 98		O 4
3197-045	C58	H112		O 5
2701-012	C59	H 86		O26
2003-026	C59	H 90		O 4
2003-029	C59	H 92		O 4
2003-030	C59	H 94		O 4
1599-027	C60	H 50		O20
1599-026	C60	H 50		O21
3197-046	C60	H116		O 5
3197-047	C62	H122		O 3
3197-044	C63	H122		O 5
2003-027	C64	H 98		O 4
2003-028	C69	H106		O 4

A 5
B 6
C 7
D 8
E 9
F 0
G 1
H 2
I 3
J 4